International Maritime Conference 2012

(Pacific 2012)

Darling Harbour, Australia
30 January - 2 February 2012

ISBN: 978-1-62276-120-3

Printed from e-media with permission by:

Curran Associates, Inc.
57 Morehouse Lane
Red Hook, NY 12571

Some format issues inherent in the e-media version may also appear in this print version.

Printed by Curran Associates, Inc. (2012)

For permission requests, please contact The Royal Institution of Naval Architects]
at the address below.

The Royal Institution of Naval Architects
10 Upper Belgrave Street
London SW1X 8BQ UK

Phone: 44 207 235 4622
Fax: 44 207 259 5912

publications@rina.org.uk

Additional copies of this publication are available from:

Curran Associates, Inc.
57 Morehouse Lane
Red Hook, NY 12571 USA
Phone: 845-758-0400
Fax: 845-758-2634
Email: curran@proceedings.com
Web: www.proceedings.com

TABLE OF CONTENTS

FULL PAPERS

REFEREED FULL PAPERS

PACIFIC 2012
International Maritime Conference

31 January - 2 February 2012
Sydney Convention and Exhibition Centre
Darling Harbour, Australia

Welcome to the proceedings from the Pacific 2012 International Maritime Conference.

Pacific 2012 IMC is held in association with the Pacific 2012 International Maritime Exposition and RAN Sea Power Conference 2012.

Pacific 2012 IMC provides a unique opportunity for people involved in maritime and naval affairs around the world to discuss the latest developments in design, naval architecture, engineering, science and technology. With the concurrent Maritime Exposition, the event will provide a meeting place for industry representatives to exchange ideas and to establish personal and business contacts.

CD-ROM Sponsor

This CD-ROM is proudly sponsored by UGL

Organising Committee

Chair
John Jeremy, The Royal Institution of Naval Architects

Members
Stuart Cannon, The Royal Institution of Naval Architects
Bill Bixley, The Institute of Marine Engineering, Science and Technology
Adrian Broadbent, The Royal Institution of Naval Architects
Catriona McAuliffe, Engineers Australia
Tauhid Rahman, The Institute of Marine Engineering, Science and Technology

Program Committee

Chair
Adrian Broadbent

Members
Craig Boulton
Rob Gehling
Gangadhara Prusty
Tauhid Rahman
Martin Renilson

Address for Communications

Please contact the Conference Managers at the address below for further information

Pacific 2012 IMC Managers
arinex pty limited
GPO Box 128
SYDNEY NSW 2001 AUSTRALIA
Telephone (61) 2 9265 0700
Fax: (61) 2 9267 5443
Email: pacific2012imc@arinex.com.au

Hosts

The Royal Institution of Naval Architects

The Institute of Marine Engineering, Science and Technology

Engineers Australia

MODULARITY: NAVAL SHIP MODERNIZATION WITH REDUCED COST AND INCREASED FLEXIBILITY

Marc DeBlasio, Peter Gentile, Chris Piaszczyk
Northrop Grumman Corporation, USA
marc.deblasio@ngc.com

Abstract

The traditional model of designing warships places severe limits on the roles that the vessel can undertake. A modular concept, where a common hull is host to a range of mission-specific modules, provides significantly greater flexibility. As the ship's required roles and missions change over time, or technology evolves, modules can be replaced, enhanced or modified to suit the mission, without having to significantly alter the basic hull sub-system. This can drive down the overall fleet costs, improve interoperability and facilitate joint and allied operations. This paper reviews the modularity concept in general, its importance and benefits in the current context of quickly evolving asymmetric threats. The discussion is then illustrated with two examples from the efforts our organization is presently involved in: the U.S. Navy Littoral Combat Ship (LCS) Mission Package Integration (MPI) and the NATO Modularity study – NIAG SG-150.

Contents

Modularity Definition, Its Importance and Benefits

In the general sense, modularity implies an architectural approach reducing the complexity of the system design by employing interchangeable components or parts that can be mixed and matched to perform desired tasks or missions by linking them via interfaces that adhere to a set of common standards. Successful implementation of modularity requires strict adherence to the systems engineering process and systems engineering principles. Key to the success of modular system design is open architecture (a foundation for an Open Business Model) that consists of standardized interfaces and standardized functionalities.

Common examples of modularity include many applications, from toys to housing and software. A very successful example of modularity application in the commercial world is the ISO standard 20-foot equivalent unit (TEU) cargo container. Another well known example is the High-Mobility Multipurpose Wheeled Vehicle which can be outfitted in 15 configurations such as cargo/troop carriers, weapons carriers, ambulances and shelter carriers by swapping the back end modules that share a common engine, chassis and transmission, with 44 interchangeable parts used in more than one position. With fewer number of part types, the maintenance mechanics require fewer training hours. Reduced number of supply lines reduces life-cycle costs and lowers the total operating costs.

Recently, some very significant modularization projects have been undertaken by several of the world's navies. Examples include the LCS by the U.S., BAM's by Spain, SEA 1180 by Australia, StanFlex by Denmark. Traditional military ships are built for single or multi-missions and can have a lifespan of 35+ years. Recent emergence of asymmetric warfare dictates constantly evolving counter strategies requiring new mission capabilities. A modularized ship's warfighting capability can be rapidly reconfigured for different missions in a short time. As a result, a navy does not have to re-design or upgrade the ship's Hull, Mechanical & Electrical (HM&E) systems for different missions, system upgrades, or additions, therefore reducing life-cycle costs. Computer networking technologies used to provide situational awareness in maritime security operations naturally lend themselves to modular design. Unmanned systems in all domains (air, surface, subsurface) that have the potential to reduce crew manning requirements can also be modularized.

Exchangeable mission modules can provide improved, focused mission capabilities for every specific mission (Figure 1). While small ships have an advantage in the littorals with shallower drafts and higher maneuverability, large ships offer more deck space. Naturally, large ships are more suitable for large scale operations such as humanitarian relief or harbor protection. Potential for modularity application is huge. Current considerations already include a long list of candidate missions:

- Mine Warfare
- Surface Warfare
- Antisubmarine warfare
- Humanitarian Assistance/Disaster Relief
- Counter-Piracy/maritime security operations
- Harbor defense
- Immigration control/border protection
- Emergency medical
- Electronic Warfare
- Irregular Warfare
- Special Operations

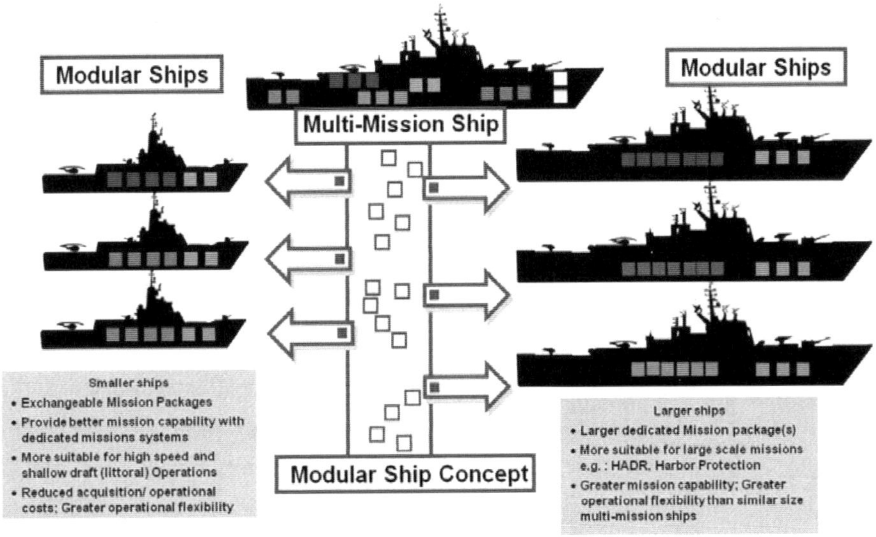

Figure 1: Modular Mission Packages for Small and Large Ships

Example Mission Packages/Modules/Platforms

Our organization is currently involved in two major modularization efforts: LCS MPI for the U.S. Navy and the NIAG SG-150 Modularity Study for NATO.

Modularity for LCS

The Littoral Combat Ship (LCS) will be a fast, agile and networked surface combatant optimized for operating in the littorals. Fundamental to the modularity of LCS is the use of interchangeable Mission Packages (MPs). LCS capability will be built up incrementally, taking advantage of its modular open systems architecture. Each MP provides warfighting capability for one of the following three focused mission areas:

- Mine Countermeasures (MCM) – from detection to neutralization of mine threats
- Surface Warfare (SUW) – prosecution of small boat threats and maritime security missions
- Anti-Submarine Warfare (ASW) – countering diesel submarine threats in the littoral shallow waters and their associated deep water approaches

Additional focused mission area packages are under evaluation for LCS. The LCS MPs will offer the combatant commanders continuously improving mission capabilities with new technologies that will be added to the MPs incrementally as they reach a level of maturity necessary for fielding. Specifically defined functionalities and standard interfaces provide an open architecture environment that enables future rapid technology insertion. Warfighting analysis of the experiences with the current generation of the MPs will be the primary tool for determining which technologies to pursue.

The modularity of the LCS MPs, illustrated in Figure 2, is hierarchical and can be described in three layers:

- Mission Systems = Vehicle, Communication, Sensor, or Weapon System
- MM = Mission Systems + Support Equipment + Software
- MP = MMs + Mission Crew Detachments + Support Aircraft

Each Mission Package (MP) is a grouping of Mission Modules (MM), mission crew detachments, and aircraft all focused on a single, specific mission area. Mission Modules (MM) combine Mission Systems (MS), support equipment, and Mission Package Application Software (MPAS). A Mission System is a broad, generic term for a vehicle, a sensor, or a weapon system that is used to accomplish a particular military task.

Mission Systems are packaged inside standard ten- or twenty-foot International Organization for Standardization (ISO) support containers (SC), or on ISO compliant flat racks and vehicle cradles. This standardized packaging allows efficient use of common means of transportation, use of standard port equipment for loading and unloading to and from the LCS seaframe and mating with LCS interface requirements documented in the Interface Control Document (ICD) for the Littoral Combat Ship (LCS).

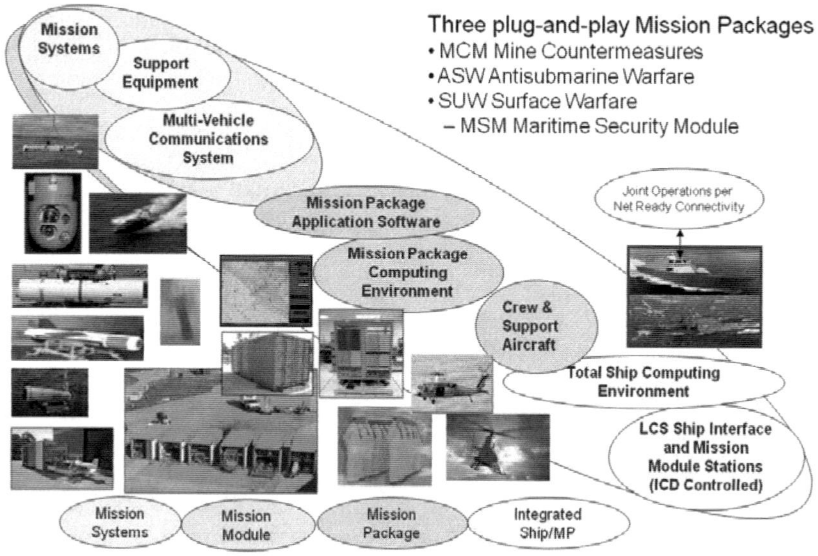

Figure 2: The LCS Hierarchical Levels of Modularity

The LCS support containers are basic (or base) ISO containers adapted for each mission system they carry with specifically designed internal mounting and support hardware. Base containers are modified with structural reinforcements and outfitted with mission systems kits. The mission systems kits themselves are made up of a variety of components such as shelving and drawers which are used to outfit the container for a particular application. Container modifications also feature bi-fold cargo doors and accommodation for common electrical and communications panels and environmental control units as required. A lighting configuration consisting of red and white fluorescent lighting, and smoke and heat detectors are also included with capability to interface with the ship's damage control system through the communications panel.

Figure 3: Operational Mission Package Support Facility (MPSF)

The LCS MP TEU mounting zones inside each seaframe mission bay are configured to accommodate the SCs in accordance with the LCS ICD specifications. The mission zones located topside are specifically configured to host certain weapons modules, such as the Gun Mission Module (GMM). These weapons modules disassemble for shipment inside ISO containers because using standard containers simplifies shipping, storage, availability of correct handling equipment, and container movement from shore to ship and ship to shore.

LCS Seaframe (SF) – Mission Package (MP) system reconfiguration will occur in homeport or overseas, using pre-positioned MPs that have been transported into theater by air, land, or sea and staged near the LCS operating area. The MP management, maintenance and other sustainment operations will be supported in a Mission Package Support Facility (MPSF). A prototype MPSF, currently located in Port Hueneme (NSWC PHD) is shown in Figure 3. Future MPSF facilities will include additional locations where U.S. Navy ships are maintained or visit on a daily basis and that can be relied on to support LCS MP swap-out or shore operations. Options for using unprepared ports in wartime are also under consideration.

Mission Package Application Software (MPAS) consists of the focused mission area specialized software tools used in conjunction with the Mission Systems hardware. Mission Package Computing Environment (MPCE) hardware and software and Multiple Vehicle Communications System (MVCS) hardware and software are both common infrastructure modules which support all MM configurations. They are permanently installed on seaframe. The MPCE provides the MPs with a standard network interface to the Total Ship Computing Environment (TSCE) on the LCS seaframe. MVCS establishes a standard interface for unmanned vehicle data links and network communication services that maximize the usefulness of radio equipment available on board the LCS seaframe.

Figure 4: Northrop Grumman € Prototype to Production Components Fabrication Integration and Installation

Northrop Grumman provides the PEO LCS with Package Production and Assembly (PP&A) services that include production, fabrication, assembly verification and process improvement. In its Mission Package Integration (MPI) PP&A role, Northrop Grumman also contributes to prototype to production conversion, requirements verification, design standardization; configuration managed baseline, obsolescence check point, as well as process and build/assembly efficiency (Figure 4).

Mission systems are provided as GFE to the PP&A contractor (Northrop Grumman) for integration into complete MPs. Northrop Grumman produces additional integration elements (customized containers and other equipment), integrates all components and supports deliveries of the complete MPs to the LCS ILS for storage and embarkation on the Seaframes for specific mission needs.

NIAG SG -150 Study

The main objective of the NIAG SG-150 study was to evaluate, based on the analysis of the scenario(s) and the modularity requirements, the architecture implications of designing, installing and operating modular mission area equipment packages that provide specific mission capabilities on LPD or Joint Support ship, or equivalent platforms. The study was to further assess modular mission equipment packages and interchangeability requirements for three missions: Humanitarian Assistance and Disaster Relief (HADR), Counter Piracy (CP) and Harbor Defense (HD). Subsequently, the study team was to prepare, for each, an outline mission module design that addresses the module equipment content, ship interface requirements and on-board, off-board ship transfer requirements.

The major goals of the NIAG SG-150 study effort were to identify ways to demonstrate the benefits of modularity through a rigorous systems engineering process.

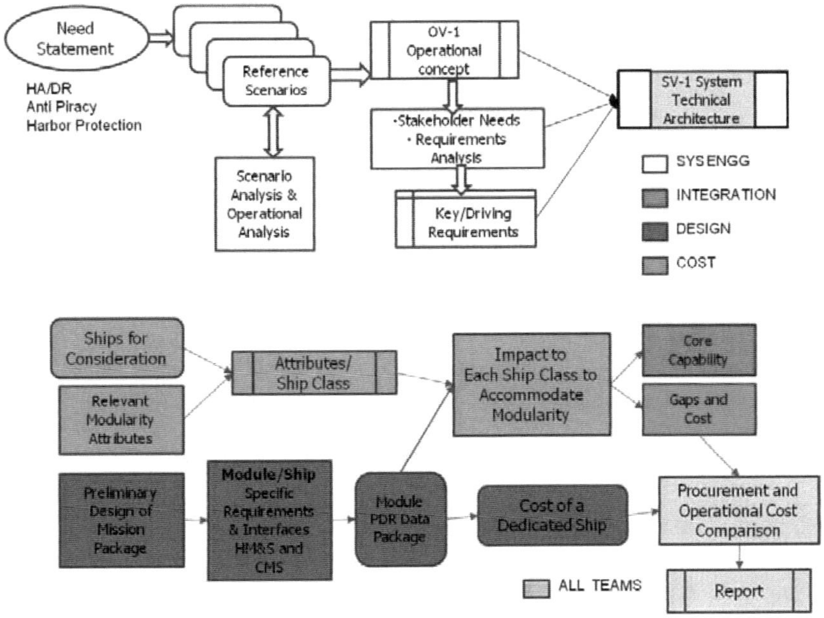

Figure 5: NIAG SG -150 Study: Process

The team assembled to perform the NATO SG-150 study included participants from France, Germany, Italy, The Netherlands, Norway, Spain, Sweden, Turkey, UK, and the U.S. The process followed by this international team is illustrated in Figure 5. The team was organized into four groups: systems engineering, integration, design and cost. The contributions of the individual groups are indicated with color mapping. The point of departure was a need statement generated by the study sponsor. The systems engineering group developed several reference scenarios plus an operational concept, and analyzed a list of stakeholder needs for key driving requirements. The systems engineering and integration groups collaborated to develop a system technical architecture. The integration group produced a selection of ship platforms for examination of attributes for analysis of impact to each class to accommodate modularity, producing a list of core capabilities and gaps. The design group performed preliminary designs of mission packages, developed module and ship specific requirements and interfaces for Hull, Mechanical and Electrical (HM&E) and produced a module PDR data package that served as a starting point for the cost group's estimates of procurement and operational cost comparison analyses.

The ship classes studied and their major attributes are shown in Figure 6. MCG-6 study guidance to SG -150 required analysis of accommodations and compatibility assessments to include a broad contingent of NATO LPDs and support ships. A review of NATO LPDs and support ships was undertaken. The list included 11 candidate ships as shown. The study was limited to public domain information and ITAR-cleared data. Some initially reviewed ships did not meet key requirements, e.g. a wet deck to support ashore operations in sub-optimal port conditions.

NATO NIAG SG150 - Mission Modularity Study							
Mission Module Host Ship Attribute Data Sheet							
Configuration Data							
Last Updated	31st March 2011	Revision G					
Nation	Ship Class or Ship Name	Ship Type	No of ships in Class	Complete	Populated By	Date	Notes
Denmark	Absalon Class	Flexible Support	2	Partial	S Jones	2010-09-03	from public domain info only
France	Foudre	LPD	2	Partial	S Jones	2010-09-03	from public domain info only
Italy	San Giorgio Class	LPD	2	Yes	Cdr Enrico	2010-07-30	
Netherlands	hr.ms. Rotterdam	LPD	1	Yes	M Janssen & R Hoogenboom	2010-08-27 & 2010-09-02	
Netherlands	hr.ms. Johan De Witt	LPD	1	Yes	R Hoogenboom	2010-09-02	
Netherlands	Karel Doorman	JSS	In Build	Yes	R Hoogenboom	2010-09-02	
Spain	Galicia	LPD	2	Yes	I Mosquera	2010-07-14	
UK	Albion Class	LPD	2	Yes	D Sharp	2010-09-03	
UK	Bay Class	LPD	4	Partial	R Boulby	2010-10-27	
USA	LPD17 Class	LPD	9	Partial	B Geldart	2010-11-04	public releasable info only
USA	LSD41	LPD	12	Partial	B Geldart	2010-11-04	public releasable info only

Foudre Bay Class Bay Class LPD17 LSD41

Figure 6: NIAG SG -150 Study: Vessels Analyzed

Using the HADR mission example, employing MMs to provide an on demand HADR capability enables a scaled response consistent with the magnitude of the crisis. Figure 7 illustrates the OV-1 developed for the Humanitarian Assistance and Disaster Relief (HADR) shore operations. This OV-1 assumes a large scale disaster that requires international assistance. As a result, the required mission package includes a rather large number of assets and will be best supported by a fleet of LPD vessels. It was assumed that the HADR fleet in this operation also includes a hospital ship fully equipped with medical mission modules. Accordingly, the medical facilities deployed to the shore consist of a patient triage center and a local surgical center for minor procedures, while the serious cases are evacuated to the off-shore hospital ship via a medivac helicopter. An alternative concept could consist of hospital mission modules that are transported to the shore. The HADR mission package portrayed in this concept also includes several modules for electrical power generation (wind, solar, diesel) that would be used according to the local conditions. Water purification and water storage modules support the temporary housing modules.

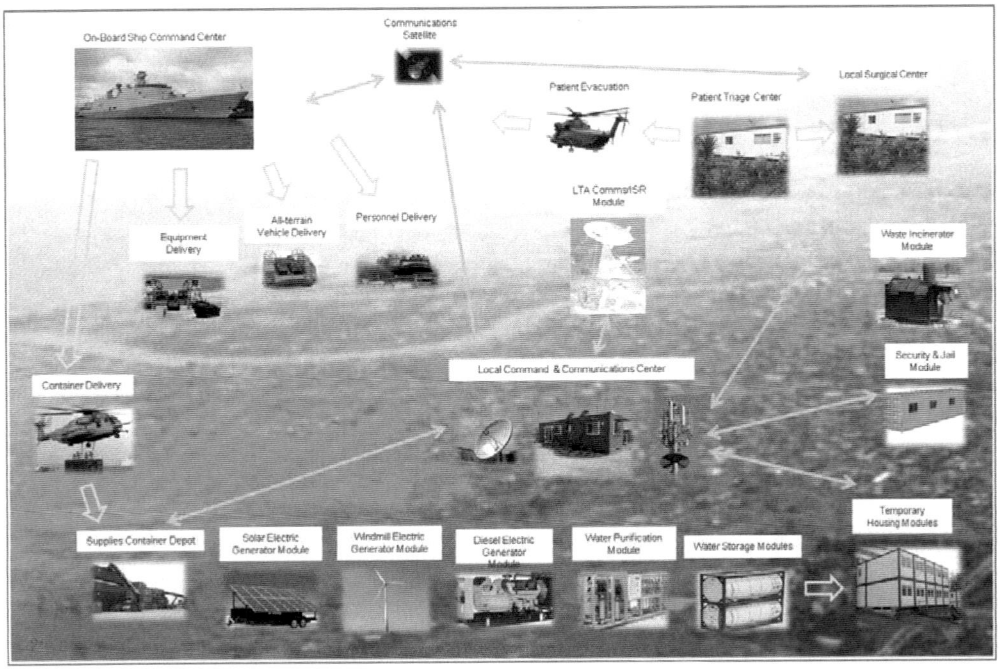

Figure 7: NIAG SG -150 Study: OV-1 High Level Operational Concepts Graphic Humanitarian Assistance and Disaster Relief (HADR) Shore Operations

Additional missions considered in the study included Counter Piracy (CP) and Harbor Protection (HP). The NATO CP mission package provides a scalable capability intended for cooperation with other maritime forces and anti-piracy networks. The study considered a modular capability for insertion into existing ships to enable reconnaissance of shipping lanes, surveillance of potential activities and gathering of intelligence. The Harbor Protection (HP) mission addressed threats to harbors associated with NATO and the interests it protects. The intent is to fill gaps in the existing port and harbor security infrastructures and mitigate asymmetric threats. The proposed HP mission package modules include a mix of sensors for reconnaissance and surveillance, anti-threat systems and countermeasures, security apparatus, tactical communications and C4I with connection to intelligence and situation awareness networks.

Interface compatibility categories ranges from structural and mechanical hardware, through environmental, electrical, pneumatic/hydraulic, and electronic data to human. The level of complexity is relatively low for hardware, while increasing rapidly for information systems featuring high density of content data transfers across electronic media systems. Figure 8 shows a summary of the MM interface characteristics according to the module type.

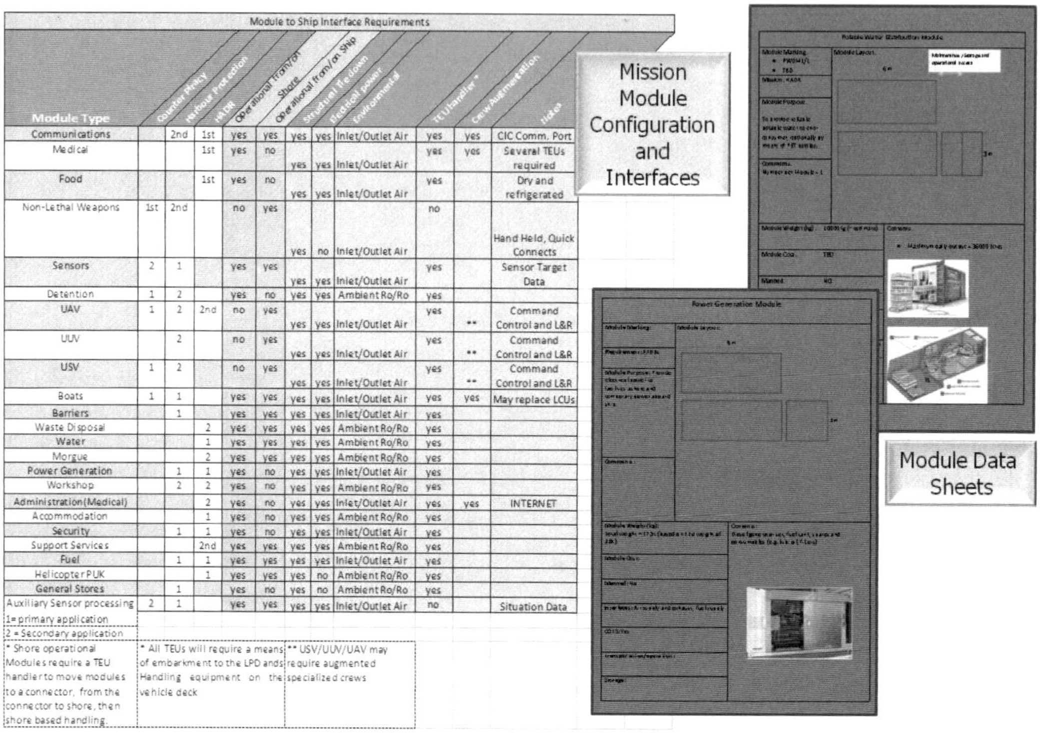

Module to Ship Interface Requirements											
Module Type	Counter Piracy	Harbour Protection	HADR	Operational from/on Shore	Operational from/on Ship	Structural Tie down	Electrical power	Environmental	TEU handler *	Crew Augmentation	Notes
Communications		2nd	1st	yes	yes	yes	yes	Inlet/Outlet Air	yes	yes	CIC Comm. Port
Medical			1st	yes	no	yes	yes	Inlet/Outlet Air	yes	yes	Several TEUs required
Food			1st	yes	no	yes	yes	Inlet/Outlet Air	yes		Dry and refrigerated
Non-Lethal Weapons	1st	2nd		no	yes	yes	no	Inlet/Outlet Air	no		Hand Held, Quick Connects
Sensors	2	1		yes	yes	yes	yes	Inlet/Outlet Air	yes		Sensor Target Data
Detention	1	2		yes	no	yes	yes	Ambient Ro/Ro	yes		
UAV	1	2	2nd	no	yes	yes	yes	Inlet/Outlet Air	yes	**	Command Control and L&R
UUV		2		no	yes	yes	yes	Inlet/Outlet Air	yes	**	Command Control and L&R
USV	1	2		no	yes	yes	yes	Inlet/Outlet Air	yes	**	Command Control and L&R
Boats	1	1		yes	yes	yes	yes	Inlet/Outlet Air	yes	yes	May replace LCUs
Barriers		1		yes	yes	yes	yes	Inlet/Outlet Air	yes		
Waste Disposal			2	yes	yes	yes	yes	Ambient Ro/Ro	yes		
Water			1	yes	yes	yes	yes	Ambient Ro/Ro	yes		
Morgue			2	yes	yes	yes	yes	Ambient Ro/Ro	yes		
Power Generation		1	1	yes	no	yes	yes	Inlet/Outlet Air	yes		
Workshop		2	2	yes	no	yes	yes	Ambient Ro/Ro	yes		
Administration(Medical)			2	yes	no	yes	yes	Inlet/Outlet Air	yes	yes	INTERNET
Accommodation			1	yes	no	yes	yes	Ambient Ro/Ro	yes		
Security		1	1	yes	no	yes	yes	Inlet/Outlet Air	yes		
Support Services			2nd	yes	yes	yes	yes	Ambient Ro/Ro	yes		
Fuel		1	1	yes	yes	yes	yes	Inlet/Outlet Air	yes		
Helicopter PUK			1	yes	yes	yes	no	Ambient Ro/Ro	yes		
General Stores		1		yes	no	yes	no	Ambient Ro/Ro	yes		
Auxiliary Sensor processing	2	1		yes	yes	yes	yes	Inlet/Outlet Air	no		Situation Data

1= primary application
2 = Secondary application

* Shore operational Modules require a TEU handler to move modules to a connector, from the connector to shore, then shore based handling.

* All TEUs will require a means of embarkment to the LPD ands Handling equipment on the vehicle deck

** USV/UUV/UAV may require augmented specialized crews

Figure 8 NIAG SG -150 Study: Defining the Mission Modules

In summary, the NIAG SG-150 study objectives were achieved and the team engagement and cooperation was excellent. The feasibility of modularized flexible capabilities that can be rapidly deployed on target host vessels for the three identified mission areas was demonstrated. The study concluded that nations can best stretch their ship building budgets by trading fixed mission with modular mission capabilities. The modularity concept can be supported by multiple nations and companies with NATO leadership.

Ship Modularity Summary

Modularity provides mission flexibility for the spectrum(s) of conflict and the phases of war. Modular mission packages can leverage the Navy's (or a navy's) ongoing investments in networks, platforms, systems, and sensors. Recent progress in remotely operated and robotic off-board vehicles provides ample opportunities for modularity implementation. The modularity concept can be applied to many surface platforms. The possible concept for a standard interface between the seaframe and the mission package has been proven and is applicable to large and small ships. Operational cost benefits of modularity are being realized. Through NATO International Standards (STANAG), implementation of ship modularity for a wide range of missions is being realized.

HSV *WestPac Express* - Ten Years Service to USMC

James M (Jim) Black
Austal, Henderson WA, Australia
jimb@austal.com

INTRODUCTION

The High-Speed Vessel *WestPac Express*, a 101m Austal built aluminium catamaran, has been in continuous service with the United States Marine Corps (USMC) since 2001, based in Naha Military Port, Okinawa, Japan and operating throughout the Western Pacific Region.

In the ten years of service the *WestPac Express* has maintained an impeccable record for reliability, availability and safety of operation, allowing the Okinawa based Third Marine Expeditionary Force (IIIMEF) to rely on the ship for virtually all of their off-island operations and exercises in the region and significantly reducing their reliance on the availability of US Transportation Command aircraft. The ship has carried the Marines and their equipment throughout the region, from Hokkaido in the North to the Philippines in the South, from Thailand in the West to Guam in the East.

This paper reviews the ten-year operation and its support requirements. This was an operation of many "firsts" and of developing and bedding down effective long term solutions born out of initial improvisations as both Austal and USMC learned "on the run" how to operate a high-speed commercial ferry as a regional military theatre support asset. Operational statistics are presented and lessons learned are discussed.

The applicability to future military programs is considered, such as the US Joint High-Speed Vessel (JHSV), with seven vessels already under contract at Austal's US shipyard in Mobile, Alabama and the first (USNS *Spearhead*) due for delivery in 2012.

BACKGROUND

On 2 July 2001 HSV *WestPac Express* set sail from her builders, Austal, in Henderson, Western Australia for the port of Naha on the island of Okinawa in the East China Sea, approximately mid way between Japan and Taiwan, and home to the US Marine Corps' Third Marine Expeditionary Force (IIIMEF). She arrived there on 14 July with an international crew on board and under the Panamanian flag, to commence a six week "proof of concept" time charter. The arrival marked the culmination of several months of intensive preparation following the signing of the contract with III MEF, in which Austal had to convert the vessel to make her more suitable for the intended operation; register and flag her; engage a crew; and obtain all the necessary trading certification, insurances and other documentation. It was with a great sense of anticipation that the vessel was seen on her way. Although the initial charter was for a period of only six weeks there were high expectations this was just the start of something with huge potential for Austal's future.

HSV *WestPac Express* - first touch down at her new home in Naha Military Port, July 2001

Austal seized this opportunity and took on this "proof of concept" contract to demonstrate to the US military the utility of high-speed craft (HSC) in a theatre support role. The first vessel in this role had been HMAS *Jervis Bay,* used very successfully by the Australian military for theatre support operations in East Timor a few years previously; however *WestPac Express* was the first HSC engaged to directly support United States forces. Even the terminology used to describe such vessels had to be developed and agreed and during her years of operation *WestPac Express* has been variously termed a High-Speed Craft, a High Speed Connector, a Theatre Support Vessel and now she is commonly accepted to be an HSV(High-Speed Vessel), which term has now been officially adopted for the joint US Army and Navy JHSV; Joint High-Speed Vessel, the first of which has just been launched by Austal from its US-based shipyard in Mobile, Alabama, something of which everyone associated with *WestPac Express* is justifiably proud.

There are many US military bases on the island of Okinawa, including those of IIIMEF, and although about 10,000 Marines are stationed on the island, almost all regular training exercises are conducted off-island at regular training locations on mainland Japan, Korea, Guam, the Philippines and Thailand.

Training is a vital element in retaining battle readiness, but for the IIIMEF this had been characterised by high costs of over US$20m per year to engage the US Air Mobility Command strategic lift aircraft to mobilise Marines and their equipment. These deployments typically required aircraft for 14 to 16 days to fully deploy one battalion as each aircraft could only carry a limited amount of troops. This extended deployment also meant that long periods of time were wasted waiting for aircraft with associated risks of individual aircraft delays further disrupting the timetable and wasting valuable time waiting to get to, or return from, exercises. Changing to an HSV ship-based approach meant that the Marines were able to travel with their vehicles and equipment, instead of flying separately and hoping to meet up with their equipment at the required time and place.

Another important aspect of III MEF's training deployments were their impact on the host population of Okinawa. The IIIMEF bases of Camp Schwab and Camp Hansen are located in the north of the island while the Kadena Air Base is in the south, close to the main population centre of Naha. Thus deployments for training required the movement of large numbers of troops and vehicles through the centre of Naha and other populated areas causing disruption to traffic and inconvenience to the host population and high levels of aircraft noise. US military transportation officials referred to these deployments as horror stories, with Marines often sitting on their sea bags for two weeks at a time waiting for enough planes or ships to get them to their off-island exercises.

In her role of transporting troops to exercises, *WestPac Express* has provided IIIMEF with a cost effective method of deployment and one that avoids the disruption on the island since the vessel loads and discharges at Kin Red pier (an old landing craft ramp) only one kilometre from Camp Hansen in the north of the island. In the words of Lt General Wallace C. Gregson (at the time the Commander Marine Forces Japan) in an interview with Stars and Stripes magazine: "*In our opinion, we found a way to deploy for training that is faster, better and cheaper, it makes operational sense, it make economic sense ... this is a transformational success story*"

THE SHIP

WestPac Express is an Austal Auto Express 101m aluminium ROPAX catamaran, a sistership to *Euroferrys Pacifica* which, for the past eleven years, has been successfully plying the Straits of Gibraltar in year-round commercial service.

Principal Dimensions

Length (overall)	101.0 m
Length (waterline)	86.2 m
Beam (moulded)	26.6 m
Depth (moulded)	9.4 m
Draft (max.)	4.3 m

Speed

33 knots with cargo and passengers, at 90% MCR
36 knots in light ship condition, at 90% MCR

Range & Fuel Consumption

6580 litres/hr @ 90% MCR
1240 nm at 90% MCR (with cargo and passengers)
1460 nm at 90% MCR (in light ship condition)

Deadweight

Maximum total deadweight	792 t
Maximum military deadweight (pax + cargo)	645 t

Crew and Complement

Officers & Ratings	15
Cadets or other crew	2
Passengers	max 900

Total COI Complement	917

Survey

Class: GL ✠100 A5, HSC-B OC3, High Speed Pax/Ro-Ro Type, ✠MC, HSC-B Aut
Certification: International Code of Safety for High Speed Craft (HSC Code)
Flag United States of America, Hailing Port Mobile, AL

Propulsion

Main Engines	4 x Caterpillar 3618 @ 7,200 kW at 1,050 rpm
Gearboxes	4x Reintjes VLJ 6831
Water Jets	4 x Kamewa 125 SII
High-speed Rudder	2 x Austal T Max
Motion Control System	Austal, 2 x DC-foil (fwd), 2 x Interceptor (aft)

Electrical Equipment

Generator Sets	4x Caterpillar 3408, each 313 kW(E) @ 1500rpm
Primary	3 wire 400V 50Hz AC 3-phase
Secondary	4 wire earthed neutral 230V 50Hz AC and 110V 60Hz AC
Shore Power	400 Amps 3-phase

Tankage

Marine Gas Oil	368,000 litres
Fresh Water	20,000 litres
Black and Grey Water	8,000 litres

Vehicle/Cargo Deck _Fully enclosed

Main deck area	1,944 m^2
Mezzanine deck area	724 m^2
Main deck clear height	5.1 m, 4.6 m & 2.7 m
Mezzanine deck clear height	2.0 m
Main deck inboard of mezzanines	15.0 t/ dual axle, 12.0 t/single axle
Main deck below mezzanines	2.0 t/single axle
Mezzanine decks	1.0 t/single axle
(HMMWV types can be loaded on mezzanine decks)	
Stern RoRo Ramp	19.2 m long for 40 t vehicles
Bow RoRo Ramp	11.8 m long for 40 t vehicles

The main enhancements made to the ship to fit her for military service were:

- Replacement of the normal RORO short stern ramp with a 19 long metre bi-fold unit that allows operation from low wharfs and shore ramps
- Upgrading of the twin anchoring system to enable "med mooring" evolutions
- Increased tankage
- Additional hotel systems
- Installation of crew accommodation, troop embarkation office, sick bay and galley

The arrangement changes on the passenger deck are such that that can be removed should the ship require to be reconfigured for commercial service in the future.

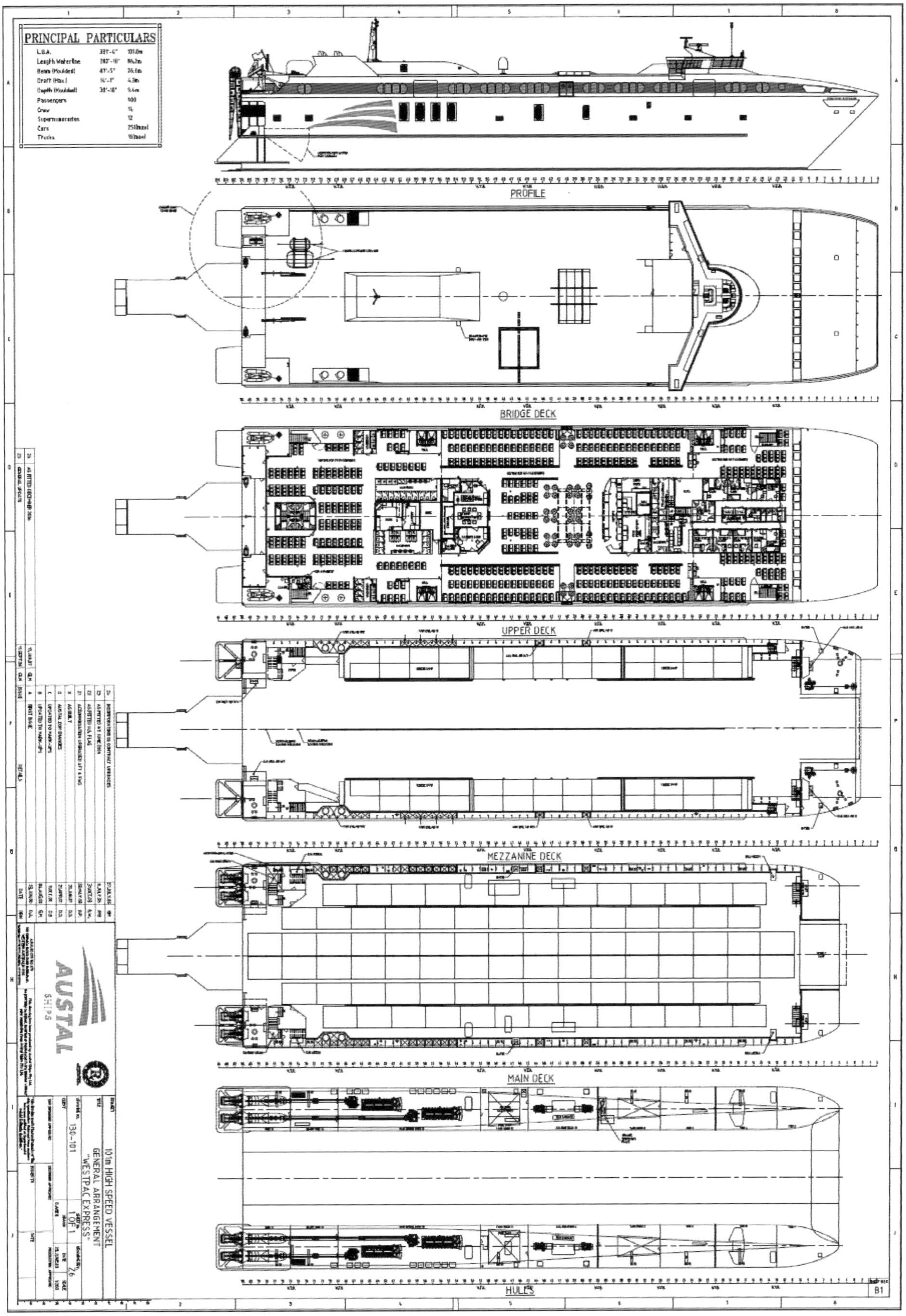
PRINCIPAL PARTICULARS
L.O.A.
Draft (Max.)
4.3m
9.4m
Passengers
Crew
12
Cars
Trucks
PROFILE
BRIDGE DECK
UPPER DECK
MEZZANINE DECK
MAIN DECK
HULLS
AUSTAL
SHIPS
101m HIGH SPEED VESSEL
GENERAL ARRANGEMENT
"WESTPAC EXPRESS"
130-101
1 OF 1
Z6
B1

***WestPac Express* General Arrangement**

Following the initial, and very successful, proof of concept charter, a five-year charter was negotiated with Military Sealift Command (MSC), on behalf of III MEF, for the same service, but with the ship required to be converted to US flag within a specified timc frame. That was an interesting undertaking in itself, but the ship was successfully reflagged in September 2003 and has sailed under the Stars and Stripes ever since. The story of that reflagging, with its own saga of many "firsts" was presented at the Pacific 2004 IMC (Ref. 1). Since then the charter has been further extended with the vessel celebrating her ten year anniversary in July 2011.

THE OPERATION

Since 2002, the ship has been US crewed by Hornblower Marine Services (now HMS Global Maritime) under contract to Austal. The crew consists of master, chief engineer, 3 x deck officers, 3 x engineer officers, 3 x deck ratings, 3 x engineer ratings, steward, and up to two cadets. The ship works a "blue/gold" crewing system, six weeks on, six weeks off. The watch system is nominally 4 hours on, 8 hours off but typical voyages of 24 to 36 hours with port evolutions in between tend to mean that the work load is significantly heavier than would first appear; the current crewing arrangement allowing for operation within STCW limits but with little margin. Typical annual OPEMPO is 25 to 30%, with the ship running up to about 2500 hours per year. For a sustained OPTEMPO above about 40% it is considered that additional crewing would be required to remain within STCW watchkeeping limits.

Base porting a vessel of this type in the East China Sea / Western Pacific region brings its own special challenges, not the least of which are that operations are required to continue year round, including throughout the typhoon season; but with good forecasting in recent years and the HSV's speed capability, interruptions to schedule have been kept to a minimum.

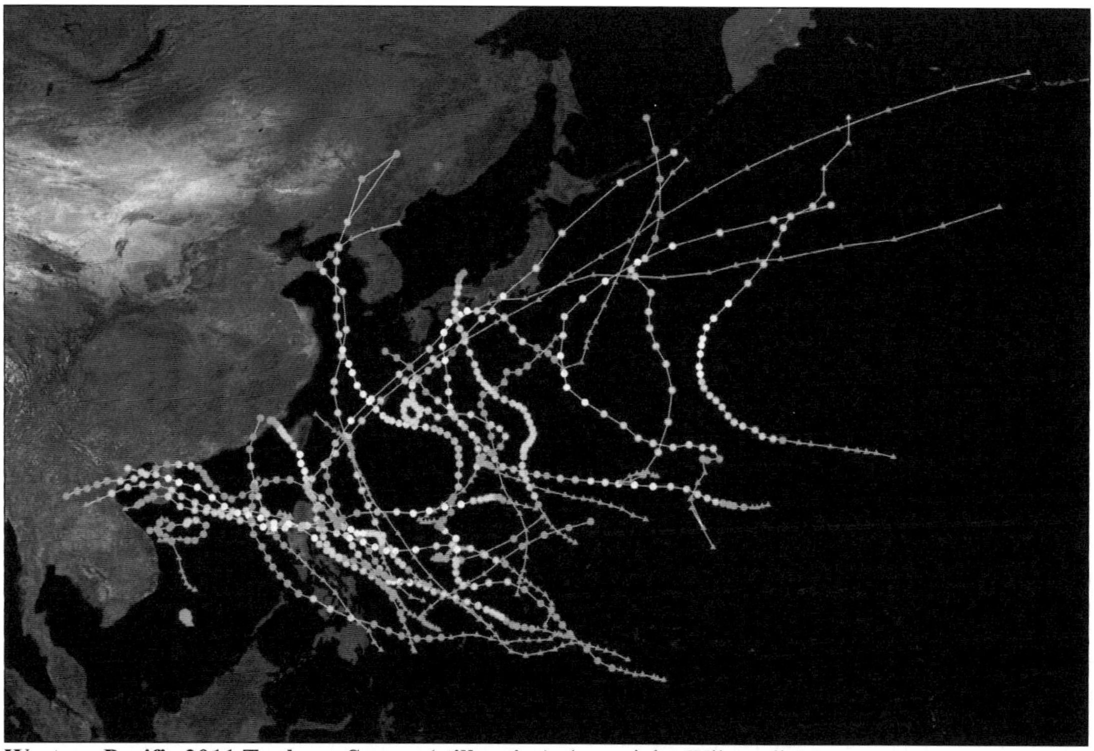

Western Pacific 2011 Typhoon Season (still active) (copyright Wikepedia)

Since commencing operations in 2001, the ship has sailed about 550,000 miles, carried 177,000 troops and 160,000 tonnes of their equipment; the main engines have run for 24,000 hours, the gensets for 46,000 hours and have between them consumed 112 million litres of fuel. Ports visited include Naha, Kin Red and White Beach on Okinawa, Yokohama, Yokosuka, Sasebo, Iwakuni, Sendai, Oita, Tsuneishi, Tomokomai and Kushiro in Japan, Busan, Pohang, Pyongteak and Chinhae in Korea, Subic Bay, Masbate and Tabaco in the Philippines, Chuk Samet in Thailand and Kota Kinabalu in Malaysia. There have also been occasional voyages out to Guam, Saipan and Tinian.

Loading at Kin Red

Unloading in Korea

In addition to regular deployments for scheduled exercises throughout the region, *WestPac Express* has taken part in foreign military training exercises, civilian evacuation exercises, anti-terrorist exercises and humanitarian relief operations.

A difficult med moor at Tabaco, PI ....

.... but a great thank you for a job well done!

Given the variety of operational scenarios, the huge variation in port facilities encountered and the varying capabilities and requirements of each Marine Corps unit embarked, competent, capable and "can-do" masters, officers and crew have been a key ingredient in the quiet success of this ship. As have the ship's shallow draft, good speed, good manoeuvrability, full width bridge, full controls at the wing stations, large robust stern ramp, twin anchoring system and powerful hydraulic capstans. A good example of this combined capability was the recent safely conducted "in-stream" refuelling from another MSC ship off Thailand, for which a "Bravo Zulu" was received.

***WestPac Express* refueling from USNS *Jack Lummus* February 2011**

THE SUPPORT

One of the biggest challenges over the years has been effective support of *WestPac Express*. Being a purely commercial time charter to Military Sealift Command (MSC) on behalf of IIIMEF, the only source of income is the daily hire rate plus specified reimbursables. Any day, or part day, that the ship is not available for service means that she is off hire and the income stops. Unfortunately the same cannot be said for the expenditure: crew still need to be paid and fed, the ship requires to be insured and maintained, fuel and port bills and bareboat charter fees still need to be paid; the list can seem endless! All of this certainly helps to concentrate the mind, and has led to some innovative solutions being implemented.

First, and most importantly, in what will soon be eleven years now, the ship has never left her area of operations. Much as it would be nice to be able to bring her back to Austal in Western Australia for major maintenance periods, the additional time off hire and the associated fuel bill could not be justified, so all support has been provided "in theatre". Using the conventional terms of Organisational Level Maintenance (OLM), Intermediate Level Maintenance (ILM), and Depot Level Maintenance (DLM), in the simplest of terms: OLM is carried out by the ship's crew on a continuous basis, ILM is carried out while on hire but alongside (usually Naha or Yokohama) between deployments, by the crew, Austal, local and international contractors as required, and DLM is carried out annually, again by the crew, Austal, local and international contractors while off hire and drydocked at a shipyard in the region.

One of the keys to successful support has been a good supportive local ship's agent with an effective network of corresponding agencies throughout the area of operations; effectively a one-stop shop for all crew handling, port handling and local logistics matters. Another key is a capable local maintenance contractor, or group of contractors, who can provide the quick turn-round unscheduled maintenance support that always seems to be required despite the best of planning intentions; and probably always will be as long as there are ships on the sea! And of course it goes without saying that a competent engineering superintendent and competent engineering officers who take pride in their ship are essential ingredients.

Maintainability

Another very important aspect of effective support is having a ship that is fitted with robust "commercial off the shelf" (COTS) equipment built in to simple systems, preferably well documented. *WestPac Express* is blessed with such relatively simple systems and equipment; mostly quite capable of doing the job required of them in a "no frills" way. The main engines, while by no means trouble free, have proven capable of complete in-situ maintenance; top ends were overhauled in 2006 and complete rebuilds were partly done in 2010 and are being completed at the 2011 DLM. The ship even changed out a cracked crankshaft during the 2010 three-week scheduled DLM period, rigging it in vertically through the engineroom maintenance access hatch.

Typical of the way maintenance is carried out was the purchasing of nine reconditioned cylinder heads from the OEM (there are 4 x 18 cylinder main engines). The cylinder heads were changed out one bank at a time during available ILM periods, using the ship's crew, OEM local dealer, and a local contractor. Each removed set was then locally overhauled and ready for the ship's return. The job has been completed and the last set of removed heads sent back to the OEM for return of the core deposit payment.

Traceability

The above mentioned cylinder head job highlights the need for effective asset tracking, sometimes down to sub-component level. Typical, for these main engines, would be the need to track cylinder heads, injectors, turbochargers, pumps and starters. *WestPac Express* uses a computer-based maintenance management system (CMMS) that has the merit of being simple, practical, easy to use and class-approved; however it is really not capable of this level of asset tracking and so is supported by a suite of spreadsheets that has grown over time. While this provides adequate asset control on this ship, and probably needs no more for most single ship operations, this would not be sustainable across a fleet and a more sophisticated CMMS would be required for that purpose.

Communications

A closely related aspect to the CMMS is IT, particularly when it comes to effective communications, data transfer and secure storage. By having IT-competent crew managers and superintendent, we have kept abreast of the ever changing technology, and use such capabilities as Remote Desktop, SharePoint, DropBox, and Skype to keep in touch and ensure that we are all looking in to the same data, be it the ship in the Far East, the crew managers in the USA, the superintendent in Sweden or the program manager in Australia (and to keep communications costs down). We also now have a reasonably sophisticated virtual server system on board to manage and safely store all the data that is an essential part of any business these days.

The use of an effective, configurable, ship-wide control, monitoring and alarm system has also been important (we use the Austal-developed MarineLink system). The ability to view event history, remotely diagnose, and even remotely monitor have been great assets; the chief engineer can now even monitor and control using his iPad from his hotel room!

Obsolescence

Thinking of IT leads naturally to thoughts of obsolescence; no great surprises here, many of the electronics aboard have become unsupportable and have had to be changed out in the last few years, although we have minimised this by keeping in close touch with suppliers and ensuring that we keep or have access to superseded components for as long as possible. This, combined with the "tramping" nature of this operation has meant that we have had to be far more self reliant than a normal ferry operation and we now carry over a million dollars worth of consumables, spare parts, special tools and equipment, either on board or ashore at our base port in Naha.

Corrosion and Cracking

No discussion of HSV support would really be complete without touching on the emotive issue of corrosion, something which has been reasonably controllable on *WestPac Express*, and with the experience now gained is considered eminently controllable going forward. What has certainly been highlighted is the necessity for understanding of the causes of corrosion, having good installation practices, and having good maintenance practices.

There have been three main types of corrosion on this ship, at least two of which are substantially preventable:

- Poorly designed or installed zinc coated steel framing of structural fire protection panels has allowed the steel to come in contact with the aluminium in a number of areas where salt water has also been present; first the zinc acts as the anode until it is gone, then the current reverses and the aluminium acts as an anode to the steel, with inevitable results. Improved design detailing and installation control can completely eliminate this problem.
- Years of operating in tropical climes with variable fuel suppliers has led to the inevitable water and contaminants in the bottom of the main fuel tanks. Although normal fuel stripping practices were carried out, in the earlier years we did not annually empty and clean the tanks and the result has been a build up of contaminants and some pitting of the bottom shell plates, particularly behind the stitch-welded stiffeners. Some pits had extended to at least 1/3 of the shell plate thickness and these plates have recently been renewed. An appropriate annual tank cleaning regime should effectively eliminate this problem.
- The third one is the pitting of cooling water overboard spigots; the combination of fast flowing hot salt water and a mix of materials across the pipework and valves inevitably leads to a finite life for these overboards. The jury is still out on the most effective prevention or cure, but meanwhile we carry spares and examine these overboards at each docking, replacing on condition about every five years or so.

The waterjet intake ducts are not included above because we have no significant problem there; good coating practice, accurate monitoring of the impressed current cathodic protection (ICCP) system, and touch up at docking has kept this area under good control.

Cracking has been left to the last as it is really not a problem on this ship. We have no, repeat no, cracking at or near the waterjets, only two equipment foundation cracks in ten years, and very minor, mostly non-structural cracks in the slamming affected areas up forward; not bad for a ship that operates on the open ocean most of her life!

THE FUTURE

HSC to ....

.... HSV to ....

.... JHSV

Ten years of operation of *WestPac Express*, and other commercially-built HSVs in US military service has well and truly proven the capability and versatility of these vessels in logistic support roles; this has been acknowledged and has led to the development of the JHSV project which is now coming to fruition with the launch of the first of class USNS *Spearhead* in September 2011 from Austal's shipyard in Mobile, Alabama.

The in-theatre support concept that we have successfully pioneered points the way to support of this fleet around the world. The adaption of efficient commercial build and support concepts to non-warfighting vessels in military fleets is now well established, and more can be learned from such ongoing operations; *WestPac Express* has maintained an operational readiness of over 99% of her required 350 days annual availability over the past ten years and is all set to continue this record well in to the future.

REFERENCES

1. Taylor, G, Black, J, "Austal High-Speed Ferry Joins the US Fleet!", Pacific 2004 IMC

Jim Black is currently Technical Manager, Service at Austal, Western Australia, having held several previous key positions there including project manager, quality manager and service manager; he is a naval architect and chartered professional engineer (BSc, CEng, FRINA, CPEng, MIEAust) who has now clocked up nearly 40 years' experience in the maritime industry. He has been Program Manager for the *WestPac Express* charter operation since its initial inception in 2001 and subsequently through several successful re-charters.

Ship Recoverability: How Important is the Human Factors Input?

Michelle R. Grech[1], David Drohan[2], Roberts Warren[1] and Grant I. Gamble[1]

[1]Maritime Platforms Division, Defence Science and Technology Organisation
[2]Navy Engineering Division, Department of Defence

michelle.grech@dsto.defence.gov.au
david.drohan@defence.gov.au
warren.roberts@dsto.defence.gov.au
grant.gamble@dsto.defence.gov.au

ABSTRACT
Today's naval shipboard technological advances include substantial improvements in navigation, communication, weapons and information systems together with improvements in ship survivability which is a critical functional area for the survival of a ship and crew safety. One aspect of ship survivability is recoverability which involves identifying and controlling damage to the ship, and restoring and maintaining degraded functionality. Although, advanced automation shows great promise for reducing the amount of crew required for ship recoverability operations, undue consideration and recognition of the human systems integration aspect could potentially lead to an increase in crew workload potentially leading to system failure. It is hence important that the design of future ships take into account both technological advancements and the application of human factors analyses to ensure that the full potential of automated and 'smart systems' can be successfully achieved. Simply automating systems may not be adequate for crew to accomplish their mission successfully; hence it is critical that the development of a system considers how the crew will be affected by the introduction of technology.

This paper will outline some of the work that DSTO and Navy are doing to address the human factors issues in the development of ship recoverability systems including escape and evacuation with the integration of various automated systems. This will also incorporate a description of some of the modelling tools being used to effectively simulate human performance during ship recoverability and escape and evacuation events under various conditions.

INTRODUCTION

Ship Recoverability

Recoverability involves identifying, managing, and suppressing damage sustained to a naval ship. Damage can be due to equipment failure, environmental factors (e.g. weather), accidents (collisions, groundings, engine-room fire, human error etc.) and weapons strike. Damage may involve multiple compartments, large areas of the ship and /or equipment. It can stem from internal sources, such as flammable liquid leaks (e.g. fuel leaks in the engine room) a weapons mishap, external sources such as mine strikes, missile hits, gunfire or groundings and collisions. Any of these situations can cause structural damage to the hull or superstructure, flooding, personnel injury, explosions and/or ensuing fire. When such situations arise, damage control (DC) equipment and personnel would be deployed to assess and report on the damage and be directed to undertake appropriate action. Although an infrequent occurrence, ship damage must be taken very seriously as it can compromise the safety of the crew and the ship, hence on naval ships there is a strong emphasis on DC training and procedures to mitigate and manage such risks.

DC training and procedures cover fire fighting, repairing of hull damage and piping, and/or shoring up watertight bulkheads. Most naval personnel are trained to perform some aspect of DC functions. In the event that the damage sustained cannot be repaired or the damage compromises the safety and survivability of the vessel beyond the recoverability stage then mustering, escape and evacuation is undertaken. The focus at this stage is on remaining afloat to enable the safe evacuation of crew and passengers (such as troops).

Human Systems Integration

Human-Systems Integration (HSI) is aimed at incorporating human factors and systems engineering tools and methods into the development and acquisition of systems and equipment to improve system performance and safety throughout the whole capability life cycle [1]. HSI places a great deal of emphasis on systems integration which includes the interfaces between the physical and human systems (both operational and maintenance personnel). HSI is divided into a framework incorporating a number of domains dealing with varying human factor and system engineering issues. The number of domains cited in the literature varies from a core of five, covering human factors engineering, crewing, personnel, training and system safety with additional domains, such as survivability, habitability, etc. being included to ensure focus on other specific areas of interest [1].

Human Systems Integration and Ship Recoverability

Records indicate that more than fifty percent of ships lost in the Falklands war were due to the inability of the crew to control the spread of damage [2]. In a more recent incident, in October 2000, the US Navy DDG 67, USS Cole was hit by a terrorist bomb while refuelling in the port of Aden in Yemen [3]. The ship sustained major structural damage, with 17 of its crew dead and several injured. In the aftermath of the event the response by the crew to damage sustained was described as '*less than adequate*' [3]. These are only a couple of examples from the multitude that indicate the critical requirement to include HSI within complex organisational structures and operations.

Runnerstrom (2003) indicates that the application of HSI to the development of naval systems in general and DC systems in particular is critical [4]. He argues the need for HSI by describing the poor state of current ship capabilities for dealing with abnormal situations and proposes HSI as a promising way to help improve ship capabilities while reducing crew levels. Advances in DC technologies and the increasing degree of automation show great promise for reducing the number of crew required for DC operations. However, not implementing an HSI approach to new acquisitions could potentially lead to, for example, an increase in crew workload which in turn could lead to system failure. It is hence important that HSI analyses are considered and applied to ensure that the full potential of any automated and/or 'smart' systems can be successfully integrated for optimum safety and capability of the platform.

IMPACT OF NEW TECHNOLOGY

Global Trends in Damage Control

A review of global trends in the area of ship recoverability and HSI carried out by DSTO highlighted the fact that most of the research focus is being dictated to some extent by future technological developments in areas such as remote monitoring systems, automatic control elements and new systems/software developments. Considerable research is being conducted in order to achieve better and more reliable onboard DC systems that would not only match current capabilities, but would actually enhance DC through automation. The general trend in navies around the world is to fit more automated DC systems onboard with a common goal to achieving further reductions in crew sizes [5-7]. Some of the more prominent research and technological developments such as monitoring sensor technologies, fire fighting and suppressant systems, smart valves, decision aids and support systems and DC communication networks show great potential for optimising crew numbers [8-12]. However, human factors research in 'out-of-the-loop' studies in domains such as aviation and process control, have demonstrated that this may lead to a negative impact if not designed with an understanding of the ways in which humans interact with these technologies. Hence, if not properly designed with a thorough HSI analysis, these systems have the potential to actually increase operator workload and human error [13]. A major challenge would be to ensure that automation employed in future DC systems capitalises on human capabilities and respects human limitations.

Ship Movement and Evacuation Modelling Tools

Ship movement and evacuation modelling tools are used to evaluate the behaviour and performance of personnel under emergency conditions. Computer models have been used to simulate the evacuation of people from buildings and aircraft in a variety of scenarios and have exposed a number of safety issues. Such models have enabled a more detailed understanding of the underlying processes involved in emergency, particularly escape and evacuation events, allowing designers, engineers and safety managers to put appropriate measures in place to reduce risks.

A number of maritime specific modelling tools have been developed which are able to model ship based movement and escape and evacuation [14]. Some of the more sophisticated tools are capable of representing populations as a collection of unique interacting individuals with the ability to represent the space in which the individuals interact (i.e. ship layout). Simulation outputs incorporate aspects such as time to assemble, levels of congestion, time to undertake specific tasks, distance travelled by individuals in achieving goals and the likely number of injuries and fatalities resulting from, for example, a fire. The more widely used and sophisticated ship specific models include AENEAS (developed by Germanischer Lloyd), EVI (developed by the University of Strathclyde), maritimeEXODUS (developed by the University of Greenwich) and ODIGO (developed by Bureau Veritas) [14].

The following sections will focus on human movement and evacuation simulation and will describe some of the research currently being undertaken in this area.

MODELLING HUMAN MOVEMENT ON SHIPS

MaritimeEXODUS Simulation Tool

DSTO and the Royal Australian Navy (RAN) Engineering Division have acquired maritimeEXODUS (*mE*) with the specific aim of evaluating and validating this tool with human performance data derived from RAN ships or purpose designed experiments utilising RAN training facilities. *mE* simulates the evacuation and circulation of large numbers of people within a variety of complex enclosures and has been used to evaluate escape and evacuation scenarios of competing ship designs [15]. Current movement times in *mE* are based on the results of experiments involving people moving along open spaces, through doors and hatches, and up and down stairs and ladders. These experiments provide a distribution of timings and *mE* selects timings from these distributions, resulting in a probabilistic outcome. At this stage *mE* seems to be the preferred option for modelling human movement on naval ships, with the UK also developing most of its human movement research and development modelling based on this tool.

Background on Human Movement Simulation Research

The UK Royal Navy (RN) has conducted a large body of work in the area of escape and evacuation with most of the work being contracted to the University of Greenwich. Initial work in this area looked at reviewing software modelling tools based on a number of criteria (e.g. human traffic flow; emergency lighting, effects of ship listing/motion, etc.). This led to human factor trials using RN training facilities and personnel to collect human performance data while participants undertook shipboard specific scenario type tasks at various heel angles (e.g. open watertight doors, etc.). The intention was to have a set of RN personnel data available that could later be used for modelling purposes within *mE* [16]. This trial was considered to be the start of an extensive program of work planned to capture a full data set representing the performance of naval personnel during damage control and escape and evacuation on RN ships. Many developments and modifications to incorporate the naval environment and some aspects of human performance have since been integrated into the original *mE* model [16].

This work instigated the RAN's Engineering Division to conduct human movement simulation studies in this area, also using *mE,* by initially comparing various simulated escape and evacuation times between civilian and naval personnel [17]. The ship layouts were imported into *mE* and nodes placed to indicate available floor space and transit areas. As shown in figure 1, the *mE* model is based on grids of free space nodes (an individual can place themselves on one node on the grids). Connections between nodes allow individuals to move between them. Each individual is modelled separately, together with some of their attributes (gender, age, etc) with their movement based on a set of rules. Twelve simulations were run for each of the four separate scenarios studied, for repeatability and reliability purposes. As hypothesised the simulated outputs confirmed that naval crew were faster in responding to the escape procedure and in reaching the mustering station than the civilian crew in both conditions (ship's structure with civilian and naval fittings).

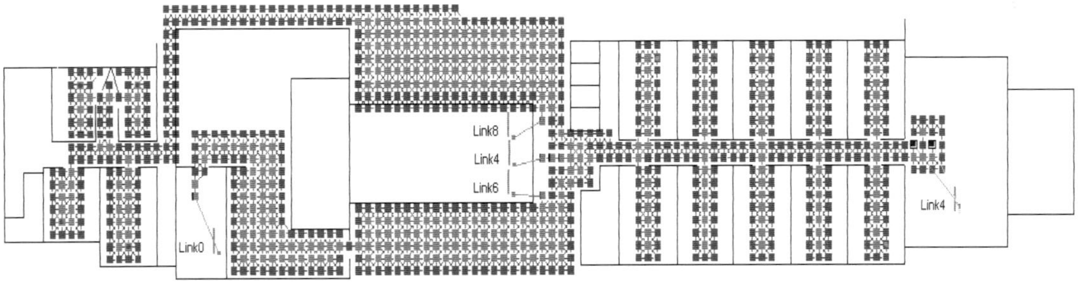

Figure 1: Part of the ship layout used in the RAN simulations using maritimeEXODUS showing upper deck only.

DSTO Human Movement Simulation Research

As indicated a program of work has been developed by DSTO aiming to develop a capability to evaluate ship recoverability operations onboard existing vessels as well as assess competing vessel designs and effects of vessel configurations on crew movement. In addition this work program enables the evaluation of the introduction of potentially new DC technologies [18] using human movement simulation and modelling tools, to examine their effectiveness in improving crew DC activities and procedures for the whole of ship. Furthermore the program also facilitates the development of relevant guidelines in the area and enables the analysis of the effectiveness of naval platforms to go from, for example, one 'state of readiness' to another.

An initial analysis conducted by DSTO included the use of experimental data from a study carried out by the National Research Laboratory (NRL) in the United States [19]. A comparison between the NRL experimental results and the DSTO simulated outputs using mE was undertaken. The same conditions as the experimental results were assigned to the simulation, which involved modelling personnel movement of a damage control team to move from a damage control base to a pipe damage location. The NRL collected data onboard the ex USS Shadwell in which timings were recorded for the team over a 20 m escape route (figure 2).

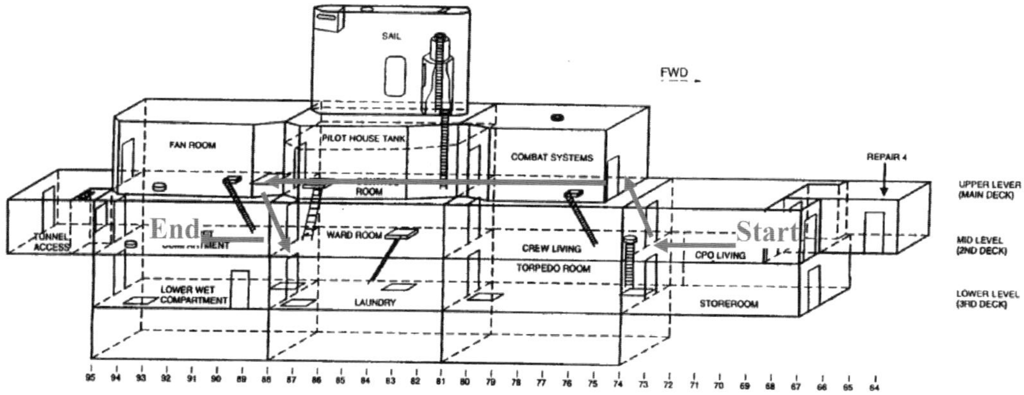

Figure 2: Damage control route through ex USS Shadwell showing start and end [19].

A model of the route layout of the ex USS Shadwell (figure 3) was built by DSTO using *mE* for the comparison between the simulated *mE* data and the experimental human movement data recorded during the trial.

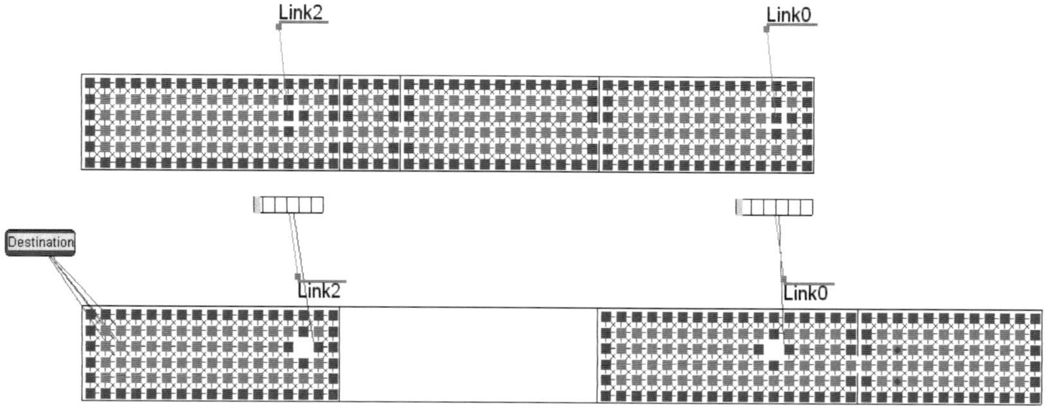

Figure 3: maritimeEXODUS model of the route layout.

The distribution of traverse times between the *mE* outputs and the recorded times on the ex USS Shadwell as shown in figure 4 indicate satisfactory agreement. Although some differences are evident (single peak for maritimeEXODUS versus double peak for NRL results) the findings for the timings fall within a comparable range. Overall the results demonstrate that this capability is useful for modelling personnel movement for simple damage control scenarios where various ship configurations, bottlenecks and the influence of new damage control technologies on the recoverability process can be assessed.

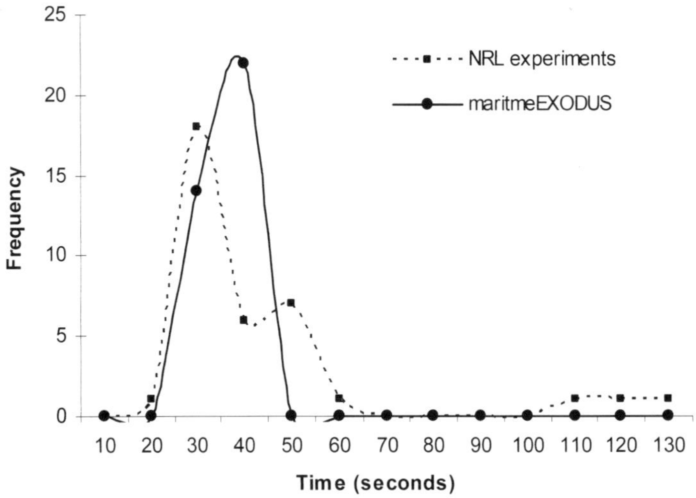

Figure 4: Distribution of timings for the recorded NRL experiment and *mE* outputs.

Limitations Identified for Naval Application

An important outcome from these initial studies is that it highlighted a number of limitations in the application of this simulation tool within a naval environment. Specifically the tool needs further development to realistically simulate DC operations as at the moment, its primarily focus and design, is on escape and evacuation events. This is especially important for complex movement and collection of specific tools and equipment (such as life jackets or DC tools) as part of DC activities. The simulation also currently does not replicate human decision behaviour and 'forces' crew to traverse the shortest possible route to muster, or are directed through a pre-determined route with the use of way-points. This was evident in the RAN Engineering Division study. This can be non-representative and unrealistic especially if the shortest route is damaged (or blocked) or fire and smoke is present within that route. The model also does not account for movement of personnel wearing protective equipment such as breathing apparatus, and for personnel carrying damage control equipment (damage control tools), or timber for leak stopping and shoring activities. In addition to these limitations, a critical shortfall is the lack of appropriate human performance and behaviour data representative of the population under study (RAN personnel). Furthermore, civilian scenarios used have been identified as being non representative of the responses expected from trained naval personnel, with current evacuation scenarios assuming an even vessel heel/trim during the event.

As a result DSTO aims to develop a capability for realistically modelling DC and escape and evacuation events on RAN ships. This entails the collection of RAN human performance data for input into *mE* to support the assessment of damage control and escape and evacuation on RAN platforms.

Pilot Study: Validation and Methodology Development

As part of this program of work a pilot study was conducted by DSTO in November 2011. This pilot study aimed to collect a set of human movement data which, as a first step, will be used to contribute to the verification and validation of the *mE* tool. This trial was also be used to test and refine the experimental methodology in preparation for future full scale ship trials.

This pilot trial was carried out at the RAN training facility HMAS Cerberus which is part of the RAN School of Survivability and Ship Safety. The facility incorporates fixed configurations of two ship structures (representative of a ship's upper deck area) designed for teaching students fire fighting techniques and flooding containment from ruptured compartments. Both configurations have identical layouts with one of the structure's being flat (0 degrees angle of heel) and the other angled at a permanent 7 degree heel. Data was collected while participants conducted identical shipboard evacuation runs (e.g. mustering, opening/closing watertight doors, collection of a life jackets etc) on both ship configurations.

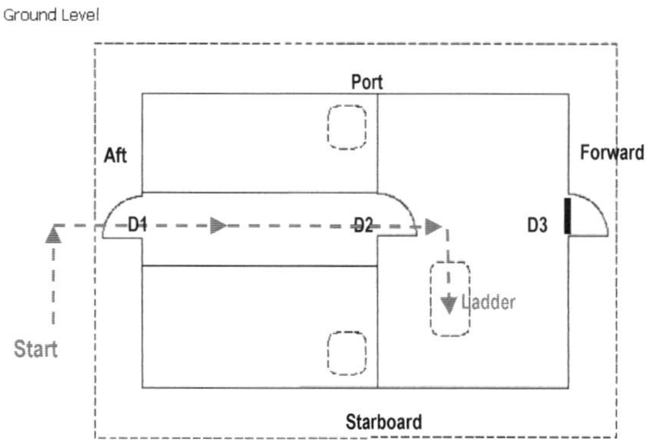

Figure 5a: Phase 1 ground level route.

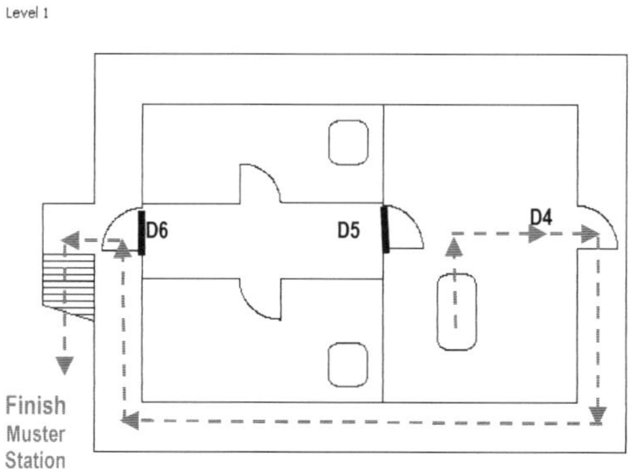

Figure 5b: Phase 1 level 1 route.

Figures 5a and 5b shows the structure layout plan that was used to run each experimental phase. The dotted line shows the designated route from start to finish (mustering station) that each participant was required to traverse. The run commenced from the ground level on the aft side, outside of the structure, designated as 'start' and shown in Figure 5a. Participants entered the structure from watertight door D1 and traversed through watertight door D2. Participants were required to go up the vertical ladder to level 1 of the structure. Figure 5b shows the level 1 plan of structure layout. Participants then exited the structure onto a platform through watertight door D4. Participants were required to go around the platform on level 1 towards the aft part of the structure and finally down the stairways. All runs ended at the mustering station which was clearly marked for all participants to see.

The data collected was mainly quantitative in nature using stopwatches, Infra-red timing tags (RFID Centre, UK) and video footage. Participants were requested to wear actiwatches and some were fitted with physiological monitors used to monitor a participant's change in movement. Participants were also be requested to complete a short questionnaire prior to the experiment to collect general demographic data.

The simulation tool *mE* was used to establish whether the simulation outputs match the data set collected during the pilot trial. This pilot study also included the influence of opening/closing water tight doors and the collection of life jackets with and without the influence of ship heel. Each phase was run through the simulation a number of times to ensure reliability and consistency in the results. Outcomes of this trial are currently being analysed.

CONCLUSION

As highlighted in this paper, it is evident that shipboard automation currently being developed is increasingly comprehensive, covering almost all DC tasks from detecting damage to determining and executing the best strategies for controlling that damage. Future RAN ships will probably incorporate a combination of some of these capabilities depending on cost and crew size, with some of the tasks traditionally carried out by crew may be made redundant through the use of such technologies. This means that decision aids, volume sensors and distributed systems will be used more extensively to detect, monitor and control damage. So although developments in DC technology show great potential for optimising crew numbers, simply automating onboard systems may not provide the crew with what is needed to accomplish their mission, and hence crew numbers may not be sufficient to successfully deal with ship recoverability requirements. Hence, it is critical that the development of a system considers how the crew will be affected by the introduction of new technology. For example, although increased automation may reduce the need for humans to carry out certain tasks, it may in turn require increased training of the crew to operate the automated system, and to interpret the data.

This paper has also provided an update on some of the work that DSTO and Navy are doing to address the human factors issues in the development of human movement and evacuation modelling tools. Such tools have the potential to inform how new DC technologies may potentially affect DC activities and processes and escape and evacuation events. DSTO is seeking to achieve a number of objectives in the development of this capability, however as indicated this tool needs to be extended so that it can provide a modelling environment for a range of crew evolutions carried out during ship recoverability under various conditions. This will need to include the transiting of crew for DC purposes carrying all kinds of DC equipment. Hence any new developments in DC technologies being carried out by DSTO, such as for example the pipe jumper, need to be accounted within the modelling and simulation tool. This will ensure that the whole of platform effectiveness is being evaluated with the piecemeal introduction of new DC technology.

An outcome from this research program will be to provide input into the development of a set of guidelines, DC policies and procedures for ship evolutions and escape and evacuation. This will enable the assessment of new acquisitions based on a number of evacuation and DC scenarios using validated modelling tools.

REFERENCES

1. Elischer, P. & Grech, M. R. (2008). *Human Systems Integration: How does it fit into warship design?* Paper presented at the Pacific 2008: International Maritime Conference, Sydney.
2. Department of Navy (1983) *Lessons of the Falklands*. 830915079, [Summary report] Washington D.C., U. S. Department of Navy.
3. DoD USS Cole Commission (2001) *USS Cole (DDG 67) Commission Report*. US Department of Defence.
4. Runnerstrom, E. (2003) Human Systems Integration and shipboard damage control, *Naval Engineers* 71-79.
5. Bulitko, V. and Wilkins, D. C. (1999) Automated instructor assistant for ship damage control. *Proceedings of the Eleventh Conference on Innovative Applications of Artificial Intelligence*, Orlando
6. Downs, R., Runnerstrom, E., Farley, J. P. and Williams, F. W. (2002) *Damage control automation for reduced manning (DC-ARM) supervisory control system software* NRL/MR/6180--02-8609, [Summary Final Report] Washington, DC, Naval Research Laboratory.
7. Kuzma, H., Gorton, J., O' Mara, J., Kelleher, P. & Rhoades, D. (2003) Advanced damage control system concepts. In: *13th International Ship Control Systems Symposium.*
8. Pehrsson-Rose, S. L., Minor, C. P., Steinhurst, D. A., Owrutsky, J. C., Lynch, J. A., Gottuk, D. T., et al. (2006) Volume sensor for damage assessment and situational awareness. *Fire Safety Journal* 41 301-310.

9. Darwin, R. L. and Williams, F. W. (2000) The development of water mist fire protection systems for U.S. navy ships. *Journal of American Society of Naval Engineers* 112 (49).
10. Hiltz, J. A. and Daniels, J. (2006) *Damage control technology: A literature review*. TM 2006-045, [Technical Memorandum] Halifax, Defence R&D Canada – Atlantic
11. Hagenaars, C. H. A. (2003) An overview of the advisory functions on RNLN's ADCF. In: *13th International Ship Control Systems Symposium.*
12. Hughes, J., Hinze, S., Roberts, M., Meyer, J., Farley, J. P., Pham, H. V. et al. (2010) *Advanced damage control (DC) communication* 6180/0057, [Final Technology Report] Washington, D.C., Naval Research Laboratory.
13. Bainbridge, L. (1983) The ironies of automation. *Automatica* 19 775-779.
14. Dongkon, L., Hongtae, K., Jin-Hyoung, P. and Beom-Jin, P. (2003) The current status and future issues in human evacuation from ships. *Safety Science* 41 (2003) 861-876.
15. Galea, E.R., Lawrence, P.J., Filippidis, L., Blackshields, D. and Cooney, D. (2008), *maritimeEXODUS V4.1 User Guide and Technical Manual*, Fire Safety Engineering Group, University of Greenwich, UK.
16. Jenkson, J. (2004) *RN Escape & Evacuation data collection trials: A detailed plan for conducting a set of human performance data collection field trials.* QAL/PI167/02/2.5/04/02-7, Surrey, Quintec Associates Limited.
17. Elliot T. (2011) Comparing the simulated escape times and procedures of civilian and navy crews using maritimeEXODUS. Report A017177, Department of Defence, Canberra.
18. Robinson, J. Gamble, I. G., Suendermann, B., Grech, M.R. and Roberts, W.S. (2012), *Damage control tools and technologies for Naval Platforms.* Paper to be presented at the Pacific 2012 conference.
19. Williams, F.W., Hill, S.A., Campbell, C. and Kuzma, H. (1997) *Results of Damage Control Tactics and Equipment Tests - Phase III: Cold Water Tests*, NRL/MR/6180-97-7919, Naval Research Laboratory, Washington DC.

Landing Craft - Choosing the Right Tool for the Job

Nick Johnson, Jeremy Atkins
BMT Defence Services, UK
NJohnson@bmtdsl.co.uk

ABSTRACT

Far from being 'simple ships' [Ref 1] the design of landing craft poses some unique and often taxing problems to naval architects and engineers. The increasing speed and safety aspirations for the next generation of landing craft further exacerbate these issues. This paper illustrates how small changes in the needs of the end user can have a significant effect on the design, leading to the need for radical changes, if cost effective and appropriate technologies are to be incorporated.

The paper identifies how the end user's requirement for offload of vehicles and troops by sea, dictates the cross-over point between the choice of a standard or more exotic hullform for a landing craft. The paper also discusses the impact on the design as a whole, in particular on propulsion equipment selection. The background to this investigation is a BMT concept design for a higher speed LCM.

Designers must ensure that they have the knowledge to properly understand the solution space. Without this they cannot be in a position to advise on how the best and most cost effective balance between desired capability, operating concept and vessel design may be achieved.

INTRODUCTION

Outline

Landing craft may seem relatively simple vehicles given their role, but a closer look at the environment in which they operate and the evolutions they perform shows that the designer has a number of complex challenges to overcome. These challenges are amplified given the design's sensitivity to requirement change. The paper explores the impact of this sensitivity in the context of the need to apply cost effective and appropriate technologies to landing craft design, with the objective of ensuring that the requirement sponsor understands how to ensure that they obtain the right tool for the job.

Context & Background

To provide context it is necessary to briefly describe the characteristics and role of the landing craft for which the arguments in this paper apply. The paper discusses issues associated with landing craft of the size normally labelled LCM (landing craft medium) or LCU (landing craft utility): approximately 25 to 30 metres in length with a loaded displacement ranging from around 120 to 240 tonnes. It is these craft that are often transported by and/or are interoperable with a large amphibious ship with a well dock such as an LPD (landing platform dock) or LHD (landing helicopter dock).

Typically, an LCM is used to transport vehicles and troops from the host ship to land and vice versa. Historically, the host ship has operated relatively close to the land in question (say 5 to 10 nautical miles). Many navies now seek to adopt a concept of operations that offers increased operational and tactical surprise whilst reducing the vulnerability of the host ship to inshore threats. This inevitably leads to a requirement to launch assaults using landing craft, from further offshore.

REQUIREMENTS AND CONSTRAINTS

The Need for Speed

When considering the procurement of a new landing craft to meet a new user requirement, or when developing options with existing landing craft to meet a change in the user's requirement, the designer must ensure that the basis of the performance requirement (for example payload capacity or speed) is properly understood.

Consider an example of an amphibious assault being conducted using a fixed number of landing craft within a fixed period of time. If a requirement exists to launch the assault from further offshore there is a direct requirement to either increase the speed or payload capacity of the landing craft. It is in this situation where a traditional landing craft design may no longer be appropriate. Determining the point at which this occurs and how this is influenced by the user requirements is an important part of ensuring that a cost effective solution can be obtained.

The relationship between speed, payload capacity and range for a single landing craft is summarised in Figure 1.

Total payload to be transferred:	M
Total period to transfer payload:	T
Payload transfer distance:	d
Payload capacity of craft:	m
Time to load and unload:	c
Number of transfers required:	$n = M/m$
Period for one return trip:	$t = T/n$
Mean speed for return trip:	$V = 2d/(t-c)$
Kedge distance:	d_k
Kedge speed:	v_k
Time to transit surf using kedge:	$t_k = d_k/v_k$
Maximum speed ratio (light/deep):	r
Maximum speed light:	$V_{lmax} = r.V_{dmax}$
Maximum speed deep:	$V_{dmax} = (d-d_k).(r+1)/(r.(t-c-2t_k))$

Figure 1 – Terminology / Speed and Range Relationship

The relationships in Figure 1 assume the following:

- The time to accelerate to maximum speed and also that to return to a standstill is negligible;
- The variance in turnaround time is small – i.e. term 'c' remains fairly constant across the range of payloads and landing craft designs;
- Other factors such as the kedge speed and distance are constant.

For an example total payload ('M') and transfer period ('T'), the relationship between speed and range is illustrated in Figure 2.

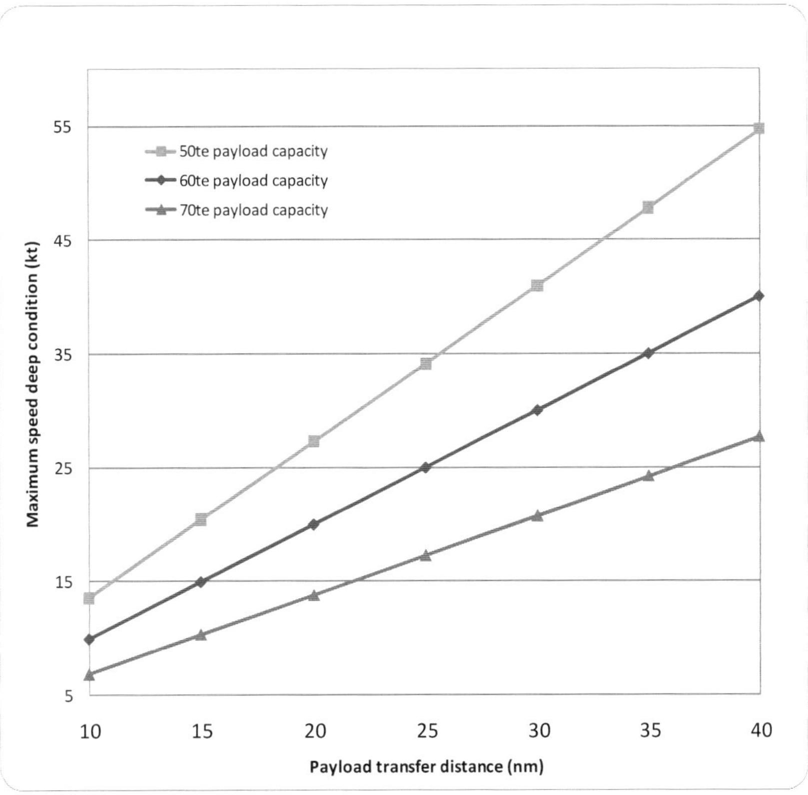

Figure 2 –Example Speed and Range Relationship

As one would expect, it is possible to see even from this crude example that the speed requirement is strongly influenced by both the distance through which the payload must be transferred and the payload capacity. Given the number of transfers required must be an integer ('n' in Figure 1), payload capacity should be tailored to ensure that the speed requirement is minimised. This is easier said than done as the influence of the propulsion system on the general arrangement is significant. Finding the right balance between payload capacity and speed can be challenging when considered in the context of the whole design.

Safety Issues

With regard to stability safety, some navies aspire to operate landing craft with levels of safety commensurate with other naval craft, or commercial craft of a similar nature. This brings into play alternative, more appropriate rules for damage stability than may have been applied in the past, for example, a requirement to withstand two compartment damage. When this is combined with other requirements associated with high speed operation (for example the requirement to survive bottom raking damage defined in the IMO High Speed Craft Code 2000) it will drive the arrangement of the internal subdivision and intensify the challenge for the designer.

The increasing of speed of operation requires careful consideration of the human factors aspects of the design. An example is the requirement for crew and passenger seating: provision of the latter in particular has the potential to severely constrain the payload capacity of the craft if not dealt with carefully in the early stages of the design.

Geometry Constraints

Landing craft will usually have a shorter service life than their host ships and not always will both the host ship and landing craft replacement programme coincide. Therefore, despite logic dictating that the host ships should be designed around the landing craft which they carry, invariably, the landing craft must be designed to fit existing well dock dimensions. A further requirement for interoperability across the fleet and amongst allies can effectively freeze the dimensional footprint of landing craft from one generation to the next.

Length and beam are constrained by the plan geometry of the host ship well dock. The draught and air draught constraints are more complicated and must include consideration of the relative motion of the landing craft and the host ship.

The constraints on landing craft geometry in the vertical plane are illustrated in Figure 3. The relative vertical motion between host ship and the landing craft will dictate the magnitude of the motion allowance above and below the landing craft ('c' and 'd'). These dimensions, in conjunction with the well dock height ('e') will define constraints on the draught and air-draught of the landing craft. Determining the relative motion is a highly complex problem, and it is likely that the designer will include some additional margin of safety to avoid problems which further constrains the landing craft dimensions.

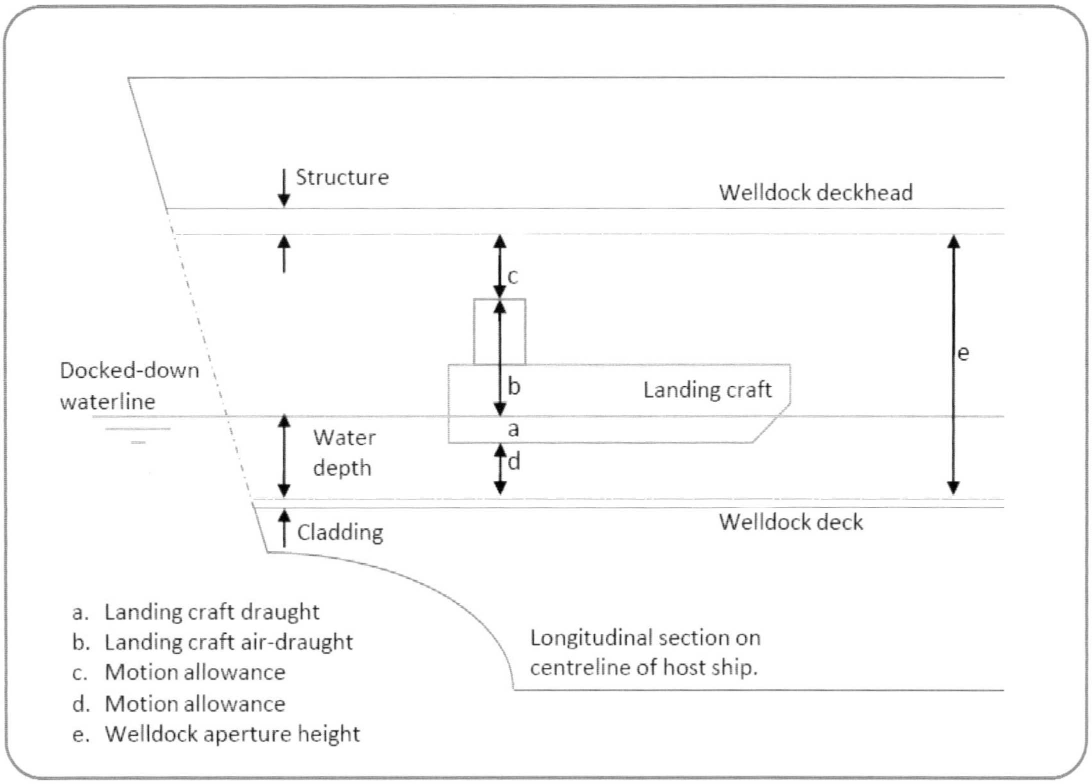

Figure 3 – Well Dock & Landing Craft Vertical Constraints

These dimensional constraints define the solution space within which the designer must strive to develop a design that meets the speed and payload requirements. In addition, mass may also be constrained by the structural properties of the well dock deck.

As the designer looks towards lower resistance and higher power solutions, the dimensional constraints will serve to hamper progress; for example the lower resistance hullforms with less waterplane area generally require a deeper draught ('a'); the internal payload area of the landing craft is influenced by not only external dimensions, but also the propulsion machinery arrangement (especially maintenance access, exhaust and air intake arrangements).

In summary, due to the constraints on landing craft geometry, the increasing speed requirement is likely to have a more significant impact on the design of the landing craft, than it would for other larger vessels.

THE IMPACT OF REQUIREMENT CHANGE ON SPEED

Hullform & Propulsion Variations

Figure 4 illustrates approximate resistance - speed relationships for three different landing craft designs in calm water. Each design can carry the same payload deadweight of around 60 tonnes; and each can operate in conjunction with the same generic host well dock equipped ship.

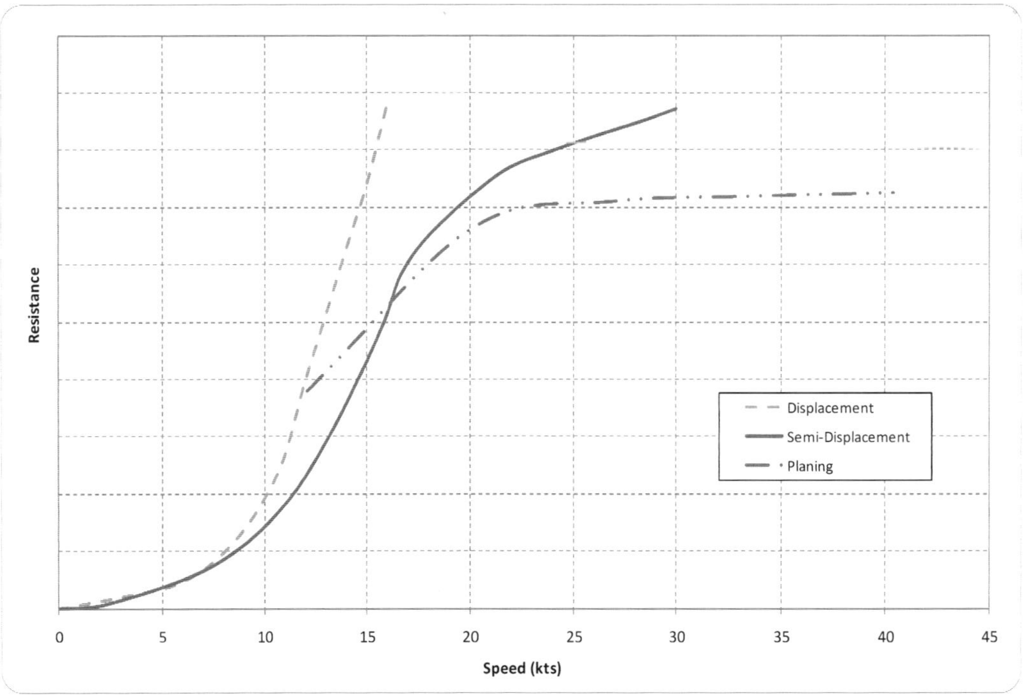

Figure 4 – Resistance/Speed Relationship for Alternative Landing Craft Hullforms

Each of the designs operates in a different regime: displacement, semi displacement and planing. The semi displacement form is represented here by a 25m long BMT LCM design that uses a derivative of the BMT Tri-Bow Monohull hullform [Ref 2]. The displacement and planing designs are representative examples of similar sized designs.

There are also differences in the power and propulsion systems for each of the designs. These differences ultimately manifest themselves as technical risk and cost. As one would expect, in broad terms, the faster you want to go, for the same capability, the likelihood is that you will pay more per knot.

Note that the curves presented in Figure 4 are for calm water. In a moderate sea state (e.g. Sea State 3 to 4) the speed loss due to waves for the planing form is likely to be significantly greater than for the semi-displacement BMT LCM design [Ref 2]. If assaults are launched from further offshore, the landing craft will become increasingly exposed to fully developed seas and it may be argued that in operational areas it will be subject to higher sea states for a greater proportion of its time at sea.

Assuming that the hullform and geometry remains constant, it is self-evident that as the speed requirement increases the thrust and installed power must also increase or the displacement must decrease. There will be a corresponding increase in both the initial procurement and through life cost. However, with no corresponding increase in the available weight or space budgets, the designer must be increasingly creative with how they install the required power and deliver the thrust.

Taking the case of the conventional displacement hullform, the power curve presented in Figure 4 demonstrates that the additional power required to achieve each additional knot of speed increases exponentially. In response to this, for small increases in the speed requirement, the designer may initially simply choose to specify larger engines and propulsors. However, there comes a point where larger engines and propulsors can no longer be accommodated in the existing hullform. This may be due to clear deck height in the machinery space or draught limiting the propulsors dimensions for example. The designer then may consider specifying a greater number of smaller engines and propulsors. There is potentially now a step change in cost as an additional entire drive train has to be procured and maintained through life. However, each case should be considered individually as there are some exceptions to the rule associated with specific engine size and manufacturer selection.

Once this avenue is exhausted the designer is left with looking for greater power density and an efficient means of delivering that power as thrust. Greater power density is likely to mean specialist engines from a high speed engine supplier. These are likely to be extremely costly as they are manufactured in small production runs and built using exotic materials to achieve the high power density. So not only does the power requirement increase exponentially, so does the cost associated with integrating that power into the craft.

It is also necessary to take into account the weight that the additional power will add to the craft; this not only exacerbates the power problem further but also will force the designer to consider reducing either payload deadweight, fuel or lightship weight in order to maintain the same loaded displacement. The impact of adjusting payload capacity on the speed requirement is demonstrated earlier in Figure 1 and Figure 2.

It is not only propulsion costs that will increase with an increasing speed requirement; other areas include the structural design. The structure will need to withstand increased loading, particularly at the forward end of the hull form and the ramp. Also, in response to the increasing mass of the propulsion system, the designer may be forced to use aluminum rather than steel as the main structural material, further increasing build and through life cost and reducing the resilience of the design.

Cost - Performance Balance

The problem obeys the law of diminishing returns. There comes a point where the investment required to achieve a little more speed, exceeds the return in performance. Figure 5 illustrates this for the conventional displacement hullform. The curve is theoretical and is provided solely as an illustration of the problem.

The curve is of design performance plotted against cost and shows that once a certain performance level is achieved, further expenditure will only return a very limited increase in capability. In terms of speed there is ultimately a physical limit on how much performance can be achieved from a certain type of hullform. In the context of the discussion regarding speed, this physical limit may be defined by the balance of installed power and payload capacity, the ability to control the craft at higher speeds or a point where the safety and comfort of the crew and passengers is jeopardised due to the seakeeping response of the landing craft.

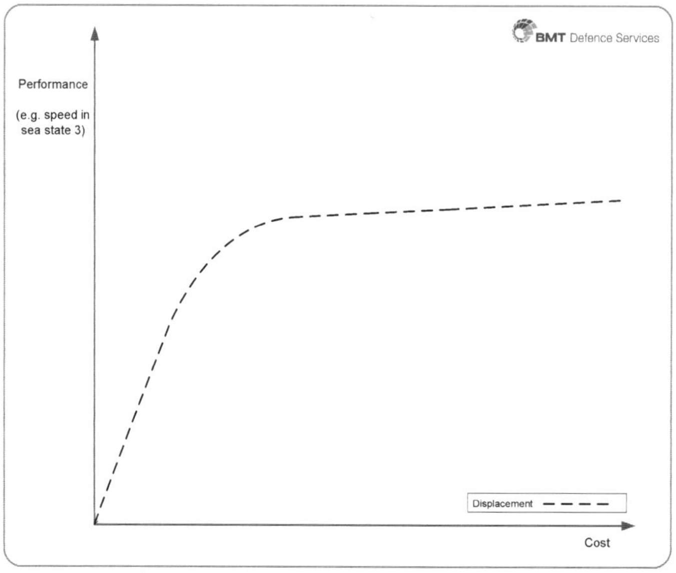

Figure 5 – Cost / Performance Curve for Displacement Hullform

Figure 6 includes the curves for the planing hullform and semi-displacement hullform. Each hullform has a unique cost/capability curve dependent on its relative effectiveness over the range of performance. The planing form curve increases at a shallower angle than the displacement form indicating a greater cost is required to achieve the same capability at the lower levels. This is due to the higher costs associated with the design and build of a more complex planing craft (for example: increased build accuracy, increased material and equipment cost etc). However, there comes a point in the level of investment where the two curves cross. Beyond this point the landing craft owner will begin to reap the benefit of the higher investment as greater capability can be achieved for lower cost. Identifying where this point is in terms of speed and then range is important for the designer.

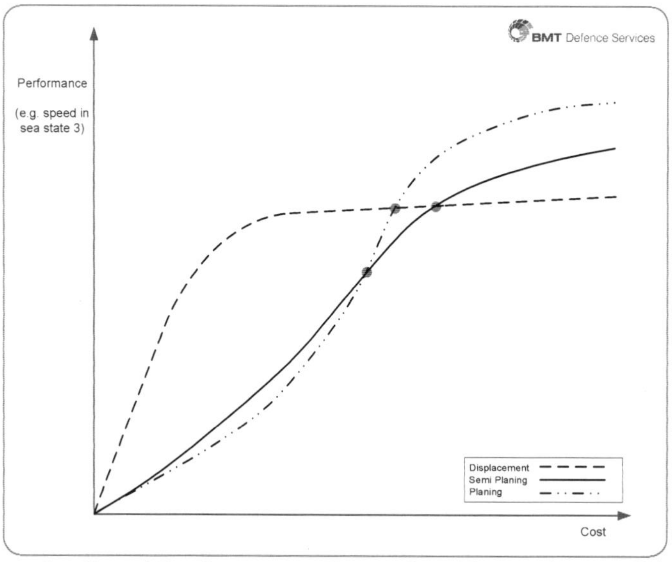

Figure 6 - Cost / Performance Curve for Alternative Hullforms

The curve for the semi-displacement hullform is a slightly different shape again. The curve rises initially at a slightly shallower angle than the planing hullform reflecting the possible additional build cost of the complex shape of the BMT Tri-Bow design. The curve then rises more rapidly to cross both the planing curve and the displacement curve before tailing off as it reaches its point of diminishing returns. This reflects the efficiency of the hullform – i.e. reducing the propulsion system requirements (in comparison to the planing form for increase speed in a moderate sea state).

The three intersections of the curves are the balance points between the various designs and dictate the cross-over point where a more exotic hullform may offer a more cost effective solution to the problem than a more traditional form, or where the converse is true.

Figure 7 illustrates an example situation where a designer may have selected a traditional displacement form for a landing craft designed to meet a particular speed requirement (the design point illustrated by the red cross on right hand side of figure). Further parametric investigation by the designer will show that a more efficient, less costly solution may exist via adoption of a more exotic hullform – something that may have been dismissed as too costly at the outset.

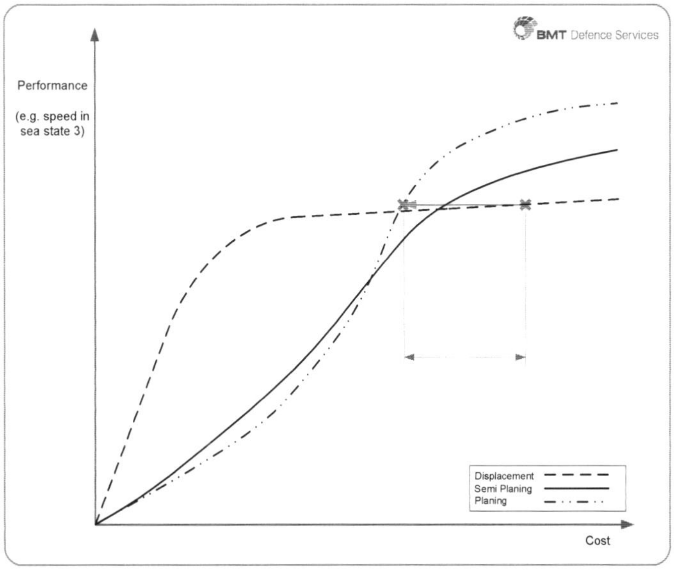

Figure 7 - Cost / Performance Curve – Example of Inefficient Solution

Figure 8 illustrates a second example situation where given the speed requirement the designer has been forced to move away from a displacement hullform and adopt a planing form for the design. However, it is clear from the plot that if the capability requirement could be reduced very slightly, the design could return to the displacement hullform option and the subsequent cost savings could be very significant.

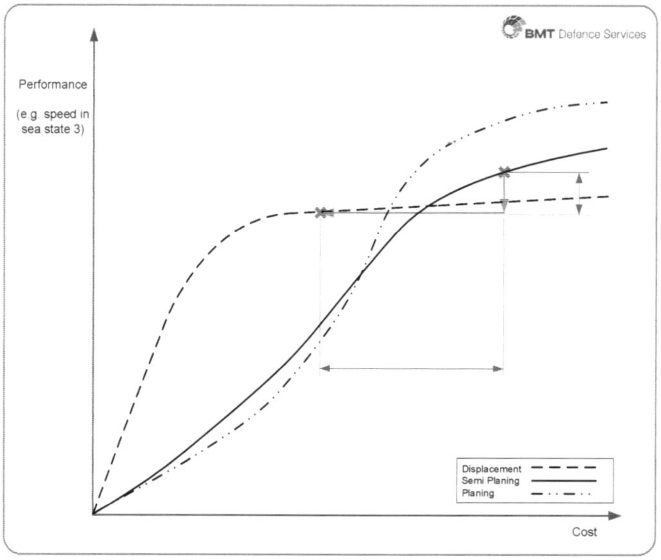

Figure 8 - Cost / Performance Curve – Example of Cost Reduction

The designer must identify this relationship at the outset of the design process and then advise the capability sponsor accordingly to ensure that opportunities to enhance performance for small additional cost, or make significant cost savings for small reductions in performance, can be exploited.

CONCLUSION

There is little margin for landing craft designs to absorb minor changes in requirement and still remain cost effective because of the broad range of constraints that govern the design of the landing craft. The sensitivity of this balance is arguably more pronounced in landing craft than in other marine vehicles.

This paper argues that it is necessary for designers to communicate the effect of requirement change on the cost and capability of the subsequent design at early stages and identify opportunity for achieving savings and enhancements. With the designer striving to properly understand the often complex problem of landing craft design, the user may well end up with the right tool for the job.

REFERENCES

1. Mackereth I, Noel-Johnson N, (2008), "Landing Craft - Simple Ships or Time for a Rethink? A Designer's Perspective", MAST 2008

2. Sime R, Dudson E, Amaratunga S, (2010), "Thinking Inside the Box – Development of a Monohull Fast Landing Craft", RINA Warship 2010 – Advanced Technologies in Naval Design and Construction

Applying Commercial Technologies to Address Electronic Obsolescence

Norman Dong, Adam Waldie
Thales Underwater Systems, Sydney, NSW, Australia
norman.dong@thalesgroup.com.au

ABSTRACT

Electronic systems are used extensively throughout a modern naval platform. In many cases these form part of a mission critical Combat Management System (CMS) or related sensor or weapon system. The underlying electronic hardware technology is evolving rapidly and this frequently leads to obsolescence problems emerging through the life of the system. The recent emergence of high performance Commercial-Off-The-Shelf (COTS) based technologies is a potentially attractive way to deal with these challenges.

In this paper we explore some of the factors to be considered and strategies to be adopted in applying commercial technologies to address obsolescence in military electronic systems.

Factors to be considered include:

- Compatibility
- Product stability
- Supply chain risks.

Strategies include:

- Learning from prior experience, both inside and outside the company
- Architectural design that facilitates component change
- Identifying and understanding the benefits of applying an open innovation model approach.

Examples are presented that illustrate these factors and application of the strategies to achieve through life cost saving and capability enhancement through exploitation of COTS electronics in military systems.

INTRODUCTION

What can be done to address obsolescence when it appears (reactive obsolescence management) or preferably before it appears (pro-active obsolescence management)? Obsolescence becomes an issue due to the pace of change, particularly for commercial components. (Cf. Moore's law)

Electronic obsolescence is an inevitable issue for all electronic systems to contend with. Within the defence and military domain this issue is exaggerated as the majority of these systems rely on equipment being in service for many years (decades) {refer Figure 1 – [9]} whilst also contending with the ever decreasing life cycles for the electronic components that are incorporated into these systems {refer Figure 2 – [10]}.

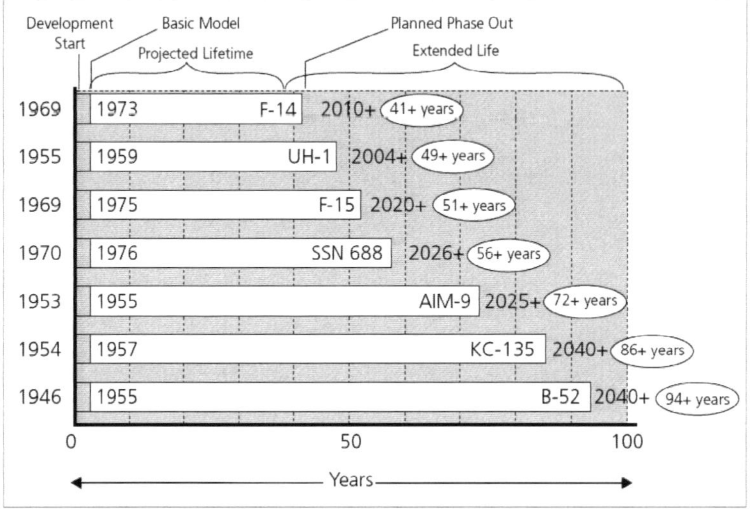

Figure 1 - Life Cycle of Aircraft and Weapon Systems

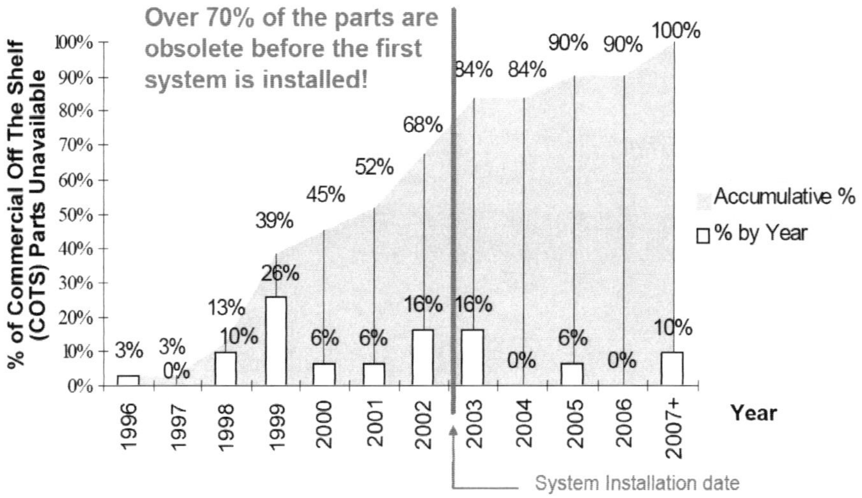

Figure 2 - Percentage of Commercial Off The Shelf (COTS) parts that are out of production (un-procurable) versus the first 10 years of a surface ship sonar system's lifecycle. (Courtesy of NAVSURFWARCENDIV Crane)

To retain key capability provided by the system, organisations need to consider how to keep the equipment operational. Our approach is to use commercial technologies to address electronic obsolescence.

For example, the US ARCI (Acoustic Rapid COTS Insertion) and APB (Advanced Processing Builds) programs were instigated for submarine systems to enable the rapid replacement of existing submarine sonar systems with COTS hardware. This was done independently of the software development to reduce costs with developing new algorithms. The results indicated a dramatic increase in the submarine sonar processing capability along with a significant decrease in the procurement costs. {refer Figure 3 - [8]}

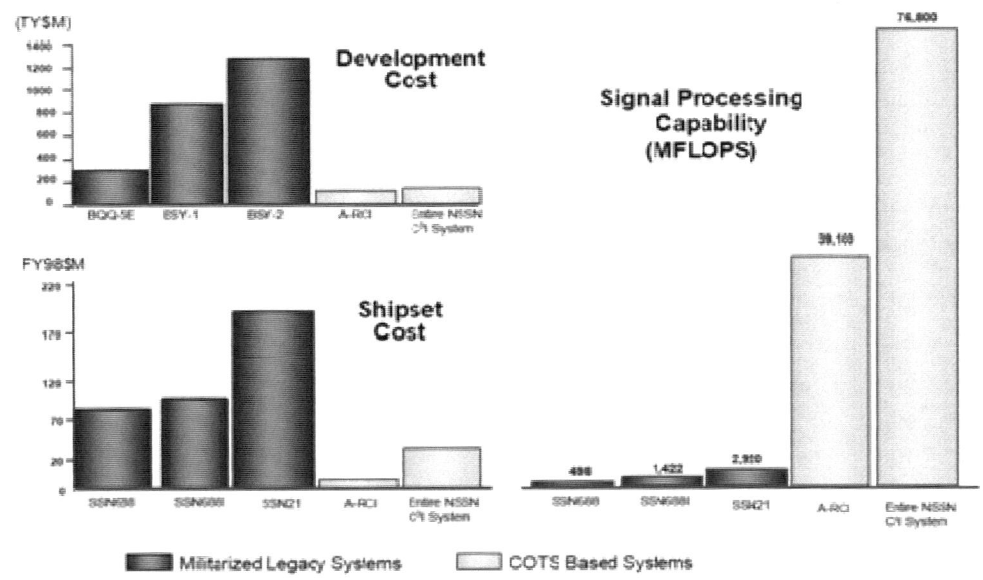

Figure 3 - Game changer - ARCI Influence

BENEFITS OF COMMERCIAL TECHNOLOGIES

Commercial technologies offer an attractive option for companies when developing systems. For example, the data presented in Figure 4 [5] indicates a significant cost saving over in-house developed technologies through the adoption of open innovation concepts. To build systems from these existing technologies provides a flexible option to the designer and builder to assist with meeting the desired outcome. By considering technologies which are already proven in various markets, the immediate challenge of realising the system can be addressed with a significantly reduced time to design. These components have been utilised in other systems bringing with it a sense of stability as well as a market willing to sustain its continued production by expanding the user base of the component product(s) and adding to its commercial viability.

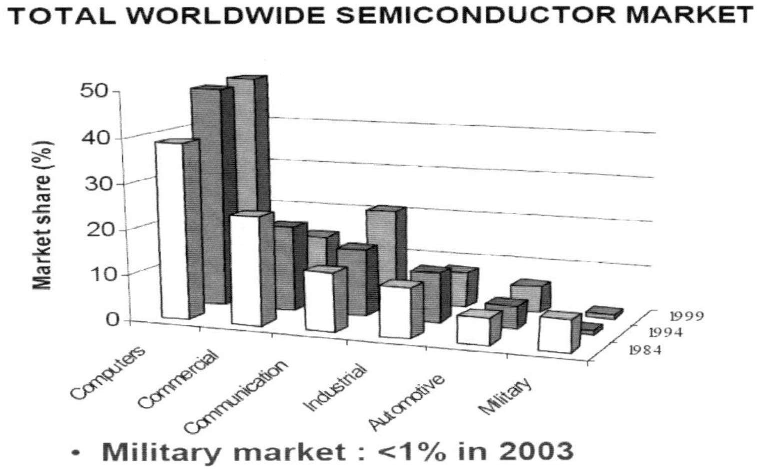

Figure 4 - Market place trends for Semiconductor Influence

The problem of keeping a system functioning where obsolescence of electronic components is forecast is a good place to explore the following question. Can the advantages of applying commercial technologies be leveraged to help address the electronic obsolescence of systems in service?

On the surface, the application of commercial technologies appears to provide many benefits. There are also many other factors that need to be considered when dealing with in-service equipment, particularly in the military domain. For example:

- Safety - RoHS (Restriction of Hazardous Substances) and LSZH (Low Smoke Zero Halogen) compliance
- Operational technical regulations
- Compatibility – functional and environmental
- Component end of life forecasts
- Supplier Company reputation (counterfeiting) – how long will it be around, is there an equivalent product from other suppliers?
- What is the field rating of the components, where else are they used and is it used in a similar situation as the one currently intended?

STRATEGIES

Buying Life-of-Type supplies as spares to enable through life maintenance of the system is a traditional approach to obsolescence issues. However, this strategy has some shortcomings that detract from its intended purpose, including:

- Significant upfront expenditure for stock-piling of the spares
- Determining which types of spares to stock-pile and how many to purchase
- On-going costs of storage facility space and associated management
- Prolonged storage leading to deterioration of equipment (e.g. Equipment needing batteries)
- Capability stagnation, only maintaining the status quo.

Other obsolescence mitigation strategies could include:

- Reclaim – salvaging of parts from other equipment
- Component redesign
- Part substitution.

Each of these will also have its unique problems, similar to the ones listed previously for the designer to address. Parts substitution is the approach considered in this paper for treating obsolete sub-systems through the application of commercial technologies.

PARTS SUBSTITUTION

Introduction

Abstracting the design into larger functional components based on an open architecture inherently allows components to be replaceable. An open system with published interfaces provides a means of addressing obsolescence through the flexibility of the design. Similar to the ARCI concepts, separation of the hardware and software layers permits easier management of maintaining as well as evolving the system.

Collins Submarine Sonar

The Scylla sonar system developed for the Collins class submarine was originally conceived in the 1980s and the design was based on custom built electronics utilising discrete COTS components (i.e. microprocessors, micro controllers and other ICs populating custom board layouts with custom bus interconnects). An example is the Thales Mangouste processor which used a proprietary coding language to perform calculations on custom designed hardware.

As one would expect, the Scylla sonar is now due for a technical refresh which is best achieved through parts substitution. Our preferred approach is to replace functionality by using generic processing machines (COTS processors) and applying open programming languages to implement the same algorithms as the existing system. The algorithm is the definer of the output, how we get there should not be the constraining parameter. This same analogy should apply to systems where the functional specification defines the system, not the implementation. The implementation is an instantiation of the design. Other implementations are possible as long as the qualification leads to the successful demonstration of the functional requirements.

Sustainment versus Capability

These aspects are inter-twined, with the solutions to treat obsolescence providing an opportunity to enable a financially beneficial outcome. The effort required to sustain capability and prepare for upgrade or include enhancement is effectively the same. However, by removing the need to separately manage both a sustainment group and a capability group leads to a more efficient use of resources. This approach can keep costs down while retaining the same or better capability through the enabling of freer thinking thereby allowing innovative solutions to be developed.

The preferred approach is to leverage open innovation partners and third party providers of sub-systems to provide a view of what is available and what will be available as potential solution enablers. It is also important to utilise university and academia to develop niche technologies to help develop the new and innovative systems. These systems or the process of developing the systems can be used to assist with the treatment of obsolescence.

The utilisation of products and technologies within a global company's world-wide product portfolio can also help to create viable solutions through product evolution by applying open innovation concepts. Solutions which started out as fully proprietary have moved to pseudo-COTS solutions and then onto full COTS component solutions. There are still some areas that require tailored equipment (e.g. sonar sensors); however, these are minimised effectively to the interfaces of the proprietary devices which themselves utilise standard (COTS) interfaces.

While addressing obsolescence issues, it is sensible to review the functional capability to enable the customer to retain operational capability advantage in the face of an ever evolving threat.

Treating obsolescence using technologies that are generic can mean that systems are capable of more than the original intended output. This provides an enhanced capability which is available for the user. In the software domain, versioning offers enhanced features that are accessible through use of keys (serial numbers etc.) already existing on the same CD/DVD distribution. A similar approach for hardware features utilises hardware dongles to release equipment capability. In both cases, the same core product (software or hardware) is used to provide the different levels of functionality.

The model of providing only software and separating the hardware is a well-known approach used by the PC world. The user can choose variations with operating systems, types of hardware components and software applications. A system integrator can be used to work on behalf of the customer to enable the custom build of equipment to suit their needs. These concepts have flowed into the military domain as seen with the ARCI and APB projects.

This type of approach can allow for the minimisation of vendor or product lock-in if implemented with this goal in mind. Evolution is performed in a measured manner (cf. ARCI, APB and Technology Insertion; going from Intel i860 to PowerPC, PowerPC + Sharc, PowerPC + Altivec through to Intel architecture; utilising Intel performance libraries {SSE, SSE2 etc.} providing a flexibility in design and modular abstraction). This model can provide an economic approach to enhance the system with inclusions of upgrades and new features to improve the existing capabilities.

Collins Submarine Sonar Revisited

Treating obsolescence in an aging system that has been in service since 1996 (first of class commission year) and initially developed in the early 80s is a challenge. Allowing obsolescence treatment to address the capability and not just the components permits a sensible approach in treating these aging systems. By permitting the original OEM developed lowest replaceable units (LRU) to be thought of as a conglomeration of LRUs that can then be replaced by another conglomeration of COTS equipment is an initial step. The ability to look beyond simple component replacement allows a more flexible solution space to be explored by the system integrators as well as the system owners. These concepts open up the options of maintaining functionality by enabling enhanced capability.

Inclusion of new algorithms to increase capability and at the same time address obsolescence is an approach demonstrated by Thales Australia with the submarine Flank Array Adaptive beamformer processor. This approach re-uses the equipment cabinet envelope and its associated hotel services as a means of pro-actively demonstrating a method of addressing obsolescence through the insertion of commercial compact-PCI processors and 19 inch rack equipment. Minimising the footprint and impact on external systems enabled a rapid implementation and installation approach. Utilising the standards associated with COTS equipment provides defined and consistent interfaces from both electronic and non-electronic aspects. The standard interfaces assist in providing an open systems approach to the design and realisation of the system.

In terms of cost benefits, an estimated 17% saving per year has been calculated based on the application of COTS equipment to treat the obsolescence and sustainment of this system.

A similar approach has been demonstrated with the Spherion B Sonar on the ANZAC ships. In conjunction with DSTO, Thales has addressed the inboard processing obsolescence utilising commercial technologies whilst also providing advanced capabilities.

OTHER APPROACHES TO OBSOLESCENCE

Virtualisation

The advent of virtualisation has brought another method of prolonging system life spans. The abstraction of hardware from the software enables possible leverage of the virtualisation technologies. Keeping the same software and replacing the underlying hardware without losing capability can be used as a cost effective way to treat obsolescence. However, the issues regarding performance and the reliance of the virtual machine to correctly emulate the original equipment needs consideration, particularly the cost to validate. Trust of the virtual environment entails not only the supporting virtual machine, but also the compatibility with the chosen hardware. Emulation on one type of hardware may not provide the same results on another hardware platform.

Virtualisation has proven to be a very successful strategy for maintaining test equipment that is no longer supported. The virtualisation of obsolete hardware has enabled continued use of legacy test software in support of the aging systems.

Software

Transformation of data into the digital domain and applying signal processing algorithms implemented by software replaces traditional discrete hardware components. Flexibility inherent in software allows the same system to be reconfigured to perform different functions. This changes the focus for system evolution moving from hardware to software. This is balanced by the cost of developing the software components required to replace the existing hardware. However, this approach has the promise of treatment of obsolescence by avoiding it. E.g. Software defined radio (SDR) – the use of software to replace traditionally hardware implemented functions such as filters and mixers. This approach is used in the US military Joint Tactical Radio System (JTRS) program.

FPGA (Field Programmable Gate Arrays)

SOC – System On Chip, this is a hardware approach to address obsolescence. FPGAs are used to enable unique interfacing to legacy systems for building specific solutions to enable the longevity of existing equipment. This programmable hardware has been successfully used in a number of Collins sonar sub-systems including evolutions of the legacy equipment.

These strategies address functional obsolescence with technologies that enable the technical obsolescence to be handled at the same time. Both of these approaches together enable a flexible mechanism that seeks to eliminate the adverse impact of obsolescence.

PARALLEL FUNCTIONALITY

Figure 5 ([7]) provides a generic overview of the reliability of a typical electronic system over its life cycle. Introduced sub-systems used to address obsolescence are compared to existing in-service equipment which is in a later stage of deployment within the bathtub curve.

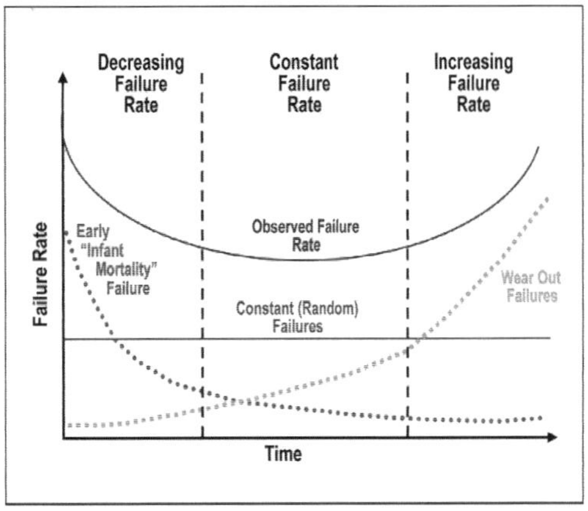

Figure 5 – Reliability "Bathtub" Curve

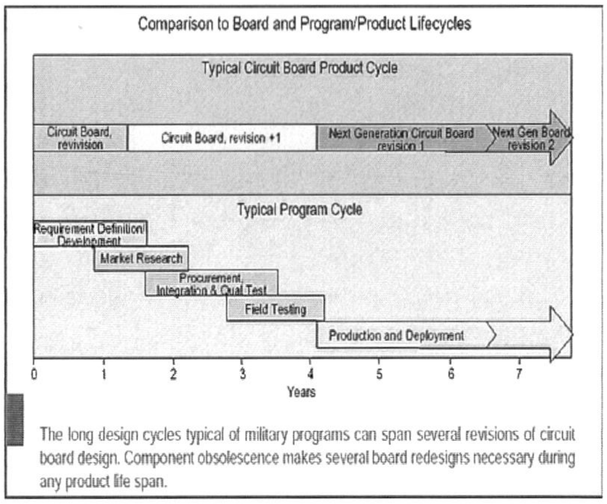

Figure 6 - Circuit Board Revision Vs. Product Lifespan

The insertion of COTS to replace obsolescent equipment must contend with the "infant mortality" inherent with newly introduced equipment. The loss of capability due to the failure rate associated with new equipment is a real possibility, which may well result in the perception that the treatment was a failure.

An alternative approach is to install the new COTS equipment while still retaining the obsolescent system. This allows both systems to operate in parallel and obviates the issue of losing functionality should the COTS insertion fail.

This risk management approach ensures that the capability is always available using the existing system as the baseline. Testing can be done without loss of capability to ensure equivalence before making the replacement permanent. When obsolescence occurs there is a proven alternative. This is not a just-in-time style approach, but a redundancy approach preparing a "hot" spare to be available whilst also enabling the replacement equipment to be "bedded in".

Qualification of the alternate implementation can be done over a period of time while the existing system is still operational. There is less reliance on a single set of trials to validate the system and ensures that the desired capability is maintained. The parallel operation provides the evidence required to show that the implementation is performing correctly.

This approach ties in with deployment of systems, particularly in military programs. Figure 6 illustrates the life-cycle of typical circuit boards and the need to utilise a number of revisions during the lifespan of the system. By allowing systems to be tested in parallel, updates and revisions of components can be inserted and proven to minimise risk exposure to fielding a non-compliant system.

SECONDARY CONSIDERATIONS

Secondary systems such as the support equipment used to maintain the main system, e.g. ATE (Automatic Test Equipment) need to be considered as part of the obsolescence treatment. The ability to keep a system in operation depends on reliable test equipment to help diagnose faults and determine corrective actions. This process will aid in the maintenance of the main system and enable obsolescence problems to be delayed by enabling the existing system to remain in service longer. Treating the obsolescence of existing ATEs provides an extension to the in-service period of the equipment. The updated additions to systems that treat the obsolescence concerns also require a reliable mechanism to prolong its service through identification of COTS ATEs.

There are other non-electronic issues which need to be considered when discussing obsolescence of systems. This includes the impact of changes required to treat the immediate obsolescence of local sub-systems on the remainder of the existing system (e.g. power, cooling and space). The support infrastructure to the electronics is an important consideration which encompasses not only the functional aspects but also the economic and service impacts. Reusing the power and cooling services as well as maintaining the installation envelope enables a modular approach to introducing new subsystems to treat the obsolescence issue.

Changes to individual sub-systems are not isolated, they inevitably affect other systems. Amongst other things, this impacts the documentation and subsequently the training (or re-training) of maintenance staff. Keeping the same supporting infrastructure in place and designing the system to be able to utilise the existing support equipment reduces the initial costs to allow a scheduled and consistent economic approach; i.e. keeping the form and fit of equipment within the same equipment envelope provides a means of isolating the impact of the changes. This modular approach combines well with the open architecture concepts to enable the controlled addition or replacement of sub-systems. Minimising the changes required to be made and maximising its effect is a desired outcome for an elegant solution to treating obsolescence from both a cost and technical view point.

CONCLUSION

Commercial technologies provide a viable means of treating electronic obsolescence through parts substitution.

From the sustainment point-of-view, there is a need to keep the system operational to enable the tools to be available for the end user. The capability point-of-view has the need to provide the tools for the end user to perform the job. The main goal for the user is to keep systems that are providing the required capability operational for as long as needed. For a supplier and/or designer, the goal is to deliver a system to provide the required capability and provide the through-life support of the system.

Sustainment and capability can both be addressed using a common approach which provides shared technical outcomes and positive cost benefits.

Knowing the functional requirements and understanding the operational needs allows systems to be considered in a modular manner. By seeing beyond the immediate and embracing the ability to evolve will permit more flexible solutions to be considered for the treatment of obsolescence.

System wide impacts are key issues to resolve in determining the suitability of obsolescence treatment. Reducing the impact on other sub-systems helps to contain the amount of re-testing required. By using a hot spare approach for the replacing of subsystems when utilising COTS equipment enables management of risks in a proactive manner and ensures capability is not lost during the system's qualification period.

The application of commercial technologies can bring with it a wealth of possibilities. It is sensible to consider these available features in providing innovative solutions to address electronic obsolescence issues.

REFERENCES (Bibliography):

1 Avnet Marketing whitepaper (n.d.) "*Authorized Component Distributors Help Military/Aerospace OEMs Plan for Obsolescence*"
Accessed from http://www.em.avnet.com/ctf_shared/pgw/df2df2usa/avnet-PGW-hirel-bestdefense.pdf

2 Brandt, J. (2004), "*Extension of Life Companies: An Obsolescence Solution*"
Accessed from http://www.cotsjournalonline.com/articles/view/100035

3 Buratti, M. and Del Brusco, D. (2000) "*The Obsolescence Management Based on a 'Pro-Active' Approach in Conjunction with a 'Pre-Planned' Technology Insertion Route*"

4 Chesbrough, H. (2003), "*Open Innovation: The New Imperative for Creating and Profiting from Technology*", Harvard Business School Press.

5 Cretenet, A. (14/9/2005), "*Obsolescence of Electronics - Potential Impact on Subsea Production Systems Availability*"
Accessed from: http://events.sut.org.uk/past_events/2005/050914/AlainCretenet.pdf

6 Howard, M.A. (2002) "*Component Obsolescence - It's Not Just for Electronics Anymore*"

7 O'Dowd, R.J. (2010), "*A Survey of Electronics Obsolescence and Reliability*", DSTO-TR-2437.

8 Schuster, J. (n.d.), "*Recent Progress in Submarine Sonar ... and Future Challenges in ASW*"
Accessed from: http://acoustics.mit.edu/dyerparty/Ira Dyer talk rev 1.pdf

9 Shekar, N.K. (2008), "*Obsolescence Management - The need for a proactive, concurrent approach*"

10 Singh, P. and Sandborn, P. (2006), "*Obsolescence Driven Design Refresh Planning For Sustainment Dominated Systems*"

11 Dr Zarnich, R. (21/2/2006), "*ASW ENTERPRISE OPEN ARCHITECTURE Briefing*"
Accessed from : https://acc.dau.mil/adl/en-US/44946/file/13047/Presentation_7_ASW OA Overview.pdf

Emerging Fibre Optic Sensor Technology For Maritime Security

Paul Baker, Tim Cain
Thales Underwater Systems, Thales Australia
paul.baker@thalesgroup.com.au

ABSTRACT

Fibre Laser based sensor technology is well suited to provide a robust solution for lightweight, ultra-thin, hydrophone arrays that can be rapidly deployed from a range of platforms to provide underwater acoustic surveillance. As a seabed array it is relevant to both maritime security and force protection. Additionally, it may be exploited for surface ship and submarine fixed and towed sonar arrays.

This technology brings a number of advantages. Optical signals from the hydrophones can propagate along several kilometres of optical fibre without significant attenuation, thus enabling remote deployment of arrays from the receive electronics. It enables use of smaller winch systems, with smaller and simpler platform interface requirements, and places less demand on the overall platform power and weight budgets.

Recent work has focussed on fibre laser based electro-optic hydrophone technology. In this paper we outline how the application of emerging fibre optic based sensor technology to underwater acoustic systems provides significant benefits and capabilities not previously possible with the current start-of the-art piezoelectric based sensors.

INTRODUCTION

Maritime Security has become an increasing concern in recent years for military, government and commercial organisations alike. The threat is multi-faceted and complex, and there are no easy solutions. Whether, the asset to be protected is a Naval vessel, an oil rig, a port or a national border, the threat could be airborne, on the surface or under the water. No single detection system can address all scenarios.

The above water threat is traditionally addressed by radar, CCTV, thermal imaging cameras or a combination of all three. The underwater threat is commonly addressed by active sonar such as a Diver Detection Sonar (DDS) and Seabed Passive Low Frequency Sonar, which is used at choke points such as the US Navy SOSAS system deployed in the North Atlantic during the cold war. In the case of the underwater solutions, active sonar is expensive and requires a significant power source, the seabed passive sonar arrays based on conventional piezo-electric sensor technology require supporting electronics to be underwater and large copper cables which suffer power and signal attenuation.

The small size and high sensitivity of Fibre Laser Sensor (FLS) technology is ideally suited for use in lightweight, ultra-thin, seabed hydrophone arrays that can be rapidly deployed from a range of platforms to provide underwater acoustic surveillance. The technology is relevant to maritime security and force protection. A further advantage of the technology is that optical signals can propagate along several kilometres of optical fibre without significant attenuation, and can therefore make use of small diameter, light weight fibre optic cables. This opens the possibility of remotely deploying an array significant distances from the dry end photo-electronics receiver system.

In the Naval domain there is an increasing emphasis on the littorals as the location for future conflicts. A rapidly deployable fibre optic seabed array has the potential for fulfilling a force multiplier role as an early warning 'tripwire' system for task-force protection. This may either be for protection of the fleet or the rear guard defence of Special Forces put ashore by submarine.

Tripwire systems also have a role to play in border security, due to the potential to remotely deploy very long arrays of multiple sensors. The advantages of a passive sonar array over an active system include the fact that they do not advertise their presence, they require much less power and infrastructure to operate and they can be deployed some distance away from the task force under protection and therefore provide an early alert to an intrusion.

A BRIEF HISTORY

The field of Fibre Laser Sensor Technology was pioneered in Australia by Defence Science and Technology Organisation, (DSTO), [1]–[5]. Thales Australia collaborated with DSTO in this area of research under the Interactive Project Agreement (IPT) during 2003-2005. This research was followed by the very successful Capability Technology Demonstrator (CTD) Program in 2005-2008. This CTD leveraged extensive existing experience in conventional piezo-electric transducer, array design and materials science. The resultant Fibre Laser hydrophone design had good frequency response, depth compensation and signal to noise ratio (SNR). The array design was small in diameter (20mm) and very robust to suit the maritime environment. This eight channel array was successfully rapidly deployed and recovered at sea and demonstrated the feasibility and advantages of seabed arrays based on hydrophones which utilise Fibre Laser Sensor technology.

Work has continued on the seabed array technology, focusing on a cross deck-able or portable sonar surveillance array product development. This work is expected to conclude in 2012, with an at sea trial of a sixteen channel FLS based array with a further reduction in diameter and much increased acoustic sensitivity. The inboard photo-electronics system has been improved, industrialised and packaged.

Parallel work is being done to develop a Fibre Optic Towed Array (FOTA) under internal research and development funding and the FOTA CTD contract which was awarded in 2010. The design is very much in its early stage, however, together with more than 30 years of experience in conventional towed array technology, the knowledge gained during previous FLS array developments has provided a good starting position for a future fibre optic based Towed Array.

FIBRE OPTIC ACOUSTIC ARRAY TECHNOLOGY

Fibre Laser Sensor

A Fibre Laser Sensor can be created by applying a Distributed Feedback (DFB) grating structure with a phase shift at its midpoint to a short length (55mm) of Erbium doped fibre having a diameter of 125μm. Such a structure forms a laser cavity that, when pumped by shorter wavelength optical radiation (980nm or 1480nm), yields a narrow line width, single longitudinal and single polarization mode output having a good Signal to Noise Ratio (SNR).

The wavelength produced by the fibre laser is determined by the optical cavity (Figure 1). The acoustic sensing mechanism of a fibre laser device is provided by the fact that it will shift its operating wavelength in response to longitudinal mechanical strain, hence change in optical cavity length on the fibre.

The advantages of a DFB laser include its high reliability and low power consumption and the fact that it easily lends itself to multiplexing.

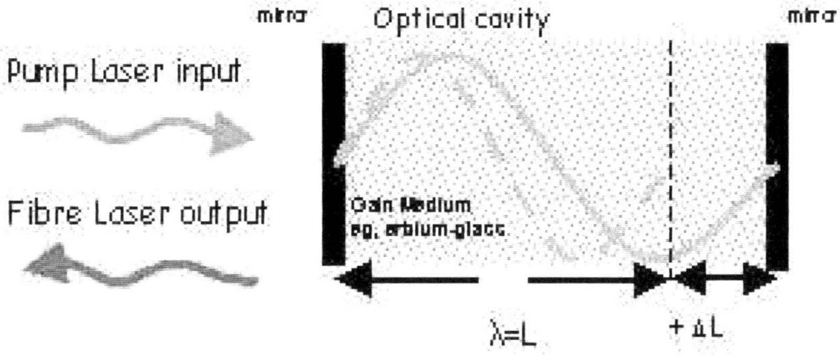

Figure 1: Fibre Laser Optical Cavity

The key to harnessing the sensitivity of the Fibre Laser Sensor, but at the same time isolating the laser from other environmental (non-acoustic) perturbations, is in the design of the packaging of the laser, both of the hydrophone itself and the mechanical structure of the array of multiple Fibre Optic Hydrophones.

Fibre Optic Hydrophone

The current Fibre Optic Hydrophone design addresses all key requirements for conventional piezo-ceramic based underwater acoustic hydrophones including depth compensation, flat frequency response, temperature compensation and reduced vibration sensitivity. Recent improvements in the design have been realised which provide much improved hydrophone sensitivity than previously reported [7]. The hydrophone sensitivity currently achievable (Figure 2) confirms the technology as competitive to conventional piezo-ceramic element hydrophones for a range of acoustic array applications.

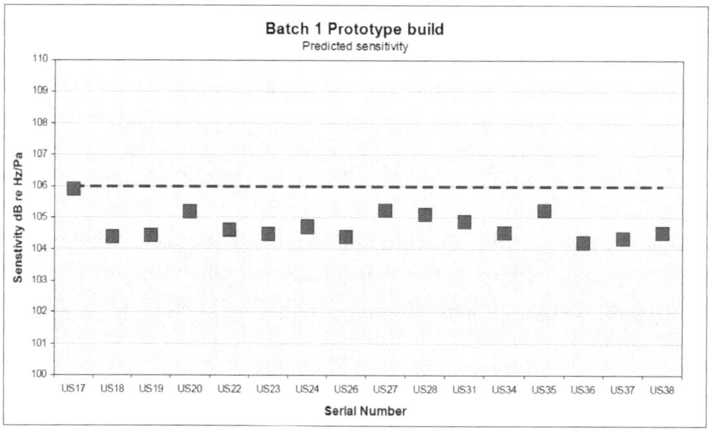

Figure 2: Sensitivity measurements, 16 Channel Seabed Array

Fibre Optic Hydrophone Array

It is possible to assemble an array of multiple Fibre Optic hydrophones onto a single optical fibre, each with a different nominal wavelength and each independently modulated by an acoustic pressure perturbation.

A sixteen channel high strength, small diameter seabed array has been assembled in a continuous oil filled hose and attached to a 1.5km ruggedized fibre optic lead-in cable via a purpose designed high strength underwater fibre optic connector developed to meet the specific requirements of this demanding application.

The Array includes mechanical features to securely mount the fibre laser hydrophones, provide axial strength and stiffness and radial crush resistance. The fibre is managed to prevent it from experiencing strain during deployment, operation and recovery, (Figure 3).

Figure 3: 16 Channel Seabed Array on its deployment drum

Inboard System

A functional view of the optical system architecture is shown in Figure 4. The system comprises a Wet End sub-system which includes the lead-in cable and Array of FLS hydrophones plus the Dry End sub-system which includes the opto-electronics module and the processor-display on a laptop.

Optical signals from the array pass through an interferometer module through the opto-receiver module for demodulation. The demodulated sonar data are then processed and beam formed and displayed on the sonar display.

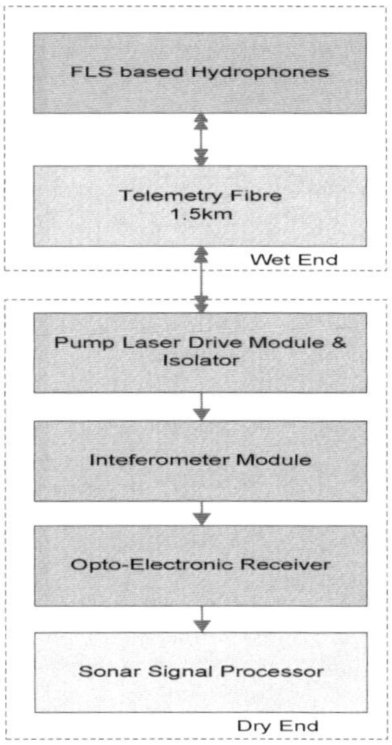

Figure 4: FO Seabed Array System Architecture

The inboard system can be powered from a 12V battery or the 240VAC mains. It draws very little power (around 24W for the 16 channels). It has been ruggedised and will fit neatly into a small suitcase (Figure 5), making it easily transportable, cross deck-able and with the potential to be packaged into a medium sized UUV.

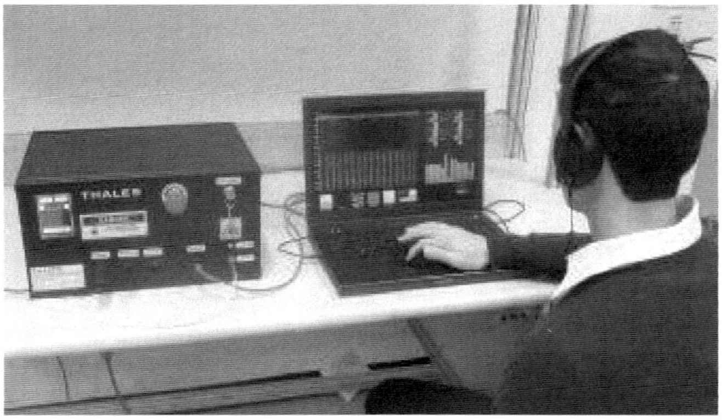

Figure 5: Inboard System, photo-electronics unit and processor with display

Fibre Optic Array Deployment System

A deployment system has been developed which protects the minimum bend radius of the assembled array and allows rapid deployment and recovery from a RHIB (Figure 6). The deployment system can be transferred from platform to platform as a result of a quick release mechanism. For the sea trials conducted in Jervis Bay, NSW in 2008, the deployment system was designed such that it could be rapidly fitted to a standard RAN, 7.2 metre RHIB with no modifications required to the boat. To enable the RHIB to maintain its standard ocean going set-up, the deployment system was fitted to the stern for array deployment and then transferred to the bow to allow recovery at the end of the operational phase of the demonstration. A deployment method was devised and applied which ensured the array was rapidly laid relatively straight on the seabed (Figure 7), with no diver intervention.

Due to the very small diameter of the fibre optic cable, the deployment system demonstrated in Jervis Bay could stow and deploy up to 5km of cable from a RHIB platform, and provide a remote surveillance capability, something which is inconceivable using existing technologies.

Figure 6: FO Seabed Array deployment System in use in Jervis Bay, NSW

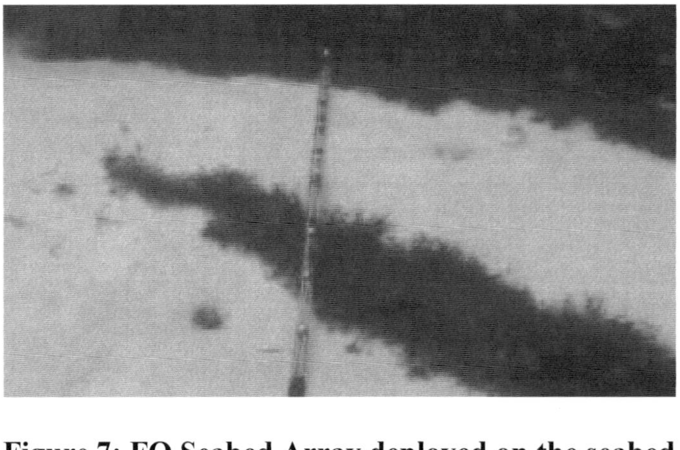

Figure 7: FO Seabed Array deployed on the seabed

BENEFITS OVER CONVENTIONAL ARRAY TECHNOLOGIES

The key advantages of an optical system architecture (especially Fibre Laser based sensors and cables) for acoustic arrays over conventional piezo-electric analogue or digital arrays are:

- No electrical power distribution is required in the wet end thus removing the need for thick power cables;

- No preamplifiers or electronics are required in the wet end thus improving reliability and thereby reducing in-service costs and allowing smaller diameter arrays to be manufactured;

- Low attenuation of optical signals enabling the use of lightweight, thin fibre optic link cable several kilometres in length.

- Much smaller power, size and weight footprint on the deployment platform, thus opening up a greater range of platform options including Rigid Hull Inflatable Boats (RHIB), Unmanned Surface Vessels (USV) and Unmanned Underwater Vehicles (UUV).

As an example of the impact this thin and lightweight technology conveys, it can be compared to the current technology used in towed array systems, (Figure 8).

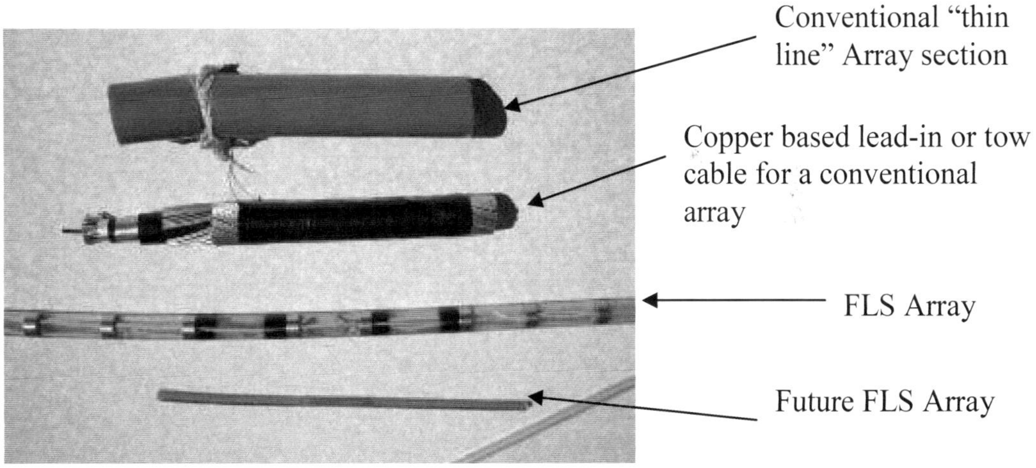

Figure 8: Array technologies, physical comparison

A typical towed array winch, fitted to a 3000 tonne frigate, weighs about 2.5 tonnes and has a 1.8m diameter drum. The towed array is typically 40mm or more in diameter and can weigh 1000kg. The deployment system designed for use with the fibre laser array comprises a 20kg winch and a 0.75m diameter drum that can be fitted to a 7.2 metre long RHIB (Figure 6). With the array section and lead-in cable, the entire system weighs 40kg and can be carried by two people, (Figure 9).

Figure 9: Winch technologies, physical comparison

MARITIME SECURITY APPLICATIONS

The underwater domain of Maritime Security has been addressed by increasingly sophisticated active and passive sonar systems. All these systems are currently based on conventional piezo-electric hydrophone technology.

Due to the thin, lightweight and low power nature of the fibre laser based acoustic array technology, many opportunities are opened up for acoustic array applications in the Maritime Security domain that previously were unfeasible using conventional array technology. These include:

- Rapidly deployable seabed arrays from small vessels;
- Towed arrays for smaller surface ships and submarines, providing small winch systems, smaller and simpler platform interfaces and less power and weight;
- Submarine hull mounted arrays as an alternative to the current outboard electronics approach;
- Arrays for Unmanned Underwater Vehicles (UUVs) and Unmanned Surface Vessels (USVs) that demand low power and small size.

Probably the most obvious and most practical application in the near term is force protection – where the array would be deployed around a high value asset while in a potentially hostile harbour in a trip wire configuration.

Force/Infrastructure Protection

The USS COLE bombing in October 2000 highlighted the vulnerability of Naval ships when in port to asymmetric threats such as small boats. Also, divers can either swim or use diver delivery vehicles to approach Naval vessels almost undetected. There is currently a number of existing active sonar systems in the market for diver detection giving detection out to about 800m. However, there may be operational scenarios whereby a passive approach may be preferable, for instance if a larger perimeter detection range is desired or if keeping the diver unaltered to his detection is preferable, or where transmitting on active sonar sources may be banned from operation in some ports. Passive diver detection utilising FLS hydrophone technology is currently being evaluated.

Similarly, offshore oil and gas infrastructure can be protected with the use of a trip wire style FLS sea bed array. Another potential application includes perimeter protection of archaeological or sensitive ecological sites from unauthorised visitors.

Border Security

The FLS based array technology lends itself to the border protection application. By acting as a trip wire laid on the seabed a very large distance can be surveyed around the clock. The array can be permanently deployed at known choke points or on transit paths for smugglers, pirates or illegal entry or departure routes of people and vessels.

The fibre optic cable can be brought back to land for signal processing and monitoring many kilometers from the array or connected to a moored buoy with RF or satellite communications links.

Harbour Defence

Entrances to ports and harbours can easily be acoustically monitored both for surface and underwater movements using an FLS seabed array. The acoustic signature of vessels can be correlated with radar and/or video or thermal camera images.

UUV Surveillance Array

UUV's are increasingly being used as a covert means to survey seabeds for mines either ahead of a fleet or to monitor mine laying activities. The FLS based small diameter array with its low drag, low power demand and compact photo-electronics system could add useful additional surveillance capability to the UUV, enabling it to also monitor shipping and submarine movements.

CONCLUSIONS

FLS based acoustic arrays have proven advantages over conventional piezo-electric based arrays. These advantages include it's inherently small size and low power needs, which have lead to the development of small footprint arrays, handling equipment and an inboard system. These advantages lend themselves very well to a number of possible applications in the Maritime

Security domain. The most promising of which include Force Protection, Border Security, Harbour Defence and UUV Surveillance Arrays.

A small diameter seabed array with good sensitivity has been demonstrated in 2008, and a small diameter surveillance array with more sensors will be demonstrated in the near future. A towed array prototype is expected to be demonstrated in 2013.

ACKNOWLEDGEMENTS

This work was carried out as part of the Fibre Laser Sensor Capability Technology Demonstrator Program, Thales' Research and Development funding programs and the Fibre Optic Towed Array Capability Technology Demonstrator Project and has benefited from close collaboration with DSTO.

The Thales team involved in the development of the FLS hydrophones, array, deployment and inboard systems included the following key personnel: Ian Bedwell, David Jones, Barry O'Keefe, David Mann, Daniel Mercer, Harley Boggis, Peter Harvey and Fabio Souto.

REFERENCES

1. Foster.S, Tikhomirov.A, Hardy.G, Milnes.M, Van Vetzen.J, (2004), "Ultra-thin all photonic hydrophone array research at DSTO", Proc. 33rd meeting TTCP Mar TP9 Newport.

2. Foster.S, Tikhomirov.A, Milnes.M, Van Vetzen.J, Hardy.G, (2005) "A fibre laser hydrophone", Proc. SPIE **5855** (OFS 17, Brugge), pp627-630.

3. Tikhomirov.A, Foster.S, Milnes.M, Van Vetzen.J and Hardy.G, (2003), "Acoustic and vibrational response of a DFB fibre laser sensor", COIN/ACOFT Conference Proceedings, R. Tucker and T. Aoyama (ed.), pp440-443, Melbourne, Australia.

4. Goodman.S, Tikhomirov.A and Foster.S, (2008), "Pressure compensated distributed feedback fibre laser hydrophone", Proceedings of the 19th International Conference on Optical Fiber Sensors (OFS-19), Perth, Australia.

5. Foster.S, Tikhomirov.A, Englund.M, Inglis.H, Edvell.G and Milnes.M, (2006), ``A 16 Channel Fibre Laser Sensor Array", Proceedings of the 18th International Conference on Optical Fiber Sensors (OFS-18)}, Cancun, Mexico.

6. Bedwell.I, Jones.D, (2010), "Fibre Laser Hydrophone Performance" IEEE OCEANS2010

7. Cain.T, Bedwell.I, Jones.D, Hodder.B, Mercer.D, Boggis.H, Foster.S, Tikhomirov.A, Goodman.S, (2010), "Fibre Laser Sensors for Underwater Acoustic Surveillance Arrays" Mast 2010

Defining Infra-Red Signature Requirements for Naval Acquisitions

Leo de Yong , Rodney Borg, Joanne Culpepper and Vivienne Wheaton
Maritime Platforms Division,
Defence Science and Technology Organisation
Melbourne Australia
Email: leo.deyong@defence.gov.au

Infra-red signature requirements often need to be specified for new naval acquisition projects. However, specifying an infra-red signature requirement for a warship is complicated by the definition of a signature and the influence of the physical environment which has a major effect on the signature. The latter leads to problems for the ship builder/designer in proving compliance of the ship with the requirements, often leads to the ship not meeting the requirements and raises the issue of where the responsibility lies if the ship signature does not meet the requirements.

In the past, specifying infra-red requirements has often been limited to approaches that have been less than ideal. These have included specifying the maximum temperature or radiance of selected elements of the ship design (e.g. the stack, exhaust plume) or of the total ship itself, specifying the details of the technologies that may be used to manage the infra-red signature (stack suppression systems, insulation, coatings) or specifying the signature in relation to an existing ship (a percentage reduction). In all these approaches, direct measurement of systems or modelling or a combination of measurement and modelling are used to prove compliance with the signature requirement.

This paper will examine some of the approaches used by navies today to define the infra-red signatures of naval platforms and discuss their limitations, and benefits.

1. INTRODUCTION

The procurement of a new ship by any navy today often requires some measure of the susceptibility of the ship to a range of threats to be defined leading to a requirement for the ship to meet these measures. These threats may be simple, such as conventional guns or may be sophisticated, such as anti ship guided missiles. Many of the sophisticated threats rely on one or more sensors to guide it to the target with the sensors usually operating in one or more electromagnetic and other wavebands.

The ship designer will usually attempt to meet the signature requirements for as many of the signatures of the ship as possible usually within cost constraints.

In the past, the concentration has been on the radar cross section, the acoustic and the magnetic signature of ships. The techniques and the design rules that are needed to achieve low signatures in these wavebands are relatively well known. Coupled with this, the measurement of these signatures and the compliance with a specified requirement has usually been relatively easy to prove through a series of field trials.

However, the infra-red (IR) signature of a ship is becoming more important in today's operational environments. Threat weapons that operate in the IR wavebands are becoming more prolific and have the added advantage that they are passive and provide little warning of their approach. They are also becoming more sophisticated with advanced seekers that operate in both radar and IR bands, or operate in two IR bands or operate in two electromagnetic bands such as visible and one IR band. The result is that IR signature management is becoming more important for ship survivability and the ability to define IR signature requirements for a new ship is also becoming more important.

The problem with specifying the IR signature requirements of a ship is firstly, the definition of what the IR signature is. The second problem is that the signature is dependent on the environment that the ship is exposed to and how to ensure that the requirements can be tested and the signature requirements can be achieved. This paper discusses both these problems and proposes some solutions.

2. DEFINITIONS OF INFRA-RED SIGNATURES

A signature is not a physical or prescriptive property of a ship but is what may be called a quantitative record of the presence of the ship in an environment. As such it is clearly not a uniquely defined quantity. However, there are many quantitative measures of a ship "signature" which are distinct representations of what may be termed the ship signature.

2.1. Contrast Signature

The contrast signature describes the difference between an image of a scene that contains a platform and one of the same scene that does not. There are several contrast signatures that may be used such as the absolute contrast signature, the far field contrast signature or the RMS contrast signature. The most likely definition used by a shipbuilder would be the absolute contrast signature which is universally accepted as the common standard by NATO and many other countries. This signature is a summation of the modulus of the difference between pixel platform intensities and the background intensities.

2.2. Beacon Signature

It may be important that the point with the highest detected intensity is defined and that its level be specified as it may be important and act as a "beacon" which will attract the attention of a viewing system. This may be referred to as the beacon signature of the platform and is particularly important when dealing with platforms which have strong exhaust plumes, exhaust hotspots or solar glints.

2.3. Information Signature

An imaged scene contains information about several discernable elements within the image. Examples include the outline or the profile and the pattern provided by the image. This information signature is used usually to provide details about the platform once it is detected.

In defining the signature of a new acquisition the contrast signature is the most commonly used approach and the absolute contrast signature is the universal accepted methodology used by many nations. Note that this does not mean that the other signatures are neglected but that they are rarely defined for acquisition purposes.

3. ENVIRONMENTAL EFFECTS

To demonstrate the influence that environmental conditions can have on the IR signature, the IR signature of a generic frigate was calculated using ShipIR (also known as NTCS, the Naval Threat/Countermeasure Simulator) [1]. In Figure 1 the average contrast radiance of the generic frigate is shown as a function of the time of day and at various ranges to the viewer. The average contrast radiance was calculated by averaging the absolute contrast radiance of 360 viewing directions around the platform. The contrast radiance reduces rapidly with range due to atmospheric attenuation. At 1 km the contrast radiance varies between 1000 W/sr at around 09:30 to a maximum of 1800 W/sr at 15:00. To demonstrate the changes that occur around the platform, Figure 2 shows the absolute contrast signature polar plots for the generic frigate at 1 km at three times of the day. These plots show how the solar loading changes the spatial distribution of the IR signature over the course of the day. The results of these simulations show that environmental factors contribute significantly to the IR signature of a platform. Therefore it is important to consider the impact of environmental factors in devising IR signature requirements, modeling IR signatures, measuring IR signatures and analyzing/interpreting the results of simulations and/or trials.

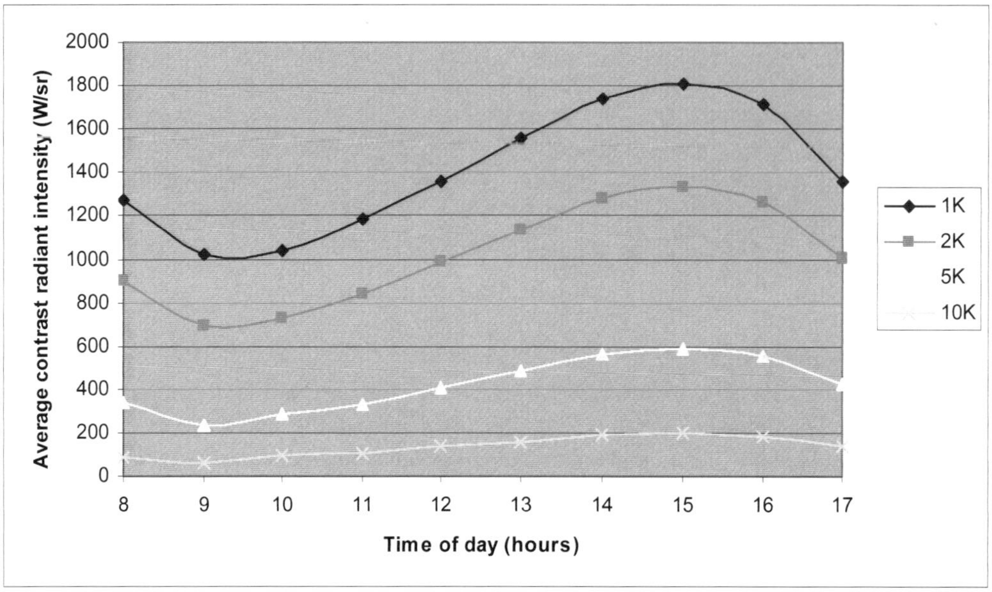

Figure 1: Average contrast radiant intensity of a generic frigate at different times of the day and at different ranges to the viewer.

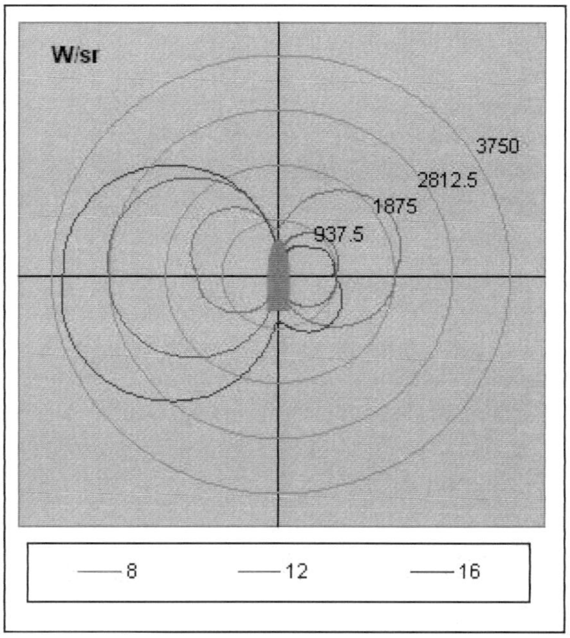

Figure 2: Polar plots of the absolute contrast radiance of the generic platform at 1 km at three times of the day.

4. DEFINING IR SIGNATURE REQUIREMENTS

4.1. Defining the Construction Parameters

The IR signature of a ship can be specified in terms of its construction parameters [2]. This may involve specifying the type of paint and its emissivity, and specifying the thermal insulation requirements for compartments that are traditionally hot (e.g. the engine room). Other signature suppression devices can also be specified such as eductor/diffusers for uptake/exhaust cooling or water wash down systems for superstructure temperature control. The level of specification can be generic or specific make/models can be stipulated. These parameters are easy to verify for compliance at acceptance and the complications of the environmental effect on IR signature are removed.

The definition of the construction parameters however, puts the risk or the responsibility back to the ship owner for the IR signature meeting the requirements. It also means that the ship owner needs to have a fundamental knowledge and capability to understand and to evaluate the consequence on the IR signature of the various construction parameters. This involves detailed operational analysis (OA) which would specify the operating envelope of the ship in terms of the physical environments. Coupled to this would be computer modelling of the ship in the range of OA defined environments with the defined construction parameters. From this analysis, the change in the IR signature for the range of construction parameters may be evaluated and the construction parameters may be defined to meet the required IR signature. Modelling of the IR signature of the ship is not a significant issue as there are a number of computer codes that are validated and are routinely used to predict the IR signature of ships. However, a limitation of this approach is the knowledge of the operating environments for the OA analysis and future changes in the operating environments. Another potential issue is that this approach assumes that all the signature related construction parameters are well known. Thus, the definer of the construction parameters must be knowledgeable and well versed in the performance and use of the parameters. Specification of construction parameters also stifles innovation since novel or emerging approaches to IR signature control may not be in the list of construction parameters specified. An advantage of this approach is that the IR signature of the ship does not need to be released to the shipbuilder/designer, only the relevant construction parameters. This helps to retain the security/sensitivity aspects of the IR signature. The ship also does not have to undergo at sea IR signature acceptance trials as the use of and checking of the construction parameters can be done during construction or alongside.

4.2. Specifying the IR Signature Requirements

As noted above, the IR signature requirements for a ship are usually specified in terms of the maximum allowable temperature for various parts of the ship (such as the stack), the contrast radiance of the ship in relation to its background or a combination of both. The problem with this approach is in the assessment of the signature compliance. Compliance usually involves a sea trial but the atmospheric conditions encountered in the sea trial (temperature, relative humidity, wind speed, wind direction etc.) are highly unlikely to correspond to the conditions on which the signature requirements were originally defined. This then leads to a situation where the IR signature compliance cannot be evaluated. The consequence of this is a ship that may be unable to meet its IR signature requirements and a designer/builder who could be unable to meet their responsibility to meet the IR signature requirements or choose to negotiate that it may not be their fault if the ship does not achieve the requirements. A further problem is that shipbuilders are not highly skilled in understanding or interpreting IR signature requirements in terms of contrast radiance.

4.3. Defining the Detection Probability

Knowing the IR signature of a ship in itself has limited operational value. It is actually the relationship between the signature and the ability of a threat to detect the ship in an environment that is of more operational use. Therefore, the probability of detection of the ship in a defined IR band becomes an important relationship. As such, the IR signature of the ship is not directly specified to the shipbuilder/designer but rather the required probability of detection of the ship in a range of defined scenarios by one or more defined IR sensors.

A number of computer models have been developed to determine the probability of detection of a target in a scene from real or synthetic imagery [3, 4] . These models consist of two main components: a target detection metric and a sensor model. The target detection metric defines target saliency or how much the target stands out from its background. It is calculated by analysing statistical properties of the image to determine the contrast variation between the target and its immediate background. This contrast is effectively a temperature difference that can be related to the sensor response in the senor model to produce a probability of detection for a target in a particular scene.

Two of the most widely used probability of detection models are the US developed ACQUIRE model and the German developed CAMAELEON IR model. The ACQUIRE and CAMAELEON IR models both calculate a thermal contrast metric for a target in a scene using image statistics. This metric is then transformed into either a probability of detection or a detection range. This detection range is calculated at a 50% probability of detection. The target thermal contrast is corrected for atmospheric transmission, and associated with either a generic or specific sensor response model to determine how well a sensor will see the target at range. Also factored into these models is the target size. The detection range estimate is determined using an empirical factor, which is based on human observer experimentation.

The ACQUIRE and CAMAELEON IR models have been developed primarily for the land environment. As such these models do not necessarily perform well for targets in maritime scenes. In fact, ACQUIRE has been shown to overestimate the probability of detection compared with human observer performance of small maritime targets in FLIR imagery [5, 6]. This is due to the interaction of the target with its background. Targets in a blue-water environment tend to be more similar to their backgrounds than land targets relative to their dimensions. This is illustrated in IR imagery shown in Figure 3. It also demonstrates that the two environments have significantly different background characteristics. Thus the interaction between the target and background for a blue-water maritime environment must be modelled differently to a land environment. Further model development is required.

One of the problems with specifying a probability of detection as a requirement is that the shipbuilder/designer is highly unlikely to have any detailed knowledge of how to determine it and has even less knowledge of the environments that the ship will operate in. So the ship owner must then provide extensive support to the shipbuilder/designer to achieve the requisite probability of detection.

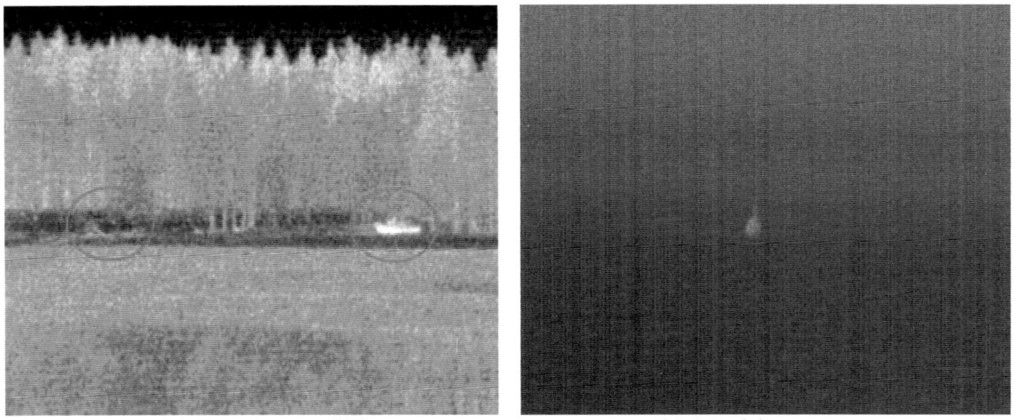

(a) Land target[7] (b) Maritime target

Figure 3: Typical IR imagery

With access to a validated probability of detection model for maritime environments an approach to requirements development is possible. If the maritime probability of detection model follows a similar approach to the models developed for land it will be based on contrast metrics. It is then possible for contrast radiant intensity levels, as discussed previously, to be calculated based on desired probability of detection performance in specified environments. However, identifying the desired probability of detection level is a non-trivial exercise, and must be done with consideration of the mission and the specific threats.

4.4. IR Signature Modelling and Prediction

There are several modelling codes that are used to predict the IR signature of ships that have had significant validation conducted over the last 5-10 years. There are several approaches that may be used with modelling as their basis.

4.4.1. The Standard IR Target (SIRT)

One approach that has been used is based on the definition of the IR signature as a percentage of a standard IR target (SIRT). The SIRT can be an existing ship or it may be an idealised ship. In essence the SIRT represents the baseline IR signature on which the requirement is based. For example, the requirement may be for the IR signature of the ship to be 70% of the SIRT when solar loaded and 50% when not solar loaded. The percentages would also be split into the various wavebands of interest (such as 3-5 μm and 8-12 μm). The advantage of specifying the requirement in this manner is that it is independent of the environment. Another advantage is that the compliance of the ship can be monitored throughout the acquisition process. The IR signature of the proposed design can be modeled and compared to the SIRT results even before the ship is built. This can highlight issues early in the acquisition cycle and allows risks to be identified prior to key decision making deadlines.

However, there can be issues with testing compliance using this approach. Compliance testing would involve a trial, or series of trials, under a limited set of conditions. It is clearly not practical to test the ship under all conditions. Since the IR signature is highly dependent on the trial conditions, the testing results could be at the extremes of possible signatures. It is known that certain environmental conditions will result in a higher signature (so called worst possible case) whereas other conditions will give a much lower signature (so called best possible case). Compliance testing would involve comparing the trial results against the SIRT result computed for the same environmental conditions as the trial. The ability to achieve compliance can be affected by whether the trial conditions were worst or best case. Under worst case conditions, the IR signature suppression methods employed are likely to produce larger percentage reductions when compared to the best case conditions. Therefore under worst case conditions it will be easier to comply with the requirements and under best case conditions it will be more difficult to comply. Another disadvantage is that the trial conditions may be nothing like the conditions that are important from an operational perspective. This could raise questions as to the relevance of the testing process and the implications for the required capability.

This method also relies on the ability to develop realistic IR signature requirements. Typically the requirements can be estimated by performing modeling on a similar platform. Essentially the SIRT model is modified to incorporate state-of-the-art IR signature suppression technology. The signature is then calculated for the desired environments and then increased by some predetermined amount to provide an engineering margin. The final requirements are then expressed as a percentage of these increased values such that they should be achievable if the appropriate IR suppression technology, or better, is adopted.

4.4.2. Requirements Test Set.

An alternative approach that addresses some of the disadvantages of the pure SIRT method involves defining a range of operational scenarios. In this method a set of environments, to be referred to as the Requirements Test Set (RTS), are devised based on operational requirements. The RTS is designed to cover a range of environments that are deemed to be realistic from an operational perspective. A unique set of IR signature requirements for each of the environments in the RTS is devised in a similar manner to the SIRT method. However, the approach to testing and compliance is slightly different. The completed ship undergoes a series of IR signature measurement trials under different conditions (perhaps a couple of daytime and one nighttime at different ranges and different speeds). The data from the trials is used to create a validated ship model. This validated model is then used to compute the IR signature of the ship in the environments of the RTS. Compliance is then assessed by comparisons between the computed signature and the IR signature requirements of each environment in the RTS.

4.4.3. Integrated Ship, Decoy and Missile Performance

A disadvantage of the previous methods is that the defined requirements ignore the effects of the on-board decoys, as well as the performance of the red threat missiles. Clearly from an operational and capability perspective, the overall performance of the platform (including decoy and threat missile behaviour) should be integral to the IR signatures requirements definition. This can be achieved by adopting an integrated approach to the signature definition phase. This approach could be framed in a very similar way to the RTS described in Section 4.4.2 except that the IR signature requirements would be determined using a holistic model. The holistic, or integrated, model would include the predicted IR signature of the SIRT, the IR performance of the decoys and the IR seeker behaviour of relevant red threats. The IR signature of the SIRT would be adjusted (by altering the level of IR suppression) until the desired performance against the threat in areas of operational interest was achieved. This then sets the IR signature requirements of the platform in a set of environments (viz. the RTS) that are used to assess the delivered platform against. Whilst this method encompasses all of the important factors that determine the susceptibility of a platform to IR guided threats, it does require a great deal of knowledge and understanding of the relevant systems. For example, detailed knowledge of the specific missile seeker behaviour is required. This includes information pertaining to physical performance (e.g. spatial resolution, spectral sensitivity, operating waveband etc) as well as the underlying guidance algorithm (e.g. imaging versus hot spot seeker).

5. CONCLUSION

A number of alternative approaches to defining IR signature requirements for naval acquisition projects have been presented. The relative advantages and disadvantages of the various options have been described.

1. *User Manual for the Naval Threat/Countermeasure Simulator (NTCS), Version 2.7*, (2000), Davis Engineering Limited: Ottawa, Ontario.
2. Neele, F., Wilson, M. and Youern, K., (2005), "'Stealth' technology: proposed new method of interpretation of infrared ship signature requirements", *Naval Forces* **26** (6), 32-40.
3. Culpepper, J. B., (2011), *Parametric techniques for predicting target detection performance in greyscale scenes,* DSTO-GD-0636, [General Document] Defence Science and Technology Organisation.
4. Hecker, R., (1992), Camaeleon--camouflage assessment by evaluation of local energy, spatial frequency and orientation., *SPIE Opt. Eng.* **37** (7), 2022-2028.
5. Bijl, P., (1996), *Acquisition of sea targets. Part I: Observer performance and 'ACQUIRE' model predictions for Air-to-Surface FLIR imagery,* [Technical Report] Soesterberg, The Netherlands, TNO Human Factors Research Institute
6. Bijl, P. a. V., J. M. , (1998), Validation of a new triangle orientation discrimination method and ACQUIRE model predictions using observer performance data for ship targets, *Optical Engineering* **37** (7), 1984–1994.
7. Saab. *MCS – Mobile Camouflage System Webpage*. [Accessed; Available from: http://www.saabgroup.com/en/Land/Force_Protection/Signature_management/Mobile_Camouflage/MCS_Mobile_Camouflage_System/Features/.

Ultimate Strength Calculations of an Aluminium Patrol Boat

T Magoga and C Flockhart
Defence Science and Technology Organisation, Australia
Teresa.Magoga@dsto.defence.gov.au

ABSTRACT

The use of aluminium alloys in the construction of naval vessels has increased considerably over the past several years. This is particularly evident in the case of high speed patrol vessels for which the requirement for reduced structural weight is critical. The Defence Science and Technology Organisation (DSTO) is currently undertaking a range of research programs to further its capabilities in structural integrity measurement and prediction of aluminium naval vessels. As part of this research effort, an improved understanding of the parameters of influence on the ultimate strength of welded aluminium ship structures and access to validated rapid assessment tools have been identified as important prerequisites to allow DSTO to support operational, through-life and acquisition requirements of the Royal Australian Navy (RAN).

This paper investigates the use of the state-of-the-art rapid assessment procedure ISFEM (Intelligent Supersize Finite Element Method) to examine the ultimate hull-girder strength in vertical bending of a generic aluminium midship section with scantlings typical of a patrol vessel. The ISFEM procedure is available in ALPS/HULL module of the MAESTRO software code. To verify the ISFEM procedure, the midship section of a steel-construction vessel is compared with the results generated using the UltSAS (Ultimate Strength Assessment) program. The effect of ultimate strength of the generic aluminium midship section due to initial plate and stiffener distortion, welding residual stress and softening in the heat affected zone is then investigated using the ISFEM model.

INTRODUCTION

The hull of a ship is designed to withstand the vertical bending moment loads resulting from lifetime extreme operational conditions. The global strength and stiffness of a ship's hull-girder are governed by the strength of the individual plate and stiffener elements assembled to comprise the overall hull girder structure. The global strength of the vessel is highly dependent on the behaviour of these structural units under various load combinations, particularly their instability characteristics in compression and elasto-plastic response. To prevent hull collapse due to inadequate strength under applied loads, it is necessary to accurately and reliably calculate the ultimate strength of the hull [1, 2].

In light of the increasing use of marine-grade aluminium as the hull material of naval vessels, efforts to improve understanding of their collapse behaviour have increased. The mechanical properties and damage response of both the welded and base material of aluminium alloys are different and less understood than that of steel. High speed vessels are also operating in increasingly harsher environments. As such, increased knowledge of the parameters that influence the ultimate strength of aluminium ship structures is required [2, 3].

Methodologies which account for local failures of structural units in the determination of the global ultimate strength of a hull require knowledge of the effects of weld-induced imperfections on the large deflection elastic-plastic response of stiffened and unstiffened plate structures [2-4]. Imperfections introduced during welding include softening in the heat affected zone (HAZ), residual stresses, and initial distortions of plates and stiffeners. Furthermore, as there is a significant amount of uncertainty in the magnitude and shape of fabrication-induced distortions, it is expedient to use measurements of imperfections in representative models [1]. To meet these needs, a database of the statistics of initial imperfections in welds of aluminium stiffened plates for marine applications has been produced by means of extensive laboratory measurement of full scale metal inert gas (MIG) welded panels [3]. The ultimate strength of unstiffened aluminium plates used in high-speed vessels has been characterised [2]. A parametric study of the influence of geometric imperfections, material properties and internal stress fields in aluminium plates using a non-linear finite element method (FEM) was performed.

Theoretical, numerical and experimental studies on the strength assessment of aluminium multi-hull ship structures have been performed in [5]. Using ALPS/HULL, an analysis of the ability of a hypothetical aluminium catamaran to withstand the maximum vertical bending moment due to extreme environmental conditions was performed. The applicability of the ALPS/HULL program for analysis of aluminium ship hulls was confirmed via comparison of results from ALPS/HULL to other progressive collapse methods of a 1/3 scale steel frigate hull and to a steel double-skin tanker model with a uni-directional girder system.

DSTO is furthering its capabilities in ship structural integrity measurement and prediction. To support operational, through-life and acquisition requirements of the Royal Australian Navy (RAN), access to efficient assessment tools for naval structures is critical. DSTO's knowledge and understanding in the areas of fatigue management and, in general, the response of naval aluminium structures to the wave environment is being increased.

DSTO has previously used progressive collapse methods to assess the strength of intact and damaged steel naval vessels. Results from ultimate bending strength calculations and analysis of flooding and sloshing pressure loads on watertight bulkheads determined the loading on the vessel. This data was then used to evaluate the probability of structural failure [6]. Such research enables support to Navy in terms of structural and operational decision-making.

This paper examines the ultimate strength of a generic aluminium patrol boat with material and scantlings typical of a high-speed vessel. The section is assumed to be fully MIG welded. The parametric study investigates the influence of initial plate and stiffener distortion, plate and stiffener residual strength, HAZ breadth, and HAZ yield stress ratio on the ultimate strength of a boat using the progressive collapse procedure ISFEM.

ULTIMATE STRENGTH ASSESSMENT METHODS

The bending moment response of a vessel can be described in relation to an applied curvature (1/r), where r is the radius of curvature (see Figure 1). As the curvature is applied to the structure the stress is initially in the linear, elastic region. After the highest bending load in the linear region, or first yield moment M_y, is reached the response of the vessel is in the elastic-plastic region. Beyond M_y there is typically buckling in compression or yielding in tension of structural items furthest from the neutral axis. This non-linear behaviour decreases the load carrying capacity of the structure. Failed parts of the structure progressively transfer load to the non-failed items, and with increasing curvature, the whole structure eventually reaches the ultimate limit state M_u. Further increases in bending moment cannot be sustained. The process of load transfer from failed to non-failed items and successive failure leading to M_u is referred to as progressive collapse. The structure retains residual strength post-collapse. The rate at which the load carrying capacity reduces with increasing curvature beyond M_u is a measure of the energy absorption or "brittleness" of the structure.

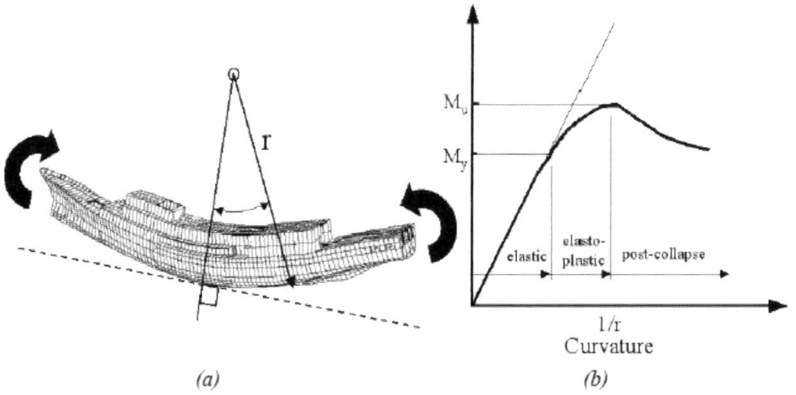

Figure 1: a) Ship Structure under sagging bending moment and b) bending moment versus curvature plot where My = moment at initial yield and Mu = moment at ultimate strength [6]

Intelligent Supersize Finite Element Method (ISFEM)

ISFEM is a numerically efficient FEM procedure implemented in the ALPS/HULL module of the MAESTRO software. ISFEM allows for the interaction of local failures of structural parts to calculate the progressive collapse of a hull cross-section. Combinations of vertical bending, longitudinal bending, vertical shear force, horizontal shear force, torsion, initial imperfections and external hull pressure can be applied to the model [1, 7].

ISFEM analysis combines features of both the conventional nonlinear FEM approach and the Idealised Structural Unit Method (ISUM), the latter also referred to as the load-shortening curve method [8]. ALPS/HULL ISFEM analysis involves discretisation of the hull section into "supersized" 4-noded plate elements (hull plate and girder webs) and 2-noded beam elements (stiffeners and general beams). The length of ISFEM elements typically spans an entire bay. The breadth of hull plate ISFEM elements typically spans the width between stiffeners. As in the case of conventional nonlinear FEM, tangent stiffness matrix formulations are developed which describe the large-deflection, elastic-plastic response of individual ISFEM elements. The element stiffness matrices are combined to form the global tangent stiffness matrix for the structure. Standard load incrementation and convergence iteration schemes are used to solve the nonlinear response. However, unlike conventional FEM different ISFEM element tangent stiffness matrices are developed to capture the pre-collapse and the failure mode dependant post-collapse behaviour of the element. A total of six unstiffened and stiffened plate failure modes are considered [9]. The tangent stiffened matrix formulations within ALPS/HULL have been developed to include the effects of weld-induced imperfections as parameters of influence. Furthermore, the ALPS/HULL ISFEM approach utilises closed-form analytical failure mode criteria to establish the transition from pre to post collapse response for individual ISFEM elements.

The computational efficiency of the ISFEM approach compared with conventional FEM is due to the spatially large elements with fewer degrees of freedom as well as the use of analytical large elements, which avoided convergence problems near element load peaks [9].

Ultimate Strength Assessment (UltSAS) Code

The UltSAS program uses the load shortening curve method of ultimate strength determination. UltSAS interrogates a database of load shortening curves from a series of pre-computed finite element analyses of structural units. UltSAS is extensively validated against scaled hull sections [10].

VERIFICATION OF ISFEM MODELLING PROCEDURE

To verify the ISFEM procedure implemented for an aluminium vessel, the ultimate strength of the midship section of a frigate sized vessel constructed of steel is first compared with the results generated using UltSAS. The UltSAS and ALPS/HULL models are idealised to have no initial imperfections. The UltSAS model was previously utilised in ultimate strength assessments of cross-sections along the length of the vessel [6, 11].

A one-bay ALPS/HULL model of the frigate sized midship is shown in Figure 2a. To create an ISFEM model of the midship section, a MAESTRO model is first created which is comprised of "strakes". Strakes enable quick definition of the plate thickness and stiffener layout of a panel. The hull girder collapse model is then generated automatically from the MAESTRO model using mesh refinement analysis type ALPS/HULL, which converts the strakes to ISFEM "supersize" elements [7].

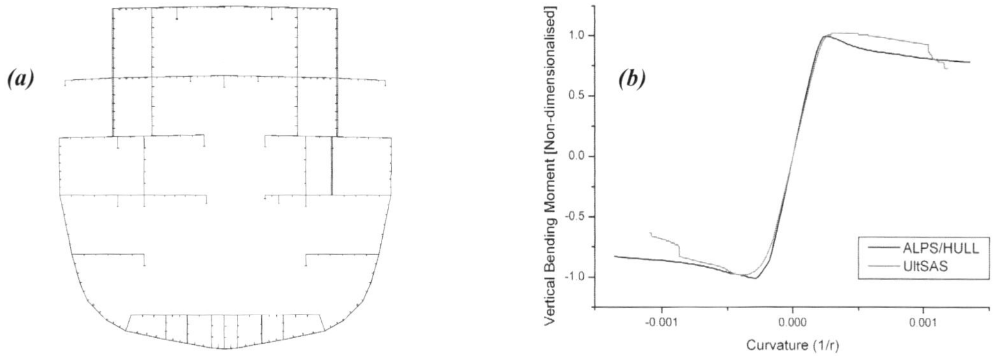

Figure 2: a) A finite element model of the midship of a frigate sized vessel and b) comparison of ALPS/HULL and UltSAS vertical bending moment response of frigate sized midship

Figure 2b compares the curvature versus bending moment of the frigate sized midship generated using ALPS/HULL and UltSAS. The sign convention employed is a positive bending moment in hogging and negative in sagging. Table 1 lists the maximum non-dimensionalised hogging and sagging bending moments.

Table 1: Parameters for the UltSAS versus ALPS/HULL comparison

Parameter	ALPS/HULL	UltSAS	Difference [%]
Height of neutral axis from baseline	1.00	0.986	1.36%
Ultimate hogging vertical bending moment*	1.00	1.02	2.41%
Ultimate sagging vertical bending moment*	-0.992	-0.977	-1.58%

The comparison between ALPS/HULL and UltSAS shows a good correlation between the two methods. UltSAS produces a slightly larger maximum hogging moment and a slightly

* Reference bending moment = ultimate hogging vertical bending moment from ALPS/HULL

smaller sagging bending moment compared to ALPS/HULL. ALPS/HULL predicts the flexural rigidity, or gradient of the initial linear section of the curve, to be slightly stiffer than UltSAS. These discrepancies can be accounted for by the different formulations of the methods. ALPS/HULL and UltSAS use different types of idealised structural units and calculate the pre-yielding response of the structure differently. However, overall there is good agreement between the results. As such, there is confidence that the implementation of ALPS/HULL presented is acceptable.

ULTIMATE STRENGTH CALCULATIONS

The sensitivity of the ultimate strength of the aluminium patrol boat midship to weld-induced initial imperfections can be estimated by completing a matrix of runs in ALPS/HULL. ALPS/HULL allows values for the following initial imperfections to be defined for all plates and stiffeners in the model; maximum plate deflection, plate residual stress, maximum stiffener deflection, stiffener residual stress, HAZ breadth, and HAZ yield stress ratio.

Figure 3 shows the model of a one-bay section of a generic aluminium patrol boat with typical scantlings. The present study limits itself to investigating aluminium alloys AA5083- H116 which is used in plating and AA6082-T6 which is used in extruded beams and stiffeners.

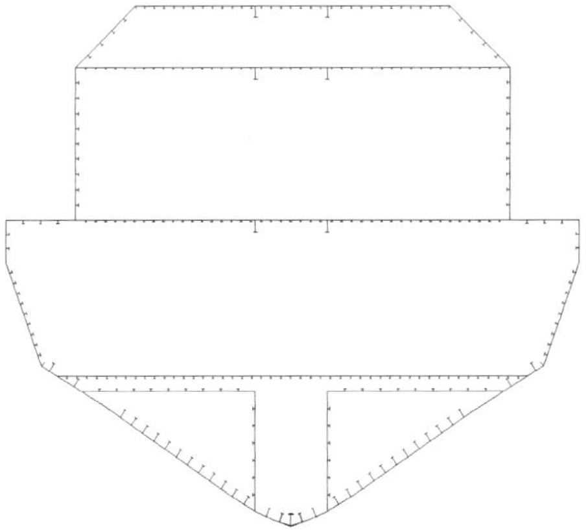

Figure 3: FE model of the midship of a generic patrol boat

An investigation of fabrication and weld-induced imperfections in stiffened aluminium panels [3] suggested slight, average and severe reference levels to characterise these distortions. These reference levels are used in this study, as presented in Table 2. As one parameter is varied, all other parameters are held at their respective average value.

Table 2: Increments of Initial Imperfections

Parameter	Symbol	Slight	Average	Severe
Plate maximum deflection / thickness	w_{opl}/t	$0.018\beta^2$	$0.096\beta^2$	$0.252\beta^2$
Plate residual stress / yield stress of unwelded plate	σ_{rcx}/σ_{yp}	-0.110	-0.161	-0.216
Stiffener maximum deflection / length	w_{oc}/a	0.00016	0.0018	0.0056
Stiffener residual stress / yield stress of unwelded stiffener	σ_{rcx}/σ_{ys}	-0.078	-0.137	-0.195
HAZ breadth	$b_f' = b_s'$	11.3 mm	23.1 mm	29.9 mm
HAZ yield stress / yield stress of unwelded plate or stiffener	σ_{HAZ}/σ_y	0.906	0.777	0.437

The maximum plate deflection is dependent on plate slenderness ratio, β. The slenderness ratio of a plate is a function of thickness, width, elastic modulus and yield stress of the material. The slenderness ratio of the bottom plates, which have the smallest slenderness ratio, was used for this input.

Figure 4 illustrates the HAZ and residual stress distribution. For simplicity, the levels of the HAZ yield stress ratio for AA5083 are used, even though both AA5083 and AA6082 are used in the model as they are the typical aluminium alloys used for plating and extrusions. In addition, in the present format of ALPS/HULL, the same value of the HAZ breadth and yield stress ratio for plating and stiffeners are defined.

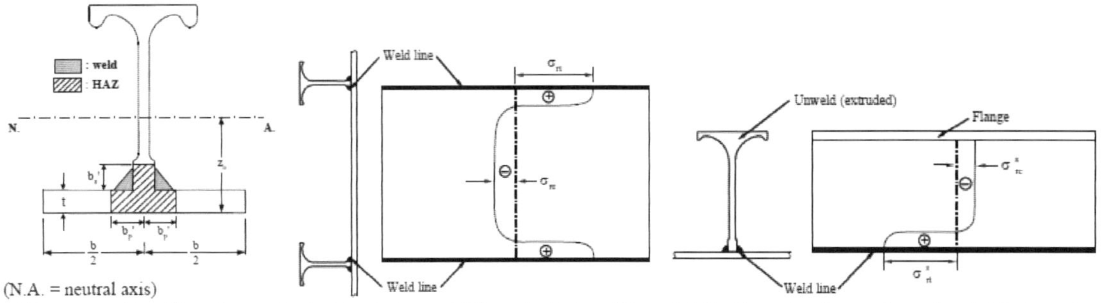

Figure 4: Profile of a typical aluminium stiffened plate with weld (left) and schematics of weld-induced residual stresses (centre and right) [3]

Results

Figure 5 compares the ultimate vertical bending response of the patrol boat midship section (see Figure 3) for the case where no imperfections and average imperfection levels for all imperfection types are defined in the model. Figure 5 also shows the fully plastic bending moment and the first yield bending moment at the top of the superstructure, as the section modulus at the top of the cross-section is smaller than that at the keel. Figure 6 shows the ultimate vertical bending response with varying imperfection levels. All results for maximum hogging and sagging bending moments in Figure 5 and Figure 6 are non-dimensionalised by the first yield bending moment.

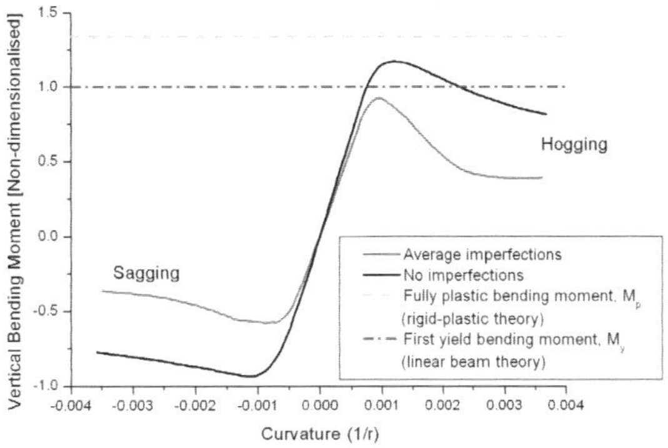

Figure 5: Curvature versus vertical bending response with and without initial imperfections

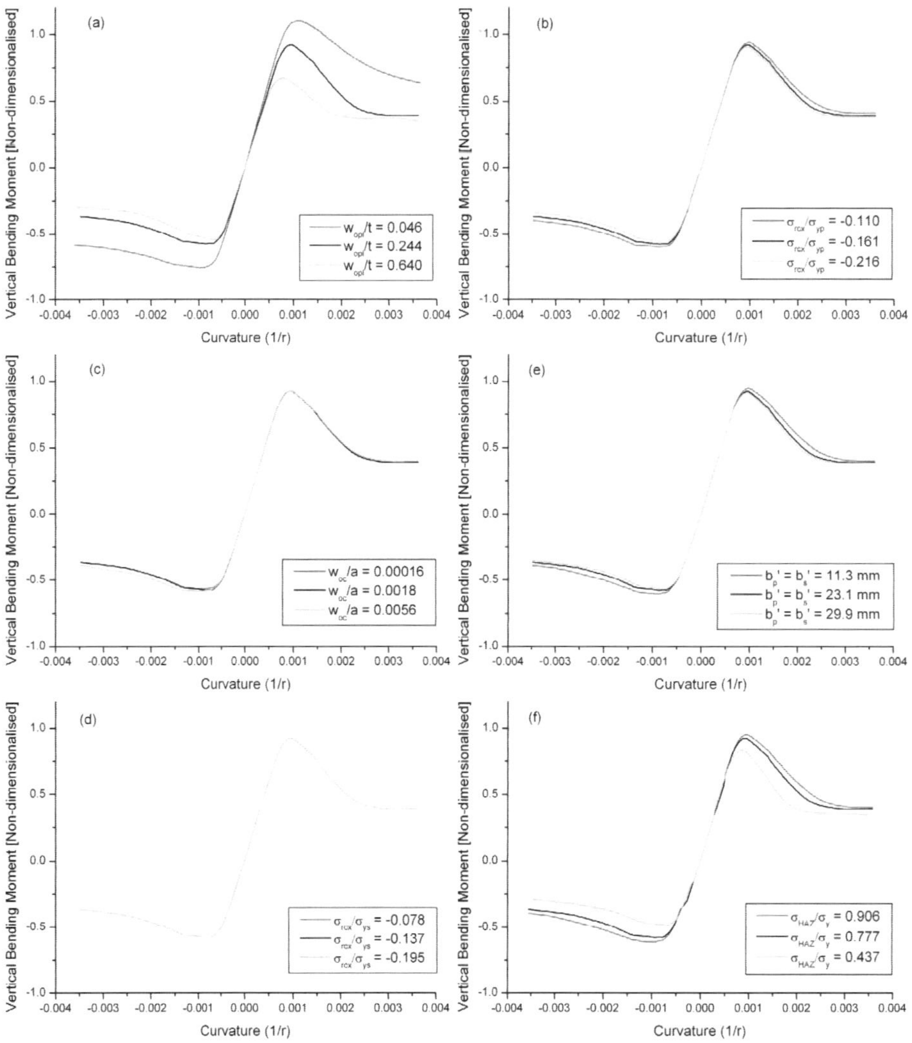

Figure 6: Curvature versus vertical bending response of patrol boat midship for varying levels of a) initial plate deflection, b) plate residual stress, c) initial stiffener deflection, d) stiffener residual stress, e) HAZ breadth and, f) HAZ yield stress ratio

Discussion

The extent and type of an imperfection, and the failure mode of an individual structural element, influence the collapse moment of the midship structure.

Figure 5 compares the bending response of the patrol boat midship for the cases where no imperfections and average imperfection levels are defined in the model. The ultimate limit state of the "imperfect" structure in the hogging condition is 80% of that of the "perfect" structure. In sagging, the ultimate limit state of the imperfect structure is 60% of that of the perfect structure. Furthermore, the introduction of imperfections resulted in reduced flexural rigidity and first yield moment of the structure. The extent of the reduction in strength emphasises the importance of modelling a representative aluminium structure with imperfections, as the effect of fabrication-induced defects can be substantial.

It is also observed from Figure 5 that for the perfect structure the value of the ultimate moment normalised by the first yield moment in the sagging condition, M_{u_s}/M_{y_s}, is less than one. In hogging M_{u_h}/M_{y_h} is greater than one. For conventional, large vessels it would be expected that both hogging and sagging M_u/M_y values are greater than one. However, a patrol boat is a relatively small vessel with a design dominated by a stress distribution induced by localised loading and slamming rather than by global bending. The top of the superstructure of the patrol boat also features thin plating and small stiffeners, so that buckling may occur at a stress level significantly less than the first yield stress of the overall structure. It should also be noted that it is not expected that the vessel will experience a vertical bending moment greater than the first yield moment.

The severe level of plate initial deflection has the greatest influence on the reduction of the ultimate strength of the aluminium patrol boat midship, as shown in Figure 6a. For increasing values of initial plate deflection, the flexural rigidity, yield moment and ultimate limit state decrease. A geometrically imperfect plate has some capacity to respond elastically to the increasing load, though this decreases with increased initial deflection amplitude which is illustrated by the reduction in flexural rigidity. This is because individual plates will reach their yield stress at different bending moment levels.

Plate distortion tends to be larger when the slenderness ratio is relatively large. Given that the plates in the patrol boat midship have different slenderness ratios and thickness, the effect of the plate deflection parameter on the load when the plate fails varies for each plate. After the yield moment has been reached, some plates behave in the plastic regime. This occurs locally, and the overall structure behaves elasto-plastically pre-collapse. In the post-collapse region the plates feature gross plasticity, though the midship can still sustain a reduced vertical bending moment.

As the ultimate strength of the patrol boat midship is most sensitive to the initial plate deflection, selection of an appropriate imperfection level is important. The slenderness ratio taken in this analysis is conservative, with the consequence that the maximum deflection of relatively slender plates is under-predicted.

Figure 6b shows how the level of plate residual stress changes the vertical bending response of the midship section. In this case, for increasing values of plate residual stress, the yield moment and ultimate limit state decrease, but the flexural rigidity remains constant. A geometrically imperfect plate has some capacity to respond elastically to the increasing load, though this decreases with increased initial deflection amplitude which is illustrated by the reduction is flexural rigidity. In contrast to the effect of initial plate deflection, the reduction of plate yield stress is a global defect resulting in a smaller capacity of the overall structure to respond elasto-plastically pre-collapse.

The effect of increasing initial stiffener deflection and stiffener residual stress on the ultimate limit state of the midship is negligible, as shown in Figure 6c and d respectively. As the stiffeners of the patrol boat are relatively small and carry a relatively small proportion of the load, and the failure modes of the midship are not dominated by stiffener failure modes, stiffener imperfections have minimal consequence on the overall strength for the vessel studied. This may not hold for steel ship structures with characteristic scantlings.

The impact of increasing HAZ breadth and HAZ yield stress ratio on the reduction of the ultimate strength of the patrol boat midship is substantial, as illustrated in Figure 6e and f respectively. In welded aluminium structures the HAZ is softened and the yield stress reduced. As fabrication of a patrol boat includes a significant amount of welding due to the high number of stiffeners, HAZ related imperfections play a large role in the reduction of overall strength. Fracture of the plate under tension or collapse under compression will occur in the HAZ prior to failure of the non-welded material. The reduced yield stress affects the ability of the overall structure to carry the bending moment elastically as the yielded regions spread. In addition, the structure has a relatively small pre-collapse plastic region due to the global nature of the imperfection.

Future Work

Although ALPS/HULL has the distinct advantage over non-linear FE procedures that it is considerably time efficient, the reduced ability to tailor a model to represent an actual build becomes an issue when a more accurate or detailed analysis is required. For example a vessel featuring both deck plating with integrated stiffeners, and bottom plating with welded stiffeners, cannot be modelled with ALPS/HULL at this time, resulting in an unrealistic decrease in strength of the deck plating.

The softening characteristics in the HAZ of an aluminium structure depend on the alloying materials. Though both AA5083 and AA6082 are used in the model because they are the typical alloys used for plating and extrusions, the reference levels of AA5083 are applied to the whole model. Furthermore, in the present version of ALPS/HULL, the same value of the HAZ breadth and yield stress ratio for both plating and stiffeners is defined. However, welding may produce different defects in plating and stiffeners. For a better representation of the build, the reference levels for HAZ defects of both AA5083 and AA6082 products should be used and different HAZ-related imperfections in plates and stiffeners can be defined.

In order to conduct a more accurate and detailed ultimate strength assessment of the patrol boat midship, a nonlinear FE analysis can be performed. In nonlinear FE modelling and analysis, it may be possible to define unique imperfections of individual panels and stiffeners. However, this procedure is much more time intensive than ALPS/HULL. As such, an ultimate strength analysis of the midship involves a trade-off between accuracy and time.

The work presented in this paper can be expanded by conducting a probabilistic study on the effects of initial imperfections on the ultimate strength of the patrol boat midship. This would require a large number of ALPS/HULL cases to be run, which could be facilitated by execution of ALPS/HULL in batch. This is currently not possible but has been suggested to the developers of MAESTRO, DRS Defense Solutions.

CONCLUSION

With increasing use of aluminium alloys in the construction of naval vessels, as well as increasingly harsher operational conditions, the need to develop better knowledge of the behaviour of aluminium naval structures and the factors that influence their strength has arisen. Research into the applicability of rapid structural integrity assessment tools is critical to enable DSTO to support the operational and through-life requirements of the RAN.

The importance of the extent and type of a weld-induced defect on the capacity of the midship structure to sustain an increasing vertical bending moment can be estimated by completing a matrix of runs using the ISFEM procedure ALPS/HULL. Using suggested reference levels for initial imperfections in welded aluminium plates and stiffeners, the contribution of failed structural items to the overall collapse of the midship section under a vertical bending moment is shown.

Plate initial deflection is shown to be the dominant parameter in the reduction of the ultimate strength of the particular aluminium patrol boat studied, though plate residual strength and HAZ-related imperfections also contribute to a large extent. In contrast, stiffener imperfections have little impact on the strength reduction of the patrol boat midship because of their size and relatively minor capacity to resist bending.

Due to the sensitivity of the ultimate strength of the aluminium patrol boat to plate and HAZ imperfections, confidence in the levels of these defects in structural assessment is necessary. Measured data of initial imperfections of the specific aluminium alloy used in a vessel's construction material is needed.

ACKNOWLEDGEMENTS

The assistance of Professor Jeom Paik of Pusan University, South Korea and Dr Stuart Cannon of DSTO, Australia is appreciated.

REFERENCES

1. Hughes, O. and Paik, J. (2010) *Ship Structural Analysis and Design*. Jersey City, The Society of Naval Architects and Marine Engineers
2. Benson, S., Downes, J. and Dow, R. (2011) Ultimate Strength Characteristics of Aluminium Plates for High-speed Vessels. *Ships and Offshore Structures* **6** (1-2) 67-80
3. Paik, J., Thayamballi, A., Ryu, J., Jang, J., Seo, J., Park, S., Seo, S., Renaud, C., Cojeen, H. and Kim, N. (2006) The Statistics of Weld Induced Initial Imperfections in Aluminium Stiffened Plate Structures for Marine Applications. *Transactions of the Royal Institution of Naval Architects Part A: International Journal of Maritime Engineering* **148** (4) 19-62
4. Paik, J., Branner, K., Choo, Y., Czujko, J., Fujikubo, M., Gordo, J., Parmentier, G., Iaccarino, R., O'Neil, S., Pasqualino, I., Wang, D., Wang, X. and Zhang, S. (2009) TCIII.1: Ultimate Strength. In: *Proceedings of the 17th International Ship and Offshore Structures Congress,* Seoul, South Korea
5. Paik, J., Hughes, O., Hess, P. and Renaud, C. (2006) Ultimate Limit State Design Technology for Aluminum Multi-hull Ship Structures. In: *SNAME Maritime Technology Conference and Expo and Ship Production Symposium,* Houston, United States
6. Gudze, M., Flockhart, C. and Cannon, S. (2006) Short Term Damage Assessment in Naval Ship Structures. In: *Pacific 2006 International Maritime Conference,* Sydney, Australia
7. *MAESTRO Manual*. (2011). Stevensville. DRS Defense Solutions.
8. Smith, C. (1977) Influence of Local Compressive Failure on Ultimate Longitudinal Strength of Ship's Hull. In: *International Symposium on Practical Design in Shipbuilding,* Tokyo, Japan
9. Paik, J. and Thayamballi, A. (2003) *Ultimate Limit State of Steel-Plated Structures*. Chichester, Wiley
10. Adamchak, J. (1997) *ULTSTR: The Revised Program for Estimating the Collpase of Ship HULLs or HULL Components Under Longitudinal Bending or Axial Compression*. Naval Surface Warfare Centre, Carderock Division
11. Morris, B. (2011) *Personal communication.*

A Comparison of Submarine Fin Geometry on the Performance of a Generic Submarine

Gregory J. Seil
Sinclair Knight Merz Pty. Ltd., Australia
GSeil@globalskm.com.au

Brendon Anderson
Defence Science & Technology Organisation, Australia
Brendon.Anderson@dsto.defence.gov.au

INTRODUCTION

The fin (sail or fairwater) is a significant feature in any submarine design, particularly on a diesel-electric boat (SSK) where the size of the fin presents a significant proportion of the boat length. The presence of a fin on a submarine increases its resistance and can introduce complex wake characteristics such as hull-junction vortices which can travel down the casing and enter the propeller plane affecting its performance and noise generation. The size and location of the fin also influences the submarine manoeuvring characteristics, particularly as tip vortices from the fin form and interact with the body vortices formed from cross-flows [1].

The challenge for a submarine designer is to meet the internal functional requirements while optimising the hydrodynamic impact of the shape. For a given height the fin should be "thin", relative to the diameter of the pressure hull, to reduce drag and wake, however this is in conflict with maximizing the useable volume. Joubert [1] notes that fins on United States submarines tend to be like streamlined stub wings, while many Russian designs favour longer squatter shapes. Some Russian fins, such as those on the Alfa and the Akula [2], will have a larger volume to surface-area ratio ie. a greater volume per unit surface area.

Gorski [3] describes the benefits of an "advanced sail" for improved effectiveness in littoral regions. This fin is a canopy-like shape, with a factor of four increase in volume over a conventional airfoil-shaped fin. Gorski [3] investigated the design of the "advanced sail" using Computational Fluid Dynamics (CFD) and experimental testing where the focus was on minimizing the drag increase of design variants of the fin whilst reducing wake distortion.

This paper investigates the effect of changing the volume of a submarine fin on drag and the wake into the propulsor. The paper also studies the effect of blending the fin into the casing.

HULL AND FIN GEOMETRY

The geometry of the DSTO generic submarine pressure hull, casing and aft control surfaces was developed by Joubert [4]. Three fins were studied, with each fin representing an evolution of the previous design. Figure 1 compares profile of each fin. Views of the three fins integrated into the geometry of the submarine are shown in Figure 2. Fin 1 was an airfoil section with constant chord length. Table 1 lists the change in surface area and volume of Fin 2 and Fin 3 <u>*relative*</u> to Fin 1. All three fins have the same height for a consistent comparison.

Fin 2 has longer chord than Fin 1 at the root and shorter chord at the tip, with swept leading and trailing edges. The section thickness at the root of Fin 2 is substantially larger than Fin 1, but the same as that of Fin 1 at the tip. Fin 2 has more than double the volume of Fin 1 and adds an extra 2.0 % to the volume of the submarine.

Fin 3 represents a modification of the basic Fin 2 geometry to reduce drag and improve wake uniformity. The modifications include a leading edge extension of large radius and rounding of the forward fin geometry near the tip. It can be seen from Table 1 that Fin 3 has slightly greater volume than Fin 2.

An important point to note is that both Fin 2 and Fin 3 produce essentially no change in the total surface area of the submarine, relative to Fin 1. This at first appears counter-intuitive but is actually logical as Figure 3 illustrates. At *relevant* locations along the chord of the fin, the length (L_3) and hence surface area of the side of Fin 2 or Fin 3, will be less than the *combined* length and surface area of the side of Fin 1 (L_1) and the corresponding casing (L_2) under Fin 2 or Fin 3. It is therefore possible to substantially increase fin volume, whilst maintaining or even decreasing wetted area.

	Change relative to Fin 1	
	Fin 2	Fin 3
Fin Volume	109.1 %	110.2 %
Submarine Volume	2.0 %	2.0 %
Fin Surface Area	29.7 %	30.0 %
Submarine Surface Area	-0.003 %	-0.007 %

Table 1 Change in volume and surface area relative to Fin 1

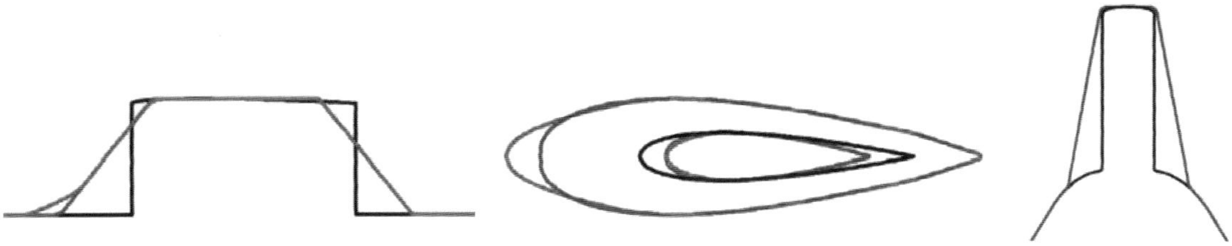

Figure 1 Comparison of fin profiles (Fin 1 – black, Fin 2 - blue, Fin 3 – red)

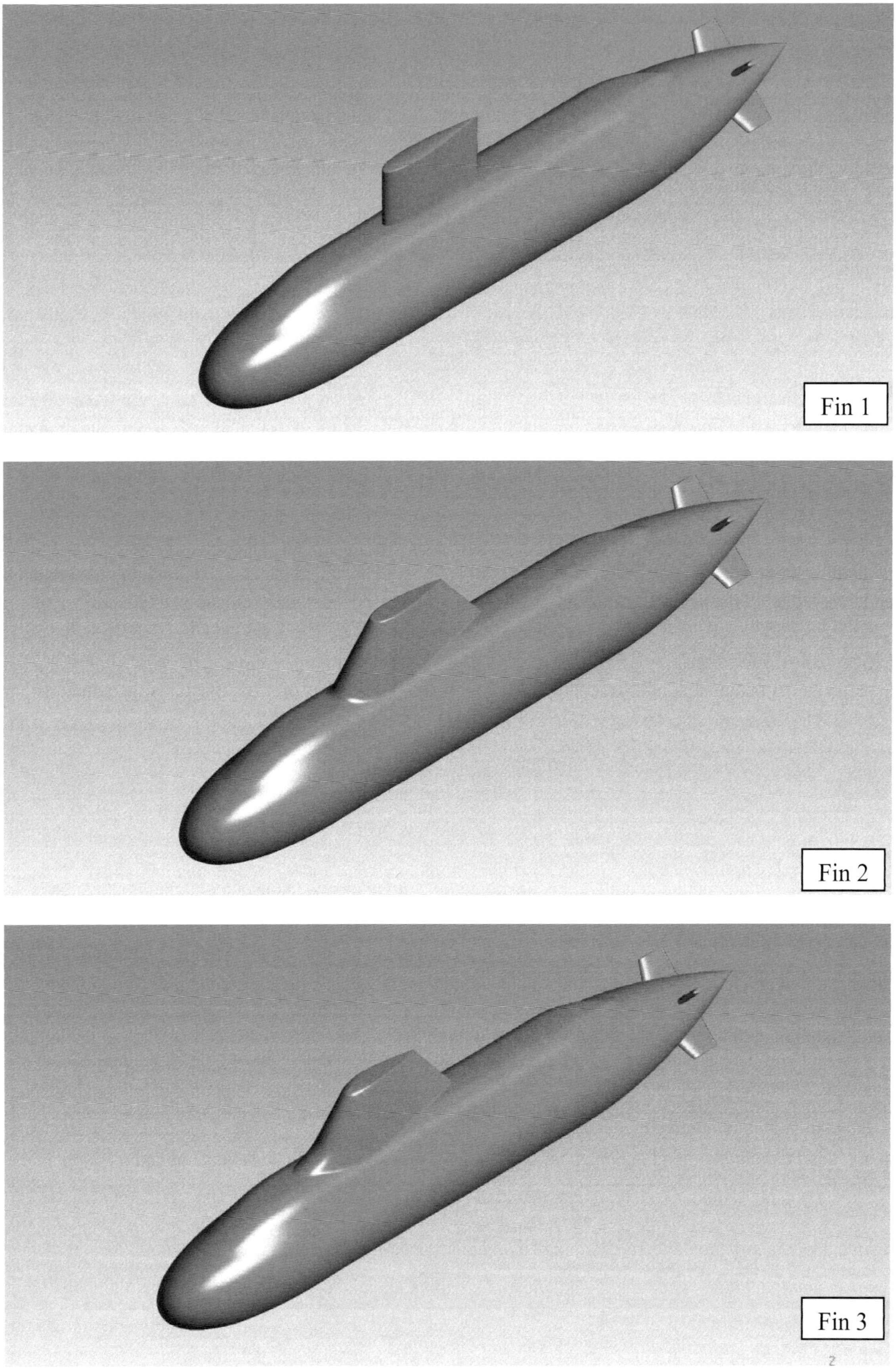

Figure 2 Comparison of submarine geometry

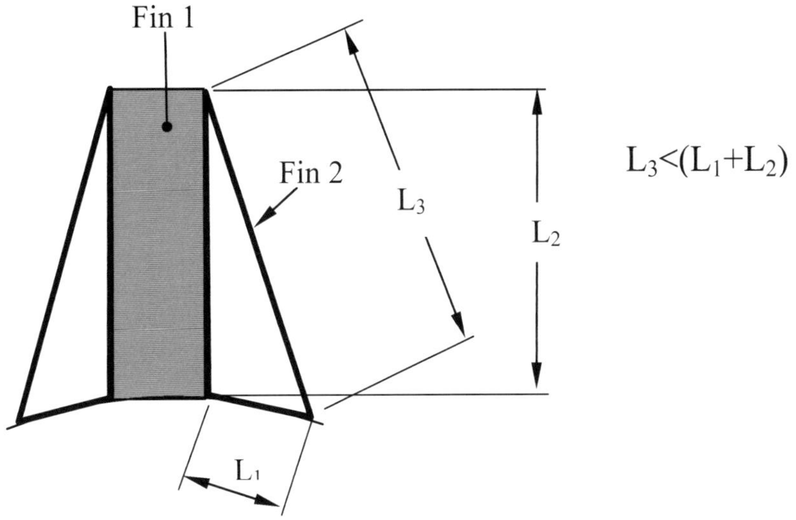

Figure 3 Effect of geometric shape on wetted surface area

COMPUTATIONAL MODELLING

The submarine was modelled for a deeply submerged straight-ahead condition. All calculations were performed for a *model-scale* Reynolds number of 1.0×10^7 using the CFD software ANSYS FLUENT 13. CFD modelling was based on the solution of the Reynolds-averaged Navier Stokes (RANS) equations with two-equation turbulence modelling.

The modelled submarine was placed in a rectangular volume, representing the surrounding water. The modelled flow domain extends 1.0L upstream and 2.0L downstream of the submarine, where L is the length of the modelled submarine. It also extends 1.5 L laterally in the horizontal and vertical directions from the centreline of the submarine. Using the assumption of flow symmetry, only one-half of the total flow domain was modelled. The inflow and outflow boundaries were modelled as a velocity and a pressure boundary, respectively. The far-field boundaries were modelled as "symmetry" boundaries, equivalent to a "slip" wall.

The flow domain was meshed using the mesh generation software Gridgen V15.16 from Pointwise Inc.. A fully structured mesh was produced in order to provide high quality resolution of boundary layer development over the hull and wake propagation downstream of the fin. The mesh of the half-model contained 9,872,475 cells for the hull with Fin 1. The SST k-ω turbulence model of Menter [5] and Menter et. al. [6] with wall functions was used for each calculation. The QUICK scheme [7] was used to provide interpolation of convective terms in the discretised transport equations for momentum, turbulent kinetic energy (k) and the specific dissipation rate (ω).

RESULTS

Resistance and Powering

The change in resistance of the submarine with Fin 2 and Fin 3 is shown in Table 2. At the same speed, the change is resistance is equal to the change in power. The values presented in Table 2 are based on model-scale dimensional values, to provide a consistent comparison based on the same overall length of the hull. The forces have been broken down into forces arising from integration of the pressure field acting on the surface of the submarine (pressure force) and those arising from skin friction (viscous force). Fin 2 increases the resistance of the submarine by 2.2%, whereas with Fin 3 the resistance of the submarine is actually decreased by close to 0.9%. This is a significant result, because it shows that despite the substantial increase in fin volume, it is possible to actually decrease resistance by careful design.

	Change relative to Fin 1	
	Fin 2	Fin 3
Pressure force	20.3 %	3.1 %
Skin friction force	-0.4 %	-1.5 %
Total force	2.2 %	-0.9 %

Table 2 Change in submarine resistance and power relative to Fin 1

Both Fin 2 and Fin 3 increase the pressure drag force acting on the submarine configuration, with the greatest increase being for Fin 2. The drag due to skin friction is decreased with Fin 2 and Fin 3, with the greatest decrease being for Fin 3. For both of these fins, the magnitude of the decrease in the skin frictional force is greater than the reduction in wetted area of the whole submarine shown in Table 1.

Flow around Hull

The development of the boundary layer over each submarine configuration is compared in Figure 4. This is shown by the contours of velocity magnitude plotted on vertical planes at various locations along the length of the hull and also downstream of the hull. The red contours represent free-stream velocity and the blue contours low velocity. The surface of each submarine has been coloured by the static pressure, with red contours representing high pressure and blue contours low pressure. The hydrostatic term associated with depth of submergence is neglected. The development of the wake from the fin and aft control surfaces is clearly evident in each plot.

The development of the absolute value of the helicity (magnitude of the streamwise vorticity ie. vortex development in the streamwise direction) of the flow over each submarine fin configuration is compared in Figure 5. This is shown by the contours plotted on vertical planes at various locations along the length of the hull and also downstream of the hull. The helicity is defined as,

$$H = \left(\nabla \times \tilde{U}\right) \cdot \tilde{U} \tag{1}$$

where $\tilde{U}$ is the velocity vector. The red contours represent high values of helicity magnitude and the blue contours low values. The surface of each submarine has been coloured by the static pressure, using the same scale as that of Figure 4.

Whilst Fin 1 produces a fin root vortex ("necklace" vortex), Fin 2 produces a much stronger vortex, which is evident both in terms of its helicity and also in the more localised thickening of the hull boundary layer at the vortex location. Since this vortex distorts the hull wake at the propulsor plane and creates stronger secondary flow, this has clear implications for signature management. Fin 3 produces a necklace vortex and boundary layer similar to Fin 1.

Fin 2 and Fin 3 produce stronger fin tip vortices. Under straight ahead conditions this will not affect flow into the propulsor, but can still be addressed through careful design. In Fin 3, the rounding of the tip of the fin at the leading edge of Fin 2 eliminates the low pressure spike evident at this location in Fin 2 and therefore reduces the possibility cavitation at high speeds and small depth of submergence.

Nominal Wake

The wake produced by each submarine configuration at the first vertical plane downstream of the aft control surfaces, indicated in Figure 4, has been plotted in terms of *axial* wake fraction and compared in Figure 6. The axial wake of the hull without fin has been plotted for reference. The axial wake fraction (w) is defined as,

$$w = 1 - \frac{U_x}{U_s} \tag{1}$$

where U_X is the axial velocity and U_S the speed of the submarine. In the figure, red contours represent high values of wake fraction and blue contours low values.

The wake from the aft control surfaces is evident in each plot of Figure 6. Fin 2 produces a more distorted "hook" shaped wake than Fin 1. The wake of Fin 3 is similar to the wake of Fin 1 up to a radius corresponding to half the radial extent of the wake from the aft control surfaces. Both Fin 2 and Fin 3 produce a greater wake from the fin. While conclusions can be drawn by observing visual differences in nominal wake from Figure 6, it is important to quantify the magnitude of these differences to avoid any ambiguity in interpretation.

For an axisymmetric body at zero degrees incidence, we will expect non-uniform wake velocities in the radial direction, but uniform velocities in the circumferential direction. Radial non-uniformity can be compensated for in propulsor design, whilst the latter poses design challenges as the propulsor blading rotates through a circumferentially varying inflow at a given radius. Therefore circumferential non-uniformity is of particular interest when quantifying wake "distortion".

The average wake fraction at a given radius is simply,

$$\overline{w} = \frac{1}{2\pi}\int_0^{2\pi} w\,d\theta \approx \frac{1}{n}\sum_{i=1}^{n} w_i \tag{2}$$

for n equal spaced measurement points around the circumference. It is assumed here that n is large enough to adequately represent the wake profile at given radius. A distortion coefficient (D_C) can then be defined at each radius based on the standard deviation of w about its mean,

$$D_C = \sqrt{\frac{\sum_{i=1}^{n}(w_i - \overline{w})^2}{n-1}} \tag{3}$$

for n equally spaced points around the circumference.

The radial variation of distortion coefficient calculated for each fin type has been plotted in Figure 7. The left hand plot shows contours of w (for Fin 1) superimposed on a grid of circumferential lines of different radial distance to provide a reference for the right-hand plot. In the figure the vertical and radial dimensions have been non-dimensionalised by the maximum outer radius (R_{Max}) of the pressure hull.

Figure 7 shows that Fin 2 produces the greatest wake distortion for $r/R_{Max}<1.55$, which is consistent with what can be observed from the wake plots of Figure 6. For $r/R_{Max}<1.0$, Fin 3 produces similar wake distortion to Fin 1, although with slightly more distortion for $r/R_{Max}<0.6$.

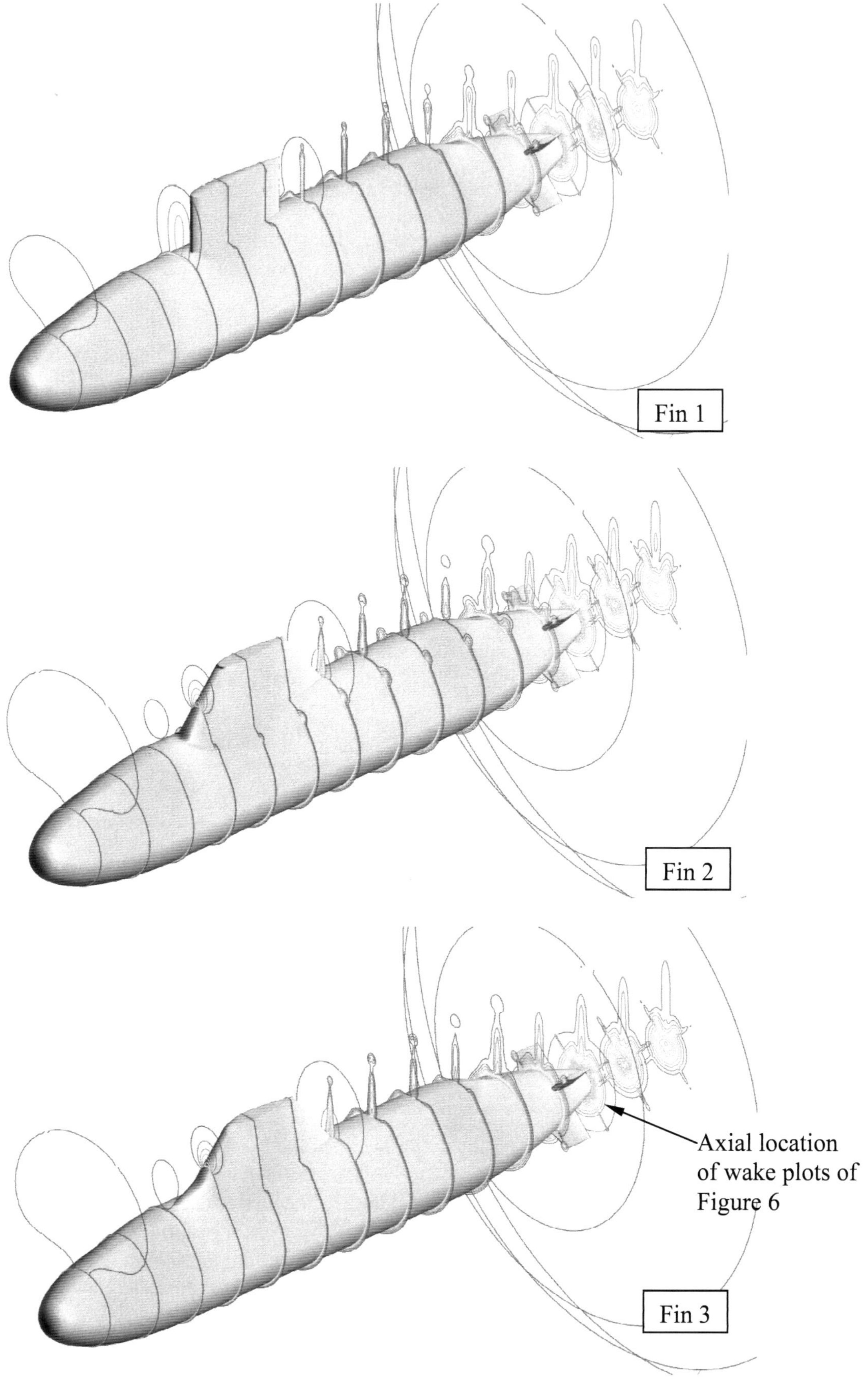

Figure 4 Submarine pressure distribution and boundary layer development

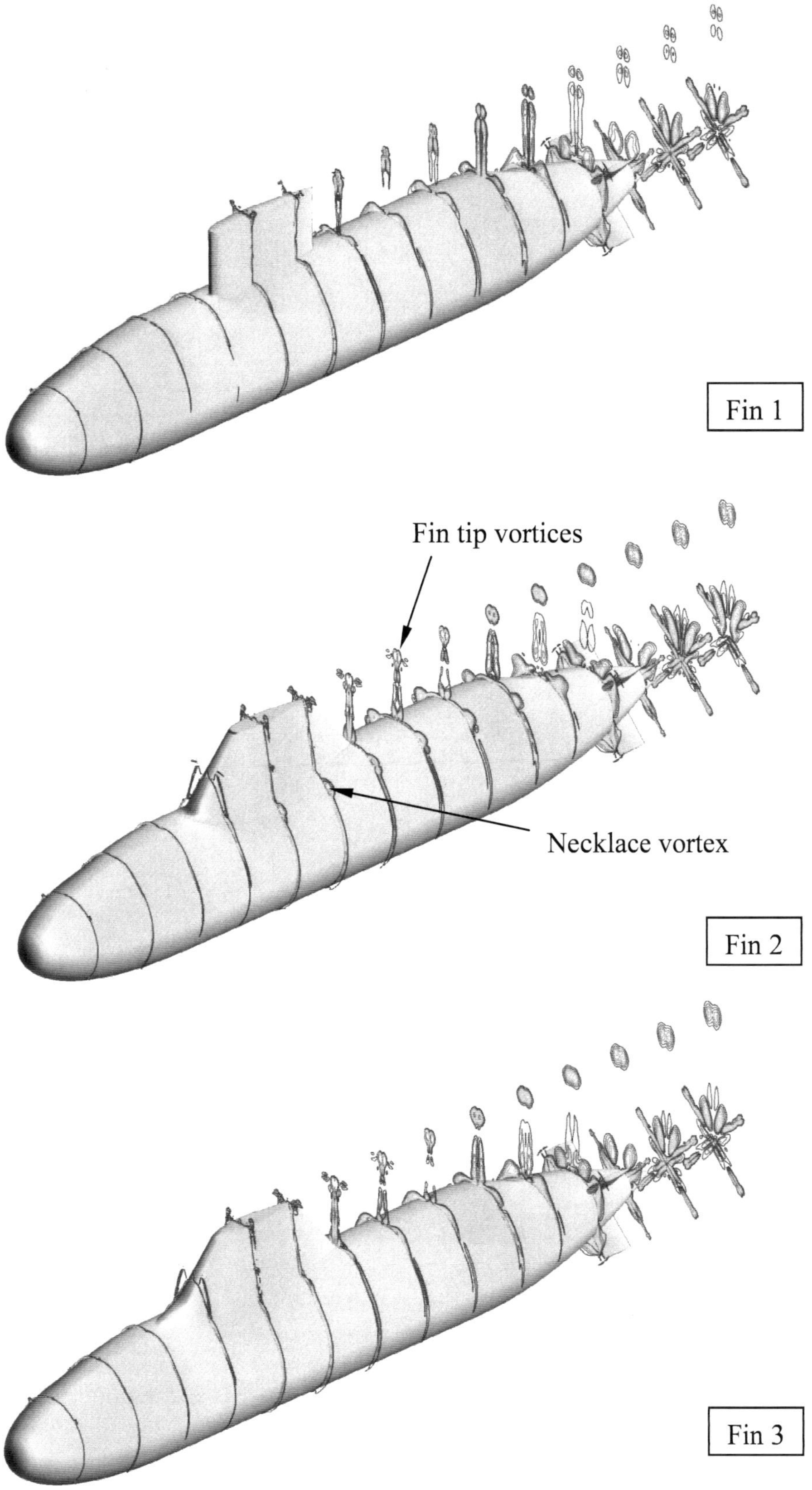

Figure 5 Submarine pressure distribution and fin root and tip vortex development

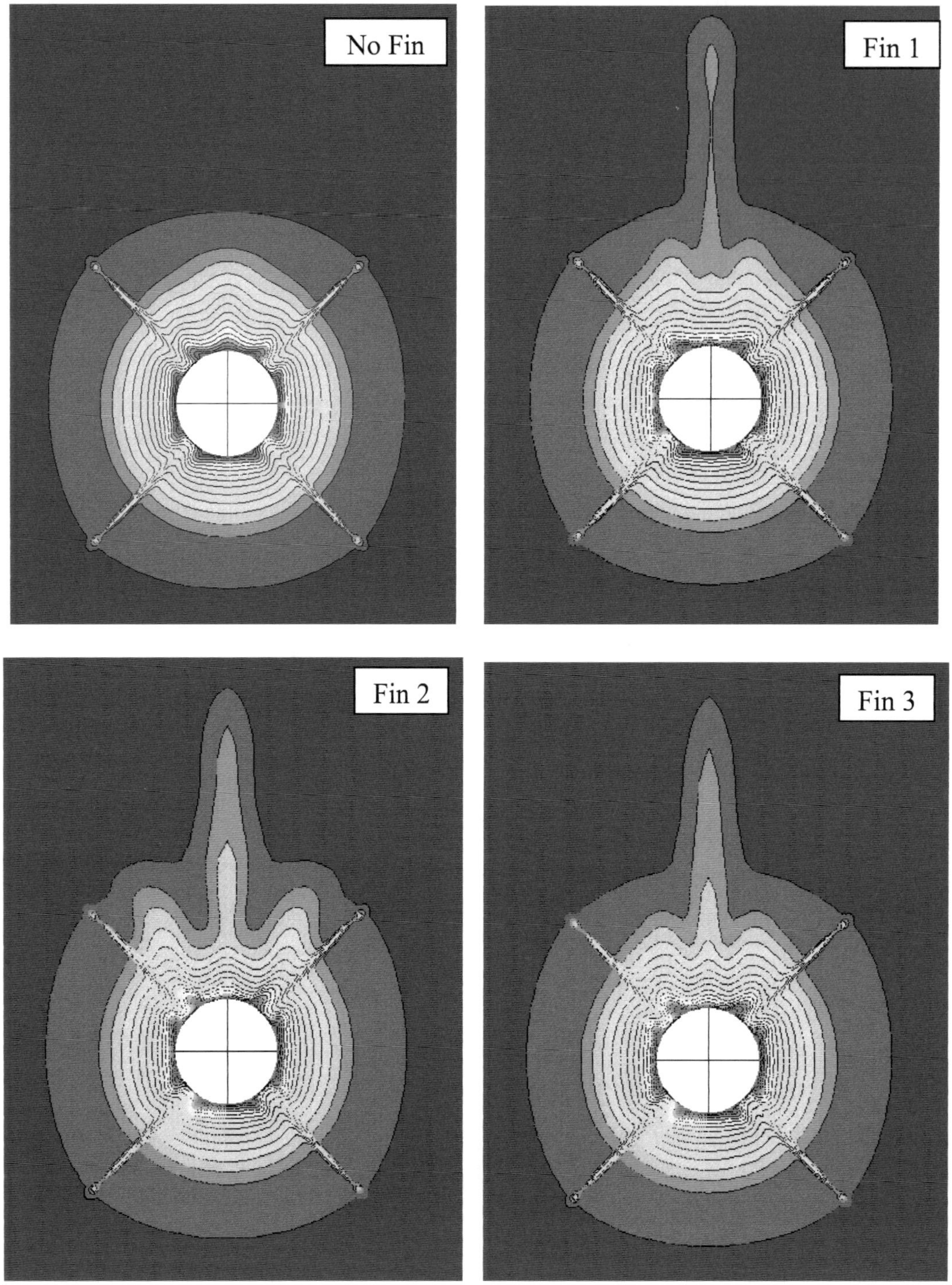

Figure 6 Distribution of nominal axial wake fraction downstream of aft control surfaces

CONCLUSIONS

The study presented in this paper shows that it is possible to increase the volume of a submarine fin above that of a "thin" fin, whilst actually decreasing drag and incurring a slight increase in wake non-uniformity at the location of the propulsor. Fin 3 is by no means optimum and there exists the possibility to further increase the volume through careful shaping whilst maintaining similar drag and wake properties.

This paper only considers the design of different fins on the resistance of a generic submarine. The effect of the fin design on the dynamic stability of the submarine whilst surfacing, due to the volume of entrained water, has not been considered. In addition to the resistance, it is important to consider the effect of fin design on the manoeuvring performance of the submarine, in particular the swept leading and trailing edges. The manoeuvring performance has not been presented as the resistance is the focus of this paper.

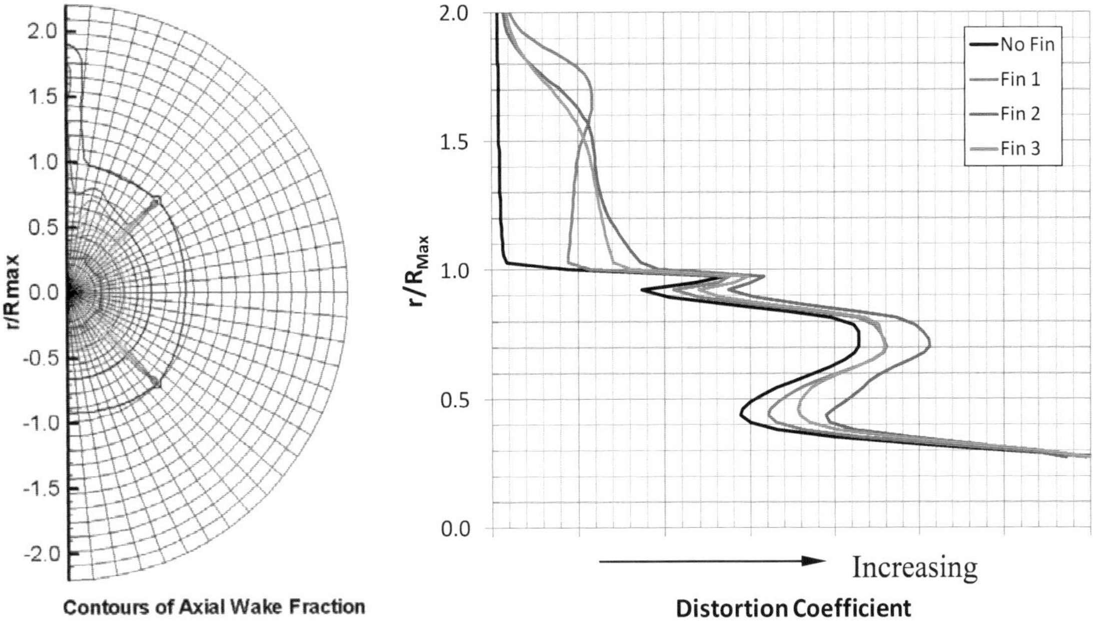

Figure 7 Radial variation of wake distortion

ACKNOWLEDGEMENTS

The authors would like to thank Sinclair Knight Merz Pty Ltd and Defence Science and Technology Organisation for supporting this research and the publication of this paper.

REFERENCES

1. Joubert, P.N., (2004), "Some Aspects of Submarine Design Part 1. Hydrodynamics", DSTO Technical Report, DSTO-TR-1622.
2. Polmar, N. and Moore, K. J., (2004), *Cold War Submarines*, Potomac Books, Washington DC.
3. Gorski, J.J., (2002), "Use of RANS Calculations in the Design of a Submarine Sail", RTO AVT Symposium on Reduction of Military Vehicle Acquisition Time and Cost through Advanced Modelling and Virtual Simulation, *RTO-MP-089*, Paris, France, 22-25 April 2002.
4. Joubert, P.N., (2006), "Some Aspects of Submarine Design Part 2. Shape of a Submarine 2026", DSTO Technical Report, DSTO-TR-1920.
5. Menter, F. R., (1994), "Two-Equation Eddy-Viscosity Turbulence Models for Engineering Applications", *AIAA Journal*, Vol. 32, No. 8, pp. 1598-1605, August 1994.
6. Menter, F R., Kuntz, M. and Langtry, R., (2003), "Ten Years of Industrial Experience with the SST Turbulence Model", In Hanjalic, K., Nagano, Y. and Tummers, M., editors, *Turbulence, Heat and Mass Transfer 4*, pages 625-632. Begell House Inc., 2003.
7. ANSYS Inc. (2010), *ANSYS FLUENT 13.0 Theory Guide.*

Sonar Acoustic Transducers Technology: A Priority Industry Capability For Australia

Tim Cain, Paul Baker
Thales Australia

ABSTRACT

The 2009 Defence White Paper defined Acoustic Technologies and Systems as a Priority Industry Capability (PIC). A PIC is a capability that confers an essential strategic advantage by being resident in Australia and, if not available, would significantly undermine defence self-reliance and ADF operational capability.

A key element of the Acoustic Technologies and Systems PIC is the design and manufacture of piezoelectric ceramic sensor components that are at the heart of underwater acoustic systems.

Whereas the inboard processing and display technologies required for sonar system implementation have become commoditised and are generally widely available in industry, the acoustic sensor technologies remain a niche capability that is frequently coveted and protected by those leading nations who hold this capability advantage.

Fortunately, Australia is one of the nations with an indigenous acoustic sensor design and manufacturing capability. This provides the ADF with assured access to the latest and best sensor technologies, to support in-country product development to achieve the performance necessary for the RAN's sonar operating requirements and environmental conditions.

In the face of emerging regional powers seeking to access strategic capabilities and driving the need to support an increased ASW effort, it is important now more than ever to safeguard and grow Australia's strategic advantage conferred by this indigenous acoustic technology and systems capability.

In this paper we outline the technologies and processes involved in the design and manufacture of acoustic sensors and identify key actions required to nurture this PIC for the continued self-reliance of the ADF operational capability.

INTRODUCTION

"Minerals of unknown composition and doubtful purity are exposed to immeasurable heat long enough to carry unknown reactions partly to completion, forming the heterogeneous non-stoichiometric materials known as ceramics." So humorously observed S.W. Bradstreet in 1958 when describing ceramics properties.

These comments are directly applicable to the family of piezoelectric ceramics as well. The control of ceramic properties from batch to batch is a major challenge for all manufacturers no matter how much care and attention is paid to the material properties of the individual constituents of modern piezoelectric ceramics, or to the manufacturing processes employed to combine them. Indeed there have been military standards written to define piezoelectric ceramic characteristics into specific "Types" but this has only served to highlight the significant differences in the resulting materials made to the same "standard" by different manufacturers. Despite the variability in material properties, material scientists and expert sonar transducer designers are able to utilise piezoelectric ceramics to produce rugged, reliable high performance hydrophones and underwater projectors.

PIEZOELECTRIC CERAMIC TECHNOLOGY IN SONAR ACOUSTIC TRANSDUCERS

Sound travels through water better than any other type of radiation– for instance RF and light. Sound is created by a number of sources, such as the ship or submarine's motion through the sea, the machinery onboard, and the rotation of propellers, engines, and pumps. Sound can travel long distances and if there is an array of pressure sensors and the associated electronics and signal processing, one can detect amplitude, frequency, direction and therefore derive important information such as the range and speed of a vessel or "contact".

The state of the art in sonar pressure sensors or hydrophones is the piezoelectric ceramic sensor, based on the piezoelectric effect discovered in 1880. Piezoelectric effect is the ability of certain crystals to develop an electric charge when acted upon by a force, in this case sound pressure. When large piezoelectric effects were discovered in ceramics just after World War Two, they became the standard for underwater pressure sensors.

The piezoelectric ceramic materials used in today's sonars are typically based on Barium Titanate or Lead Zirconate Titanate (PZT) formulations. These piezoelectric ceramic elements can be manufactured in a wide variety of shapes and sizes to suit a number of sonar applications, from small high sensitivity sonobuoy hydrophones to high power transducers used in variable depth sonar transmitters. A selection of piezoelectric ceramics manufactured in Australia utilising these formulations and a selection of completed sonar transducers is shown in Figure 1.

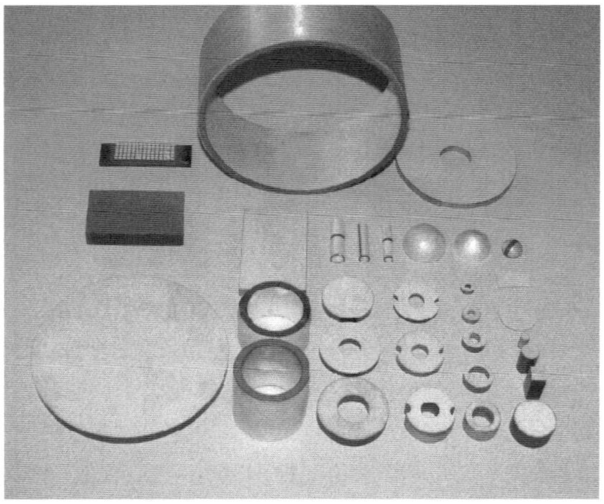

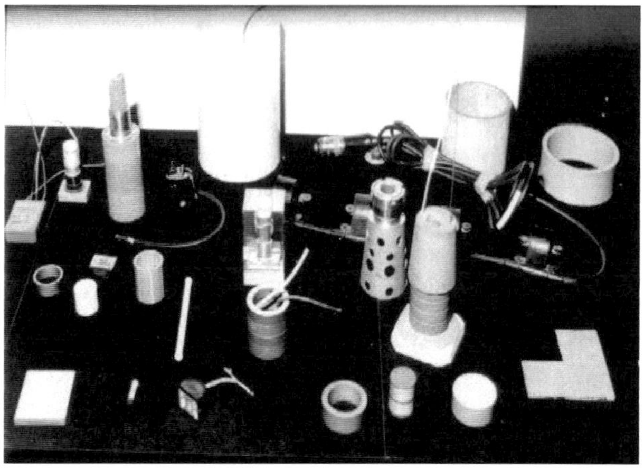

Figure1; left: piezoelectric ceramic parts; right: assembled sonar transducers

A BRIEF HISTORY – PIEZOELECTRIC CERAMIC MANUFACTURE IN AUSTRALIA

The ceramics facility at Thales Australia has its origins in the capacitor manufacturing division of the old Ducon company after World War Two. After some early collaboration with the Erie Capacitor Company of the USA, Ducon began ceramic capacitor manufacture at Villawood, NSW, in about 1950 and employed Mr Hans Ruckert as chief ceramicist. Mr Ruckert developed a full range of ceramic material for capacitor use. The range included piezoelectric barium titanate compositions.

Although PZT had been developed in the 1950s, it was closely associated with the Brush Clevite Company of the USA and their patents ensured an effective stranglehold on the market particularly as the US Navy used the materials as the basis for their own standards. This situation resulted in the Clevite tradename for their product, "PZT", becoming the generic name for the whole class of materials.

In the mid-1960s Hans Ruckert began looking at ways to circumvent the Clevite patents. Ruckert took his lead from developments in Japan and eventually patented a series of compositions based on PZT with the addition of lead nickel niobate in 1969. These materials have formed the basis of Thales's product line since that time. In the 1980s and 1990s the benefits of these materials gained worldwide recognition and they are now produced by many companies, especially for high performance actuators.

The 1970s saw the demise of the electrical components industry in Australia. The ceramics facility survived because of its involvement in the Barra Sonobuoy program. This program enabled the facility to be moved to the Meadowbank site and during the 1980s the facility was expanded and modernised.

Following the retirement of Hans Ruckert in 1979, the facility was managed by Dr Gavin Tulloch who greatly expanded the product base and commenced transducer manufacture for a number of RAN projects. New materials were actively investigated and a significant composite material program was set up.

In 1996 the ceramics facility was moved to its third home at the Rydalmere site. At this time, the facility expanded its product range to enable better matching to customer requirements (such as the production of a Type III high power ceramic). Design and manufacture hydrophones for the commercial seismic market commenced and the facility started supplying larger quantities of ceramic to export markets.

In 2002 Thales Australia celebrated the manufacture of its 1000th batch of piezoceramic. This milestone represented over 100 tonnes of piezoceramic powder and reaffirmed the support for a fully Australian developed technology.

AN EXAMPLE OF SONAR ACOUSTIC TRANSDUCER INDUSTRY CAPABILITY

Piezoelectric ceramic properties are complex and non-linear. Not only do they depend on the characteristics of the individual compounds used in the formulation and the manufacturing processes employed to cause them to react but the properties change with temperature and age. It requires deep expert knowledge in material science, physics and electronics to develop underwater acoustic devices.

Fortunately, Australia is one of the nations with an indigenous acoustic sensor design and manufacturing capability. This provides the ADF with assured access to the latest and best sensor technologies, to support in country product development to achieve the performance necessary for the RAN's sonar operating requirements and environmental conditions. The Thales Australia Acoustics Engineering Department provides:

- Innovative and timely solutions for internal and external transducer customers;
- Mastery of PZT materials – specification, formulation, prototypes, production and test;
- Transducer Design and prototyping capability – utilising tools such as Finite Element Analysis (FEA) modelling (refer Figure 2), and
- Acoustic Testing of Transducers and Arrays – using a 700,000 litre Acoustic Test Tank Facility and pressure vessels of various sizes;

This engineering capability is coupled with a manufacturing team employing techniques such as Design For Manufacture and Assembly (DFMA) to provide the only vertically integrated transducer design and manufacturing department in Australia able to supply low cost, mass manufacture acoustic sensors. Many of the sonar arrays fitted to current RAN submarines and surface ships were manufactured locally in the Rydalmere NSW facility. Sonar sensors manufactured in Australia also go into some of the most advanced high performance sonar system programs in Europe.

Recent developments in single crystal piezoelectric ceramics are providing prototype sensors with phenomenal performance in terms of sensitivity and bandwidth. This technology will ultimately lead to a revolution in underwater acoustic sensors that Australian industry must understand, master and be able to take advantage of where necessary.

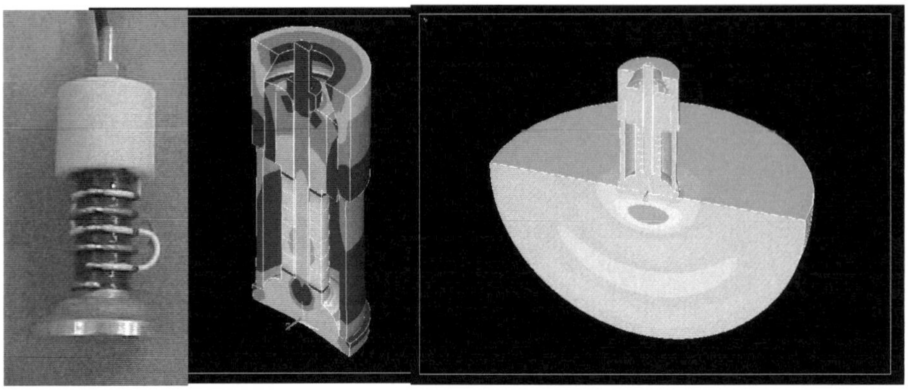

Figure 2; An example of transducer modelling and prototyping capability

It should now be self-evident that acoustic sensor technologies remain a niche capability that is frequently coveted and protected by other leading nations who hold this technology advantage.

A significant amount of this capability resides in the people. The scientists and engineers skilled in the art of piezoelectric ceramic formulations or sonar transducer design are quite rare. Their skills are not easily acquired in the market but are developed over time "on the job". There is a need to specifically target recruitment of graduate materials scientists to familiarise them with piezoelectric ceramic development and associated processes and techniques. This has presented a significant challenge during the current resources boom, competing with mining companies for the talented graduates of Australia's universities. To be successful, recruitment requires development of relationships with Australian universities (such as UNSW School of Material Science and Engineering) and the "positive advertising" of the interesting technical challenges that await the nascent Material Scientist.

Some of those technical challenges go straight to the ability to support the RAN in the unique waters that surround Australia and so give them the capability edge. The RAN has extensive experience with the performance of minehunting sonars in tropical waters and therefore developed an extremely rigorous performance specification during the acquisition of the Minehunter Coastal (MHC) platform. The 2093 Mine Hunting Sonar, developed initially for operation in northern hemisphere waters, was selected as the main sonar. After detailed comparison and analysis of the underwater acoustic operational environment, sonar performance improvements were identified to compensate for the increase in high frequency absorption due to the elevated water temperatures experienced off northern Australia.

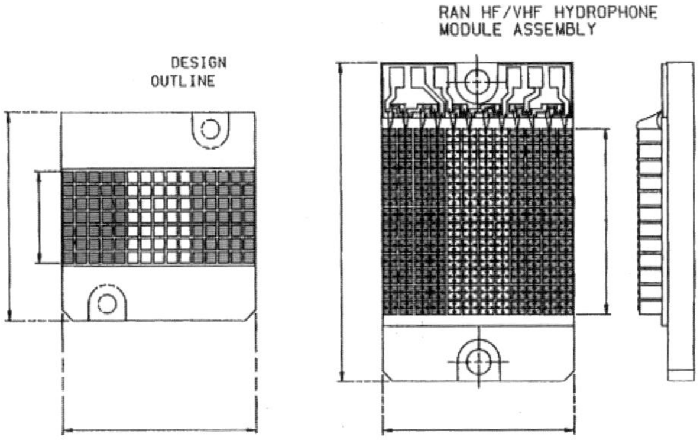

Figure 3; Doubling the height of hydrophone modules to improve tropical water sonar performance

Some examples of the array design adaptations developed to meet the sonar performance requirements have been:

- Doubling the height of hydrophone modules (refer Figure 3) to increase the Directivity Index (DI)
- Moving the hydrophone resonance including changing filter the characteristics of preamplifier
- Dividing each channel in two and connecting them in series for better noise matching to the FET preamplifier
- Employing low capacitance cabling to reduce losses
- Changing the curvature of the arrays for improved beam patterns.

Without these changes there would have been a reduced operational mine hunting capability in the northern ports and harbours of Darwin, Cairns and Brisbane.

PRIORITY INDUSTRY CAPABILITY (PIC)

A PIC is a capability that confers an essential strategic advantage by being resident in Australia and, if not available, would significantly undermine defence self-reliance and ADF operational capability. A key element of the Acoustic Technologies and Systems PIC is the design and manufacture of piezoelectric ceramic sensor components which are at the heart of underwater acoustic systems.

The next ten to fifteen years will see an explosion in the number of next generation conventional submarines appearing in our region. The Indian and Pacific Oceans will see expanded operations of these quiet submarines as they seek to operate beyond their coastlines and project a presence closer to Australian shores. Enabling technologies such as Air Independent Propulsion (AIP) and improved high density battery storage will provide already existing stealthy diesel submarines with extended underwater endurance, making them more difficult to detect.

The ability of submarines potentially to operate undetected cannot be underestimated. The force multiplier effect of submarines is well known and the number of assets necessary to perform Anti-Submarine Warfare (ASW) is disproportionately high. In addition, these submarines pose a significant threat to both Australia's lines of communications with allies and its international shipping routes through the Indonesian archipelago and beyond. Next generation conventional submarines proliferating in the region will be fitted with the latest export sonar systems from major sonar system providers. In the face of emerging regional powers seeking to access this strategic capability and driving the need to support an increased ASW effort, it is important now more than ever to safeguard and grow Australia's strategic advantage conferred by this indigenous acoustic technologies and systems capability.

Initiatives such as the Capability Technology Demonstrator (CTD) program greatly assist in the retention of technical skills and fuel innovation in underwater acoustic technologies and systems. An example of this is the Fibre Laser Sensor, initially developed by DSTO, and then developed further by Thales Australia into an electro-optic hydrophone, the first of its type in the world. This technology has the potential to revolutionise sonar systems in terms of the reduction in supporting infrastructure (e.g. copper cabling, preamplifiers) and provide building blocks for future sonar systems capability that is impossible with current technology. All this has been developed in Australia. All this would be impossible to achieve without a local acoustic technologies and systems capability. Without it, Australia would be reliant on imported technologies and subject to associated supply limitations and restrictions.

The CTD program has had a healthy number of sonar related topics over recent years but more could be done to encourage acoustic sensor development, for it is here where the major obstacle to accessing overseas technology resides. A sonar system is only is as good as its acoustic sensor. Low noise, high sensitivity, multiple sensor, large aperture arrays are the crown jewels of established navies around the world. If Australia is to maintain its regional capability edge, it must be able to have access to next generation sensors or alternatively have the capacity to develop its own equivalent technologies. It is with some pride to note here that some Australian companies have already demonstrated their transducer design skills and expertise in developing high performance acoustic sensors, particularly 1-3 composite sensor technology, to the point where sonar houses supporting established navies have engaged Australian transducer designers seeking innovative technical solutions for their next generation sonar sensors. CTD topics targeted at the evolution of RAN in-service sensor fits, where the fit and form remains the same but the function is improved, would provide a low cost capability upgrade that could be realised in the near term.

Another action in support of indigenous underwater acoustic technologies and systems is to ensure large sonar system acquisition programs include a significant Australian Industry Involvement (AII). Recent sonar system acquisitions by DMO regrettably contain no Australian transducer technology, no Australian piezoelectric ceramic and manufacturing input for the first time in more than twenty years. Achieving significant Australian content in the underwater acoustic technologies and systems in future acquisition programs would provide practical support for the Underwater Acoustic Technologies and Systems PIC.

Finally, to showcase Australian underwater acoustic technologies and system technical skills to our allies, Defence Export Division co-ordination and funding support of Australian company trade delegations to overseas sonar transducer workshops should be welcomed.

CONCLUSION

If Australia is to continue to have access to high performance underwater acoustic technologies and systems, it is necessary for the Government to acknowledge and actively support Australian industries engaged in the areas of development and through-life support of underwater acoustic systems.

The first step in doing this has been achieved in identifying Underwater Acoustic Technologies as a Priority Industry Capability. The next steps are considerably more complex, especially defining the mechanisms by which Government and Defence can engage with industry to maintain the capability in a competitive "value for money" framework.

This paper has proposed some ideas as to how this next stage might be accomplished including:

- Capability Technology Demonstrator topics directed at the evolution of in-service acoustic sensors

- Australian Industry Involvement required in major sonar system acoustic sensor acquisitions

- Defence Export co-ordination and funding of an "acoustic technologies and systems" trade delegation to appropriate countries.

This will go some way towards providing the conditions under which the best underwater acoustic technologies necessary for the defence of Australia might be realised.

ACKNOWLEDMENTS

The authors would like to thank Dr Peter Bryant of Thales Australia for his contribution to the section concerning the history of piezoelectric ceramics manufacture in Australia.

REFERENCES

S.W. Bradstreet — Bulletin of the American Ceramic Society 37, 510, 1958.

O. B. Wilson — Introduction to Theory and Design of Sonar Transducers

Morgan Electric — Ferroelectricity Primer, Technical Publication TP-217

P. Bryant History of 1000 Ceramic Batches, Thales Technical Note, 2005

B. Pacey Under the Air Sea Gap - Australia's ASW Challenge, Kokoda Paper 16, 2011

Department of Defence http://www.defence.gov.au/dmo/id/pic/PIC_Factsheet.pdf

Department of Defence Building Australian Defence Capability: A Policy for a Smarter and more Agile Defence Industry Base

Driving Efficiency into Maritime Defence Programs

Rear Admiral Ric Cheadle (Rtd) CB DL

Director Newton Europe
ric.cheadle@newtoneurope.com

INTRODUCTION

Maritime Defence programs in the United Kingdom have, over the last decade, attracted the attention of bodies such as the National Audit Office, Public Accounts Committee and House of Commons Defence Committee for their inability to deliver on time and to the agreed budget. The reason behind the headlines are many, varied and complex and while this paper does not seek to explain these, it remains an evident truth that original estimates on time and cost have invariably proven optimistic. Over the last decade several actions have been put in place to improve performance. These include the introduction, in 2006 of the Maritime Industrial Strategy to secure a more balanced program and limit the 'peaks and troughs' within the maritime build sector and more recently, as a result of the Gray Report, a much greater level of 'due diligence' to ensure that programs are genuinely affordable before significant investment is made. Such moves might be described as 'top down' or programmatic. This paper attempts to complement these by looking at a range of issues, through a number of worked examples, where efficiencies have been achieved by 'bottom up', approaches. These equally mitigate the risks associated with time and cost overrun but can often be overlooked by project teams who lack specific expertise in engineering problem solving or by Management who fail to recognise their potential for influencing the bottom line.

EFFICIENCY

In its simplest form, efficiency can be expressed as the difference between actual delivered performance and the highest level of performance than can practically be achieved. This definition is somewhat subjective; while it is straightforward to measure the former, it is less easy to agree on the definition of 'practical'. What might be considered practical to one operation may not be to another; efficiency might be improved by investing in new plant, but this may not be cost effective against the current order book's value. For the purposes of this paper, and the application of the principles which will be outlined in the case studies later, a more straightforward approach is taken. Practical is defined as the limit that can be reached, and if reached, confers measurable benefit, without the need for capital investment.

So, if an industrial operation currently takes 10 weeks to perform and the efficiency of it could be improved to take 8 weeks, *and* by so doing, this would be beneficial *and* this can be done without capital investment, then the current operation is 80% efficient. Of course, the same approach can be taken with costs, or a combination of both time and cost. These conditions are important. Improving the efficiency of an operation may not realise a direct cost saving but may, for example, free up labour. If this labour cannot be productively employed, the overall efficiency has not been improved. Equally, improving the time taken for a non-critical path activity may not reduce the overall time for the program. It is therefore important when considering where and when to act that this simple 'truth' is tested, as improving efficiency is, itself, a resource consuming activity and unless it makes a difference, it may not return the investment. There is, of course, a school of thought that endorses 'efficiency for the sake of it' as it can induce a culture of efficiency. While not criticising this approach, the examples cited here follow the 'rule' defined above.

WHY, WHEN & WHO?

Why?

Organisations can seek to invest in improving their efficiency for at least three reasons:

- Insurance: in order to put themselves in a favourable position to win future work. It is not uncommon in defence contracting for organisations to be required to benchmark their efficiency against other, similar businesses and therefore there is merit in being accredited with being efficient.

- Assurance: to provide comfort that a given program can be completed to time and cost prior to the production activity. Typically this is driven by the need to mitigate risk identified at the planning stages of the project.

- Ensurance: to improve float or recover critical path slippage and/or limit cost overrun. Additionally, some contracts demand year-on-year performance improvement.

When?

Too often the need for improvement only becomes an issue well past the optimum time to plan for its delivery. Perhaps the most obvious reasons for this are that:

- it is not until the business is visibly failing to maintain program or budget that alarm bells ring at the right level in the Company, and

- as many organisations either believe themselves to be efficient (or reasonably efficient) or simply do not know how efficient they are (or could be), the concept of simply improving efficiency is often dismissed without adequate analysis.

Who?

Organisations tend to deploy internal resource for the first two of these and external resource, as a last resort, for the third. Commonly, external expertise is drafted in only then when the 'home team' has failed to deliver the required results or the business is in trouble completing an order. While using the 'home team' is quite understandable, it is not uncommon to find that the level of genuine experience of delivering engineering efficiency in home grown teams is limited and their ambitions (in terms of the improvements they can make) equally limited. Why this is so is touched on later in the paper.

It is of interest to see how, in terms of efficiency, Defence compares with other manufacturing sectors. The data for Fig 1 is derived from actual projects undertaken in the last 10 years by Newton and uses the above definition of efficiency. Two features are of note; firstly Defence lags the others and secondly, the quantum of available improvement is greatest. Many reasons are cited for Defence's position; operational and technological complexity, lack of direct competition, complex contractual mechanisms and most commonly the excuse that 'Defence is different'. While these may be true, there are few occasions where they are valid reasons, or genuine obstacles to the ability to improve efficiency.

Fig 1 – Comparative Efficiency

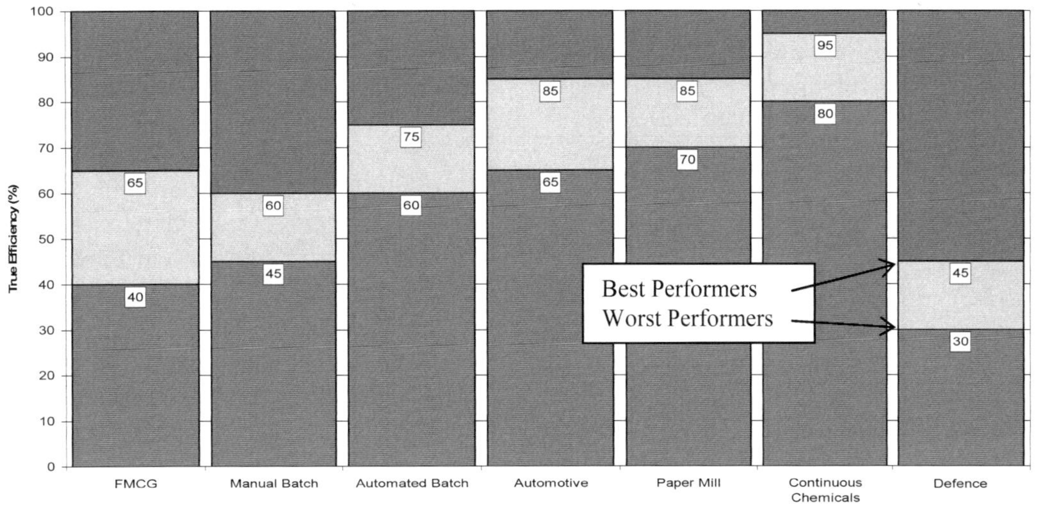

CASE STUDIES

The paper sets out below, three case studies, all real examples drawn from work in the Defence environment. The detail in each simply provides evidence that step changes in efficiency are achievable. The importance of these examples is to make two points: firstly, that double digit improvements in performance can routinely be achieved if approached systematically and secondly, there are a number of generic reasons which can prevent such efficiencies being realised and thereby limit the efficiency with which organisations operate.

Case Study 1

An organisation involved with the manufacture of steel fabrications for one of the UK's maritime programs was faced with the problem of unsatisfactory delivery rate. Fifteen of those directly responsible for the production process were asked to list the three predominant reasons for the failure. Of a theoretical maximum of 45 reasons, 42 were listed and of the 15 sets, not one identified an issue within their direct responsibility to rectify. This is not as extreme an example as it may sound. When organisations are struggling to maintain program, individuals are reluctant to 'self identify' as it can be taken as a sign of personal inefficiency and seldom do organisations feel they can afford to take 'time out' to truly identify the causes of the problem. Most commonly, anecdotal explanations are pursued or inanimate causes identified such as the age or inadequacy of the plant.

In the case of a known issue such as the inadequate production rate above, the Management was readily able to measure the current production rate against the planned, promised, production rate and therefore were in no doubt that they needed to improve. What they did not understand was where they were in relation to the practically possible production rate, nor how to reach it. While many of the 42 proffered ideas were valid, there was no mechanism to quantify the individual contribution each might be able to make and no systematic analysis to demonstrate whether any of the 42 was the principal causes of unsatisfactory performance.

So how was this particular situation remedied? The issue was the production rate of steel panels required for the construction of the bow section of the Type 45 Destroyer program. The panels were relatively simple assemblies; steel plate re-enforced with a variety of stiffener bars. The production process was a simple 'pulse' line with each panel moving down the production line having the stiffener bars welded to it at specific welding stations with stiffener bars of various lengths delivered to the line. While it is difficult to comprehend how there might be as many as 42 possible reasons for these relatively assemblies to miss their production target, the measured production rate was some 60% below the required rate.

Fig 2 – Pareto of Type 45 Panel Line

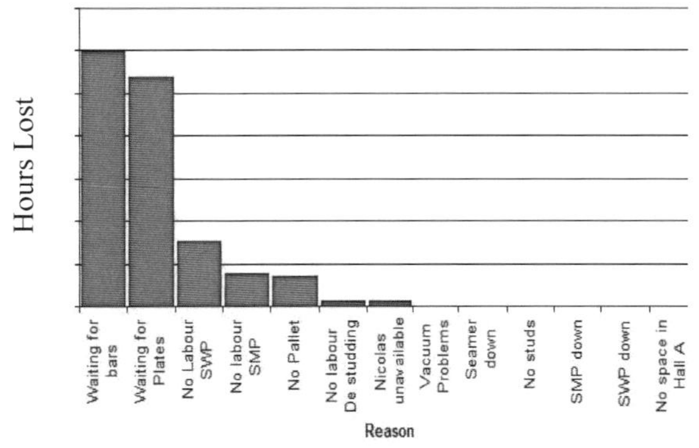

After a short assessment period, dominated by observing the production line at very close quarters, a range of issues was identified and their impact on performance calculated and arranged, by significance, in the pareto curve shown in Fig 2. It is of interest to note, that of the 42 'solutions' only 6 appeared in the analysis of the top 20 issues and none were in the 'top five'.

Once the genuine reasons for poor production rate had been identified and scaled, specific, very targeted correction plans could be drawn up and relatively quickly implemented. Within 3 months the production rate had increased by over 120% (as shown in Fig 3). To sustain the improvement, a very simple tracking system was developed, placed in the middle of the production floor which showed all of the components for the next 5 day's worth of production, whether the components were to hand and if not, named the individual responsible for getting them there and the time at which they were needed. This plan was very visual, all those on the shop floor 'owned' it and updated it.

What is interesting in this example was why the Management had not been able to implement these relatively simple improvements without external assistance. They understood the leap in efficiency required, they were very familiar with the technology of this type of production and it was very much in their interest to solve it. A number of factors were identified and these are not uncommon elsewhere:

Fig 3 – Panel Line Improvement – Hours to produce a panel

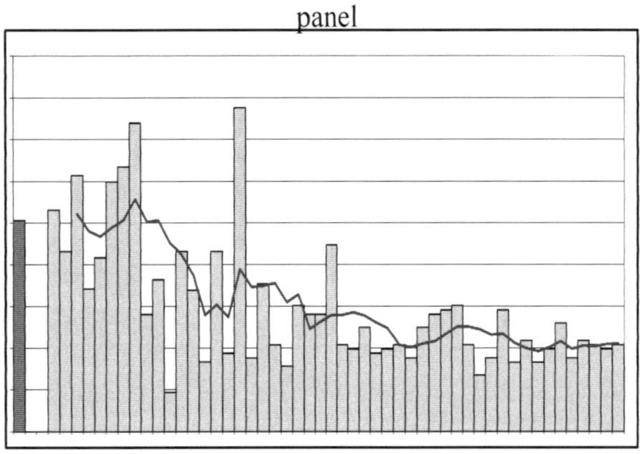

- Management chose to rely on the shop floor's supervisors' expertise. These individuals were largely 'stove-piped' and therefore did not have the breadth of vision, nor autonomy, to look across the whole production process, and none were particularly expert it the business of improving efficiency.
- The improvement program depended on accurate, unambiguous and unbiased information about the current state of the assembly process. No-one had the necessary time, nor expertise, to gather, analyse and present this in a logical way that then allowed action to be taken.

- Interestingly, the Management was reluctant to get too close to the shop floor activity possibly in order to avoid exposing their own lack of knowledge of the production process. Many had reached the managerial layer without gaining experience on the shop floor.

- Agendas were running; 'we need a better IT system', 'we've tried everything already', 'we are as efficient as it is possible to be', 'we need more overtime / staff' although none of these explanations was found to be valid. They simply needed to be better organized.

Case Study 2

An organisation responsible for refitting warships was facing the risk of not completing on time. The nature of the contract (Target Cost Incentive Fee) meant that a comparatively small degree of slippage would result in a significant reduction in profit. They were midway through the refit before the gap between the actual risk adjusted completion date and the planned completion date became wide enough to demand that corrective action (other than everyone just 'doing better') was taken.

In this example, the issues were more complex because more individual processes were involved and the overall refit plan contained a number of inter-dependent activities. Nevertheless, the root cause of slippage was clear; those managing the refit did not have the information they needed in the form required to make the critical project decisions required.

The project had a weekly information pack which took the project control team, the individual project managers and the supporting teams, hours to prepare. The report produced, however was largely qualitative rather than quantitative and more a collection of opinions of progress (or the lack of it) unsupported by auditable data. The lack of evidence together with no logic based prioritisation mechanism meant that problems could be intentionally hidden or could be easily missed. By the time the problems came to light the damage was done, often with compounded consequences.

When a new issue arose, the natural response was put an additional page in the weekly report to cover that specific problem, regardless of whether it had any impact on the critical path. The weekly report blossomed into over one hundred hand generated pages, bogging the delivery team down and eating into the time they needed to deliver the program.

So what was the solution? An assessment was completed with the project team to identify what key pieces of information were required. This identified that although the vast majority of data required to drive the right decisions was already collected within the existing processes and IT systems, in many cases it was being processed locally on spreadsheets for personal use or local decision making. Different departments had different levels of IT literacy and provided statistics in different and often contradictory formats. In most cases the data was just too complex to process and never made it off the screen.

The decision was taken to develop a concise set of metrics, using the data collected in the assessment, to give a very clear picture of project performance and highlight where insufficient progress of the supporting processes was being made.

In a matter of weeks the resulting metrics were agreed, and with the support of a capable IT programmer, delivered on a web based interface which enabled the whole management team to view live project performance. The data improved the decision making up and down the organisation and provided a simple unambiguous single point of truth which in turn, allowed proper prioritisation across the project.

The metrics developed were not complicated, were all quantitative and verifiable and were generated from existing processes. Amongst them were:

- The measured performance of the critical path against plan together with the biggest problem on the critical path and what, specifically, is being done about it.

- The measurable progress in the remaining bulk of work and the paths that have the potential to become critical and what, specifically, is being done about them.

- The performance of the supporting processes (supply chain, design change etc), which have the potential to impact the critical and subcritical paths and what is being done about them.

The resulting management recovery plan allowed the Project Manager to redirect resources and restructuring of the delivery team. Process outputs rose by between 80% and 200%.

The significant lessons from this case were:

- There was no shortage of data! In fact, there was so much, the Project Manger had little chance of spotting the vital piece of information which would make the difference between success and failure. The more the project slipped, the thicker the data pack became.

- A great deal of potentially vital information was held, by individuals, on 'local' IT systems, in formats which did not allow it to be readily shared or allowed others to help resolve the issues.

- Much of the information was qualitative rather than quantitative. Far too often, the 'we've got that one in hand' could not be substantiated by measurable progress.

- Today's fires were being fought; tomorrow's fires were being ignored.

- The Project Manager needed help to concentrate on those issues that were driving slippage rather than the myriad of issues being brought to his door. By developing appropriate metrics, he was able to use his time solving the right issues rather than other peoples' favorite ones.

Case Study 3

An organisation charged with undertaking systems testing at the end of build was struggling to achieve the stringent cleanliness standards required on the test form. The system was on the critical path and so the overall program was slipping as a consequence. The team undertaking the flushing had exhausted all obvious avenues with no measurable impact. A relatively quick assessment using structured problem solving techniques identified the two causes of the increased flushing durations. Firstly the flushing rigs themselves had decayed over time and were actually adding to the contaminates within the system. Cleaning the rigs and reducing the size of storage tanks for flushing water reduced flushing durations, bringing them back in line with previously achieved times. However, more interestingly, it was also identified that the cleanliness specified was actually higher than the laboratory observed quality of the flushing medium and significantly cleaner than that specified for in-service conditions. The only reason that flushes were passing at all was that the test form required only one successful result in order to pass. So against any statistical basis, the

Fig 4 – Flushing Improvement

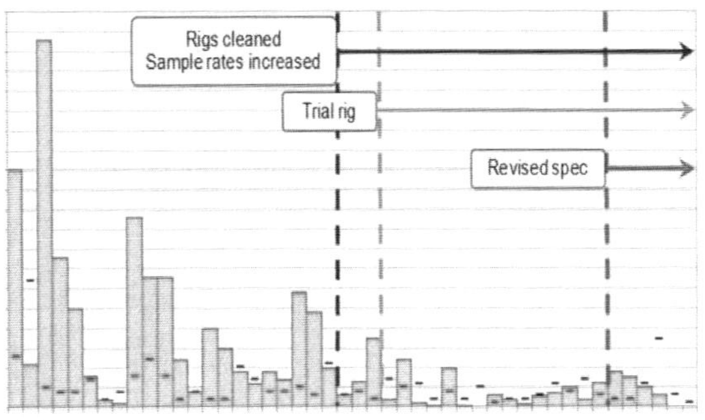

flushes were not reaching the specified level, very occasionally, one sample might just prove 'clear' and the system would be signed off as 'clean'. Not only was the flushing duration vastly exceeding that planned (waiting for the 'one in a hundred' result), but there was actually no justifiable reason for confidence that the specification had, in fact, been met even when the system 'passed'. How was the problem solved? Working with the equipment manufacturer to understanding the actual system requirements allowed a new specification to be developed, agreed by the end customer and then used by the flushing team both improved flushing durations whilst improving the robustness and confidence of the cleanliness checks. One of the challenges here was the relationship between the client, the equipment manufacturer and the end customer, all of whom had strong, largely unsupported views on the cause of the problem and the solutions being proposed; each confident that the problem was being caused by a combination of the other two. Being a 'trusted third party' able to work across all parties with an objective and factual approach enabled buy in from all stakeholder groups

Here, the interesting point was what prevented the team doing the flushing from spotting the solution themselves.

- The data was not being collected and analysed in a way that allowed the nature of the problem to be identified. Had it been so, the team would have seen that, statistically they were routinely plateauing out at a cleanliness level below the specification and only recording successes when, by chance, a sample exceeded the specification.

- The 'wisdom' of the shop floor was that 'sometimes, it just takes that long' and this explanation was reluctantly accepted by the Project team without challenge.

- Within the team undertaking the flushing, there was no understanding of why the specification was set in a particular way, nor any authority to challenge it or to suggest an alternative approach.

CONCLUSIONS

The above examples are but three of many but usefully illustrate a number of lessons about delivering efficiency improvements. In summary, these are:

- The majority of production failures stem from realising too late into the project that intended progress is not being made. While efficiency throughout an organisation is a worthy objective, it is vital that the business ensures that they are *efficient at those activities that are likely to dominate the critical path.* For such activities it may be a worthwhile investment to insure against failure by improving the efficiency of these activities alone before the contract starts. For example, at the end of build and refit, setting to work dominates the critical path, and system flushing is invariably on it. So a determined and pre-emptive program to look at measures that could be taken to massively reduce the time of activities like this would seem to be a worthwhile investment. It may not be at the exciting end of the spectrum but neither is week-for-week slippage at the end of the build or refit.

- While many organisations have their own 'Tiger Teams' who are charged with improving efficiency (usually by removing cost) these are seldom manned by people who are genuinely skilled at undertaking the root cause analysis required, have unfettered ability and the motivation to manage the downstream activity and the experience of techniques and processes that have worked successfully elsewhere. Such individuals must be able to rise above any inter-departmental barriers, demonstrate with confidence how their proposed solutions will work and achieve rapid results to instill confidence.

- To drive efficiency, reliable, unambiguous and auditable data is vital. Without reliable baseline data showing actual performance, improvements are impossible to quantify. It is then vital to really understand the root causes of poor performance and then to prioritise these for action. Too often, potential efficiencies are listed, not prioritised and tackled in order of ease rather than bottom line impact.

- If time is the issue, it is important to understand activities that prescribe the critical path. Human nature tends towards solving 'easy / quick win' problems but the only problems that should be tackled are those on the critical path. Failure to do this will leave the original slippage unresolved.

- It is common for inefficiency to be blamed on failings with plant and machinery but experience shows that only very infrequently is plant operating at maximum capacity. Much more frequently the causes of inefficiency are associated with logistics; getting the right part to the right place at the right time together with the right guidance to the operator of what is required.

- Experience shows that most processes can be improved (without capital investment) by between 15% - 25% within 6 months. This is often in stark contrast to internally set efficiency targets, loosely overlaid across the business seeking 5% efficiency 'year-on-year'. For any organisation seeking to improve, it is vital that only those processes that genuinely improve the bottom line should be undertaken. Too many initiatives confuse and demoralise and invariable fail to provide the feedback that the 'pain is worth the gain'. Equally, long drawn out exercises lose their impact.

- It is vital that efficiency gains are seen to be 'wins' for every level within the organisation. Workers on the production line enjoy demonstrating that they can work efficiently and being involved in the business of improvement. The 'Efficiency team' needs to be integrated into the process such that improvements are not done 'to' the workforce, but 'with' the workforce.

Radiation Characteristics of Turbulent Flow Noise from an Underwater Vehicle, and a Feasible Prediction Method

L Chen, I R MacGillivray and Y K Tso,
Maritime Platforms Division, Defence Science and Technology Organisation
506 Lorimer Street, Fishermans Bend, VIC 3207, Australia
li.chen@dsto.defence.gov.au

ABSTRACT

Noise induced by turbulent flow has a crucial impact on submarine stealth. The generation of flow noise by a submarine is complex, and involves a wide range of physics. Radiation by flow-noise sources near an acoustically compact structure can be estimated using a semi-analytical solution with the help of a Computational Fluid Dynamics (CFD) approach. However, for a very low Mach number flow over a non-compact structure, the interactions between the structure and the acoustic waves play an important role and, as a result, can only be analysed using numerical methods. In this study, the radiation characteristics of turbulence-induced noise from a submarine is analysed numerically. A model based on the combination of CFD and the conventional acoustic analogy is developed to predict the trailing-edge noise of a finite span hydrofoil. Turbulence-induced broadband noise generated by a NACA0012 trailing-edge at $Re = 2.68\times10^6$ is estimated using the model. The predicted results are in good agreement with the published experimental data.

INTRODUCTION

Noise generated by turbulent flow has a critical impact on submarine stealth. An understanding of the characteristics of such noise is therefore one of the most important aspects in submarine operations. Flow-induced sources from a complex structure such as a submarine may include noise from the rotation of the propeller, turbulence induced noise caused by the turbulent eddies scattering from a discontinuity of geometries (such as a trailing edge), cavity noise due to the interactions between turbulent flow and acoustic waves, and tonal noise induced by the coherent bluff-body-vortex shedding over appendages such as a sail, mast or control surface. Turbulence induced noises are mainly broadband in nature. There are three classes of prediction methods that can be used for turbulence broadband noise: 1) a computationally expensive Direct Numerical Simulation (DNS), 2) a semi-analytical solution and 3) a hybrid method (the acoustic analogy combined with a computationally economic CFD). Application of DNS to any practical problem is not feasible now and this will most likely be the case for some time to come. A semi-analytical method can be used very effectively in comparative experimental and design studies, but the absolute sound level can only be determined for some very simple cases. The hybrid method is designed to predict the flow noise for arbitrary geometry and is therefore feasible for the noise prediction of a complex objects such as a submarine. However, the main obstacles faced by the hybrid method are the expensive processes involved in flow simulation studies and the requirement for the further development of wave propagation models. For a time-dependent solution of turbulent submarine flow, the required computational resources are very high, even without considering the numerical solution of acoustic wave propagation. Although it is the only practically achievable method usable to predict flow-induced noise from a full scale submarine or a structure of complex topology, it is still severely limited by currently available computational power if a time-dependent solution is involved. An alternative method which requires much fewer computational resources is to combine the flow solution of turbulence statistics with a numerical solution of wave propagation. Some results of a hybrid method is presented in this study.

The prediction of the flow noise of a full-scale submarine is challenging because it is a complex multi-physics problem. For example, turbulence noise radiation from a large smooth flat plate is broadband and quadrupole in nature (Black 1986). Adding appendages to the plate complicates the noise generation mechanism. It is well known that the turbulence noise radiation from a trailing or leading edge is caused by scattering of broadband turbulence noise near the edge and is dipole in nature (Black 1986 and Howe 1998). But the effect of combining large surfaces with trailing edges is not well understood. The understanding of such interaction is important for the development of flow noise models for structures with complex geometry. The characteristics of turbulence noise radiated from an isolated hydrofoil and a bare-hull SUBOFF (a generic submarine) were studied by Chen & MacGillivray (2011). In this study, the turbulent flow and associated acoustic source for the research unmanned underwater vehicle, Wayamba, developed at Maritime Platforms Division (MPD), DSTO, is simulated to demonstrate the nature of the turbulence noise sources for a structure of complex geometry. A detailed study of the turbulence noise sources for an underwater vessel such as Wayamba is beyond the scope of this paper. The current study investigates the turbulence noise radiation characteristics for an underwater vessel, with control surfaces, due to a model source. It is believed that the results from this study will be applicable to a submarine. Further, as a brief example, a hybrid method is used to predict the broadband noise generated by the trailing edge of a NACA0012 hydrofoil, based on the combination of turbulence statistics and the conventional acoustic analogy. The application of this hybrid method to an underwater vessel requires further study.

PREDICTION OF FLOW OVER AN UNDERWATER VEHICLE

To illustrate the complexity of the flow field around a submarine, the flow field around Wayamba (the MPD unmanned underwater vehicle), travelling at approximately 12 knots, has been computed as a demonstration as it shares many common fluid dynamic features with a submarine. The diagram of Wayamba is shown in Figure 1. The main hull of Wayamba has dimensions $4.33\ \text{m} \times 1.44\ \text{m} \times 0.50\ \text{m}$ $(L \times W \times H)$, with an added sail of elliptic cross-section and four fins shaped as two pairs of NACA0009 hydrofoils. The computational domain was $5L \times 1.4L \times 1.1L$. The outlet boundary was located at distance $3L$ away from the end of the body, and the inlet boundary condition was implemented at distance L upstream of the nose of the body. Symmetrical boundary conditions were used at the side boundaries located at $0.5L$ distance from the body. The flow was simulated using the steady state RANS (Reynolds Averaged Navier-Stokes) model of FLUENT (a commercial CFD package) with a total of around 5.6 million grid points. The turbulence was modelled by the $k-\omega$ model.

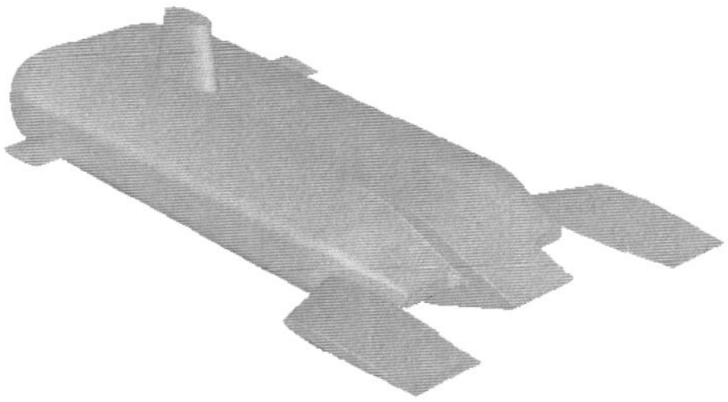

Figure 1. Diagram of Wayamba.

It can be seen from the flow path lines in Figure 2 that the flow is complex, with recirculation occurring behind the sail and aft where the flow is highly unsteady or turbulent. The turbulence kinetic energy distributions at different locations, as shown in Figure 3, are highly concentrated on the aft of the main body and the sail, as well as the trailing edges of the control fins. The turbulence noise source distributions at the same locations to those in Figure 3 are depicted in Figure 4, based on Lilly's broadband noise model in FLUENT which is extracted from the RANS solution. The two sets of results are in good agreement.

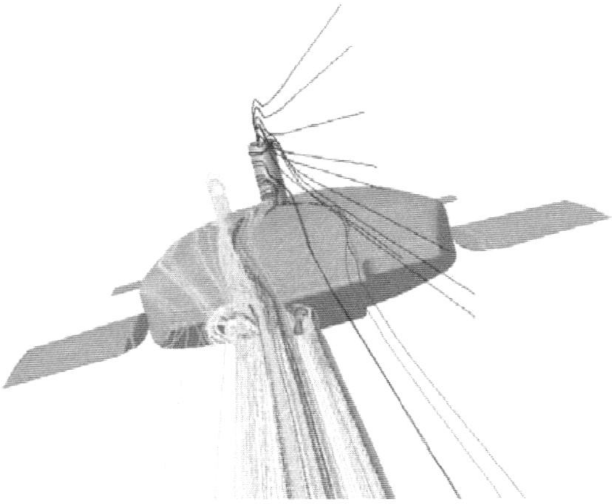

Figure 2. Path lines for flow over Wayamba at free stream velocity U = 6 m/s.

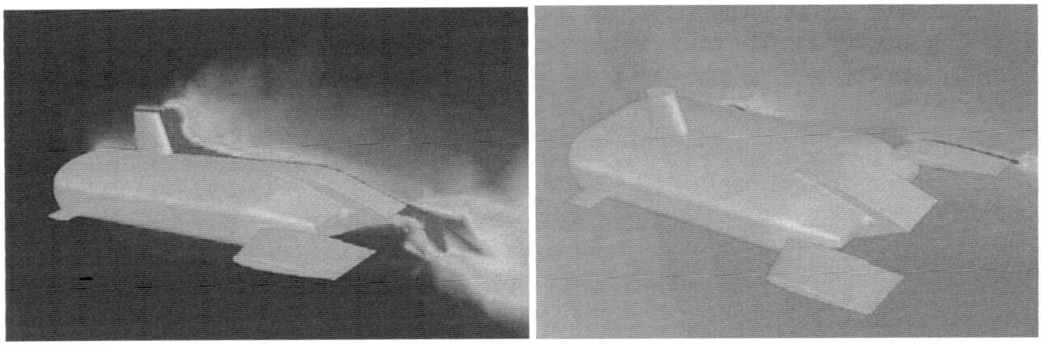

Figure 3. Turbulent kinetic energy distributions around Wayamba travelling at $U = 6$ m/s on a plane at the middle section (left), and through the fins (right).

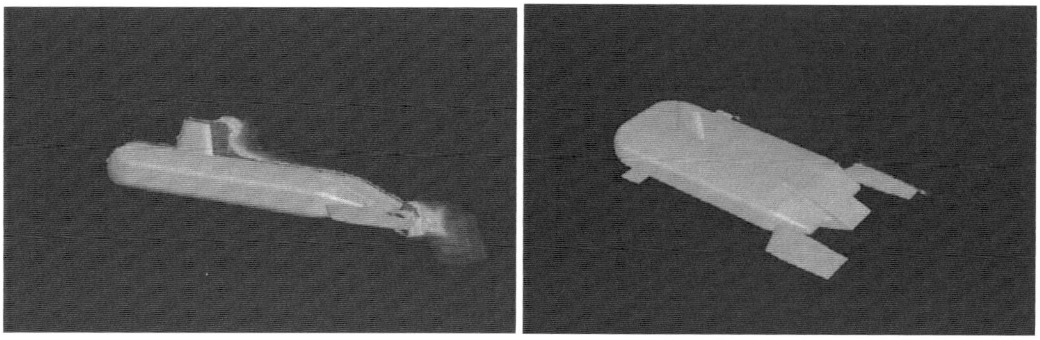

Figure 4. Lilly's total turbulence noise source distributions around Wayamba travelling at $U = 6$ m/s (contour range $10^2 \sim 10^9$) on a plane at the middle section (left), and through the fins (right).

TURBULENCE NOISE RADIATION FROM A SUBMARINE

According to Lighthill (1952, 1954), the sound pressure p radiated by a turbulent flow has a quadrupole nature and is described by the inhomogeneous wave equation

$$\frac{\partial^2 p}{\partial t^2} - c^2 \frac{\partial^2 p}{\partial x_i^2} = \frac{\partial^2 T_{ij}}{\partial x_i \partial x_j}. \quad (1)$$

The Lighthill stress tensor, T_{ij}, is given by

$$T_{ij} = \rho u_i u_j + \left(p - c^2 \rho\right)\delta_{ij} + \tau_{ij}, \quad (2)$$

where ρ, u, c, and τ are the fluid density, fluid flow speed, speed of sound, and shear stress tensor, respectively. An analytical solution to Eq. (1) can be found using a free space Green's function for some simple problems with the absence of solid surfaces, such as spinning vortex sound. However, to find an analytical solution to Eq. (1) for stationary or moving surfaces of arbitrary geometry is not trivial. Kim *et al.* (2006) studied noise generated by flow over a hump using DNS and confirmed that Lighthill's source term in Eq. (1) was well correlated with the surface pressure fluctuation. Gloerfelt (2010) reached a similar conclusion from his Large Eddy Simulation (LES) solutions on turbulent boundary layer flows. The implication of these studies is significant as they provide a very good foundation to develop a practical CFD solution for Eq. (1) based on the surface pressure fluctuations. However, the ability of this method to account for the scattering of sound from non-compact surfaces depends on how the pressure p is defined. Combining a CFD solution and a boundary-element based wave equation solver (such as SYSNOISE) will be a practical method to study flow-induced sound for underwater applications of a complex geometry where the noise source is provided by CFD and the associated sound field is solved by FEM/BEM.

According to Grighton *et al.* (1992), unsteady turbulent flows are acoustically equivalent to a particular quadrupole distribution radiating into a uniform acoustic medium. That is, the turbulent flow is comprised of many energetic eddies, and each energetic eddy has random phase relative to any other eddy. Inside each eddy the motion is well correlated, but uncorrelated with the motion inside other eddies. To get the sound field generated by a single eddy, one needs to add the density fields generated by points within the eddy linearly, for they are all in-phase. To get the total intensity, one can sum the intensities from all separated eddies. For the generation of flow noise, the fine-scale structure of turbulence is largely irrelevant and only one length parameter is needed, representing the scale on which the turbulent energy is mainly concentrated. The flow simulations reveal that the turbulence induced source is compact over a wide range of frequencies. Therefore, it is very useful to study the characteristics of radiated turbulence noise for a moving underwater structure using a quadrupole source. The detailed prediction of a flow-induced noise from an underwater vessel will require more study.

The radiation characteristics of a generic submarine of dimensions $61.8\text{m} \times 7\text{m}$ ($L \times D$), with control surfaces, is studied using a model quadrupole source of specified orientation and unit strength. The locations of the model quadrupole shown in Figure 5 are chosen according to the results of our CFD simulations to study the effect of the source location on the directivity of the radiated field.

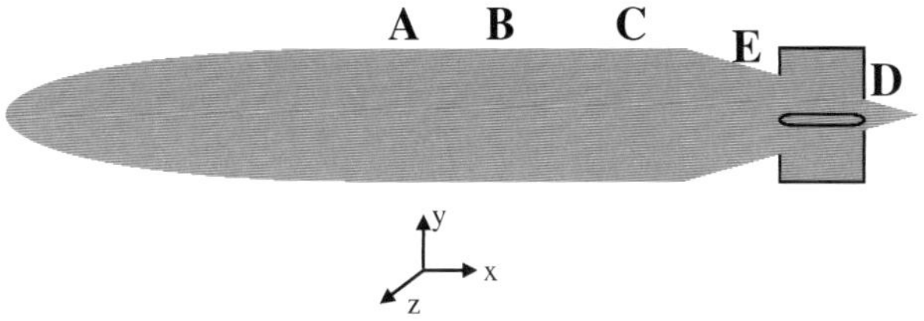

Figure 5. Locations of quadrupole sources around the generic submarine.

Figure 6 shows the sound radiated from a model quadrupole at location A. An aspect angle of 0° is the +x direction and 90° is the +y direction. All directivity plots shown are for the $z = 0$ plane. It can be seen that at all frequencies the scattered field from this object is very weak, and it shows a quadrupole pattern. At 50 and 200 Hz the submarine is non-compact in the axial direction and the scattering from the surface becomes destructive, due to the nature of the model quadrupole, leading to a radiation field weaker than the incident. It is interesting to note that at a higher frequency of 1000 Hz, the underwater vessel becomes non-compact in all directions, and the scattered radiation field is enhanced above the source at an angle between 20° and 160°. However, the enhancement is weak.

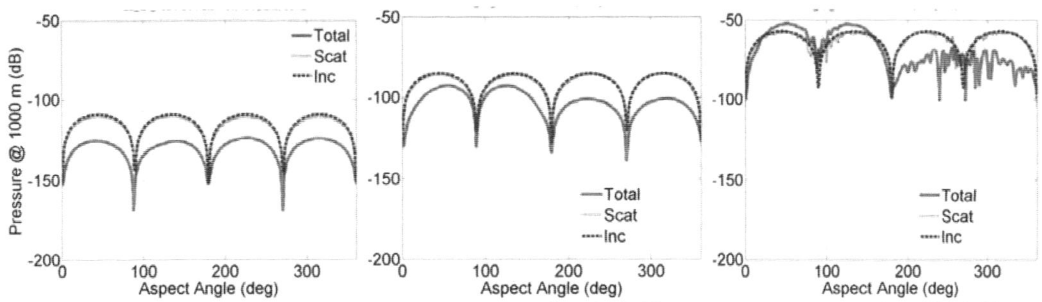

(a) $f = 50$ Hz (b) $f = 200$ Hz (c) $f = 1000$ Hz

Figure 6. Directivities at different frequencies with the model source at location A.

Moving the source to location B, the combination of dipole scattering and quadrupole incidence at low frequencies (Figures 7a and 7b) produces a radiation field of a dipole pattern that is weaker than the incident due to the destructive nature of the scattered field. At 1000 Hz (Figure 7c), the directivity pattern is similar to that of the source at location A, and the radiation is enhanced slightly by scattering below an aspect angle of 160°. The source located at C, which is closer to the aft and control surfaces, generates a weakly enhanced radiation field of a dipole nature at 50 Hz and 200 Hz (see Figures 8a and 8b). With an increase in frequency to 1000 Hz, the scattered field becomes complex, and reveals some features of scattering from a leading edge of control surfaces (Chen & MacGillivray, 2011). Compared with the sources at locations A and B, the enhancement of the incident field is increased. With the sources located close to the front and rear of the control surfaces, as shown in Figures 9 and 10 respectively, the radiation fields of a dipole pattern are significantly enhanced. This is particularly so at low frequencies, as shown at 50 Hz and 200 Hz, and scattering is the dominant radiation mechanism. At 1000 Hz the directivity pattern of the scattered field shows maxima downstream for the source located near the leading edge of the control surface (Figure 9c), and shows maxima upstream for a source located near the trailing edge (Figure 10c). The directivity plots shown are for the $z = 0$ plane, and it should be noted that the directivity on the $x = 0$ plane will be different. For instance, significantly enhanced radiation fields of a dipole pattern are obtained for sources located at the aft of the submarine for all frequencies studied. For sources located on the hull at 1000 Hz, the radiation can be enhanced significantly around the azimuth (not shown), and the directivity has a pattern of order higher than a quadrupole.

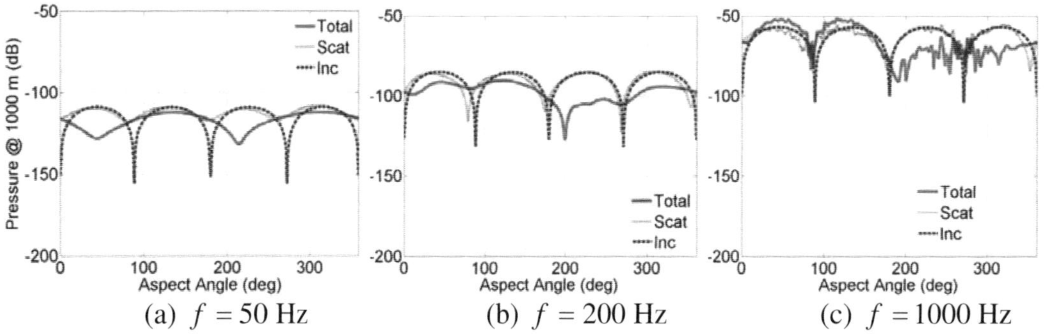

Figure 7. Directivities at different frequencies with the model source at location B.

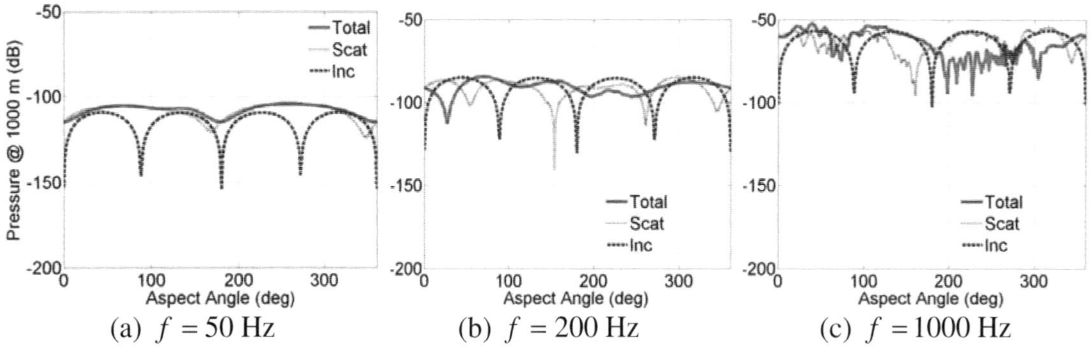

Figure 8. Directivities at different frequencies with the model source at location C.

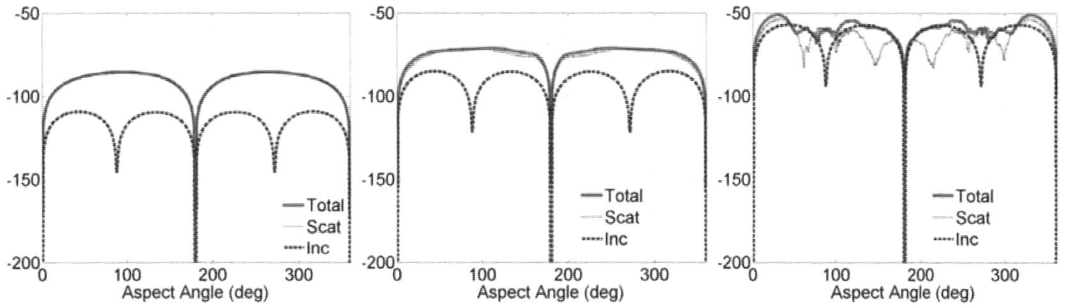

(a) $f = 50$ Hz (b) $f = 200$ Hz (c) $f = 1000$ Hz

Figure 9. Directivities at different frequencies with the model source at location E.

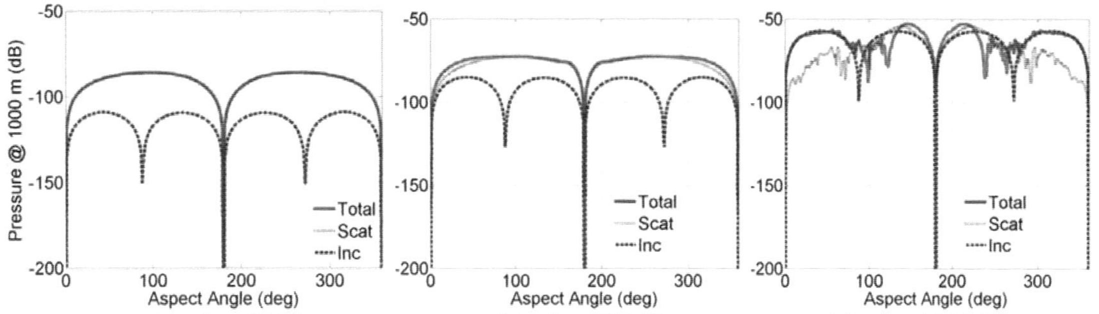

(a) $f = 50$ Hz (b) $f = 200$ Hz (c) $f = 1000$ Hz

Figure 10. Directivities at different frequencies with the model source at location D.

PREDICTION OF TRAILING EDGE NOISE BASED ON RANS

It can be seen from the results that the closer the source is to the control surfaces, the more efficient the radiation, especially at low frequencies. This is consistent with many experimental observations that the flow-induced noise generated by a submarine-shaped object with control surfaces is concentrated at the aft section (e.g. Harddle and Shudrzyk 1969). Given the characteristics of the noise source distribution over such a vessel, scattering of flow noise from edges is likely to be the dominant mechanism of flow-noise radiation.

It follows that to predict the fluid-induced noise generated by a submarine it is very important to model the sound radiation due to trailing edges of appendages accurately. According to Howe (1998), the trailing-edge noise for a hydrofoil of a finite span can be calculated by

$$\Phi(\mathbf{x},\omega) \approx \frac{U_c L \sin^2(\theta/2)\sin\psi}{2\pi^2 c_0 |\mathbf{x}|^2} l_3 \Phi_{pp}(\omega) \left| \frac{G(\mathbf{x},\mathbf{y},\omega)}{G_1(\mathbf{x},\mathbf{y},\omega)_{half}} \right|, \quad (3)$$

in which U_c, L and l_3 are the convection speed, wetted length of trailing edge, and turbulence correlation length scale in the spanwise direction, respectively. $\Phi_{pp}(\omega)$ is a point surface pressure spectrum, which depends on the mean shear and turbulence statistics and can be modelled using a fast steady-state RANS solution based on Chase (1986). θ, ψ and $\mathbf{x}$ are source angle, observation angle and observation coordinate respectively. G is a frequency dependent Green's function and G_1 is the first-order expansion of the G which accounts for multi-scattering from the edges of the hydrofoil. The spanwise turbulence correlation scale is approximated according to Amiet (1971).

The trailing edge noise for the NACA0012 hydrofoil studied experimentally by Brook and Hodgeson (1981) is predicted using the above model. The turbulent flow at $Re = 2.8 \times 10^6$ is simulated using the $k - \omega$ model of FLUENT. Details of the numerical modelling can be found in Do, Chen & Tu (2010). The predicted surface pressure spectrum compares well with the experimental data of Brook and Hodgeson (1981) and with other empirical models. The predicted radiated noise at a distance 1 m above the trailing edge for sharp and round trailing edges is compared with the experimental data of Brook and Hodgson and illustrated in Figure 11. A semi-infinite plate solution is also presented in the figure. The prediction for the sharp edge generally agrees better with the experimental data. The over-predicted noise level at low frequencies needs further investigation, although the predicted surface pressure is in good agreement with the experimental data.

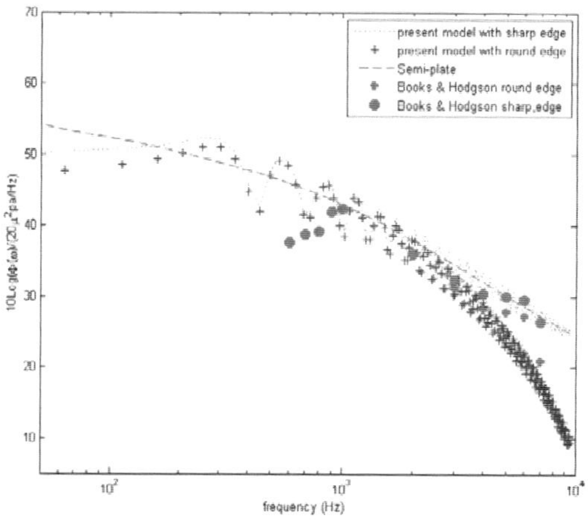

Figure 11. Comparison between predicted trailing edge noise and experimental data of Brook and Hodgeson (1981). Semi-plate refers to the modelling for a semi-infinite plate.

CONCLUSION

For highly turbulent flow around a submarine-shaped underwater vessel, the flow-induced noise sources of highest intensity will concentrate around the appendages and the aft section. Sources on the hull at frequencies lower than about 200 Hz do not radiate efficiently due to diffraction, resulting in quadrupole directivity. However, the radiation by sources at higher frequencies can be enhanced in some directions. The directivity of the radiation field depends strongly on the frequency. The closer the source is to control surfaces the more efficient the radiation, especially at low frequencies, with the directivity of the radiation field being dipole in nature. Due to enhancement of flow noise as a result of scattering at the trailing edge of appendages the turbulence at those regions of a submarine are the main contributors to noise radiation.

Trailing edge noise for a NACA0012 hydrofoil was predicted using a computationally efficient RANS solution of CFD and Howe's radiation model. The results are in good agreement with the experimental data.

REFERENCES

Amiet, R 1976, "Noise due to turbulent flow past a trailing edge", *Journal of Sound and Vibration*, 41(4), 407-420.

Blake, WK 1986, *Mechanics of Flow-Induced Sound and Vibration*, Academic Press (London).

Brook, TF and Hodgeson, TH, 1981, "Trailing edge noise prediction using measured surface pressure", *Journal of Sound and Vibration*, 78, 69-117.

Chase, D 1987, "The character of the turbulent wall pressure spectrum at subconvective wavenumbers and a suggested comprehensive model", *Journal of Sound and Vibration*, 112(1), pp. 127-147.

Chen, L & MacGillivray, I, 2011, "Characteristics of turbulence noise radiated from hydrofoils and bare-hull Suboff", *Proceeding of ACOUSTIC2011*, paper No. 105, ACOUSTIC 2011, Australia Acoustic Society, Gold Coast Australia, Nov. 2011.

Do, T, Chen, L & Tu, JY 2010, "Numerical study of turbulent trailing edge flows with base cavity effects using URANS", *Journal of Fluids and Structures,* 26, pp. 1155-1173.

Gloerfelt, X 2010, "The link between wall pressure spectra and radiated sound from turbulent boundary layers", *AIAA paper 2010-3904*.

Grighton, DG, Dowling, AP, Ffowcs Williams, JE, Heckl, M & Leppington, FG 1992, *Modern Methods in Analytical Acoustics, Lecture notes.* Springer-Verlag London.

Haddle, GP and Skudrzyk, EJ, 1969, "Physics of flow noise", *Journal of the Acoustic Society of America*, 46(1), 130-157.

Howe, MS 1998, *Acoustics of Fluid-Structure Interactions*, Cambridge University Press.

Kim, J & Sung, HJ 2006, "Wall pressure fluctuations and flow-induced noise in a turbulent boundary layer over a bump", *J. Fluid Mech.* 558, pp. 79-102.

Lighthill, MJ 1952, "On sound generated aerodynamically, Part I: general theory", *Proc. Roy. Soc. Lond.* A211, pp. 564-587.

Lighthill, MJ 1954, "On sound generated aerodynamically, Part II: turbulence as a source of sound", *Proc. Roy. Soc. Lond.* A222, pp. 1-32.

Analysis of Platform Configurations using the Integrated Platform System Model

S. Harrison, J. Rodgers, J. Wharington and S. Demediuk
Maritime Platforms Division, DSTO
simon.harrison@defence.gov.au

ABSTRACT

Analysis of high level capability of maritime platforms is difficult and time consuming when there is a need to assess multiple designs by different manufacturers and different configurations or option sets of these designs.

This paper presents the Integrated Platform System Model (IPSM) developed by the Naval Architecture and Platform Systems Analysis group in DSTO's Maritime Platforms Division and describes its use to model multiple platform designs. An example analysis of a generic submarine model demonstrates that the IPSM provides a capability to manage these evaluations within a consistent and robust framework (which is described in a companion paper) at the integrated platform level.

The IPSM uses ModelicaML graphical modelling language to represent the system. ModelicaML documents the models in a visual format from which source code can be generated and then executed as a Modelica computer simulation. The object oriented nature of ModelicaML makes efficient modelling possible through reusable building blocks. From these reusable building blocks, platforms with different configurations (including both platform design configurations and mission profiles) are created and modelled. This allows fast turnaround for providing advice relating to the capability and feasibility of specific platform design configurations.

The IPSM may be used in the evaluation of Military-Off-The-Shelf (MOTS) and derivatives as well as unique designs. Development of a robust verification and validation framework is ongoing, and information from platform manufacturers and other empirical data are being used for validation.

1. INTRODUCTION

The Australian Defence White Paper [1] indicates that future development of the Australian Defence Force (ADF) will include several major naval acquisition projects to enhance Australia's maritime capabilities in the years to come. The design phase of such projects will require significant amounts of time and effort. Developing a capability for assessing maritime platform configurations (both surface and undersea) at the integrated platform level in a timely manner will provide respective Defence Materiel Organisation (DMO) project offices with valuable input to requirements development and translation of capability objectives to design characteristics. By providing a better understanding of integrated platform characteristics this capability will also support initial trade-off decisions relating to the high level capability of these future platforms.

Analysing high level capability of maritime platforms is difficult and time consuming when there is a need to assess multiple design configurations and option sets offered by different manufacturers. This paper presents the Integrated Platform System Model (IPSM) developed by the Naval Architecture and Platform Systems Analysis (NAPSA) group from DSTO's Maritime Platforms Division (MPD). The IPSM provides a capability to model platform designs within a consistent and robust framework (this framework is described in future reports).

In this paper we present a declassified generic submarine model configuration, which was implemented and analysed to demonstrate the capability of the IPSM to make these assessments.

1.1 Approach

A Model Based Systems Engineering (MBSE) approach [2] has been adopted by NAPSA for developing the IPSM and surrounding framework. A key element of this approach is the use of a standardised method for graphically representing models. As a derivative of the Systems Modelling Language (SysML), the Modelica Modelling Language (ModelicaML) provides additional rules that when combined provide a standard method for graphical representation of all model elements including: class diagrams, behaviour diagrams, composite diagrams, state machines and requirements. These additional rules provide ModelicaML with the ability to automatically generate Modelica source code [3] so models can be executed as part of a Modelica simulation. Finally, the adoption of an MBSE approach in developing the IPSM has provided the group with an effective medium for communication by bringing the group's methodologies inline with those of project stakeholders.

The object oriented nature of ModelicaML makes efficient modelling possible through the use of reusable building blocks from which platform configurations are created and modelled. This also allows for highly reusable models to be created. Furthermore, factors influencing platform configurations such as operational constraints, sub-system characteristics and requirements may be modified and reflected in platform configuration models with minimum effort. This provides flexibility and efficiency in the creation and modelling of multiple platform configurations.

Creating models at the integrated platform level provides insight into the cumulative effect of multiple sub-systems operating together and hence the high level capability of the system. ModelicaML provides a vehicle for both the creation of individual sub-system models and their implementation within larger models as desired.

2. THE INTEGRATED PLATFORM SYSTEM MODEL

The capability provided by the IPSM is intended to be used for making assessments and exploring the solution space of both surface and undersea platform configurations. Initially this work has been focussed on making assessments of submarine platform configurations which has driven the creation of three core building blocks that combine to describe the behaviour of a conventionally powered nominal submarine platform. These core blocks are: Power and Energy; Propulsion and Drag; and Design.

In this section the component building blocks used to build a generic submarine platform are described in detail.

2.1 Power and Energy

Modern conventional submarines are powered by a combination of diesel generators, battery systems and in some cases an Air Independent Propulsion (AIP) system typically in the form of a Stirling Engine, Hydrogen Fuel Cell or MESMA (Module d'Energie Sous-Marine Autonome) system. An example Power and Energy block has been developed for use in the nominal submarine configuration which contains three fuel systems. This block is shown in Figure 1 with the main components of Diesel, AIP and Battery represented as a ModelicaML composite diagram.

The Power and Energy block (shown in Figure 1) provides an example of the high degree of reusability within models due to the object oriented nature [3] of ModelicaML. The EnergyStore class is used to represent a generic fuel tank, which is then instantiated three times as a specific block – once for each fuel system (shown in the top row of Figure 1). The parameters within the block are tailored in each case so that the specific characteristics of each fuel tank are modelled as desired.

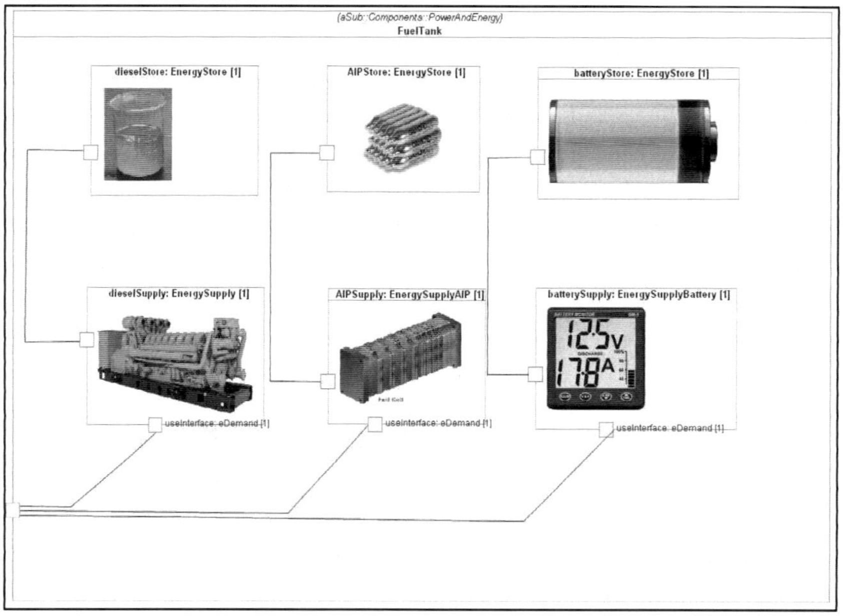

Figure 1: ModelicaML composite diagram of the Power and Energy system for the generic submarine configuration

Similar to EnergyStore the EnergySupply class is generic and is designed to govern the rate of energy (fuel) being drawn from (or entering in the case of a charging battery) the respective EnergyStore blocks. An EnergySupply object is then instantiated and specialised to create a specific block for each of the three fuel systems (shown in the bottom row of Figure 1). This enables specific system efficiencies to be incorporated into their respective blocks in the form of derating curves. The coefficients in the derating curves are derived from manufacturer specifications or complementary empirical research and can be adjusted to represent the desired characteristic for each fuel system.

Within the ModelicaML representation of the power and energy system flow variables allow the model to solve for unconstrained variables in the model. This is best illustrated by example. Assume a mission state where the nominal submarine platform is conducting a snorkelling transit. The submarine has a target speed of 12 knots and the diesel engines are providing power to the drive train at their maximum capacity of 5MW. The rate of energy entering or being drawn from the battery system is left unconstrained. If the power required to drive the submarine at 12 knots is less than 5MW then residual power is passed to the battery system to charge the batteries. If the power required is greater than what is being provided by the diesel engines (>5MW) then the model will draw power from the battery system. Alternatively the speed of the platform or the power output of the diesel system may be left unconstrained and the charging rate of the battery system defined. Modifying model behaviours such as these do not require changes to the model building blocks. The flexibility provided by this method adds significantly to the ability to explore the various sizing trade-off solution spaces efficiently.

2.2 Propulsion and Drag

The nominal Propulsion and Drag block developed for integration in the generic submarine platform governs the translation from fuel energy to drive train power. Drag force is described by a modification to a simple quadratic drag relation [4]. Additionally, propeller, drive shaft and permanent magnet motor efficiencies are also accounted for by this block.

2.3 Design

Information relating to the sizing of the generic submarine is provided to the rest of the model by the last of the core blocks, the Design block. These sizing parameters are specified as hard constraints and include: total submerged displacement of the platform, fuel fractions, fuel reserve levels and hotel loads for various operating states. Future versions of the IPSM will allow for these design variables to be left unconstrained and solved for by the model.

2.4 Other Blocks

Additional blocks calculate and record data used in post processing analysis of performance. Included are a Kinematics block evaluating distance travelled, indiscretion ratio and mission endurance. High level assessments related to the habitability of the modelled platform are performed in a Habitability block.

2.5 Mission State Machines

Exercising a platform configuration through a mission profile provides a means for testing the ability of the platform to meet a set of operational constraints. The platform is created from component building blocks and paired with a mission profile to define an experiment for simulation. Mission profiles exist in the IPSM as a ModelicaML state machine through which a platform configuration is exercised. Currently non-deterministic events such as unplanned dives or emergency course variations are not accounted for. Parameters in mission profiles are modified in a similar manner to the tailoring of parameters in component building blocks such that the relative length of the mission phases and therefore the overall length of the mission can be modified even after the experiment model has been compiled. Constraints defining mission success are added to the mission profile before the simulation is executed.

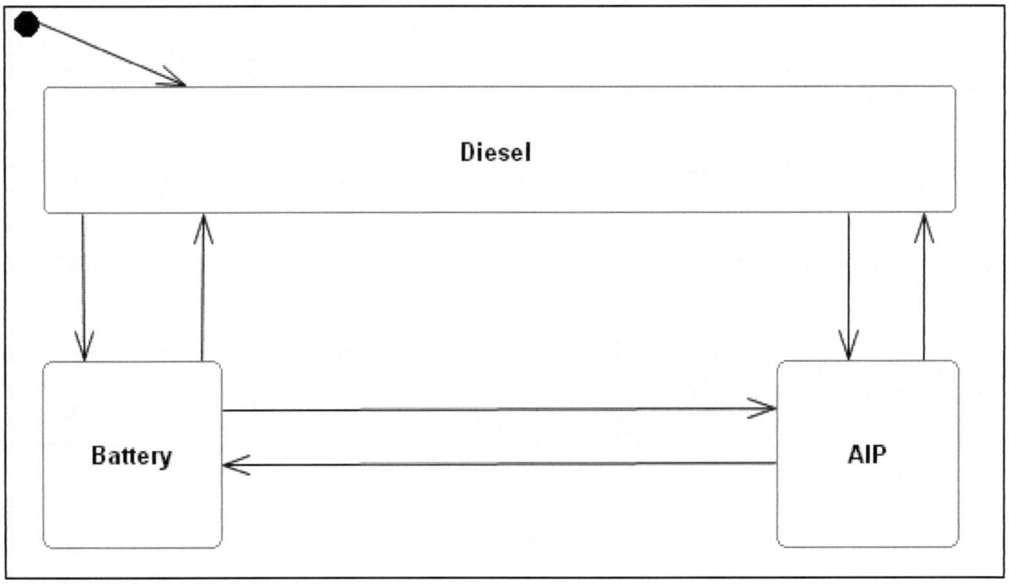

Figure 2: Snorting Cycle Loop, ModelicaML State Machine

Figure 2 shows an extract of the ModelicaML state machine representing the mission profile, created for use with the generic submarine configuration. In this example, three states are defined that determine which of the three available fuel systems the platform will use. Links between the states are gated by conditional expressions to control transitions and therefore model behaviour. The system will move to the Battery state by preference and as the battery charge level drops to a lower reserve limit, the ModelicaML state machine will change to the use of either the AIP or Diesel fuel systems according to the conditional state transitions and continues for the length of the multi day mission.

3. IMPLEMENTATION AND ANALYSIS OF A GENERIC SUBMARINE

Implementation of a platform configuration in the IPSM requires component building blocks describing a platform configuration to be selected and associated links between those blocks to be formed. The means to establish these links is provided through the use of ModelicaML composite diagrams..

This section provides an overview of how component building block models are brought together to form the generic submarine platform configuration. The completed configuration model is then simulated and some brief analysis of the simulation results is conducted.

3.1 Creating Links and Associations

Figure 3 shows the ModelicaML composite diagram describing the generic submarine platform. The platform consists of five building blocks which include:

- Power and Energy
- Propulsion and Drag
- Design
- Habitability
- Kinematics

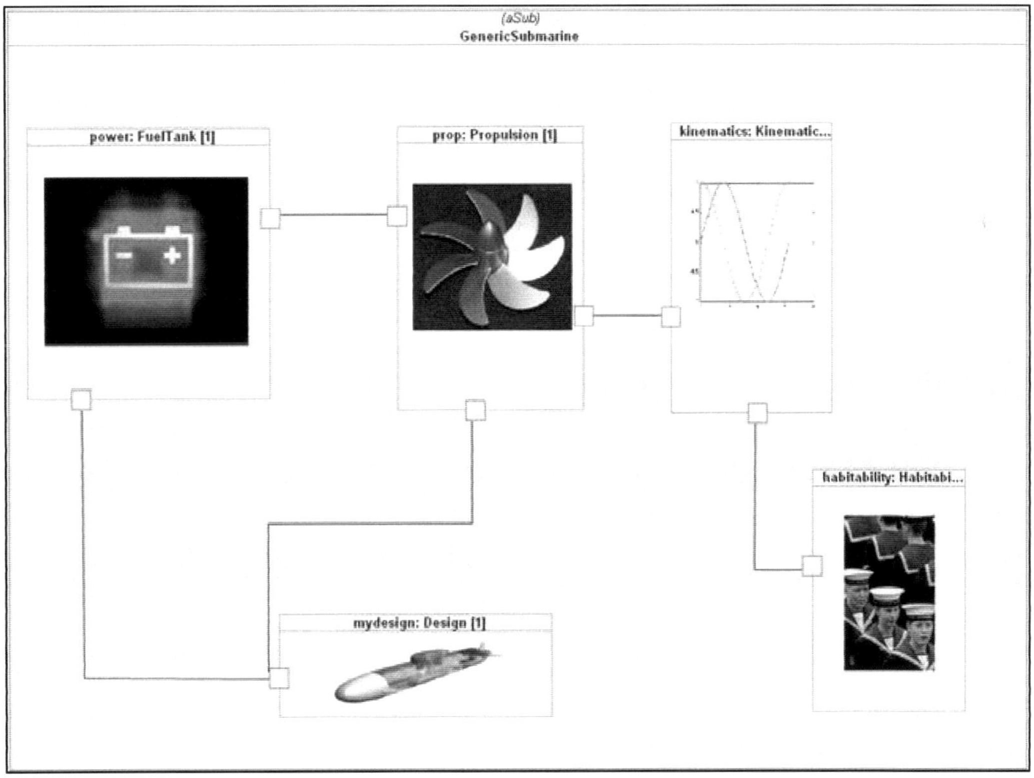

Figure 3: ModelicaML composite diagram representing the generic submarine platform

Links between blocks provide a means of passing variable values to other blocks due to causal or implicit interdependencies. The links also provide a valuable means of exposing these interdependencies within the Generic Submarine model to the user.[1]

3.2 Mission Profiles

The completed platform configuration is paired with a mission profile in the form of a ModelicaML state machine. The generic submarine was exercised through a mission profile (see Figure 4) consisting of two open water transits separated by a constrained area transit in which the use of diesel engines was forbidden. This accounted for the transit to the mission area where the generic submarine spent a period of time on patrol, followed by a similar transit home.

The mission constructed for demonstration was deemed to have failed if the generic submarine completely depleted any fuel stores (diesel, AIP or battery) during the mission or if a threshold maximum indiscretion ratio specified for each mission phase was exceeded.

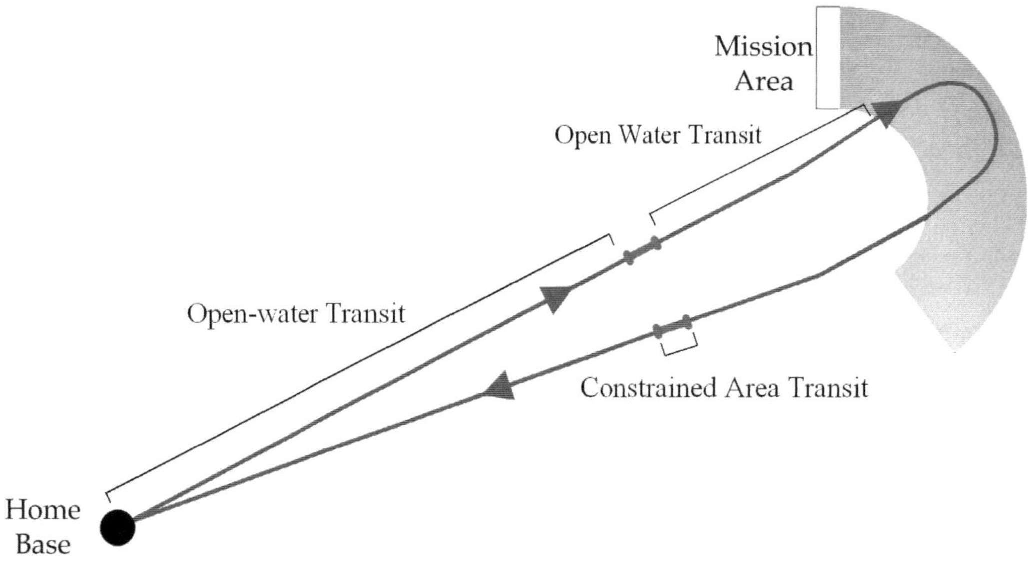

Figure 4: Example mission profile

3.3 Simulations and Results

Simulations were conducted with the use of OpenModelica software which compiled generated Modelica source code into an executable model. Simulations were run for long enough to allow the submarine to complete the example mission.

[1] This method of first constructing subsystem building blocks and selecting and assembling these blocks into a platform configuration was found to offer several engineering benefits: it aided management of complexity by separating architecture from implementation; it promoted re-use by enabling construction of models from existing components; and it promoted incremental model refinement through the ability to specialise object classes.

3.4 Fuel Level Analysis

Fuel level analysis allows observation of the generic submarine model in each phase of the mission profile selected for simulation. Mission success is constrained by the requirement that none of the three fuels available are exhausted at any stage during the mission; this constraint was satisfied by the generic submarine and is illustrated in Figure's 5 and 6.

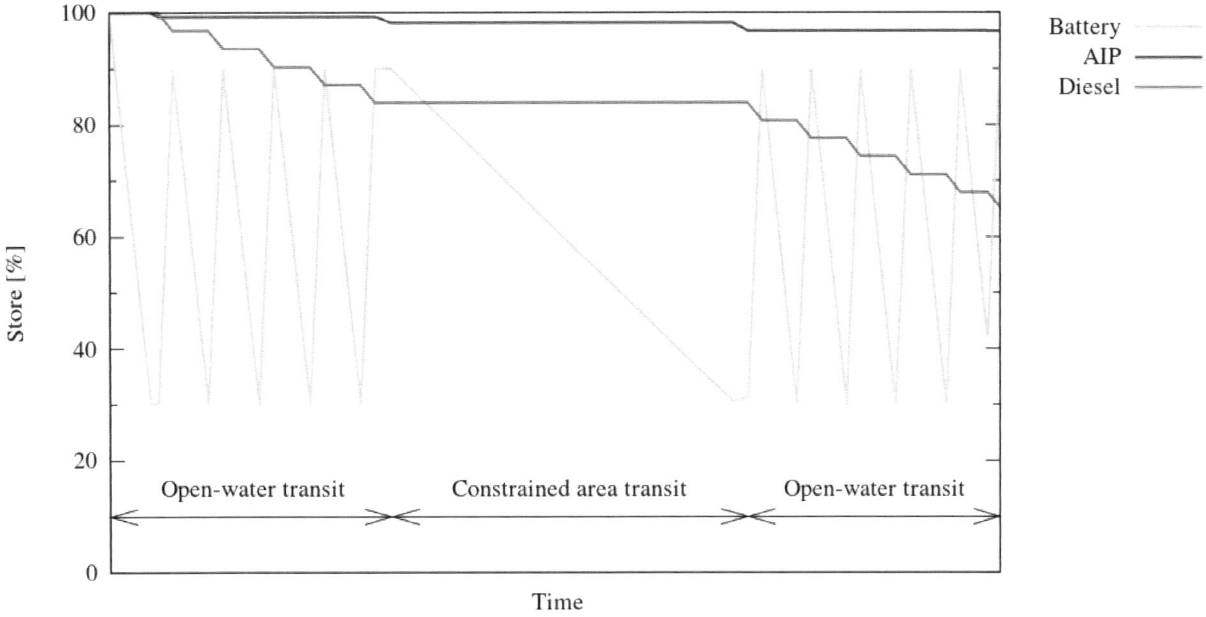

Figure 5: Fuel Levels Trace for transit to Mission Area

Figure 5 describes the fuel use during transit phases between home base and the mission area. The return transit (mission area to home base) produced similar results and is not displayed here. The AIP system was required at the end of the first battery discharge during the first open-water transit due to the constraint that the diesel engines may not be used during daylight hours so as to avoid detection from the exposure of the snorkel mast.

Figure 6 shows the behaviour of the generic submarine's fuel systems during the patrol of the mission area. The submarine was constrained to not use the diesel system during the mission patrol. This constraint was satisfied due to the adequate capacity of the battery and AIP systems. The majority of the AIP fuel capacity was used during the mission patrol phase.

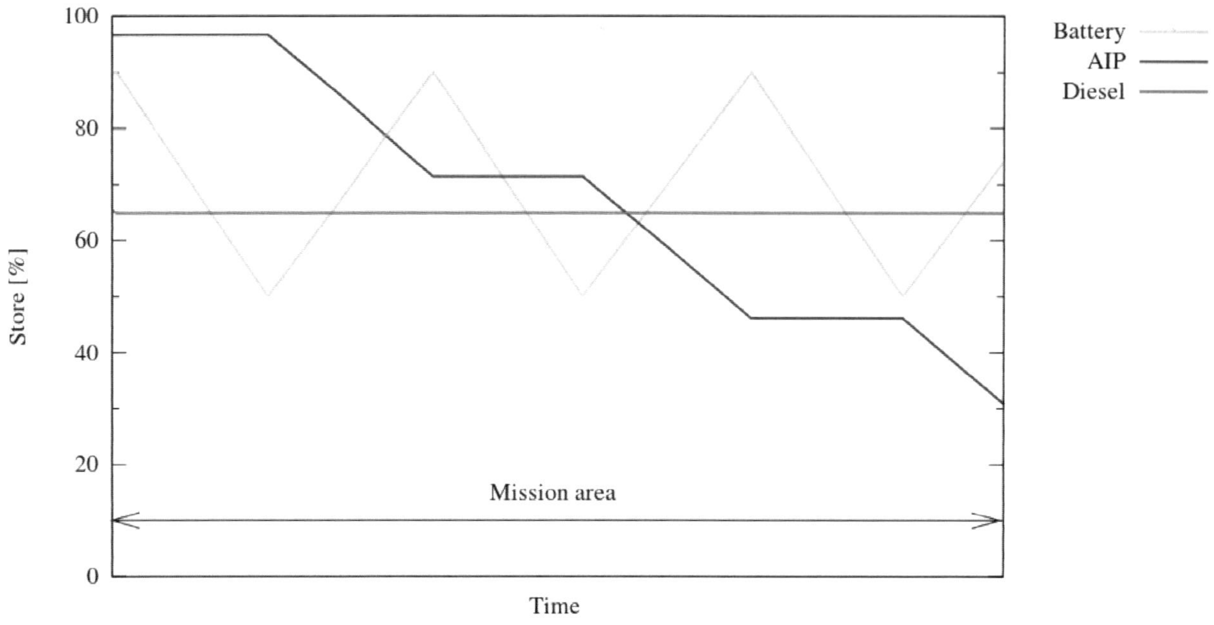

Figure 6: Fuel Levels Trace describing fuel use in the mission area

Whilst the analysis provided here does not relate to any specific submarine or operation insight is provided into the types of capability assessments that lie within the capabilities of the IPSM itself. The IPSM provides a vehicle for making assessments of high level capability of both submarine and surface platforms and exploring sizing trade-off solution spaces. The overarching advantage of this modelling method is the ability to perform these assessments in an efficient manner in terms of time and effort required. A similar analysis of crew sustainment aspects (i.e. Provisions, oxygen etc) will provide assurance that these aspects of the platform design would not constrain mission success. The capability to perform such analysis is a current and major focus of IPSM development.

4. FUTURE WORK AND CONCLUSIONS

4.1 Future Work

Development of a supporting infrastructure for the IPSM is an active area of work. This infrastructure will ultimately provide a means for making the use of IPSM more accessible to project stakeholders and significantly reduce the maintenance effort required as further development work is conducted.

Further development of IPSM models will occur in two areas: continued maturing of current sub system building block models and the development of new building block representations of systems. Future versions of the IPSM will have a repository of blocks to choose from when configuring platforms. Incorporating system models with varying levels of fidelity into platform configuration models will provide a means for assessing specific aspects of integrated performance of a given platform whilst minimising modelling effort required. Currently a major focus is the development and integration of robust and accurate habitability assessments. This development will enhance the fidelity of the assessments made by the IPSM and broaden the scope of the integrated performance assessments in the future.

The IPSM provides a capability to make integrated performance assessments of design configurations as well as assessments of the ability of specific configurations to meet the operational requirements of different mission profiles. Furthermore this tool provides a capability to explore the feasibility of solution spaces for configurations by taking a top down view from requirements through to design and sizing trade-offs. Assessments made in these two areas will provide the Royal Australian Navy and DMO project offices with advice on both specific design configurations and the effect that changes in requirements has on the feasible solution space of various design trade-offs.

4.2 Conclusions

The example analysis presented here demonstrates the ability of the IPSM to construct submarine design configurations for simulation. In a manner similar to tailoring component building block parameters to represent specific systems, the mission profile parameters are selected to facilitate adjustment of the mission phases. Infrastructure supporting IPSM is currently being developed to facilitate incremental and Monte Carlo type analysis of specific design parameters and is further improving the efficiency of the modelling process.

A greater overhead in staff labour was experienced in developing the initial component building block Modelica models, when compared with traditional ad-hoc model programming. However, the advantages of presenting models in a visual and exposed format combined with the efficiency in implementing different design configurations through the reuse of component building blocks, offset these overheads. The NAPSA group anticipate that the modelling approach adopted will provide even greater benefits, in both implementation and interpretation, as the models become more complex. Flexibility offered through the reuse and tailoring of component building blocks and mission profiles added a significant amount of efficiency to the modelling process.

5. REFERENCES

1. Department of Defence, 2009, *Defending Australia in the Asia Pacific Century: Force 2030, Defence White Paper 2009*, ISBN: 978-0-642-29702-0, Commonwealth of Australia.
2. Estefan, J.A., *Survey of Model-Based Systems Engineering (MBSE) Methodologies*, INCOSE MBSE Focus Group, 2007
3. Fritzon, Peter, *Principles of Object-Oriented Modelling and Simulation with Modelica 2.1*, Wiley-IEEE Press, 2004
4. Burcher, R., Rydill, L., *Concepts in Submarine Design*, Cambridge University Press, 1995

Tanking Control

C Gates
BMT Design & Technology, Melbourne, Vic, Australia
cgates@bmtdt.com.au

ABSTRACT

The Royal Australian Navy tanker HMAS Success has recently undergone modifications to achieve International Maritime Organisation compliance. The complex modifications involved the installation of a second hull, internal to the ship, to reduce the risk of cargo spillage in the event of a serious incident. New control stations were introduced, and existing stations upgraded, to provide Human-Machine-Interfaces for the new system. The control stations interfaced with existing systems as well as the new systems, many in hazardous areas.

The engineering design process for the new control system is discussed, including the key challenges encountered. Subsystems include tank level monitoring sensors, PLC systems, touch input displays, pump motor control, ventilation systems, communication systems and alarm and monitoring devices. Challenges faced were time constraints, interfacing an Ethernet communications protocol to a PLC controller, and dealing with obsolete equipment and hazardous areas.

Conversion works to HMAS Success were undertaken, from December 2010 to April 2011, by Singapore Technology Marine (ST Marine). The designer was actively involved in set-to-work activities for the control system during the conversion phase. In late May 2011, the ship returned to Australia where minor additions were incorporated into the control system to meet emergent requirements.

INTRODUCTION

Background

The design task for the new ballast control system on HMAS SUCCESS was commenced well after the rest of the double hulling project, with an inevitable tight schedule. It was performed by an engineer who had just commenced working for BMT due to the unexpected loss of critical personnel, and the task had been inadequately scoped. Consequently, the design engineer was confronted with some of the key reasons for project failure.

Purpose

The purpose of this document is to summarise the approach used to deal with the perceived project risks and to provide details of the technologies and main components used in the control system.

DESIGN TASK

HMAS SUCCESS is a Durance Class replenishment oiler that (previously) did not comply with IMO double hull requirements. In addition to structural and mechanical changes (to provide a double hull) there was a need to provide a ballast control system (for new ballast pumps and pipe work) and a monitoring system for tank level indication.

The new control and monitoring system needed to:

1. Be compatible with ship operations and maintenance routines;
2. Provide local (pump room) and remote (control room) control of two new ballast pumps;
3. Provide operators with indication of the:
    a. tank levels, temperatures and pressures,
    b. position of ballast system butterfly valves,
    c. position of cargo system overflow valves, and
    d. pressure of ballast system transducers.

Control functionality included limiting operation of the two ballast pumps to particular valve alignment arrangements, automatic shut-down of pumps in the event of certain conditions, and the provision of key-locked system overrides. Additional requirements included actuation of overflow valves dependent on control switch position, visual and audible alarms and communications between control and pump rooms. A picture of one the newly installed ballast pump motors is shown in Figure 1.

Figure 1 - Installed Ballast Pump Motor

Some of the equipment was located in hazardous areas, requiring special treatment.

DESIGN APPROACH

The basic design approach relied on improving the planning and organising phase of the task to facilitate goal achievement (see Ref. 1.). The task work-flow proceeded from the work breakdown structure by identifying and obtaining all identified prerequisites for the individual design activities. A buffer of work was created from the individual activities with met prerequisites, and these activities were completed in turn. When an impediment was encountered with one activity, another task was chosen from the buffer allowing work to continue while the impediment was resolved. There was a strong focus on key design issues and the design task was tailored to do only what needed to be done. The approach intended to limit the impacts (from the perceived risks) on project cost, schedule and quality.

The design process commenced with an analysis of customer needs to establish what the system had to do (functional requirements) and how well it had to do it (performance requirements). These requirements were documented along with the identified method of verification, and traceability back to the customer's needs was recorded for each. Verification of system requirements was by analysis, demonstration, test or inspection. The requirements were formally presented to the customer for review and acceptance.

Next, the functional and performance requirements were assessed to define system functions and interfaces. A review of available technologies was conducted to create a list of feasible options that were available to meet these functions. The options were evaluated to determine the best solutions. Options were evaluated for:

1. Technical feasibility, in terms of people, processes and products;
2. Technical risk and the levels of maturity/uncertainty;
3. Prior acceptance within the Royal Australian Navy (standardisation);
4. Reliability (that the option would meet expectations, now and in the future);
5. Flexibility (that the option could still be used with changes in scope);
6. Quality;
7. Availability; and
8. Cost.

The determined best solutions were combined to create the system design including subsystems, individual components and interconnections. The design process took into account system interfaces and applicable constraints. Documentation of the design followed and included a design description and installation details. Instruction details included functional overviews, arrangement, location and connection information as well as parts lists, label lists and data sheets. Additionally, a suite of test procedures was created for verification of component, subsystem and system performance.

The design task was critical path (for the double hulling project) and, towards the end of the task, elements of the design were managed using methods normally employed for software development (Ref. 2.). These methods start by identifying outstanding activities (the backlog). Rather than using a traditional work breakdown structure, the backlog is used to decide on task allocation and project coordination. As work proceeds the backlog is continually assessed and tasking is assigned accordingly. Impediments are eliminated as a management function.

Prior to commencing the task a search was conducted to identify (at a high level) the main reasons for project failure. Commonly, these were found to be lack of communications and information overload. Consequently, efforts were made to establish good communications with members of the customer's project office, suppliers and Navy regulators. Informal discussions were also had with ship system operators and maintainers. Additionally, (in an effort to reduce information overload) the design team concentrated on functional and performance requirements. Constraints were treated by exception unless they specifically related to the function or performance of the system.

The design task provided an opportunity to improve document templates and mainline component lists for future tasks. It taught the importance of being flexible and the need to continually assess and reallocate work to meet project needs.

COMPONENTS

Choosing high quality, reliable components that are (technically) well supported and documented was a key part in successfully delivering the design task. The evaluation process considered mature and proven maritime solutions as priority criteria. Generally, these products are well supported by subject matter experts from within the supplier's organisation.

The value added by these subject matter experts assisted greatly in mitigating risks. Note: The use of Commercial-Off-The-Shelf (COTS) equipment had been approved (by the customer) as a method of reducing project costs.

VEGA products were chosen for tank monitoring. VEGA Grieshaber KG is a world-leader in the supply of level, switching and pressure instrumentation and their measurement technology covers an extremely wide range of applications.

VEGAPULS radar sensors were initially investigated for level measurement, but the manufacturer's recommendations for a smooth bore sounding tube would have been difficult to implement. Instead, VEGAWELL 52 pressure transmitters were chosen for the task. The VEGAWELL 52 transmitters met the operational requirements for tank level monitoring and have been proven to be robust and reliable in a maritime environment. They offer quick and precise monitoring characteristics and long-term stability. The VEGAWELL 52 transmitters were able to be retrofitted relatively easily into tanks and were connected to VEGASCAN 693 signal conditioning instruments in a multi-drop arrangement. The multi-drop arrangement utilises the HART communication protocol and was limited to five transmitters (instead of fifteen) per signal conditioning instrument, due to hazardous area limitations. Use of the HART communication protocol meant temperature information could also be captured from the transmitter's integral PT100 temperature sensor. VEGABOX 02 breather housings were used as the interface between VEGAWELL 52 pressure transmitters and VEGASCAN 693 signal conditioning units.

VEGABAR 52 pressure transmitters were chosen for pressure monitoring. VEGABAR 52 pressure transmitters are well suited to monitoring gas pressures in product tanks. They provide precise monitoring and can withstand pressure shocks caused by rough seas. Like the VEGAWELL 52 transmitters, the VEGABAR 52 pressure transmitters were connected to VEGASCAN 693 signal conditioning instruments in a multi-drop arrangement, for pressure and temperature monitoring. The VEGASCAN 693 signal conditioning instruments also served as power supply units for VEGAWELL 52 and VEGABAR 52 transmitters and provided an Ethernet interface for connection to the logic controller. PACTware software is available for set up of VEGA products and provides a convenient method to adjust, test and record sensor parameters. VEGA products used are shown in Figure 2.

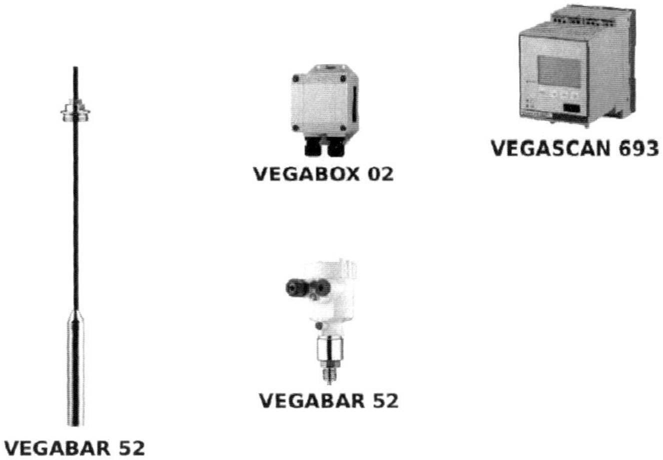

Figure 2 - VEGA Products

A distributed Siemens SIMATIC S7-300 Programmable Logic Controller (PLC) was chosen for control of the system. Siemens automation systems provide a future-proof solution and are engineered to meet the needs of the maritime industry. The chosen controller arrangement provides an efficient, flexible and compact solution. Local and remote Control and Monitoring (C&M) is provided by an S7-315 Central Processing Unit (CPU) with digital and analogue ET 200M input/output (I/O) modules (for remote C&M) and ET200S I/O modules (for local C&M) in a decentralised arrangement. ET200 S I/O modules were used for local operation as they are certified for use in hazardous areas (Zone 2). The local (pump room) and remote (control room) modules are housed in steel control panels and interconnection is via a PROFIBUS field bus. The SIMATIC product range is shown in Figure 3.

Figure 3 - SIMATIC Product Range

The PLC is the core of the system and is connected to VEGASCAN 693 signal conditioning instruments, ballast system pressure transducers and butterfly valves, cargo system overflow valves, bar-graphs (for primary tank level indication) and ballast pump motor contactors. Generally, I/O signals are 24-volt DC and 4 to 20 milliamps with the exception of the VEGASCAN 693 instruments, which are via Industrial Ethernet. Seven VEGASCAN 693 instruments have been used and are connected to the PLC via a SCALANCE X-200 Industrial Ethernet Switch. The VEGASCAN 693 instruments and PLC use different communication protocols, and conversion from MODBUS (VEGASCAN) to PROFIBUS (PLC) is made via software in the PLC.

Two SIMATIC MP 377 INOX 15" touch panels have been used for Human Machine Interfaces. One is installed in the pump room (for local C&M) and the other is installed in the control room (for remote C&M). The resolution of the MP 377 panel is 1024 x 768 pixels and it has a high-contrast thin film transistor (TFT) display and a stainless steel front. The MP 377 touch panel was chosen for its high ingress protection rating (IP66 installed) and suitability for hazardous areas (Zone 2).

Primary level indication is provided by LED bar-graphs, which replaced obsolete units. HI-Q119 bar-graphs, from Precision Instrument Company, were chosen as replacements units. The bar-graphs incorporate programmable intelligent controllers and are rugged and reliable. The HI-Q119 bar-graphs were selected because they were available in a similar size to the obsolete units and for their compliance with MIL standards for shock, vibration and electro-magnetic interference (EMI). The bar-graphs connect to 4 to 20 milliamp (analogue) outputs from the PLC. An installed HI-Q119 bar graph is shown in Figure 4.

Figure 4 - Installed HI-Q119 Bar Graph

Schneider LC1D contactors are used for direct-on-line (DOL) starting of ballast pump motors. Start and stop signalling is enabled through touch panel inputs and PLC digital I/O. The motors are protected by GV3P motor circuit breakers and LRD thermal overloads that are designed for protection of motors used in explosive atmospheres. Figure 5 shows the motor switchgear during panel fabrication.

Figure 5 - Schneider Motor Switchgear during Panel Fabrication

The installed system worked well after some minor changes that were made during set-to-work activities. BMT would like to thank the customer's project team for involving the designer in set-to-work activities. This improved the installer's understanding of the design and allowed for the required minor changes to be made quickly and simply.

CONCLUSION

The new ballast tank design task for HMAS SUCCESS, and associated project risks, has reinforced:

1. The need for a systematic design approach; and
2. The value of using high quality, reliable components that are supported by subject matter experts.

The approach used for this design task has been summarised in this paper, along with an overview of the major components used.

REFERENCES

1. Ballard G B, 2000, The Last Planner System of Production Control, University of Birmingham
2. Schwaber K and Beedle M, 2002, Agile Software Development with Scrum, Prentice Hall

The Application of Foresight Planning to Australian Defence Force Capability Acquisition

Anthony Woolley and John Wharington
Defence Science and Technology Organisation, Australia
anthony.woolley@dsto.defence.gov.au
john.wharington@dsto.defence.gov.au

ABSTRACT
When the Australian Defence Force (ADF) identifies a capability gap an acquisition process commences, supported by science and technology (S&T) guidance. Though the S&T support requirements are governed by the needs of the acquisition project, the S&T planning process would benefit from the introduction of a framework to improve the robustness and transparency of decision making with regards to the allocation of S&T resources. This paper presents an overview of a proposed framework, encompassing the Foresight Planning methodology, to assist in the identification of critical design issues; technology readiness; and research plans for critical technology areas in support of ADF capability acquisition. The framework facilitates establishment of S&T requirements thereby assisting the development of an S&T Plan.

1. INTRODUCTION

In 2003, the Defence Procurement Review (DPR) was established to initiate change in the Australian Defence Force (ADF) acquisition process [1]. Resulting from this review, the Defence Science and Technology Organisation (DSTO) became responsible for reviewing all technical risk assessments and managing science and technology (S&T) plans for ADF capability development and approval processes. Guidance has since been published relating to the development of and performing technical risk assessment, as well as a template for writing S&T Plans that define S&T requirements and support provided by the DSTO [2, 3]. However the existing procedure concerning the provision of the S&T support is aimed at a relatively high level and could benefit from the introduction of a process to assist in the construction of S&T Plans. Attempting to develop and apply such a formal procedure in a recent acquisition project, the DSTO developed a conceptual framework, utilising Foresight Planning techniques, to systematically identify and prioritise relevant S&T.

Developing an S&T Plan is akin to strategic and long range planning and numerous tools, such as 'SWOT Analysis', 'Portfolio Analyses' and 'Balanced Scorecard', are available for use in developing those plans. However, these tools do not constitute a formal framework. Instead, Foresight Planning, with a lengthy historical application in examining innovation policy; considering alternative futures; generating a vision of possible futures; and planning and developing actions to achieve desired goals [4, 5, 6], provides the necessary methodology to enable the development of strategic and long range plans. Noting that while Foresight Planning methods are not without their limitations, Foresight Planning has had successful application in government, commercial and defence applications [7, 8, 9]. The Foresight Planning methodology, therefore, appears to provide an ideal basis for use in developing S&T Plans in support of ADF capability acquisition.

This paper presents an overview of an S&T planning framework, as shown in Figure 1, designed to assist the development of S&T Plans for capability acquisition. Examples of the planning products produced by the framework for a fictitious ADF maritime project are included to assist in the explanation[1]. A full description of the framework will be available in a forthcoming report [10].

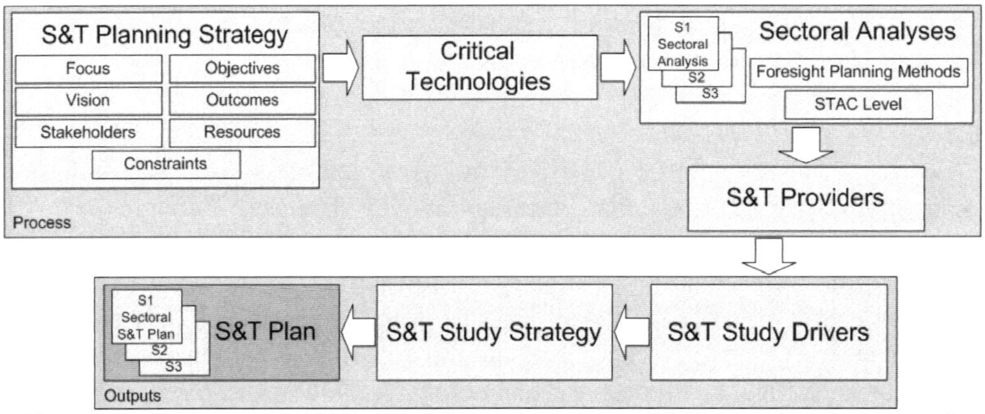

Figure 1: Conceptual ADF capability acquisition S&T planning framework.

The application of Foresight Planning within the S&T planning framework described here differs little from that published in the general literature; however the following aspects are considered somewhat unique: the methodology for identifying critical technologies; the methodology for identifying suitable providers of S&T support; and the establishment of a new metric related to the level the S&T advice required during various acquisition phases, known as the S&T Advice Capability (STAC) Levels. Outputs from the framework will contribute to the identification of S&T study drivers, enabling development of an S&T study strategy and the S&T Plan. The framework will benefit ADF capability acquisition by enabling: technology options assessment; the identification and prioritisation of critical technical risks; and the analysis and prioritisation of system integration issues. The aim of the framework is to provide a traceable method to ensure objective decision-making during ADF capability acquisition. It is anticipated that such systematic S&T planning will assist in reducing project risk, project costs and scheduling delays.

2. SCIENCE AND TECHNOLOGY PLANNING STRATEGY

The first step in the S&T planning strategy is to identify, and document, the goals of the S&T planning for acquisition analysis, as defined in the following sub-sections. This approach follows from the general Foresight Planning methodology documented in the literature [6]. Articulating these definitions ensures that high-level S&T guidance may be aligned with stakeholder requirements.

2.1 Focus

The focus defines the core problem to clearly establish expectations [6]. For example, assume that the Royal Australian Navy (RAN) is acquiring a new capability referred to as Project SEA X,

[1] While it is recognised that there are many different fundamental inputs to capability, the S&T planning framework shown here focuses primarily on the technology issues related to the major systems and their acquisition, sustainment and upgrade, along with consideration of whether an appropriate S&T advice provider can be identified or needs to be developed/established to support the project.

a potential focus statement would be: *The focus of the S&T planning strategy for Project SEA X is to identify technology and technology trends, including opportunities and threats, spanning the life of the capability. Issues of consideration include: technology influencers and developments; in-country capabilities to provide S&T advice at various levels of expertise; support and test facilities; priority setting for national S&T policy; and enhancing the commercial competitiveness of Australia.*

2.2 Vision

The vision is an imagined representation of the future [6]. For Project SEA X planning, the vision statement might be: *The vision of the S&T planning strategy for Project SEA X is to identify the future morphology of the capability, and its systems, to assist development of the long term S&T Plan.* The vision enables identification of the level of technological advancement; identification of capabilities; and development of scenarios, such as the threat environment.

2.3 Stakeholders

Stakeholders include participants and anyone interested in, or affected by, the outcomes [6]. Stakeholders are specific to the acquisition project but will include the ADF, DSTO and the Defence Materiel Organisation (DMO).

2.4 Constraints

Constraints impede delivery of objectives. Documenting constraints ensures stakeholders do not have unrealistic expectations of the acquisition project. Constraints include project implementation details; Government policies; program timelines; acquisition, build and sustainment strategies[2]; consideration of which fundamental inputs to capability are within planning scope; program costs; through-life capability objectives; survivability requirements; and staffing constraints. For example, during the capability life-cycle of Project SEA X, it might be assumed that up to 8% of Australian Defence Organisation (ADO) staff will depart the organisation, requiring recruitment, redeployment and/or re-training to sustain the Project SEA X workforce. Another example constraint may relate to the decision to acquire military-off-the-shelf (MOTS) systems for Project SEA X. As a result, the balance of S&T activities may shift away from technology selection and enhancement, and instead lean towards understanding the integrated performance of the system in the ADF context.

2.5 Objectives

Objectives represent high-level questions to be answered; desired documentation; the degree of involvement by stakeholders; and the duration of the S&T Plan [6]. Objectives, guided by the focus, incorporate information needs and benefits. The objectives for S&T planning would be to:

1. determine strategic technologies and research and development (R&D) priority areas to enable identification of capability related critical technologies that provide for improved: capability edge; survivability; habitability; service life; and through-life costs;
2. develop an S&T planning vision for the capability to identify technology trends and suitable technologies that may be incorporated during the capability life-cycle. Identification of

[2] It should be noted that a lack of an agreed acquisition strategy will result in uncertainty in defining S&T program deliverables and the expected delivery schedule. While this will not in itself interrupt the S&T planning activities, it is possible that S&T resources will be wasted by focussing on S&T issues that are subsequently found to be outside the project scope or are inconsistent with the timing of key decisions.

technology threats is also considered. Threats include disruptive technology, that is technology that renders other technology obsolete or those that counteract technology included in the capability design; and
3. facilitate development of S&T policies for the life-of-type, governed by Objectives 1 and 2.

2.6 Outcomes

Outcomes are derived from the objectives and contribute to the ADF acquiring a capability utilising state-of-the-art, or at least up-to-date, technology and facilities [6]. Outcomes from S&T planning in support of ADF capability acquisition include:

1. recommendations for S&T policy for the project;
2. recommendations for technology areas for inclusion within the capability;
3. critical technology areas and cross impact within the capability;
4. a vision for the future direction for critical technology areas within the project; and
5. identification of key providers of S&T support and advice for the capability.

3. CRITICAL TECHNOLOGIES

The purpose of the S&T planning framework is to identify technology areas that influence the acquisition process, relating to specific capability or sustainment requirements, vulnerabilities or systems. To achieve this, technology areas relevant to the capability must be identified[3]. S&T prioritisation is aided by the definition of criticality measures that relate the technology areas to performance, risk and cost of the provision of capability through-life. These definitions are developed by the Foresight Planning practitioner and vary widely in the literature [12], with a set of example measures being presented in Table 1. The next step is to determine the significance of each technology area to capability acquisition, thereby enabling identification of critical technologies [6, 10]. Accordingly, the measures are assigned values by Foresight Planning practitioners working with subject matter experts.

Table 1: Measures of Significance (Project SEA X – Major Systems Technologies).

Measure	Definition
Capability	Measures the technology area's effect on capability including direct influences via the coverage of the technology area on function, as well as indirect influences via functional dependencies.
Cross Impact	Measures the indirect effect that the technology area has on dependent functions. This considers how the technology area is expected to influence the integrated design via its design drivers.
Integration	Measures the degree to which the technology area is subject to influence by other technology areas that have shared functions and design drivers. This can be interpreted as the sensitivity of the technology area to competing demands or conflicting requirements by other systems.
Sustainment	Measures the degree to which the technology area affects capability sustainment indicating whether: the technology area is vital to ensure cost effective through-life sustainment; or that the technology area is purely driven by the capability itself.
Risk	Measures the risk associated with the technology area, comprising of the *Cross Impact* and *Integration* measures, and allowing for technology readiness levels [10, 13].
Critical	Critical technology areas are those deemed significant according to having a high ranking of *Capability*, *Sustainment* or *Risk*.
Cost-effective Edge	Measures the degree to which significant developments in the technology area, that directly results in enhanced capability, are affordable within the scope of funding by the acquisition project. Even if this yields a low score, developments in those technology areas may still become available due to large investment by industry or the greater scientific community.

[3] This may through the use of such techniques as Work Domain Analysis [11].

Assessment of whether a technology area is considered *Critical* is proportional to the *Capability*, *Sustainment* and *Risk* measures. When assessing the technology areas against the measures of significance, it is important to consider:

1. technology evolution – including refresh rates, permanence and obsolescence issues. Table 2 presents technology evolution characteristics relevant for critical technology analysis; and
2. design influencers – a major factor in determining the timing of design specification and technology de-risking is if the technologies have strong, widely varying demands on major design parameters. Table 3 defines design influence characteristics for an example system.

Table 2: Technology evolution characteristics (Project SEA X – Major Systems Technologies).

Technology Characteristic	Definition
Refresh Rate (static) (measured in years)	The rate at which static aspects of the technology undergo major evolution in form, function, major interfaces and system demands. Static aspects are typically built-in to the capability and not upgraded except, possibly, during capability upgrades.
Refresh Rate (upgrade) (measured in years)	The rate at which non-static aspects of the technology undergo major evolution in form or function. This relates to items that can be upgraded readily, including software upgrades and modular subsystem replacements.
Permanence	Indicates if the influencers on the design are permanent for the life of the capability.
S&T Cost-effectiveness	Refer to definition for *Cost-effective Edge* in Table 1.
Novelty for Australia	Identifies if the technology has been implemented in a relevant ADF environment.
Obsolescence Issues	Identifies if the technology area is subject to lack of access to technical expertise, equipment suppliers and replacement parts after several technology refresh rate cycles. This can include technology involving replaceable components, non-standard equipment, evolving interface standards and small marketplaces.
New Facilities Required	Identifies if new facilities requiring significant investment or establishment lead-time (of approximately five years) are required in order to provide advice at a specified level.

Table 3: Design influence characteristics (Project SEA X – Major Systems Technologies).

Design Influence	Definition
Power and Cooling	Identifies if the technology relates to power, 'hotel load', cooling and thermal management.
Life Support	Identifies if the technology governs crew levels, accommodation and life support.
Weight and Moment	Identifies if the technology significantly influences weight and moment, stability, margins and ballasting requirements.
Size, Layout and Form	Considers items involving demands on the hull-form, displacement, deck layout and major equipment layout.
Networking and Bus	Considers items involving demands on the data bus, sensor distribution and cabling.

Technology evolution characteristics may be determined in consultation with subject matter experts, or for greater fidelity, by the use of technology trend analysis techniques such as bibliometrics [14]. Design influence characteristics are expected to be tailored to the particular system by subject matter experts, based on knowledge of factors involved in fundamental design constraints. Since the scope of influence by the acquisition process on the fundamental design of commercial-off-the-shelf (COTS) and MOTS options is limited, the influence characteristics may be reduced to a small set for COTS/MOTS projects.

Table 4 presents technology evolution characteristics and design influences for the major system being acquired in the Project SEA X example. Included is a simple decision timeframe, defined in Table 5, to identify the required timing of planning decisions. Input for this table is derived from subject matter experts and the application of Foresight Planning methods (presented in Section 4). The table contains two technology areas relevant to Project SEA X. Consider *Hull Materials*. Here the refresh rate is 30 years and it is a permanent feature for the capability; it is also an early decision requirement with a need for facilities to perform test and evaluation. Hull material cannot be upgraded during the life of the capability and therefore the refresh rate must be evaluated against the major system's life-of-type requirement. Construction of the major system cannot commence until hull material is evaluated; a delay in selecting appropriate material will affect the acquisition strategy schedule.

Table 4: Technology evolution characteristics and design influences analysis (examples for Project SEA X).

Technology Area	Technology Evolution							Design Influence					Decision Timeframe				
	Refresh rate, static (years)	Refresh rate, upgrade (years)	Permanence	S&T cost effectiveness	Novelty for Australia	Obsolescence issues	New facilities required	Power, Cooling	Life Support	Weight and Moment	Size, Layout and Form	Networking and Bus	Early	Delay	At Convenience	Urgent Facilities Required	Plan Upgrade
Sensors (Sonar)	8	4	✓	✓	✓			✓		✓	✓	✓	✓				
Hull Materials	30	30	✓	✓			✓			✓	✓		✓			✓	

Table 5: Technology decision timeframes.

Decision Timeframe	Definition
Early	Decisions that need to be made early and influence aspects of the capability design, including technology with long refresh rates and development lead-times.
Delay	Decisions that are appropriate to make at the latest possible stage to exploit ongoing evolution or maturity. They have lower levels of design interdependency with other technologies/systems or the interdependency can be managed.
At Convenience	Short lead-time and low interdependency decisions that can be made at convenience.
Urgent Facilities Required	It is necessary to establish facilities for technologies with refresh rates of five years or more. Assumes: five years to establish facilities; that technology development will require half to one refresh rate cycle to mature any advancement suitable for capability inclusion.
Plan Upgrade	Identifies technology areas subject to obsolescence that are early design items. This avoids degraded sustainment of capability due to technology areas requiring midlife upgrade.

4. SECTORAL ANALYSES

Sectoral analysis consists of studies conducted by the industry in critical technology areas, requiring input from subject matter experts. Sectoral analysis assists identification of:

- technology drivers – key features and characteristics, including disruptive technologies; local industry manufacturing skills; and the level of knowledge;
- timing – technology evolution and obsolescence issues; refresh rates; potential key technology advancements during the life of the capability; and
- providing S&T advice – ability of provider sectors to impart advice in relevant technology areas, measured using the STAC level (described in the following sub-section); existing or new facilities required to support the technology area; and intellectual property.

4.1 Science and Technology Advice Capability Level

STAC levels identify the ability of a provider sector to supply specific technology advice [10]. STAC levels, defined in Table 6, may influence aspirations to develop long term knowledge and in-country capability. Minimum STAC levels for each technology area and acquisition phase are specific to, and determined by the needs of, respective acquisition projects. In particular, the acquisition of COTS or MOTS systems involving mature technologies would potentially allow for lower minimum STAC levels than would bespoke designs.

Table 6: STAC level definitions.

STAC Level	Definition
0	Insufficient knowledge to provide meaningful advice.
1	Awareness of the technology and knowledge of advancements in the technology area is being maintained.
2	Along with STAC 1, there is understanding of the technology such that: principles of the technology can be explained; there is ability to operate it; and/or there is ability to understand the science behind it.
3	Along with STAC 1 and 2, there is ability to: specify the requirements of the technology; understand the capability of the technology; and understanding of technology requirements.
4	Along with STAC 1, 2 and 3, there is ability to design/perform innovative R&D in the technology area.

4.2 Foresight Planning Methods

Numerous Foresight Planning methods may be utilised to assist the identification of technology trends, integration issues and priority technology areas that influence key design decisions; capability edge; and capability risk required in the sectoral analyses. Foresight Planning methods are well documented in the literature [6, 15] and their application in the development of S&T Plans is not unique. Tools also exist to assist the selection of appropriate Foresight Planning methods [6]. The methods identified as being most suitable in relation to S&T planning and sectoral analyses include [6, 15]:

- Expert Panels;
- Environmental Scanning;
- Morphological Analysis;
- Structural Analysis.
- Cross Impact Matrices;
- Trend Extrapolation;
- Technology Roadmap;
- Portfolio;
- Delphi Survey;
- Backcasting; and

The reader is directed to the literature for specific details in relation to these techniques [6, 15].

5. SCIENCE AND TECHNOLOGY PROVIDERS

Resources to support S&T activities originate from a number of provider sectors, each having unique strengths, characteristics and constraints. Categories of provider sectors to be considered for ADF capability acquisition are presented in Table 8, with the suitability of provider sectors to supply S&T resources determined by considering the constraints defined in Table 9. Requirements analysis of critical technologies, such as performing sectoral analyses and application of Foresight Planning methods, will provide the necessary information for each constraint in Table 9. Whilst it is expected a complete set of provider constraints would require analysis of procurement and security policy, the relationship used to assess Provider Sector Applicability for Project SEA X was defined as:

$$\text{Provider Sector Applicability} = \frac{STACAppl.(Free + Critical.C_w + Risk.R_w + CommercialValue.CV_w + ThroughLifeSupport.TLS_w + Security.Sec_w)}{TechAppl}$$

where:

- *STACAppl* – a Boolean value indicating if the technology area has a STAC level requirement;
- *Free*, *Critical*, *Risk*, *Commercial Value*, *Through Life Support* and *Security* – criticality measures derived from the analysis described in Section 3 and defined Table 9;
- C_w, R_w, CV_w, TLS_w and Sec_w – weightings (subjectively assigned) indicating applicability of respective constraints to the provider sector; and
- *TechAppl* – number of technology constraints satisfied for each critical technology area.

For each critical technology, the result is a unitless, unbounded value for each provider sector and the greater the value, the more suitable the provider. Where the value is low, the provider is deemed less suitable but may be available if the technology applicability constraints are managed. So, in Table 10, continuing with the Project SEA X example, hull material for this major system was deemed to be critical, with no significant technological risk but having commercial value and through-life support requirements. Therefore there were three applicable technology constraints (that is, *TechAppl* = 3). Analysts for Project SEA X had also set a minimum STAC level requirement for the technology. Application of the aforementioned equation yielded the values presented in Table 10 for the applicability of each provider sector. Here, DSTO was identified as being the most suitable S&T provider, however other providers, particularly Australian industry and academia, may also provide suitable S&T support.

Table 8: Provider Sectors capable of providing S&T resources (Project SEA X).

Provider Sector	Definition
DSTO	The Defence Science and Technology Organisation
National Industry	Australian based industry and multinational industry with an established local presence in Australia.
National Academia	Australian academic and research organisations
Multinational Industry	Foreign industry (to Australia) with a multinational presence but not in Australia
International Academia	International academia and government funded research organisations
International Industry	International industry, with no local presence in Australia
International Defence	International defence industries
International Government	International government departments and in-country research organisations equivalent to the DSTO.
Multinational Government	Multinational technical and research organisations, for example: The Technical Co-operation Program (TTCP) and the North Atlantic Treat Organisation (NATO).

Table 9: Technology applicability constraints to Provider Sector suitability (Project SEA X).

Constraint	Definition
Critical	Identifies if the technology is established as of critical significance (see Section 3).
Risk	Identifies if application of the technology is associated with significant potential risk (see Table 1).
Commercial Value	Identifies significant commercial value for application outside the acquisition project, to identify areas where investment by the acquisition project may have national wealth building benefits. It is also attractive for industry with non-Defence customers; or export potential is likely to be more cost-effective and robust.
Through-life Support	Identifies if access to advice concerning the technology area is expected throughout the capability's life. Through-life applicability is assumed to be directly related to whether the technology area is significant with respect to sustainment.
Security	Identifies if the technology area includes designs, systems, specifications, trade secrets and security-sensitive elements requiring management of special access arrangements.
(Free)	Items are unbounded by technology area provider applicability constraints.

Table 10: Alignment of Provider Sectors with Technology Areas (examples for Project SEA X).

	Technology Applicability						Provider Applicability								
Technology Area	Critical	Risk	Commercial value	Through-life Support	Security	(Free)	DSTO	National Industry	National Academia	Multinational Industry	International Academia	International Industry	International Defence	International Government	Multinational Government
Sensors (Sonar)	✓	✓					2.0	2.0	0.0	1.2	0.0	0.0	0.0	0.0	0.0
Hull Materials	✓		✓	✓			1.5	1.3	1.3	0.8	0.3	0.2	0.4	0.0	0.6

6. SCIENCE AND TECHNOLOGY STUDY DRIVERS

S&T study drivers distil the results from the S&T planning analysis, focussing on: early design decisions with high bearing on the acquisition strategy; ability to provide cost effective capability edge; and identifying areas of high risk. Study drivers identify requirements and technology provider sectors, for each critical technology area, which may, for example, enable leverage from Australia's allies or identify technology areas best suited for in-country development. Five study driver categories, defined in Table 11, may be used to document the relevant information.

Table 11: Definitions of S&T study driver categories.

Category	Definition
Vision	A summary of goals to be achieved by the acquisition project's S&T effort in the technology area.
Significance	Summarises the significance the technology area has on the acquisition project in terms of: capability; sustainment; and risk. It will identify if the technology area compels early design decision making.
Advice Capability	Assesses immediate and ongoing STAC requirements and actual capabilities within the DSTO, national industry and academia. Capability gaps are identified and highlighted for action to build-up. Where the actual capabilities are deficient, potential S&T advice providers are identified by sector. Where there is excess capability, potential areas for trade-off are identified.
New Facilities	Summarises new facilities, including experimental facilities and large-scale modelling and simulation projects, identified as being required for providing S&T advice for the specification, design and assessment acquisition phases. Facilities required to be established in the short term are highlighted. Facilities may not need to be built if access to an equivalent facility elsewhere can be secured.
Options	Available technology options that are expected to be topics of research and assessment.

7. SCIENCE AND TECHNOLOGY STUDY STRATEGY

The S&T study strategy is an evolving document developed by subject matter experts and Foresight Planning practitioners in response to the S&T study drivers that is refined as capability requirements are defined. The study strategy assists development of the S&T Plan and presents the agreed high level objectives, constraints and rationale for the acquisition program:

1. constraints, such as acquisition project timelines; and resource allocation.
2. aspirations, such as the benefits to Australia; and identifying a capability effectiveness model, including capability/performance trade-offs; and
3. rationale, the logic to determine STAC levels and Technology Readiness Levels (TRLs) [13].

8. SCIENCE AND TECHNOLOGY PLAN

The S&T Plan, developed by subject matter experts in collaboration with acquisition project stakeholders, recommends the short and medium term development of facilities and advice capability to assist in registering and managing high risk areas. It summarises the S&T strategy to be implemented by individual sectoral S&T Plans. Sectoral S&T Plans correspond to individual critical technology areas, outlining the significance and vision for technology development and technology options and present specific objectives of the sectoral S&T activities.

9. CONCLUSION

The proposed S&T planning framework is designed to provide a formal, systematic approach for the identification of critical technologies that require investigation in the short to medium term of a capability acquisition project. Sectoral analysis, and the application of the Foresight Planning methods, enables identification of study drivers for these critical technology areas, as well as highlighting costs, facilities, resource requirements, the scope of work and associated goals contributing to the study strategy. It is from the study strategy that the capability acquisition S&T Plan will evolve, documenting S&T analyses to be performed in each of the critical technology areas. The S&T planning framework for ADF acquisition provides a comprehensive, documented and traceable S&T process for key decision-making points during capability acquisition. While this paper provided examples of S&T planning products for the major systems technology aspects of a fictitious Project SEA X, the framework would benefit from further research examining the sensitivity of the methods, the choice of metrics and the subjective inputs provided by participants and demonstration of the framework for a more complete list of the fundamental inputs to ADF capability.

10. REFERENCES

1. Kinnaird, M., Early, L. and Schofield, B., 2003, *Defence Procurement Review*, Commonwealth of Australia
2. O'Neill, J., Thakur, N. and Duus, A., 2007, *Technical Risk Assessment: a Practitioner's Guide*, DSTO-GD-0493, Defence Science and Technology Organisation, Commonwealth of Australia
3. Projects and Requirements Division, 2010, *Technical Risk Assessment Handbook*, Defence Science and Technology Organisation, Commonwealth of Australia
4. Becker, P., 2002, *Corporate Foresight in Europe: A First Overview*, EUR 20921, European Commission
5. Cagnin, C. and Scapolo, F., 2007, *Technical Report on a Foresight Training Course*, EUR 22737 EN, Institute for Prospective Technological Studies, Joint Research Centre, European Commission
6. European Commission, *European Foresight*, URL - http://forera.jrc.ec.europa.eu/, Institute for Prospective Technological Studies, Joint Research Centre, European Commission, sighted: 23 August 2011
7. Department of the Navy, 2004, *The Navy Unmanned Undersea Vehicle (UUV) Master Plan*, Department of the Navy, United States of America
8. Anderson, P. D., Jørgensen, B. H. and Rasmussen, B. 2001, *Sensor Technology Foresight*, Risø-R-1292(EN), Risø National Laboratory, Roskilde
9. UNIDO, *United Nations Industrial Development Organization Technology Foresight*, URL - http://www.unido.org/index.php?id=o5216, sighted: 21 October 2011
10. Woolley, A. and Wharington, J., 2012, *A Framework to Support S&T Planning for Royal Australian Navy Capability Acquisition,* DSTO-TR-XXXX, Defence Science and Technology Organisation, Commonwealth of Australia [In Review]
11. Naikur, N., Hopcroft, R. and Moylan, A., 2005, *Work Domain Analysis: Theoretical Concepts and Methodology*, DSTO-TR-1665, Defence Science and Technology Organisation, Commonwealth of Australia
12. Klusacek, K., 2003, *Selection of Research Priorities – Method of Critical Technologies*, UNIDO – Training Course "Technology Foresight for Practitioners", 6-10 October 2003, Prague, Czech Republic
13. Mankins, J. C., 1995, *Technology Readiness Level: A White Paper*, Advanced Concepts Office, Office of Space Access and Technology, National Aeronautics and Space Administration
14. Abercrombie, R. and Udoeyop, A., 2011, *A Study of Scientometric Methods to Identify Emerging Technologies*, International Society for Scientometrics and Informetrics 2011 Conference, 4-7 July 2011, Durban
15. Cooper, R. G. and Edgett, S. J., 2001-2008, *Portfolio Management for New Products: Picking the Winners*, Reference Paper #11, Product Development Institute Inc. and Stage-Gate Inc.

Assessment of Governor Control Parameter Settings of a Submarine Diesel Engine

Peter Hield and Michael Newman
Defence Science and Technology Organisation, Melbourne, Vic, Australia
peter.hield@dsto.defence.gov.au

ABSTRACT

Modern conventional submarines use diesel generators to provide power for propulsion and the hotel load. They are run at a nominally constant engine speed against a constant imposed load, but are subject to back pressure disturbances due to the underwater exhaust. Under these conditions the engine speed governor attempts to maintain a constant speed (often using a PI or PID controller) by regulating the fuel flow. Poor control can cause fluctuating exhaust gas temperatures, leading to increased wear and reduced reliability.

This paper investigates the performance benefits that can be obtained through tuning of the governor control parameters. While more complicated control systems may be feasible to improve performance, this paper focuses on the comparison of poor versus optimally tuned governors using an industrial standard control topology. A linearised model of a turbocharged submarine diesel is developed using the simulation package Matlab/Simulink. Along with other analysis techniques, this model will be applied to find the PI gain boundaries as limited by the stability criteria and allowable engine parameter ranges defined for this application. The performance variation of the submarine diesel within these boundaries will then be investigated and discussed.

INTRODUCTION

Diesel generators typically operate at a constant engine speed and power a constant electrical load. One method of maintaining the desired engine operating point is to use a governor, which regulates the fuel flow in order to maintain a constant engine speed. A governor of this type is often all that is necessary for full control of the engine. For example, an increase in the load will cause a drop in engine speed. In response, the governor will increase the fuel flow rate until the speed is brought back to the set point. A common governor implementation is via a proportional-integral (PI) or proportional-integral-derivative (PID) controller. This paper only considers the simpler and more common PI based control structure, with the impact of an added derivative component left for future work.

Under submarine snort conditions, the underwater exhaust system imposes a high exhaust back pressure on the diesel engine, due to a combination of the long exhaust ducting and the head of water above the exhaust outlet (see Figure 1). Sea surface waves passing over the submarine cause the water depth to vary, causing a corresponding fluctuation in the engine exhaust back pressure. This can result in fluctuations in the engine speed, air flow rate, fuel flow rate and cylinder temperature as the governor struggles to maintain constant speed [1-5]. Under high sea state conditions the engine power output must be reduced to limit the maximum cylinder temperatures, as excessive temperatures cause increased wear, reduced reliability and reduced engine life [1, 6].

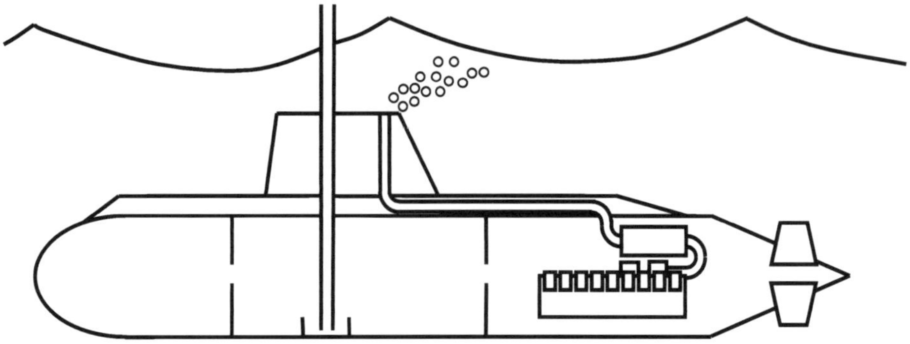

Figure 1 Diagram of a snorting submarine

From a control systems viewpoint, the engine and governor form a standard closed loop control system, with the back pressure acting as a disturbance input to the system (see Figure 2). This paper examines the effect of varying the proportional and integral gains on the ability of the controller to reject the disturbance and maintain a constant engine speed, while investigating the impact of the control parameters on selected engine parameters.

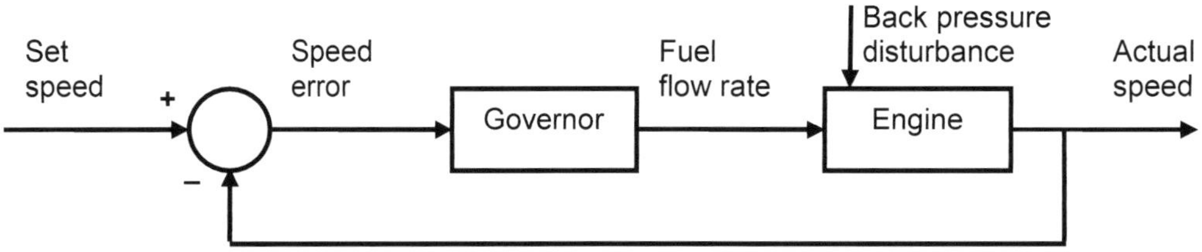

Figure 2 Diagram of the engine control system

THE ENGINE MODEL

A mean value model of a submarine diesel engine was created using Matlab/Simulink, and validated against the more detailed Ricardo Wave engine model developed by DSTO in prior work [1]. The engine control system consists of a speed governor which monitors the engine speed and varies the fuel flow with the aim of matching the engine speed to the set speed. The speed governor is implemented using a PI controller. An exhaust back pressure disturbance was applied to simulate the engine operating conditions for a snorting submarine. Simulink's control system design tools were then used to linearise the model around a typical operating point, allowing linear control techniques to be used.

The mean value engine model is based on those of Biteus [7], Hopka *et al.* [8] and Karmiggelt [9]. The model consist of a series of components (compressor, intercooler, combustion chamber, turbine) connected via control volumes representing the inlet and exhaust manifolds. The engine and turbocharger shafts are modelled using Newton's second law. Special attention was paid to correct modelling of the effect of the high back pressures. The engine model is shown in Figure 3.

The engine modelled is an 11 litre six cylinder inline turbocharged diesel engine. Although this is not an existing submarine engine, this engine topology is commonly used for submarine engines. The chosen operating point has a load torque of 1000 Nm and an engine speed of 1800 RPM, and the engine is subjected to a mean back pressure of 145 kPa with a fluctuating pressure of amplitude 62.5 kPa and period 7.4 s superimposed on top, representative of a submarine snorting in sea state 3 [1, 10]. The engine speed is controlled using a PI controller with the equation

$$\dot{m}_f = Pe + I\int edt,$$

where $\dot{m}_f$ is the fuel flow rate, e is the engine speed error, P is the proportional gain and I is the integral gain.

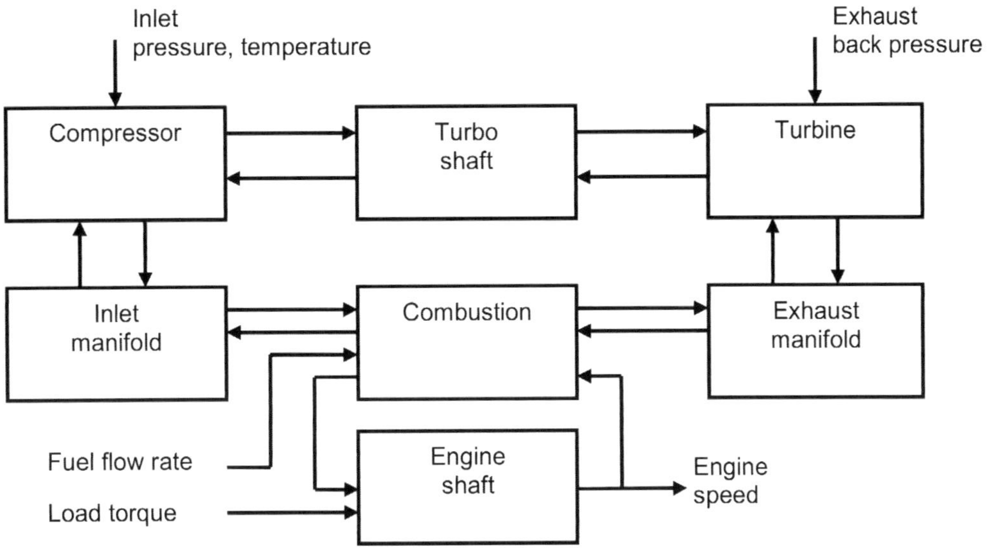

Figure 3 The engine model

RESULTS AND DISCUSSION

The model was run for a wide range of values of P and I, and the system performance was evaluated at each point. Figure 4 shows a typical engine response to imposed back pressure fluctuations for one controller setting. Several measures of system stability and performance were calculated, and the results are discussed in this section.

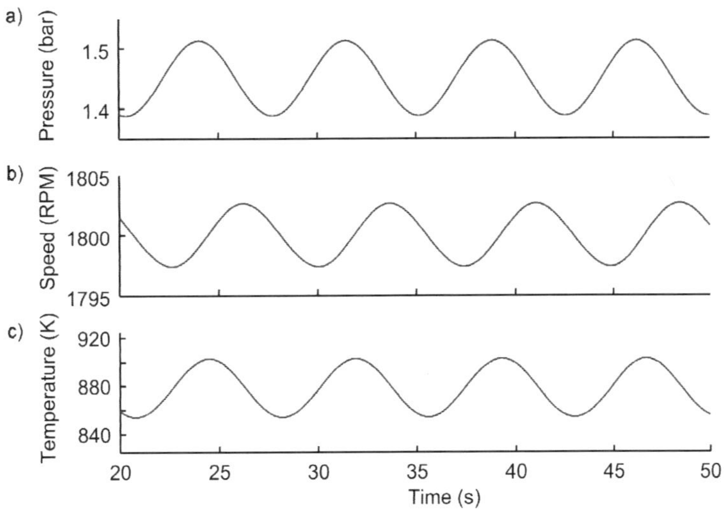

Figure 4 Typical results from the mean value engine model, showing a) the exhaust back pressure (bar), b) the engine speed (RPM) and c) the cylinder exit temperature (K), for $P = 0.0037$ and $I = 0.032$

System Stability

Two common measures of relative stability of a linear system are the phase and gain margins. The phase margin is defined as 180° plus the phase angle of the open loop transfer function at the gain crossover frequency (the frequency at which the gain is unity). The gain margin is the reciprocal of the gain of the open loop transfer function at the frequency where the transfer function phase angle is -180° [11, 12]. For stability, both the phase margin and the gain margin (expressed in dB) must be positive. Figure 5 shows the variation of the system phase margin as a function of P and I, with the variation in gain margin over the same range shown in Figure 6. The model predicts stable engine operation over a very wide range of both P and I, which may be varied by several orders of magnitude (note the logarithmic scales used for the P and I axes).

When both P and I were set to low values it was not possible to run the model. This is because the controller gains are too low for the system to respond to disturbances sufficiently quickly to keep the engine operating parameters within an acceptable range. A step change in the engine load, for example, would cause the engine to stall as the engine speed would drop too far before the fuel flow increased to compensate for the increased load. This region has been marked as having insufficient gain. For low I but larger values of P, the model predicts stable operation up to a threshold of $P \approx 0.2$, above which the phase margin rapidly decreases and unstable behaviour is predicted. Within this band of P values, I can be reduced to zero without affecting the stability of the system. For low P but larger values of I, stable behaviour is predicted, which gradually becomes unstable as I is increased further. The value of I at which instability sets in increases as P increases. At the centre of Figure 5 is a large region representing the values of the controller gains P and I for which the model predicts that the engine will respond to back pressure disturbances in a stable manner.

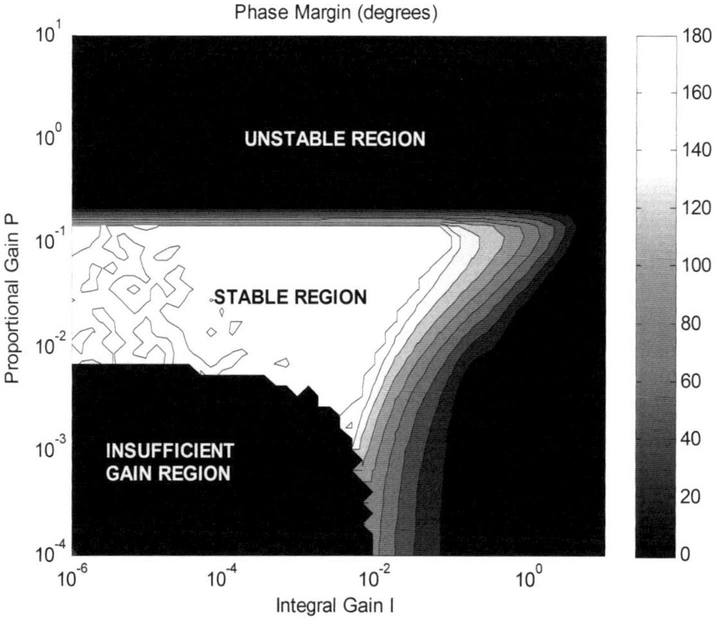

Figure 5 The system phase margin stability indicator as a function of P and I

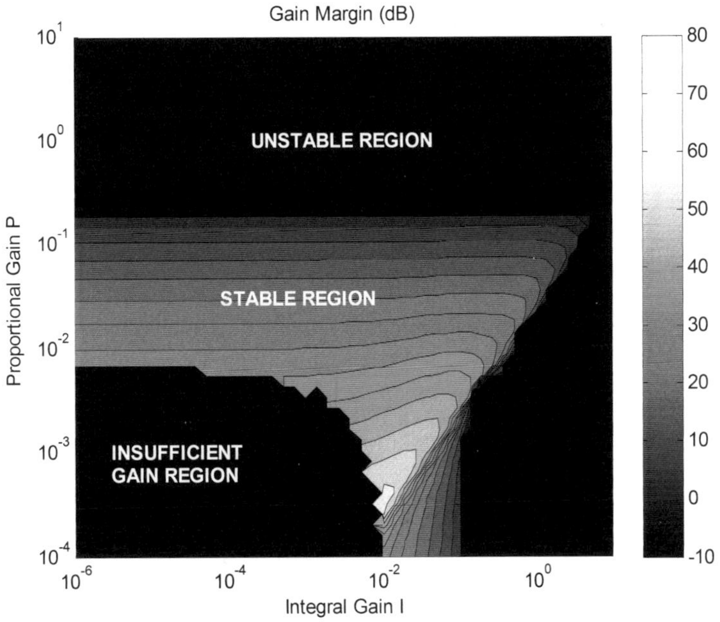

Figure 6 The system gain margin stability indicator as a function of *P* and *I*

The threshold *P* value appears to be independent of the value of *I*. The unstable behaviour in the region is due to the delays inherent in the fuel system. For a six cylinder 4-stroke engine operating at 1800 RPM, one cylinder fires every $1/90^{th}$ of a second. Thus, even if the controller detects an error in the engine speed, the fuel system is incapable of responding until the next cylinder fires, thus introducing a delay of up to $1/90^{th}$ of a second. The speed error may grow during this time, and a large proportional gain will produce a large corrective action when the cylinder does fire, potentially leading to an overcorrection and thus unstable behaviour. The threshold *I* value is due to the delay introduced by the integration of the error signal. The significance of this delay increases as *I* increases relative to *P*, and thus *P* cannot be reduced to zero for any value of *I*.

The theoretical stability point of the linearised system is independent of the sea state conditions, as the varying exhaust back-pressure is treated as a disturbance into the system and not a change to the closed loop control system model structure. Therefore, the results above are presented independent of the sea state condition. However, while the system may be stable for a given operating point, for large disturbances on the system (such as varying exhaust back pressure under high sea state conditions) the response on the engine parameter may still be undesirable (or even unacceptable). Under excessive disturbance conditions the validity of the linearised system may also be affected. For these reasons the transient response of the system to disturbances as the PI control gains are varied also need to be investigated.

Sea State 3 Disturbance Response

The peak-peak speed variation for sea state 3 conditions is shown in Figure 7. Contours are plotted over the stable region of the *P-I* space. As the task of the engine governor is to maintain a constant speed, the amplitude of the speed variation provides the primary measure of the controller performance. This is close to zero over most of the stable region, although there is a ridge of severe speed fluctuations along the line of $P \approx 10I$. The amplitude of the speed fluctuations reduces gradually as *P* and *I* are moved away from this line, in some areas requiring a variation of P or I by up to two orders of magnitude before the amplitude is reduced to zero.

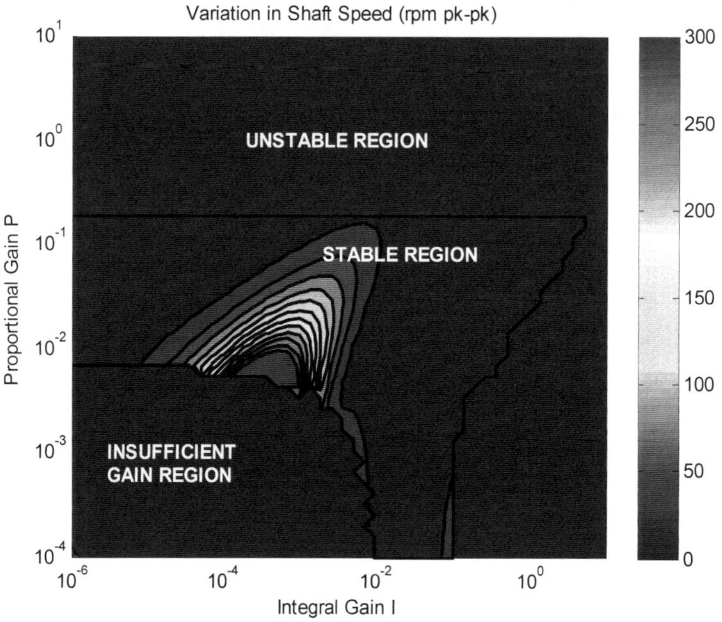

Figure 7 The peak-peak variation in shaft speed (RPM) as a function of P and I, for sea state 3 conditions

A second performance metric is the peak-peak variation of the cylinder exit temperature (see Figure 8). This follows a similar pattern to the engine speed fluctuations, although adjusting I has progressively less effect to the left of the line $P \approx 10I$. Also, the peak-peak variation of the cylinder exit temperature remains significant over the whole range of possible values of P and I.

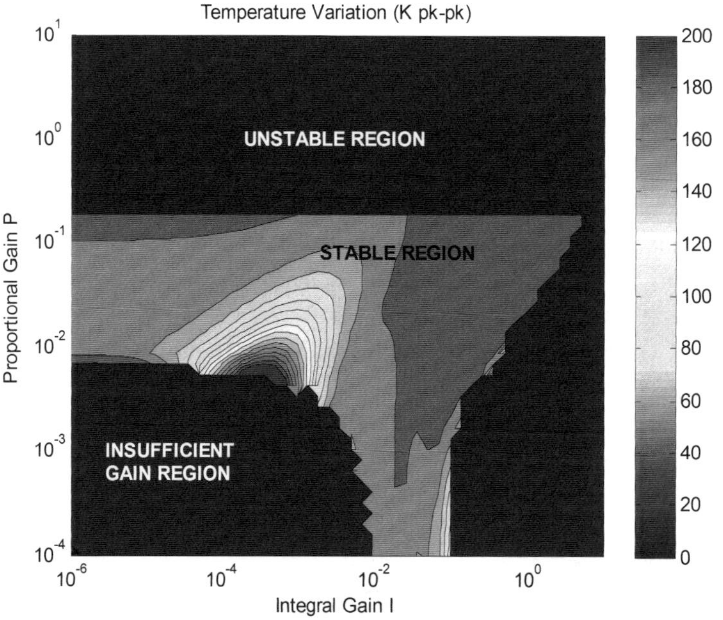

Figure 8 The peak-peak variation in the cylinder exit temperature (K) as a function of P and I, for sea state 3 conditions

Thus, with a speed governor as the sole means of controlling an engine subject to fluctuations in the back pressure, it is possible to reduce but not completely eliminate the cylinder exit temperature fluctuations.

The maximum cylinder exit temperature is shown in Figure 9. This performance metric follows a pattern significantly different from the fluctuations in engine speed and cylinder exit temperature, and can be split into two distinct regions. To the left of the line $P \approx 10I$, the maximum cylinder exit temperature is independent of I, but strongly dependent on P, decreasing with increasing P. To the right of this line, the maximum cylinder exit temperature is approximately constant at its minimum value, and is independent of both P and I. There is only a small region close to the line $P \approx 10I$ where the value of I has a significant effect.

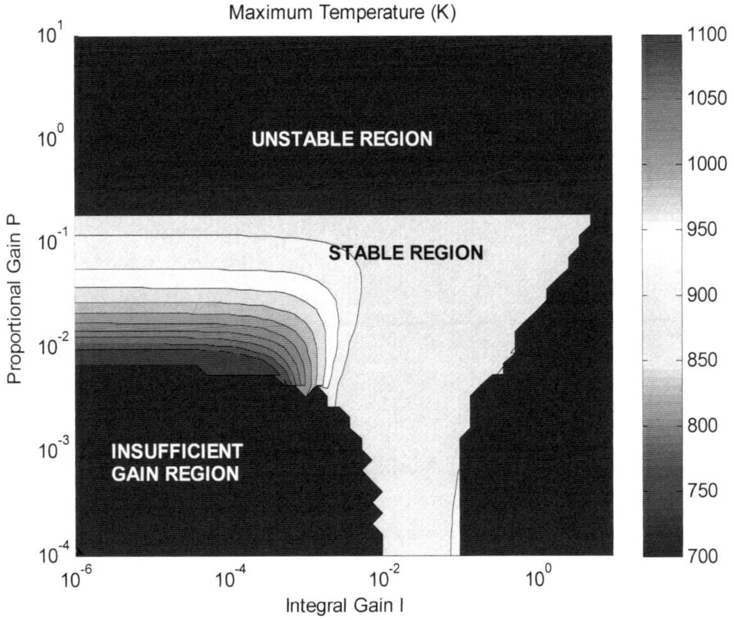

Figure 9 The maximum cylinder exit temperature (°K) as a function of P and I, for sea state 3 conditions

Other performance measures are also possible, and, although not shown here, the peak-peak variation in the turbocharger shaft speed was also considered. This showed very similar behaviour to the peak-peak variation in the engine speed, although the minimum value was significant, at ~20,000 RPM peak-peak. This is due to the pressure ratio across the turbine being directly affected by the back pressure disturbance, and not directly controlled by the engine speed governor.

The stability and performance parameters can be used to impose further limits on the range of acceptable values of P and I. This allows trade offs to be examined and the optimal values selected when tuning the controller gains. Although a system is stable for positive phase and gain margins, it is usual to allow a safety factor to keep the operating point well away from the stability boundaries and to compensate for modelling errors. In this case, a phase margin of 45° and a gain margin of 20 dB have been selected. Regions for which the phase and gain margins are below these values are highlighted in Figure 10, in light green and light blue, respectively. Similar boundaries can also be placed on the performance measures. The region for which the maximum temperature is more than 20 K above the optimal operating point has been highlighted in red on Figure 10 as a region to avoid when tuning the controller. Similarly, the region for which the peak-peak speed variation exceeds 25 RPM is highlighted in green, and the region in which the peak-peak temperature variation is more than 20 K higher than the optimal operating point is highlighted in orange. This leaves the area highlighted in dark blue as the preferred region for setting the *P* and *I* control gains. This is quite a large area, covering almost two orders of magnitude for *I* and three orders of magnitude for *P*. Figure 7 to Figure 9 show the controller performance improves as both *P* and *I* are increased, although for sufficiently high values of *P* and *I*, further increases have only a minimal effect. Figure 5 and Figure 10 show that the limiting factor on the maximum values of *P* and *I* is the system stability.

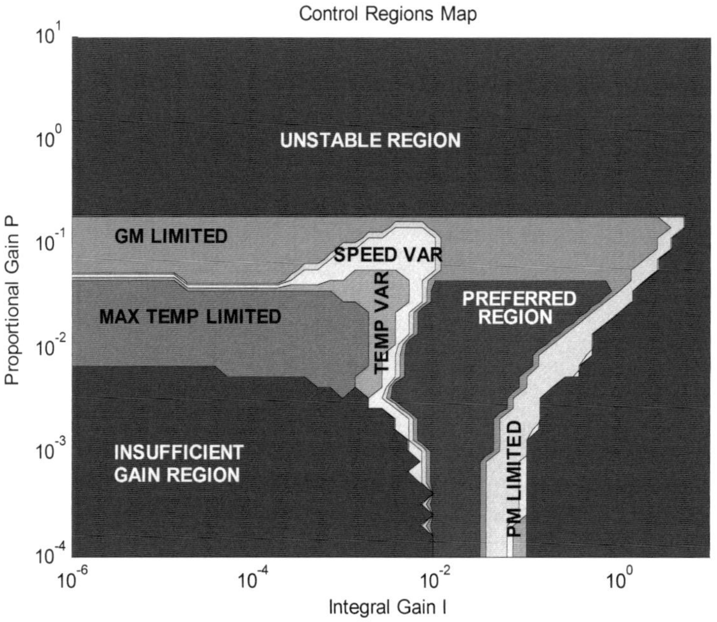

Figure 10 The limiting stability and performance metrics over the region of allowable values of *P* and *I*, for sea state 3 conditions

Sea State 6 Disturbance Response

Sea state 6 was investigated using the same approach as applied in the previous section for sea state 3. The resulting preferred region (blue) for selection of control parameters shown in Figure 11 is clearly smaller than presented in Figure 10. As previously discussed, there is no change in the stability region as this limit is independent from the disturbance. However, with the increase in applied disturbances, the resulting responses in speed and temperature variations have also naturally increased, which results in the more constrained preferred operating region.

While occurrences of snorting in sea state 6 conditions is not likely to be common, it is still a plausible condition that the submarine will have to be able to support. For simplicity of design, implementation and operation it is not desirable to have different control gains for different operating conditions. Luckily, in this case the preferred region for the higher sea state condition is simply a sub-set of the lower sea state condition. Therefore, by selecting control gains within the preferred region of the worse case sea state, the conditions for all required sea states can be met.

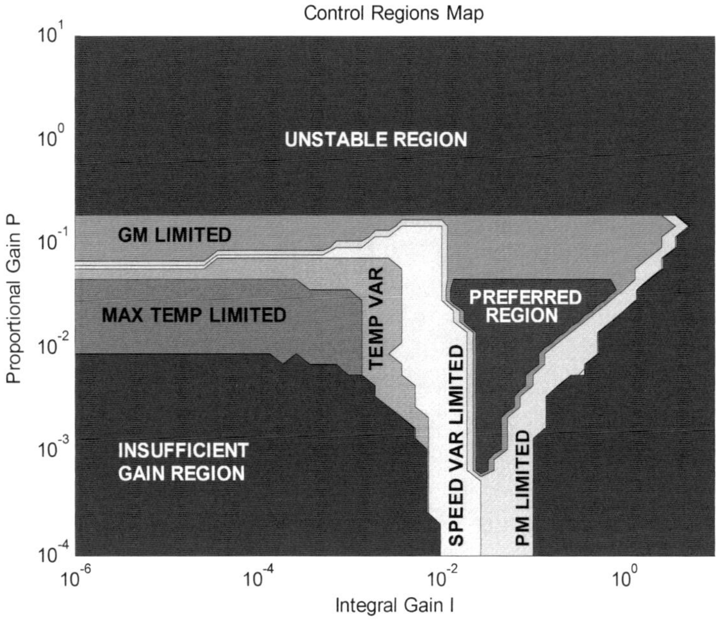

Figure 11 The limiting stability and performance metrics over the region of allowable values of *P* and *I*, for sea state 6 conditions

CONCLUSIONS

Matlab/Simulink has been used to create a mean value model of a diesel engine, which was then validated against a more detailed Ricardo Wave engine model. The engine control system consists of a speed governor which monitors the engine speed and varies the fuel flow with the aim of matching the engine speed to the set speed. The speed governor is implemented using a PI controller. An exhaust back pressure disturbance was applied to simulate the engine operating conditions for a snorting submarine.

The model was used to examine the engine behaviour for different values of the controller gains *P* and *I*, with a back pressure disturbance representative of sea state 3. Stable engine behaviour was predicted over a very wide range of values of both *P* and *I*, and both can be adjusted by several orders of magnitude while still allowing the engine to operate successfully. However, there were significant variations in engine performance over this range and with a poor choice of the controller gains, large fluctuations in the engine speed, turbocharger shaft speed and cylinder exit temperature will occur, along with excessively high maximum cylinder exit temperatures. The results show that a good choice of *P* and *I* can reduce the engine speed fluctuations effectively to zero, and significantly reduce the maximum cylinder exit temperature. For the engine modelled in this paper, this is achieved by increasing both gains as much as possible, while being mindful that increasing the gains too far will lead to instability. Optimising the controller gains to minimise the engine speed fluctuations also minimises the cylinder exit temperature fluctuations, the turbocharger shaft speed variations and the maximum cylinder exit temperature.

While it is possible to eliminate the fluctuations in engine speed, with a control strategy that only targets the engine speed it is not possible to eliminate the fluctuations in cylinder exit temperature.

The key conclusions from this work are therefore as follows:

1. for a speed governed engine, with the control system implemented using a PI controller, the engine exhibits stable behaviour over very wide ranges of both the proportional and integral gains;
2. it is possible to tune the controller in such a way as to effectively eliminate engine speed fluctuations due to back pressure disturbances;
3. tuning the control system to focus on minimising speed fluctuations may also minimise the cylinder exit temperature fluctuations (for the engine described in this paper, there is a large region for which the speed variations are minimised, but the cylinder exit temperature fluctuations are only minimised over a small part of this region); and
4. even at the optimal control point, a PI controller based on speed error alone still results in significant temperature variations due to exhaust back-pressure disturbances. This effect cannot be simply tuned out. A different control structure is required for reduced engine temperature variations.

REFERENCES

1. Hield, P. A. (2011) *The effect of back pressure on the operation of a diesel engine*. DSTO-TR-2531, [Technical Report] Melbourne, Vic., Defence Science and Technology Organisation (Australia)
2. van den Pol, E. (1990) Aspects of Submarines, Part IV: The submarine and the Diesel Engine. *SenW 57STE* (JAARGANG NR 5)
3. Goodwin, G., McGrath, J. and Bowden, D. (2008) Performance of turbocharged diesel engines in ocean-going submarines. In: *Pacific 2008, International Maritime Conference,* Sydney, IMarEST
4. Kirkman, E. T. F. and Hopper, R. A. (1990) Turbocharging for submarines - A special case. In: *International Conference on Turbochargers and Turbocharging*, IMechE
5. Mann, J. (2011) Twin-turbocharged diesel performance under snorkelling conditions. In: *UDT 2011,* London
6. Challen, B. and Baranescu, R. (1999) *Diesel engine reference book.* 2nd ed. Bath, Butterworth-Heinemann
7. Biteus, J. (2004) *Mean value engine model of a heavy duty diesel engine*. LITH-ISY-R-2666, Linkoping, Sweden, Department of Electrical Engineering, Linkopings Universitet
8. Hopka, M., et al. (2003) Identification of a mean value model of a modern diesel engine for control design. In: *2003 ASME International Mechanical Engineering Congress,* Washington, D. C.
9. Karmiggelt, R. (1998) *Mean value modelling of a s.i. engine*. Report Number 98.041, Eindhoven, Eindhoven University of Technology
10. Department of Defence (2003) *DEF (AUST) 5000 ADF Material Requirement Set, Volume 3, Part 4 - Seakeeping*. Canberra, RAN
11. Ogata, K. (1997) *Modern control engineering*. Upper Saddle River, NJ, Prentice-Hall
12. DiStephano, J., Stubberud, A. and Williams, I. (1976) *Theory and problems of feedback and control systems*. Los Angeles, CA, McGraw Hill

Diesel Engine Back Pressure Controller

Owen Tregenza, Peter Hield
Defence Science and Technology Organisation, Australia

ABSTRACT

A submarine diesel engine operates in an adverse environment with a low intake pressure and a high and fluctuating back pressure. This causes high and fluctuating exhaust gas temperatures, which may lead to increased wear and reduced reliability.

This report proposes the use of a system of butterfly valves as an accurate method of reproducing submarine inlet and back pressures, for use with the Defence Science and Technology Organisation's Subscale Submarine Research Diesel Engine and Hedemora V6 Submarine Research Diesel Engine test facilities. A model based on butterfly valve performance characteristics, sea state conditions, engine capacity and operating conditions was developed in Simulink. The valve control system must follow the required pressure against time profile. The development of a target pressure is dependant on exhaust mass flow and valve plate angle. Fluctuations in back pressure cause fluctuations in engine pumping work and as a consequence, fluctuations in exhaust mass flow. The control system must be robust to such fluctuations.

INTRODUCTION

Submarine Environment

The Netherlands invented the first snorkel system for a submarine, and in 1939 the first operational submarines were equipped with a complete system [1]. Whilst this offered an operational advantage, allowing the submarine to recharge its batteries relatively unseen below the waterline, this had significant implications for the operation of the diesel engine.

Low intake pressure and high fluctuating back pressure are a result of the sea state the submarine experiences whilst snorting. With an increasing sea state, both wave height and wave period increase [2]. The back pressure on a submarine engine is due to a combination of flow losses through long exhaust ducting and the pressure head due to the underwater exhaust [3]. The intake pressure can also be reduced due to wave interaction causing the snort mast to close. Fluctuations in intake pressure and back pressure lead to increased fuel consumption and increased exhaust gas and cylinder temperatures [4], potentially leading to increased engine wear and reduced reliability.

The sea state scale gives an indication of the fluctuations in back pressure due to surface waves [3]. By estimation of a submarine snort exhaust depth, it is possible to build an exhaust back pressure versus time profile based on the sea state condition. Typically the mean component of the back pressure is approximately 1.45 bar absolute [3]. Figure 1 shows pressure versus time profiles generated using the estimation of the mean back pressure and sea states '2 max', '4 mean' and '6 mean' [2]. Waves by nature are trochoidal in form, however for simplicity, sinusoidal waves are often assumed [5]. This concept can extend to the fluctuations in exhaust back pressure.

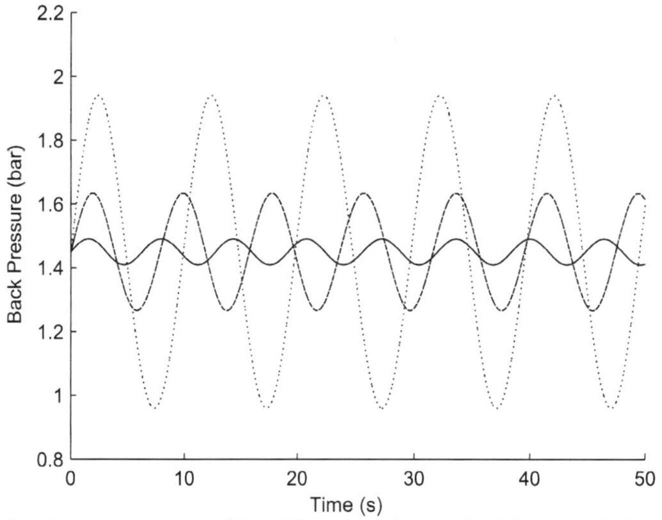

Figure 1: Exhaust back pressure profiles '2 max' (——), '4 mean' (----) and '6 mean' (········).

BACK PRESSURE SIMULATION

Manufacturers of submarine diesel engines need to consider the influence of back pressure on engine performance as the difference between exhaust back pressure and intake pressure has a significant impact on the maximum engine power [6]. Submarine engine test and development programs therefore need to incorporate exhaust back pressure and inlet pressure variations when evaluating engine performance. However, the extent of information on the techniques used to simulate a snorting submarine is limited, with literature only mentioning engine test cells equipped with 'special features' such as 'exhaust flaps' [6] or 'valves monitored electronically' [7] to simulate exhaust back pressure variations.

Two methods suggested in the literature were considered for this report. Periodically immersing the exhaust under sea waves or a water column [4], or a system of butterfly valves or exhaust flaps situated in the exhaust line [6-8].

Periodically immersing an exhaust to sea waves or a column of water would be 'rather complicated' [4], requiring a water filled tower at least fifteen meters in height, and the ability to vary the water depth by up to ten metres every thirty seconds to replicate sea state '6 mean' conditions (see Figure 1). In particular the generation of a transient sinusoidal back pressure corresponding to wave influence would be difficult [4]. Studies using butterfly valves as the primary mechanism for pressure development show success in experimentation [6-8]. Exhaust back pressures from about 0.2 to 2.0 bar were successfully developed using a butterfly valve in the exhaust line of a twelve cylinder aircraft engine [8], as well as the submarine engine test cells discussed in [6] and [7]. This range covers the approximate exhaust back pressures of a submarine in snort.

BUTTERFLY VALVE EQUATIONS

Valve Flow Coefficient

Butterfly valves give little resistance to flow when fully open but are sensitive to flow control when open between fifteen and seventy degrees [9]. The valve flow coefficient is an 'inherent characteristic' related to the valve geometry, and is used to establish flow capacity [10] as a function of valve plate position. It represents the capacity of water that will flow though a valve with a unit pressure drop at a specific opening position. The valve coefficient, C_v, is an empirical relationship and is given by the equation [10]

$$C_v = Q\sqrt{\frac{\Delta P_0}{\Delta P}\frac{\rho}{\rho_0}}, \tag{1}$$

where Q is the volume flow, ΔP_0 is the reference differential pressure, ΔP is the operating differential pressure, ρ_0 is the density of the reference fluid and ρ is the density of the operating fluid. Figure 2 shows C_v as a function of valve position for the general class of butterfly valves outlined in [11].

Calculating Back Pressure

Valve manufacturers publish extensive data on valve flow coefficients, appropriate sizing and pressure drop calculation. Equations (2) and (3) are for valve sizing for volume flow rate for 'ANSI Class 600' butterfly valves with gas flow [11].

$$\Delta P < P_1/2, \quad Q = 272C_v\sqrt{\frac{\Delta P(P_1+P_2)}{G(273+T)}}, \tag{2}$$

and

$$\Delta P \geq P_1/2, \quad Q = \frac{236C_vP_1}{\sqrt{G(273+T)}}, \tag{3}$$

where Q is the volume flow rate across the valve $(\text{m}^3/\text{h for liquid or N m}^3/\text{h for gas})$, P_1 is the pressure upstream of the valve and P_2 is the pressure downstream (Pa), ΔP is the pressure differential (P_1-P_2, in Pa), G is the specific gravity of the fluid and T is the temperature of the fluid $(^\circ\text{C})$. Equation (2) and (3) are for un-choked and choked gas flow conditions respectively. To adjust for the degree of valve opening, the example valve data shown in Figure 2 can be multiplied by the published valve flow coefficient.

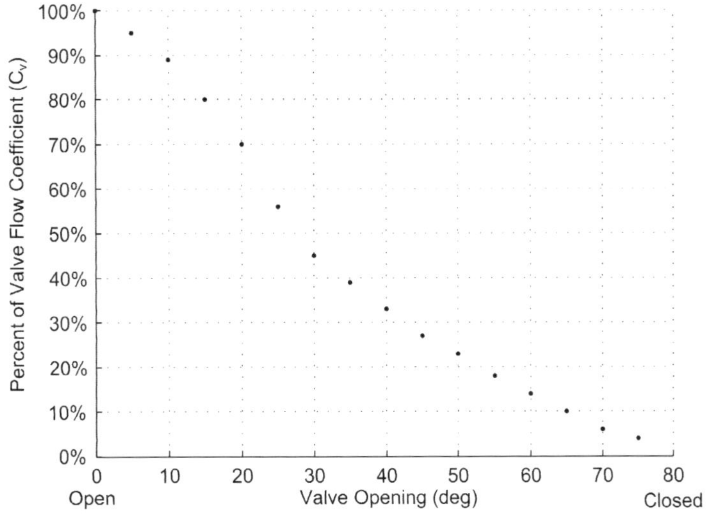

Figure 2: Percent of valve flow coefficient (C_v) as a function of valve opening.

Equations (2) and (3) can be rearranged to obtain the upstream pressure:

$$\Delta P < P_1/2, \quad P_1 = \sqrt{G(273+T)\left(\frac{Q}{272C_v}\right)^2 + P_2^2}\,, \tag{4}$$

and

$$\Delta P \geq P_1/2, \quad \frac{Q\sqrt{G(273+T)}}{236C_v}\,. \tag{5}$$

P_1 represents the pressure upstream of the butterfly valve. Equations (4) and (5) can be used for calculating developed back pressure.

For further information on valve flow coefficients and flow equations refer to American National Standard 'ANSI/ISA-75.01.01-2002 (60524-2-1 Mod)' in 'Flow Equations for Sizing Control Valves' [12].

MATLAB AND SIMULINK MODEL

Back Pressure Development

The capacity and operating speed range of a diesel engine was used as an indication of the volume flow across the butterfly valve. The developed back pressure was calculated as a function of valve opening. Figure 3 shows the developed back pressure as a function of valve opening for a four inch diameter valve with a valve flow coefficient of 255 at engine speeds of 1000, 1500 and 2000 revolutions per minute.

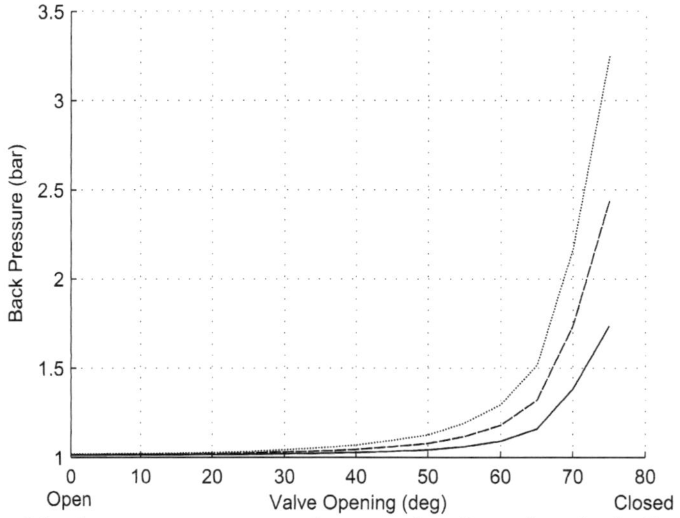

Figure 3: Developed back pressure versus valve opening for a four inch diameter valve with a valve flow coefficient (C_v) of 255. 1000 RPM (——), 1500 RPM (----), 2000 RPM (·········).

Comparison of Figure 1 and Figure 3 shows that the back pressures of sea state '2 max' and '4 mean' can be produced within an opening range of twenty five to seventy degrees. The back pressures of sea state '6 mean' can be produced within an opening range of twenty five to seventy five degrees. Figure 3 shows the highly non linear relationship between valve opening and back pressure.

Valve Position Control

A model was developed in Simulink to simulate the back pressure response of an actuated butterfly valve. Diesel engine capacity and operating speed, the valve characteristics from [11] and DC servo actuator characteristics were incorporated into the model. DC servo actuators are common in control valve applications and are suitable for continuous closed loop control [13]. The actuator characteristics were modelled based on torque, where the motor must overcome the maximum dynamic torque seen by the valve. The empirical equation given in [10]

$$T_D = C_t \Delta P, \tag{6}$$

provides a method for calculating the dynamic torque (T_D) as a function of the pressure drop across the valve (ΔP). C_t is a coefficient based on valve size and allowable maximum rotation. Manufacturers also supply charts for actuator sizing for maximum dynamic torque as a function of the maximum differential pressure and valve diameter (e.g. [10]).

The valve positioning mechanism was simulated using the principle of 'bang-bang' (or 'hysterisis') control. The sign of the error between the set and measured position determines the required direction of rotation. The actuator applies a constant torque in the indicated direction to achieve the new set position. Figure 4 shows the general structure of the position controller.

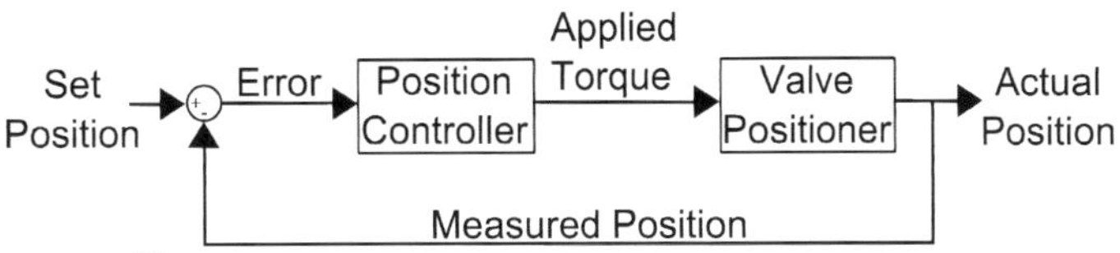

Figure 4: Position controller.

Pressure Drop Control

A proportional-integral-derivative controller (PID) was used to control the required valve position, given the error between the set point and measured back pressures, and outputting the required position change. Figure 5 shows the overall control structure.

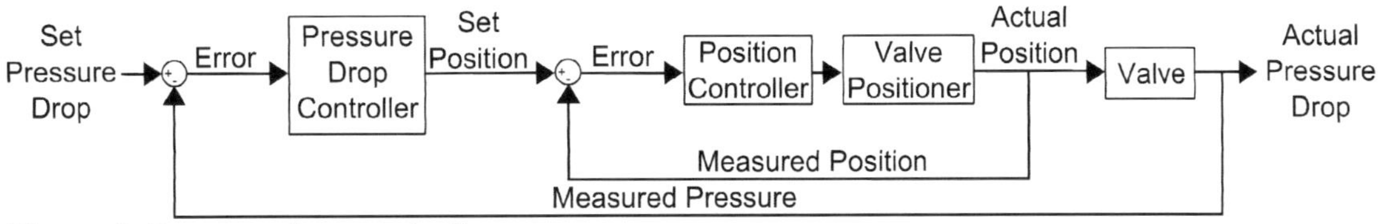

Figure 5: Pressure drop controller.

The proportional (K_P), integral (K_I) and derivative (K_D) terms of the PID controller were tuned heuristically until an adequate response was achieved. As a general rule, increasing K_P provides a faster but more oscillatory response. Increasing K_I provides a faster response and significantly removes steady state error. Increasing K_D provides a slower but more stable response [14, 15].

DISCUSSION AND RESULTS

Back Pressure Development

Figures 6 to 9 show the response of the model given a set point back pressure. A constant set point of 1.45 bar and the sinusoidal pressure versus time profiles, '2 max', '4 mean' and '6 mean', discussed in the introduction, were used.

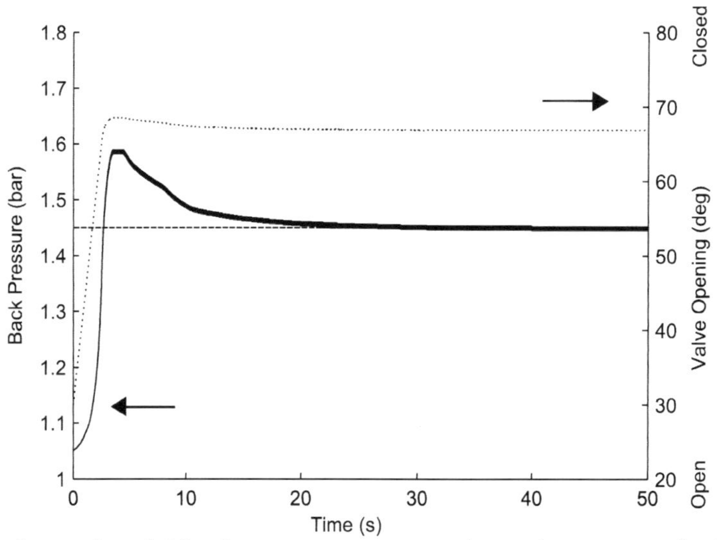

Figure 6: Comparison of model back pressure response (——) to a set point back pressure of 1.45 bar (----) and the valve opening (·······) in response to the set point back pressure.

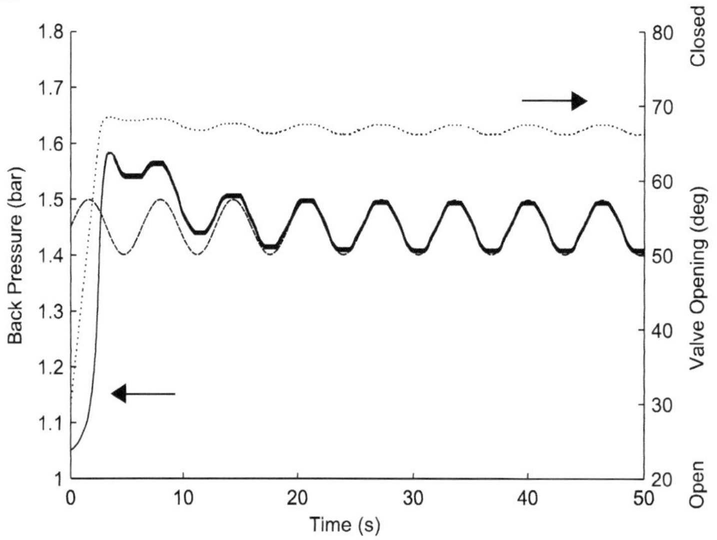

Figure 7: Comparison of model back pressure response (——) to the sinusoidal set point back pressure '2 max' (----) and the valve opening (·······) in response to the set point back pressure.

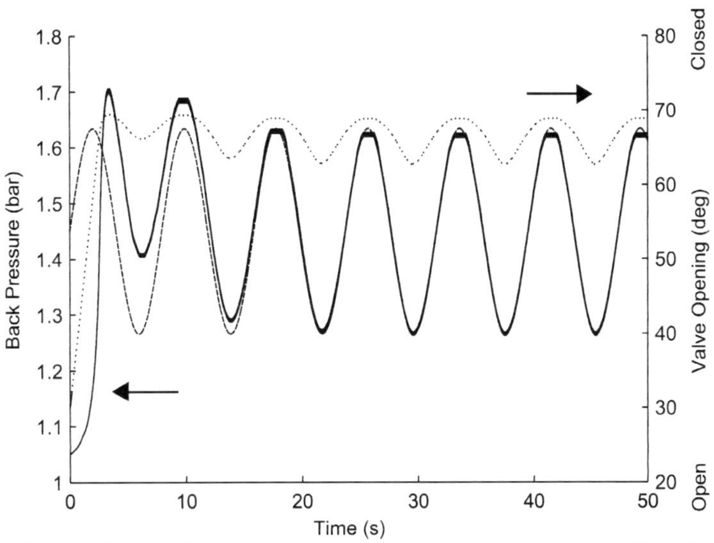

Figure 8: Comparison of model back pressure response (——) to the sinusoidal set point back pressure '4 mean' (----) and the valve opening (·········) in response to the set point back pressure.

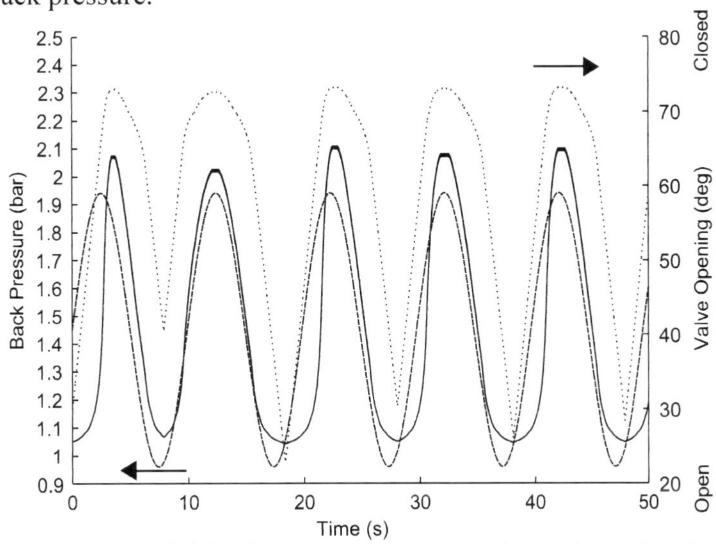

Figure 9: Comparison of model back pressure response (——) to the sinusoidal set point back pressure '6 mean'(----) and the valve opening (·········) in response to the set point back pressure.

For the constant, '2 max' and '4 mean' cases, the response reached the set point after approximately fifteen seconds with a ten percent initial over shoot. The '4 mean' response in Figure 8 also shows a small flattening at each peak. This is a result of the limited actuator speed, the quick transition from increasing to decreasing set point back pressure and the valve opening region. For regions beyond sixty degrees a relatively small increase in valve opening results in a rapid increase in back pressure. No change in pressure will result if the required increase in valve opening is of the order of the hysteresis band, and hence a 'flattening' will be observed.

Although following with approximately the same frequency, the response for the '6 mean' case did not entirely match the set point. An over shoot can be seen at each peak and an under shoot at each trough. The over shoot is a result of the valve opening extending beyond seventy degrees, a region where a small increase in valve position results in a dramatic increase in back pressure. The opposite occurs for the zero to forty degree region, where only a small increase in back pressure occurs with a considerable increase in valve opening. The lag is due to a combination of this valve back pressure relationship, and the limiting speed of the actuator. This behaviour demonstrates the highly non linear relationship between valve opening and back pressure and the difficulties it can pose for designing a butterfly valve pressure drop controller.

Additional minor adjustments to the PID controller terms would yield a quicker response with less overshoot for a particular sea state. However, due to the non linear relationship between valve opening and back pressure, different operating regions will require different controller terms. It was for this purpose that the controller was tuned to produce an adequate response over a wide range of set points to enable the system to simulate a wide range of sea states. Whilst the back pressure response of the Simulink model was adequate, further controller development may minimise the effect that the highly non linear relationship between valve opening and back pressure has on the system. In particular, those behaviours observed for the states '4 mean' and '6 mean'. The development of a back pressure profile map, similar to that seen in Figure 3, could be used to provide feed forward control for the position controller, reducing dependence on the pressure drop controller. Such a map could be easily generated by plotting the back pressure profile for a range of engine operating speeds and valve openings.

Engine Speed Disturbance

Fluctuations in the exhaust back pressure on an engine cause fluctuations in the engine speed. In turn, fluctuations in engine speed affect the exhaust volume flow through across the butterfly valve. An increase or decrease in exhaust volume flow causes a corresponding increase or decrease in the pressure drop across the valve. As a result, the developed back pressure changes. Sinusoidal speed variations, with a mean speed of 1500 RPM, amplitude of 50 RPM and a period of five seconds were introduced. This represents an approximate response of an engine to an imposed and varying back pressure [3]. Figure 10 shows the influence the varying engine speed has on the developed back pressure. The valve position fluctuates to compensate for the fluctuations in exhaust volume flow rate. The back pressure is maintained within two percent of the set point back pressure with small fluctuations due to the limited rotational speed of the valve actuator.

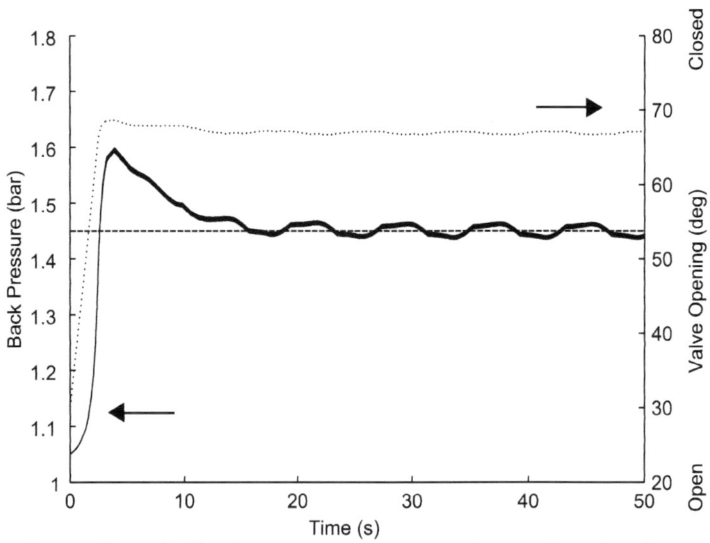

Figure 10: Comparison of model back pressure response (——) and valve opening (········) to a set point back pressure of 1.45 bar (----) for a varying engine speed.

For simplicity, the behaviour of the engine speed and applied back pressure are not related in the Simulink model. In reality, applied back pressure and engine speed are dependant [3]. An approximation of this influence is to simulate the fluctuations in engine speed exactly out of phase with the fluctuations in set point back pressure. It is expected that an increase in back pressure results in an increase in engine load and hence a corresponding decrease in engine speed. Figure 11 shows the influence a varying engine speed has on the developed back pressure for a sinusoidal set point. These examples serve to illustrate controller effectiveness in maintaining a set point back pressure given disturbances in engine speed.

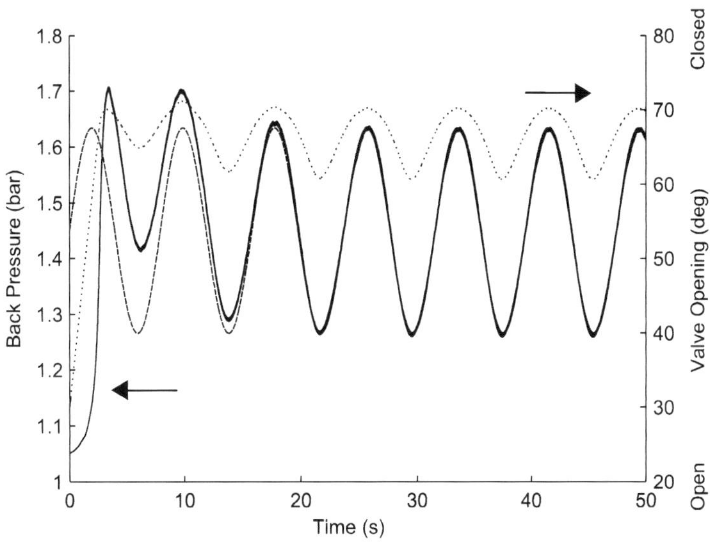

Figure 11: Comparison of model back pressure response (——) and valve opening (·········) to the sinusoidal set point back pressure '4 mean' (----) for a varying engine speed.

CONCLUSIONS

This report proposes the use of butterfly valves as an accurate method for reproducing and controlling exhaust back pressures. The characteristics of a four inch diameter valve with a valve flow coefficient of 255 were modelled, giving calculated back pressures in the range of those seen by a submarine in snort. A PID controller was used to effectively control the pressure drop across the butterfly valve, enabling it to track a set back pressure profile. The tolerance of the controller to speed and hence volume flow disturbances was observed. This is essential for any butterfly valve based engine back pressure controller due to the interaction between the pressure drop across the valve and the engine operating point. Additional controller development may yield a better back pressure response. In particular the use of a 'back pressure profile map' providing feed forward control, reducing pressure drop controller dependence. The development of this advanced controller is left for future work. The results have shown that an actuated butterfly valve can be successfully applied to land based submarine engine test facilities.

This model serves as a proof of concept. The following points need to be considered during implementation of a back pressure system of this type:

- Valve selection should be based on an approximate exhaust volume flow and hence be made with the engine size and operating speed in mind. Equations calculating the pressure drop are empirical. The related coefficients are determined by valve manufacturers in experimentation. The relationship between valve opening and developed back pressure can be estimated. The set point back pressures should be within the developed back pressure range for a reasonable range of valve positions. The influence of changing engine speed also needs to be considered. Increased speed may push the developed range away from the set point. If this is the case, resizing the valve may be required.

- The valve opening range should be between fully open and approximately seventy degrees. The highly non linear relationship between back pressure and valve opening has implications for control and the tuning of PID controller terms. In addition to the previous point, the set point back pressures should be within the 'valve opening range' to avoid the dramatic increase in the developed back pressure beyond a valve opening angle of approximately seventy degrees.

- Actuator sizing is made according to the maximum valve torque. Estimations can be made given the maximum pressure differential across the valve. The maximum rotational speed of the actuator also needs to be considered. This will influence the response time to changes in set point back pressure and disturbances in exhaust mass flow.

REFERENCES

1. van den Pol, E. (1987) The Simulation of the Dynamic Pressure Variations Inside Snorkelling Submarines
2. Australia. Department of Defence (2003) *DEF (AUST) 5000 ADF Material Requirement Set, Volume 3, Part 4 - Sea Keeping.* Canberra. RAN.
3. Hield, P. (2011) *The Effect of Back Pressure on the Operation of a Diesel Engine.* Defence Science and Technology Organisation, *Technical Report DSTO-TR-2531*
4. Tauzia, X., Chesse, P. and Maiboom, A. (2008) Simulation Study of a Ship's Engine Behaviour Running with a Periodically Immersed Exhaust. *Proceedings of the Institution of Mechanical Engineers*
5. Tupper, E. (2004) *Introduction to Naval Architecture.* Oxford, Elsevier
6. Rohm, H. (2005) Development and Testing of MTU Submarine Propulsion Units. In: *Submarine and ASW* Asia
7. Herrmann, R. (2001) Combined Supercharging System: The Best Fit for Submarine Engines. In: *Proceedings of the 2001 Spring Technical Conference of the ASME International Combustion Engine Division,* Philadelphia, ASME
8. Desmon, L. and Doyle, R. (1947) *Effect of Exhaust Pressure on the Performance of a 12-Cylinder Liquid-Cooled Engine.* National Advisory Committee for Aeronautics
9. Smith, P. and Zappe, R. (2004) *Valve Selection Handbook. Engineering Fundamentals for Selecting the Right Valve Design for Every Industrial Flow Application* 5th ed. Oxford, Elsevier
10. Fisher (2005) *Control Valve Handbook.* U.S.A. Fisher Controls International.
11. Korea Unicom Valve Co. Ltd. (1996) *High-Seal (GTD/FSD/MTD) High Performance Butterfly Valves ISO 9001 Certified.* Seol.
12. Standard, A. N. (2002) *ANSI/ISA-75.01.01-2002 (60524-2-1 Mod).* In *Flow Equations for Sizing Control Valves.*
13. McMillan, G. and Considine, D. (1999) *Process/Industrial Instruments and Controls Handbook.* U.S.A. McGraw-Hill.
14. Ang, K., Chong, G. and Li, Y. (2005) PID Control System Analysis, Design, and Technology. *IEEE Transactions on Control Systems Technology* **13** (4) 559-576
15. Franklin, G., Powell, D. and Emami-Naeini, A. (2007) *Feedback Control of Dynamic Systems.* 5 ed. New Jersey, Prentice Hall

OPERATIONAL ADVANTAGES AND BENEFITS OF "GREEN" SUBMARINES

Eric FUSIL,
Whole warship Architecture and Deterrence division
DCNS
Lorient, France
Eric.Fusil@dcnsgroup.com

INTRODUCTION

Taking into account the environment is an increasing concern for industry. Defence blends into this trend. More particularly, DCNS, leader group in naval defence in Europe, is on the frontline to reduce and optimize the environmental impacts of its activities and products. The ISO 14001 certification of the group and its eco-design approach of ships attest of this strategy and its results. For several years, more and more environmental criteria have been included in the design and construction of DCNS' surface armed ships. In 2008, ECOSHIP® was the first eco-designed ship by DCNS and a world premiere that shows it is possible to fully integrate the reduction of the impact of a military ship on its environment without affecting the required performances, while only using available technologies.

The life cycle assessment is a compulsory first step to have a precise and complete overview of the impacts of a ship on its environment, and then to intend to reduce them. This analysis seeks to quantify precisely the impact of a vessel by considering its entire life: from conception to dismantling. Therefore, it allows to target the most effective areas of improvement and to quantify the benefits.

The first part of the study presents the development and results of the first environmental analysis conducted on a specific submarine: the Scorpene submarine. From these results, the paper will develop the alternative architectures and technologies liable to reduce the environmental impact along life, while maintaining all operational performances. The combination of these technologies and architectures resulted in a concept submarine, SEPIA, to check the viability of the selected choices. Finally, it will be shown how green choices resulted in improved operational performances.

1. WORLD RESOURCES DEPLETION AND COMMITMENT IN ENVIRONMENTAL MANAGEMENT AND PRESERVATION

1.1 A world statement: increasing environmental impacts

Even if figures can change, they all eventually express the same trend: natural resources (fossil fuels or metals) are depleting, and we may run out of several raw materials in a close future. To produce an idea of the currently known resources depletion rates, the numbers of years of reserve are described in Fig. 1.

Metal	Stock (Proven store)	Main application	Stress at the current rate of consumption
Silver	270 to 383 kt	Jewelry, electricity	13 years
Gold	42 000 t	Jewelry, electricity	17 yrs
Zinc	180 Mt	Steel galvanizing	17 yrs
Tin	6,1 Mt	Welding	20 yrs
Lead	79 Mt	Battery, ammunition	22 yrs
Copper	490 Mt	Electricity	31 yrs
Nickel	67 Mt	Stainless steel	40 yrs
Platinum	1300 t	Jewelry	56 yrs
Iron	150 Gt	Steel industry	79 yrs
Aluminium	25 Gt (bauxite)	Transportation	131 yrs

Figure 1. Metal world stock

Besides, the oil reserves are estimated for 40 years of consumption, based on the hypothesis of known oilfields.

Materials	Stock (Proven store)	Current rate of consumption	Stress at the current rate of consumption
Coal	578 Gtoe	3,2 Gtoe/yr	183 yrs
Natural Gas	185 Gtoe	2,9 Gtoe/yr	64 yrs
Uranium	30 Gtoe	0,62 Gtoe/yr	48 yrs
Oil	172 Gtoe	3,9 Gtoe/yr	44 yrs

Source : BP statistical review of word energy 2008

Figure 2. Energy world stock

Therefore, even if some spots remain to be found, thorough use is a necessity.

Besides, industry does not only use natural resources, but also releases waste, CO2 and greenhouse gases. On average, nine tons of waste and two tons of greenhouse gases are generated by a ton of a manufactured product: the need for a clean and respectful industry is prominent.

Along with this statement, regulations and standards are constantly evolving in order to ensure a slowdown of natural resources depletion, and to establish a respectful behaviour towards environment.

Naval sector is responsible for 15% of worldwide CO2 emissions due to the transport and 41% of nitrogen oxides (NOx), and is therefore submitted to severe environmental standards (for example, the International Convention for the Prevention of Pollution from Ships, known as "MARPOL").

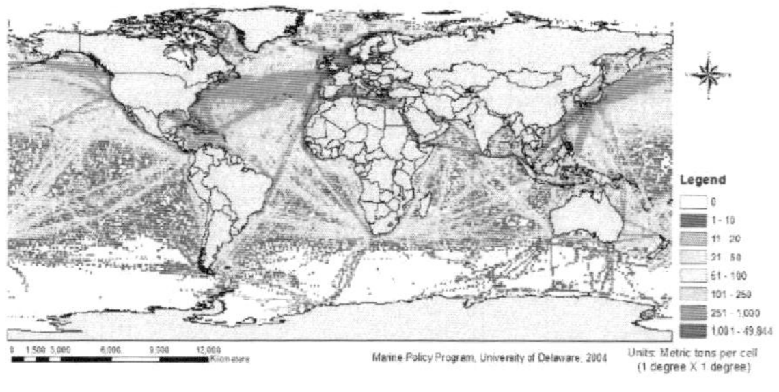

Figure 3. NOx emissions from shipping in 1997 [1]

1.2 ISO 14001 certification : life cycle assessment and eco-design

Integrating environmental considerations at the beginning of design allows architects to take every phase of a ship's life into account. Furthermore, by considering environment since the very beginning of design, it is possible to reduce efficiently the impact of a ship. The

ISO14001 certification permits an easier integration of environmental considerations inside the existing design process. It also controls rigorously eco-design activities.

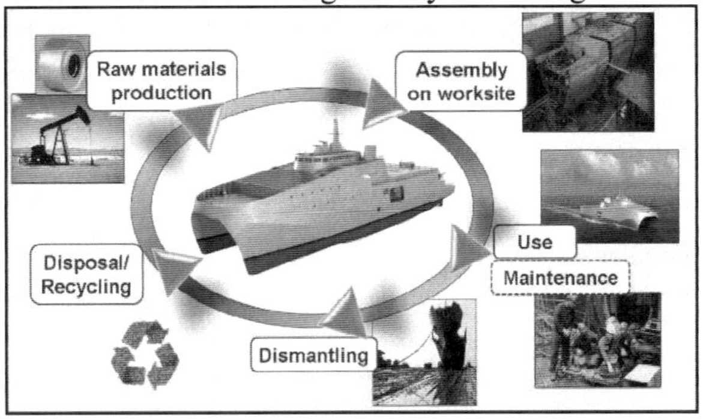

Figure 4. The different phases in a ship's life

2. SCORPENE SUBMARINES ENVIRONMENTAL IMPACT ANALYSIS

2.1 Overview of the process for environmental impact analysis: life cycle assessment

A 'Life Cycle Assessment' (LCA) is the investigation and evaluation of the environmental impacts of a given product during its whole life cycle.

- This analysis includes every single phase of a product's life called "cradle-to-grave":approach: the assessment of raw material production, manufacture, distribution, use, dismantling as well as the energy and the materials consumed for transportation, manufacture, etc. The sum of all those steps - or phases - is the life cycle of the product.

2.2 Life Cycle Assessment of the Scorpene submarine

The Scorpene is the most recent DCNS-designed conventional (diesel) submarine. She was chosen by the Malaysian, Chilean, Indian, and Brazilian Navies. In the case of the chosen configuration of Scorpene, she is a 67 meters-long submarine, weighing 1700 tons. Her maximal diving depth is over 350 meters and her autonomy goes past 70 days for a crew of 31 sailors and her combat system is tailored to the customer's needs, whatever the supplier (DCNS or overseas companies).

Figure 5. The Scorpene submarine during trials

Thanks to the precise knowledge of the mass quotation and the manufacturing processes used for the Scorpene submarine, DCNS' specialists were able to model almost completely the submarine for the analysis.
Indeed, the breakdown product structure represents more than 96% of the submarine, which is a tremendous score; keeping in mind it means knowing precisely the chemical composition of each considered system. Moreover, the data concerning manufacturing are recent and accurate, which allows us to analyse realistic results.

2.3 SCORPENE: results and analysis

The analysis can offer several calculation methods. We chose here to use the Ecopoint-99 method, which allows us to consider several impacts, such as land use, resources depletion, impact on human health, etc (see Figure 6.).

The results present the impact in six stages: materials and manufacturing, active duty and maintenance, dismantling and afterlife.

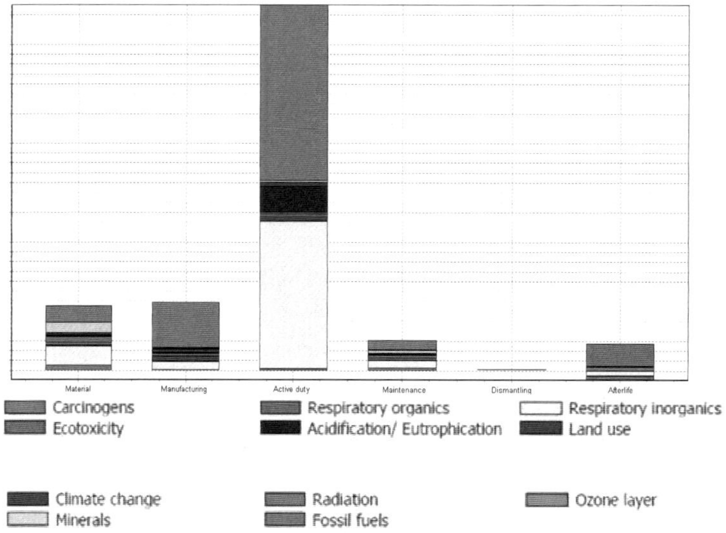

Figure 6. LCA's global results, comparing the impact of each phase

It is not surprising to see that the impact of the active duty is bigger than the others phases. Indeed, manufacturing lasts approximately six years while active duty is meant to last at least thirty five years.

The most represented impact is obviously the depletion of fossil fuels. It is obviously because of the fuel consumption during active duty, but also because the extraction of many metals requires large amounts of fuel.

Searching for solutions.

The analysis must be used as a starting point for an efficient abatement of the global impact of the submarine. The consequences of a change can be numerous, virtuous or not and sometimes really hard to detect. For example: choosing a more efficient antifouling reduces the resistance force, therefore a less powerful propulsion is required, which diminishes the amount of fuel required, which allows the architect to resize the ship, etc. But a specific choice can have at first excellent consequences, but be eventually globally more impacting. It means that every choice must be analysed to determine its global impact.

The best solutions must be levers: they have to induce a chain of positive consequences. In the case of a surface warship or a submarine, the biggest impact is the fuel consumption during active duty, and the manufacture of the hull. For instance, by lightening the structure of a surface warship, the impact of the hull will diminish, and in the same time, the propulsion will be resized, and the fuel consumption will decrease.

As for manufacture, maintenance is more demanding for a submarines, more particularly for batteries (more than 150 tons), which need to be changed every six years. It means that improvements made in the materials and manufacture parts can be really influent.

The analysis can also reveal hidden sources of impact. For instance, when examining more carefully the hull and its structures, the employed alloy shows that a rather low rate of nickel (less that 5%) generates more than 30% of the impact.

Eventually, looking for solutions consists in eliminating vicious circles, and integrating virtuous circles.

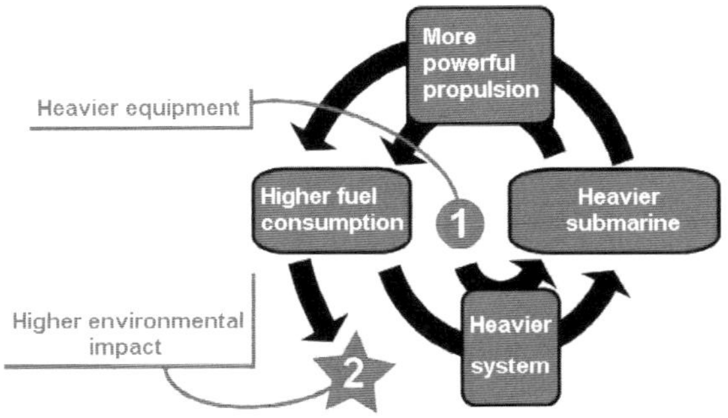

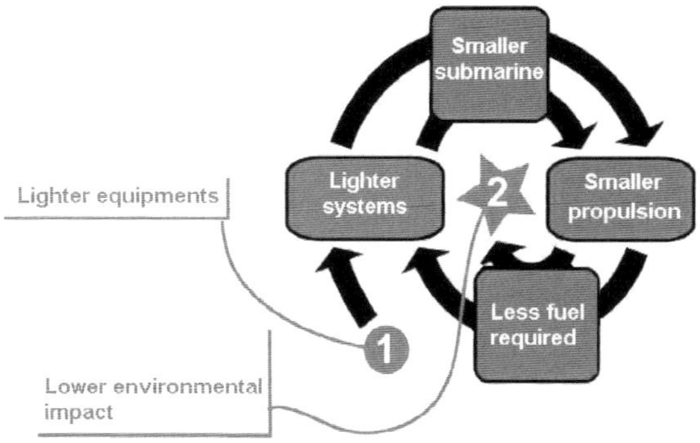

Some bitter disappointment may fall on careless thinkers: let's imagine one has a terrific idea to reduce the quantity of steel used for the hull, thus its mass. What a great reduction in environment impact! Unfortunately, one may reduce the hull mass, if no other factor enables to reduce the volume of the submarine, the volume of displaced water remain unchanged, so is the resistance to movement and the required propulsion. Only the ballast capacity will have been increased to keep the submarine balanced when submerged…
A submarine naval architect has one major goal: to get a balanced submarine, which means:
The total weight (systems + ballast) must be equal to the Archimedes thrust, e.g. buoyancy,
Thus lighter equipment only means more ballast, the submarine's drag is not reduced.
The very challenge is to diminish both weight AND size of the equipments to enter the spiral leading to a smaller, thus less demanding submarine, by reducing the drag.

3. THE ARCHITECTURAL AND TECHNOLOGICAL SOLUTIONS TO REDUCE THE ENVIRONMENTAL IMPACT OF FUTURE SUBMARINES

3.1 The lessons from the SCORPENE life cycle analysis

As detailed previously, the major effects impacting the environment are:

- The diesel-oil consumption in active duty,
- The choice of materials and manufacturing in shipyard.

To reduce the diesel-oil consumption without modifying the speed profile and the number of days at sea per year (it would be cheating…), the naval architect has to:

- Reduce the drag. To reduce the drag, few means exist but diminish the displacement (e.g. volume) of the submarine or, and, obtain a better hydrodynamic yield.
- Reduce consumption onboard. For that purpose, the naval architect has to diminish the Hotel Load (e.g. the power consumption of the equipment different from the electrical propulsion), obtain the best efficiency on the propulsion, from the batteries (we are talking about a diesel-electric submarine) up to the propeller.

3.2 Ways of improvements

Different categories were defined to improve overall performance:

- Energy savings,
- Reduction of pollutants from the submarine into the sea,
- Choice of less impacting materials on the environment,

Nevertheless, these solutions are available, with a technological readiness level of at least 6 on a scale from 1 (scientific research coming to practical R&T) up to 9 (sea proven solution-on-the-shelf): prototypes in their real environment have been tested.

3.3. Energy savings solutions

Reduced drag and propeller yield

With an optimized hydrodynamic shape, a reduced drag is obtained thanks to a silicone painting.

The advantage is the use of a biocide free antifouling system, with scarcer maintenance and a reduced drag (between 2 to 9 % depending on the level of maintenance on the painted hull).

Besides, a pump-jet system gives higher propulsive performances compared to classical propellers.

Figure 7. Pump-jet technology on DCNS submarines

Advantages: with a reduced drag and a more efficient propeller, the mechanical thrust to produce is lower than for the reference submarine.

Hybrid propulsion system

Due to the speed profile (see Figure 8.) of the submarine, one can see that:

- The main electric motor is designed for the maximum speed but,
- That the submarine spends most of its time at medium speed.

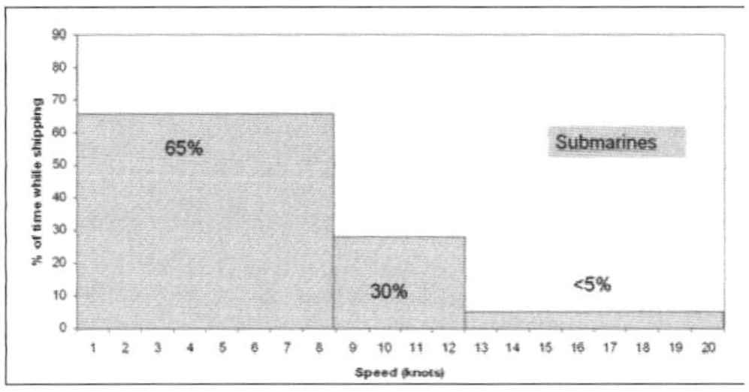

Figure 8. Compared speed profiles between surface ships and submarines

Thus, the main electric motor works at a lower yield than its maximum most of its time.

The idea is to separate two problematics:

- The maximum speed to guarantee during rare moments in the submarine life,
- The most frequent speed where the main electric motor is to work at maximum yield.

The architecture could be based on hybrid propulsion with:

- A low speed motor up to medium speed,
- A high speed motor, for high speed.

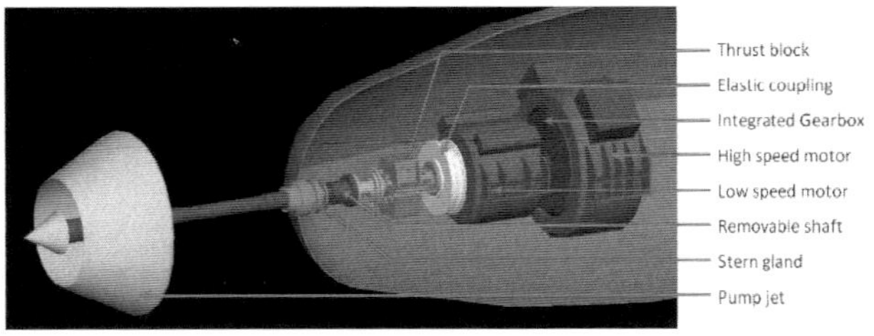

Figure 9. Hybrid propulsion architecture

The benefits are immediately seen through the following figure:

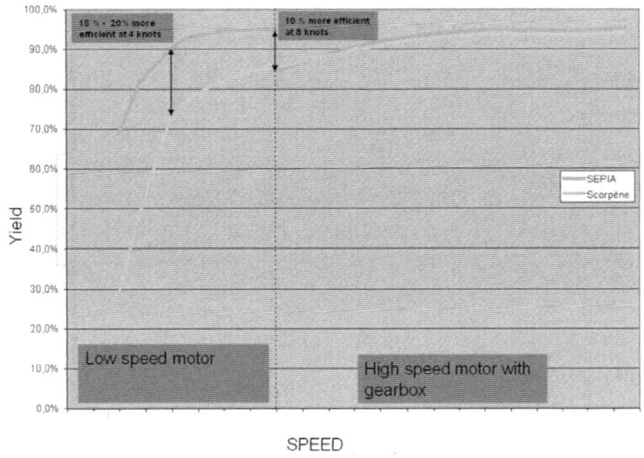

Figure 10. Yields of classical and hybrid propulsion system

The smaller motor works with an increased efficiency of 15-20% more than the single motor solution.
Advantages: energy saving and reduced fuel consumption due to the running at higher yield.

The assessment is about 160 tons of diesel oil per year.

Lithium ion batteries

With a higher energy density than lead acid batteries, the lithium-ion batteries are promising. Nevertheless, lithium is more impacting than lead-acid technology. This should doom such a choice…

Fortunately, the life expectancy of lithium ion batteries is higher than lead acid technology. This fact, added with the lighter weight of lithium ion batteries. It means that on a life expectancy of 35 years for a submarine, the investment is profitable.

Lighting system

From the initial light bulbs, the fluoro compact system enabled to divide by 3 the part of lighting in the ship hotel load.
With the LED lighting coming on the market, another reduction should be attained, but only of a few percent.
Nonetheless, the development of this popular technology makes it a relevant one for future submarines.

3.4 Polluting emission reductions solutions

Sewage waters and waste releases

With a purifying system, the sewage water from the life support systems can be purified with a membrane filtering station.
This has two consequences:

- The water can be either reinjected in the systems or released at sea in a clean state,
- The sludges and residues are stored onboard and can be neatly removed at quay.

In terms of architecture, this means an increased volume and dedicated tanks onboard the submarine. It is estimated that 15% of extended volume is required to perform such functions.

Solid garbage
For solid waste, a grinder / crusher coupled with a sterilizing system enables to store solid waste in a non-refrigerated atmosphere, for instance, in the galley. This avoids to send solid waste in the sea in ballasted plastic bags through a garbage ejector.

Gaseous emissions
To reduce NOx and SOx emission, the submarine situation has to be considered. The submarine uses its diesel engine to recharge the batteries before diving or to ship from harbour to its diving point. In the first case, the submarine is snorkelling, whereas it is on surface otherwise.
When snorkelling, the exhaust gases are released by overpowering the counterpressure of the seawater. Beyond that difficulty, it has the great advantage of trapping the SOx underwater, due to the salinity of seawater.
To treat the case of surface running of the diesel engines, the scrubber solution is efficient, with gases going through a saline solution. The existing surface exhaust on submarines can then be fitted with such system, without creating a specific by pass.
Nonetheless, the NOx pass through water. To reduce their volume, an adapted Exhaust Gases Recirculation (EGR) system is an effective solution.

3.5 Less environment impacting materials

Purely based on the environmental cost of some materials, it appears that titanium alloy ducts are an interesting alternative due to their ability to last for the life expectancy of the submarine, whereas other alloys may require to be changed during a major overhaul.
In the same philosophy, PER ducts may be less environment impacting in the long term.
As for electric cable, copper is three times more impacting than aluminium for the environment. Aluminium cables are an interesting alternative.

3.5 The combination : the SEPIA submarine

A technology or architecture must be assessed in a combination, meaning a designed submarine, to gain any value. Otherwise, a nice looking solution on a marketer's catalogue may prove to be of much reduced interest than expected, once integrated (hopefully successfully) in a whole.

The exercise of re-thinking was performed on that basis, and lead to a concept submarine: SEPIA (Submarine with Environmental Performance Improvements Along-life), stemming from the cuttlefish (*Sepia sp.*), which is able to melt in its surrounding environment.

Figure 11. General view of the SEPIA submarine

- <u>Surface displacement : </u>1570 t
- <u>Beam : </u>7,6 m
- <u>LOA</u> : 54 m

3.6 Global results

Based on these assumptions, we assess that we could diminish the environmental impact up to 35 % compared to a reference submarine. This is a very good result for it preserves the performances of the reference submarine.

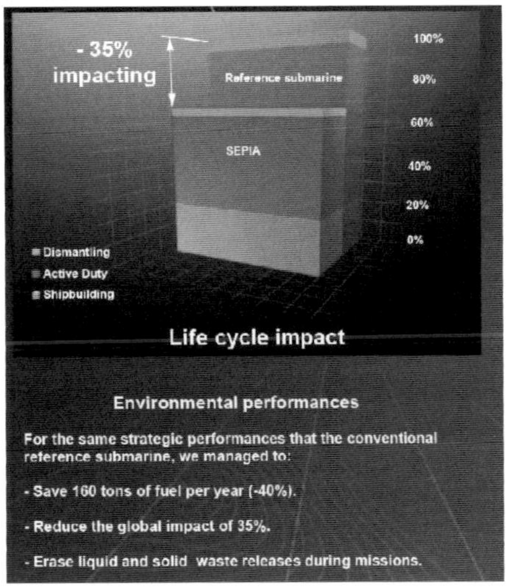

Figure 12. Assessment of global environmental reduction with listed solutions

4. THE INCREASED OPERATIONAL PERFORMANCES

The overall energy efficiency of the boat is increased, whether it is on output, storage and consumption (propulsion and hotel load). Thus, for a given stored volume of diesel-oil, the submarine can have an increased range, compared to currently existing submarines. Moreover, the indiscretion ratio of the submarine will be drastically reduced.

With a reduced need in diesel oil storage for a given range, the size of the submarine can be reduced, meaning less hull structures. As a consequence, the magnetic signature (linked to the mass of ferromagnetic, i.e. metallic, structures) will be drastically reduced against maritime patrol aircrafts.

Moreover, with a smaller submarine, the hotel load is reduced, and so the number and power of auxiliary equipment: the consequence is reduced noise, thus increased acoustic discretion and improved acoustic signature.
Besides, sewage processing, and other waste treatments will lead to excellent stealth. This is vital for the submarines of the 21st century due to the uprising capabilities of chemical sensors in terms of sensitivity to artificial products, enabling future submarines to "smell" and detect the wake of their preys as a shark for a bleeding fish.

It is remarkable to note that some of the solutions evoked in this paper are just the engineers' "good old sense" that enables to do the most with the fewest. For instance, the hybrid propulsion did exist on Agosta class submarines, but interestingly for other technological reasons (see cruise electric motor A and main electric motor B on Figure 13.).

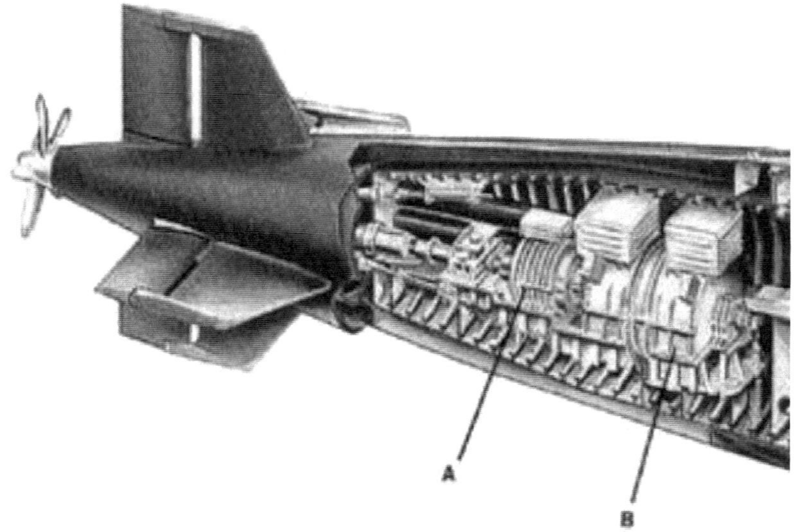

Figure 13. Agosta class submarine with two electric motors

5. CONCLUSION

. Far from being a burden or a trend, eco design is a valuable opportunity for engineers to go further in the continuous questioning over their choices of architectures and technologies. If some constraints come along with the eco-design, it is but another variable entering the equation naval architects have to solve to produce the best compromise ever.
Different axes have been identified: energy saving, polluting emissions reduction and waste management and less impacting materials. For each of them, some solutions exist and can be implemented. Some already are.
Even for the military, the environmental approach has collateral benefits with improved performances for their primary missions (such as extended autonomy and increased stealth), and a more efficient submarine from the architectural point of view.
This paper demonstrates that it is possible to improve the environmental impact of a submarine of about 30% while keeping the overall performances. We state that the research of less impacting solutions is a potent means for efficient engineering solutions and we hope the military will seize this opportunity in the near future.

REFERENCES

[1] Corbett, J.J., and Wang, C., 2004. Unpublished image of 1997 NOx emissions distribution from ships, http://coast.cms.udel.edu/ShipGISData/ShipNitrogenData/ShippingEmissionsNox_1997.jpg, University of Delaware

Propulsion system options for a modern multi-role patrol vessel.

Michael O'Connor
Rolls Royce plc
Michael.O'Connor2@rolls-royce.com

SYNOPSIS:
Much discussion is currently underway throughout the naval community as to an optimum vessel design to encompass multi-roles. In general such roles include Mine countermeasure surveillance, hydrographic and oceanographic missions and also provide all the necessary functions of an offshore patrol vessel. Increased flexibility, multi-role capability and longer design life will need to be accompanied by reduced through life costs with a greater emphasis on whole platform fuel efficiency, a low maintenance burden, reduced manning and minimal environmental impact whilst on deployment and alongside. Factors such as top speed, range, bollard pull, manoeuvrability and station keeping are becoming important operational requirements for a single platform. Such a multirole application certainly places challenges for a single propulsion system to meet a variety of different operational modes.
This paper looks at various propulsion system options for such a multi-role vessel, but also aims at highlighting the impact propulsion system selection will have on ship design and deck system layout. Of particular importance is the modular nature of the deck systems of such a vessel and their ability to Launch and Recover multiple assets.
Trade-off studies have been conducted in assessing multi-role vessel concepts with the aim of highlighting some optimum propulsion system solutions for a vessel with a broad range of operating requirements. Propulsion solutions considered include a twin shaft mechanical systems, hybrid electro-mechanical propulsion options and the use of Azimuth thrusters for such a platform.

ASSUMPTIONS:

In assessing as appropriate propulsion system for such a platform it is most important to get the vessel operational profile correct. It is assumed that such a profile will be a combination of current operational profiles of the three classes of ships. That being Mine Counter measure vessels, hydrographic/oceanographic research vessels and patrol craft. Some typical operational profiles of the 3 ship types are contained below. It is also assumed however that there may be some variation in the use of such platforms, for instance both mine countermeasure and hydrographic/oceanographic roles will be employing more launch and recovery technology using UXV's. This may mean station keeping and dynamic positioning is necessary. However as a baseline study combining the operational profiles of the each of the 3 ship types was used.
The operational profile is possibly the single most important element to obtain when designing a relevant propulsion system. This will provide an overview of vessel functionality and the main ship operational drivers effecting the selection of the propulsion system. It will also provide relevance when conducting trade off studies of through life costing and any cost vs capability analysis.
Some common operational profiles for the various ship types are contained below and also the principle particulars of the ship class operational in the Australian Navy in which the vessel is intended to replace.

Traditional Mine countermeasure operational profile: Speed vs Percentage of time spent

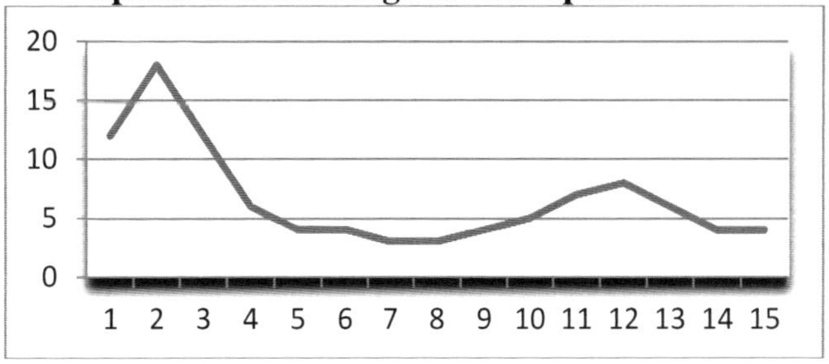

HMAS Huon Class particulars:
Length: 52.5 m
Beam: 9.9 m
Draft: 4.87 m
Displacement: 720 t
Range: 12 knots
Top speed: 14 knots

Oceanographic/hydrographic operational profile:

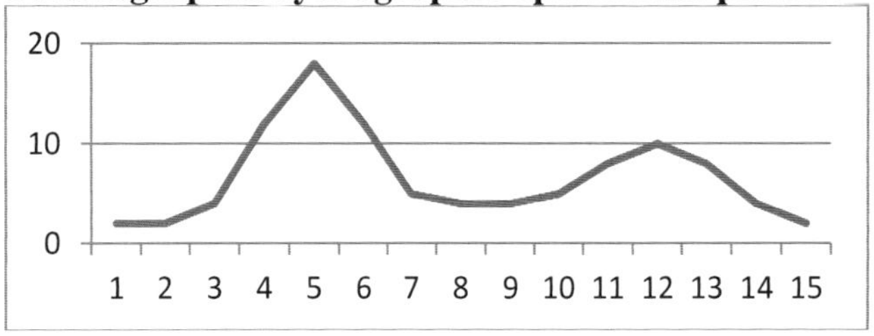

HMAS Leeuwin Class Particulars:
Length:71.2 m
Beam:15.2 m
Draft:4.3 m
Displacement: 2550 tons
Top speed: 12-14 knots

Patrol Craft Operational profile

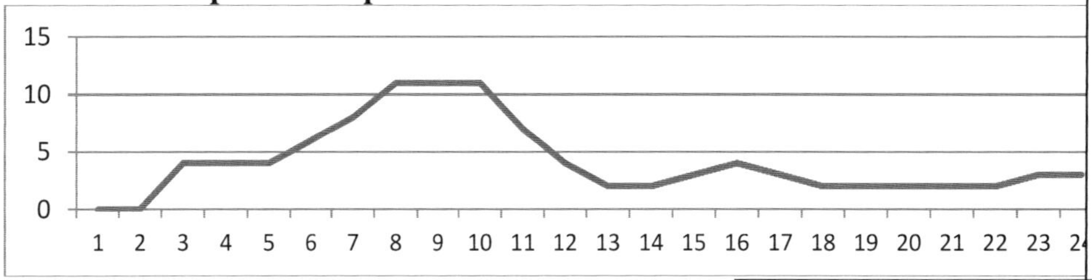

Armidale class patrol boats:
Length:56.8
Beam:9.5m
Draft:2.25
Displacement: 270 t
Speed: 25 kn

Combined average operational profiles

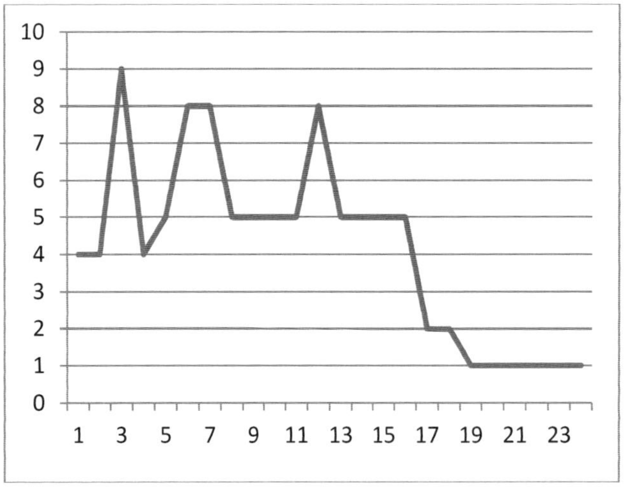

Note: The adjacent operating profile is considered reasonably accurate in performing key mission critical tasks of the 3 specified ship types. However the major difference may be the top speed requirement as initial requirements on the offshore combat ship are in the range of 27 knots.This is not considered realistic when considering the hull form needed to perform vessel missions, as a wide stern mission bay is needed with a large portion of time spent at low speeds towing, loitering or station keeping. A wide stern is desirable with a lower length to beam ratio than that of a vessel designed for 27 knots. A top speed of 24 knots is considered more realistic and is used in the study.

It is also shown that the vessel will need to be highly manouverable, spend extended time loitering, possibly stationkeeping, be able to tow bodies at low speeds and have a large mission bay for launch and recovery operations. As a minimum a Dynamic Positioning DP1 notation is considered appropriate. All these factors are considered when assessing various propulsion systems.

Vessel principal particulars used as part of the propulsion study include:

Length: 90m
Beam: 14 m
Draft: 4m
Propeller Diameter: 3.5m
Top speed: 24 kn
DP1 Notation

Note: These are indicative dimensions used to obtain vessel resistance data and develop an accurate power speed curve. Such a hull form is considered reliable and accurate in performing preliminary propulsion calculations and is considered appropriate for the ships intended capability.

Range

This is very important for designing the propulsion system, as the optimum propulsion system efficiency would need to be achieved at this speed range. This will effect fuel load carried, vessel hullform and propeller design. It should also be considered in terms of vessel hotel load also and the difference in range speed and top speed is important. As it will determine the percentage of MCR of the diesel or electric motor at transit speed compared to top speed. If this is uneconomical or operating the diesels at below 20% MCR then alternative options will be needed to achieve top speed.
It is assumed this vessel will be operating within a task group so a transit speed and range would be around 16 knots. This is operating the 2 main engines at about 25-30% MCR if main engines are designed to operate at a maximum speed of 24 knots.
At speeds below 16 knots trailing a shaft would be necessary and this provides vessel propulsion inefficiencies.
For twin screw conventional propulsion systems to extend the range speed from 14-20 knots it is advisable to use a smaller sized engine and operate it at a more efficient operating point. For top speed the use of an electric (PTI) for boost mode and increasing pitch on the propeller while operating the diesel engine also through a reduction gearbox as shown below.

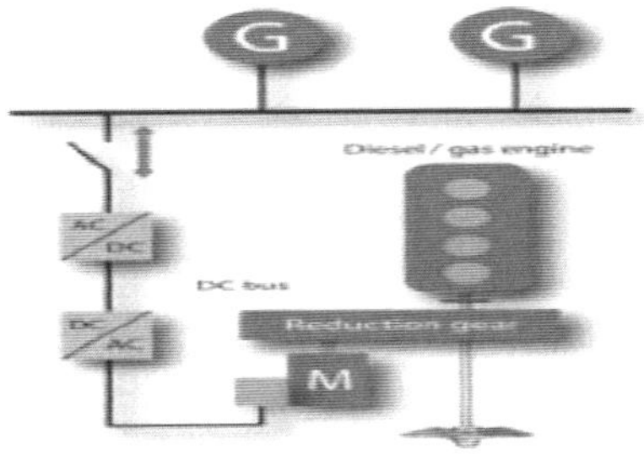

The use of Active Front End frequency convertor technology allow you to reduce the speed of the main engine to match the propeller rpm needed, whilst providing ships power through a PTO shaft generator.

For azimuthrusters the top speed and power is reduced as the maximum power input for a single azimuthruster is 5MW, then operating at 16 knots is running the electric motors at 50% MCR. However some losses will be apparent through the use of frequency convertors and electric generators. Where substantial fuel savings of Azimuthrusters can be achieved is when lower speed more efficient diesel generators can be used compared to high speed generators, as the installed electric power will be more. However this has not been taken into account in this study as this will have an impact on vessel weight and further reduce the ship speed when using Azimuthrusters.

Bollard Pull/thrust:

This is term is widely misused as Bollard pull is generally referred to tugs with maximum thrust at zero speed. However this vessel will need to tow bodies at low speed, so some thrust at low speed is needed. It is not considered this vessel will be undertaking salvage operations so Bollard pull is not an important requirement. However high thrust at low speeds is, and the use of electric motors to achieve high thrust at low speeds may provide some efficiency gains if using Controllable pitch propellers. This however is dependent on the speed and towing thrust needed.

Station keeping/Dynamic Positioning:

This is the ability of the ship to maintain a position with sufficient redundancy. The use of azimuthrusters is widely used for DP vessels as it provides superior directional thrust for all heading angles and it negates the need to have additional tunnel thrusters at the stern of the vessel. Additional tunnel thrusters at the stern will not only increase the number of installed thrusters and associated Electrical load, but also increase the ship resistance underway with additional power needed to meet top speed requirements. An aft tunnel thruster will also alter the design of the aft hull form, as a stern skeg is necessary to fit the tunnel thruster. Careful design is needed here as the tunnel in the aft skeg can effect the in flow of water into the propellers and reduce propeller efficiency, and also A-brackets and shaft lines may effect the inflow into the tunnel and reduce the side thrust of the tunnel thrusters.
It has been assumed that as a minimum this vessel will need the DP1 notation so a stern thruster has been used as part of the analysis for all propulsion systems except the Azimuthrusters.

Noise:

This is an important consideration when conducting MCM and oceanographic/Hydrographic operations. The important requirement is to ensure underwater radiated noise does not interfere with any towed body or UXV measuring equipment. It is important to understand the noise levels and the operating range to restrict underwater noise. This will allow appropriate modelling of the propulsion system to ensure noise levels are restricted to acceptable limts within the appropriate operating range. However this is not a simple black and white issue and involves detailed system analysis, usually from a noise specialist. Each case is also different as noise source from propeller, gearbox, motors, generators, auxiliary equipment and the hull interaction designed for a specific ship mission will always be ship specific.
Some areas or measures to use to meet Acoustic noise criteria include

Propellers:

This is very dependent on the detail propeller design and the wake field. A propeller with offloaded propeller tips is very important in order to reduce the effect of tip vortices. This is a propeller design that is focused on a low load distribution at the tip of the blade. This will also ensure a large clearance between the propeller tip and the hull is not necessary and propeller and hull interference in the water field will be minimal and not affect underwater noise.
Placing the propellers in front of any appendages like the Azipull is arranged also allows good undisturbed flow to the propeller.

Azimuthing thruster gears:

Yes, azimuthing thruster bevel gears are a contributor to underwater noise. However through careful selection of gear ratios, gear type and gear design Rolls Royce can provide azimuthing thruster bevel gears to meet acoustic noise criteria. Some design features and solutions below are used on our bevel gears to reduce gear noise into the water include:

- The use of high precision gears
- Better surface finish on gears
- Use of suitable lubrication
- Apply high rigidity gears
- Apply sufficient backlash of gears
- Obtain good tooth contact

The use of traditional helical gears also will require consideration in design to meet noise criteria, however there noise propogation is not considered as critical due to the larger distance the noise needs to travel through the hull structure to the water.

Structure:

Natural frequencies in the hull structure which coincide with the natural frequency (resonance) of any noise source need to be avoided. Occurrence of such abnormal phenomena should be controlled at the detail design phase through vibration calculations.

Diesel generators and main engines:

High airborne noise usually requires acoustic insulation of bulkheads, hull sides and ceiling within the engine room. Such insulation is not considered necessary within the azimuthruster room. However this should be part of any noise calculation.
Engines are usually double resilient mounted to prevent the alternator from passing any structural borne noise into the water.
The exhaust, fuel oil, lubrication oil and cooling water systems for the diesel generators may require resilient mounts and/or accumulators or silencers to avoid the flow of acoustic energy through these parts.
shaft coupling and bearings will need to be appropriately designed to ensure resonant frequencies of equipment are controlled.

Auxiliary machinery:

Resilient mounting is usually necessary of machinery operated in the research condition. Ventilation fans may need silencers and resilient mounting. Steering gear pumps can be acoustically isolated from the structure and frequency controlled. Rolls Royce provide both acoustically quite electric steering for both Azipulls and also rotary vane steering gear.

Underwater appendages:

These should have smooth trailing edges and good hydrodynamic design. Of particular note is the hydrodynamic design of the AZIPull with a hydrodynamically optimised skeg and housing profile. The design of any stern skeg with a tunnel for a thruster will need careful consideration .

AC Motor:

AC motors should be subject to low noise design. This may include slot passage frequencies, structural design and electric control systems made to smooth out excitation. Resilient mounting may also be necessary.

Rolls Royce can provide assistance in the areas above where we have equipment. This includes diesel generators,diesel engines, shaftlines and bearings, AC motors, Azipull propellers and conventional propellers.
However it is further stressed that the amount of additional design measures taken for individual equipment will be dependent on noise modelling and noise analysis by a specialist of the entire ship system.

SHIP INTEGRATION:

Propulsion system selection should never be done in isolation to the Electrical system design and also the greater ship design process. Various propulsion systems will have an effect on the ship build, ship design, ship functionality and the integration of the electrical system.

Hull Integration:

The hull form chosen will effect propulsion selection. Azimuthrusters can be easily integrated into most hull forms as they are modular and can easily be installed parallel to hull lines. Azimuthrusters are usually installed in simple hull forms which often reduces ship building costs. Conventional shaftlines will need some further careful consideration due to the rake angle needed, shaft brackets to ensure efficiency is maintained and also the design of a skeg with stern thrusters. Conventioanl shaft lines will also require a rudder and steering gear further aft of the propeller which will effect the aft arrangement and the size of the intended mission bay for launch and recovery equipment.

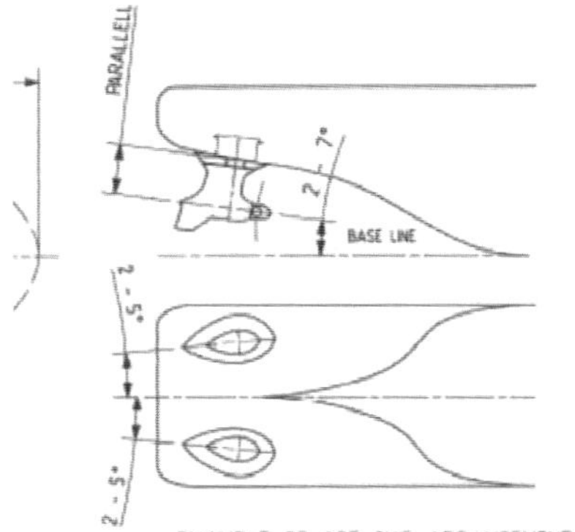

EXAMPLE OF AFT SHIP ARRANGEMENT
SEEN FROM THE SIDE
AND FROM BELOW

Typical open shaft stern arrangement with centreline skeg and tunnel thruster

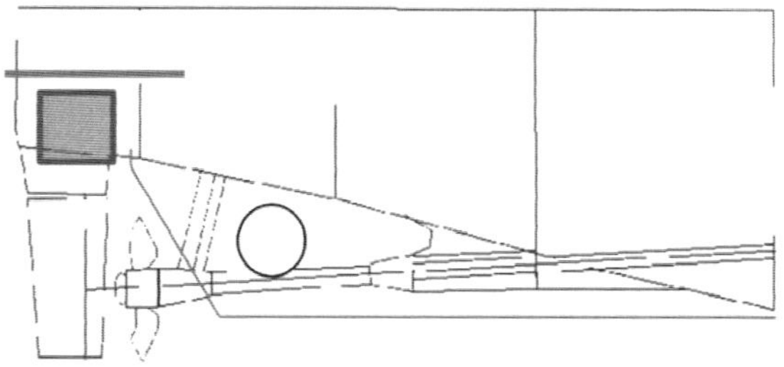

Launch and Recovery and Mission Bay:

This is an important consideration and will effect the propulsion layout. It is envisaged that the launch and recovery capability of the vessel will need to be flexible and modular. A large mission bay at the stern of the vessel may be necessary to provide launch and recovery for towed bodies, unmanned UXV's and also manned ships boats for boarding party operations. The use of azimuthrusters provide additional space in the stern of the vessel for a large mission bay as the machinery space can be shortened and moved fwd of the vessel due to the all electric architecture and a steering gear room further aft of the propellers is not necessary, in comparison to conventional open shaft propellers.

Electrical System:

The electrical hotel load and propulsion load should be modelled over the vessels operational profile. With the use of various launch and recovery equipment and the probable need for station keeping and DP capability it is assumed the hotel load at low speed will be high, and at higher speeds the propulsion load will be higher with minimal hotel load.
It is wise to provide a ratio of hotel load to propulsion load over the vessels operational profile to indicate the main drivers for vessel power and propulsion and provide an optimised design.
The higher the vessels hotel load to propulsion load over the operational profile indicates an all electric architecture is advantageous with the use of Azimuthrusters. In particular if this is at low speeds stationkeeping.

PROPULSION SYSTEMS

Twin shaft – single engine Mechanical Propulsion system:

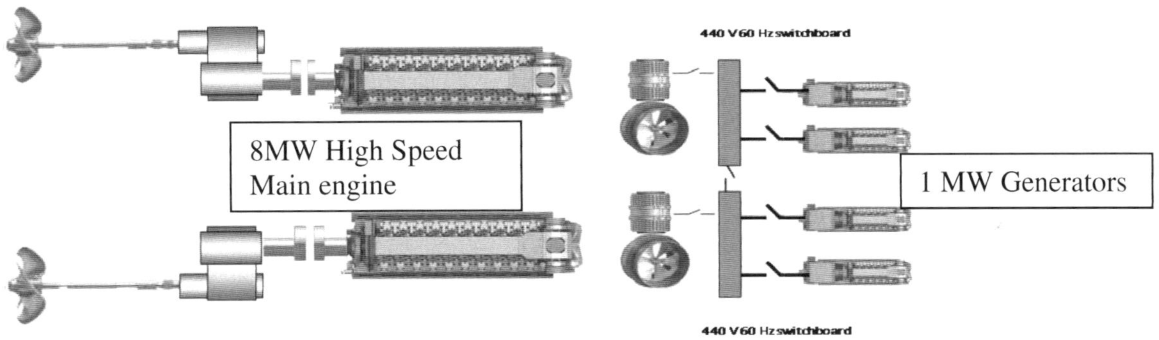

The simple twin shaft mechanical propulsion system is a low risk low cost installation, however is completely inappropriate for this ship type. The amount of time spent at low speeds will mean that main engines are spending some 60% of the time at below 20% MCR. This is even while trailing a shaft up to 14 knots. The propulsion system is soley designed to meet the top speed requirement of 24 knots, of which is only a small component of the vessels operational profile.

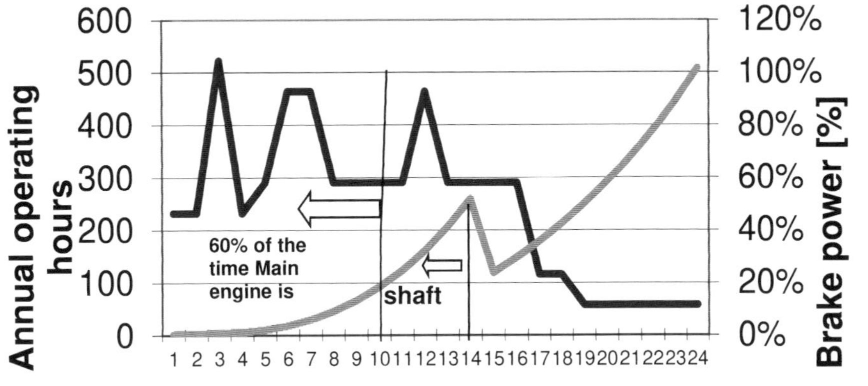

Twin Shaft – twin engine mechanical

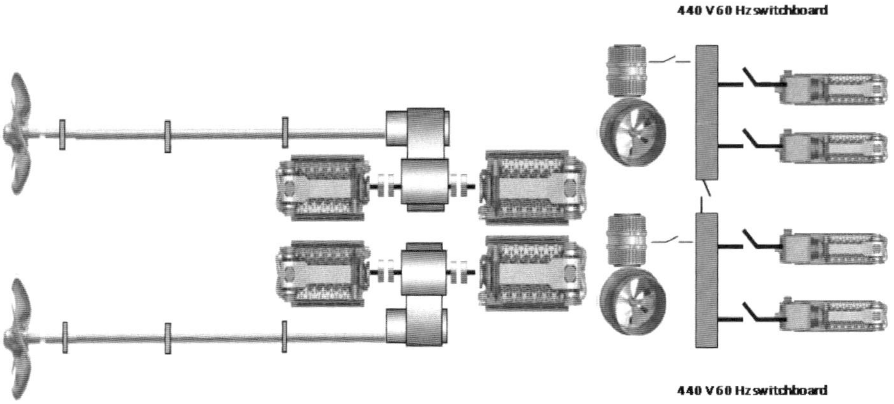

This is also a low cost low risk installation similar to the twin shaft single engine mechanical solution. It is an appropriate and common propulsion system for patrol craft where a high range is desirable at transit speed and there is a higher top speed requirement. However for lower speeds you will still need to trail a shaft. If the vessel is performing stationkeeping or even towing a body on 1 shaft this solution is not appropriate. Also the engine running hours are excessive for this solution and the installed number of engines is high compared with the other solutions. This will lead to a high maintenance and spares cost and additional maintenance burden on crew, possibly leading to an increased crew size.

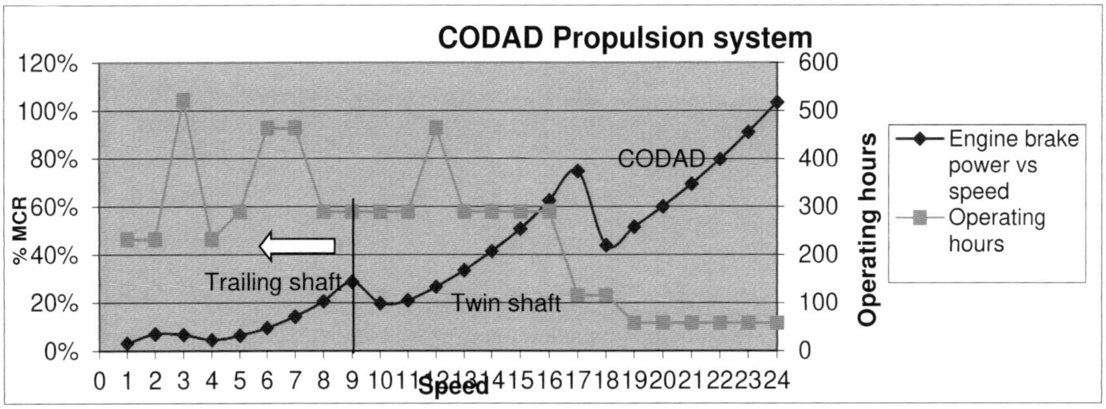

Hybrid Propulsion system with HSG:

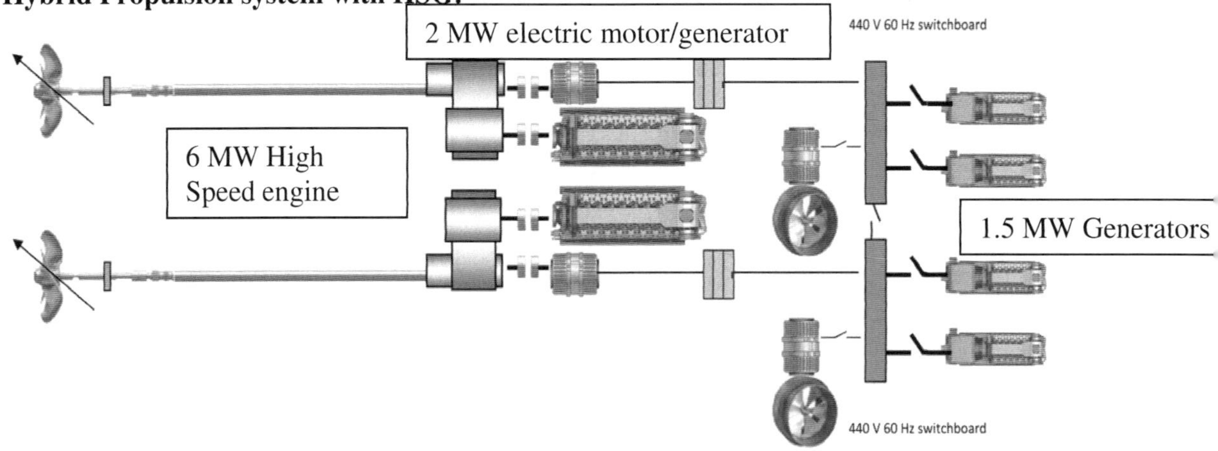

This is a very flexible propulsion system and can be adapted for a range of operating profiles. Its initial cost is slightly higher than the mechanical propulsion systems, however as the electric motors can operate as generators, there is scope to reduce the number of installed generators needed. With the use of HSG substantial fuel savings can be obtained with a low number of engine running hours, as generators can be shut down for speeds of 6 to 21 knots.
Top speed is achieved through a PTI electric motor used as a boost motor combined with the diesel engines. Thus the size of the main engine is more optimised for transit speeds and there is no need to trail a shaft.

Azimuthrusters Integrated For Electric Propulsion:

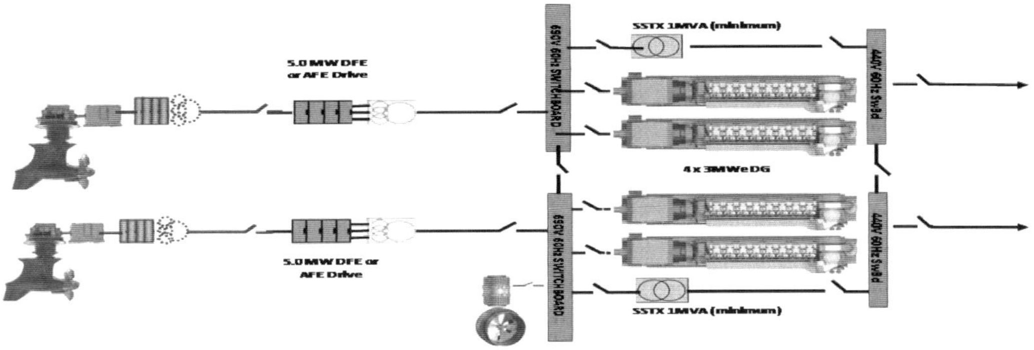

This is a very appropriate propulsion system for this ship type. The use of azimuthrusters will greatly enhance vessel manouverability, particularly at low speeds. With the use of the Rolls Royce Azipull, where the propeller is at the front of the azimuthruster housing greater propeller efficiencies can be achieved, and particularly at higher speeds. The azipull can be used in vessels up to 24 knots, however with the expected vessel hullform the maximum power obtainable is 5MW, which will provide a speed of 20 knots. Thus top speed may need to be compromised if the Azipull is used on this vessel type.
The azipull allows for a simpler hullform, particularly at the stern and provides a greater mission bay for launch and recovery operations.
The all electric architecture also allows for a lower number of installed diesel engines and the lowest number of vessel engine operating hours. This provides a lower ship maintenance and spares burden and provides opportunities for a reduction in manpower.

Procurement cost

Initial cost is always an important consideration and any study should assess procurement cost of the propulsion system,the associated electrical system and any effect the propulsion system may have on the ship design which are cost drivers.

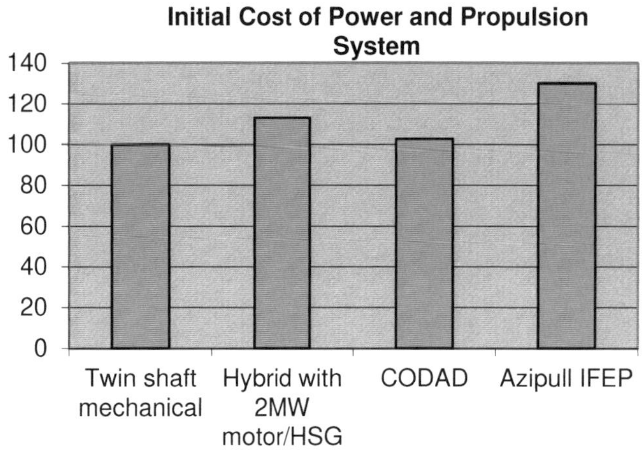

Note: The main cost driver with the Azipull is the additional installed power needed to operate electric motors in an Integrated Electric Propulsion Architecture. What is not considered is the Azipull allows a reduced ship building cost with a much simpler hull form at the stern of the vessel, a smaller engine room with a larger mission bay and less installed side thrusters or retractable thrusters used in Dynamic Positioning

Through Life Cost:

By obtaining an accurate operating profile of the ship in service various propulsion systems can be modelled and compared over the vessels operating profile. This can give an indication of fuel consumption over the life of the vessel and also engine Time Between Overhaul which will effect spares and maintenance costing on the main engines. The engine running hours may also give an indication of crew loading and the number of crew needed to operate and maintain engines.

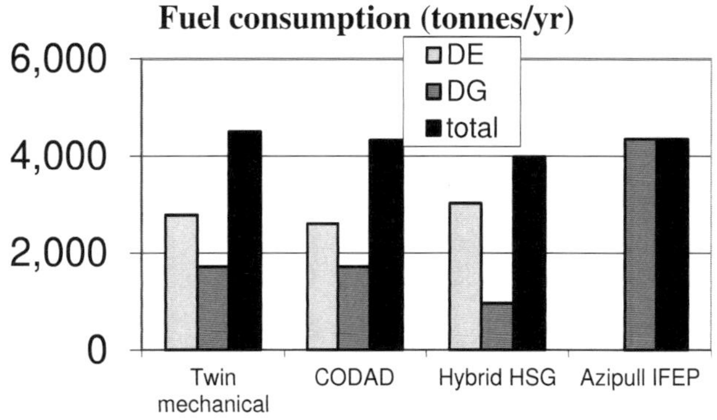

Note: Fuel consumption calculations were conducted while the vessel is underway. If the vessel is spending considerable time stationkeeping, then fuel consumption will be considerably less for the Azipull, as less power is needed to stationkeep and their are less zero pitch losses when stationkeeping on electric motors rather than diesels.

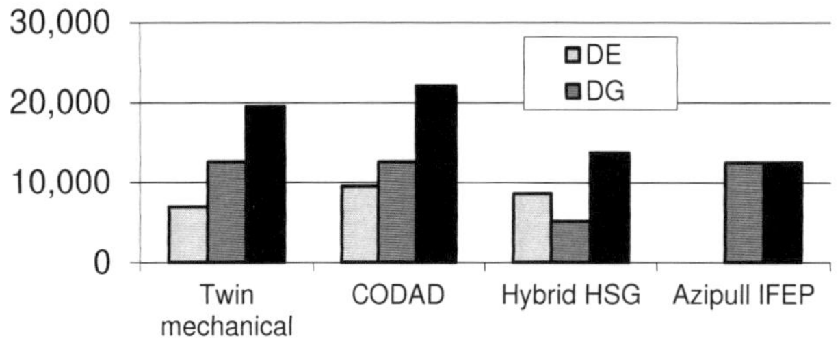

Note: Much better utilisation of vessel diesel engines can be obtained when using the Azipull in an all electric architecture, with less installed prime movers, or the Hybrid solution using HSG.

CONCLUSIONS:

- **Simple mechanical propulsion systems are not appropriate for this vessel type**
- **Hybrid electro-mechanical propulsion systems utilising Hybrid shaft generator technology can meet all vessel capability requirements. However any DP requirement will change the vessel hull form at the stern and increase the number of thrusters needed and installed electrical power.**
- **Azimuthrusters provide high vessel manouverability are very appropropriate for this vessel type and in particular for DP vessels with stationkeeping requirements.**
- **The rolls Royce azipull in particular can obtain higher vessel speeds than traditional azimuthrusters, however with its power limitations and the expected hull form the maximum obtainable speed is estimated to be 20kn**
- **Azimuthrusters provide ship construction, installation and design benefits providing a simple hull form, being modular in nature for easy removal and replacement and allowing for a large mission bay for launch and recovery activities**
- **A greater understanding of the vessels intended operational profile at this stage in the project is important in ship design and power and propulsion assessment.**

FURTHER STUDIES:

Optimising SSK Transit Performance through Hullform, Transit Mode and Diesel Powerplant Selection

Jacob Mann AIMechE MEng
BMT Defence Services, UK

Tom Gibbs AMRINA MEng
BMT Defence Services, UK

Summary

This paper presents the results of a study into conventional submarine (SSK) transit performance through optimising the hullform, transit mode and diesel powerplant selection. The diesel powerplant analysis has been carried out using BMT DyDEM [1] (an in-house dynamic diesel engine model), which allows diesel performance to be simulated for a range of conditions and engine variants. The advantages and disadvantages of the combinations are assessed and the resulting sensitivity of overall submarine performance characteristics to future requirements presented. This paper is therefore of interest to submarine designers exploring concept options for future submarines.

Introduction

There is an operational need to minimise transit duration for SSK submarines, to ensure propulsion plant efficiency and to maximise the time on station in operational theatres. This will deliver operational effectiveness, minimise overall fleet number requirements and ensure future submarines represent value for money. For larger submarines, these needs are particularly challenging for the submarine designer to balance when coupled with relatively high radius of operation requirements (ROO).

Current diesel-electric SSKs typically conduct transits in either surfaced or near surface 'snorting' modes. This paper considers a range of diesel powerplant, submarine hullform and transit mode options to meet high transit range requirements, both covert and overt. The performance of diesel engine variants, e.g. supercharged versus turbocharged, available for the hullform and transit mode options are investigated. The subsequent wider whole submarine design impacts such as total fuel requirements, indiscretion ratio and transit time are therefore identified. This paper follows on the work presented in [2].

History

Before the 1950s, submarine design was dominated by the surfaced running craft which were designed to transit on the surface, but with a capability to submerge for short periods of time, driven by poor energy storage capability and lack of snort capability. These craft therefore had hullforms optimised for the surface, with high Length/Diameter ratios to minimise wave-making resistance, and saddle tanks to provide adequate stability whilst on the surface.

With the advent of nuclear propulsion and improved energy storage technology, submarine hull forms have reverted to the tear drop hull style utilised by the early Holland craft, and of course epitomised by the AGSS Albacore. A tear drop hull shape, with a L/D ratio closer to 7 and minimal appendages and structures, is designed to minimise the submerged resistance in order to maximise the propulsive efficiency. With this style of hullform, submerged performance is maximised at the compromise of surfaced capability.

The current SEA-1000 program in Australia is looking at replacing the current fleet of Collins class SSKs with a more capable, longer range and greater endurance SSK platform [3]. Maximising the transit performance is therefore extremely important if these aims are to be met.

OPERATIONAL SCENARIOS

AREA OF INTEREST, RANGE AND TRANSIT TIME

An indicative modern blue water SSK might be required to have a ROO of 5000nm [4] . At a transit speed of 12 knots, over two weeks would be required to reach the furthest point of the operational area.

SPEED

SSK design speed is governed by the need for an optimised quiet speed that allows the maximum time submerged on batteries or AIP, and a high top speed for transit and evasion.

A submerged transit is composed of two elements: part-snorting and part-fully submerged on batteries. The snorting speed cannot be too great as it is limited by the need to reduce visible wake signature and prevent damage to masts. The power left available to recharge the batteries is reduced if the speed is too high and this has a detrimental effect on the indiscretion ratio [4].

STUDY

For the purposes of this study, the submarine configurations will be evaluated over a nominal 3,000nm ROO. It is assumed that there is a need to travel this distance in the shortest possible time, so the submarine surface speeds have been set as a balance between the available installed power and the hull resistance.

SURFACE TRANSIT

It may be desirable to travel part of the transit distance on the surface (e.g. to minimise transit time or shallow water environments), with the submarine operating under submerged/snorting operation for the remainder of the journey.

To investigate the effect a partial surfaced transit had (e.g. the submarine may wish to submerge for the last 200nm of the transit as it reaches the operational area), the percentage of the transit that is spent on the surface was varied to identify the impact on range and fuel consumption.

This study evaluates the impact of varying submarine hullform, powerplant, and transit mode options on a number of key criteria:

- Transit Time

- Indiscretion Ratio
- Fuel Consumption

SSK HULLFORMS

A number of hullforms have been developed to allow analysis of transit performance. A BMT SSK design, Vidar-36 [5], consisting of a conventional modern hull has been selected to provide a benchmark against which an SSK with a surfaced form can be compared. The Vidar-36 hullform can be seen in

Figure 1.

Figure 1 - BMT Vidar-36

BMT VIDAR-36

BMT Vidar-36© is a 3600te capable conventional submarine design offering multiple-mission capabilities hosted in an ocean-going platform. The key characteristics of Vidar-36 are shown in Table 1.

Vidar-36 Characteristics	
Length	79m (overall)
Beam	8.4m (maximum)
Submerged Displacement	3,600 te

Table 1 - Vidar-36 Characteristics

SURFACE FORM SSK

This is a submarine hullform that has been designed for a transit speed of 12-13 kts on the surface. The shape of the boat is designed to reduce wave making resistance, and shares similarities with traditional surface ship hulls. Previous hullforms designed for transit on the surface include the very successful Oberon and Foxtrot Class SSKs.

For the purposes of this study, to eliminate the additional variable of submarine size, the surfaced hull form has been scaled to match the form displacement of Vidar. This allows a comparison to be made between the different hull forms on a level basis.

Figure 2 - Surfaced Hull Form

TRANSIT MODES

To evaluate the different hullform and diesel configurations, a model has been run to vary the distance spent at the surface as a percentage of the overall transit distance. The remainder of the transit is conducted submerged (with snort to recharge batteries). The submerged speed has also been varied to show the effect on transit time, fuel consumption and indiscretion ratio.

DIESEL ENGINES

Traditionally diesel generators (DGs) onboard submarines have been naturally aspirated or supercharged. Recently however, a number of turbocharged diesels have been installed on modern SSKs and SSNs. The benefits of turbocharging and supercharging are increased power density and reduced fuel consumption, however there is some conjecture on the stability and performance on the turbocharged engine under snorting conditions [1], whereby the engine may experience inlet depression and high back pressure.

For this study supercharged and turbocharged engine configurations have been investigated for each hullform and transit mode. It may transpire that a turbocharged engine would perform better in a submarine platform optimised for surface transit, due to the relatively small time period spent snorting.

The diesel powerplants have been evaluated on their snort and surfaced performance using a proprietary BMT in-house tool, BMT DyDEM. The available power and specific fuel consumption will be used in the analysis.

DIESEL POWERPLANT PERFORMANCE & CHARACTERISTICS

The performance of both the supercharged and turbocharged diesel powerplants under submarine installed conditions (both surfaced and snorting) was evaluated using BMT DyDEM and is shown in Table 2.

Pressure losses due to exhaust runs and intake routes have been taken into account. Snort conditions assume an exhaust depth of 5m and Sea State 4.

	1400kW Turbocharged Diesel	1400kW Supercharged Diesel

	Available Power (kW)	S.F.C.	Available Power (kW)	S.F.C.
Submarine Surfaced	1400	205	1400	220
Submarine Snorting	1000	212	1200	231
Percentage change	-29%	+3%	-14%	+7%

Table 2 - Diesel Powerplant Performance

ENGINE CONFIGURATION

The powerplant configuration evaluated consists of 2 diesel engines, in either turbocharged or supercharged options, sized to produce a total of 2800kWe at atmospheric conditions.

The reduction in power and fuel efficiency whilst snorting is proportionally less for the turbocharged configuration, however, the stability of the turbocharger under snort conditions may cause problems [**Error! Bookmark not defined.**].

HULL FORM POWER REQUIREMENTS

The resistance and hence effective power curves for the evaluated hullforms at submerged and surfaced conditions are shown in Figure 3. The difference between the surfaced resistances for a modern SSK design (Vidar) and a surfaced hullform can clearly be seen.

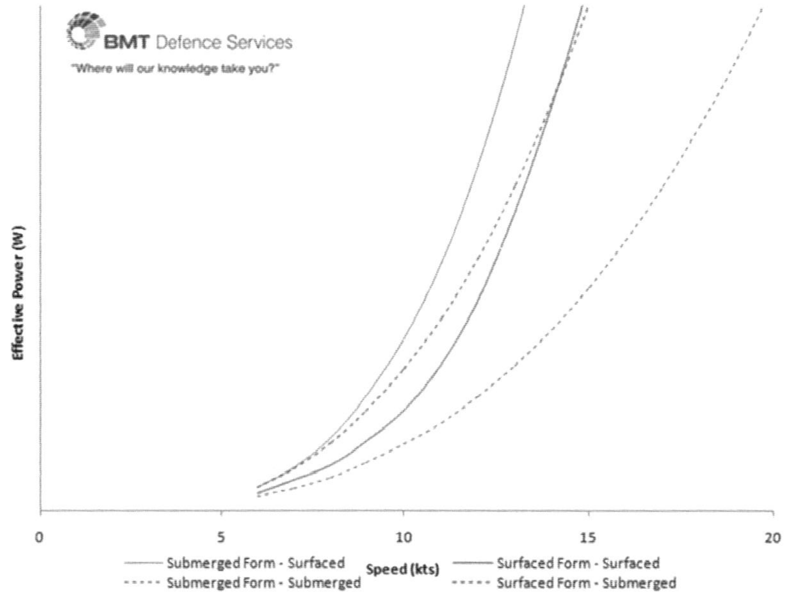

Figure 3 - Hullform Effective Power Requirements

At a certain speed (around 14kts), the surfaced resistance of the surface hullform becomes greater than the submerged resistance. This is due to wave making effects becoming dominant.

ASSUMPTIONS

For the purposes of this study the following assumptions have been made:

- Radius of Operations: 3000nm
- Snort mast exhaust depth: 5m
- Battery capacity: 600 x 8800Ah lead acid cells.
- Transit hotel load: 150kW
- Snort hotel load: 260kW
- Sea state: 4

The snort speed and submerged speed have been set as a balance between the available propulsive power from the DG's and the submarine resistance in that mode. The propulsive power is the available DG power minus the hotel or snort load and the power supplied to the battery for charging.

RESULTS

Varying the submerged speed and the percentage of the transit distance spent on the surface allows a contour map of the results to be plotted showing the transit time and the total transit indiscretion ratio.

To compare results, the performance of each configuration over the 3000nm transit for a fixed fuel consumption of 120 tonnes will be evaluated.

BMT VIDAR-36

TURBOCHARGED PERFORMANCE

With the available power ratings from turbocharged diesels, the maximum snort speed is 4kts and the maximum surface speed that allows fast battery recharge is 10kts.

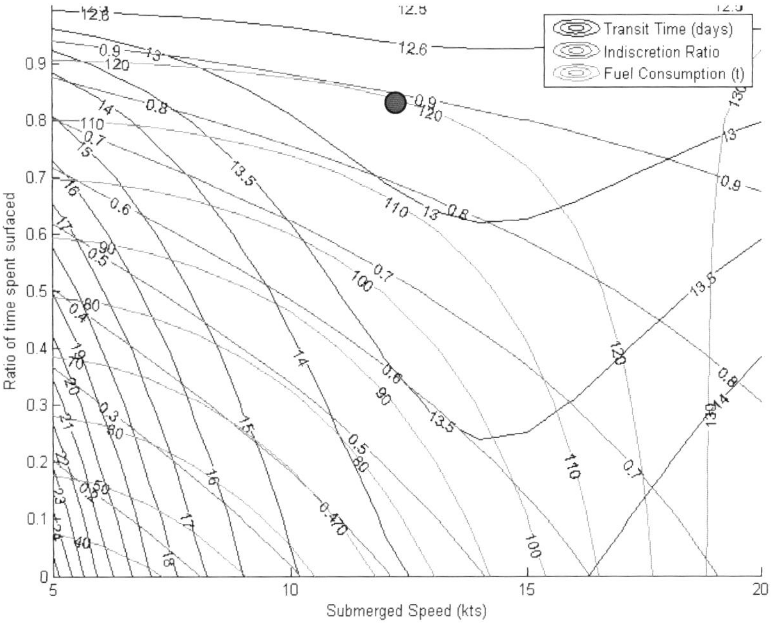

Figure 4 - Transit Time, Indiscretion Ratio and Fuel Consumption for Vidar-36 with Turbocharged Diesels

Figure 4 shows the transit performance of the Vidar-36 hullform configured with turbocharged diesels. As turbocharged performance suffers significantly at snorting conditions, reducing the proportion of the transit distance spent at surface conditions results in an increase in transit time. However due to the submerged resistance being significantly less, the fuel consumption decreases. At submerged speeds greater than 14kts, it becomes beneficial to increase the proportion of the transit distance spent on the surface.Figure 4 This is due to the time spent snorting becoming dominant in submerged operation.

At 120 tonnes fuel consumption, the minimum transit time is shown on the graph as approximately 12.7 days.

If fuel consumption is ignored, the minimum transit time of approximately 12.5 days for this configuration can be found at a surfaced transit proportion of 100%.

The minimum indiscretion ratio and minimum fuel consumption can be found by decreasing the percentage of the transit spent surfaced and by decreasing the submerged speed. However, this has a significant effect on transit time.

SUPERCHARGED PERFORMANCE

With the available power ratings from supercharged diesels, the maximum snort speed is 5kts and the maximum surface speed is 10kts.

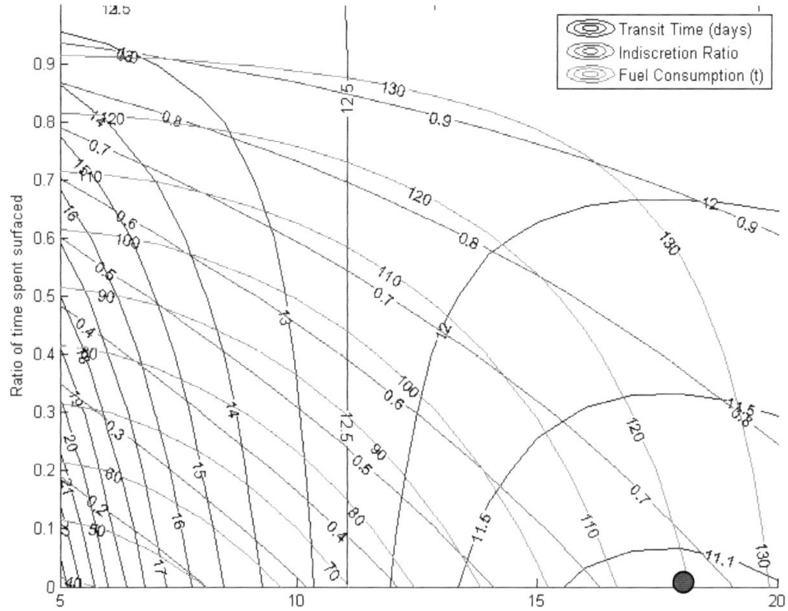

Figure 5 - Transit Time, Indiscretion Ratio and Fuel Consumption for Vidar-36 with Supercharged Diesels

Figure 5 shows the transit performance of the Vidar-36 hullform when configured with supercharged diesels. As the supercharged diesel suffer from less performance drop than the turbocharged configuration at snort conditions, the minimum transit time at 120 tonnes fuel consumption can be found at a surfaced percentage of 0% and a submerged speed of approximately 18kts.

The submerged speed has increased, however, the fuel consumption has risen due to the reduced fuel efficiency.

BMT SURFACE HULLFORM

TURBOCHARGED PERFORMANCE

With the available power ratings from turbocharged diesels, the maximum snort speed is 4kts and the maximum surface speed is 12kts.

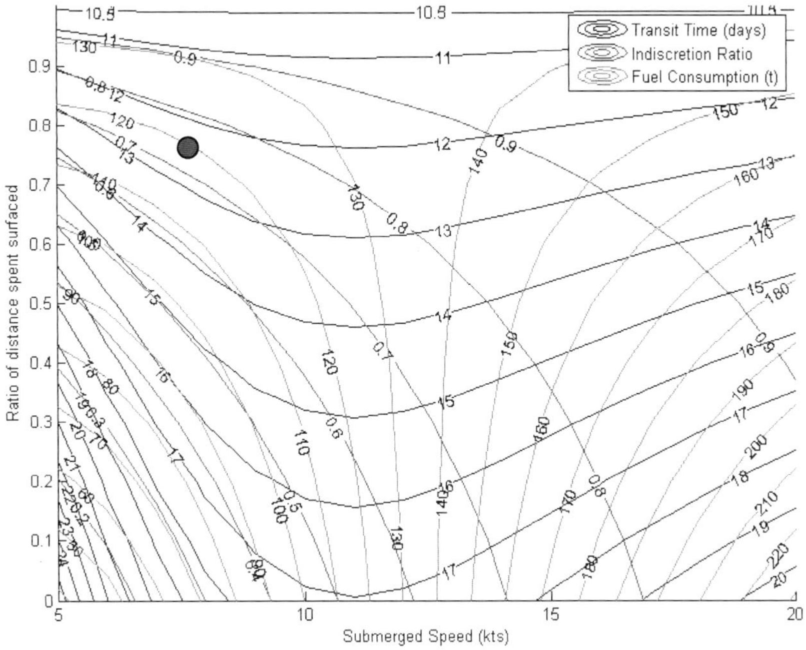

Figure 6 - Transit Time, Indiscretion Ratio and Fuel Consumption for Surface SSK with Turbocharged Diesels

Figure 6 shows the transit performance of the surface form SSK when configured with turbocharged diesels. At a fuel consumption of 120 tonnes, the minimum transit time is approximately 12.5 days.

The transit time for the surface form SSK is much more sensitive to distance spent submerged than the Vidar-36 hullform, however, the minimum transit time has decreased to under 11 days. This time occurs when the submarine is operating 100% on the surface.

SUPERCHARGED PERFORMANCE

With the available power ratings from supercharged diesels, the maximum snort speed is 6kts and the maximum surface speed is 12kts.

Figure 7 shows the transit performance of the surface form SSK when configured with supercharged diesels. At a fuel consumption of 120 tonnes, the minimum transit time is approximately 13 days.

The surface transit performance has degraded due to the reduction in fuel efficiency. Submerged performance has improved due to the increase in available power from the supercharged configuration.

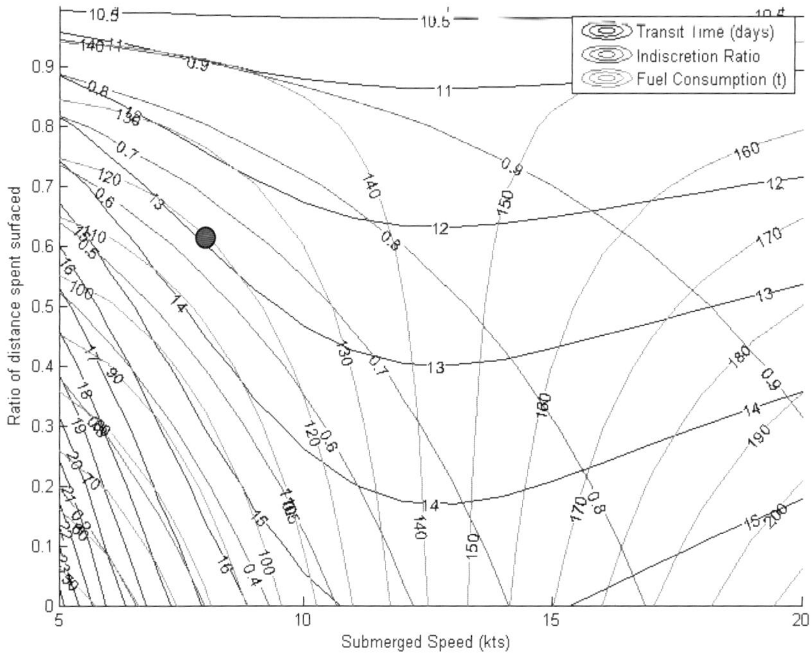

Figure 7 - Transit Time, Indiscretion Ratio and Fuel Consumption for Surface SSK with Supercharged Diesels

SUMMARY

For a fuel consumption of 120 tonnes, the attainable transit times and indiscretion ratios for each configuration are shown in Table 3.

	BMT Vidar-36		Surface Form	
	Turbocharged	**Supercharged**	**Turbocharged**	**Supercharged**
Minimising Transit Time	Transit time: 12.7 IDR: 0.9	Transit time: 11.1 IDR: 0.67	Transit time: 12.5 IDR: 0.75	Transit time: 13 IDR: 0.65
Minimising Indiscretion Ratio	Transit time: 14.5 IDR: 0.65	Transit time: 11.1 IDR: 0.67	Transit time: 17 IDR: 0.55	Transit time: 15 IDR: 0.52

Table 3 - Summary of performance at a constant transit and constant fuel consumption

The minimum transit times for each configuration are shown in Table 4.

	BMT Vidar-36		Surface Form	
	Turbocharged	**Supercharged**	**Turbocharged**	**Supercharged**
Minimum Transit Time	Transit Time: 12.5 IDR: 1 Fuel: 125	Transit Time: 11 IDR: 0.65 Fuel: 120	Transit Time: 10.5 IDR: 1 Fuel: 135	Transit Time: 10.5 IDR: 1 Fuel: 145

Table 4 - Summary of minimum transit times

CONCLUSIONS

This study has used the BMT Vidar design to compare the transit performance against a hullform more suited to surface operation. Two different diesel configurations have been investigated and the trade-offs compared.

The Vidar-36 hullform, is shown to be more efficient at greater percentages of the transit distance spent submerged, due to the much reduced resistance when compared to its surface performance. However, there is a limit to maximum submerged speed, at which no greater benefit to transit time or fuel is obtained. A supercharged diesel powerplant configuration would seem to offer better transit performance than a turbocharged configuration.

The surface form SSK, has the ability to transit the 3000nm distance in a reduced time when compared to Vidar, however, this is at an increased fuel consumption. Both engine configurations offer the same transit times at fully surfaced running, however the turbocharged configuration offers a better fuel consumption.

This study has shown that different engine configurations may have benefits with different operating modes and hullforms.

Any submarine design is a balance between the available technology and the operational requirements, it is therefore necessary to carefully consider how the submarine will be used. A surface form SSK could be useful if a need is identified to transit large distances in the minimum amount of time, with a trade-off in fuel consumption and indiscretion ratio. The modern SSK hullform has advantages in submerged performance, and as battery technologies improve, leading to higher power densities, along with an increase in diesel powerplant performance, the transit performance will subsequently improve.

ACKNOWLEDGEMENTS

The opinions expressed in this paper are those of the authors and not those of BMT Defence Services, other companies, the UK MoD or any of its agencies. The authors would like to thank their colleagues at BMT Defence Services for their kind help and support whilst writing this paper.

REFERENCES

1 Mann, J. "Twin-Turbocharged Diesel Performance Under Snorkelling Conditions" UDT 2011.

2 Buckingham, J et al. "Multi-Engine Submarine Power Supplies: The Operating Case" Pacific 2010.

3 Australian Government, Department of Defence, Defence capability plan. http://www.defence.gov.au/dmo/id/dcp/html/sea/Sea1000.html.

4 Prof Joubert, P. "Some Aspects of Submarine Design Part 1. Hydrodynamics" DSTO.

5 Binns, S. "Meeting the Current Challenge of Designing High Capability SSKs" Pacific 2008.

Multi-role replenishment/support vessels: Is this a cost effective solution?

Michael O'Connor
Principal Engineer - Propulsion systems: Rolls Royce plc
Michael.O'Connor2@rolls-royce.com

Leif Helle Becker
Senior Ship Designer: Rolls Royce Marine As
leif.helle.becker@rolls-royce.com

SYNOPSIS

The Naval Tanker and Amphibious vessel are traditionally two mission specific ship types. However are there real cost effective benefits in combining two important military assets into a single platform? Many navies are beginning to think so, as demonstrated in the Dutch Joint Support Ship and similar programs in Norway, Sweden, Canada and New Zealand to name a few. Rolls Royce has recent experience in designing this new ship type which combines Replenishment at Sea capability with the Roll on - Roll off capability needed for amphibious support. Other capabilities of such a vessel include aviation, hospital & humanitarian support, command and control, ammunition storage & even dynamic positioning has been considered. Some key design drivers include fuel carrying capacity, fuel type, aviation capability, number of Roll on – Roll off lane meters needed, hospital size, intended ships company, provision for landing craft and ammunition storage. The ships operational profile is also critical in designing an optimum propulsion system. In particular top speed requirements will influence hull form selection which will drive vessel particulars, installed power and ultimately propulsion selection. Future MARPOL environmental regulations and vessel environmental operational requirements are also considered key design drivers. Of great importance is the integration of various deck systems to achieve a harmonious ship design that is both cost effective but also provides a smooth logistical flow of stores throughout the vessel, ensuring good operability of ship functions. What are the cost drivers? Combining two naval platforms into one must be justifiable. A thorough analysis is conducted.

1. SHIP CAPABILITY REQUIREMENTS:

A Naval Replenishment at Sea (RAS) ship is traditionally a purpose built design whose primary capability is to replenish other task group naval vessels with fuel, ammunition, food and stores whilst underway.

RAS vessels contain survivability requirements in areas of susceptibility, vulnerability and recoverability in line with the vessels intended level of operational threat. This is usually to operate as part of a task force group but to replenish task group assets away from the threat level. These vessels are not intended to operate directly in a theatre of war and as such self defense capability is usually minimal. Weapons carried on replenishment vessels are usually small caliber guns, machine guns and sometimes a Close in Weapons System (CWIS).

Vulnerability & recoverability characteristics include naval damage stability compliance, extended firefighting equipment and damage control measures, system redundancy, some shock rating of equipment, fragmentation protection and protection from airborne CBRN threats. Vulnerability and recoverability rules affect the ship design, and the RAS vessel is a distinctly different design than a commercial tanker.

Figure 1: Principal particulars of Naval RAS Tankers

HMAS Success
Displacement: 18000 t
Length:157.2m
Beam: 21.2 m
Draft: 8.65
Speed: 19 knots
Crew: 220

Berlin Class
Displacement: 20240 t
Length: 173.7m
Beam: 24 m
Draft: 7.6 m
Speed: 20 kn
Crew: 139 crew, 94 additional

UK Wave Class
Displacement: 31500t
Fuel Capacity: 13000 m^3 F76, 3000 m^3 F44
Length: 196 m
Beam: 27.8 m
Draft: 9.97 m
Speed: 18 kn
Crew: 80 RFA crew, 22 RN Crew

Patino Class Spanish Navy Tanker
Displacement: 17000
Fuel Capacity: 8000 m^3 F76, 2000 m^3 F44
Length: 166 m
Beam: 22
Draft: 8 m
Speed: 21 kn
Crew: 154

IMO Compliant Double Hull Tanker →

Cantabria Spanish Navy Tanker
Displacement: 19600
Fuel Capacity: 8000 m^3 F76, 2000 m^3 F44
Length: 174 m
Beam: 23
Draft: 8m
Speed: 21 kn
Crew: 154

It is important to note naval replenishment vessels are adopting commercial IMO MARPOL compliance. This includes the need for a double hull surrounding the outside perimeter of the fuel tanks for extra environmental protection in the case of hull damage. Navy flagged vessels can claim an exemption for this rule, however most navies have decided to comply with this requirement. To comply with this requirement and maintain vessel fuel carrying capacity and speed, the vessel will have an increased beam and length (and subsequent steel weight). This is shown in Figure 1 above where the Spanish Cantabria class tanker has maintained vessel fuel carrying capacity and speed, but is larger in order to comply with MARPOL regulations when compared with the previous Patino class. This will lead to a more expensive ship.

Amphibious warfare is the use of naval firepower, logistics and strategy to project military power ashore and this involves numerous classes of ships. The amphibious ship can be a purpose designed vessel for the carriage of military vehicles or equipment to a port facility or it can be a high value military asset capable of commanding a military operation, and transporting equipment and personnel into an operational theatre of war. Force projection capabilities of the amphibious ship impacts its costs particularly when considering the vessels survivability requirements when operating in a high threat environment. Capabilities include communications and combat information centers (to command a military operation) acoustic and magnetic reduction measures, electronic warfare systems & decoys, a Close In Weapon System (CIWS) and sometimes missile launchers. Amphibious vessels also often include a shock capability to maintain equipment functionality after damage and dual redundant mission critical equipment. Such equipment and capabilities are over and above the requirements of a RAS ship and as such are a significant cost consideration.

Figure 2: Principal particulars of Naval amphibious vessels.

HMNZ Canterbury
Displacement: 9000t
Length: 131 m
Beam: 23.4 m
Draft: 5.4m
Speed: 19 kn
Complement: 53 Naval, 10 air force, 7 army, 35 trainees

HMAS Kanimbla LPA
Displacement: 8534t
Length: 159.2 m
Beam: 21.2 m
Draft: 5.3m
Speed: 22 kn
Complement: 23 Naval officers, 2 army officers, 197 sailors, 18 soldiers, 400 embarked forces

HMAS Tobruk LSH:
Displacement: 5800t
Length: 126 m
Beam: 18m
Draft: 4.9m
Speed: 17 kn
Complement: 145 Naval 7 army, 300-550 soldiers

HMS Albion: LPD
Displacement: 18500t
Length: 176m
Beam: 28.9 m
Draft: 7.1 m
Speed: 18 kn
Complement: 325

LSD HMAS Choules
Displacement: 16160t
Length: 176.6m
Beam: 26.4 m
Draft: 5.8 m
Speed: 18 kn
Complement: 60-67 crew, 356 troop – 700 overload

Juon Carlos1 (L61) : LHD
Displacement: 27079 t
Length: 230.82 m
Beam: 32 m
Draft: 6.9 m
Speed: 21 kn
Complement: 243 ships crew, 172 Air force, 913 soldiers

Mistral class Helicopter carrier
Displacement: 21300 t
Length: 199 m
Beam: 32 m
Draft: 6.3 m
Speed: 18.8 kn
Complement: 160 ships crew, up to 900 soldiers

The primary capability of amphibious vessels is to carry and transfer military vehicles from sea to the shore. A large vehicle deck is necessary for the carriage and transfer of army vehicles and battle tanks. Amphibious vessels designed to enter the operational theatre of war require a much larger accommodation for troops, a combat information center and additional survivability measures.

Multi-functionality should not be considered a new term in military vessel design. Inherently Naval Replenishment ships and amphibious vessels are both multi-functional naval assets. This is illustrated in the capability plot shown below in figure 3.

Figure 3: An overview of Capability requirements of the RAS ship vs the Amphibious vessel

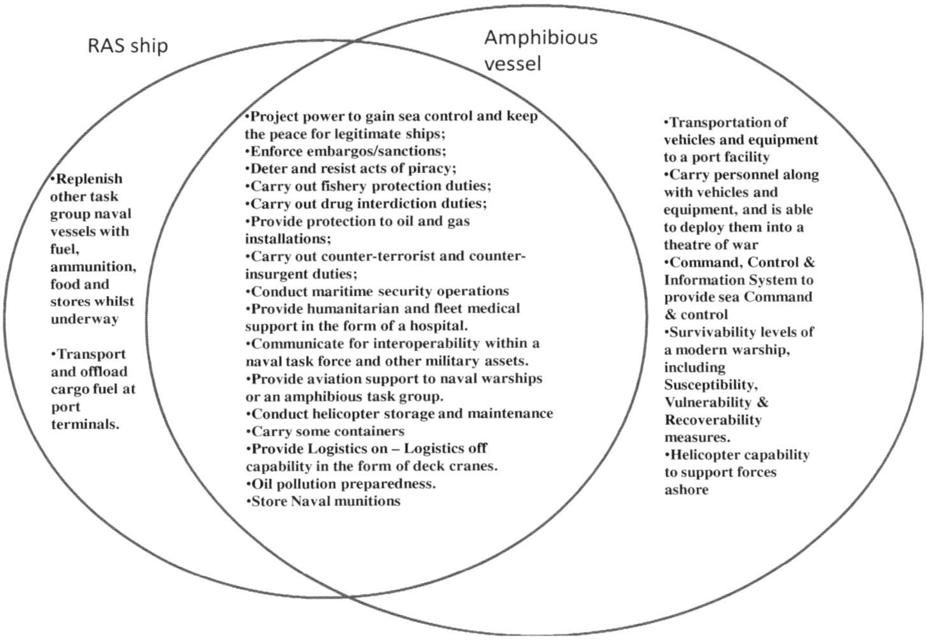

2. DESIGN CHALLENGES

2.1 Design challenges and the commercial sector:

The main design challenges of a multi-functional ship type is merging Replenishment at sea capability with Roll-on/Roll-off capability (contains a vehicle deck with unloading & loading features). In the commercial sector the Ro-Ro vessel and the tanker are specialized and quite separate designs. A commercial tanker is purpose designed for the carriage of the specified cargo from A to B, with a small crew and the cargo tank area occupies the majority of ship length. Whilst a Ro/Ro vessel is also purpose designed often with a larger accommodation block and is dominated by the vehicle deck. Whilst there are some lessons to be taken from both tankers and Ro-Ro in terms of oil storage and lane meters respectively the replenishment ship and the amphibious vessels requirements demand specific design features whether in being for purpose ships or in multi-functional designs.

2.2 Hull & Arrangement:

Figure 4 shows a typical Naval Replenishment ship. Design drivers for ship length, beam & weight (which are a good indicator of ship construction costs) are the ships helicopter hanger

and flight deck, the size of the accommodation and possibly hospital, ammunition stores and the amount of liquid fuel carried. The naval intact and damage stability requirements affect the principal particulars (and affect cost) compared to a commercial hull form converted to the naval sector. Naval damage stability compliance will affect the bulkhead subdivision spacing, but most importantly will cause an increase in beam to comply with the minimum heel angle in the damage condition and to ensure V-lines and any red-risk lines in the damage condition are below an acceptable level. This is an important difference between the naval and commercial sectors. Naval vessels are designed to remain afloat in the damaged condition at an acceptable heel angle with the damaged waterline below the damage control deck to allow the crew an adequate platform to save the ship. Commercial vessels are designed for collision damage with stability rules to allow crew a safe platform for evacuation in a timely manner.
Figure 5 is an example of combining capability of the Replenishment at Sea and the amphibious vessel. The accommodation block has increased to account for the increase in crew size, hospital capacity, command and communications capability and the carriage of any army troops. The main deck forward of the accommodation can be used as deck space for container storage, landing craft, helicopter landing, ships boats, vehicles and RAS rigs. It should be noted this area is premium space.
A vehicle deck running the length of the vessel above the fuel tanks allows vehicles with containers to be loaded through a stern ramp and stored on the vehicle deck. If the vessel requires Logistics On-Logistics Off (LO-LO) capability then the vehicle deck can run the length of the vessel and containers or equipment can be lifted through a deck hatch to the RAS deck and transferred ashore via a crane. Thereby negating the need to transfer via the Ro-Ro ramps and remaining independent of the shore infrastructure.
Fuel cargo space is below the vehicle deck and separated from it by a cofferdam. This is acceptable with F76 and F44 fuel as the flash point of the fuels is above 60 degrees Celsius. Deck space on the RAS deck however is dependent of the number of, and space taken by, the RAS rigs. RAS rigs should be as close as practically possible to amidships longitudinally to provide a safe platform for underway replenishment. If two RAS rigs are needed per side then it is expected the length of the vessel will increase proportionally to maintain the same deck area for cargo capacity. Upper deck ship length space is a premium.

Thus ship length drivers include:

- Number of RAS Rigs needed.
- Lo-Lo off capability on the main deck.
- Accommodation & hospital capacity.
- Flight deck and hanger capacity.
- Ammunition stores

Ship size (weight and internal volume) drivers include:

- Fuel carrying capacity.
- Accommodation space and hospital.
- Cargo lane metres

Figure 4: Typical Naval Replenishment at Sea Tanker

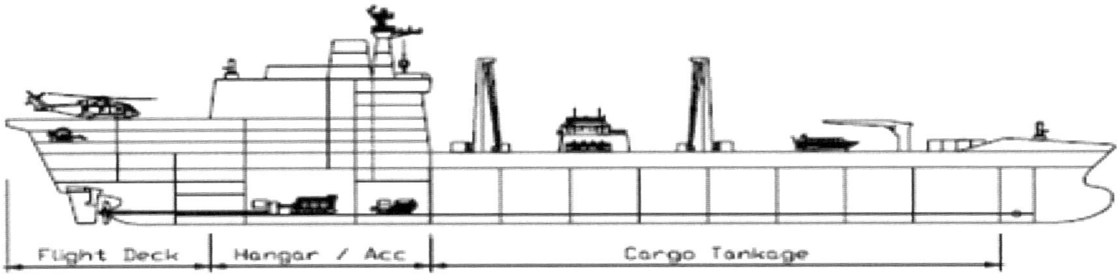

Figure 5: Combining Naval Replenishment capability with Ro/Ro capability

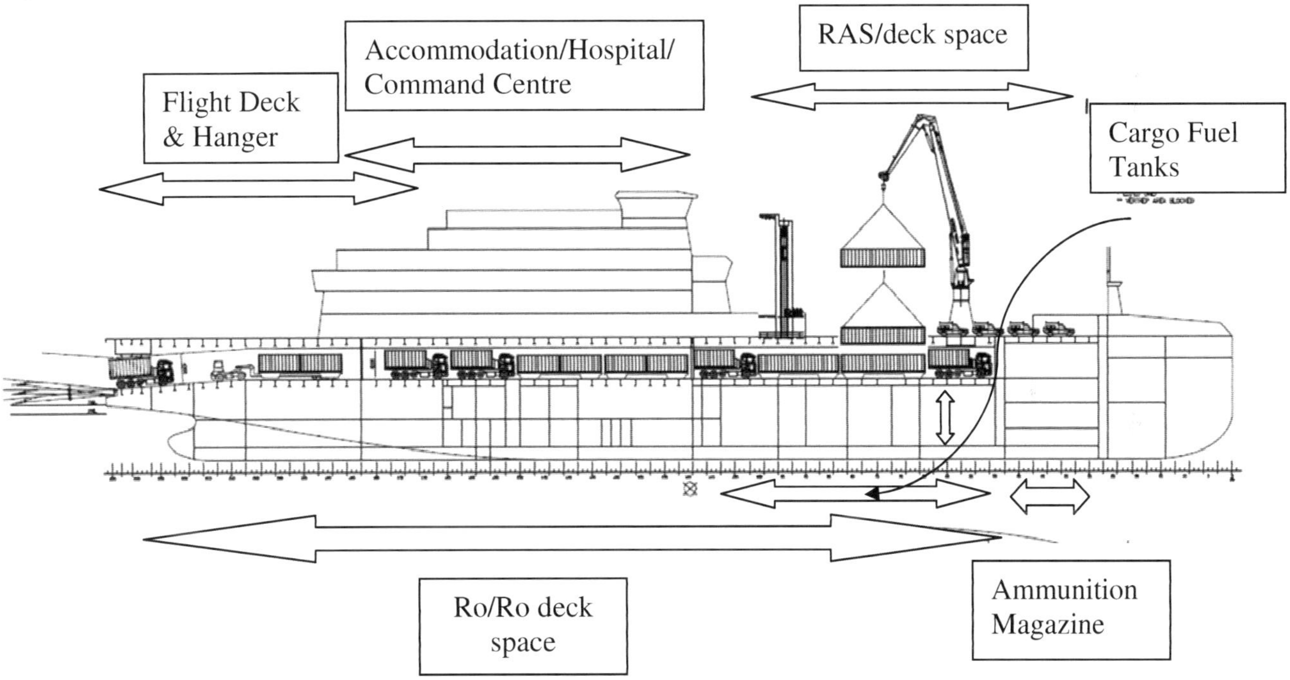

Figure 6: Logistics on – Logistics off (LO-LO) capability

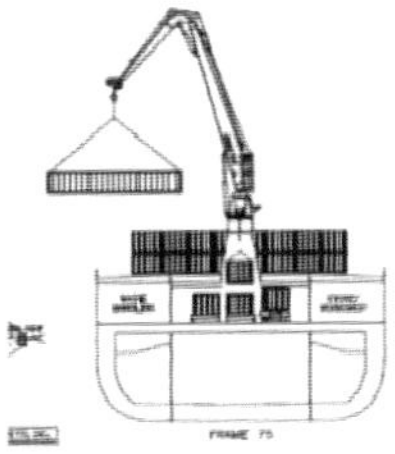

Figure 7: RAS deck space

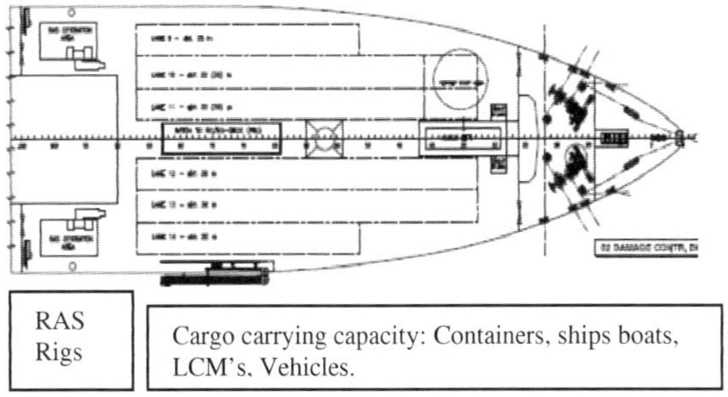

3. VESSEL CAPABILITY:

3.1 Hull form:

A hull form which combines both Naval RAS & Amphibious warfare will need a long sustained range (usually at a transit speed of 16 knots or greater), have good sea keeping capabilities and be directionally stable for performing RAS operations. Amphibious vessel hull forms and RAS vessel hull forms have been similar in the past with RAS vessels having a slightly higher block coefficient. (approx. 0.5 for amphibious vessels to approx. 0.6 for RAS vessels)
This increase in block coefficient in order to carry cargo fuel means the vessel length to beam ratio may need to increase to maintain the same speed. Thus the top speed requirement is also a vessel length driver.

Commercial Ro-Ro hull forms are suitable for long efficient transit speeds, a good top speed and having good sea keeping characteristics to minimize speed loss in heavy seas. Conventional tanker hull forms and designs do not provide the capability required to meet this speed requirement due to a much fuller hull form (higher block coefficient). There are no or few conventional tanker designs today designed for speeds above 16 knots. Commercial tanker hull forms tend to have less directional stability due to their fuller hull form.
(In general a Ro-Ro vessel is considered to be a volume critical vessel with higher top/service speed, while a tanker is a lower speed and deadweight critical vessel. A typical RAS vessel is rather a volume critical vessel even though having'tanker' functionality.)

3.2 Propulsion:

The multi-functional vessel will need a highly flexible propulsion system to be suitable for the various ship functions. A high speed of 18-20 knots, an economical transit speed of around 16 knots, a RAS speed of between 10-16 knots, a loiter speed of 6-12 knots and, possibly, a dynamic positioning operation where the vessel is station keeping while conducting boarding operations, pollution prevention, fire fighting or stores transfer to military or civilian installations. To get real cost savings it is important to have a propulsion system integrated with the ships electrical system. For this reason power and propulsion systems are moving towards Hybrid or All electric architecture. A hybrid electro-mechanical propulsion system is a cost-effective and flexible solution for a multi-functional logistics vessel combining both RAS and Amphibious ship roles.

3.3 Electrical & Auxiliary System:

Naval vessels contain a high component of ship-board auxiliary systems. The RAS vessel contains RAS rigs and a cargo pumping system and the amphibious vessel will contain a large accommodation block, possibly a large hospital, weapons and decoys and a high associated content of ship board auxiliary systems. Combining such ship functionalities gives importance to the integration of the ship electrical system and auxiliary systems. Particularly when designing to operate in both tropical and cold climates.

3.4 Command systems:

C4I is a major cost driver and is a distinct difference between a RAS ship and an Amphibious vessel. The need for a combat information centre and specific naval communications equipment to command and provide interoperability with naval task group assets adds complexity and cost and when combined with electronic warfare and radar systems the cost increase can be in the order of approx. 20%.

4. INTEGRATION OF SYSTEMS

Integration of military capabilities is a key consideration when combining such vessel functionality to achieve a functional and harmonious design and also obtain real cost savings. Putting together amphibious and replenishment capabilities will add crew size, increase vessel size and weight and drive up acquisition cost. What is important however is that the factor of increased vessel size, weight & capability is not directly proportional to acquisition cost by a factor of two or greater. The amount of cost savings will be in the amount of real harmonious design integration in the initial phase of the design.

Integration includes general arrangements that can provide an acceptable vessel size and volumetric capacity for all ship functions and allow them to operate in a safe an efficient manner. It also includes the integration of ship systems to ensure installed power is kept to a minimum and that through life cost savings can be maximised.

5. CONCLUSIONS:

Designing a new class of ship for multifunctionality which includes both underway replenishment and amphibious capability can provide real acquisition cost savings in the ship building process. What needs to be defined however very early in the design stage is the type of amphibious capability to be combined with RAS. Combining an amphibious vessel which is not designed to enter a high threat environment with a replenishment capability allows for a consistent vessel survivability requirement, a limited accommodation increase and no combat information centre and hence good cost savings. Trying to turn a RAS ship into an amphibious vessel intended to be projected into a theatre of war will add significant costs to the purchase price. It is not intended to discourage this ship type; however a sense of reality of the amount of cost savings should be understood in naval project teams as they apply for government budgetary approvals.

The level of ship "survivability" requirements in combination with the mission of the vessel is crucial to define early in a project.

Through life cost savings however can be achieved when combining functionalities. Especially in the possibility of a reduction in crew size as operations are performed on a single platform.

References:

1. Partridge R, North M, Scott A; "Hybrid electric propulsion for Naval auxiliaries", Rolls Royce plc, INEC 2008

2. Cooper S L, Burger D P G, McDonald T P; *"Concepts for a fleet tanker: An exploration into options and pricing"*, UK Ministry of Defence,

3. Bricknell D, Skarda R, Vedlog P; *"A family of underway replenishment ships"*, Rolls Royce plc, WMTC 2009

Submarine Power And Propulsion: Balancing The Energy Elements

John Buckingham CEng FIMechE

BMT Defence Services Ltd, Bath, United Kingdom

SUMMARY

The challenge for designers of submarines with an ocean going capability is to achieve a power and propulsion system design which allows the ambitious transit distances to be achieved whilst still allowing for stealthy operations in the mission area. Using the BMT concept submarine design, Vidar-36, this paper shows the consequences of changes to the size of the main battery, the main engine, embarked fuel and the AIP capability within prescribed limits.

The way in which the different combinations of these energy elements affect the operating parameters, such as the achievable range, for a range of snort and submerged speeds is shown.

The performance benefits of modern batteries and modern diesel-engined generating sets (DG sets) are also explored. This information will also allow those who set platform requirements to better appreciate the conflicting challenges which need to be overcome to achieve a balanced submarine design.

CONTEXT

Scope

If diesel electric (SSK) submarines are to conduct operations of longer and more distant duration they will need to be able to go faster during the transit to allow the maximum time on mission. To achieve this, the power and propulsion machinery needs to have the right design parameters to allow operations within the prescribed operating envelope together with a high availability and low upkeep burden.

The machinery in a submarine comprises a number of key energy elements: energy consumption; power generation; energy storage; propulsion drives and these have been addressed in previous BMT papers in 2008 [Ref. 1] and 2010 [Ref 2)]. This paper considers the subtle issues and consequences relating to the variation of these different elements.

Current SSK Arrangements

The typical modern SSK has two gensets to allow for a high battery charge rate and to allow for a degree of redundancy. In a cramped arrangement there is limited access for upkeep. In the Australian Collins class this led to the use of horizontal crankshaft removal in the Hedemora engines [Ref 3]. The Collins class is one of the few SSK with three gensets: apart from redundancy and the need for a compact genset arrangement to avoid increasing the submarine length, the provision of three sets supports the enduring need to ensure that there is always one genset available to allow for battery recharge.

Current Engines

Engine Range

In modern Western submarines, the DG sets principally employ the following engines

Engine Make & Model	Power kW (bhP)	Speed rpm

MTU 396: Surfaced : Submerged	600 to 1350 600 to 1250	1800
Pielstick PA4	590 to 2215 (800 to 3000)	1500
Paxman Valenta	1300	1800

Table 1 - Submarine Snort Engines

The stated power ratings are estimates for normally aspirated operations. When operating in the snort condition, their rating will typically be reduced by 25% from their surface ship rating. As an example the MTU 12V 396TE54 is rated at 1,200kWe for the surface ship TE54 designation [Ref. 4, pg 52] but when employed in the submarine version it is derated to 1,000kW on the surface and 900kW when snorting [Ref.5].

MTU 396

MTU have produced many diesel engines for submarines both with and without super-charging [Ref. 5]. For many years their offering has been the 396 which has been available with a turbo-charger design since the 1980's after MTU had a public test-bed demonstration of its performance in 1982. Over 210 MTU 396 gensets are in use in SSK applications. They come from a broad pedigree which has also been developed for a range of specialist marine applications including low-magnetic applications.

Pielstick PA4

The 12 cylinder Pielstick PA4 engine is used on French submarines [Ref. 6] and those exported from France to other navies. The best known models are the PA4-185 and the 200-VG models. The PA4 engines are fitted with a variable geometry combustion chamber and are now supported by MAN.

Paxman Valenta

Paxman engines have been on UK submarines since the U and V class of World War II. The last design to enter service before Paxman was acquired by MAN Diesels was the Valenta 16SZ which has a mechanically-driven super-charger on HMS Upholder [Ref. 7 and 8], now HMCS Victoria, and the remainder of that class. The design is reviewed in some detail by Sessions et al [Ref. 9]. His paper shows how supercharging both increases the rated power of a naturally aspirated engine and lowers the specific fuel consumption (Sfc) by over 20%. Mann reported on this aspect and the issues with snorting [Ref 10].

MTU 12V4000U83

In 2011 MTU announced their concept design for a new submarine engine; the submarine version of the 12V4000, the U83 would be rated at 1300kWb. The expected performance of the new engine was compared with the performance expected from two MTU 12V396SE84 gensets using information from Ref. 5 and due to a lack of some information, BMT's own assumptions and estimates.

Feature	**MTU 16V396SE84**	**MTU 12V4000U83**[1]	**Comments**[2]
Rated snort power	1,200kW	1,300 kW	8.3% benefit

1 Estimated from given 12V4000M23S data

2. Comments show the benefits of the 4000 over the 396.

Feature	MTU 16V396SE84	MTU 12V4000U83[1]	Comments[2]
Length – DG set	4.94m	4.58m	7% benefit
Length - Engine	3.320m	2.52m	7% benefit
Width	1.60m	1.55m	3% benefit
Height	2.85m	2.60m	9% benefit
Weight – Engine	8,800kg	9,737kg	11% worse
Weight – DG set	13,050 kg	13,987 kg	7% worse
Unit Footprint	7.9m2	7.1m2	10% benefit
Unit Volume	22.5 cubes	18.5 cubes	18% benefit
Power density	53.3 kW/cube	70.4 kW/cube	32% benefit
Standard Sfc: surface (Open)	214 g/kWh	205 g/kWh	4.2% benefit
Standard Sfc: snorting	280 g/kWh [3]	2683 g/kWh3	4.2% benefit

Table 2 - Engine Comparisons

Table 2 indicates the potential benefits of the new engine in terms of volume (i.e. due to power density).

Platform

The BMT Vidar-36 submarine design [Refs. 11 & 12] is the platform for this assessment. BMT Defence Services developed the Vidar-36 through its Inspira™ process and the design serves as a virtual technology test-bed for many methods and design ideas. The baseline Vidar-36 design is a conventional SSK which comprises the following machinery features

- A displacement of 3,600 tonnes with tankage for 176 tonnes of burnable fuel;
- A brushless DC (active stator) propulsion motor rated at 5,300kW;
- A 7m long AIP plug methanol-fuelled, liquid oxygen supplied PEM fuel cell for a submerged endurance of 14 to 28 days (21 standard);
- Two 1200kW diesel engines (i.e. two MTU 16V396SE84(L)) installation for snort power generation;
- Top sprint speed of 20 knots.

Range

The key operating requirement is range, where this figure is taken to mean the furthest distance from port possible at the standard operating conditions. A long range will permit operations far from base-port, as well as extended duration in the mission area once the submarine has arrived. The transit to the theatre will use snorting and batteries to avoid depleting the AIP's fuels and oxidant. This assumes that a surface transit cannot be undertaken due to the surface threat.

Self-noise and indiscretion on such occasions is still critical but to arrive in the mission area in a realistic timescale, transit speeds of over 10 knots are therefore to be expected. A reasonable future transit requirement, the standard operating conditions, would be a range of

[3] BMT estimates

over 3,000nm at 11 knots submerged transit speed with 6 knots snort speed. This is to achieve a patrol duration of ~60 days.

Vidar Mission Details

Once in mission area, the submarine travels at four to five knots. The hotel load is assumed to be 150kW with a snort electrical load of 260kW. The endurance when snorting is a key performance criterion for a platform which may have to travel long distances to the mission area. The operation of the Vidar-36 in snort operations for battery recharge between submerged transits is presented below. Figure 1 shows the anticipated range (i.e., or half-range) and patrol duration performance on submerged transit speed (y-axis) and snort speed (x-axis) for such an arrangement.

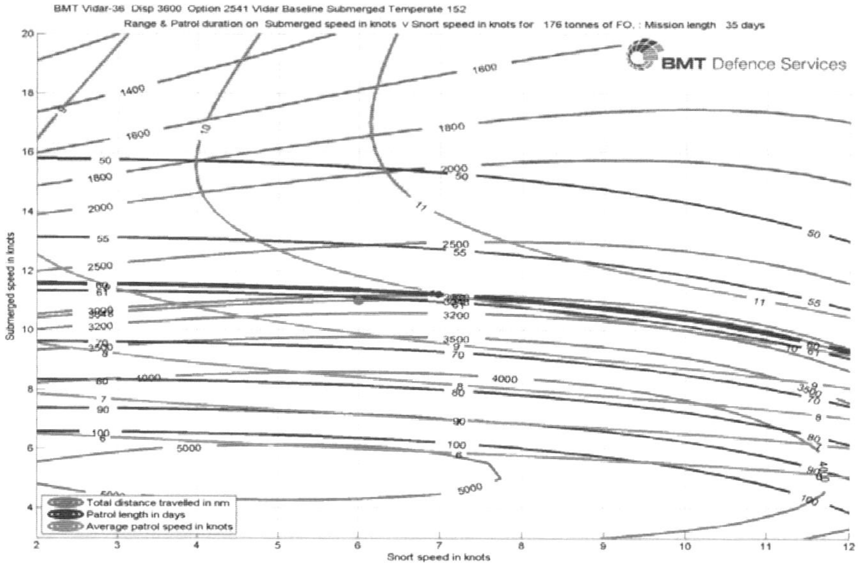

Figure 1: Range and Patrol Length Performance

Figure 1 shows how the provision of 176 tonnes burnable fuel permits a range of 3,048 nm to be achieved at the standard operating conditions shown as a red circle. The total patrol length is 61 days (blue lines).

TRANSIT SPEED V MISSION DURATION

Requirements

To be able to operate at ranges far from the home port, and to stay on mission for the longest possible time is a standard challenge addressed by those in the operational analysis world [Ref. 13]. However such issues need to be addressed in the early stages of design if the vessel is to be suitable for such operations and indeed if the designed operating tempo is to be balanced with the intended number of vessels.

This study has identified the achievable range for a range of submerged transit speeds and mission durations. In this case, the snort speed has been fixed at 6 knots and the duration of snort periods in transit are not limited due to the lack of an active threat outside the mission area.

Range and Patrol Duration

For the total amount of available fuel burn, the range of the submarine can be deduced from the mission duration and the submerged transit speed. The variation of range with these parameters is shown in Figure 2.

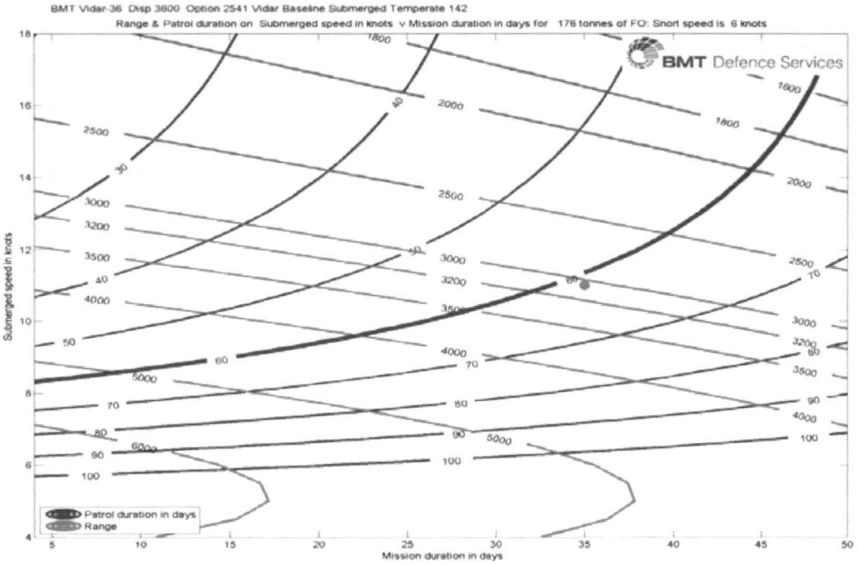

Figure 2: Range & Patrol Duration v Mission Duration

Figure 2 shows how the range (in red) increases with slower submerged transit speed and shorter mission durations. The effect of the square resistance curve is seen in the rapid falling away of the range distance for the higher submerged transit speed. Figure 2 also shows how the patrol duration lengths (in blue) decrease with increasing submerged transit speed.

The red dot shows the design point of 11 knots submerged for 35 days which gives a patrol duration of ~60days and a range of ~3000 nm.

DESIGN VARIATIONS

BMT sought to explore the impact of changes to the design of the submarine in terms of:

a. Changes to the length of the AIP plug,

b. Trades between the number of batteries and the fuel stored onboard.

c. The use of lithium ion batteries instead of standard lead acid units;

d. And the adoption of a modern DG set.

AIP PLUG AND FUEL/BATTERY TRADES

The length of the AIP plug is 7m in the baseline. This was varied between 5m and 9m to identify the impact on range from the change to resistance.

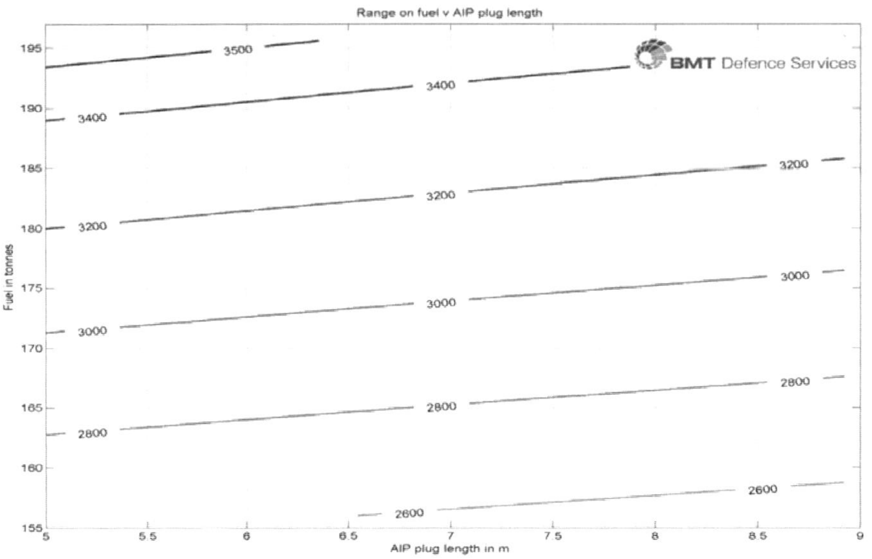

Figure 3: Range on Fuel v AIP plug length

Figure 3 shows how the 2m variation in AIP plug length has a 2% impact on the range of the submarine's operations compared to the +/- 20 % change in batteries/fuel which has a +/- 14% impact.

If 2m of the AIP plug is set aside for additional fuel storage with the remaining 5m retained for a reduced AIP fit, the standard AIP capacity falls to 15 days but the extra fuel provides a 7% increase to the submarine's range.

LITHIUM ION BATTERIES

The Gaia 500Ah lithium ion cell being developed by LTC is provided in modules of 23 cells (Ref 14) . The module is 1450mm high by 500mm x 290mm and weighs ~385kg. It stores 40kWh of usable energy with a maximum continuous current of 320A at a voltage of 82.8V.

The baseline Vidar-36 design has 600 lead acid cells each rated at 8800Ah. These are arranged in two strings each of 300 cells to yield a usable voltage of over 700 Volts. Due to the limitation in the rated current of the lithium ion battery module and also to take advantage of its significant voltage, the 330 lithium ion modules are arranged in 33 strings of ten with a system voltage of 828V.

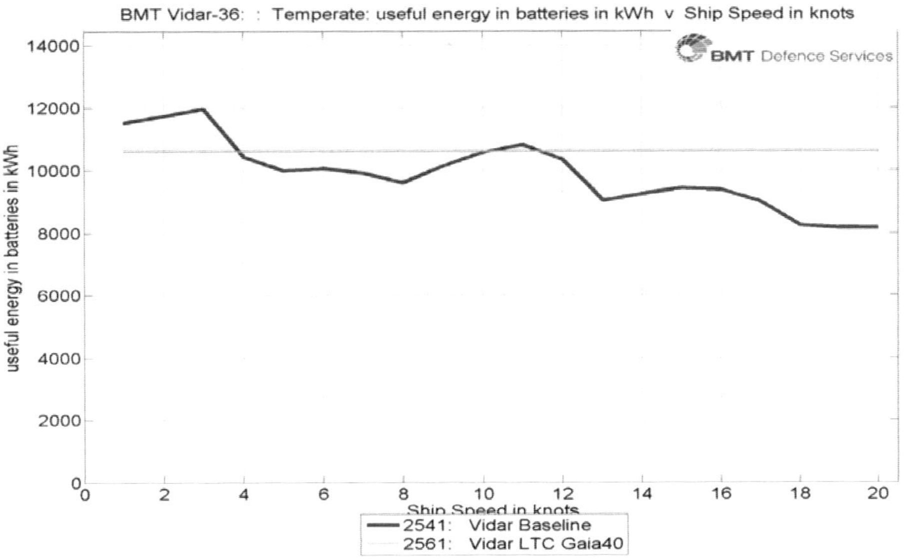

Figure 4: Useful Energy Stored in Ship's Battery

The lead acid battery is limited in the energy it can supply when delivering high current, i.e. high power at high submarine speeds; it is able to deliver more energy with lower currents.

To ensure the lithium-ion battery fit has the same energy delivery at all submarine speeds, it is matched to the lead acid battery at the lower ship speed as shown in Figure 4. This is better demonstrated in Figure 5 which shows the same discharge duration at low powers but a significantly higher duration at higher powers.

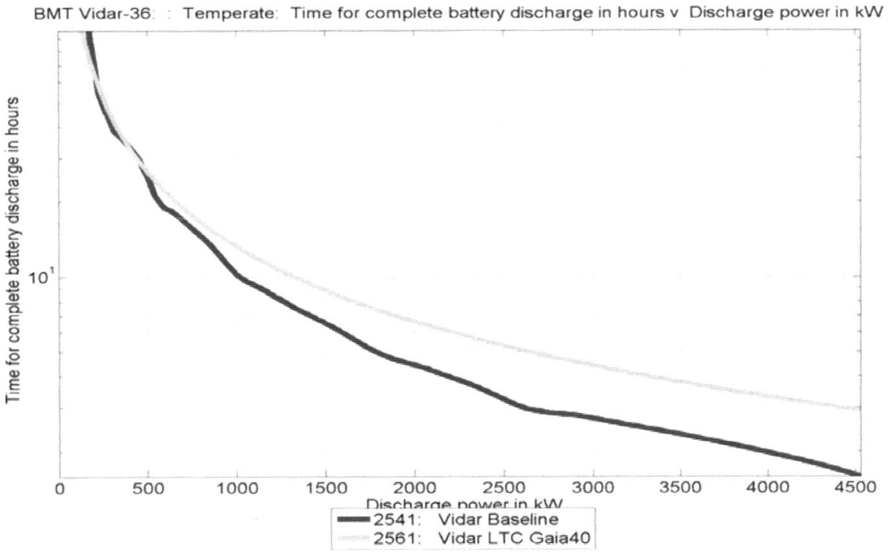

Figure 5: Time for Complete Battery Discharge in Hours

This extended battery capacity at higher speeds does not affect the range of the submarine which is defined by the original energy stored onboard when leaving port, i.e. the amount of usable fuel. Neither does this improved battery allow for the limits of indiscretion ratio to be altered as the ratio between time submerged and time to snort is the same for a given speed, the rate of energy consumption and the rate of battery re-charge are not changed.

However the ability to stay submerged for longer periods of time with a larger battery charge and with a greater scope to go at full speed for longer is a major operational and tactical benefit. The lithium ion batteries provide extra energy and therefore the potential for increased time and/or speed and travel distance between snorts.

This extra time is translated into significant range benefits at higher speeds as shown in Figure 6. At 20 knots, the baseline distance of 36nm is increased to ~50nm.

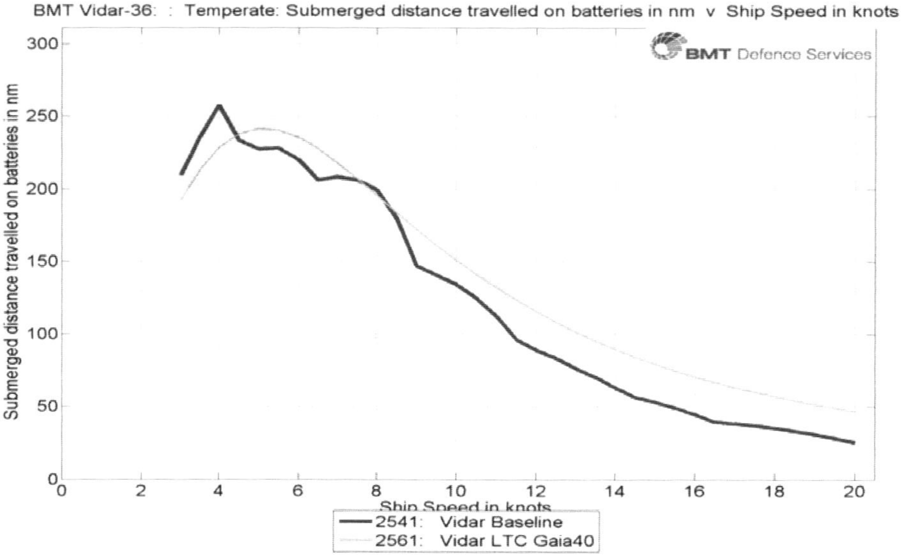

Figure 6: Submerged distance v ship speed

Figure 7 shows how the use a greater part of the stored battery energy can lead to longer snort times to fully recharge the battery. However in practice the larger and more flexible battery capacity gives the submarine commander more options on when to charge and for how long: he has less range anxiety!

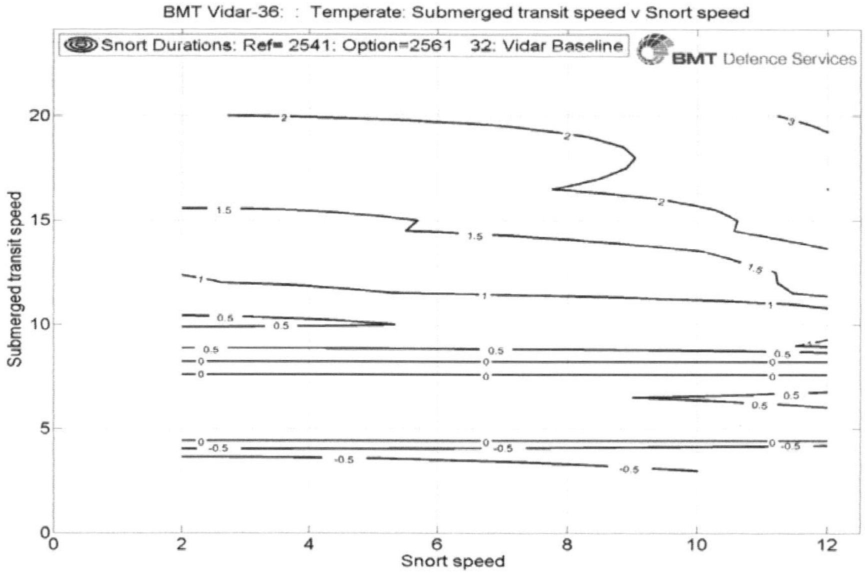

Figure 7: Increased Snort Time v Snort Speed

MODERN DG SET

Figure 8 shows the increased range offered by a DG set with the more powerful and more efficient MTU12V4000U83 engine.

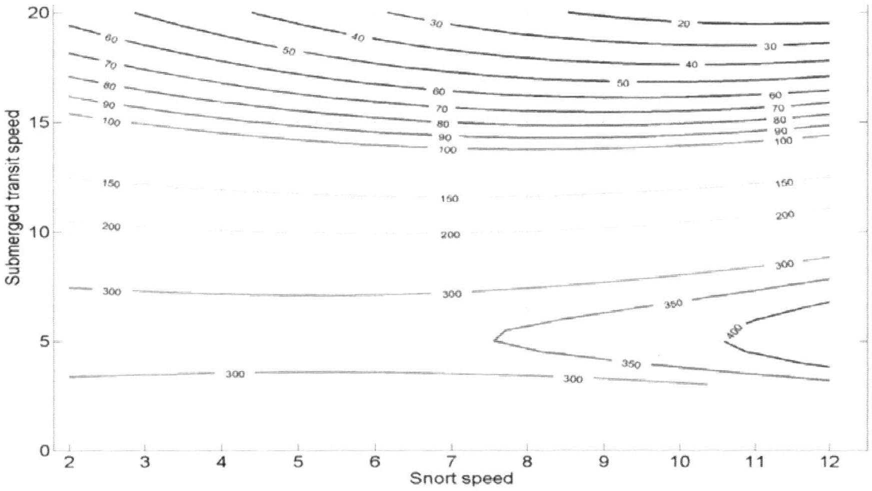

Figure 8: Range increase with MTU12V4000U83

Figure 8 shows how the new engine with its better Sfc allows the fuel to take the submarine further. The greater rated engine power allows the batteries to be re-charged more quickly with an IDR benefit of 3% at the standard operating condition.

CONCLUSIONS

The study has used the BMT Vidar-36 submarine design to explore the consequences of the use of different AIP plug sizes and a mix of battery and fuel provision as well as the use of lithium ion batteries and modern DG sets.

The considerations for mission and patrol times as well as operational range have been reviewed and the method of analysis has been demonstrated.

The changes to the AIP plug sizes had a small (2%) impact on submarine resistance and powering but the achievable operating range is sensitive to changes in DG set designs and the fuel provisions.

Lithium ion batteries offer a significant benefit in terms of the allowable times between battery re-charge and also allow for longer periods at top speed. A modern more efficient and more powerful engine provides a greater range and shorter snort times.

ACKNOWLEDGEMENTS

The opinions expressed in this paper are those of the author and not those of BMT Defence Services Limited, other companies, the UK MoD or any of its agencies. The author wishes to acknowledge the co-operation and able assistance of the BMT staff in the writing of this paper.

REFERENCES

1 J. Buckingham et al. (2008) "Submarine Power and Propulsion - Application of Technology to Deliver Customer Benefit" UDT 2008 Glasgow, Scotland.

2 Buckingham, J E et al. “Multi-engine Submarine Power Supplies: The Operating Case” Pacific 2012.

3 Yule, P et al “The Collins Class Submarine Story”. 2008. Cambridge University Press.

4 MTU Marine Sales Program. Edition 1/09.

5 MTU document . “Diesel engines in submarines” Jan-2008.

6 Magnet, J-L. "New Concept for Submarine Diesel Engines"

7 Bullard, Cdr J P RN et al. “The Type 2400 ’Upholder’ Class submarine”. JNE Volume. 28, No. 3.

8 Wrobel P G “Design of the Type 2400 Patrol Class Submarine”. RINA. 1984.

9 Sessions, Lt Cdr et al. “The mechanically supercharged Valenta diesel engine”. JNE Volume 29, No 3.

10 Mann, J “W. “Twin-Turbocharged Diesel Performance Under Snorkelling Conditions” UDT Europe 2011.

11 Binns, S. “The High Capability Conventional Submarine Solution Space.”. BMT Defence Services, RINA Warships June 2008: Naval submarines.

12 Binns, S. “Meeting the Current Challenge of Designing High Capability SSKs” Pacific 2008.

13 Torresan, H D and R A Jarvis. (2008) “Modelling of Conventional Submarine Power and Propulsion Systems for Future Capability Assessment”.. DSTO. SIA 2008. Sydney 2008.

14 Buchen, A. “Development And Integration Of Lithium Ion Batteries For Submarines”. LTC Gaia. UDT Europe 2008.

Exploring Options To Reduce Fuel Consumption

J Buckingham, CEng, FIMechE

BMT Defence Services Limited, Bath, United Kingdom

johnb@bmtdsl.co.uk

INTRODUCTION

Fuel Conservation

For any vessel, the ability to reduce fuel consumption at sea leads to:

- Reduced fuel purchase costs;
- A reduced logistical cost of fuel acquisition;
- Reduced emissions of all types.

For a naval replenishment tanker, reduced fuel consumption allows for longer lines of logistics support and the ability to operate at a range of speeds and an upper sustainable speed which matches fleet operations.

Increasingly the benefits of reduced engine emissions both for climate change and pollution legislation reasons are also a strong influence. Therefore, with the continuing upward trend of fuel costs, all those means by which fuel consumption can be reduced require consideration.

Those means which seek to reduce the fuel demand may also allow operational benefits possibly through an increased top speed (subject to adequate qualifications for machinery loading).

This paper presents a range of energy saving technologies in the context of the Royal Australian Navy's replenishment tanker, HMAS SIRIUS. The study seeks to show, from an independent viewpoint, how such technologies may be employed to achieve real financial, environmental and possibly operational benefits. The consequences for machinery performance is identified along with a consideration of the parasitic loads such features can require. The estimated acquisition and implementation costs are compared with the benefits to the ship's resultant in-service costs to show how the choice of an energy saving solution needs to be considered in conjunction with the ship's overall design, its machinery and its operating role.

Scope

Since the widespread use of propeller driven steel ships in the past 100 years, there has been a sustained but sometimes unsteady effort to identify technologies which offer to improve the energy efficiency of marine propulsion. As each ship and its operations are different and each technology has a specific design point, clearly not all technologies will suit each ship.

The current emerging requirement for improved energy efficiency has three main drivers: the increasing cost of energy, its impact on the environment and to a lesser extent, energy security. BMT has assessed the benefits of a set of relevant emerged and emerging technologies in the context of HMAS SIRIUS.

Context

BMT conducts power and propulsion studies for a wide range of commercial and naval clients. Often this is to improve propulsion efficiency and to promote energy conservation. BMT actively seeks to monitor those technologies which offer potential for these aims and maintains a database of over 150 technologies of varying degree of maturity. For each project, these are applied to a given platform and its operating profile to identify those which have the most merit, i.e. those with the greatest savings in terms of fuel and emissions at an appropriate acquisition cost.

For shipping as a whole, the price of fuel and its usage is likely to continue to rise due to the triple pressures of:

- Public and regulatory pressure on environmental emissions
- Peak oil is/will lead to higher fuel prices
- Rising world demand for energy.

With the world's known available fossil fuel reserves apparently dwindling, future prices will rise in real terms due to the anticipated future growth in real demand. However, rising fuel prices are nothing new: Figure 1 shows the situation in the 1970's which underlines how events such as wars and natural disasters can affect the prices of commodities.

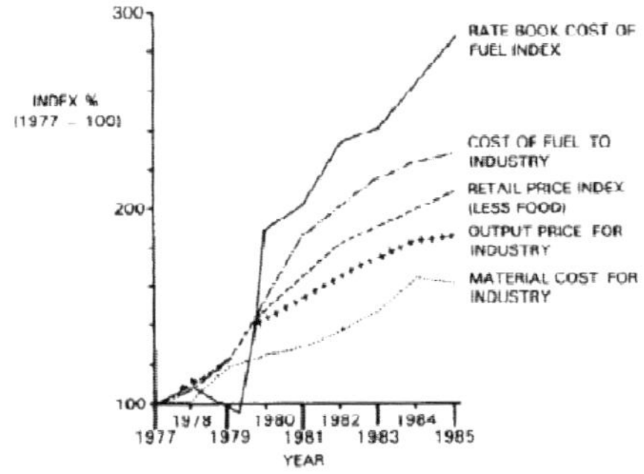

FIG. 1—COMPARISON OF A 'RATE BOOK' COST OF FUEL INDEX WITH SEVERAL IMPORTANT INDICES

Figure 1- Past Fuel Price Increases

ENERGY DEMANDS

For various reasons, nations have been struggling to meet the carbon reduction targets set by various to meet international agreements. Closure of inefficient coal-fired power stations has removed some large contributors but as the accountable emissions are identified and reduced, the contribution from shipping will come under greater scrutiny and pressure.

Alternative fuels such as LNG may be much discussed, but for military operations liquid fuel and specifically oil will still be the most important fuel in 2020, especially in transport due to the limited substitution options.

IMO

The International Maritime Organisation (IMO) has tabled measures to be introduced to limit or reduce shipping's emissions of CO_2. The IMO has laid out design guidance to enable designers to reduce marine-based Green House gases (GHG) (MEPC 58) which employs the concepts of:

- Energy Efficiency Design Index (EEDI);
- Energy Efficiency Operational Index (EEOI).

The EEDI seeks to simplify the machinery arrangements onboard a ship and uses a relatively simple equations to generate the mass of CO_2 emitted per tonne.nm (g. CO_2/tonne-nm) travelled. The EEDI is subject to much debate as it makes swingeing assumptions about the way in which power is generated and used onboard. The EEDI continues to be a disputed index for designers: it is contested that it cannot form the basis of a legally binding design objective as it does not consider the commercial aspects of ship operations.

The EEOI uses actual data from ship operations to identify the same metric. This index is clearly more accurate and as there are so many new ships due to the large production of the past ten years it is these which may benefit most from retro-fitting opportunities.

Although sea shipping is relatively "climate friendly" compared to other modes of transport when measured with the efficency indices, in absolute consumption terms, it is a considerable consumer of fuel, much of it is of poor quality and it is a significant source of air pollutants of all types.

In Europe, GHG emissions from land-based sources have decreased in the last 20 years, but emissions of GHG from sea shipping have been rising due to the increase in the global trading of goods. With unchanged legislation, it has been claimed fuel usage and emissions from international shipping will "more than double by 2020" within the European Union. Even if this is not so and the fleet stays the same, its fractional contribution will grow as other forms of transport become more efficient.

CARBON DIOXIDE EMISSIONS

In 2007, worldwide shipping CO_2 emissions were estimated to be of the order of 3.3% of the total with 2.7% of this being international shipping [Ref 1]. The shipping industry is increasingly expected to share in the burden of reducing global emissions of CO_2. Therefore, the shipping community needs to act now to take responsibility for their share of the GHG burden.

Shipboard

Fuel tankers due to their large size, consume significant quantities of fuel and so HMAS SIRIUS is used in this study as a basis for exploring the feasibility of a set of energy conserving technologies.

MACHINERY

Improvements to inboard machinery may offer the following benefits:

- Reduced Fuel Usage;
- Increased endurance between refuelling;
- Reduced machinery maintenance burden;
- Potential for reduced heat stress in the machinery spaces;
- Actual cost savings if the efficiencies outweigh the price increases;
- Helps save the environment.

However to use the source energy more efficiently, extra and/or more complex machinery is often required. There is therefore a real ship and upfront financial impact with a more efficient plant. With regard to the ship, the basic impacts are footprint, insertion and removal routes, maintenance access as well as the added complexity: interfaces, controls and interfaces to the ships control systems.

The new plant may also place additional demands on the ships services and a need for better heat management. Thus these demands may detract from the full benefits of the original energy saving measure.

Specifically additional weight from the machinery and its supporting structure increases displacement and hence resistance, unless the bunker fuel is reduced accordingly. There may also be stability aspects due to the location and the weight of the plant.

The opportunity for changes to existing ships is also a challenge as scheduled refits are planned to be short for economic reasons. The changes to the baseline machinery also lead to:

- Operating and support (upkeep) changes
- An increase in the manning and training burden.

All these issues affect the cost-benefit balance and so many of these financial impacts need to be factored into the assessment.

MAKING IT PAY

The cost of change and its impact on the return on investment is always an issue when decisions on whether to proceed with the introduction of an energy conservation technology. The size of the benefit needs to be significant therefore to make the effort worthwhile although unfortunately the need for a low-risk short-payback period often dominates the decision making assessment. The increased cost of fuel is making these payback periods shorter than ever.

Cost assessments of manning and training needs are complex but the main financial benefit is reduced fuel consumption. The efforts to simply reduce emissions measures do allow for a green image to be presented and until carbon taxes are enforced, it is fuel reduction that saves the money and allows for the savings to recover the outlay costs. The main criterion used in the study was the balance of through life fuel savings to the estimated changes to the acquisition costs.

HMAS SIRIUS

Context

BMT Defence Services do not have access to the whole set of machinery and resistance data related to the HMAS SIRIUS. With the limited available data on the MT DELOS as she was once known and the HMAS SIRIUS and also using first principles, BMT has developed a definition of the ship to allow a Power and Propulsion assessment to be undertaken. To ensure the assumed data is clear to the reader, it has been shown in italics.

Description

The HMAS SIRIUS is a converted product/chemical tanker formerly known as the MT DELOS. The original vessel was rated at product tanker of 37,000 tonnes deadweight with a service speed of 15 knots at the design draft of 9.0m. The ship is 175.93m overall with a beam of 31m.

The estimated resistance of the hullform was derived using Holtrop and Mennen [Ref 2[with suitable assumptions for appendages. The single four-bladed fixed pitch propeller of diameter 5.8m. It is assumed to have a blade area ratio of 0.6025 and a pitch-to-diameter ratio of 0.735. The Wageningen 4B blade series was used to model the propeller behaviour.

The single main engine is a MAN 6S50MC which is rated at 11,640 bhp at 127 rev/min, 100%. Engine power is 9,890 bhp at the normal service speed with a shaft speed of 120 rev/min.

There are three Diesel Generating sets (DG sets) to supply the ships electrical power demand of *800*kWe seagoing, and in port discharging *2,000kW* (max discharge rate).

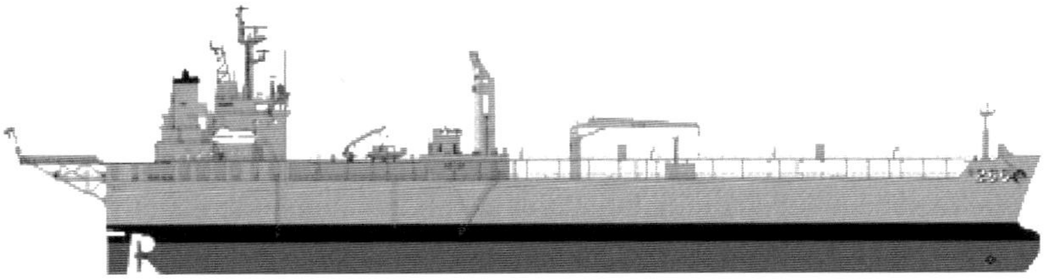

Figure 2 - Replenishment Tanker: HMAS SIRIUS [1]

Operating Profile

The assumed ship's time-speed operating profile is shown in Figure 1. This profile considers the principal operating speed but also the small time spent operating in confined waters and when performing operations with craft of different sizes and speeds.

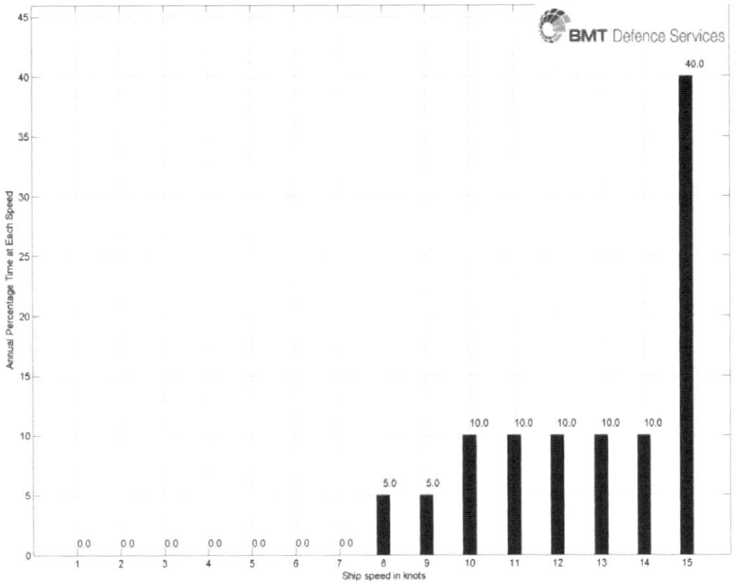

Figure 3: Operating Profile

TECHNOLOGIES

Whole Set

A set of over 100 technologies were reviewed to identify a sub-set which were most applicable to the ship under study. The benefits for each technology were identified using first principles energy-based assessments together with a consideration of the feasibility of retro-fit in terms of ship impact and the likely increases in the upkeep burden. The set technologies identified in Table 2 were assessed against the operating profile.

---

1. Copyright Australian DoD

Drag Reduction Or Propulsion Assist	Drives & Power Sources
Air Lubrication	Exhaust Gas Waste Heat Economiser
Skysails	Organic Rankine Cycle
Wing-sails	Wind Turbines
Flettner Rotor	Photo-voltaic solar panels
Mewis Duct	Larger Engine

Table 2 - Technology Merit Set

Some of these technologies are described in detail below.

Air Lubrication

Although a full air layer approach to air lubrication is becoming a popular area for study, the more limited use of air in the principle of Microbubble Drag Reduction (MBR) is a more conservative approach. MBR involves the injection of very small air bubbles into the boundary layer of ships to reduce friction drag. Since 1973 [Ref. 3], extensive studies on drag reduction using micro-bubbles have allowed their effectiveness to be assessed [Ref. 4]. The detailed research and methods used were addressed in a previous BMT paper [Ref. 5]. Also known as an Air Lubrication System (ALS) it has been fitted to ships from MHI and Maersk have their own WingedAir Induction Pipe (WAIP) system which has limited success [Ref 6)].

Figure 5 shows how the resistance varies with speed after microbubbles have been introduced. The rate of air supply is proportional to the ship's speed and the application area by a factor 1e-4 in accordance with the guidance of Kodama [Ref. 7].

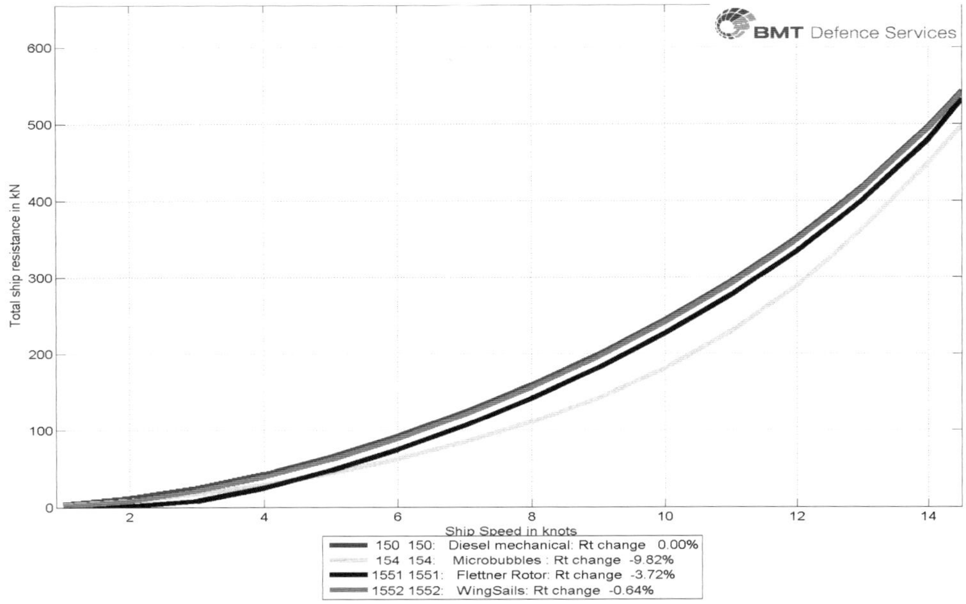

Figure 4: Resistance Reduction

Figure 4 shows the benefit of microbubbles for an air supply provided at a pressure of ~1 barg. The power for the air compressor would go some way to offset resistance power saving. Therefore a microbubble solution is best suited to a ship which has a surplus of compressed air.

Flettner Rotors

Flettner rotors use the Magnus effect to provide a lateral thrust from a spinning vertical cylinder placed in a moving air stream. The effect was first used in the 1920's to propel a ship [Ref. 8]. WindAgain Ltd of Singapore now offer modern versions. BMT designed three cylinders, 15m tall, 1.1m diameter to act as Flettner rotors. Their performance calculated from first principles for a range of wind speeds to make an estimate of their effect. The wind spectra and full azimuth of wind directions was considered for each ship speed.

Photo-Voltaic Solar Panels

Photo-voltaic solar panels have already been fitted to ships to provide auxiliary power. This is an area of fast-moving technical development with the latest efficiency figures exceeding 13%. A 10% efficiency figure with a pessimistic annual solar radiation of 900kWh/sq.m/year has been used to derive the average power supply from the onboard installation using 90% of the free weather deck area: in practice the super-structure roof could also be covered.

Wind Turbines

To quantify the scope for harvesting useful energy from the wind that strikes the port and starboard sides, a 2m by 2m wind turbine design was developed. The unit provides a modest average power of 0.5kW when wind speed variations and ship's heading factors have been factored in. This is considered to be a conservative yield with plenty of scope for further efficiency improvements. *** units are located each side.

Sky Sails

Sky Sails from the German company of the same name are simple, light and relatively low expense. Real benefits have already been realised but the performance is clearly dependent on the ship's heading relative to the wind and for ship's operating fixed routes this compromises the benefits.

At 15 knots with the wind aft the savings can be as much as 10%. For faster vessels the ship speed may exceed the wind speed and the benefits will be much lower. A 400m^2 design with a 300m tether has been modelled using weather data for the North Atlantic.

Wing Sails

The current Wing Sails market offering is from Shadotec/WingSail PLC. BMT has used their own variant of a wing-sail for this study: it comprises 3 blades each 15m high. A model of this has identified the kind of performance that might be achieved with North Atlantic wind spectra for a range of ship speeds and the whole circle of headings. The wind speed and ship heading was used to identify the average thrust from the wing-sails for each ship speed. Installation capability is limited by the necessary clearance height between the bridge top and the lowest bridge it has to pass under.

Exhaust Gas Waste Heat Recovery

Exhaust gas waste heat recovery (EGWHR), is an energy recovery method whereby the heat in the engine exhaust is used to create steam which can then be used for:

- Space heating;
- A feed to a steam turbine generator to produce electrical power.
- A feed to an absorption chiller to reduce the load on the chilled water plant.

An EGWHR can allow for the effective use of fuel into the engine to be increased by 12% or so. This technology is widely used in ships such as the Queen Mary 2 and in offshore applications. As of the order of 35% of the engine's waste heat is in the exhaust gas this is an obvious source of recoverable energy.

The MAN Thermo Efficiency System [Ref. 9] is one example of a modern means of saving energy from a two-stroke design. A fraction of the exhaust gas is bypassed from the turbochargers and fed directly to a turbine driving an alternator. This turbine is combined on a shaft with the steam turbine with a reduction gear and over-speed clutch between the high speed exhaust gas turbine and the slower steam turbine. Energy savings have been estimated using first principles and considering the effectiveness of heat exchangers and the efficiency of the Rankine cycle.

For the SIRIUS the MAN 5S60MC engine exhaust gas temperature is low at 245°C. This provides a wet steam issue from the steam generator. This steam enters a fuel-fired super-heater to raise it to 8 bar, 170°C before it enters the steam turbo-generator rated at 1,800kWe.

At slow ship speeds the exhaust mass flow is low the operation of the steam generator facility may not be feasible or indeed viable. As the ship's speed increases, the increasing load on the engines allows more heat to be available for steam and hence electrical power generation. The power generated is a bonus even though the Rankine cycle is only about 30% efficient.

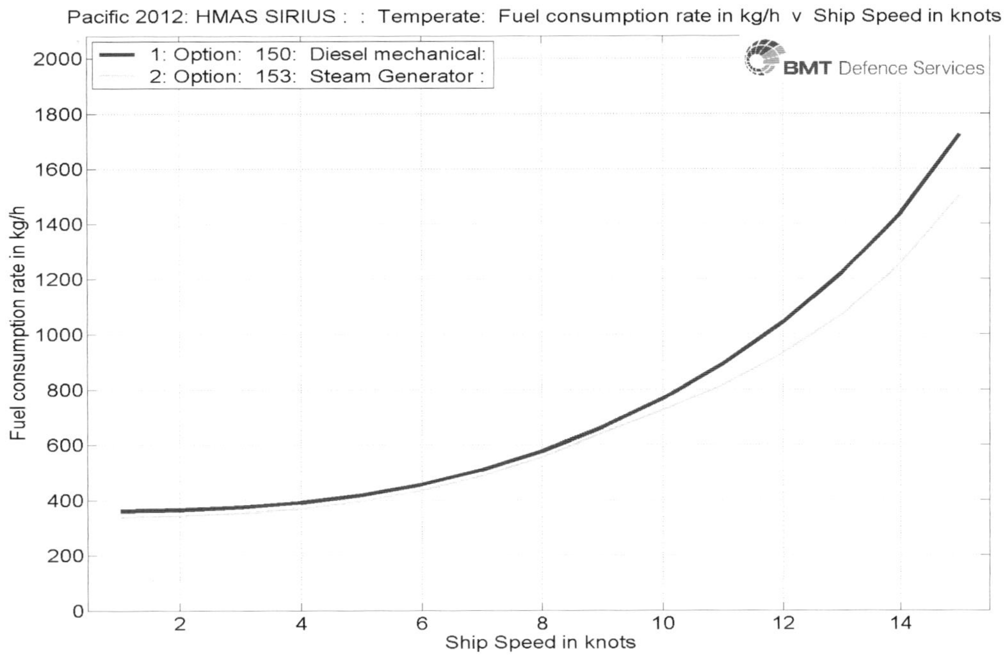

Figure 5: Steam Generator Fuel Benefit

Organic Rankine Cycle

An Organic Rankine Cycle (ORC) driven by the exhaust gases may allow for higher efficiencies depending on the design parameters. The heat transfer to the ORC system could be via thermal oil. Again thermal oil operates at 300 degrees C so an additional heater would be required.

TECHNOLOGY COMPARISONS

Figure 6 shows how the steam generator with the turbo-genset offer significant fuel savings with the SkySails and the Microbubble feature also offering a tangible benefit.

If these features are all added as a composite design then it is estimated that over 20% fuel savings can be achieved.

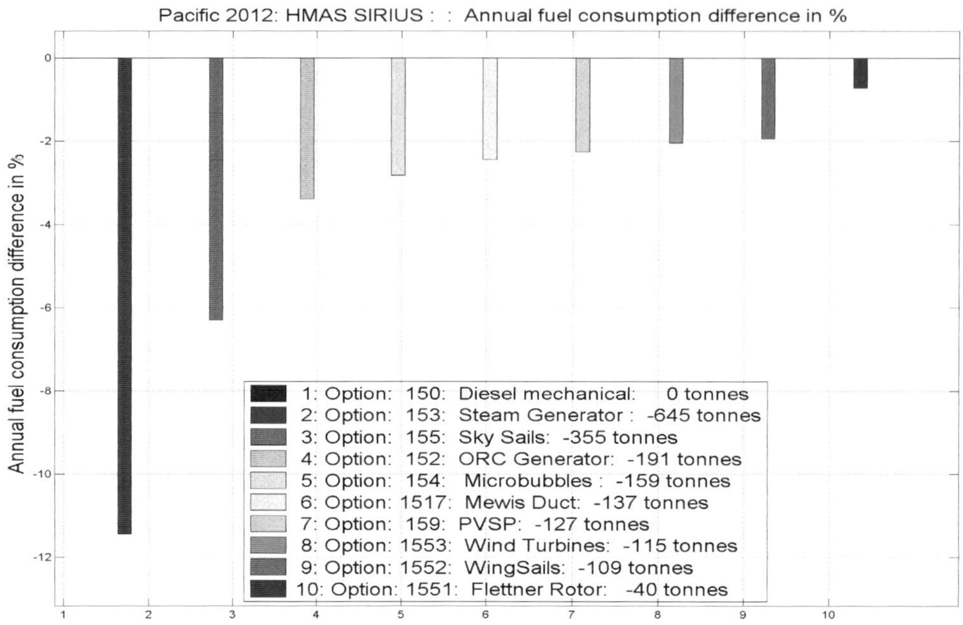

Figure 6: Annual Fuel Consumption Benefits

Figure 7. shows the estimated balance between the cost of acquiring and inserting the new technologies versus the accrued financial benefits through life. The large fuel saving of steam turbo-genset make it attractive despite it high initial costs which are recovered within a few years assuming no change in the fuel price. The Mewis Duct is also a cost-effective solution.

Other technologies such as microbubbles are marginally not cost-effective but future changes to fuel costs would affect this outcome.

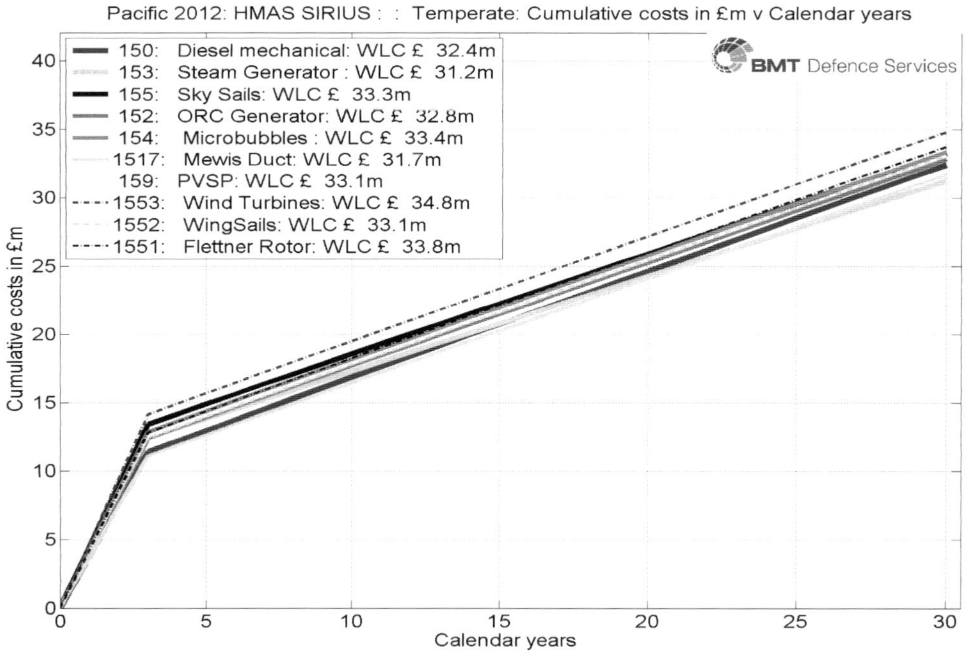

Figure 7:. Baselined Whole Life Costs

HIGHER TOP SPEED

BMT are aware of comments made about top speed of the vessel, since selection of the MT DELOS and then the introduction into service of the HMAS SIRIUS. A higher upper sustained speed was originally sought by the RAN but during the process of ship selection a decision was made to obtain a slower ship. A review of the technologies considered for this vessel indicate that the drag reduction and propulsion assist features do not offer any significant benefit to achieving a higher upper sustained speed.

BMT then considered the scope for replacing the current engine with one that is two cylinders longer. The ship's general arrangement drawings give an approximate indication of the size of the engine room and an engine with two extra cylinders was considered to be feasible in terms of engine length. No assessment was made of the scope for the propeller shaft to absorb a higher power as such detailed information was not available to BMT. Such a new engine would allow a revised top speed of about 16.5 knots.

CONCLUSIONS

The greatest scope for fuel saving onboard a ship is usually with the Power and Propulsion system: here there is more scope to render the main engine and the propellers more efficient to yield significant fuel savings.

The adoption of an exhaust gas waste heat recovery steam generator together with SkySails and Microbubbles offers scope for significant fuel savings and a reduced carbon footprint.

However each candidate technology needs to be considered with the specific operating profile of the ship and the method used by BMT allows changes of ships loads and usage to be assessed quickly to yield information on the comparative benefits.

ACKNOWLEDGEMENTS

It is BMT's intention to include copyright with the submitted paper. However BMT acknowledge Pacific 2012's request for permission to publish the paper with their proceedings. The kind permission and time granted to the author by BMT is acknowledged with thanks. All ideas, opinions and errors herein are those of the author.

REFERENCES

1 IMO, 2009. *Second IMO GHG Study 2009*. International Maritime Organization (IMO) London, UK.

2 Holtrop and Mennen

3 McCormick, M.E., Bhattacharyya, R., 1973, "Drag Reduction of a Submersible Hull by Electrolysis", Naval Engineers Journal, Vol.85, No.2, pp. 11-16.

4 Fontaine, A.A., et al., "The Influence of the Type of Gas on the Reduction of Skin Friction Drag by Microbubble Injection," Experiments in Fluids, Springer-Verlag, vol. 13, No. 2/3, pp. 128-136, 1992.

5 Buckingham J E. "Energy Conservation: Matching Technologies to the Platform". MAST Sep. 2009. Stockholm.

6 "Maersk's Bubble Bursts" The Naval Architect . April 2011.

7 Kodama, Y et al "Practical application of microbubbles to ships - large scale model experiments and a new full scale experiment".. 6th International Symposium on Smart Control of Turbulence, Tokyo, March 2005.

8 Seybold, G.B. "A Sailing Ship Without Sails: New Wonder of the Seas" . Popular Science Monthly (February 1925). http://www.rexresearch.com/flettner/flettner.htm

9 MAN Diesels: Thermo Efficiency System (TES) for reduction of Fuel Consumption and CO2 emissions. 2005.

http://www.manbw.com/files/news/filesof5055/P3339161.pdf

AUTHOR'S BIOGRAPHY

John Buckingham is the Chief Mechanical Engineer at BMT Defence Services, Bath, UK. He joined BMT in 1992 after 8 years with Vickers Shipbuilding and Engineering Limited, Bath. John has a BSc in Engineering with French and an MSc by research in hydraulics and simulation; both at the University of Bath.

John leads the development of tools, information and methods for mechanical engineering aspects within the Company and has published papers on a range of subjects from waterjet modelling, biofuels and rail-gun cooling through to the analysis of warship and submarine power and propulsion systems. He has recently been working on the identification and assessment of technologies for energy efficient ships.

Sensitivity of Submarine Performance to Power and Energy System Parameters

Robert A. Jarvis
Defence Science and Technology Organisation, Melbourne, VIC, Australia
robert.jarvis@dsto.defence.gov.au

1. INTRODUCTION

The performance of any naval platform can be measured using a number of different metrics to quantify the platform's capabilities when operating in a specific manner. While metrics can be constructed for a variety of platforms and can cover many aspects of performance, this paper will focus on submarine metrics that depend heavily on the power and energy system of the submarine. Examples include range and endurance in particular modes of operation, indiscretion ratio, and the maximum possible time that can be spent in an area of operation.

The Defence Science and Technology Organisation (DSTO) have developed a set of modelling tools used to investigate platform power and energy systems. This paper outlines the application of a global sensitivity analysis to the latest version of the DSTO Submarine Power and Energy System Static Model, first introduced in [1]. A global sensitivity analysis explores the entire range of acceptable parameter values at once, without being confined to a localised exploration around a specific design point [2]. Therefore, this paper presents one method of performing a wide-ranging investigation of the link between performance metrics and power and energy system parameters, while avoiding the restriction of defining a 'starting-point' submarine around which to modify the parameters.

1.1. Scope

The scope encompasses high-level power and energy system parameters, and compares their level of influence on a selection of performance metrics when the parameters are varied within a generic range of values. The aim is to investigate sensitivity in a generic and global sense, where no specifications have yet been made to constrain the boundaries of a submarine design. Hence, the parameter ranges are kept as wide as possible while attempting to stay within the bounds of reason. Sensitivity to submerged displacement is incorporated into the analysis, but the parameter is varied independently, with no attempt made to size the submarine based on the size of the power and energy system.

1.2. Sensitivity Analysis

Sensitivity analysis seeks to describe how variation in the outputs of a model can be linked to variation in the different inputs of the model [3]. Assuming that the model used is a reasonable representation of the power and energy system of a submarine, the sensitivity analysis undertaken on the inputs and outputs of the model can be considered indicative of the actual link between the variations in the power and energy system parameters (inputs) and the performance metrics (outputs) of a real submarine.

The two objectives (also called 'settings') of the sensitivity analysis in this paper were [2]:

1. **Factor fixing** – determining which parameters are non-influential on a performance metric, and hence could be fixed at any value *within their specified range* without affecting the value of the metric
2. **Factor prioritisation** – ranking and comparison of the level of influence of the remaining parameters of notable influence for that performance metric

Using a variance-based method of sensitivity analysis, the two sensitivity indices of interest in addressing the above objectives are the 'first-order index' (S_i) and the 'total-effect index' (S_{Ti}) for each input parameter (X_i) with respect to each output (Y). These are defined as [2]:

$$S_i = \frac{V_{X_i}\left(E_{X_{\sim i}}\left(Y \mid X_i\right)\right)}{V(Y)}, \tag{1}$$

$$S_{Ti} = \frac{E_{X_{\sim i}}\left(V_{X_i}\left(Y \mid X_{\sim i}\right)\right)}{V(Y)} = 1 - \frac{V_{X_{\sim i}}\left(E_{X_i}\left(Y \mid X_{\sim i}\right)\right)}{V(Y)}, \tag{2}$$

where $X_{\sim i}$ represents all input parameters except for X_i; $Y \mid X_i$ and $Y \mid X_{\sim i}$ are the values of output Y conditional on the input parameters X_i or $X_{\sim i}$ being fixed; and $E(\cdot)$ and $V(\cdot)$ are respectively the mean and variance taken over the set of input parameters that is indicated by their subscripts, or over all results if no subscript is used. The values of both indices fall in the range from 0 to 1, with a higher value suggesting greater sensitivity and influence.

If a scatter-plot was constructed of output Y against input parameter X_i, the amount of defined 'shape' in the cluster of points could be considered to indicate level of influence of the input parameter. The first-order sensitivity index of (1) is one measure of this shape, which looks at how the mean value of the output Y at any X_i varies across the range of the input parameter, but does not take into account interaction effects with other input parameters. The total-effect sensitivity index of (2) attempts to capture the full influence of a parameter by measuring the variance in the output Y, taken over the range of values that parameter X_i can assume, when all other inputs except the parameter of interest are fixed, and with the variance then averaged over all possible values to which those other parameters can be fixed. By its design, the total-effect index takes into account interactions between input parameters that are missed by the first-order index.

The total-effect sensitivity index is used to address the factor fixing objective, as parameters with low values of S_{Ti} can be considered non-influential on the performance metric of interest. The remaining parameters can then be ranked and compared in their level of influence using their first-order sensitivity indices, with higher values of S_i indicating greater influence. However, while (1) and (2) are the formal definitions of the sensitivity indices, use of these equations is computationally expensive [2]. The indices can instead be accurately approximated with a streamlined process, proposed by Saltelli [2], which is used in this paper.

1.3. Submarine Power and Energy System Model

In order to investigate the sensitivity of submarine performance metrics to changes in platform power and energy parameters, a submarine propulsion system model was required. For this investigation, DSTO's established Submarine Power and Energy System Static Model was used [1]. The model is called 'static' as it does not include transient events or time-varying quantities. Instead it represents the power and energy system at a high level, modelling only basic steady-state energy and fluid flow through the system. The model is load-driven, with each component in turn demanding an input power or mass flow to meet its output requirement according to an empirical equation or lookup table that relates the output to the input. Information flow originates at the control system, which monitors and sends signals from and to each of the components, providing control of the operational state of the submarine using configurable state-machine logic.

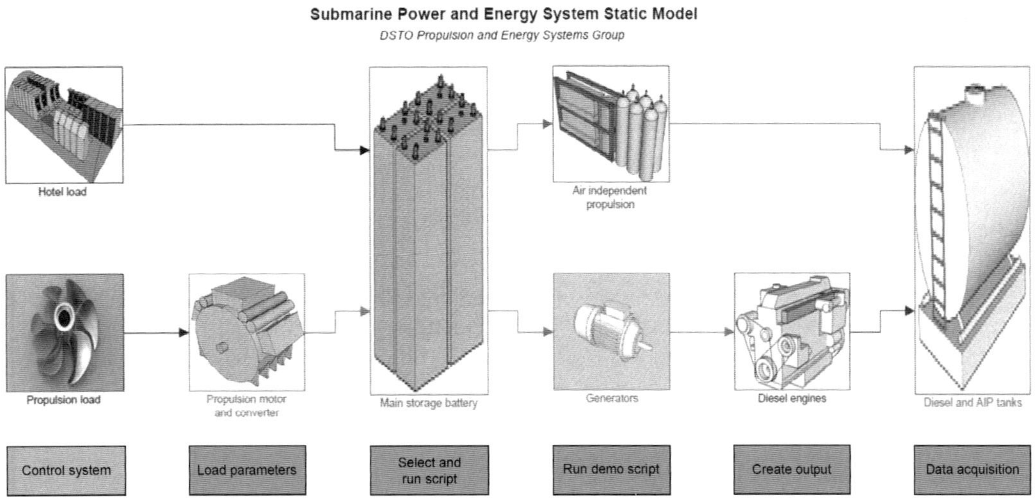

Figure 1 DSTO Submarine Power and Energy System Static Model

2. INPUTS: POWER AND ENERGY SYSTEM PARAMETERS

The input power and energy system parameters used in this analysis are listed in Table 1, along with the generic range of values it was estimated that they could reasonably assume. The sensitivity analysis used a uniform distribution of each parameter within its range, and the value of each parameter was assigned independently of the other parameters.

Table 1 Power and energy system parameters and their range of variation

Element	Parameter	Range
Whole submarine	Submerged displacement	1500 – 5000 t
	Mission endurance	40 – 80 days
	Hotel load	50 – 350 kW
Propulsion motor	Efficiency at very low output power	70% – 85%
	Efficiency at low output power	90% – 98%
	Efficiency at rated output power	90% – 98%
Main storage battery (MSB)	Energy capacity at very low discharge power	5000 – 50 000 kW·h
	Fraction of capacity at very high discharge power	10% – 95%
	Charging efficiency	75% – 99%
Air independent propulsion (AIP)	Rated output power	0 – 600 kW
	Max. system efficiency	50% – 90%
	Efficiency loss at very low output power	25% – 45%
	Efficiency loss at rated output power	0% – 25%
Diesel generator sets	Total combined output power	1500 – 7500 kWe
	Specific consumption of diesel	220 – 300 $g{\cdot}kWe^{-1}{\cdot}h^{-1}$
Tanks	Mass of diesel fuel	100 – 600 t
	Mass of AIP hydrogen fuel	1 – 8 t

In addition to the bounds defined in Table 1, and using the generic guidelines from [4], the submarine was assumed to always have a maximum speed of 20 knots, and a maximum snorting speed of 10 knots. The propulsion motor's rated output power was always scaled to meet the maximum speed requirement. Mission endurance is the maximum time the submarine can be at-sea. The air independent propulsion (AIP) system was assumed to have an efficiency curve shaped similar to that of a proton exchange membrane fuel cell, to consume hydrogen and oxygen, and to produce only water as a product with oxygen and product water tanks sufficient for complete consumption of the AIP hydrogen fuel. The diesel-generator sets were modelled as one complete system (i.e. no distinction made for individual engines or generators) that consumed diesel to produce electrical power. The diesel fuel and the energy capacity of the main storage battery (MSB) were the usable quantities when at-sea (i.e. after the application of diesel operational reserves and MSB state-of-charge limitations). The efficiency curves of the propulsion motor and the AIP system were constructed from three points, with values between these points determined via linear interpolation. The efficiencies of the motor operating points were varied absolutely, along with the AIP system maximum efficiency, while the efficiencies of the other points were determined as losses relative to the system maximum.

The usable energy capacity of any battery reduces as discharge power is increased, due to internal losses that increase with power. In this paper, the submarine's MSB energy capacity was assumed to decrease exponentially with discharge power. The baseline energy capacity was set by the parameter that represents the MSB energy capacity at very low power. The 'flatness' or 'steepness' of the exponential decrease was then controlled by varying the MSB parameter that represents the fraction of the low power capacity that can be recovered at a very high discharge power. These two points were then fitted with an exponential function to determine the energy capacity at discharge powers between the two extremes.

3. OUTPUTS: SUBMARINE PERFORMANCE METRICS

This paper used performance metrics for a number of different modes of submarine operation, as defined in Table 2. The first three modes and the unconstrained platform range were limited only by energy resources, while constrained range and time in operational area were limited both by energy and by the time limit imposed by the mission endurance [1].

Table 2   Modes of submarine operation

MSB only	Submarine operating powered solely by the MSB
MSB-snort cycling	Submarine alternately fully discharging the MSB, then fully recharging the MSB by snorting
AIP and MSB	Submarine operating with power sourced preferentially from the AIP, with any additional power coming from the MSB either because the load is greater than the max. power of the AIP, or because the AIP resources have been consumed. Operation terminates when the MSB is fully discharged.
Platform range	Submarine going as far as possible by MSB-snort cycling and using AIP, travelling at the best speeds to maximise range, either unconstrained or constrained by the time limit of the mission endurance
Time in operational area	Submarine transiting to operational area by MSB-snort cycling, staying as long as possible (first consuming AIP resources, then MSB-snort cycling, all at patrol speed), then transiting back to base

The investigated performance metrics associated with each mode of operation are presented in Table 3. Generic patrol and transit speeds of 4 knots and 10 knots were selected based on [5] and [6] respectively. As stipulated previously, 20 knots was chosen as a generic maximum speed using the guidelines in [4]. Indiscretion ratio was the fraction of time spent snorting during one complete MSB-snort cycle, with the minimum achievable indiscretion ratio being a function of the average speed over the cycle.

Table 3   Performance metrics for each mode of operation

MSB only	**MSB-snort cycling**	**AIP and MSB**
Max. range, any speed	Min. indiscretion ratio, patrol ave speed	Max. range, any speed
Endurance, patrol speed	Min. indiscretion ratio, transit ave speed	Endurance, patrol speed
Endurance, transit speed		
Endurance, max. speed		
Platform range	**Time in operational area**	
Unconstrained	Near operational area	
Constrained by mission endurance	Far operational area	

4. RESULTS AND DISCUSSION

The following results demonstrate the application of a variance-based global sensitivity analysis to the DSTO Submarine Power and Energy System Static Model, using the parameters and performance metrics defined in the previous sections. The point at which a parameter is judged non-influential or is considered to have only a minor or marginal influence varies between metrics, since it comes from a contextual comparison of all the sensitivity indices for the metric of interest. For a particular metric, all non-influential parameters have approximately the same low values of S_i and S_{Ti}, while those with a minor influence have slightly larger values that are just enough to make them stand above the non-influential parameters. Parameters of significant influence have values much greater than all others, while marginally influential parameters fall somewhere in-between. For the full set of calculated sensitivity indices refer to Table 4 in the conclusion.

4.1. MSB Only Performance Metrics

Table 4 shows that only four of the seventeen parameters have a notable influence on the MSB only performance metrics as shown in Figure 2. The raw modelling results indicated that maximum range is usually achieved around the patrol speed, which appears to be confirmed by the similarity shown in the sensitivity results of the maximum range and endurance at patrol speed metrics.

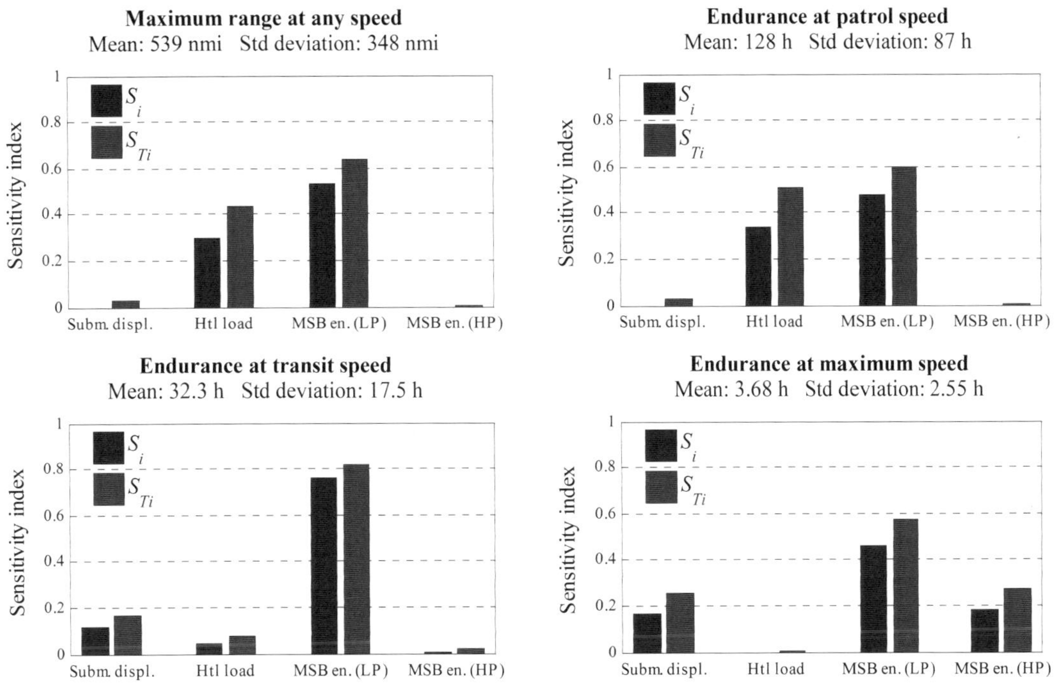

All other power and energy system parameters were found to be non-influential, each with $S_{Ti} < 0.01$.

Figure 2 Results for the MSB only performance metrics. The parameters of notable influence are submerged displacement (Subm. displ.), hotel load (Htl load), MSB energy capacity at low power (MSB en. (LP)), and the fraction of MSB capacity at high power (MSB en. (HP)).

The most influential parameter on all four of these metrics is the MSB energy capacity at low power, which is the parameter that defines the baseline capacity of the MSB. It remains influential at high speeds (high discharge powers) because the energy capacity at high power was defined as a fraction of the low power capacity. These results indicate that the first consideration when attempting to increase range and endurance when operating solely on the MSB should be to increase the energy capacity of the MSB, either by specifying a larger MSB or by moving to a battery chemistry with higher energy density. Additionally, the flatness or steepness of the decline in capacity as discharge power is increased, which is a function of the battery's chemistry and design, is an important consideration for maximising endurance during high speed sprints, but will have little effect on the low speed performance.

The parameters of next greatest influence are those that define the total system load: submerged displacement and hotel load. Propulsion load is modelled as being proportional to submerged displacement to the power of 2/3 (based on an equation from [5], and similar to another found in [4]). Over the parameter ranges used in this paper (Table 1), at patrol speed the hotel load is usually a few times the size of the propulsion load, while at transit speed the situation is reversed. At maximum speed, the hotel load is always dwarfed by the propulsion load – usually by an order of magnitude or more. This is reflected in the declining influence of hotel load and increasing influence of submerged displacement as speed is increased, although the influence of the submerged displacement is dampened by its 2/3 power relationship to propulsion load. The results suggest that reducing hotel load at low speeds and reducing submerged displacement at high speeds should be secondary considerations after maximising MSB energy capacity when attempting to increase these performance metrics.

4.2. MSB-Snort Cycling Performance Metrics

Table 4 shows that only five of the seventeen parameters have a notable influence on the MSB-snort cycling performance metrics as shown in Figure 3.

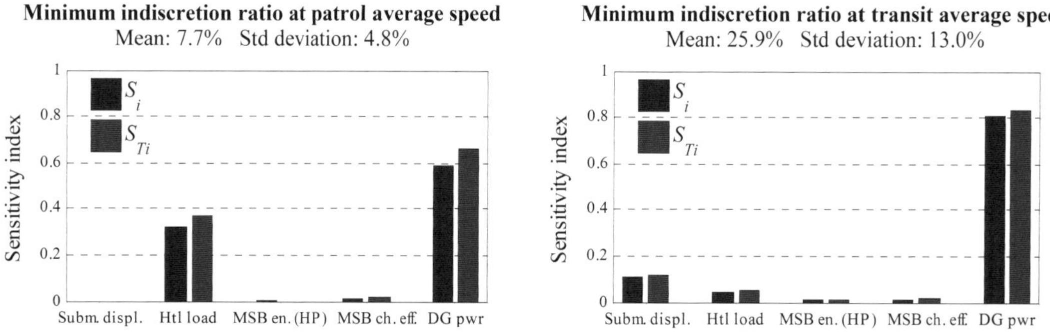

All other power and energy system parameters were found to be non-influential, each with $S_{Ti} < 0.01$.

Figure 3 Results for the MSB-snort cycling performance metrics. The parameters of notable influence are submerged displacement (Subm. displ.), hotel load (Htl load), fraction of MSB energy capacity at high power (MSB en. (HP)), MSB charging efficiency (MSB ch. eff.), and the total combined power of the diesel-generator sets (DG pwr).

As shown in Figure 3, the results imply that the most direct approach to reduce indiscretion ratio is to increase the power from the diesel-generator sets, either by installing more sets or by increasing the power delivered by each set. Some improvement can be made by reducing the total system load, which is best achieved by reducing hotel load at low speeds, or by reducing submerged displacement at higher speeds. Only minor reductions are possible by making changes to the MSB, with the absolute energy capacity of the MSB being unimportant, as demonstrated by the non-influence of MSB energy capacity at low power, which is the parameter that controls the baseline MSB energy capacity. MSB charging efficiency and the fraction of capacity usable at high power both play a minor role, since both reduce the quantity of energy that can be removed from the MSB on discharge relative to the energy consumed during snorting.

Fundamentally, indiscretion ratio is a measure of how quickly energy can be returned to the MSB when snorting relative to how quickly energy is removed during discharge. By definition as a ratio, the quantity of energy transferred to and from the MSB plays no role. Therefore, indiscretion ratio is most affected by changes to the rates of energy transfer, which are altered by changing the power of the diesel-generator sets or by changing the hotel and/or propulsion loads. Over the defined parameter range for diesel-generator power (Table 1), the maximum value is five times larger than the minimum, while the maximum hotel load value is seven times larger than its minimum. When combined with the propulsion load, the maximum variation they can create in the total system load is reduced to a factor of about five at patrol speed, and a lot less at higher speeds. Submerged displacement creates even less variation, changing the total system load by a maximum factor of about two at transit speed, and significantly less at lower speeds. Therefore, when averaged across the sample space of parameters, the influence of the power of the diesel-generator sets appears greater than the influence of the hotel load and submerged displacement.

4.3. AIP and MSB Performance Metrics

Table 4 shows that only four of the seventeen parameters have a notable influence on the AIP and MSB performance metrics as shown in Figure 4. As with the MSB only metrics, the results are very similar for the maximum range and the endurance at patrol speed metrics, again suggesting that maximum range is usually achieved when operating near patrol speed.

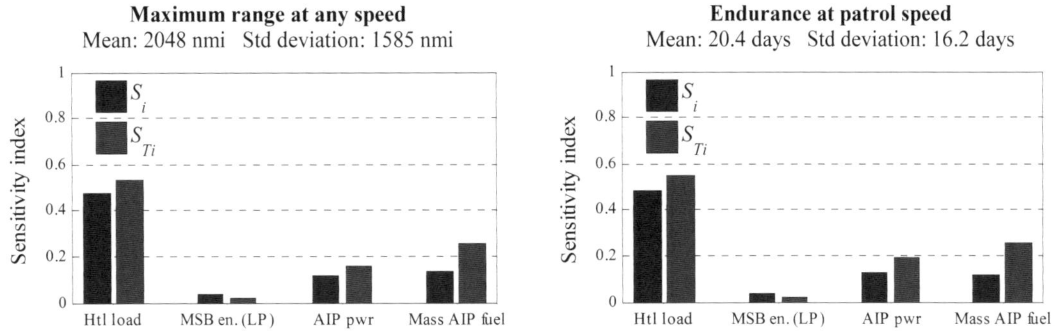

All other power and energy system parameters were found to be non-influential, each with $S_{Ti} < 0.01$.

Figure 4 Results for the AIP and MSB performance metrics. The parameters of notable influence are hotel load (Htl load), MSB energy capacity at low power (MSB en. (LP)), AIP system rated power (AIP pwr), and mass of AIP fuel (Mass AIP fuel).

To understand the large influence of the hotel load, recall from Table 2 that any power that cannot be supplied by the AIP comes from the MSB – either because the AIP system has consumed its energy resources, or because the total system load is too great for the AIP alone. Then once the MSB is fully discharged, this operational mode terminates. Additionally, these are low speed metrics, and it has been previously discussed how the hotel load tends to drive the system load at low speed. Therefore, the parameters of hotel load and AIP power together define the point at which excess power must be supplied by the MSB. If the power draw on the MSB is large, this leads to a rapid discharge, and greatly restricts range and endurance. If the hotel load is small, or the AIP power is large, the MSB does not discharge until the AIP resources are consumed, and the full capability of the AIP system can be exploited. In between these extremes, the hotel load is still the primary parameter that determines the rate of energy consumption. The relative influence of AIP fuel to that of MSB energy capacity at low power is indicative of the extra range and endurance that can be achieved by additionally utilising AIP in comparison to just using the MSB alone. To illustrate, refer to Figure 2, where the mean maximum range and mean endurance at patrol speed using only MSB are just 26% of the corresponding means for the metrics that use both AIP and MSB in Figure 4.

The results suggest that a critical consideration when maximising the underwater performance of a submarine using both AIP and MSB is to minimise the hotel load in order to avoid drawing power from the MSB for as long as possible. This can also be mitigated by increasing the power output of the AIP system. Further gains can be made by increasing the AIP fuel, but only minor gains will be made by increasing the energy capacity of the MSB.

4.4. Platform Range Performance Metrics

Table 4 shows that only five of the seventeen parameters have a notable influence on the platform range performance metrics as shown in Figure 5.

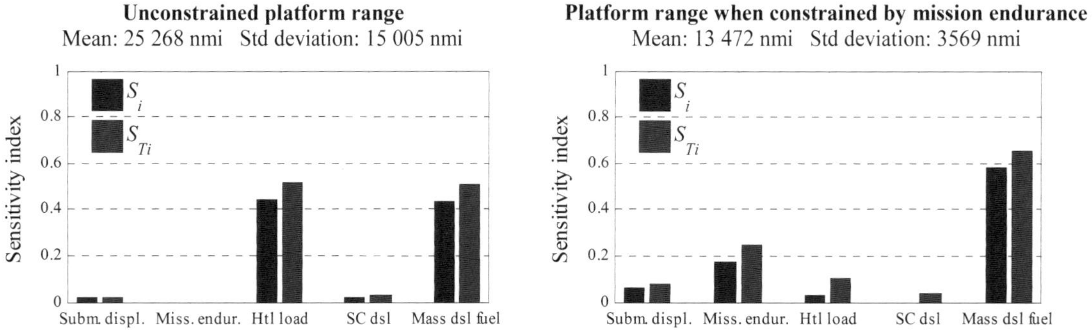

All other power and energy system parameters were found to be non-influential, each with $S_{Ti} < 0.03$ for unconstrained range and $S_{Ti} < 0.05$ for constrained range.

Figure 5 Results for the platform range performance metrics. The parameters of notable influence are submerged displacement (Subm. displ.), mission endurance (Miss. endur.), hotel load (Htl load), specific consumption of diesel by the diesel-generator sets (SC dsl), and mass of diesel fuel (Mass dsl fuel).

When unconstrained, the submarine travels at its most energy-efficient speed for maximising range, no matter how long such a journey would take. As discussed previously, the speed that maximises range tends to be close to patrol speed for a generic submarine, and at these low speeds hotel load is generally influential, while submerged displacement has only a minor influence. Therefore, the two most influential parameters are the mass of diesel fuel, which is the main store of energy on the submarine, and the hotel load, which is the primary factor in determining the rate of energy consumption. The specific consumption of diesel by the diesel-generator sets has a minor influence due to the small allowable range that is realistic for this paramater, and therefore tends to be dominated by variations in other parameters with wider ranges that are used in this study (Table 1). Once the mission endurance constraint is applied, range is maximised by travelling at faster speeds that are less energy-efficient, which slightly increases the influence of submerged displacement and drastically reduces the influence of hotel load. However, diesel fuel quantity remains the most influential parameter, because larger quantities of diesel allow for faster transits that consume more fuel. The results imply that platform range is best maximised by carrying as much usable diesel fuel as possible. Then unconstrained range can be further increased by reducing hotel load, while constrained range can be improved by increasing mission endurance.

4.5. Time in Operational Area Performance Metrics

Table 4 shows that only four of the seventeen parameters have a notable influence on the time in operational area performance metrics shown in Figure 6.

Maximising time in the operational area balances the need to transit quickly, leaving plenty of time but consuming a lot of diesel, and the need to transit slowly, leaving plenty of diesel but consuming a lot of time [1]. Hence, mission endurance and diesel fuel are the two most influential parameters. The influence of diesel fuel is reduced in the near case, because when the operational area is close, the total transit diesel consumption will be low even for a fast transit. However, a large quantity of diesel will always be consumed getting to and from the far operational area, making diesel fuel just as influential as mission endurance. Therefore, to maximise these metrics, it is most important to maximise both mission endurance and diesel fuel, although diesel can be sacrificed if deployment will primarily be close to home. Improvements from reductions in hotel load and submerged displacement will only be minor.

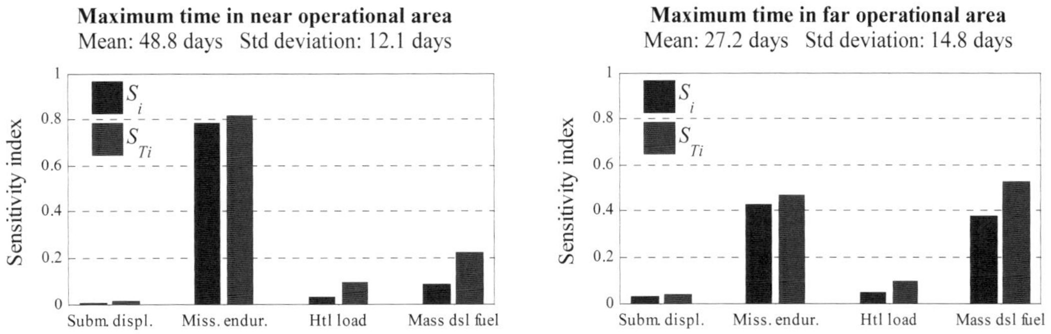

All other power and energy system parameters were found to be non-influential, each with $S_{Ti} < 0.03$.

Figure 6 Results for the time in operational area performance metrics. The parameters of notable influence are submerged displacement (Subm. displ.), mission endurance (Miss. endur.), hotel load (Htl load), and mass of diesel fuel (Mass dsl fuel).

4.6. Minor to Non-Influence of Efficiency and Specific Consumption Parameters

Note that parameters that control the efficiency of the propulsion motor and the AIP system were non-influential for all performance metrics. Similarly, the MSB charging efficiency and the specific consumption of diesel fuel by the diesel-generator sets (effectively an efficiency parameter) have at most a minor influence. When the variations in these parameters are confined to the ranges of Table 1, their impact is minimal, especially relative to the much wider ranges for other parameters that directly modify the quantity of stored energy or the size of the loads. If the other parameters were more tightly restricted, e.g. due to design or specification constraints, then the efficiency parameters would likely increase in influence.

5. CONCLUSION

This paper has demonstrated a method of applying a global sensitivity analysis to a submarine power and energy system model to investigate the influence of power and energy system parameters on a number of performance metrics. The scope was to understand the influence at a high level, before the application of constraints to the design and specification of a submarine. Hence, there was a wide range of possible values for many of the parameters.

The parameters identified as being the most influential – suggesting they are good candidates for early design consideration – were those that specified the quantities of stored energy (or mission endurance), the output power generation sources, and the load placed on the system. Within the assumed parameter ranges, efficiencies were found to have minor to negligible influence on performance when other parameters could be so widely varied. Hotel load was identified as a parameter of significant influence, particularly on metrics that required good performance at low speed. Submerged displacement was regularly shown to have a minor or marginal influence, except at the highest speeds, as propulsion load is relatively insensitive to changes in submerged displacement. This suggests that the disadvantages of increasing submarine size from a power and energy viewpoint are relatively small, especially if the increase allows for other gains, e.g. increasing the size of the power and energy system.

Table 4 provides the complete set of sensitivity analysis results. These were calculated using the approximation method proposed by Saltelli, and described in [2], which can produce small negative values for input parameters when they have little influence on the output.

Table 4 Sensitivity results for all parameters and performance metrics

Power and energy system parameters		Submarine performance metrics: *MSB only*				*MSB-snort cycling*		*AIP and MSB*		*Platform range*		*Time in op. area*	
		Max. range, any speed	Endurance, patrol speed	Endurance, transit speed	Endurance, max. speed	Min. IR, patrol speed	Min. IR, transit speed	Max. range, any speed	Endurance, patrol speed	Un-constrained	Constrained, miss. endur.	Near	Far
Submerged displacement	S_i	-0.01	-0.02	0.11	0.16	0.01	0.11	0.01	0.00	0.02	0.06	0.01	0.03
	S_{Ti}	0.03	0.03	0.17	0.26	-0.01	0.12	-0.03	-0.01	0.02	0.08	0.01	0.04
Mission endurance	S_i	-0.03	-0.03	0.00	-0.02	0.00	0.00	0.01	0.00	0.00	0.17	0.78	0.43
	S_{Ti}	0.00	0.00	0.00	0.00	-0.02	0.00	-0.05	-0.03	0.00	0.24	0.82	0.47
Hotel load	S_i	0.30	0.34	0.04	-0.01	0.32	0.05	0.48	0.48	0.44	0.03	0.03	0.05
	S_{Ti}	0.44	0.50	0.08	0.00	0.37	0.05	0.53	0.55	0.52	0.10	0.09	0.10
Motor eff., v. low output power	S_i	-0.03	-0.03	0.00	-0.02	0.00	0.00	0.01	0.00	0.00	-0.02	0.00	0.01
	S_{Ti}	0.00	0.00	0.01	0.00	-0.02	0.00	-0.05	-0.03	0.01	0.02	0.00	0.01
Motor eff., low output power	S_i	-0.03	-0.03	0.00	-0.02	0.00	0.00	0.01	0.00	0.00	-0.02	0.00	0.00
	S_{Ti}	0.00	0.00	0.01	0.00	-0.02	0.00	-0.05	-0.03	0.00	0.02	0.01	0.00
Motor eff., rated output power	S_i	-0.03	-0.03	0.00	-0.01	0.00	0.00	0.01	0.00	0.00	-0.02	0.00	0.01
	S_{Ti}	0.00	0.00	0.00	0.00	-0.02	0.00	-0.05	-0.03	0.00	0.02	0.00	0.00
MSB energy cap., v. low power	S_i	0.53	0.48	0.76	0.46	0.00	0.00	0.04	0.04	0.00	-0.02	0.00	0.01
	S_{Ti}	0.64	0.60	0.81	0.57	-0.02	0.00	0.02	0.02	0.00	0.02	0.01	0.01
MSB frac. energy cap., v. high power	S_i	-0.03	-0.03	0.00	0.18	0.00	0.01	0.01	0.00	0.00	-0.02	0.00	0.01
	S_{Ti}	0.00	0.00	0.02	0.28	-0.02	0.01	-0.05	-0.03	0.01	0.04	0.01	0.02
MSB charging efficiency	S_i	-0.03	-0.03	0.00	-0.02	0.01	0.01	0.01	0.00	0.01	-0.02	-0.01	0.01
	S_{Ti}	0.00	0.00	0.00	0.00	0.02	0.02	-0.05	-0.03	0.02	0.03	0.02	0.02
AIP rated output power	S_i	-0.03	-0.03	0.00	-0.02	0.00	0.00	0.12	0.12	0.01	0.00	0.03	0.02
	S_{Ti}	0.00	0.00	0.00	0.00	-0.02	0.00	0.16	0.19	0.01	0.02	-0.01	0.01
AIP max. system efficiency	S_i	-0.03	-0.03	0.00	-0.02	0.00	0.00	0.04	0.03	0.00	-0.02	0.00	0.01
	S_{Ti}	0.00	0.00	0.00	0.00	-0.02	0.00	0.00	0.01	0.00	0.02	0.01	0.01
AIP efficiency loss, v. low power	S_i	-0.03	-0.03	0.00	-0.02	0.00	0.00	0.01	0.00	0.00	-0.02	0.00	0.01
	S_{Ti}	0.00	0.00	0.00	0.00	-0.02	0.00	-0.05	-0.03	0.00	0.02	0.00	0.00
AIP efficiency loss, rated power	S_i	-0.03	-0.03	0.00	-0.02	0.00	0.00	0.02	0.01	0.00	-0.02	0.00	0.01
	S_{Ti}	0.00	0.00	0.00	0.00	-0.02	0.00	-0.04	-0.03	0.01	0.02	0.00	0.00
Diesel-gen. total output power	S_i	-0.03	-0.03	0.00	-0.02	0.59	0.81	0.01	0.00	0.00	-0.02	0.01	0.01
	S_{Ti}	0.00	0.00	0.00	0.00	0.66	0.83	-0.05	-0.03	0.00	0.03	0.01	0.01
Diesel-gen. spec. consump. of diesel	S_i	-0.03	-0.03	0.00	-0.02	0.00	0.00	0.01	0.00	0.02	-0.01	0.01	0.02
	S_{Ti}	0.00	0.00	0.00	0.00	-0.02	0.00	-0.05	-0.03	0.03	0.05	0.01	0.02
Mass of diesel fuel	S_i	-0.03	-0.03	0.00	-0.02	0.00	0.00	0.01	0.00	0.43	0.58	0.08	0.38
	S_{Ti}	0.00	0.00	0.00	0.00	-0.02	0.00	-0.05	-0.03	0.51	0.66	0.22	0.52
Mass of AIP fuel	S_i	-0.03	-0.03	0.00	-0.02	0.00	0.00	0.13	0.12	0.00	-0.02	0.00	0.00
	S_{Ti}	0.00	0.00	0.00	0.00	-0.02	0.00	0.26	0.25	0.01	0.02	0.02	0.01

Non-influential **Minor influence** **Marginally influential** **Influential**

REFERENCES

1. Torresan, H. D., Jarvis, R. A. (2008), "Modelling of Conventional Submarine Power and Propulsion Systems for Future Capability Assessment", UDT Pacific 2008 Conference
2. Saltelli, A., et al. (2008), "Global Sensitivity Analysis: The Primer", John Wiley & Sons
3. Saltelli, A., Chan, K., Scott, E. M. (2000), "Sensitivity Analysis", John Wiley & Sons
4. Burcher, R., Rydill, L. (1994), "Concepts in Submarine Design", Cambridge Uni. Press
5. Brighton, D. R., et al. (1994), "The use of fuel cells to enhance the underwater performance of conventional diesel electric submarines", Journal of Power Sources, 51, 375-389
6. Buckingham, J., Hodge, C., Hardy, T. (2008), "Submarine Power and Propulsion – Trends and Opportunities", Pacific International Maritime Conference 2008

SOLAS Compliant Navigation Systems On Naval Vessels

D C Bradley, A Scicluna
BAE Systems Australia Defence Pty Ltd, Williamstown, VIC 3016, Australia
david.bradley4@baesystems.com, andrew.scicluna@baesystems.com

1) ABSTRACT

In addition to modern military naval vessels meeting all of their operational requirements, the vessels are also being specified to meet commercial shipping registry classifications; such as Lloyd's Registry. Registration with a shipping registry requires vessels to be compliant with commercial shipping standards. For navigation systems, commercial classification means compliance with Safety Of Life At Sea (SOLAS) Chapter V regulations.

This paper provides a background to the legal requirements of SOLAS Chapter V, navigation equipment type approval, and Australian Maritime Safety Authority (AMSA) Marine Orders. The historical non-compliance of naval vessels, use of flag state waivers and the current drivers behind the need for new warships to be registered and meet SOLAS are discussed. This paper also examines the challenges of naval ship navigation system certification including military Ships Inertial Navigation Systems (SINS) and Global Position System (GPS) approval.

A typical navigation system example is provided representing the minimum equipment requirements and the required interfaces between equipment. Additional aspects of a naval navigation design are briefly discussed covering redundancy, Emissions Security (EMSEC) and protecting classified data. The example also includes the typical additional requirements on a military naval navigation system to provide data to other ships systems, e.g. Combat System.

The paper also examines the commercial shipping trends of integrating navigation functions in Integrated Navigation Systems (INS) and with other ship's functionality in Integrated Bridge Systems (IBS). The benefits of these systems are discussed and the possible application to military naval vessels.

2) INTRODUCTION

The history of the International Convention for the Safety of Life at Sea (SOLAS) can be traced back to the sinking of the Titanic. The first version of the convention was adopted in 1914 and prescribed the required emergency equipment and safety procedures. The International Maritime Organization (IMO) was formed to bring international shipping conventions into an international framework. The first task of IMO was to update the SOLAS convention resulting in the 1960 convention. In 1974 the SOLAS convention was completely updated including a simplified process for future amendments.

Naval vessels are using commercial navigation equipment more regularly as the drive to reduce costs and provide an equivalent level of safety as commercial shipping. This brings challenges in interfacing commercial equipment to military combat and control systems. There is also a challenge of proving to regulators that military equipment provides at least the same level of safety as the commercial equivalents.

3) MARITIME COMPLIANCE REGIME

Australian Maritime Safety Authority Marine Orders 21

The Navigation Act 1912[1] provides the Chief Executive Office of Australian Maritime Safety Authority (AMSA) with the authority to make Marine Orders. The Navigation Act 1912 defines which vessels are covered by the act and the Marine Orders define the detailed technical requirements that the vessels must comply with. Marine Orders also provide the implementation of SOLAS. Marine Orders – Part 21[2] provides the requirements for Safety of Navigation and Emergency Procedures and is presently at Issue 7.

For the navigation systems, Marine Orders – Part 21[2] refers to SOLAS Chapter V[3] regulations which are discussed in the next section. Marine Orders – Part 21[2] requires that type approvals for equipment that SOLAS mandates to be fitted to the vessel. These functional requirements for type approvals are contained in IMO resolutions and circulars referenced from Marine Orders – Part 21[2] Appendix 3. The specific requirements of the IMO resolutions and circulars are discussed in the type approval section below.

SOLAS Chapter V

SOLAS Chapter V[3] identifies the requirements for the safety of navigation. The chapter is divided in to several regulations that cover all aspects of navigation. The chapters particularly relevant to the vessel's navigation system are:

- Regulation 17 – Electromagnetic compatibility;
- Regulation 18 – Approvals, surveys and performance standards of navigation systems and equipment and voyage data recorder (VDR);
- Regulation 19 – Carriage requirements for shipborne navigational systems and equipment; and
- Regulation 20 – Voyage Data Recorders (VDR).

Regulation 17 outlines the requirement for marine equipment to be designed and tested for electromagnetic compatibility taking in to account IMO Resolution A.813(19)[4].

Regulation 18 requires that mandatory navigation equipment (defined by Regulation 19 and 20) to be tested and type approved to specific IMO resolutions and circulars.

Regulation 19 identifies the navigation equipment that is required to be carried by a vessel. The type and quantity of equipment depends on the Gross Tonnage of the vessel as shown in Figure 1.

Regulation 20 covers the requirement for vessels engaged on international voyages to be fitted with a VDR (dependent on the size and type of vessel).

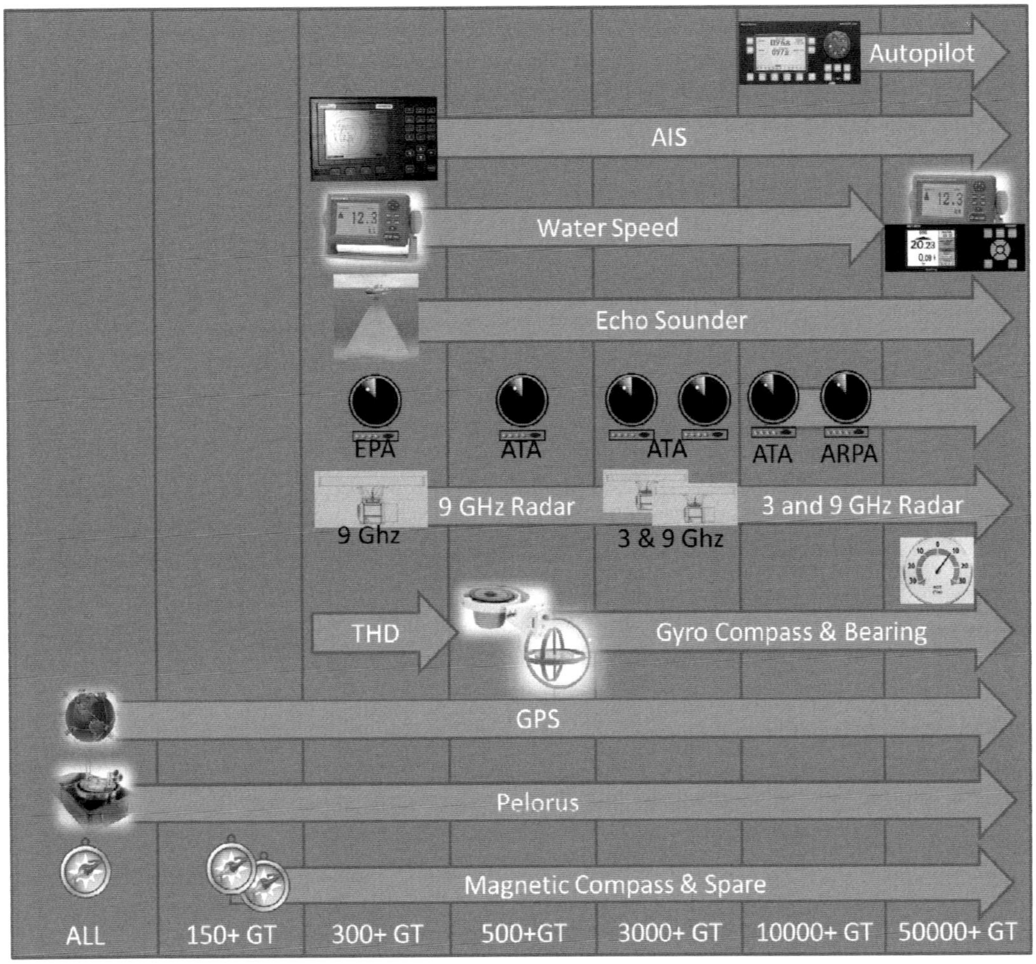

Figure 1, SOLAS Mandatory Electronic Navigation Equipment Summary

Type Approval

A type approval provides the Classification Organisation with a level of confidence that equipment complies with the required IMO resolutions and circulars. The International Electrotechnical Commission (IEC) has taken the majority of the equipment resolutions and circulars and derived test standards to allow third party independent test agencies to test navigation products in a consistent manner.

In addition to the specific functional and performance requirements of each type of equipment there are also general IMO resolutions that apply to all equipment, for example with IMO Resolution A.813(19)[4] and A.694(17)[5]. Compliance to these general requirements is shown by testing to IEC 60945[6]. IEC 60945 covers Electro-Magnetic Compatibility (EMC), environmental and safety requirements.

Warship Non-Compliance

Regulation 1 of SOLAS Chapter V[3] allows Governments to have exemptions for warships, naval auxiliaries and other non-commercial government ships from SOLAS Chapter V[3]. However these vessels are encouraged to act in a manner consistent, so far as reasonable and practicable, with the requirements of SOLAS Chapter V[3].

As a result naval vessels generally follow the procedures of SOLAS Chapter V[3] and fit similar equipment, however it is likely that naval vessels do not have all the equipment required by SOLAS Chapter V and where the equipment is of military supply it is unlikely to be type approved.

4) WARSHIP COMPLIANCE

Drivers For Future Compliance

The development of "Naval Rules" by classification societies such as Lloyd's Registry has provided a route for governments to have naval vessels assessed and classified against a set of common requirements. This provides flag state authority with an independent review of a naval ship design against 'best of class' requirements.

The Lloyd's Registry Naval Rules[7] is divided into different notations; the main notations applicable to navigation are Safety of Navigation and Communications (SNC), Superior Standard of Navigation (NAV) and Single Bridge Watchkeeper (NAV1). The SNC notation is performance based with a set of goals against each section. In general conformance against the SNC notation can be achieved by compliance with SOLAS Chapter V[3] regulations 19, 20, 23, 24, 25, 26, 27, 28, 30 and 34.

NAV and NAV1 notations are both aimed at reduced-manning bridges (single operator for NAV1) and are requirements based standards. The NAV notation, which is a requirement for the Canberra Class, provides increased equipment requirements compared to SNC notation and specifies requirements for bridge physical layout and alarm systems. NAV1 notation has additional requirements including a Bridge Warning System (BWS), the integration and management of alarms across navigation, communications and power.

Military Equipment Type Approval

The majority of the navigation system needs of a naval ship can be met with IMO type approved equipment, however there are a couple of areas where specialized military products will be required.

Global Positioning System (GPS) receivers have become the standard positioning system for many uses including ships. The civilian maritime market has a large number of type approved receivers for both the NAVSTAR and GLONASS global navigation systems that include Very High Frequency (VHF) radio receivers for the reception of differential corrections.

The military market is more specialised with the majority of GPS receivers either in a card format, e.g. GPS Receiver Application Module (GRAM); or targeted more at the portable role, e.g. Defense Advanced GPS Receiver (DAGR). Whichever type of receiver is selected, it will be required to be marinised and include displays to provide GPS position, performance and warning indicators to the navigator.

Due to the nature of military operations, GPS receivers are also connected to Controlled Reception Pattern Antennas (CRPAs) that have null-forming capabilities to reject deliberate jamming and other forms of interference. Again these products were not specifically designed for marine usage and require a degree of packaging to make them suitable for the marine environment.

These factors result in a bespoke GPS made up of equipment from different manufacturers and a certain amount of custom integration. Obtaining type approvals is an expensive and time consuming activity and is unlikely to be done; instead the ship designer/systems integrator will be required to prove equivalence to the IMO GPS resolutions, MSC.112(73)[8], and general requirements for shipborne equipment, A.694(17)[5], to both the classification society and the flag state authority. These resolutions include requirements for position performance, minimum displays and environmental performance.

The other typical piece of military navigation equipment installed on naval ships is a Ships Inertial Navigation System (SINS). SINS are required to accurately stabilize the vessel's mission systems with respect to attitude and position. For the navigation system the SINS provides an accurate source of heading that exceeds the IMO resolution performance requirements for gyro compass systems, A.424(XI)[9]. Again, it is unlikely that the SINS will be type approved and the ship designer or systems integrator will need to prove equivalence to the IMO resolution for gyro for performance and environmental requirements.

BAE Systems is presently generating equivalence documentation for GPS and SINS for the Canberra Class. The equivalence includes assessments of the GPS and SINS used on the Canberra Class against the functional requirements of the IMO resolutions, comparison of military GPS receiver performance against IMO GPS and gyro performance requirements and comparison of the military environmental standards against IEC 60945[6].

5) EXAMPLE NAVIGATION SYSTEM

Navigation System Aspects

An example of a typical modern naval navigation system with Lloyd's Register Classifications SNC and NAV notations is shown in Figure 2 for a vessel of 50 000 Gross Tonnage. Ideally, all of the equipment would be type approved to the applicable IMO resolutions, however it's common that the naval vessels will require greater accuracy and additional functionality. Typically, a naval vessel would consist of the following type approved equipment:

- 3 GHz and 9 GHz navigation radars each with Automatic Radar Plotting Aids (ARPA)
- Automatic Identification System (AIS);
- VDR;
- Electromagnetic speed log for speed through water;
- Doppler log for speed over ground;
- Echo sounder(s);
- Autopilot, either heading or track control; and
- Electronic Chart Display and Information System (ECDIS).

The positioning systems are military supply and unlikely to be type approved and would consist of:

- Military GPS (capable of using commercial C/A code and military P(Y) code); and
- SINS.

In some cases a standalone type approved Differential GPS (DGPS) may be provided on the bridge. On a naval vessel, the navigation system would also include a meteorological subsystem to supply wind and barometric information for aircraft and combat system operations.

Interfacing between the navigation components will be via National Marine Electronics Association (NMEA) 0183[10] / IEC 61162[11]. The internal interfacing includes:

- SINS providing heading and integrated GPS-SINS position to the navigation radars, ECDIS and AIS;
- Electro-magnetic log supplies speed through water to the navigation radar and ECDIS to allow drift calculations to be performed;
- AIS providing track information to the ARPA for display and correlation with radar tracks;
- Navigation radars providing ARPA and AIS tracks and radar video to the ECDIS; and
- All systems provide critical data and radar video to the VDR for recording

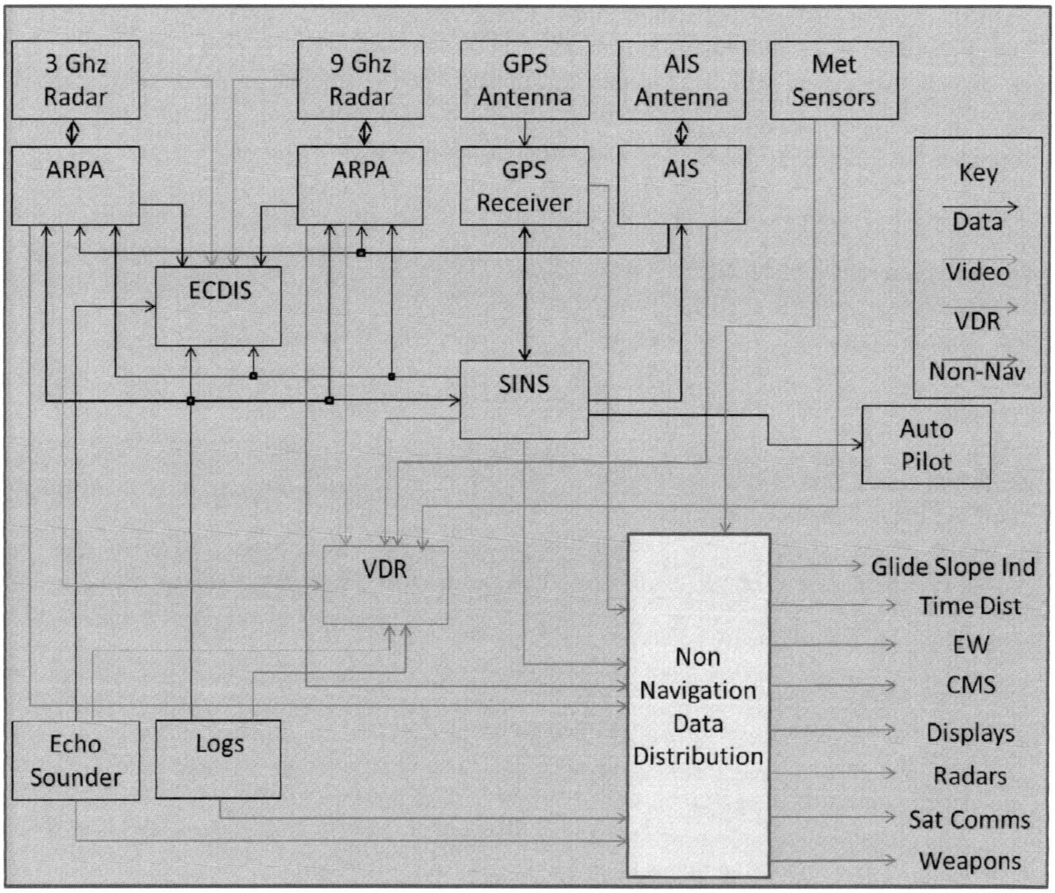

Figure 2, Example Navigation System

External Communications

External systems require various sources of navigation input as shown below in Table 1.

Source	Data	Typical Users
SINS	Integrated SINS/GPS Position, Velocities and accelerations. SINS heave, heave rate and heave accelerations SINS attitudes, attitude rates and attitude accelerations	Combat System Weapon Systems Satellite communications Radars Information Displays Aircraft (navigation alignment) Dynamic Positioning System
GPS	Position and velocities Time	Combat System Time Displays Communications equipment Dynamic Positioning System
Logs	Water Speed, Ground Speed	Combat System Steering System Dynamic Positioning Systems
Meteorological	Wind Speed (True and Relative) Barometric Pressure	Combat System Displays for air operations Information Displays Dynamic Positioning systems
Navigation Radars	Radar tracks Radar image	Combat System

Table 1, Navigation System Typical External Communications

In addition to supplying the required data any data distribution system has to convert the data in to the required format. Different external systems will require data over different bearers including analogue (voltage, current or pulses), synchro (115 V 400 Hz) and various digital formats.

Special Requirements For Naval Ships

There are some specific naval vessel requirements that impact the design of the navigation system. First is the storage of classified information on commercial equipment. Generally, current known navigation information is not classified, however route planning, and track log information is considered classified. The VDR is an obvious example of equipment that records navigation information that may be classified (ship's position) and needs to be controlled. Less obvious is that ECDIS incorporates a recording mechanism and equipment such as GPS may also have a position log. The location of the equipment that stores the data needs to be considered and the appropriate security put in place.

The second related issue is Emissions Security (EMSEC). Where classified data is displayed or transmitted along an interface the equipment and installation should meet the requirements of the Australian Government Information Security Manual[12]. This requires classified equipment and interfaces to be physically and electrically isolated from un-classified equipment. This impacts the installation design on the ship and needs to be considered early in the design.

Finally, Emission Control (EMCON) needs to be considered for navigation transmitters, specifically radars and AIS. The functionality of included power and standby controls needs to be understood to ensure that inadvertent transmissions are not possible. Where the functionality is not clear additional power switches or antenna switching in to dummy loads may be required. However, this needs to consider the impact to equipment type approvals.

6) FUTURE TRENDS

RAN Regulatory Framework

Following the recent Rizzo Report[13] and the obligations under the new Work Health and Safety (WHS) Act[14] the Royal Australian Navy (RAN) has approved the use of the North Atlantic Treaty Organistaion (NATO) Naval Ship Code (NSC)[15], described below. The initial vessel to implement the NSC will be HMAS Choules and the Naval Flag Administrator will be responsible for the development of implementation strategies for all new and existing vessels.

The NSC[15] or Allied Naval Engineering Publication (ANEP) ANEP77 was developed by ten navies and six classification societies from around the world, including Australia. The maintenance of the code is the responsibility if the International Naval Safety Association (INSA). The NSC[15] states the aim of the code is to provide a "cost effective goal based standard for naval ship safety and environmental assurance, benchmarked against statute and accepted by the global naval community and inter-government bodies". The code is divided into a number of chapters each covering different aspects of ship design. Each chapter has a set of goals that have been developed by INSA which provides an equivalent standard of safety to a commercial ship. Each chapter also has functional and performance requirements and a description of common approaches to verification of the requirements.

NSC[15] Chapter IX covers navigation and seamanship. Chapter IX regulation 0 lists the goals of the navigation systems including independent navigation, awareness of fixed and moving hazards, receiving weather forecasts, measuring and interpreting environmental data and assisting other vessels and persons in distress. Regulation 0 also covers goals on reliability and failure, requiring that essential safety functions are maintained following a single system or equipment failure. Finally, possibly one of the most important goals, is that the navigation systems' essential safety functions are not to be dependent on the ships combat system being available.

NSC[15], Chapter IX Navigation and Seamanship, Regulation 1 covers the functional and performance requirements of the navigation system. This chapter requires the ship to be designed, constructed and maintained in accordance with the requirements of SOLAS. Future revisions of the NSC should have the navigation requirements updated following working group activities in this area.

Technology Drivers

In the last ten years there have been several changes to navigation systems. These have included Integrated Bridge Systems (IBS), Integrated Navigation Systems (INS) and improvements to equipment interfacing. Initially there was no definition of what an IBS and an INS were and manufacturers could integrate functions without any specific requirements for the integrated system. Equipment interfaces are starting to move away from point to point NMEA 0183[10] serial communications to network based topologies.

The IMO resolution MSC.64(67)[16] Annex 1 defines an IBS as a system that supports at least two of the following operations: passage execution; communications; machinery control; loading, discharge and cargo control; or safety and security. The IBS should comply with the individual equipment resolutions and be as effective as individual equipment. This requires redundant system design with fall back operation for essential functions. This aligns well with the standard naval requirement for redundant systems and networks. The IMO resolution also requires IBS systems to have a transitional form of power from main to emergency source. Again, naval vessels typically have significant Uninterruptable Power Supply (UPS) requirements and an IBS fits well with naval requirements.

IMO MSC.252(83)[17] defines the purpose of an INS to "enhance the safety of navigation by providing integrated and augmented functions to avoid geographic, traffic and environmental hazards" and requires an INS to "combine, process and evaluate data from connected sensors and sources". This is very well aligned with how the combat system integrates data from multiple sensors.

The move to networked interfaces is being driven by the increased complexity in system integration and greater need for connectivity, two of which are described below. NMEA have developed NMEA 2000[18] based on Controller Area Network (CAN) system. CAN was initially designed for use in the automotive industry and has high levels of robustness. Several navigation manufacturers are distributing NMEA 0183 messages over Internet Protocol (IP) based networks. Although both systems use different network technologies, both allow data to be distributed from multiple sensors to multiple receivers using a network topology and both have significantly higher bandwidth than RS-422 serial communications. IP based networks onboard naval vessels can be used for more than just NMEA 0183[10] distribution, it offers accurate transmission of video, timing and management systems; for this reason it's likely that IP based networks will become the standard for navigation system communications on naval vessels.

7) REFERENCES

1 Australian Government, 1913, "Navigation Act 1912".
2 Australian Maritime Safety Authority, 2010, "Marine Orders Part 21 Safety of Navigation and emergency procedures", Issue 7.
3 International Maritime Organization, 1974, "International Convention for the Safety of Life at Sea (SOLAS)", as amended.
4 International Maritime Organization, 1995, Resolution A.813(19) "General Requirements for Electromagnetic Compatibility (EMC) for All Electrical and Electronic Ship's Equipment".
5 International Maritime Organization, 1991, A.694(17) "General requirements for shipborne radio equipment forming part of the global maritime distress and safety system (GMDSS) and for electronic navigational aids".
6 International Electrotechnical Commission, 2002, IEC 60945 "Maritime navigation and radio communication equipment and systems - General requirements - Methods of testing and required test results", Fourth Edition.
7 Lloyd's Register, 2011, "Rules & Regulations for the Classification of Naval Ships 2011".
8 International Maritime Organization, 2000, MSC.112(73) "Revised Performance Standards for Shipborne Global Positioning System (GPS) Receiver Equipment".
9 International Maritime Organization, 1979, A.424(XI) "Performance Standards for Gyro-Compasses".
10 National Marine Electronics Association, 2008, NMEA 0183, Version 4.0.
11 International Electrotechnical Commission, 2010, IEC 61162 "Maritime navigation and radiocommunication equipment and systems - Digital interfaces", Edition 4.0.
12 Department of Defence, 2011, "Australian Government Information Security Manual".

13 Rizzo P J, July 2011, "Plan to Reform Support Ship Repair and Management Practices".
14 Australian Government, 2011, "Work Health and Safety (WHS) Act 2010".
15 International Naval Safety Association, Naval Ship Code (ANEP-77), Edition 2.
16 International Maritime Organization, 1996, MSC.64(67) "Recommendations on new and amended performance standards".
17 International Maritime Organization, 2007, MSC.252(83), "Adoption of the Revised Performance Standards for integrated Navigation Systems (INS)"
18 National Marine Electronics Association, NMEA 2000 "Standard for Serial-Data Networking of Marine Electronic Devices", Edition 2.2.

9) BIOGRAPHIES

David Charles Bradley

David is a Combat Systems Engineering Manager at BAE Systems.

David graduated from Salford University, UK in 1990 with an Honours Degree in Electronic and Electrical Engineering.

David spent 11 years at Royal Aerospace Establishment (and subsequent DRA and DERA) conducted research and flight trials on helicopter navigation and flight control systems. For the majority of this time David specialised in Naval Aviation and the Helicopter Ship Dynamic Interface.

David spent 14 months at DSTO Fishermans Bend as part of the Anglo-Australian Memorandum Of Understanding on Research conducting simulation trials on helicopter ship operations before deciding to remain in Australia.

David joined Vision Systems (now Xtralis) in 2001 and worked as the lead systems engineer and product manager for the development of smoke detection and voice evacuation products.

David joined BAE Systems in 2009, initially working on ANZAC class update projects before joining the Australian LHD project as the lead of the navigation team.

Andrew Scicluna

Andrew is a Combat Systems Engineer at BAE Systems.

Andrew graduated from RMIT University in 2007 with a Bachelor of Electronic Engineering.

Andrew joined BAE Systems in January 2008 as a graduate engineer on ANZAC class upgrade project to conduct test and integration of the navigation data distribution system in preparation for the Anti Ship Missile Defence (ASMD) project.

Andrew joined the Australian LHD project navigation systems engineering team in 2010.

HVAC Concept Design for Small Surface Combatant

Clint Domine
Directorate Navy Platform Systems
clint.domine@defence.gov.au

1.0 OVERVIEW

The objective of the concept study is to estimate the Heating, Ventilation and Air Conditioning (HVAC) cooling load for a Small Surface Combatant. The small surface combatant used for this study was a concept approximately 80 metres in length and could accommodate a crew of 45. The parameters were chosen to reflect a realistic situation of operating in the tropics. Also the ramifications of integrating a Chemical, Biological, Radiological and Nuclear (CBRN) Citadel system were considered.

2.0 SCOPE

The scope of the study encompasses estimation of the total cooling load and required chiller water plant cooling capacity. To aid with the study, the following items were calculated using relevant heat transfer equations and HVAC design theory:

- Compartment heat gain and losses
- Total connected cooling load
- System air flow

In addition, the study will address the general requirements for implementing a CBRN citadel to provide protection from CBRN threat. The following topics will also be discussed briefly in the report:

- Citadel system design requirements
- CBRN filters and airlocks
- Cleansing station arrangements

The concept study aims to provide estimates of the Small Surface Combatant HVAC system and CBRN Citadel requirements to stimulate further discussion and research. It is not in the scope of this report to determine the specifications of individual compartment chillers, sizing of condensers, Air Filtration Units (AFU), HVAC ducting, citadel wash down facilities and total weight of the HVAC system. This report also does not discuss specific CBRN Defence arrangements and ventilation requirements for machinery spaces.

3.0 DESIGN ASSUMPTIONS

3.1 General

The total cooling load is based on the total ship peak cooling. The chilled water plant (CWP) may not be required to produce the peak cooling capacity at all times but it must be able to cope with extreme tropical environments where there is a possibility the ambient temperature will reach 40 degrees Celsius.

3.2 Design External Ambient Temperature

The external ambient temperature was selected to be representative of the operational environment of the Small Surface Combatant. It is assumed that the Small Surface Combatant will operate in littoral waters close to the equator where the climate is extreme tropical. The extreme tropical regions such as the Persian Gulf and north Australia are examples of where the Small Surface Combatant may be operating and where ambient temperatures of 40+ degrees C dry bulb can occur.

3.3 Seawater Temperatures

The seawater temperature does not affect the cooling load calculations; it does however have a profound effect on the performance of the CWP. Seawater is used as the cooling medium in the heat exchangers of the refrigerant circuit. Near surface seawater temperatures can reach as high as 35 degrees C in tropical waters. A high sea water temperature will decrease the heat transfer between the refrigerant and the seawater thus reducing refrigerant cooling capacity. This will have an adverse effect on the chiller which produces the chilled water.

3.4 Internal Design Temperature

The internal design air temperature was chosen to be +27 deg C 50% relative humidity.

3.5 Boundary Heat Load Temperature differences between adjoining internal spaces

The boundary heat load is the heat that is transferred through the bulkheads, deck and deckhead. The boundaries between adjacent compartments and external boundaries that would be susceptible to solar heat gain were established. The temperature differences between adjoining internal spaces such as a bulkhead, deckhead, or deck separating an air conditioned and non air-conditioned internal space was obtained from Table 1 of ISO-7547:2002(E). The heat transfer coefficient, U for each bulkhead, deck and deckhead was obtained using guidance values on commonly used materials as specified in ISO-7547:2002(E). It is also assumed that the temperature difference between adjacent air conditioned compartments is zero.

3.6 Equipment Heat load

The lighting heat load is due to the heat produced from the lighting in a compartment. The occupancy heat load is the heat gain due to personnel in the compartment space. The equipment heat load is the heat gain from equipment within the space such as boilers, pipes etc. The lighting heat load is generally accepted as 16 watts/square metre of deck area.

3.7 Human heat load

The human heat load due to perspiration is the following:

* Sensible = 60 watts/person
* Latent = 90 watts/person

3.8 Fresh air required per man

The overall fresh air per person required is 0.0056 kg/sec (0.00467 m^3/second). The minimum fresh air per person requirement is 0.0038 kg/sec.

3.9 Fan heat load

The fan used for distribution of the air also produces sensible heat and will vary depending on ducting arrangements and filtration back pressures. For the purpose of this design study it is assumed to be 14%.

3.10 Outdoor air load (fresh air)

The supply air entering the cooling apparatus is a mixture of fresh air and recirculated air from the compartment. The fresh air or outdoor air load is the sensible and latent heat that is contained in the fresh air supplied to the cooling apparatus of the air handling unit. It is an additional heat load that must be taken into account along with the compartment heat load. It is not in the scope of this study to calculate the bypass factors and mixture ratios which govern the amount of recirculated and fresh air entering the system. For the purpose of this design study, the fresh air percentages ranging from 50% to 100% will be assessed to provide a range of cooling loads.

Outside air load (OATH) = $1.19 \times (Fresh\,air\ percentage) \times Q_c \times (h_{outside} - h_{design})$

Where,

$Fresh\ air\ percentage$ = fresh air proportion of the airflow entering cooling apparatus

Q_c = the compartment airflow (kg/second)

$h_{outside}$ = The enthalpy of the outside air (kJ/kg)

h_{design} = The enthalpy of the design air condition (kJ/kg)

Using a psychrometric chart, the enthalpy difference for an internal ambient design temperature of 27 deg C 50% relative humidity and an ambient outdoor air temperature of 40 deg C 70% relative humidity is 69 kJ/kg.

3.11 Electronic Equipment Heat Load

In determining the cooling load, the heat gain from electronic equipment such as communications, navigation, radar, and electronic cabinets must be accounted for. It is not in the scope of this study to accurately calculate the electronic equipment heat load. Typically, the electronic equipment will have a dedicated fan coil unit and heat exchanger attached through which a refrigerant or chilled water line would pass through.

For a vessel of 80 metres in length and combining the capabilities of a Hydrographic, mine hunter, and patrol boat the electronic equipment cooling load was estimated to be 58 kW. It was also assumed that the electronic equipment cooling load was constant regardless of the air distribution arrangements. For the breakdown of the electronic cooling loads refer to Table 1 below.

Electronic Component	**Estimated load (kW)**
Hydrographic survey system	9
Radar	17.5
Communication Equipment	31.6
Estimated Total	**58.1**

Table 1: Electronic cooling load breakdown

4.0 HVAC COOLING LOAD CALCULATIONS

4.1 General

To determine the cooling capacity of the CWP to supplying the required cooling to achieve the design ambient air temperature, the total cooling load of the ship is to be calculated. The total compartment cooling load is the work that the CWP must produce to maintain the desired air temperature for all air conditioned compartments. Heat transfer equations were used to determine the total compartment heat gain.

4.2 Compartment Volumes

The volume for each compartment on the ship was calculated. Annex A contains the volumes for all compartments including air conditioned spaces, citadel compartments, and non air conditioned space such as machinery spaces.

4.3 Compartment Heat gain calculation

For each air conditioned compartment, the following sensible and latent heat loads were calculated and added to produce the compartment total heat load, Qt:

- Boundary heat load
- Lighting heat load
- Occupancy heat load
- Equipment heat load
- Fan allowance

For breakdown of compartment total heat loads please refer to Annex B.

4.4 Compartment Supply air quantities

Once the compartment heat load data was collated for all air-conditioned compartments, the individual heat loads was summed to produce the total compartment heat load. The total air quantity is the total compartment heat load divided by the enthalpy difference.

TAQ = Total Heat Gain/Enthalpy Difference

The enthalpy difference is the difference in enthalpies of the air entering and leaving the cooling apparatus. For this study the air entering the apparatus was assumed to be 29 degrees (50% Relative humidity) or a few degrees higher than the design ambient of 27 degrees. The air leaving the cooling apparatus is assumed to be the accepted value of 13 degrees C (dry bulb) 12 degrees (wet bulb). Using a psychrometric chart, the enthalpy difference between the two values is calculated as 23.5 kJ/kg. This value however will change; however, as air entering the cooling apparatus is influenced by the proportion of fresh air and the number of air changes required.

Once the total air quantity (TAQ) is calculated the apportioning of air to each air-conditioned compartment is based on the following relationship:

$$\text{Compartment Air Quantity} = TAQ \times \frac{H_{SR}}{H_{SRT}}$$

Where;
TAQ = Total Air Quantity
H_{SR} = Compartment Sensible load
H_{SRT} = Sum of Compartment Sensible Loads

4.5 Application of six minimum air changes per hour per compartment

Once the air quantity for each compartment is calculated it had to be checked against the minimum requirement of six air changes per hour. All air conditioned spaces which had air quantities less than six air changes per hour were increased to the minimum of six air changes per hour. This effectively increased the total air quantity and subsequently produced a 'revised' total air quantity.

4.6 Total Cooling Load

The total cooling load is the sum of the compartment heat gain, outside air loads (OATH) and additional loads from electronic equipment.

Total Cooling load = Grand total heat + Electronic Equipment Cooling loads

Where;
Grand total heat (GTH) = Qt + OATH
Qt = Sum of compartments heat gain (sensible and latent loads) (kW)
OATH = Outside air total heat (kW)

5.0 RESULTS - HVAC SYSTEM

5.1 General

This section outlines the estimates of total compartment heat gain, supply air requirements and total cooling load capacity for the HVAC system without a citadel. The internal ambient and external temperatures from Section 3 of this document were used as the design temperatures. The calculation process described in Section 5 was used to calculate the total cooling load capacity.

5.2 Total Compartment heat gain

The total compartment heat gain was calculated to be 94.1 kW. The sensible heat and latent heat components were 72.3 and 21.8 kW respectively.

5.3 Supply Air Requirements

The total air quantity for the HVAC system alone was initially calculated to be 4 kg/second (12015 m^3/hr) but after applying a minimum six air changes per hour per compartment it was recalculated to be 5.3 kg/second (15789 m^3/hr).

5.4 Total Cooling Load Capacity

The total cooling load capacities for fresh air proportions ranging from 50% to 100% were graphed in figure 1 below. It can be observed that with increasing fresh air proportion, the cooling load increases. This is due to the fact that with a higher fresh air percentage, a greater quantity of fresh air at a higher temperature and humidity is brought into the system which must be conditioned. Ideally, a balanced proportion of fresh and recirculated air will be chosen based on air change, fresh air per man, and power consumption requirements. If cost was not a factor, a 100% fresh air system will be desirable as air will not be recirculated reducing the amount of stale air or impurities being recirculated. A higher fresh air quantity, however, will mean the cooling coils will have to remove a greater quantity of heat and therefore power consumption will increase. Please refer to Annex C for breakdown of compartment outside air heat and grand total heat loads.

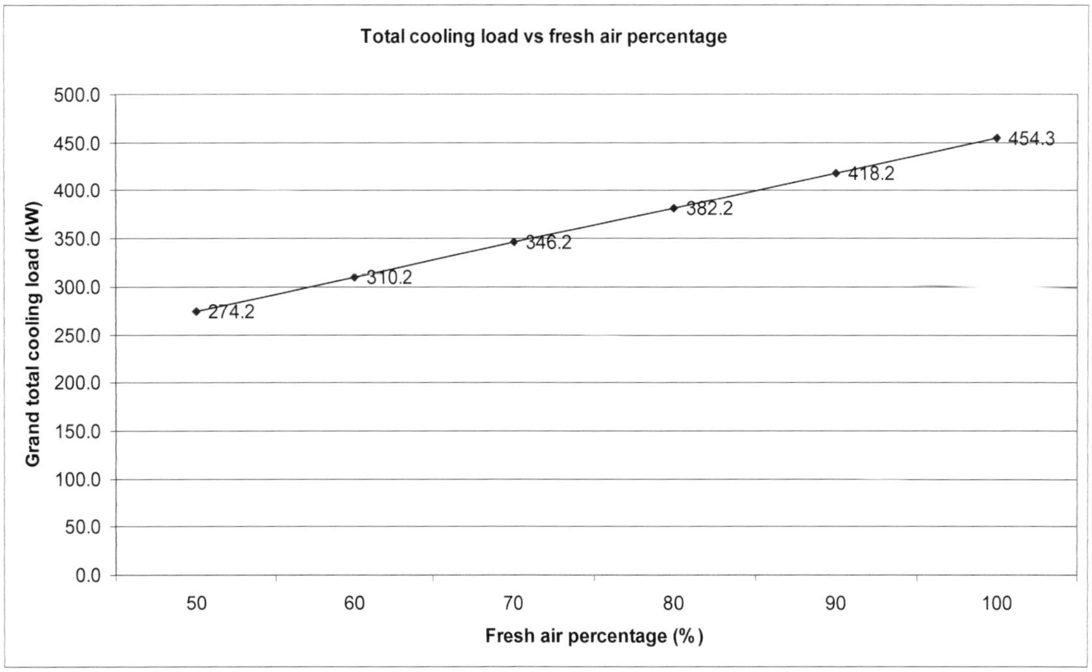

Figure 1: Total Cooling load vs. fresh air percentage

6.0 CITADEL SYSTEM DESIGN & REQUIREMENTS

6.1 General

A citadel is defined as the compartments within a vessel which form a group of interconnecting compartments enclosed by a vapour-tight boundary, within which filtered air can be circulated at a positive pressure to provide collective nuclear, biological and chemical protection. In the context of the small surface combatant the chosen citadel is a hardened COLPRO system because it provides the highest level of protection to the crew from a CBRN threat. The hardened COLPRO system is effectively a citadel that is integrated into the ship structure as opposed to an unhardened COLPRO system where by a toxic free area is achieved through the application of an inflatable CBRN liner.

The primary role of the citadel is to prevent the ingress of CRBN agents into compartments by providing an overpressure, gas tight boundary, and CBRN filters. To prevent CBRN agents from entering citadel during personnel entry and exit, cleansing stations and air locks are used for decontamination. Cleansing stations, air locks, and CBRN filters impose an additional air requirement on the vessel's HVAC system. Also to maintain the overpressure air losses due to uncontrolled leakages and breathing air replacement requires additional airflow.

This section provides an overview of the process in calculating the additional airflow required to maintain the CBRN citadel system. The following additional air flow requirement was calculated and added to the total air quantity calculated in Section 6:

- Losses due to uncontrolled leakages; and
- Breathing air replacement (Control of CO_2).

In this section the following areas are also addressed:

- Cleansing station and airlock purging;
- Cleansing station arrangements; and;
- Citadel Subdivisions; and
- CBRN Filter requirements.

6.2 Citadel Compartments

Areas generally considered part of the citadel are spaces vital to ship operations such as the bridge, communications, and crew living quarters. The number and type of citadel compartments will vary with each vessel and is constrained by factors such as compartment arrangement, available space, and budget. The table in Annex D is a list of compartments that form the Small surface combatant CBRN citadel.

6.2 Air supply/intake Requirements

For a citadel equipped ship in normal open state, the supply air is sufficient to provide the required compartment air changes and fresh air flow requirements. In a closed down state such as during wartime, supply air is diverted via dampers through pre-particulate, particulate, and vapour filters before entering the air treatment/handling units. For efficient removal of CBRN agents, each filter is normally loaded with airflow ranging between $270m^3/hr$ to 340

m^3/hr. Therefore the total supply air flow entering the citadel dictates the number of CBRN filters required to filter the air.

The citadel air flow is usually air conditioned in a citadel system is the sum of the air flow required for the HVAC system including air changes, uncontrolled losses, and breathing air volume requirements. This leads to a total air quantity of 17697 m^3 per hour (5.9 kg/second).

6.3 Cleansing station and airlock purging requirements:

6.3.1 General

The purging air flow in the airlock shall be sufficient to provide at least five air changes per minute for removal of all contamination. This is equivalent to 300 air changes per hour (5 air changes/min x 60 mins/hr = 300 air changes/hr). The purging air flow for the airlocks and cleansing stations are only filtered air so it does not need to be added to the air-conditioned supply. The purging air requirements will only have an impact on the number of fans and filtration units required.

6.3.2 Airlock purging calculation

To maintain a citadel system the Small Surface Combatant requires airlocks in the following locations:

1. 4 airlocks with 2 each in the fwd and aft cleansing stations;
2. 4 airlocks for the four machinery spaces (assuming that each machinery space is independent);
3. 2 airlocks for the contaminable spaces (hangar and paint storeroom); and
4. 1 airlock for access between the fwd and aft sub-citadels.

This leads to a total of 11 airlocks for the Small Surface Combatant citadel. Each airlock occupies a volume of approximately 1 m^3. The total swept air lock volume will be 11 x 1 m^3 = 11 m^3. The required airlock purging airflow is 11 m^3/air change x 300 air changes/hr = 3300 m^3/hr.

6.3.3 Cleansing station air change calculation

The largest volume compartment in the cleansing station generally receives a minimum of five air changes in the processing interval between men. The processing interval between personnel entering the cleansing station is specified to be no more than five minutes. The cleansing station air change requirement is five air changes per five minutes or 60 air changes per hour.

The volume of a typical cleansing station with stretcher capabilities is 39 m^3. The largest compartment in the cleansing station is typically the first stage compartment at 24 m^3. Because there are two cleansing stations on the Small Surface Combatant (Aft and Fwd cleansing stations) the total volume is 2 x 24 m^3 = 48 m^3. For 60 air changes per hour the cleansing station air change equates to 48 m^3/air change x 60 air changes/hr = 2880 m^3/hr.

6.3.4 Total purging airflow requirement

The total purging airflow requirement for cleansing station and airlocks is 3300 + 2880 = 6180 m^3/hr

6.4 Breathing Air Volume Requirements:

6.4.1 Breathing Air Volume Formula

The breathing air volume equation determines the quantity of filtered air that is required to offset the CO_2 generated from personnel breathing within a compartment. The breathing air volume varies depending on whether the person is resting, performing light work, or performing normal work. For this design exercise, the breathing air volume was assumed for personnel performing normal work loads at 1.25 m^3/hr/person.

Filtered air = $BAV \times N \times \left(\frac{A}{B_2 - B_1} \right)$

Where:

BAV = breathing air volume;
BAV resting = 0.5 m^3/hr/person;
BAV light work = 0.75 m^3/hr/person;
BAV Working = 1.25 m^3/hr/person;
N = number of persons
A = CO_2 generated during breathing (4%)
B_2 = Permitted CO_2 concentration in fully manned compartment;
B_1 = CO_2 content in fresh air (0.03%)

CO_2 permitted concentration in various types of compartments:

Operational spaces	0.15%
Berthing, resting spaces	0.25%
Dining, lounge spaces	0.25%
Workshops, offices, stores	0.45%

6.4.2 Citadel Breathing Air volume calculation

Breathing air volume was assumed worst case scenario so BAV working of 1.25 m^3/hour/person was selected. The number of persons was taken as the total complement of the platform which was 45. An average figure of 0.25% was selected for the permitted CO_2 concentration in fully manned compartment. Given these parameters the filtered air required was calculated to be:

$$1.25 \times 45 \times \left(\frac{4}{0.25 - 0.03} \right) = 1022.7\ \text{m}^3/\text{hr}$$

6.5 Uncontrolled Leakage Distribution:

The uncontrolled leakages are based on a requirement of 900 m^3/hr of filtered fresh air for every 3400 m^3 of citadel volume that is above the deep water line. It is assumed that the uncontrolled leakages were distributed evenly across the citadel areas of the ship. The breakdown of the citadel volume calculation is described in Annex D. The citadel volume, including 300 m^3 allowance for passageways, was calculated to be 3342 m^3.

$$900 \times \left(\frac{3342}{3400}\right) = 884 \text{ m}^3/\text{hr}$$

Therefore the air flow requirement due to uncontrolled leakages is calculated to be 884 m^3/hr.

6.6 Citadel Sub-division Overpressure

According to the overpressure requirements of Det Noske Veritas Part 6 Chapter 10 – *Nuclear, biological and chemical protection* there shall be a positive overpressure of at least 500 Pa between the citadel and the external environment.

7.0 DISCUSSION

From this HVAC design study, it can be observed that a chilled water plant with a cooling capacity of at least 300 kW is required to provide a design temperature of 27 degrees Celsius with 50% relative humidity. Increasing the amount of fresh air supplied to the HVAC system will increase the cooling load. At 100% fresh air system will require a cooling capacity of 454 kW.

Another critical factor that determines the cooling capacity of the chilled water plants is the temperature of the seawater and the performance of the seawater condensers. It is vital that an appropriate condenser is chosen for the chilled water plant. An undersized or inefficient condenser will result in the refrigerant not being cooled sufficiently and consequently the cooling capacity of the plant will be reduced. High temperature seawater passing through the condensers will also reduce its heat transfer potential.

8.0 REFERENCES

1. International Standards Organisation 7547:2002(E)

2. DA9 Air conditioning – load estimation, *The Australian institute of Refrigeration heating and air-conditioning and heating (Inc.)*, 1982

3. Det Noske Veritas Part 6 Chapter 10 – Nuclear, biological and chemical protection, *Rules for high speed, light craft and Naval surface craft*, 2011.

Annex A: Compartment Volumes

Name	Floor Area (m2)	Height	Width	Length	Fwd Bulk width	Aft Bulk width	Port	Stbd	Fwd Bulk (m2)	Aft Bulk (m2)	Port (m2)	Stbd Head (m2)	Com plim ent	Volume (m^3)
03 Deck														
Bridge	91.0	2.4	13.4	9.1	5.3	5.3	13.4	13.4	12.7	12.7	32.2	32.2	6.0	218.4
HVAC and Systems	15.9	2.4	3.2	5.0	3.2	3.2	2.4	2.4	7.7	7.7	5.8	5.8		38.2
Public Head	3.7	2.4	2.0	1.9	2.0	2.0	1.9	1.9	4.8	4.8	4.4	4.4		8.9
02 Deck														
HVAC and Systems	31.8	2.5	7.5	4.4	6.1	5.5	7.7	7.7	15.3	13.8	19.3	19.3		79.4
Ships Op Office	24.9	2.5	4.3	6.2	2.2	4.3	6.8	6.2	5.5	10.8	17.0	15.5	3.0	62.3
CO's Day Cabin	21.0	2.5	4.3	6.0	4.3	2.8	6.0	7.5	10.8	7.0	15.0	18.7	1.0	52.5
Pantry	5.2	2.5	1.5	3.5	1.5	1.5	3.5	3.5	3.8	3.8	8.8	8.8		13.0
CO's Sleeping Cabin	13.1	2.5	4.3	3.0	4.3	4.3	3.0	3.0	10.8	10.8	7.5	7.5	1.0	32.8
Comcen	25.3	2.5	4.3	6.3	2.3	5.8	5.0	6.8	5.8	14.5	12.5	17.0	8.0	63.3
Single Berth Officer 1	11.0	2.5	2.8	3.9	2.8	2.8	3.9	3.9	7.0	7.0	9.8	9.8	1.0	27.5
HVAC and Systems	3.8	2.5	1.6	2.3	1.6	1.6	2.4	2.4	4.0	4.0	6.0	6.0		9.5
Single Berth Officer 2	11.0	2.5	2.8	3.9	2.8	2.8	3.9	3.9	7.0	7.0	9.8	9.8	1.0	27.5
Repair Base 1	5.4	2.5	2.6	2.0	2.6	2.6	2.0	2.0	6.5	6.5	5.0	5.0		13.5
01 Deck														
Ships Magazine	12.6	1.8	4.4	2.3	4.0	5.2	2.3	2.3	7.2	9.4	4.1	4.1		22.7
Public Head	4.2	2.1	2.4	1.9	2.0	2.4	1.9	1.9	4.2	5.0	4.0	4.0		8.8
Battery Comp	3.3	2.5	1.2	2.8	0.9	1.5	2.8	2.9	2.3	3.8	7.0	7.3		8.3
Electrical Equipment Compartment	4.2	2.5	1.5	2.8	1.5	1.5	2.8	2.8	3.8	3.8	7.0	7.0		10.5
Unassigned	11.6	2.5	3.3	3.5	3.0	3.7	3.6	3.5	7.5	9.3	8.9	8.7	0.0	29.0
Engineering	16.4	2.5	4.1	4.0	3.7	4.5	4.1	4.0	9.2	11.3	10.2	10.0	2.0	41.0

Admin														
Electronics Workshop	15.7	2.5	4.6	3.4	4.5	4.6	3.4	3.4	11.3	11.5	8.5	8.5	1.0	39.3
Sickbay	22.3	2.5	4.6	5.1	4.6	4.6	5.1	5.1	11.5	11.5	12.7	12.7	1.0	55.8
Public Head	3.3	2.5	2.3	1.4	2.3	2.3	1.4	1.4	5.8	5.8	3.5	3.5		8.1
Hangar		2.5							0.0	0.0	0.0	0.0		0.0
Small Arms Stowage	9.3	2.5	2.2	4.7	2.0	2.2	5.5	4.8	5.0	5.5	13.7	12.0		23.3
Unassigned	6.5	2.5	1.9	3.5	1.9	1.9	3.5	3.5	4.6	4.6	8.7	8.7	0.0	16.3
Paint Store	5.2	2.5	1.5	3.5	1.1	1.9	3.5	3.6	2.8	4.6	8.7	8.9		13.0
Ships Office/ Pay Office	20.2	2.5	4.2	5.4	3.7	3.8	6.2	5.5	9.2	9.5	15.5	13.8	1.0	50.5
Aviation Admin Office	7.6	2.5	3.8	2.0	3.8	3.8	2.0	2.0	9.5	9.5	5.0	5.0	1.0	19.0
Avionics and Helo Workshop	24.2	2.5	4.6	6.5	4.6	2.5	6.5	8.6	11.5	6.2	16.2	21.5	1.0	60.5
Emergency Generator	10.5	2.5	2.1	5.0	2.1	2.1	5.0	5.0	5.3	5.3	12.5	12.5		26.3
1 Deck														
Naval Stores 1	12.1	3.2	6.1	2.4	5.5	6.8	2.5	2.5	17.6	21.8	8.0	8.0		38.7
Naval Stores 2	20.3	3.1	3.7	6.5	2.8	4.5	6.7	6.5	8.7	13.9	20.6	20.0		62.9
Berth (2) 1	13.7	3.1	4.8	3.2	4.5	5.2	3.2	3.2	13.9	16.1	10.0	9.8	2.0	42.5
Berth (2) 2	13.2	3.1	4.7	2.9	4.7	4.7	2.9	2.9	14.5	14.5	8.8	8.8	2.0	41.0
2 Berth PO 1	13.2	3.1	4.7	2.9	4.7	4.7	2.9	2.9	14.4	14.4	8.8	8.8	2.0	41.0
2Berth CPO 1	13.2	3.1	4.7	2.9	4.7	4.7	2.9	2.9	14.5	14.5	8.8	8.8	2.0	41.0
Pantry	7.3	3.1	3.2	2.3	3.2	3.2	2.3	2.3	9.8	9.8	7.1	7.1		22.6
Wardroom	41.6	3.1	6.9	8.8	3.7	3.6	8.8	13.4	11.6	11.2	27.3	41.5	6.0	129.0
Public Head	2.6	3.1	1.6	1.6	1.6	1.6	1.6	1.6	5.0	5.0	5.0	5.0		8.1
UUV Support and Battery Charging	11.7	3.1	3.6	5.0	1.5	3.6	5.0	7.1	4.7	11.2	15.5	22.0		36.3
Demolition Magazine	5.4	3.1	3.6	1.5	3.6	3.6	1.5	1.5	11.2	11.2	4.7	4.7		16.7
Boat Gear Store	12.2	3.1	3.6	4.5	3.6	1.6	4.5	6.5	11.2	5.0	14.0	20.2		37.8
Diving Gear Store	4.0	3.1	2.0	2.0	2.0	2.0	2.0	2.0	6.2	6.2	6.2	6.2		12.4
Repair Base 2	3.7	3.1	3.0	1.5	2.8	3.2	1.5	1.6	8.7	9.9	4.7	4.8		11.5

Damage Control Store	12.2	3.1	3.2	4.5	3.2	3.6	5.3	4.7	9.9	11.1	16.4	14.4		37.8
Berth (2) 3	14.5	3.1	5.2	3.6	3.6	5.2	4.4	3.7	11.1	16.1	13.5	11.4	2.0	45.0
2 Berth PO 2	13.4	3.1	4.7	2.9	4.7	4.7	2.9	2.9	14.5	14.5	8.8	8.8	2.0	41.7
2 Berth PO 3	13.3	3.1	4.7	2.8	4.7	4.7	2.8	2.8	14.5	14.5	8.8	8.8	2.0	41.3
2 Berth CPO 2	13.3	3.1	4.7	2.9	4.7	4.7	2.9	2.9	14.5	14.5	8.8	8.8	2.0	41.1
2 Berth Officer	13.4	3.1	4.7	2.9	4.7	4.7	2.9	2.9	14.5	14.5	8.8	8.8	2.0	41.5
Berth (2) 4	13.4	3.1	4.7	2.9	4.7	4.7	2.9	2.9	14.5	14.5	8.8	8.8	2.0	41.5
Garbage Compactor/Store	13.3	3.1	3.6	4.5	3.6	1.5	6.6	4.5	11.2	4.7	20.5	14.0		41.2
Dry Garbage	4.5	3.1	1.5	3.0	1.5	1.5	3.0	3.0	4.7	4.7	9.3	9.3		14.0
2 Deck														
General Store	16.2	2.9	4.6	6.6	2.0	5.8	6.9	6.9	5.7	16.7	19.9	19.9		47.0
Accommodation Store	7.9	2.9	3.6	3.9	2.5	3.6	4.1	3.9	7.2	10.4	11.7	11.3		22.9
Laundry	18.3	2.9	4.3	5.7	3.6	5.1	5.9	5.7	10.4	14.7	17.0	16.5		53.1
SS Mess	34.4	2.9	5.5	9.6	5.1	2.9	9.7	12.9	14.7	8.4	28.1	37.3	9.0	99.8
SS Pantry and JS Servery	8.8	2.9	2.5	3.5	2.5	2.5	3.5	3.5	7.3	7.3	10.2	10.2		25.4
Galley	24.2	2.9	6.2	4.6	6.2	6.2	4.6	4.6	17.8	17.8	13.3	13.3		70.2
Stores Cool Cold and Dry	23.0	2.9	6.2	5.6	6.2	10.9	4.2	5.6	17.8	31.5	12.2	16.2		66.7
Engineering Store	10.4	2.9	2.2	7.2	2.2	2.2	7.2	7.2	6.3	6.3	20.9	20.9		30.2
Eng Workshop	16.1	2.9	4.5	4.2	4.5	4.5	4.2	4.2	13.0	13.0	12.2	12.2	5.0	46.7
JS (2) 1	13.1	2.9	4.7	2.9	4.7	4.7	2.9	2.9	13.5	13.5	8.3	8.3	2.0	38.1
JS (4) 1	13.6	2.9	2.8	4.9	2.8	2.8	4.9	4.9	8.0	8.0	14.1	14.1	4.0	39.3
JS (4) 2	13.6	2.9	2.8	4.9	2.8	2.8	4.9	4.9	8.0	8.0	14.1	14.1	4.0	39.3
JS (8)	32.4	2.9	3.7	8.1	3.9	3.7	8.1	8.1	11.3	10.6	23.6	23.6	8.0	94.0
Unassigned	17.2	2.9	4.5	4.8	4.5	4.5	4.8	4.8	13.2	13.2	13.9	13.9	0.0	49.9
Steering Gear	12.4	2.9	3.5	3.6	3.5	3.5	3.6	3.6	10.2	10.2	10.4	10.4		36.0
General Store	6.0	2.9	2.7	3.9	2.2	2.6	4.6	4.1	6.4	7.5	13.3	11.7		17.4
Canteen and Canteen Store	14.3	2.9	4.1	5.7	2.6	3.8	7.4	5.9	7.5	11.0	21.4	17.1	1.0	41.5
JS Dining	38.0	2.9	5.5	9.6	4.8	5.2	10.3	9.7	13.9	15.0	29.8	28.1	16.0	110.2
Scullery	4.5	2.9	1.8	2.5	1.8	1.8	2.5	2.5	5.2	5.2	7.3	7.3		13.1

JS Rec Area	20.8	2.9	5.9	5.1	4.1	5.9	6.9	5.1	11.7	17.0	20.0	14.8	5.0	60.3
Dry Store	4.6	2.9	2.7	2.6	2.7	2.7	2.6	2.6	7.8	7.8	7.5	7.5		13.3
MCR and DCC Central	27.3	2.9	6.5	4.2	6.5	6.5	4.2	4.2	18.8	18.8	12.2	12.2	3.0	79.2
JS (2) 2	13.2	2.9	4.7	2.9	4.7	4.7	2.9	2.9	13.5	13.5	8.3	8.3	2.0	38.3
JS (4) 3	13.4	2.9	2.8	4.9	2.8	2.8	4.9	4.9	8.0	8.0	14.1	14.1	4.0	38.9
JS (2) 3	13.2	2.9	2.2	4.9	2.2	2.2	2.2	4.9	6.4	6.4	6.4	14.1	2.0	38.3
Baggage Store	8.6	2.9	4.0	2.1	4.0	4.0	2.1	2.1	11.6	11.6	6.2	6.2		24.9
JS (6)	20.6	2.9	2.9	7.1	2.9	2.9	7.1	7.1	8.3	8.3	20.7	20.7	6.0	59.6
Gym	17.2	2.9	3.6	4.8	3.6	3.5	4.8	4.8	10.5	10.2	13.9	13.9	4.0	49.9
Steering Gear 2	12.4	2.9	3.5	3.6	3.5	3.4	3.6	3.6	10.2	9.8	10.4	10.4		35.9
Echo Sounder/ Equipment Room	14.3	1.7	2.9	4.8	2.4	3.7	4.8	4.8	4.0	6.2	8.2	8.2		24.3
FWD Auxiliary Machinery	48.5	1.7	5.2	9.6	3.7	6.0	9.7	9.7	6.2	10.3	16.5	16.5		82.5
Generator Room	67.5	1.7	6.5	10.2	6.0	7.2	10.2	10.2	10.3	12.3	17.4	17.4		114.8
Engine Room	83.4	1.7	7.2	11.4	7.2	6.5	11.4	11.4	12.3	11.1	19.4	19.4		141.8
AFT Auxiliary Machinery Room	25.7	1.7	6.5	6.8	6.5	0.0	7.6	7.6	11.1	0.0	12.9	12.9		43.7

Annex B: Compartment Heat loads

Index	Compartment	Comp. Sensible Heat Gain Qs	Comp.Total Heat Gain Qt	Comp. Sensible Heat Ratio	#Men	Revised Comp air quantity	Comp. Air Quantity Mc	Compt. volume	6 Air changes per hour (minimum)
Units		kW	kW	SHRc	N	m^3/s	m^3/s	m^3	m^3/s
1	Bridge	15.13	16.12	0.94	6	0.57	0.57	218.40	0.36
2	Ships Operation Office	3.27	3.76	0.87	3	0.13	0.13	62.25	0.10
3	CO's Day Cabin	0.63	0.80	0.79	1	0.09	0.03	52.50	0.09
4	CO's Sleeping Cabin	0.54	0.71	0.77	1	0.05	0.03	32.75	0.05
5	Comcen	6.89	8.21	0.84	8	0.29	0.29	63.25	0.11
6	Single Berth Officer 1	0.63	0.79	0.79	1	0.05	0.03	27.50	0.05
7	Single Berth Officer 2	0.71	0.87	0.81	1	0.05	0.03	27.50	0.05
8	Unassigned 01 Deck Fwd Port	0.44	0.44	1.00	0	0.05	0.02	29.00	0.05
9	Engineering Admin	2.84	3.17	0.90	2	0.11	0.11	41.00	0.07
10	Electronics Workshop	1.63	1.79	0.91	1	0.07	0.06	39.25	0.07
11	Sickbay	2.97	3.14	0.95	1	0.11	0.11	55.75	0.09
12	Unassigned 01 Deck Fwd Stbd	0.31	0.31	1.00	0	0.03	0.01	16.25	0.03
13	Ships Office	0.87	1.03	0.84	1	0.08	0.04	50.50	0.08
14	Aviation Admin Office	0.40	0.56	0.71	1	0.03	0.02	19.00	0.03
15	Avionics and Helo Workshop	2.05	2.22	0.93	1	0.10	0.08	60.50	0.10
16	Berth (2) 1 Port	0.83	1.16	0.72	2	0.07	0.04	42.47	0.07
17	Berth (2) 2 Port	0.62	0.95	0.65	2	0.07	0.03	41.04	0.07
18	2 Berth PO 1 Port	0.59	0.92	0.64	2	0.07	0.03	41.04	0.07
19	2 Berth CPO Port	0.59	0.92	0.64	2	0.07	0.03	41.04	0.07
20	Wardroom	2.10	3.09	0.68	6	0.21	0.11	128.96	0.21
21	Berth (2) Stbd 1	0.84	1.17	0.72	2	0.07	0.04	44.95	0.07
22	2 Berth PO 1 Stbd	0.62	0.95	0.65	2	0.07	0.03	41.66	0.07
23	2 Berth PO 2 Stbd	0.59	0.92	0.64	2	0.07	0.03	41.26	0.07
24	2 Berth CPO Stbd	0.62	0.95	0.65	2	0.07	0.03	41.14	0.07
25	2 Berth Officer	0.62	0.95	0.65	2	0.07	0.03	41.54	0.07

26	Berth (2) Stbd 2	0.79	1.12	0.70	2	0.07	0.04	41.54	0.07
27	SS Mess	5.24	7.23	0.73	9	0.26	0.26	99.76	0.17
28	Eng Workshop	2.33	3.16	0.74	5	0.11	0.11	46.72	0.08
29	JS (2) Aft Port	0.83	1.16	0.71	2	0.06	0.04	38.11	0.06
30	JS (4) Aft Port 1	0.77	1.43	0.54	4	0.07	0.05	39.32	0.07
31	JS (4) Aft Port 2	0.60	1.26	0.48	4	0.07	0.04	39.32	0.07
32	JS (8)	1.74	3.06	0.57	8	0.16	0.11	93.96	0.16
33	Unassigned 2 Deck Aft Port	0.86	0.86	1.00	0	0.08	0.03	49.88	0.08
34	Canteen and Canteen Store	0.88	1.04	0.84	1	0.07	0.04	41.47	0.07
35	JS Dining	2.57	5.21	0.49	16	0.18	0.18	110.20	0.18
36	JS Rec Area	1.82	2.65	0.69	5	0.10	0.09	60.32	0.10
37	MCR and DCC Central	2.57	3.07	0.84	3	0.13	0.11	79.17	0.13
38	JS (2) Aft Stbd	0.74	1.07	0.69	2	0.06	0.04	38.31	0.06
39	JS (4) Aft Stbd	0.60	1.26	0.47	4	0.06	0.04	38.92	0.06
40	JS (2) Aft Stbd 2	0.51	0.84	0.61	2	0.06	0.03	38.31	0.06
41	JS (6)	1.13	2.12	0.53	6	0.10	0.08	59.60	0.10
42	Gym	1.03	1.69	0.61	4	0.08	0.06	49.88	0.08
	TOTAL	72.34	94.13	0.74	129.00	4.39	3.34		3.78

Annex C: Outside air total heat and Grand total heat with varying fresh air proportions

	50% Fresh Air		60% Fresh Air		70% Fresh Air		80% Fresh Air		90% Fresh Air		100% Fresh Air	
Compartment	OATH (kW)	GTH (kW)	OATH (kW)	GTH (kW)	OATH (kW)	GTH (kW)	OATH (kW)	GTH (kW)	OATH (kW)	GTH (kW)	OATH (kW)	GTH (kW)
Bridge	23.5	39.6	28.2	44.3	32.9	49.0	37.5	53.7	42.2	58.4	46.9	63.0
Ships Operation Office	5.5	9.2	6.6	10.3	7.7	11.4	8.8	12.5	9.9	13.6	11.0	14.7
CO's Day Cabin	3.6	4.4	4.3	5.1	5.0	5.8	5.7	6.5	6.5	7.3	7.2	8.0
CO's Sleeping Cabin	2.2	2.9	2.7	3.4	3.1	3.8	3.6	4.3	4.0	4.7	4.5	5.2
Comcen	12.0	20.2	14.3	22.6	16.7	25.0	19.1	27.3	21.5	29.7	23.9	32.1
Single Berth Officer 1	1.9	2.7	2.3	3.0	2.6	3.4	3.0	3.8	3.4	4.2	3.8	4.6
Single Berth Officer 2	1.9	2.8	2.3	3.1	2.6	3.5	3.0	3.9	3.4	4.3	3.8	4.6
Unassigned 01 Deck Fwd Port	2.0	2.4	2.4	2.8	2.8	3.2	3.2	3.6	3.6	4.0	4.0	4.4
Engineering Admin	4.6	7.8	5.5	8.7	6.5	9.6	7.4	10.5	8.3	11.5	9.2	12.4
Electronics Workshop	2.7	4.5	3.2	5.0	3.8	5.6	4.3	6.1	4.8	6.6	5.4	7.2
Sickbay	4.6	7.7	5.5	8.6	6.4	9.5	7.3	10.4	8.2	11.4	9.1	12.3
Unassigned 01 Deck Fwd Stbd	1.1	1.4	1.3	1.6	1.6	1.9	1.8	2.1	2.0	2.3	2.2	2.5
Ships Office	3.5	4.5	4.1	5.2	4.8	5.9	5.5	6.6	6.2	7.3	6.9	7.9
Aviation Admin Office	1.3	1.9	1.6	2.1	1.8	2.4	2.1	2.6	2.3	2.9	2.6	3.2
Avionics and Helo Workshop	4.1	6.4	5.0	7.2	5.8	8.0	6.6	8.8	7.5	9.7	8.3	10.5
Berth (2) 1 Port	2.9	4.1	3.5	4.6	4.1	5.2	4.6	5.8	5.2	6.4	5.8	7.0
Berth (2) 2 Port	2.8	3.8	3.4	4.3	3.9	4.9	4.5	5.4	5.1	6.0	5.6	6.6
2 Berth PO 1 Port	2.8	3.7	3.4	4.3	3.9	4.9	4.5	5.4	5.1	6.0	5.6	6.5
2 Berth CPO Port	2.8	3.7	3.4	4.3	3.9	4.9	4.5	5.4	5.1	6.0	5.6	6.5
Wardroom	8.8	11.9	10.6	13.7	12.4	15.4	14.1	17.2	15.9	19.0	17.6	20.7

Berth (2) Stbd 1	3.1	4.2	3.7	4.9	4.3	5.5	4.9	6.1	5.5	6.7	6.2	7.3
2 Berth PO 1 Stbd	2.9	3.8	3.4	4.4	4.0	4.9	4.6	5.5	5.1	6.1	5.7	6.7
2 Berth PO 2 Stbd	2.8	3.7	3.4	4.3	4.0	4.9	4.5	5.4	5.1	6.0	5.6	6.6
2 Berth CPO Stbd	2.8	3.8	3.4	4.3	3.9	4.9	4.5	5.5	5.1	6.0	5.6	6.6
2 Berth Officer	2.8	3.8	3.4	4.4	4.0	4.9	4.5	5.5	5.1	6.1	5.7	6.6
Berth (2) Stbd 2	2.8	4.0	3.4	4.5	4.0	5.1	4.5	5.7	5.1	6.2	5.7	6.8
SS Mess	10.5	17.7	12.6	19.8	14.7	22.0	16.8	24.1	18.9	26.2	21.0	28.3
Eng Workshop	4.6	7.7	5.5	8.7	6.4	9.6	7.3	10.5	8.3	11.4	9.2	12.3
JS (2) Aft Port	2.6	3.8	3.1	4.3	3.7	4.8	4.2	5.3	4.7	5.9	5.2	6.4
JS (4) Aft Port 1	2.7	4.1	3.2	4.7	3.8	5.2	4.3	5.7	4.8	6.3	5.4	6.8
JS (4) Aft Port 2	2.7	4.0	3.2	4.5	3.8	5.0	4.3	5.6	4.8	6.1	5.4	6.6
JS (8)	6.4	9.5	7.7	10.8	9.0	12.1	10.3	13.4	11.6	14.6	12.9	15.9
Unassigned 2 Deck Aft Port	3.4	4.3	4.1	5.0	4.8	5.6	5.5	6.3	6.1	7.0	6.8	7.7
Canteen and Canteen Store	2.8	3.9	3.4	4.4	4.0	5.0	4.5	5.6	5.1	6.1	5.7	6.7
JS Dining	7.6	12.8	9.1	14.3	10.6	15.8	12.1	17.4	13.7	18.9	15.2	20.4
JS Rec Area	4.1	6.8	5.0	7.6	5.8	8.4	6.6	9.2	7.4	10.1	8.3	10.9
MCR and DCC Central	5.4	8.5	6.5	9.6	7.6	10.7	8.7	11.7	9.8	12.8	10.8	13.9
JS (2) Aft Stbd	2.6	3.7	3.1	4.2	3.7	4.7	4.2	5.3	4.7	5.8	5.2	6.3
JS (4) Aft Stbd	2.7	3.9	3.2	4.5	3.7	5.0	4.3	5.5	4.8	6.1	5.3	6.6
JS (2) Aft Stbd 2	2.6	3.5	3.1	4.0	3.7	4.5	4.2	5.0	4.7	5.6	5.2	6.1
JS (6)	4.1	6.2	4.9	7.0	5.7	7.8	6.5	8.6	7.3	9.5	8.2	10.3
Gym	3.4	5.1	4.1	5.8	4.8	6.5	5.5	7.1	6.1	7.8	6.8	8.5
TOTAL (kW)	180.1	**274.2**	216.1	**310.2**	252.1	**346.2**	288.1	**382.2**	324.1	**418.2**	360.1	**454.3**

Annex D – CBRN Citadel Compartments

Item	Compartment	Volume m^3	Item	Compartment	Volume m^3
03 Deck			**1 Deck (cont.)**		
1	Bridge	218	33	Damage Control Store	37.8
2	HVAC and systems	38.2	34	Berth (2) 3	45.0
3	Public head	8.9	35	2 Berth PO 2	41.7
02 Deck			36	2 Berth PO 3	41.3
4	Ships operation office	62.3	37	2 Berth CPO 2	41.1
5	CO's day cabin	52.5	38	2 Berth Officer	41.5
6	Pantry	13.0	39	Berth (2) 4	41.5
7	CO's Sleeping Cabin	32.8	**2 Deck**		
8	Comcen	63.3	40	Accommodation Store	22.9
9	Single Berth officer 2	27.5	41	Laundry	53.1
10	Repair Base 1	13.5	42	SS Mess	99.8
01 Deck			43	SS Pantry and JS Servery	25.4
11	Ship's magazine	22.7	44	Galley	70.2
12	Public head	8.8	45	Stores Cool Cold and Dry	66.7
13	Battery compartment	8.3	46	Engineering Store	30.2
14	Electrical equipment compartment	10.5	47	Eng Workshop	46.7
15	Unassigned	29.0	48	JS (2) 1	38.1
16	Engineering Admin	41.0	49	JS (4) 1	39.3
17	Electronics workshop	39.3	50	JS (4) 2	39.3
18	Sickbay	55.8	51	JS (8)	94.0
19	Public head	8.1	52	Unassigned	49.9
20	Small arms stowage	23.3	53	General Store	17.4
21	Ships office/pay office	50.5	54	Canteen and Canteen Store	41.5
22	Aviation admin office	19.0	55	JS Dining	110.2
1 Deck			56	Scullery	13.1
23	Naval Stores 2	62.9	57	JS Rec Area	60.3
24	Berth (2) 1	42.5	58	Dry Store	13.3
25	Berth (2) 2	41.0	59	MCR and DCC Central	79.2
26	2 Berth PO 1	41.0	60	JS (2) 2	38.3
27	2Berth CPO 1	41.0	61	JS (4) 3	38.9
28	Pantry	22.6	62	JS (2) 3	38.3
29	Wardroom	129.0	63	Baggage Store	24.9
30	Public Head	8.1	64	JS (6)	59.6
31	UUV Support and Battery Charging	36.3	65	Gym	49.9
32	Repair Base 2	11.5			

Integration of Unmanned Air Systems for Maritime Operations

Andrew Chadwick
Frazer-Nash Consultancy Limited, UK
a.chadwick@fnc.co.uk

INTRODUCTION

The benefits of operating unmanned air systems in the maritime environment, both for military and commercial applications, is slowly being recognised. As these systems develop, they bring with them the challenge of integrating the system with maritime platforms and their operating infrastructure.

This paper discusses the opportunities that the introduction of unmanned air systems into the maritime environment bring, identifies system requirements, current air system types, plus interoperability, integration and ship safety issues. In addition, this presentation examines the potential utilisation of unmanned air systems for maritime security, in particular for anti-piracy operations.

UNMANNED AIR SYSTEM TYPES

There is an increasing growth in the development of unmanned air systems (UAS), particularly for military usage, where there is significant market expansion being recognized globally. Interestingly, however, there appears to be limited growth, and fewer systems developed, for shipboard operations, even though this presents a more amenable environment for their usage, compared to land operations.

It is a view that one of the reasons that UAS have not been adopted more readily in the maritime is the lack of understanding of the advantages of using the new technology, and concern from current maritime air operators that the introduction of unmanned systems will result in a reduction of the use of manned aircraft. This is unlikely to occur, as the key to the successful introduction of unmanned systems, of any type, in this environment lies around the overall capability, or effect, that can be achieved with unmanned systems complementing manned aircraft and providing an extension of a ships' sensors. Historically, procurement has been based around specific platforms, however the introduction of unmanned systems forces planners and procurement agencies to review their overall capability requirements and needs in more detail.

System developers have not chosen to wait for these specific requirements to be developed. Their drive is to create a market for their UAS products and then take advantage of that market. Therefore the few systems that have been developed for the maritime environment have been developed with a specific market niche in mind that they hope will act as an enabler for other markets, for example the Boeing/Insitu Scan Eagle was developed for use in support of tuna fishing but is now being used, very successfully, for military Intelligence, Surveillance Reconnaissance (ISR) work, particularly in the United States and Australia.

ISR is a key capability, combined with interoperability, in enabling military, and in particular coalition, operations. Studies in the UK and overseas have identified that a military UAS capability could be utilized to fill capability gaps providing ISR for Force Projection and Force Protection.

Other UAS system types, specific to the maritime environment, also include the Northrop Grumman Fire Scout, Scheibel S-100 Camcopter, the Bell Eagle Eye, and the Cassidian Orka 1200. Conversely, a number of Medium-Altitude-Long-Endurance (MALE) and High-Altitude-Long-Endurance (HALE) systems have been used, as off-board assets in the maritime environment in support of operations and demonstrations. These include the Northrop Grumman Global Hawk, General Atomics-Aeronautical Systems Inc. Mariner, and Israeli Aircraft Industries Heron. Indeed, the variety of options introduces a further question regarding what capability unmanned systems are required to provide, and whether there is a need for them to be based both on-board and ashore. However, for the purposes of this paper we shall focus on the use of on-board based UAS for military purposes.

SYSTEM REQUIREMENTS

Interoperability

If unmanned air systems are to be introduced into the force mix, it is important to get the balance right, in terms of integration, force requirements, power projection, interoperability and cost. UAS need to match or enhance current maritime ISR requirements, i.e. shared situational awareness - delivered through efficient and cost-effective use of joint resources, including multi-national; information collection – the acquisition of data from all-sources which is then processed and analysed; information handling – the processing, dissemination and display of gathered information, including communications for both situational awareness and command execution; battlespace management - effective and efficient direction of forces through accrued situation awareness; and network enabled capability - optimising the maritime and off-board resources by networking sensors, information handling systems, command and control systems.

In addition, of significant importance is the interoperability of UAS with other assets – nationally and internationally – both manned and unmanned (including Unmanned Surface Vehicles and Unmanned Underwater Vehicles). Interoperability enhances capability, flexibility and results in a common operating picture.

Integration

In addition to interoperability requirements, it must be possible to integrate UAS into the maritime environment. For off-board platforms, integration requirements are centred around information collection, management and dissemination, whilst on-board UAS systems have additional integration considerations.

On-board footprint is especially important on all surface vessels. For UAS, these include the air vehicle itself, the launch and recovery system(s), control system, command & control (C2) antennae, crewing, integration with on-board ship sensors, communications equipment, etc. Obviously for smaller UAS, whose role may be to supplement the capability of manned aircraft, the physical footprint of the aircraft may be small, but the associated supporting infrastructure and control system may not be.

As with manned aircraft, concepts of operation need to be developed for UAS, including the derivation of Ship-UAS Operating Limits (SUOLs) for launch and recovery of on-board based aircraft. This is especially the case for automatic take-off and landing systems (ATOLs) where there is no operator intervention except in exceptional circumstances.

Operational integration raises the issue of management of UAS in the air traffic environment. Regulation dictates that UAS must integrate seamlessly with manned systems, including the ability to detect, sense and avoid other air users (the unmanned equivalent of "see and avoid" for manned aircraft). This is particularly necessary in operations in the littoral. However it may be argued that in a more benign environment further offshore, where overflight of civilians and the likelihood of other air users being present is lower, this is not such a critical issue. There is still a need for a method to sense and avoid other air users, but it could be argued that, considering the extremely low density of air traffic, shipping, etc. the fidelity of such a system (or methodology) need not be as prescribed as for busier areas. The notion of a variable safety environment – which is operation based – may be possible, dependent upon the perceived level of risk.

The most significant issue relating to the integration of UAS into the maritime environment is the on-board Electro-Magnetic (EM) environment. A UAS must integrate safely on-board, and issues have arisen in previous trials, with the EM compatibility of UAS C2 systems Those who work in the environment will confirm, that the on-board EMC environment is not benign. It is imperative when designing UAS for maritime operations that the system - particularly the control system, antennae and supporting equipment - are EM compatible with the ship's systems. On-board trials have taken place where ship's radars have had to be powered down during the launch and recovery phase of a UAS operation, as the UAS C2 system has interfered with the ship systems. This is not a satisfactory situation in an operational scenario.

Architecture

When designing UAS for use in the maritime environment it is important that the system is designed for versatility. By their very nature, UAS should be designed such that they can respond readily to changes – these are multi-role, multi-configurable systems – hence an open architecture is preferred, which is re-configurable, permitting "plug and play" operations and enable seamless system updates, for use in a variety of roles, including coastal patrol, customs surveillance, search and rescue, harbour security, maritime security, etc.

For military operations, a UAS needs to have an architecture that facilitates a network enabled capability, optimising the maritime and off-board resources by networking sensors, information handling systems, command & control systems, plus the compatibility of systems, data formats, and data protocols which are imperative for multi-national operations providing the ability to act faster with greater agility, precision and control. However, within this, bandwidth problems also need to be resolved. There is an ever decreasing amount of available bandwidth for military operations. Interestingly, this issue also translates into the civil UAS arena in the current Single European Sky ATM Research programme (SESAR®) which is the technological and operational dimension of the Single European Sky (SES) initiative targeting future air traffic management performance needs. SESAR® aims to develop a new generation of Air Traffic Management (ATM) System in

Europe. In particular SESAR® is researching the concept of ATM as a System Wide Information Management system – a networked-enabled capability.

The architecture in which a UAS operates needs to take into account data management requirements, particularly data fusion requirements need to be considered to achieve timely analysis. In addition, and dependent upon the type of mission being flown, the architecture may need to enable data reach-back, enabling more detailed, slower-time, deeper analysis. This leads to higher volume databases, introduces the requirement for specialist analysts (either on-board or ashore), and raises the question of how much analysis could actually be automated. Automatic data analysis techniques have significantly developed in recent years reducing the resource burden, and should be exploited more. In addition higher volume of databases raises issues relating to data storage (compatible formats, storage space, historical analysis, and user interface) to be considered.

SHIP SAFETY

Naval ship safety management policy has been developed by the majority of global naval forces. Policy generally requires a safety regime to be developed for all ships and equipment to address each stage of the ship or equipment lifecycle. The level of effort and the depth of analysis given to the goal-setting approach of the safety regime should at all times be proportionate to the complexity of the ship or equipment, the perceived level of risk, time at risk and the hazard posed.

For example, in the United Kingdom, the Ministry of Defence ship safety regime is based upon a safety management system, safety case evidence and complimentary safety case reports. Specific ship-borne operations and the ship/air interface analysis are covered by the Ship Aviation Safety Report. This is a stand alone document within the Platform Safety Case which is used to provide evidence towards the Letter of Release. The Letter of Release authorises operation of a specific aircraft type on a specific ship or class of ships. As with all manned aircraft, ship-based UAS operations require a Ship Aviation Safety Report and Letter of Release.

These processes may be adopted for both military, security and civil use of unmanned aircraft.

The safety case for operating ship-borne UAS must cover safety in the role, which includes an understanding of the operational context of the use of the systems, both in peacetime and war. It also needs to consider the operational limitations of the system, such as weather limits, flight deck handling, datalink range, and SUOLs - as for any other aircraft - plus the issue of collision avoidance techniques or processes. Of note, the safety case for a UAS must also cover the entire system not just the aircraft, i.e. the air vehicle, the payload element, the datalink(s) for C2 and video/data relay, and the control system.

Human Factors also plays a significant factor in the safe operation of UAS: both for the operator (or "UAV-pilot") and maintainer, in terms of skill sets, manpower integration, teamwork, management, task allocation, shift patterns and fatigue, particularly as some unmanned aircraft have flight times of over 6 hours.

OPERATING ROLES

Typically, for maritime missions, an organic capability is required for what is termed the dull, dirty, dangerous missions that UAS are designed for. Dull roles are those that take place over long periods of time (such as surveillance tasks), dirty roles are those tasks involving operations in contaminated areas, and dangerous roles are those that take place in dangerous environments.

Example roles for consideration that a UAS can perform may include:

- Battlegroup support: maritime/littoral battlespace surveillance, targeting, imagery, command and control;
- Situational awareness;
- Persistent maritime and land ISR capability;
- Airborne communications relay (Beyond Line of Site - BLOS);
- Extension of ship's sensors;
- Working around fleet;
- Search/detect/exploit;
- Naval Fire Support/Coastal suppression;
- Littoral Manoeuvre ISR support;
- Over The Horizon - covert, Anti Surface Warfare (ASuW), offensive and defensive operations;
- Detect/Track/Identify small targets – overcoming a ships' inability to detect and track small targets, for example FIACS, determination of a FIAC's intent at range, and provide an extension of a ship's situational awareness;
- Assess engagement outcomes, for example battle damage assessment;
- Off board electronic surveillance;
- Mine detection.

In addition, in the security environment, UAS may be utilised for tasks such as:

- Port security
- Shipping lane surveillance
- Oil and Gas platform security
- Border patrol
- Coastal surveillance
- Private marine security
- Counter-piracy

COUNTER-PIRACY

The International Maritime Bureau (Reference 2) estimates that 53 ships were hijacked worldwide in 2010: 49 of them off Somalia's coast – and a record 1,181 hostages were taken. The problem appears to be growing. Overall, there were 445 pirate attacks in 2010, a 10% rise from 2009.

Maritime piracy costs the global economy between US$7 billion and US$12 billion a year, according to Oceans Beyond Piracy (Reference 3). The maritime industry is therefore forced to pass on to customers the costs in security, higher insurance premiums, ransoms and extra fuel for longer routes.

Attacks can be expected at any time, but most occur during the day, often in the early hours. They may involve two or more skiffs, that can reach a speed of 25 knots. With the help of "mother ships", GPS and other technologies, the operating range of the skiffs has been increased far outside their traditional range. Typically, an attacked vessel is approached from quarter or stern, RPGs and small arms are used to intimidate the operator to slow down and allow boarding. Light ladders are then brought alongside to climb aboard. The pirates will then try and take control of the bridge to take operational control of the vessel and move it to a pirate-friendly harbour.

Traditional counter-piracy efforts have been to react and evade if possible, or surrender and pay the steadily rising ransom amounts. Additionally, an increasing number of ships have on-board armed security teams, which are costly and incur significant liability.

Internationally, NATO, the U.S. and several European countries have sent ships to the region to protect shipping, but without wide area surveillance and early detection, the patrols have not prevented piracy from rising in the area.

Many experts agree that the use of armed security operatives is problematic under existing maritime law, although recently the British Prime Minister, David Cameron, stated that he wanted to legalise armed guards, after talks in Australia with Commonwealth leaders from the region over the escalating problem faced in waters off their shores, but armed guards would only be permitted while passing through dangerous waters, such as the Red Sea and the Gulf of Aden.

Ship owners and insurers can find other ways to protect themselves: increased training; conducting drills; and purchasing equipment that can provide advanced warning to the captain and crew, allowing them to place the vessel outside of visual range, provide communication links, and surveillance to naval forces and have more time to employ citadel procedures before pirates can board.

The importance of maintaining a continuous wide area surveillance, coupled with direct full-motion video feeds to naval vessels and law enforcement agencies, leads to a potential cost-effective solution, utilising UAS. Wide area surveillance and early detection provide additional options for crews to divert over the horizon and avoid being boarded by pirates. However, in circumstances where staying outside a pirate's vessel visual horizon is unavoidable, a UAS may act as a deterrent, or provide a record of the event. Communication, and data-link connectivity to NATO-allied navies or law enforcement agencies provides earlier intervention and interdiction within the local vicinity, and also supports regional naval assets for large-scale counter-piracy efforts. The UAS may be operated by the naval assets, the law enforcement agencies, or by the ship owners (or their representative bodies) themselves.

SUMMARY

The endless possibilities that the introduction of UAS into the maritime environment brings is slowly being recognised. The increase in operational use, and in the number of demonstrations of these systems, provides further evidence of how their capabilities may be used to fill capability gaps providing ISR for Force Projection and Force Protection, complementing manned aircraft in their agility, flexibility, and (in some cases) expendability.

Integration and interoperability are key, and embarked UAS can provide a fast moving organic capability that deploys and tasks with speed, agility, control and precision. A capability that is independent of host nation support.

There is now an expectation from the UAS community that UAS will form part of a nation's maritime air capability and that they have a role to play, both nationally and internationally. In addition, as the use of UAS proliferate, they will have more of a role to play in maritime security.

REFERENCES

1: UK Ministry Of Defence, 2006, Joint Services Publication JSP 430 : MoD Ship Safety Management, Part 2: Policy Guidance, Issue 3 Amendment 1

2: ICC International Maritime Bureau, 2011, Piracy and Armed Robbery Against Ships: Report for the Period 1 January – 30 June 2011

3: Boden A, 2010, The Economic Cost of Maritime Piracy, One Earth Future Working Paper

Submarine Energy and Propulsion Considerations

David Williams
© Deep Blue Tech Pty Ltd.
david.williams@deepbluetech.com.au

ABSTRACT
Australia requires its submarines to undertake long transits and lengthy deployments. In addition, the energy chain of a conventional submarine is subject to severe weight and volume limitations. The ability of the Future Submarine to meet transit and endurance requirements within these limitations is a challenging problem space unique to Australia. It will be a key differentiator from other diesel electric submarines.

Deep Blue Tech has developed several tools to model the performance of the submarine energy chain. This paper uses these tools to explore which aspects have the greatest impact on transit and operational area energy performance (Speed of Advance, Indiscretion Ratio etc).

This work shows the importance of optimising the whole submarine system for the entire mission and the need to include operational considerations in the requirements.

INTRODUCTION
Australia requires its submarines to undertake long transits and lengthy deployments. In addition, the energy chain of a conventional submarine (SSK) is subject to severe weight and volume limitations. The ability of the Future Submarine to meet transit and endurance requirements within these limitations is a challenging problem space unique to Australia. It will be a key differentiator from other diesel electric submarines.

Comparing the energy density of the energy chain for different types of warships illustrates the constraints inherent on the SSK energy chain.

- **Nuclear:** essentially unlimited energy store
- **Gas Turbine (and or) Diesel Surface Ship:** Energy density of propulsion & electrical system (including fuel) in the order of 3000 Wh/kg.
- **Conventional Submarine:** This is more complicated and is split up as follows:
    - Diesel Generator System and Fuel: in the order of 3000 Wh/kg
    - Air Independent Propulsion (AIP) System (if fitted): 400 – 600 Wh/kg
    - Main Battery: 40 – 60 Wh/kg (lead acid battery at low discharge rates)

A similar picture emerges when comparing the actual energy stored by each of the systems. Thus optimal use of energy on an SSK is vital and every effort should be made to maximise the overall efficiency of the energy chain i.e. make the best use of the stored energy.

Energy Chain Overview
The energy chain for an SSK with AIP is shown in Figure 1. The normal power source is the Main Battery, which provides power for all loads. At a high level the electrical load is split into the propulsion load and the remainder, called the 'hotel load'. The Main Battery is charged by the Diesel Generator system. Charging while submerged is called snorting and is when an SSK is at its most vulnerable.

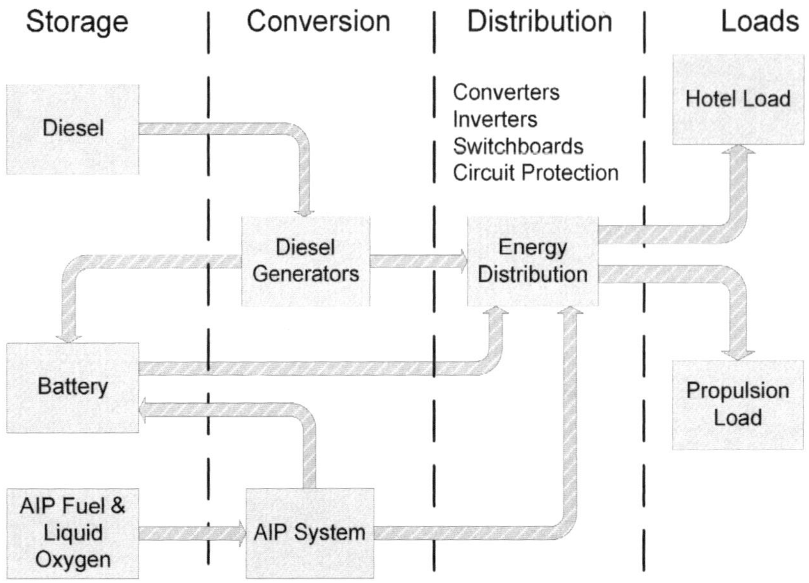

Figure 1: Submarine Energy Chain

The Main Battery is very large (>10 MWh) and has a major impact on the overall submarine layout and design. Flooded lead acid chemistry is currently used and is assumed throughout this paper. Compared to most batteries, the Main Battery is generally discharged at very low rates. The Main Propulsion System has a high torque, low speed electric motor that directly drives the propeller shaft i.e. no gears.

Many modern SSK's have an Air Independent Propulsion (AIP) system. This system provides sufficient power for both the propulsion and hotel load at low speeds without discharging the Main Battery and greatly increases low speed submerged endurance. The energy density of an AIP system is much greater than the Main Battery System, but once exhausted it cannot currently be 'recharged' by the submarine at sea during a mission.

The different components of the energy chain require significant mass and volume and thus have a major impact on the overall submarine layout and design.

TRANSIT CONSIDERATIONS

The ability of an Australian SSK to successfully undertake long transits at high speed is a key differentiator from other SSK designs. Transiting to the operational area will make up 30 – 50% of the total mission time for a long range mission and is undertaken at medium to high speeds to optimise time on station. Therefore an understanding of the requirements that make up this part of the mission is essential, especially given the energy constraints inherent in an SSK.

The overall requirement of a transit is essentially to travel from point A to point B in a given time with a particular level of stealth. During this time, the submarine will cycle between running deep on battery power and charging the battery at Periscope Depth i.e. snorting. The average speed over the transit (or a specified part of the transit) is called the Speed of Advance (SoA). One measure of submarine stealth is its Indiscretion Ratio (IR). This is simply the ratio of time spent at Periscope Depth with a mast raised compared to the total time. Many transit regimes are possible, each with different combinations of dived speeds and times, snorting speeds and percentage of battery discharged each battery cycle.

Snorting Speed

The power required to move a submarine through the water while it is deep is close to a cubic relationship with speed i.e. doubling the speed requires eight times the power. The speed power relationship when snorting is more complicated, with peaks and troughs related to submarine length; however it still has the same basic cubic shape.

One of the variables when calculating transit performance is snort speed. Whilst the shaft power increases cubically with speed, at low speeds, only a small increase in power is required (in absolute terms) for small increments in speed, especially when compared to the total power output of the Diesel Generators. For example, to go from 4 to 5 knots, a 25% increase in speed only requires 10's of additional kW. An example illustrating this for a notional 4000 tonne submarine with 4.8 MW of diesel power is in Table 1. For example, increasing the snort speed from 4 knots to 6 knots (a 50% increase in speed) only requires an increase in shaft power of 40 kW (1% of the installed diesel power). Thus there will be only a small reduction in charging rate for a significant increase in speed.

Table 1: Charging Speeds

Snort speed	Shaft power	Shaft Power Increase	Percentage of diesel power	Snorting Speed Increase
4 knots	~50 kW	0	0	0%
5 knots	~90 kW	~40 kW	1%	25%
6 knots	~150 kW	~100 kW	2%	50%
8 knots	~340 kW	~290 kW	6%	100%

Therefore it is expected that better performance (from an energy perspective) can be achieved by increasing the snorting speed. This is illustrated in Figure 2 and Figure 3. These show that even an increase from 4 to 5 knots results in a performance increase (or reduced fuel requirement for the same level of performance). There is also a point at which increasing the snort speed further reduces performance, refer Figure 2 and Figure 3.

Another aspect to consider is the wake generated by the submarine when snorting. The wake increases with speed, thus increasing the signature of the submarine. Operational considerations will determine what the allowable wake is, which in turn may place restrictions on the maximum permissible snorting speed.

The conclusions from this discussion are:

- Determining the desired snorting speed for different transit options is a balance between the performance gains of higher snorting speeds (up to a point) and the surface wake generated by the submarine.
- Even a small increase in allowable snort speed e.g. from 4 to 5 knots can result in a reasonable performance gain. Therefore time and effort should be spent in determining the optimal snort speed(s). The maximum permissible snorting speed will most likely vary depending on the area the submarine is in. Therefore a more complicated requirement set that looks at the different phases of a transit (as opposed to a single requirement e.g. 3500 NM transit at a SoA of 9 knots with a snort speed of 4 knots) will help avoid an over-conservative approach.
- Designing the masts to minimise wake at preferred snort speeds will benefit submarine transit performance.
- Any comparison of range and transit performance between different submarine designs must use appropriate and consistent snorting speeds.

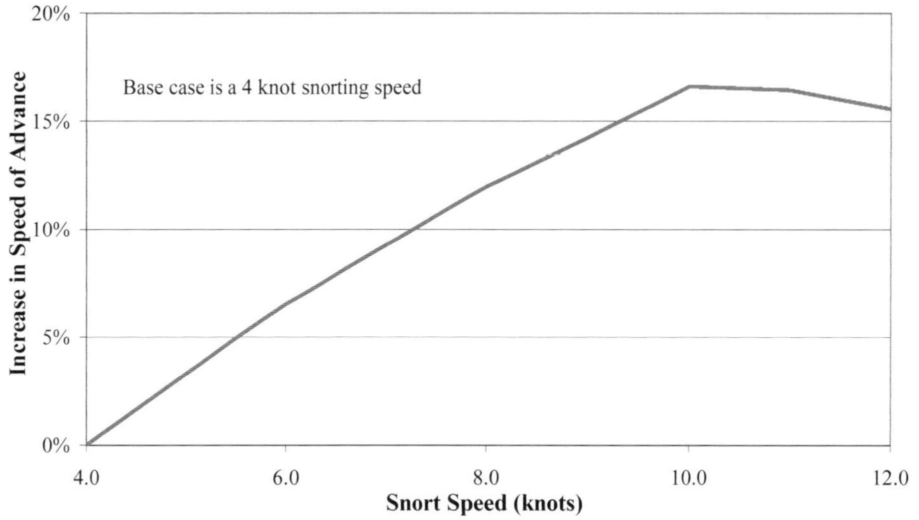

Figure 2: Speed of Advance Increase with an IR of 35%

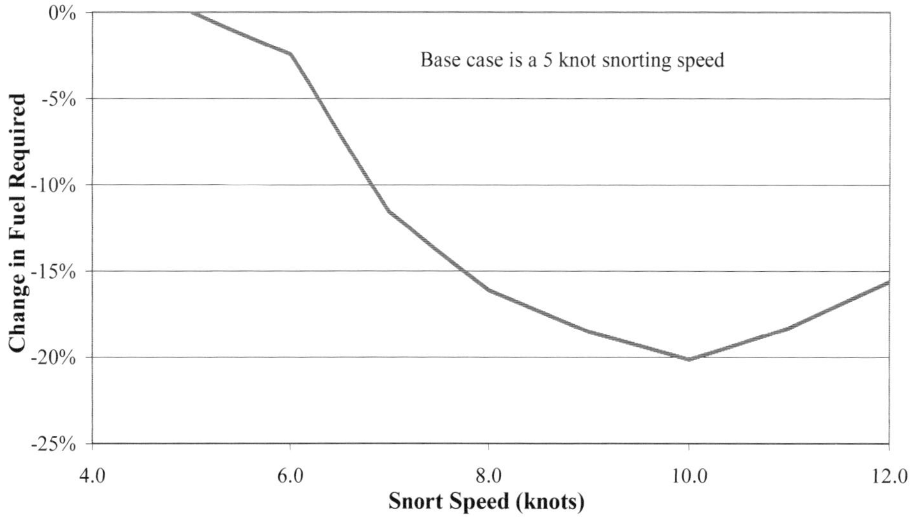

Figure 3: Change in Required Fuel for a 2500 NM Transit at 10 knot SoA

Diesel Power

The Diesel Generator System needs to be balanced with the Main Battery size and transit performance requirements. If the Diesel Generators are undersized the advantages of a larger Main Battery will not be fully realised (long charging times required) and transit performance will suffer. Conversely, the maximum charge acceptance rate of the Main Battery limits how quickly it can be charged. Therefore at some point, additional diesel power will not result in faster charging. Figure 4 to Figure 6 illustrate the impact diesel power has on transit performance. Each shaded band represents the performance between hypothetical 2000 & 4000 tonne submarines. All other factors e.g. battery mass of 350 tonnes, have been kept the same in each case[1].

These figures illustrate that diesel power has a greater impact on transit performance than displacement. For example, at least 4 MW of diesel power is required to transit with a SoA

[1] The 2000 tonne data shown is of course overly optimistic. In reality, a balanced 2000 tonne naval submarine will not have sufficient space and weight for 350 tonnes of battery and e.g. 4 MW of diesel power.

of 10 knots, refer Figure 6. The required SoA performance of a SSK will dictate a minimum level of diesel power.

These results are explained by examining the speed power curves. An approximate relationship between speed, volume and power is:

$P = K \times V^{\frac{2}{3}} \times U^3$ *Ref [1], where P = power, V = form volume, U = speed & K is a factor relating to shape.*

Therefore (assuming a similar K), doubling the form volume increases the required power by approximately 60% i.e. doubling the displacement does not double the required shaft power. Additionally, at snorting speeds the installed diesel power is much greater than the required propulsion power i.e. when snorting at 4 knots, a 2000 tonne submarine with 2 MW of diesel power will take much longer to charge its battery than a 4000 tonne submarine with 4 MW of diesel power. This leads to the results shown in Figure 4 to Figure 6. These figures also show the performance improves as the snort speed increases from 4 knots. However, generally past 8 knots performance starts to level off or reduce. This affirms the above discussion on snorting speed. The transit profiles for 2 MW of diesel power at high snorting speeds would not be used in practice and give poor performance anyway. They have been included for completeness and to show the trend.

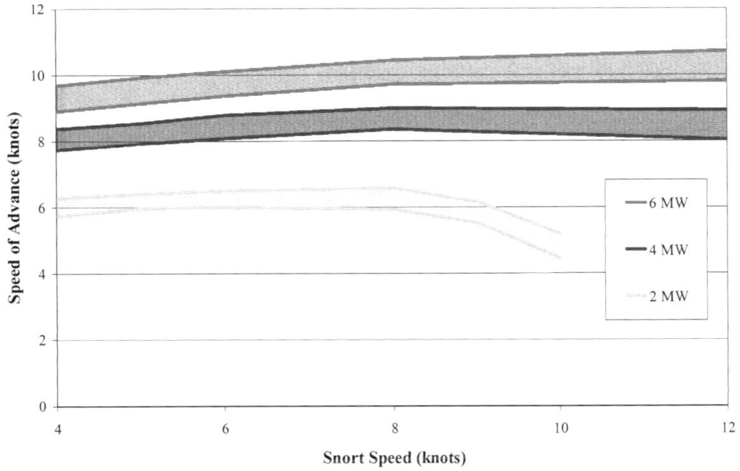

Figure 4: Speed of Advance for 25% Indiscretion Ratio

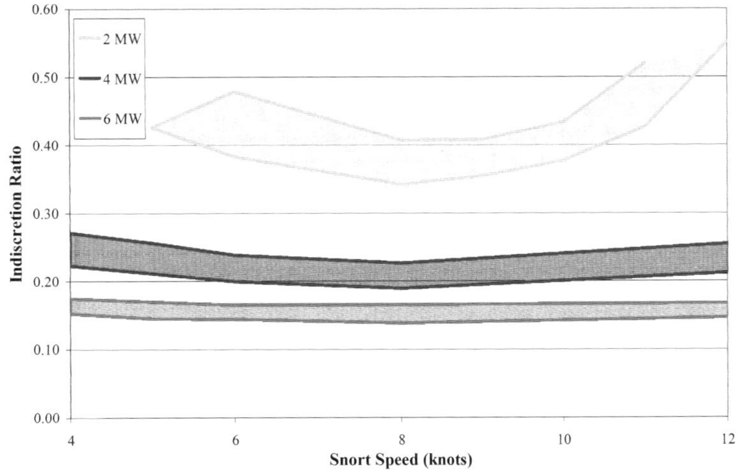

Figure 5: Indiscretion Ratio for an 8 knot Speed of Advance

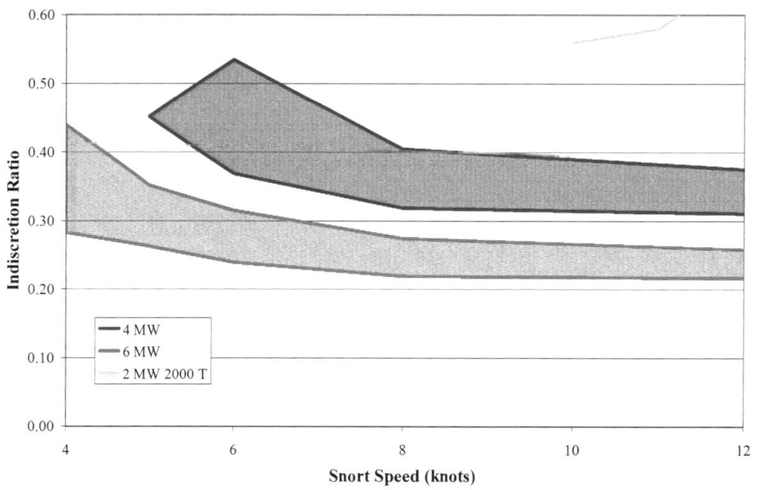

Figure 6: Indiscretion Ratio for a 10 knot Speed of Advance

MAIN BATTERY CONSIDERATIONS

Balancing the size of the Main Battery with the Diesel Generator output is discussed above. Other important considerations relate to the performance of the Main Battery itself. The Main Battery consists of many individual battery modules electrically connected into battery sections with long series strings. Each module is subject to slightly different environmental conditions (due to their location within the battery compartment). Along with manufacturing tolerances this results in performance differences between each battery module. The performance difference between modules becomes greater as the mission progresses or the battery ages, resulting in weaker modules developing. These weaker modules will reach the minimum permitted cell voltage before the other battery modules, therefore limiting the discharge of the entire battery section.

The charging regime on an SSK with lead acid batteries generally has three different stages, constant power (stage 1), constant voltage (stage 2) and constant current (stage 3). Stage 1 charging restores the bulk of the discharged energy but cannot completely charge the battery. Periodic stage 2 charging is required to equalize the battery cells and 'complete' the charge at sea. The constant voltage of stage 2 charging is required to be just under the voltage at which significant hydrogen production occurs. The required length and frequency of stage 2 charging depends upon factors such as how (e.g. depth of discharge and discharge rate) the battery has been discharged and charged previously. A stage 3 charge is required to fully equalize the battery. However stage 3 charging is both a long process and results in high levels of hydrogen being generated. Therefore it is assumed to only occur when alongside. If the required stage 2 and 3 charges are not conducted, battery life and capacity are degraded permanently.

When charging, some modules will reach their stage 2 charging voltage before the other modules. This limits the amount of stage 1 charging that is possible, which leads to reduced usable battery capacity or increased charging times.

These effects can appreciably reduce battery (and thus submarine) performance. The car industry is exploring these issues as a result of the introduction of hybrid and electric cars with 'large' (though not 'submarine large') battery packs, Ref [2], [3] & [4]. The use of battery equalisation systems to minimise the performance difference between battery cells (and modules) is being investigated.

For an SSK, the advantages of equalising battery modules are:

- an increase in the usable discharge capacity of the battery at sea i.e. improving performance
- making it easier to accurately predict and model battery performance. Therefore design uncertainty margins can be decreased, which in turn allows the Main Battery to be reduced in size.
- reducing the amount of stage 2 charging required, thus improving indiscretion ratio.

The combination of the first two factors in particular leads to a reduction in the required battery mass to achieve a given level of performance. This in turn can either reduce the submarine size or free up space for additional capability.

HOTEL LOAD CONSIDERATIONS

At a high level the overall electrical load on the submarine can be grouped into the propulsion load and the hotel load. The hotel load is the sum of the average electrical power drain of all the non-propulsion loads. The contribution of a particular piece of equipment to the hotel load is a function of its average power drain when operating and frequency of operation. Therefore a large pump that is used infrequently may contribute less to the hotel load than a small load that is always on. For example, a 10 kW pump that is used once an hour for 5 minutes will contribute 0.83 kW to the hotel load, while a processing cabinet that only requires 1 kW but is continuously on will contribute 1 kW to the hotel load.

At low speeds, the hotel load is greater than the propulsion load. Therefore it has a significant effect on low speed submerged endurance performance (either lowering the endurance or increasing the required battery, diesel fuel and (if applicable) AIP oxygen and fuel. Most of the time spent in the operational area will be at low speeds. Thus the hotel load is a major driver on the overall size and cost of a submarine. Therefore it is important to both accurately define and minimise hotel load, especially for low speed scenarios. The impact of hotel load on Main Battery size is illustrated by the example in Table 2.

Table 2: Hotel Load versus Battery Mass

Propulsion Power (kW)	Hotel Load (kW)	Total Drain (kW)	Required Battery Mass for 80 hours endurance	% Increase
70	100	170	302	0%
70	110	180	320	6%
70	120	190	338	12%
70	130	200	356	18%
70	140	210	373	24%
70	150	220	391	29%

An Australian submarine will operate in warm waters for a significant part of a deployment, refer Figure 7. Therefore the cooling systems contribute a significant portion of the overall hotel load, refer Figure 8. Optimising the cooling systems can lead to significant reductions in the overall hotel load.

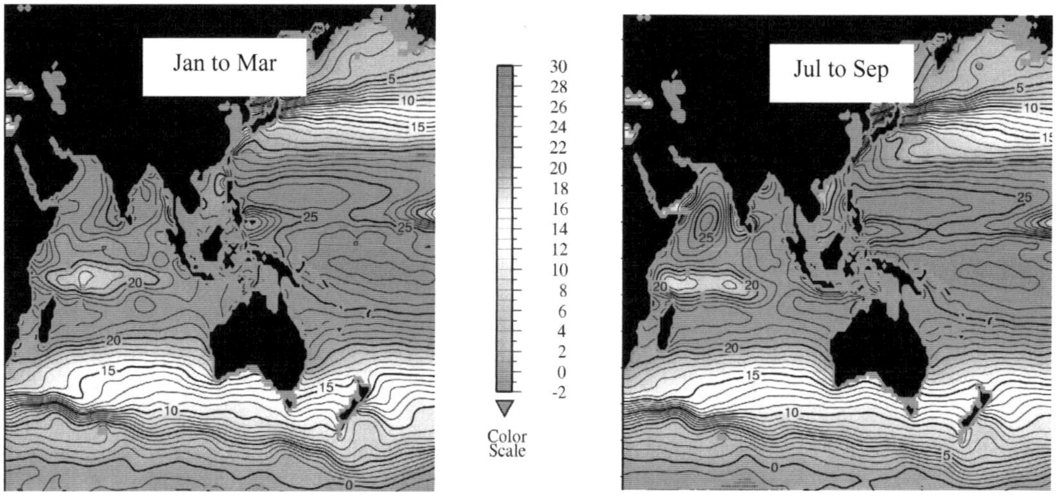

Figure 7: Average Temperature at 100m. Source: Ref [5]

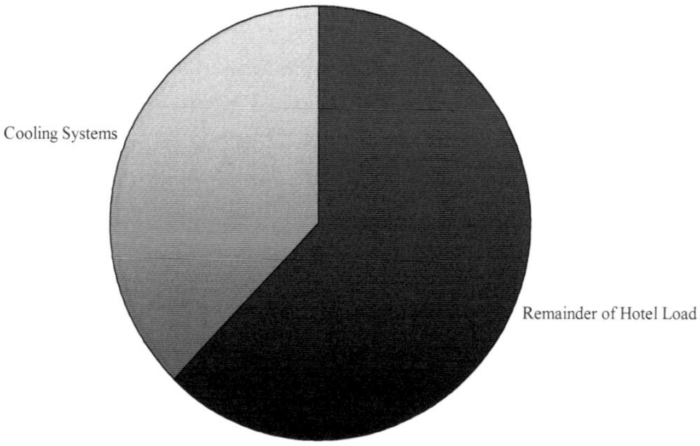

Figure 8: Cooling Systems Hotel Load Contribution: Deep on Batteries in Warm Waters

Cooling System Considerations

The interaction between the different cooling systems is illustrated in Figure 9. When deep, all heat (ignoring the small amount of conduction through the hull) is removed from the submarine to sea water via the Sea Water Cooling System. Sea water temperature has a large impact on the power required by the cooling system equipment (especially for the chiller units), refer Figure 10. Therefore the sea water temperature profiles used when designing the system have a large impact on the overall design. For instance, calculating endurance based on the maximum sea water temperature would greatly increase the hotel load of the cooling systems, thus increasing the Main Battery size. Another approach is to size the cooling equipment based on maximum (or close to) sea water temperature but calculate endurance on an 80th or 90th percentile (or similar) temperature. The reduced endurance at extreme temperatures can then be calculated and confirmed as being acceptable. It is considered that this will result in a better balanced overall submarine design.

Other methods to reduce the hotel load of the cooling systems are:

- Minimising heat loss to the submarine internal atmosphere: This is the least efficient method of removing heat from the submarine (refer Figure 9). Direct seawater or freshwater cooling of equipment should be used whenever possible (the use of seawater has safety implications that must be considered).

- Minimising the need for chiller units: These make a large contribution to the hotel load (refer Figure 10). Their impact can be reduced by;
    - Minimising the chilled water load by using fresh water whenever possible. Cooling of electronics by tepid water (slightly cooled freshwater) is an option that should be investigated in lieu of chilled water.
    - The use of modern, efficient compressor technology e.g. magnetic bearing variable speed chillers
    - Closed loop, variable speed control of the chiller units
- Closed loop, variable speed control of the overall system.

Implementation of these approaches will potentially increase the complexity, size and cost of some individual components. However this will be more than offset by the reduction in hotel load and corresponding reduction in size of the submarine energy chain components.

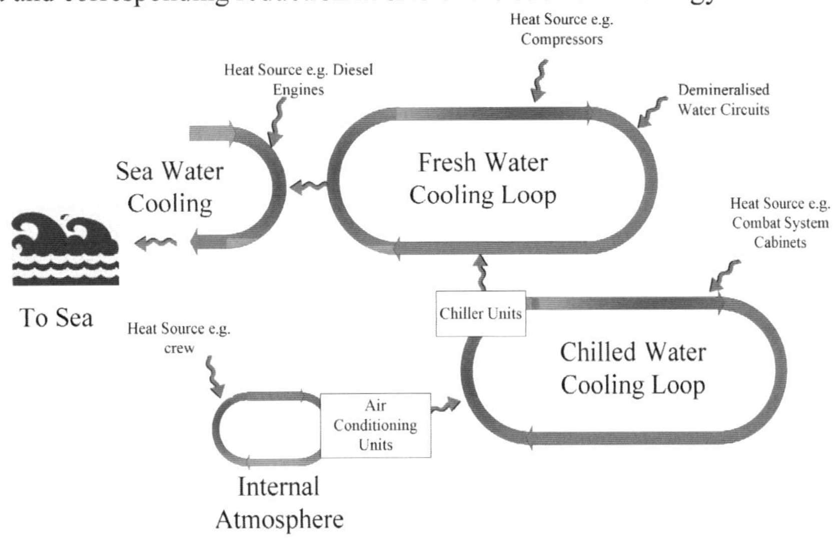

Figure 9: Cooling Systems Interactions

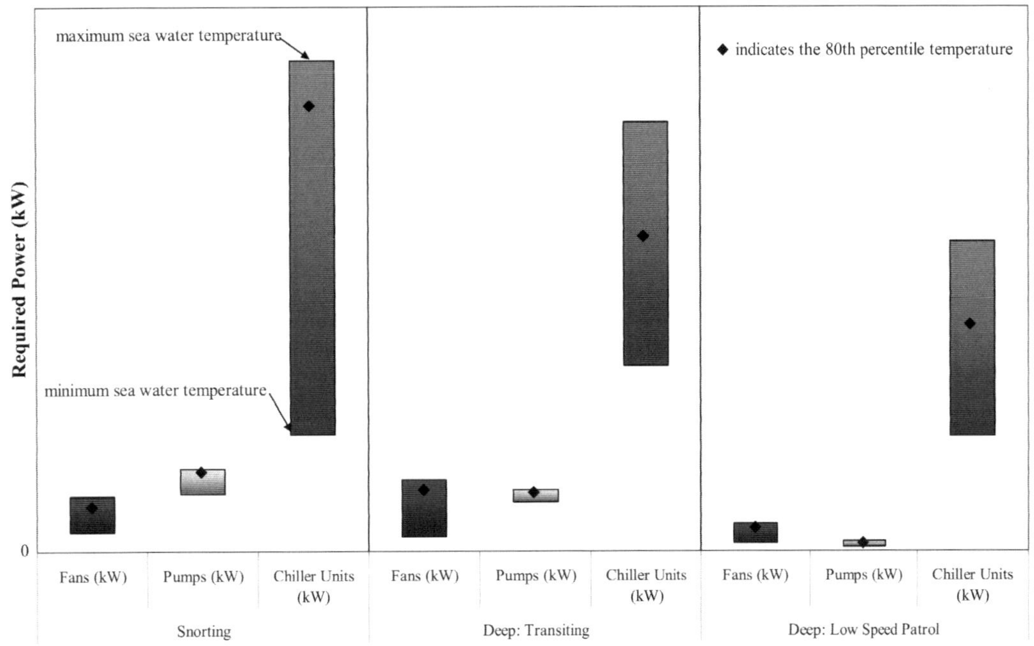

Figure 10: Effect of Sea Water Temperature on Cooling System Power Requirements

SUMMARY & CONCLUSIONS

Submarine design is a non-trivial exercise. Many different factors relating to technology e.g. the efficiency of a motor, the environment e.g. sea water temperature and operational considerations e.g. what are the limitations on snorting speed, all contribute to overall performance. This paper has illustrated the impact (either positive or negative) of:

- snorting speed
- diesel power
- battery equalisation
- hotel load and
- sea water temperature

on the energy systems and submarine performance. All of these have a large impact on overall submarine performance and cost.

In addition the energy systems of an SSK are highly constrained and interdependent, which reduces the margin for error in the design. This emphasises the need to understand the effect of each factor on cost and performance.

To understand the impact of the different factors requires knowledge in a wide range of areas such as operational considerations i.e. more than just pure design and technology expertise. Thus to arrive at a design that meets the needs of the operators within cost and any other constraints requires involvement from the customer, operators and designers starting from the initial design phases.

The examples discussed in this paper have all been generated from calculation tools developed by Deep Blue Tech. The use of these tools can allow many different scenarios and options to be quickly explored and analysed. Thus the impact of different factors can be quantified. Undertaking this exploration of the solution space early in the design phase is considered vital to arrive at an optimal solution.

REFERENCES

[1] R. Burcher & L. Rydill, *Concepts in Submarine Design.* Cambridge University Press, 1994

[2] P. Krein & R. Balog, *Life Extension Through Charge Equalisation of Lead-Acid Batteries.* In proceedings of Telecommunications Energy Conference, 2002.

[3] S. Moore & P. Schneider, *A Review of Cell Equalisation Methods for Lithium Ion and Lithium Polymer Battery Systems.* Paper for Society of Automotive Engineers Inc, 2001-01-0959, 2001.

[4] S. Williamson & A. Shafiei, *Plug-In Hybrid Electric Vehicle Charging: Current Issues and Future Challenges.* In proceedings of IEEE Vehicle Power and Propulsion Conference, 2010.

[5] www.nodc.noaa.gov accessed June 2011.

RAN Investigation into AFFF Hose Lines

I Leaver, I Raymond, D Moloney and B McNeice
Directorate of Navy Platform Systems – RAN
Ian.leaver1@defence.gov.au

ABSTRACT

AFFF hose lines, be they hose reel or inline inductors off a fire hydrant, are currently widely used within the RAN fleet. They are used in a variety of situations ranging from fighting flight deck fires to engine room fires, and anywhere else that a class B fire may occur. The methods of foam mixture production range from pre-mixing foam to the use of inductors.

To determine the effectiveness of inline inductors in RAN systems, Navy Engineering performed a series of tests to determine the level of performance of inline inductors when used with a variety of different nozzles, foams, expansion ratios and operating pressures. A low pressure (7 bar) 200 lpm inductor was used in the testing, along with a higher pressure (10-12 bar) 473 lpm inductor. Testing was performed over a range of pressures from 6 to 12 bar.

The paper discusses and highlights the effect that changing one element of a system can have on overall system performance. Extinguishment testing was undertaken on 50m^2 and 100m^2 Kerosene fires, smaller scale crude oil fires in both small and large fire pits, as well as running fires on an incline.

All extinguishment tests were performed with both US Navy MILSPEC (MIL-F-24385F) and ICAO-B foam so that direct performance comparisons could be made. RAN has moved to exclusively using MILSPEC foam due to the corrosion level limits which are specified in the MILSPEC.

INRODUCTION

Ship-based fire fighting introduces a variety of issues in addition to those faced within the typical urban environment. The issue of trying to achieve consistent hydrant pressure across a whole platform is one which is almost impossible to achieve. This, among other issues which arise during fire fighting on board ships at sea, prompted Navy Engineering to launch an investigation of how configuration changes would affect the overall performance of hose based fire fighting systems.

The RAN uses hand held hoses in a variety of situations including compartment fire fighting and flight deck fire fighting, with and without the addition of AFFF to the water. Due to this wide range of applications the investigation aimed to determine whether it was appropriate to use a single nozzle in these varying situations. The goal of this testing was to determine the effect of changing selected components on the overall performance of the system and its fire fighting effectiveness.

Both cold and hot tests were performed measuring a wide variety of variables including hydrant pressure, nozzle pressure, flow rate and throw distance. This paper will summarise the findings, focusing on the implications for ship-based fire fighting and fire hose fire fighting system design.

TEST RESULTS

Testing introduction

On board RAN ships there are both 'fixed' AFFF systems, which utilize pre-mixed AFFF solution pumped directly to the hydrant, as well as 'semi-fixed' systems, where an inline inductor is used to add the foam at the hydrant. Ideally, fixed systems are used where there is a high certainty that the use of AFFF will be necessary in a fire fighting situation, while semi-fixed systems are used where the need to use AFFF is less certain, and will vary from case to case.

Testing was performed with a variety of nozzles in order to determine the characteristics of each nozzle design. The data presented in this paper is the average of the results which were obtained during the testing. The nozzles which were tested included:

- Non-aspirating nozzle (rated at 7 bar);
- Low expansion, (a nozzle rated at 5.17 bar with a low expansion tube and a nozzle rated at 7 bar with a tube);
- Medium expansion, (an adjustable nozzle rated at 3.8 bar and 6.9 bar using a medium expansion tube);
- Combined nozzle which consisted of a 'built-in' foam tube (rated at 6.9 bar)); and
- Master stream tubes (dedicated foam tubes).

All testing was performed with one of the two inductors or with no inductor in the system, in order to determine the impact of the inductors on system performance. The details of the inductors are as follows:

- A low flow inductor:
  7 bar
  200 lpm
  Educt increments 1% 2% 3% 4% 5% 6% (continuous)
- A high flow inductor:
  4-12 bar
  473lpm (125gpm)
  Educt increments 0.25% 0.50% 1% 3% 6% (discrete)

Most tests were also performed with either one length of 15m hose or two lengths of hose, in order to determine the effect that this has on the system. Currently, depending on the situation, either one or two lengths of 15m hose are used when undertaking fire fighting. This differs from the 30m hose length standard of many land based fire fighting units.

Hydrant pressures were also varied during the testing to measure the impact on the performance of the system of varying hydrant pressures around a ship. Ultimately, this testing provided data which Navy Engineering can use to assess fire fighting systems on board RAN platforms.

Pressure Variation testing

Initial testing focused on the effect that different hydrant pressures across a platform will have on fire fighting system performance. The pressures were varied between 5 bar and 12 bar, measured at the hydrant. All testing with an inductor used the low flow inductor during this phase of testing. All testing was also done with the nozzles set to full jet in order to accurately compare the effect that the nozzle has on the performance of the overall system. It should be noted that regardless of whether the foam tubes were fitted to these nozzles or not these results would have been the same. Fitting an expansion tube to a nozzle has little impact on the back pressure created in the system.

Table 1: Pressure drop (% of input pressure) across one hose, with no inductor.

Pressure at Hydrant	**7 bar Nozzle non-aspirated**	**5.17 bar Nozzle low exp.**	**6.9 bar combined Nozzle**	**3.8 bar Nozzle med. exp.**	**6.9 bar Nozzle med. exp.**
Low (5 bar)	8%	40%	26%	36%	10%
High (Varied)	9%	33%	27%	46%	29%

As predicted, a pressure drop of approximately 0.6 bar occurred along the length of hose. The varied results show the impact of the different nozzle designs on operating pressure. Certain nozzles created significantly different amounts of back pressure, resulting in highly varying pressure drops. It is interesting to note that the 7 bar nozzle, which is providing significantly lower back pressures in both Table 1 and Table 2, is a much older nozzle than all of the other nozzle designs. This is potentially due to a lack of flow control mechanisms within this nozzle.

Table 2: Pressure drop (% of input pressure) across two hoses, with no inductor.

Pressure at Hydrant	**7 bar Nozzle non-aspirated**	**5.17 bar Nozzle low exp.**	**6.9 Bar combined Nozzle**	**3.8 Bar Nozzle med. exp.**	**6.9 Bar Nozzle med. exp.**
Low (5 bar)	18%	44%	30%	40%	4%
High (Varied)	14%	42.5%	39%	51%	---

Table 2 shows that other than this result there is a consistent increase in pressure loss across the nozzles when a second hose line is added. It is expected that with one 30m hose, rather than two 15m hoses, the results would show a slightly lower pressure drop, due to the absence of the additional coupling half way down the hose.

Table 3: Pressure drop as a percentage of hydrant pressure – one hose, inductor fitted

Pressure at Hydrant	**7 bar Nozzle non-aspirated**	**5.17 bar Nozzle low exp.**	**6.9 bar combined Nozzle**	**3.8 bar Nozzle med. exp.**	**6.9 bar Nozzle med. exp.**
Low 5 bar	46%	72%	58%	48%	14%
High 12 bar	47%	73%	60%	73%	47.5%

Table 4: Pressure drop as a percentage of hydrant pressure – two hoses, inductor fitted

Pressure at Hydrant	**7 bar Nozzle non-aspirated**	**5.17 bar Nozzle low exp.**	**6.9 bar combined Nozzle**	**3.8 bar Nozzle med. exp.**	**6.9 bar Nozzle med. exp.**
Low 5 bar	50%	74%	54%	50%	14%
High 12 bar	51%	75%	57.5%	73%	48%

Table 3 and Table 4 show the results of adding the low pressure inductor to the system. As expected, the addition of a low pressure inductor drastically reduced the maximum pressure which was available at the nozzle. Little discussion of these results is necessary, other than to observe that the inductor selection will have a significant impact on the performance of the nozzle. Care needs to be taken to ensure that a suitable inductor is selected. The higher flow inductor which was used in testing performed well for a range above its minimum operating pressure, showing that inductor design should also be considered when selecting an inductor. In a ship-based system, due to the wide range of hydrant pressures which will be available, an inductor should be selected which can support a range of operating pressures, rather than one which is designed to run at a specific pressure.

Throw distance testing

Throw distance is not typically an issue within the RAN due to the close-quarters nature of ship-based fire fighting. However, due to the larger areas involved in flight deck fire fighting, it was decided to obtain information on the impact of foam aspiration on the throw distance. Aspiration of AFFF is recommended in all circumstances where life is at risk, due to the increased effectiveness of the foam in extinguishing hydrocarbon fires when aspirated. A brief set of tests were performed and the outcomes were as predicted. The throw distance tests were performed with the nozzle positioned one meter above the ground. The outcomes of this testing are as follows:

1. Higher throw distances were achieved with non aspirated nozzles (range of up to 25m at high pressures)
2. Low expansion foam achieved significantly lower throw distances (range of up to 20m)
3. Medium expansion foam achieved the shortest throw distances (ranges up to 13m)

Testing was also done in windy conditions to highlight the effects that wind on a weather deck will have on fire fighting using AFFF. As expected, with the wind at the back of the fire fighter, the range of the foam was significantly increased, with each level of aspiration effectively achieving the range of a lesser aspirated foam. Medium expansion foam achieved the range that low expansion did without assistance from the wind, and low expansion foam the same as non-aspirated foam without the wind.

The results obtained in the throw distance experiments were as expected, however it still highlighted the requirement to take into account the operating environment when it comes to designing a fire fighting system on board a ship. The location of the nozzle (i.e. inside or out on the weather deck) has a significant impact on the performance of that system. There will be some further discussion on the effect of throw distance on extinguishment times later in this paper.

'Stickiness' testing and the insulating properties of AFFF (and foam in general)

One aspect of AFFF (or any other fire fighting foam) which can be exploited is its ability to act as an insulator to protect temperature sensitive items. Similarly, it can be used to contain the fibre release from composite structures as they are heated or burned. A series of tests were performed in order to highlight the differences between, non-aspirated, low aspirated and medium aspirated AFFF and its ability to effectively 'stick' to and cover a structure reliably.

As expected, the tests showed that the higher the aspiration, the better the foam was at sticking to the structure. It was, however, unexpected that the amount of foam which would stay stuck to the structure would increase so drastically as the aspiration level was increased.

Figure 1: Non-aspirated foam sprayed on structure

It can be seen in Figure 1 that with non aspirated foam there is very little foam which is stuck to the structure. This results in a very low level of insulation, which is providing basically no barrier and very limited containment of fibre release. Once heated, this layer would evaporate rapidly, leaving the structure unprotected.

Figure 2: Low-aspirated foam on wall

Figure 2 shows an increase in residual foam left on the structure once it has been applied. It can be seen that there is a significant increase in the amount of 'trapped' air, providing a higher level of insulation for the structure. However, it is still running off the structure rapidly, and even without the application of heat, low aspirated foam will still flow off the structure relatively quickly.

Figure 3: Medium-aspirated foam on wall

A further increase in insulation and ability to stay on the structure, Figure 3 shows the results of using a medium aspiration tube on the same nozzle which was used to create the AFFF layer in .

Figure 4: Highly aspirated foam sticks to the roof

Figure 4 highlights the ability of highly aspirated foam to provide a lasting cover on a structure. This foam was left on the structure for up to one hour with no sign of the cover degrading. Coating a structure with foam like this not only provides the structure with an insulating cover, but also gives the fire fighter a visual sign that the structure is being exposed to a high temperature. When the foam 'boils off' the structure, the fire fighter knows that there is a significant temperature being applied to the surface. This is particularly important in situations where fuels or other potentially flammable items are being stored, and in the case of the RAN, ammunition.

Furthermore, the foam also provides significant cooling effects which aid in protecting the fire fighter from the heat of the fire. Figure 5 and Figure 6 highlight this cooling effect, and outline a benefit of using foam in both confined internal fire fighting, as well as flight deck fire fighting against large aircraft fires.

Figure 5: Fire fighter temperature exposure

Figure 5 shows the temperature of the outside of a fire fighters suit being measured by a thermal imaging camera (TIC). This figure shows the temperature at 109 degrees Celsius - this temperature was indicative of the temperatures that the fire fighter's suit was being exposed to. The picture in Figure 5 was taken prior to any fire fighting actions being undertaken.

Figure 6: The cooling effects of foam

The image in Figure 6 is taken with a TIC from the perspective of the fire fighter. The image was shot over the shoulder of the fire fighter, just after the application of foam had begun. It can clearly be seen in the image that the line where the foam has been laid, and where the structure has been coated, the heat radiating back at the fire fighter has been greatly reduced. This ability to quickly knock down the heat is significant, and was not experienced to the same extent when using a lower aspirated foam. At these temperatures the foam evaporates off rapidly, so reapplication is needed in order to keep the temperatures low.

Extinguishment testing.

As well as the cold testing to determine the effects of configuration changes on the flow, throw distance, pressure loss and 'foam stickiness' a series of hot tests were also performed in order to compare an ICAO-B foam to the MILSPEC (MIL-F-24385F) currently employed by the RAN. Testing was performed in 50m2 and 100m2 test pits with Kerosene. A number of tests were also performed with crude oil. There were approximately 72 burns performed over the course of a week, with extinguishment times being the main test focus, when comparing various nozzles, inductors and foams. A number of configurations were selected based on the results from the cold testing which had already been performed. Some configurations were selected based on the expectation that they would perform well, thus verifying Navy Engineering's understanding of fire fighting system design. Some configurations were selected with the expectation that they would perform poorly, due to mismatched inductor/pressure/nozzle selection which would result in poor concentrate mixtures and low flow rates. The systems which were expected to perform poorly were used in order to highlight the differences in ICAO-B foam when compared to MILSPEC foam, given that to pass the MILSPEC test a significantly lower flow rate is used.

Table 5: Full range of Kerosene testing - control and extinguish times.

		Inductor 1				Inductor 2			
		MILS		ICAO		MILS		ICAO	
		Control	Ext	Control	Ext	Control	Ext	Control	Ext
12 bar	200 lpm Masterstream	35	50	60	120	-	-	-	-
	400 lpm MasterStream	30	40	40	60	-	32	25	30
	7 bar Nozzle non-aspirated	25	30	90	150	-	-	-	-
	5.17 bar Nozzle low exp.	45	55	31	36	-	-	-	-
	6.9 bar combined Nozzle	67	70	-	-	-	38	30	40
	3.8 bar Nozzle med. exp.	65	73	-	-	-	40	30	40
	6.9 bar Nozzle med. exp.	-	-	-	-	-	50	30	35
9bar	200 lpm Masterstream	90	100		60				
	400 lpm MasterStream	45	55	75	90	36	56	30	35
	7 bar Nozzle non-aspirated	44	48	75	105	-	-	-	-
	5.17 bar Nozzle low exp.	52	62	27	32	-	-	-	-
	6.9 bar combined Nozzle	45	90	35	45	33	40	33	35
	3.8 bar Nozzle med. exp.	-	-	-	-	30	43	50	75
	6.9 bar Nozzle med. exp.	-	-	-	-	52	92	FAIL	FAIL

Table 5 outlines the results of the testing on Kerosene. It is noted that there was a slightly different set of nozzles used for the hot testing compared to the cold. Five of the nozzles were selected based on their performance in the cold testing, be it that they performed well and therefore warranted further testing or that they underperformed and were therefore used to highlight the differences in foam performance.

Figure 7: 50m2 test pit with helicopter mock-up

It can be seen that there are two results highlighted in red. These tests were considered fails as on the 50m2 pit, shown in Figure 7, the fire was still not controlled within one minute and another hose was used to help extinguish the fire. This inductor and nozzle combination was selected specifically due to the poor concentrate mixtures which were produced during cold testing - the mixture produced was well below the 6% which the foam was designed to operate at (both the MILSPEC and the ICAO-B foams used were 6% foams).

It can be seen that under the same circumstances the MILSPEC foam, while slow, still controlled the fire within one minute and then extinguished the fire in approximately one and a half minutes.

While the MILSPEC foam outperformed the ICAO-B foam in most situations the difference was surprisingly small. None-the-less, where a system is sub-optimal, such as on a damaged ship, the MILSPEC foam is more forgiving. There was enough of a performance difference on average to highlight that not only the type of foam (i.e. AFFF, AR-AFFF, FFFP etc...), but the standard to which the foam will be qualified should be taken into consideration. While this may seem obvious, further down the track when replacing systems throughout a ship, the same thing needs to be kept in mind, to ensure that the whole system is always compatible with all other parts of the system.

Another key issue which was highlighted throughout our testing was the difference in extinguishment time due to nozzle pressure. It is important to remember that Table 5 shows the hydrant pressure rather than the nozzle pressure. It was shown consistently throughout the testing that varying the nozzle pressure of the hose produces very different results when it comes to extinguishment times. This comes back to the throw distance discussion which was mentioned previously. If a fire fighter has range in addition to good quality foam, they gain the ability to accurately direct the flow of water/foam where it is required the most. In some tests, a change of 3 bar at the hydrant had the effect of doubling the extinguishment times due to the fire fighter not being able to accurately deliver the foam to the required locations, resulting in foam boiling/evaporating off and the fire reigniting in that area. Ideally, the range of the setup which the fire fighter is using should far exceed the range required to fight the fire.

This was further highlighted when 'running fire' tests were performed. A relatively free flowing crude oil was 'spilt' in a controlled manner, in order to achieve fire sizes of approximately the same size. The crude was then lit, and when the fire had spread a certain distance the fire fighting effort began.

Figure 8: Running fire setup

Figure 8 shows the setup which was used in order to test the performance against running fires. In this set of tests it was clear that the quality of the foam mixture (i.e. as close to the designed 6% as possible) was not as important as the configuration's ability to throw what foam was achieved where it was needed. The medium expansion foams, while a much 'better' structured foam suffered due to the typically lower throw distances which could be achieved with the medium expansion nozzles. There was little performance difference between the ICAO-B foam and the MILSPEC foam in these tests. As stated above, it was clear that it was significantly more important that the fire fighter could place the foam where it was required rather than achieving a high quality thick foam blanket. Once the fire was extinguished it would be beneficial to cover any remaining fuel (or in this case oil) with a good quality foam blanket. However, in terms of extinguishment, the foam quality was not an issue.

Another key aspect is extinguishment times. While these times are important in all fire fighting situations, quickly controlling a fire and minimising ongoing damage is critical on a ship at sea, when the nearest help may be a significant time away.

Figure 9 shows the different extinguishment times between the two inductors which were tested. The higher flow 4-12 bar inductor achieved relatively consistent control times at approximately 30 seconds, while the lower flow 7 bar inductor achieved an average of approximately 48 seconds. This significant difference in control times highlights the importance of correctly selecting an inductor which will support the worst expected fire. Sizing the fire system components lower than this could result in significantly increased fire control times, allowing further damage to the ship/cargo/weapons etc…

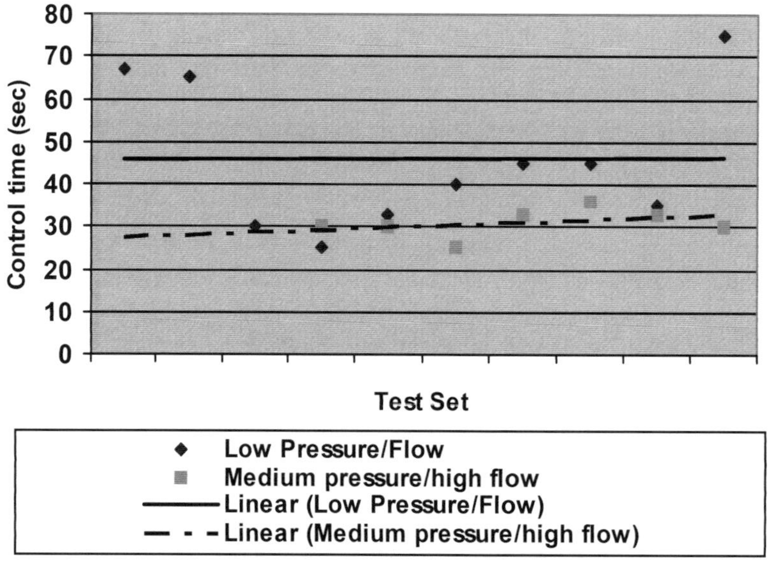

Figure 9: Fire control times, comparison of low pressure/flow and a medium pressure high flow inductor

CONCLUSION

Throughout this set of tests it became clear that in order to design a system which should achieve the best possible extinguishment times, a wide range of system elements must be considered, such as hydrant pressure, nozzle back pressure, foam to be used, aspiration level, the operating environment of the hose, throw distances required and equipment protection requirements. It also became clear that modifying any one component of a system (such as 'upgrading' the nozzle or changing fire pumps) can have a significant effect on the overall performance of the system - the system must be considered as a whole in order to achieve a functional fire fighting system. Simply replacing a 7 bar nozzle with an alternative 7 bar nozzle is not an option, because the replacement nozzles back pressure characteristics may result in 4-5 bar at the nozzle. The potential flow on effects from this include a reduced throw distance and reduced fire fighter control.

It was also clear that in order to achieve the most ideal solution, sacrifices in other areas may need to be made, e.g. a medium expansion foam may be desirable, due to the superior foam coverage and the barrier that it provides over the fuel. However, in order to achieve the required control, a low expansion foam may need to be used, with reapplication of the foam required after initial attack to ensure a sufficient barrier. A summary of the findings of this testing is as follows:

- More highly aspirated AFFF will knock down the fire quicker where the foam has been applied, however this does not always relate to a quicker extinguishment of the whole fire.
- Throw distance is an important quality of fire fighting foam, even in relatively confined locations, due to the control it gives the fire fighter.
- Inductors should only be used in situations where they match the rest of the components of the system - placing an inductor which is poorly matched to the system will only give the crew a false sense of equipment performance.
- Adding an inductor into an existing system needs to be done with caution as the implications on the pressure and other performance aspects of the system are significant.
- Nozzles cannot be hot swapped into a system without adequate investigation into the back pressure which they create; flow regulating and pressure regulating nozzles behave very differently from standard nozzles.
- There are situations where a technically inferior system configuration may outperform another configuration - there is no one size fits all solution.

A hose line based fire fighting system is one that requires all of the user requirements and operating environments to be considered. The change of any component will affect the performance of the system as a whole.

Underwater Signature Management Solutions

Samantha Davidson
Ultra Electronics PMES, United Kingdom
Email:Samantha.Davidson@ultra-pmes.com

INTRODUCTION

The electromagnetic signature management process for a new vessel design begins at the concept stage. This can include, for example, evaluating the signature effects of the choice of drive type and hull coating. The electromagnetic design team are involved throughout the vessel design process determining the signature effect of changes proposed by the Naval Architect. Often the end customer specifies the target signatures to be achieved with the contractor taking 'complete signature responsibility'.

Once the vessel is in-service the signature management process continues with configuration of onboard electromagnetic signature reduction systems and periodic signature measurement and reconfiguration to ensure the signature is maintained below signature target through its lifetime. Signature measurement can be undertaken utilising multi-influence sensors (magnetic, electric, ELFE, acoustic, seismic).

Over the last decade there has been much technological progress that has enabled the development of combined signature measurement ranges systems that gather electromagnetic, acoustic and other influence measurements on the same range system or upgrades which allow new influences such as electric field to be gathered at the same location as an existing magnetic and acoustic range.

Acoustic bandwidths up to 100kHz may be gathered simultaneously with ELFE signatures up to 3kHz. A compact sensor package that is easily deployable has many applications for both commercial and commercial applications to range ships, submarines, ROVs and divers and also key in the field of hydrocarbon exploration. Rapid system deployment and straight-forward user friendly data analysis are important features.

VESSEL SIGNATURE MANAGEMENT

Signature Management is a key element of operational capability. The primary goal of signature management technology is to reduce the likelihood of detection and thereby to increase survivability in the operational field.

There are three prime requirements for vessel signature management:

- Vessel design for low signature vessel ('stealth')
- Signature optimisation
- Threat prediction

A vessel may be designed by its construction and countermeasures with regards to stealth. For the electromagnetic influence threats the design is generally achieved using finite element and boundary element methods. Finite element methods are used to model the ferromagnetic structure and the countermeasures (degaussing coils), whilst boundary element methods are used to model electric currents flowing round the vessel which cause both electric and corrosion related magnetic (CRM) fields.

Once built the electromagnetic signature for every vessel will be different. For the ferromagnetic components, the resultant magnetic field will be dependent on the handling of the magnetic materials used in its construction and the location or locations of its construction. This signature will then change gradually over time depending on the vessel's usage and transits made. The electric signature component will initially tend to be much more similar from vessel to vessel in a class but will over time diverge as the hull paint is damaged during its particular usage.

In order to maintain a vessel's electromagnetic signature within target at the threat depth a knowledge of that vessel signature is required. This signature may be measured directly using an underwater range or alternatively inferred from knowledge of magnetic fields and electric hull potentials at or close to the vessel itself. The use of a modelling process enables signatures to be predicted at other locations other than the measurement points.

REDUCTION OF MAGNETIC SIGNATURES

Magnetic signatures can be reduced in several key ways that extends from initial design to through-life signature maintenance. The intrinsic magnetic signature of any vessel depends intrinsically and significantly on the vessel structure, material composition and design. These parameters are largely constrained by the operational requirements of the vessel. However as part of the design concept, system solutions can be put in place that will enable a high degree of signature reduction to be obtained using either internal of external solutions.

In this paper we consider the following:

- Magnetic treatment or 'deperming'
- Open loop onboard DG and ICCP system
- Multi-influence ranging
- Closed loop solutions

MAGNETIC TREATMENT SYSTEMS

A straightforward method of signature reduction that can be deployed even after the vessel design and build phases is a 'magnetic treatment' process in which a sequence of large alternating magnetic fields are applied to a vessel via an external magnetic coil. This method can of course also be in used in conjunction with a permanently fitted onboard degaussing system and in this scenario can result in both power and cost savings in relation to the onboard signature reduction system.

Magnetic treatment systems vary in scale, infrastructural complexity and cost. For many vessel types comparable performance can be achieved from any of the various common systems including the following coil system types:

- Over-run
- Drive in
- Close-wrap

The choice of facility type is therefore largely dependent on the relative priority of cost, convenience and available real estate. Close wrap systems are the only systems that can be configured in a transportable mode whereas over-run and drive in treatment systems are necessarily fixed systems.

A facility type that is convenient for many locations and has the flexibility to treat many different ship types is the redeployable close wrap facility. The system can be made more flexible by having power supplied from transportable diesel generators removing reliance on dock yard power supplies and allowing greater freedom of deployment location. A typical system is shown in Figure 1 below.

Figure 1. Actual System and Schematic Drawing of a Redeployable Magnetic Treatment System

ONBOARD SIGNATURE CONTROL SYSTEMS

Magnetic and Acoustic Influence Measurement Ranges

Onboard open loop degaussing systems comprising current carrying coils within the vessel's structure can result in optimised magnetic signatures when the coil currents and/or turns are configured using a magnetic ranging facility.

A modelling range comprises of typically 3 to 5 three axis magnetic sensors, which can be easily and speedily deployed on the seabed [Ref 1, 2]. Additional influences can alos be integrated or co-located.

A typical Transmag periodic ranging uses vessel modelling techniques that require only two DG-off runs across the range to enable the signature to be minimised (once a first of class ranging has been completed for a vessel of the same class). This means that a typical ranging would be complete in approximately one hour. In contrast the manual signature optimisation technique often takes one day for a frigate and up to three days for a new MCMV.

The main functions provided by ranging software are to enable modelling of the magnetic state of the vessel, to determine the magnetic effect of the vessel's degaussing coils and to optimise the ship's DG-on signature by calculating the optimum degaussing coil current settings.

In addition the models can be used to calculate the threat to the vessel worldwide. A range may be used to determine the optimum DG-on state for a vessel for a particular local scenario.

This may be undertaken either at a range site prior to operational deployment or indeed by using a forward area range in the operational area to confirm the optimum vessel state has been achieved and maintained. Note that in contrast closed loop systema, which are considered later in this paper, will maintain the signature within a certain tolerance band but that its performance is governed by the quality of the input data gathered on the range and the data measured onboard the vessel. Thus it may often be possible to improve on the signature in the forward area for critical operations using a range.

Redploy able Measurement Ranges

Transportable range systems are useful for forward area military measurement missions and also can be used at ship yard locations for verification of acoustic signatures of commercial platforms. Deployability is a key concept when considering sensor design. Compact sensors clearly can achieve improved man-handleability over traditional systems and can be deployed without the requirement for a large deployment vessel or crane. This provides the benefit of ease of deployment and recovery for transportable systems or as part of maintenance programme in a fixed mount system.

Figure 2 Set up of a transportable range system office and tracking basestation

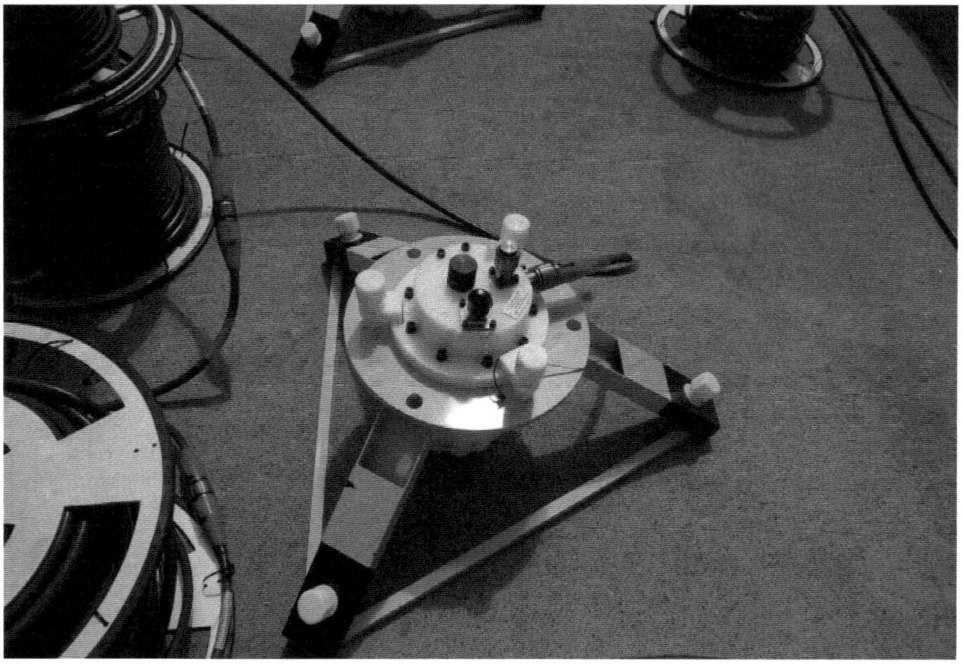
Figure 3 A Magnetic and Acoustic Influence Measurement Sensor in its Tripod

Electric Field Ranges

Historically magnetic and acoustic influences have been most commonly measured. In recent years the electric influence has come to the fore in both defence and hydrocarbon survey fields with experimental very low noise sensors being trialled [Ref. 3, 4] for detection of electric fields.

Electric field ranges have been developed using 3-axis electric field sensors which can be easily deployed in parallel to or combined with an existing magnetic and acoustic range. Each sensor comprises 3 orthogonal pairs of silver-silver chloride electrodes and has been mounted on a flat concrete mattress to create a flat seabed area around each sensor to minimise distortion of the electric field by its mounting. The low noise electrodes can provide high accuracy electric field measurements in a spherical housing of 0.25m diameter; this is a considerable improvement over previous sensor technologies.

Each sensor has a digital interface allowing Ethernet communications to an underwater network hub. The electric field range has been deployed in a line parallel to an existing magnetic and acoustic range which comprises eight multi-influence sensors. The magnetic and acoustic sensors have analogue outputs which are converted to digital signals at an underwater junction box. The two ranges are connected into a common fibre optic cable at a fibre optic interface box.

The electric range has been developed such that it can be run as a standalone range or in parallel to the existing multi-influence range such that all three influences can be gathered at the same time.The development of compact redeployable high sensitivity, high bandwidth multi-influence digital sensors has only recently become a viable proposition. Sensors are can be supplied in magnetic only, electromagnetic or combined magnetic/electric/acoustic formats.

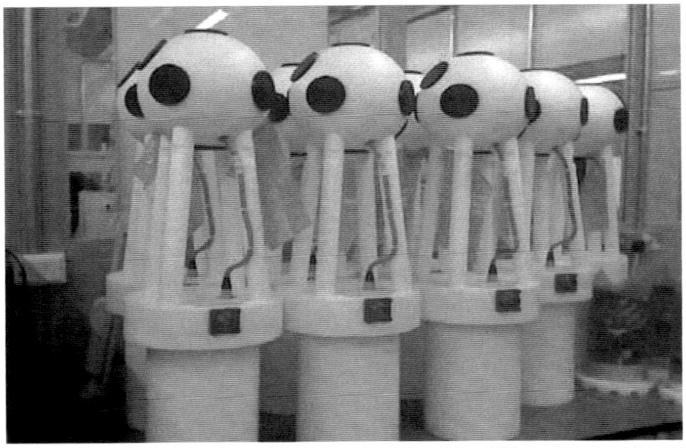

Figure 4 Compact Sensor with 3-axis electric field in Ultra's production test facility

Electric Signature Field Control

The interaction of a vessel hull with its environment, notably the sea water gives rise to corrosion where metallic areas of the hull are exposed. The corrosion of a vessel can be minimised by the suitable application and maintenance of coatings. Since coatings cannot be applied or maintained perfectly an impressed cathodic corrosion protection system, ICCP, are used in order to impress current on to the hull such that the relative potential of hull relative to the seawater is such that corrosion is rendered energetically impossible. Corrosion and the action of the anodes from the ICCP system lead to ionic current flow in the seawater and therefore there will be also be an associated electric and corrosion related magnetic, CRM field, due to the vessel in the seawater. The electric field and CRM of a vessel poses a potential threat to it from third party detection and thus it needs to be controlled and minimised.

To enable the control or optimisation of the electric field or CRM, from a vessel it is necessary to have a detailed knowledge of the relative hull to seawater electrical potential or corrosion hull state. Knowledge of the vessel's corrosion hull state is necessary for the control of the signature because the systems placed on a vessel to control the signature will also impact on the hull state and thus affect its corrosion protection. Furthermore it is the hull state that ultimately determines the vessel's electric signature and also the majority of electric signature control systems employed are the same type of systems used to prevent corrosion of the vessel.

The corrosion hull state of vessel continually changes with time due to both the physical state of the hull and its interaction with the environment, the seawater. Thus in order to control the electric signature of a vessel it is necessary to routinely characterise the corrosion hull state of a vessel. Numerical modelling techniques combined with onboard and off board ship measurement can be used to characterise the vessel hull corrosion state and have been shown to control the electric signature. Similar principles can also be applied to producing closed loops systems.

CLOSED LOOP DEGAUSSING SYSTEMS

Closed loop degaussing (CLDG) systems can be used to maintain the ferromagnetic signature of a vessel over extended periods of time. We shall not discuss possible CLDG methodologies

in any detail here other than to list several alternative methods that might be used to create such a system including:
a) Vessel State Methods
b) Compartment Modelling
c) Enhanced Source Modelling
d) Boundary Element Methods.

In the following section we will look at an example of the development of a boundary integral method for CLDG.

Boundary Integral Modelling

Boundary Integral (BI) based algorithms are finding applications in vessel modelling especially in, for example, CLDG where the measurements are made close to the steel surface. At the heart of the BI method is an algorithm which calculates the keel signature from the field measurements at the hull. This type of algorithm is based on two main principles: Green's theorem; an application of which states that if the flux of a field is known all over a closed boundary, then the field elsewhere, external to the boundary can be calculated from this provided there are no other sources of flux in the domain external to the boundary; and the definition of a scalar magnetic potential which is possible because over a short period of time the field can be said to be magnetostatic.

In practice, of course, for most practical implementations where sensors may be fitted, there will be sources external to the measurements. We may justify this by re-wording the Green's theorem statement to say that there are no sources which change significantly external to the measurement locations. In the case of CLDG, the field to be calculated in the domain is that of the threat signature and the boundary encloses the majority of the ferromagnetic material in the vessel. The flux is measured by magnetic sensors which measure the magnetic field. The number of sensors used is chosen to be both acceptable in an engineering context and also to be able to encompass the majority of the ferromagnetic material and hence represent all of the flux over the vessel. Again, in practice, this will be a compromise between coverage and practicability, and techniques must be employed to 'fill-in' those areas where sensors cannot be located.

The actual creation of the BI mesh is relatively straightforward as the geometry of the vessel is basically known and for submarines can be considered as combinations of cylindrical and spherical surfaces (Figure 5).

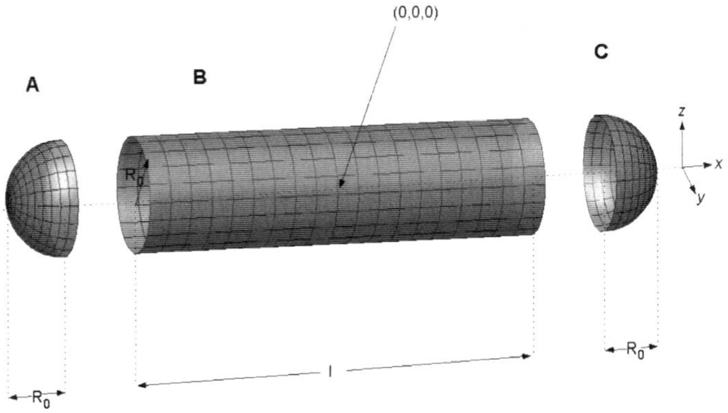

Figure 5 - Basic BI Formulation for Simple Hull

Boundary Integral Results

As stated above, there is a significant problem in the generation of data for testing algorithms designed to take on-board measurements as their input data. It is possible to carry out initial development using FE representations using models with a mesh high mesh complexity. Whilst the input data is not that of any specific vessel, sufficient variation can be applied to the material characteristics of the model to provide a suitable test for any algorithm under investigation (Figure 6).

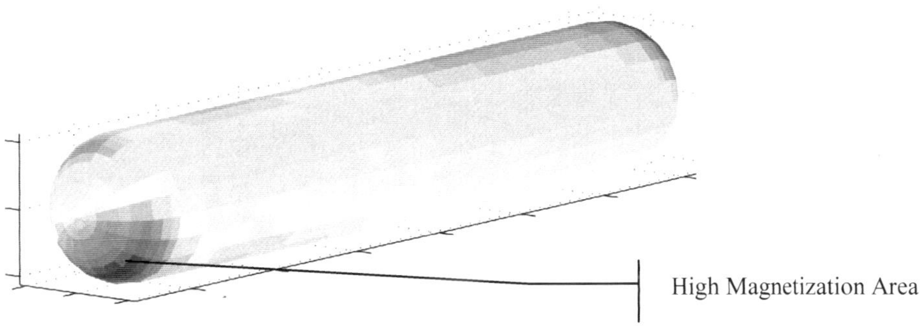

Figure 6 - Simple FE Model Magnetization Distribution Used in BE Evaluation

Using such models it is possible to quickly investigate the optimum number of sensors required, the optimum distribution of these sensors and, perhaps more importantly, the consequence for performance of the effect of not locating the sensors in the ideal locations. Results from this type of modelling (Figure 7) can demonstrate the readiness of an algorithm to be tested against physical (real) data. In this case we illustrate the x and z signature components of the BI and FEM signatures at the depth of interest beneath the keel line of the vessel where the x-axis is aligned with the length of the vessel and the z-axis is vertically aligned.

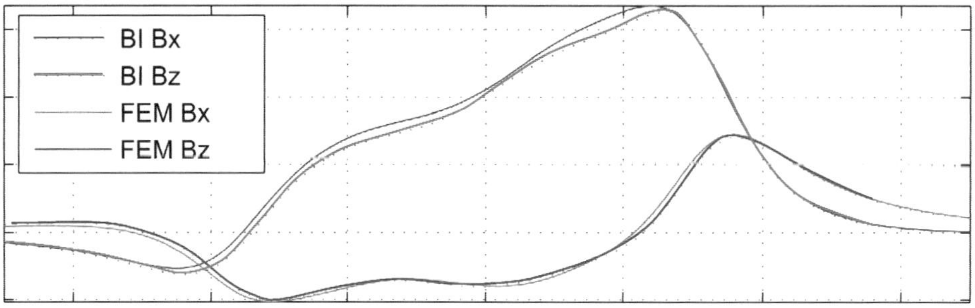

Figure 7 - Comparison of BI and FEM Keel Signatures

As has been previously stated, comparisons with real vessel signatures are more problematic. Obtaining contemporaneous close-in and range data measurements so that comparisons of the predicted and actual keel (threat) signatures can be made have been found to be difficult without suitable transportable ranges. Such data is invaluable however in comparing the magnetization distribution within real vessel steel and the FE approximation (Figure 8) and for confirming that the algorithms (in this case the BI algorithm) are effective in the real world (Figure 9). The three axes in Figure 9 represent the three signature components x, y and z for the measured signature and that predicted from the BI model at the depth of interest. This figure clearly demonstrates that the closed loop method is a valid choice as a means of magnetic signature control.

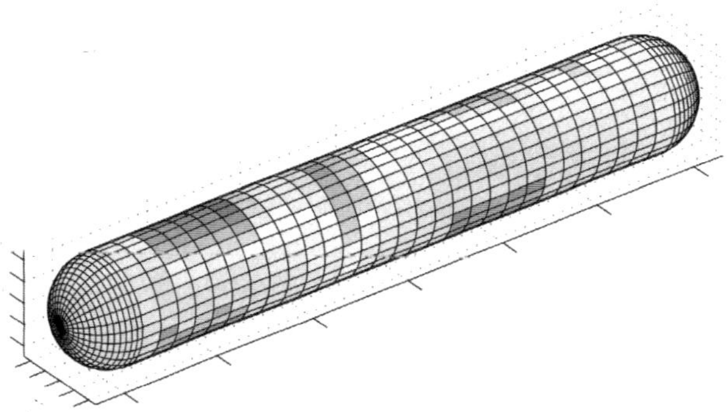

Figure 8 - Magnetization Distribution on Real Hull cf. Figure

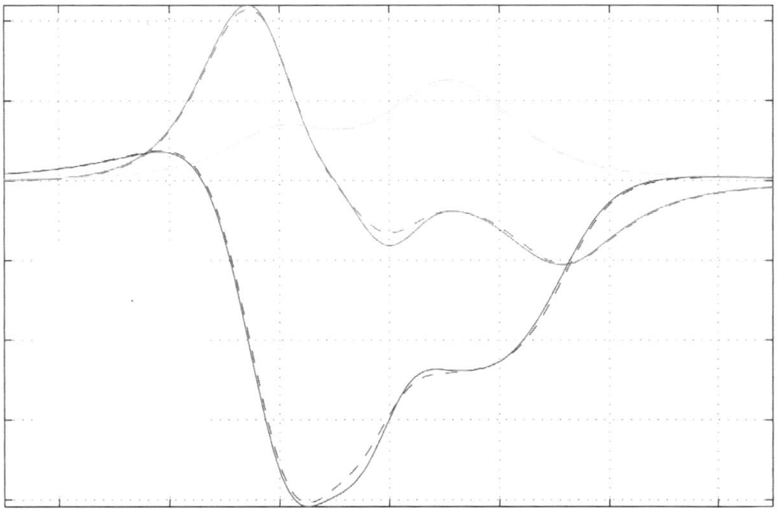

Figure 9 - Comparison of Actual (solid line) and BI Predicted (dotted line) Signatures

CONCLUSIONS

In this paper we have discussed Underwater Signature Management is an integrated process that extends from initial design to through-life signature maintenance with stages including:

- Electromagnetic concept evaluation
- Magnetic treatment via deperming process
- DG and ICCP system design and through life signature control
- Multi-influence ranging
- Closed loop solutions

We have looked at the use of Multi-influence ranges to configure both degaussing coils and ICCP systems to minimise the total electromagnetic signature. We have also shown that for closed loop systems with sensors fitted to the vessel hull, accuracies sufficient to determine the signature threat can be found using boundary integral (BI) methods. Using CLDG based algorithms ranging data can be supplemented using BI methods specifically to enable very near field measurements near the hull to be used in predicting threat depth signatures.

REFERENCES

1. S.J.Davidson et al, 1994, 'Transmag a transportable degaussing range', UDT Europe 1994.
2. S.J.Davidson et al, 2009, 'Electromagnetic Underwater Signature Management: Signature Design Techniques and Redeployable Measurement Systems', Marelec 2009.
3. H.Jones, S.J.Davidson, P.Rawlins, 2004, 'Development of a Low Noise Electric Field Sensor For Measurement and Ranging Applications', Marelec 2004.
4. Lucy MacGregor, 2004, 'Detecting and Characterising Hydrocarbon Reservoirs Using Marine Active Source Electromagnetic Sounding', Marelec 2004.

The Design of a Submarine Countermeasure Launch System Using Dynamic System Modelling

M Ratcliffe[1], N Malik[2], R Drake[2]
1 Babcock Pty Ltd., Adelaide, SA, Australia
2 Babcock, Bristol, UK.

ABSTRACT

The need for industry to provide cutting edge technology to naval customers has long been appreciated. More recently, the importance of economic drivers has also been recognised. It is no longer acceptable to seek the best solution at any price; the objective has to be to achieve a demonstrable capability edge at a cost that is sustainable economically.

Engineering tools such as Finite Element Analysis or Computational Fluid Dynamics have their place within the design process and have already been remarkably successful in reducing reliance on expensive prototypes to de-risk designs. However, these technologies may not capture interdependencies between different engineering disciplines because they are tools traditionally used by specialists in one particular field. It is at the interface between specialist fields where procedural standards may be inadequate and the impact of a seemingly minor change in design can be misunderstood.

A "model based design" approach in the hands of a Systems Engineer can facilitate understanding of the behaviour of a design far earlier in the development life cycle, significantly reducing unanticipated emergent properties and reducing or even eliminating the need for expensive physical prototypes.

Babcock has recently designed a novel submarine countermeasure launcher that is scalable for a number of different size stores and uses an inherently safe prime mover that does not require traditional pyrotechnics.

This paper presents a review of how a model based design strategy was used for the development of the countermeasure launcher.

1. INTRODUCTION

Uncertainty and risk are an inherent part of engineering development activities, and consequently there is always a certain amount of doubt whether a designed system will perform as required.

1.1. Prototypes

Prototypes are a staple within engineering development and have been used to reduce risk as part of the product design process for decades.

Typically, many iterations of prototypes are used to refine aspects of a design until it is deemed to be exhibiting the desired characteristics and meeting defined requirements. These prototypes will range from early proof of concept models to functional prototypes that attempt to replicate the final design in terms of function, space envelope, material selection, etc.

Prototypes are a reliable method of reducing (but never eliminating) uncertainty from the development process. However, there are a number of limitations with the use of prototypes. Apart from various approximations that may have been adopted for the prototype production, it may be impossible to test a prototype through the full operational envelope required of the production design. For example, in the submarine environment, a prototype can be pressure tested; however, it is unlikely that its performance in a fluid flow environment can be readily confirmed by test.

Perhaps the biggest shortcoming of prototype testing is cost, especially if multiple iterations are required. Recent advances in computer technology have reduced the need for physical prototypes in development. A number of different simulation methods have been developed by Babcock for use in the design process due to the difficulty and expense of prototype testing in the submarine environment.

1.2. Specialist analyses

Theoretical analyses – and computational analyses in particular – are typically within the purview of specialists. As the relevant technologies have matured, specialised modelling techniques have developed to cater for very detailed representations of their domain, modelled as continuous fields.

Specialised analyses can provide very precise results in their respective domains, and the effects of Moore's law have resulted in remarkably sophisticated analyses being available on desktop PCs. Specialist analysis methods include:

- Finite Element Analysis (FEA);
- Hydraulic/pneumatic circuit analysis;
- Kinematic simulations; and
- Computational Fluid Dynamics (CFD).

The specialised techniques have a shortcoming: their very specialist nature. At the interface between domains, the interactions may be quite complex. Yet this is precisely where specialist programs and specialist engineers become most uncertain. It is at these interfaces where a different approach such as Model Based Design (MBD) has enormous potential.

1.3. Model Based Design

Model Based Design (MBD) uses an overall system model within a unified development environment. It is intended to provide a faster and most cost-effective technique for the engineering development of dynamic systems. It has considerable pedigree in the automotive industry, and is used extensively in signal processing and control systems engineering, where the importance of capturing the interaction between the controller and the dynamics of the system is self-evident.

Efficiency improvements are achieved through MBD by:

- Using a common design environment across individual project teams and domains;
- Linking requirements directly into the design process;
- Facilitating scaling of the design by adopting appropriate parametric inputs;
- Enabling system level "what-ifs"; and
- Integrating prototype testing with design to identify and errors early in the design phase.

One common failing of a typical design team is the difficulty of communicating properly outside individual engineers' fields of expertise. MBD allows all members of a design team to identify interdisciplinary constraints and understand emergent issues at a system level.

2. UNDERCASING COUNTERMEASURES LAUNCHERS

Countermeasures launchers form a crucial safeguard in the defence of a submarine. A number of soft kill variants are available that seek to seduce the incoming threat away from the platform. Some modern countermeasure systems also incorporate a hard kill capability by disabling the incoming torpedo after seducing it to a safe distance away from the submarine.

Babcock has been involved in integration of the Rafael Torbuster hard kill countermeasure into submarine platforms. Torbuster is a 4th generation torpedo decoy that combines acoustic countermeasures as well as hard kill to provide active defence versus all acoustic homing torpedos.

The Torbuster system involves firing the countermeasure a safe distance from the submarine and transmitting acoustic signals to deceive and attract the torpedo that has been fired at the platform. When the offensive torpedo is close enough, the Torbuster decoy explodes inflicting enough damage to the torpedo to neutralise it.

2.1. The Babcock "Kestros" Launcher

The Kestros Launcher is a pneumatically-powered launcher that uses a self-contained high-pressure air-bottle to provide the motive power and avoids using a pyrotechnic prime mover.

The air is used to push a piston with a number of simple mechanical connections that actuate an obturator behind the store that propels the store out of the launch tube. At the end of the launch stroke the pusher seals against the forward end of the launcher and retains the residual air within the launcher, reducing the noise signature and improving covertness.

The launchers are able to operate independent of platform depth because of the inclusion of a valve which opens a pre-flood port to equalise pressure prior to discharge. The exit velocity is practically depth independent because the full firing pressure is retained in the system until depth equalisation has occurred.

Traditional countermeasures launchers have utilised pyrotechnic methods. These can never be made inherently safe due to the presence of the pyrotechnic itself and so remain a hazard even when the platform is alongside. The Kestros launcher system can be made inherently safe by being depressurised when the boat comes alongside.

Time at risk is minimised by pressurising the system when onboard and as late as possible prior to sailing. Even once pressurised, inadvertent launch of countermeasures is prevented by the a safety pin at the fore-end of the launcher, which have been designed to react the launch forces. The safety pins are removed just prior to launch of the submarine.

A CAD model of the Kestros launcher is shown in Figure 1.

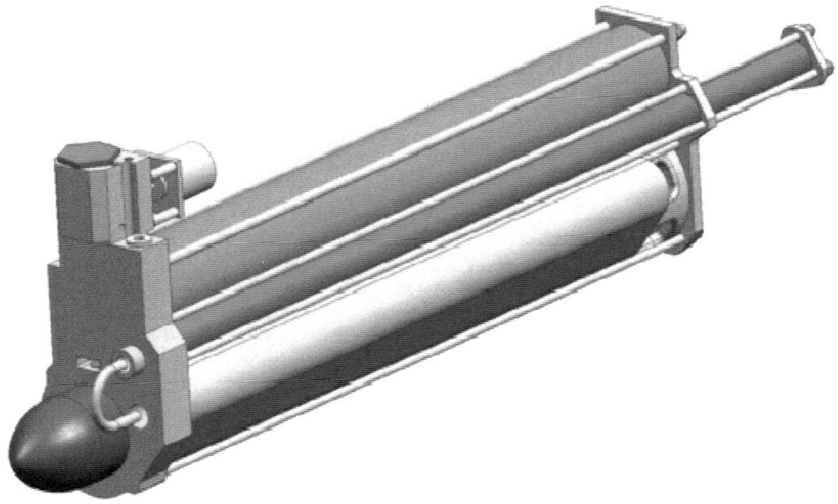

Figure 1 – The Kestros Launcher

2.2. Platform Integration

The system is scalable and 4" and 8" store versions have been produced. The 8" version has been demonstrated at sea, while the 4" version is currently in production for the Spanish S-80 Class.

The launchers are intended to be mounted into cradles (or modules) allowing several launchers to be loaded in a single operation. The cradle/module design is the one item that would probably be boat class specific, although their design can be relatively simple and the integration of the modules straightforward. Essentially, they consist of a simple frame that provides the interface to the casing and pressure hull structures. Current designs have the cradles fitted via shock mounts to strong points on the pressure hull but the cradles could equally be suspended from the casing structure instead.

Figures 2 and 3 show, respectively the Torbuster (8") and S-80 (4") launchers located within their modular cradles.

Figure 2 – Two Torbuster Launchers in Modular Cradle

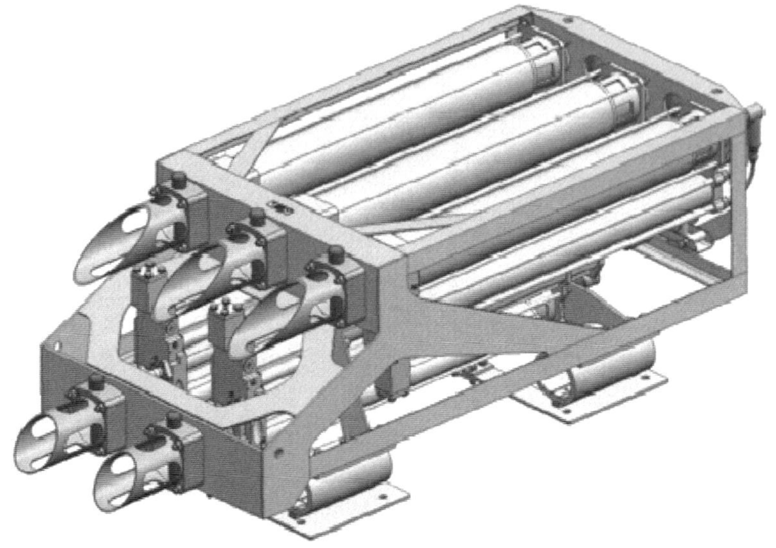

Figure 3 – Five S-80 Launchers in Modular Cradle

Current practice for loading countermeasures involves loading the stores through exit apertures. This is not ideal, as it is time consuming and requires a dedicated loading platform. The Kestros launcher allows the countermeasures to be installed into a modular arrangement (as shown in Figures 2 and 3) in a controlled workshop environment. These can then be craned onto the boat via casing apertures, resulting in a much more rapid turn around.

3. SYSTEM MODEL DEVELOPMENT

An early physical prototype indicated that enhanced understanding of the physical subsystems would be required to facilitate the continued development of the design. Scalability was an especially crucial parameter that required investigating, and the interaction of a number of different physical domains prompted the adoption of an MBD philosophy.

The system performance requirements were linked into the philosophy from the outset, ensuring that they were at the forefront throughout the design development process.

3.1. Requirements for the launcher

Locating the countermeasures beneath the casing rather than within the pressure hull has a number of advantages, including:

- Reduced number and size of pressure hull penetrations;
- Modules can readily be loaded by crane; and
- Improved system safety.

The primary requirements for the launcher system were that it should:

- Achieve a store exit velocity sufficient to ensure a safe separation from the launch platform;
- Keep launch accelerations below a specified payload g limit; and
- Complete the launch within a specified timeframe.

As the space beneath the casing is typically limited, a compact design was essential to minimise the equipment's required space envelope. This resulted in an additional primary requirement, namely to accelerate the store to launch velocity over a stroke of one store length.

The launcher system needs to be able to satisfy these requirements at all depths and for any combination of agreed boat manoeuvring accelerations and inclinations.

3.2. Requirements for the Model

Once the requirements for the launcher system itself had been established, the first step in ensuring that the model would achieve what was required of it was to determine exactly what the desired characteristics of the simulation were.

The following system characteristics and parameters were selected as key areas for the model:

- The design parameters – i.e. the model inputs;
- The launch system performance – i.e. the model outputs;
- The system interdependencies – i.e. the coupling variables; and
- The calculation subroutines – i.e. how the physics is being modelled.

As stated above, the countermeasure launcher was required to achieve a high enough launch velocity to ensure safe separation from the platform and launch within a prescribed timeframe, but at the same time, limit the peak accelerations to avoid compromising the countermeasure. These requirements need to be satisfied by the launcher system for different payloads, boat manoeuvres and boat depths. These criteria formed the outputs and inputs, respectively, to the simulation model.

In the physical countermeasure launcher, there is a particular launch sequence; certain events need to take place in the correct order. A logic subsystem was therefore included into the model.

The loads experienced by the countermeasure during launch were predicted to be a function of coupled physics within three domains: thermo-fluid, hydrodynamic and mechanical. An appropriate model for each domain was created to capture the relevant physics.

Finally, a solution module (or numerical integration routine) was also required.

In order to avoid substantial rework as the design matured, the development of the model was constrained to avoid significant changes to subsystem coupling or the flow of information. The model architecture was therefore subject to careful preparatory design as described in the following section, as well as strict change control.

3.3. Architecture Design

In order to have a manageable model that informed the developing design at all stages, it was essential that the model was configured correctly from its inception. The benefits provided by MBD are significantly reduced if the model needs to be subject to major changes of structure as the design matures.

However, any developing design will change throughout its design phase, and it is not possible to know exactly what features are needed. A flexible system architecture for the model was therefore conceived that enabled the model to develop in scope and sophistication as the design matured.

The development of the architecture was simplified because the fundamental system design was deemed unlikely to change significantly. Therefore, the top level layout within the architecture could be defined even though the theoretical approach applied to model particular system functions was initially not known.

The architecture was designed so that a "plug and play" approach could be adopted for subsystems. Each physical domain was allocated a separate module within the overall model, and each subsystem itself comprised a number of smaller subroutines. This allowed each subsystem to be developed in line with the increasing understanding of the relevant physics as the design matured. The approach also provided significant benefits in terms of using parameterised subroutines allowing for reuse of subroutine code. Figure 4 shows a schematic of the model architecture.

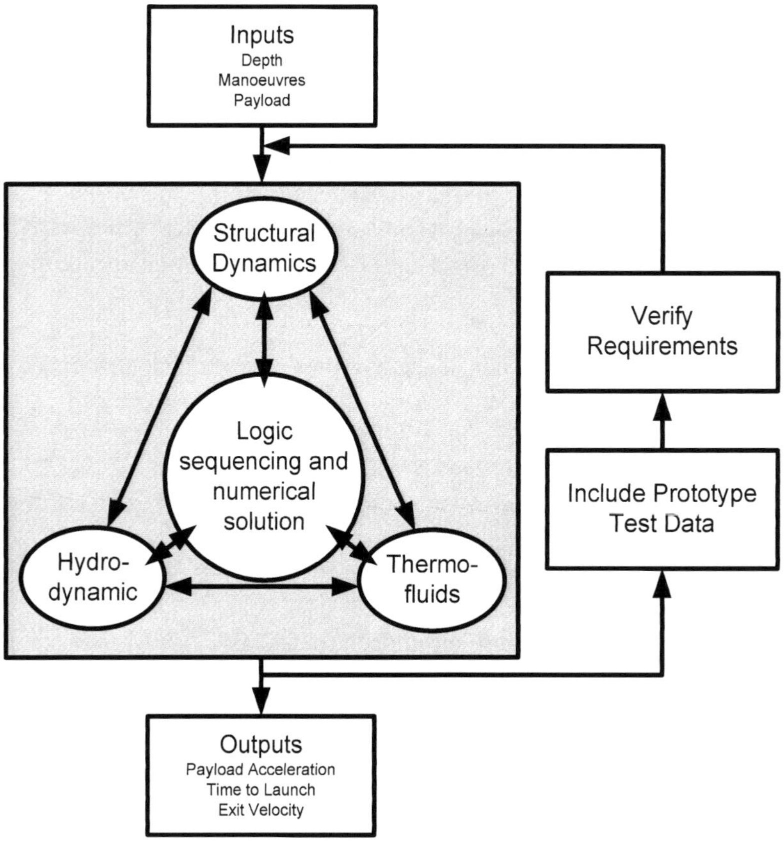

Figure 4 – Overall System Architecture

3.4. System Model Features

There are a significant number of complex physical interactions within the launcher design that could influence the performance of the completed system. The architecture described above allowed more of the physics to be captured in the model as detailed aspects of the design emerged.

Depth and propulsive forces were determined by the system model, which also included the frictional retarding loads that develop within the launcher tube as a result of the small clearances between the tube-mounted guides and the launched store.

The operation of the firing valve was also modelled as were the tube flooding processes (which involved plugging the appropriate fluid dynamic subroutines into their location within the overall architecture). This allowed the complex interaction of the flow effects and the accelerating pusher and store to be captured.

The variation of length scales and difficulty of modelling turbulence result in theoretical representations of fluid dynamics that require experimental coefficients. Provision was made in the system model to allow for experimental parameters to be included into the simulation, and these were derived through a combination of model testing and research in the published literature.

The first generation prototype was also a fertile source of empirical data for incorporation into the system model. The prototype was instrumented with an array of pressure transducers, accelerometers and a velocity recorder, and experimental data from this prototype was then fed into the system model.

3.5. Sensitivity Studies

The system model runs a standard simulation in approximately one second and was configured to enable batch runs from a controlling input deck. This allowed a large number of studies to be conducted to enable the effects of minor variations in the system parameters to be studied.

3.6. Design Support

The dynamic model was used throughout the design development to ensure that the critical requirements listed above were being satisfied. The model was used to specify loadcases for a number of the internal components which were then subjected to further detail investigation via specialist analyses.

4. STORE EJECTION CFD ANALYSIS

The discharge velocity calculated by the system model was used as an input into the extensive CFD work conducted to predict the trajectory of the countermeasures in the vicinity of the submarine. This is essential to ensure that the risk of re-contact with the submarine is minimised.

These analyses were conducted using Store Launch Simulation (SLS), Babcock's bespoke package that couples CFD analyses with dynamic modelling to determine the behaviour of stores within a flow field. The analyses included modelling the store behaviour at various submarine speeds, roll angles, turn angles/radii and depths.

A schematic of the operation of SLS is shown in Figure 5, below, while Figures 6 and 7 show a selection of outputs from the hydrodynamic modelling, showing the predicted paths of launched countermeasures for a range of boat conditions.

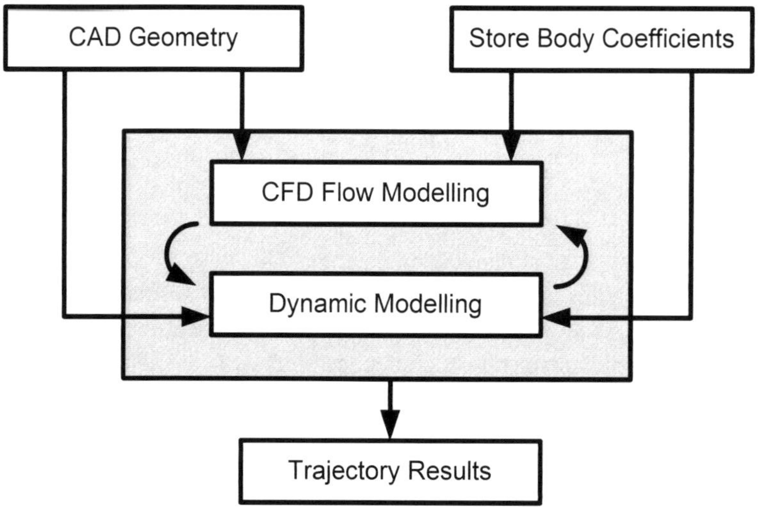

Figure 5 – Schematic of SLS

The stores considered in the SLS analysis shown in Figures 6 and 7 are negatively buoyant immediately after discharge. SLS predicts an initial drop in trajectory prior to the release of a nose weight from where the countermeasure then transits to its operating depth, while the submarine continues evasive action.

Figures 6 and 7 are example plots from SLS and depict simultaneous launches from forward and aft, port and starboard launchers which probably would not happen in practice. Nevertheless, they demonstrate the simulation power that is available to model discharge events to enhance the self-defence capabilities of a submarine.

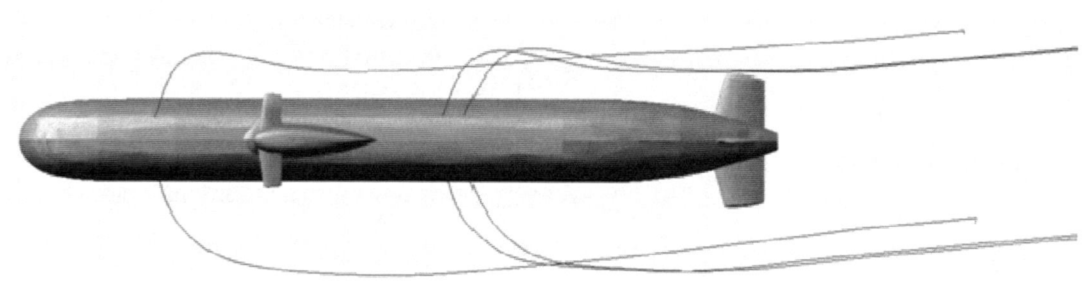

Figure 6 – Top View of SLS Trajectory Simulation

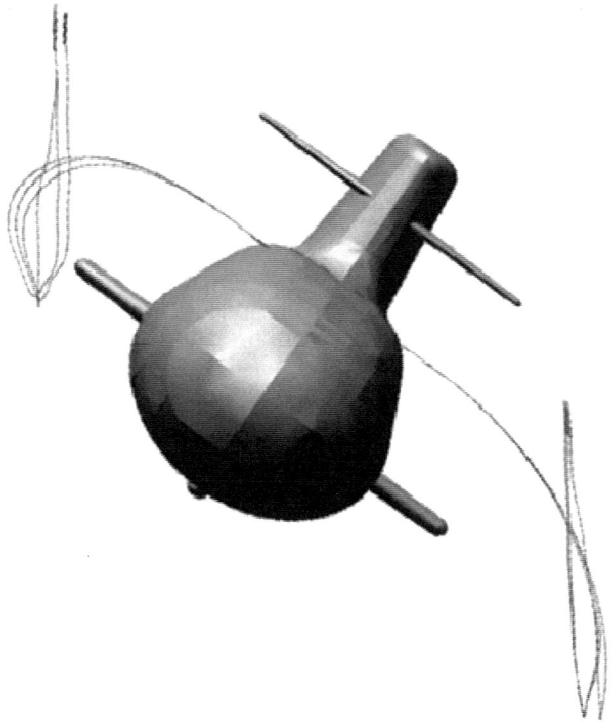

Figure 7 – Front View of SLS Trajectory Simulation

5. FUTURE DEVELOPMENTS

The scope of the countermeasure launcher integrated model could readily be extended to include SLS simulations, as well as the countermeasure control system and combat system simulator. This would result in a fully-integrated system model which would allow simulation of an entire sequence from detection of the threat through the launch of the correct countermeasure to elimination of the threat.

There is significant opportunity for applying MBD into the design of Weapon Handling and Launch Systems (WHLS, another of Babcock's core products). These are complex systems with a very large number of subsystems with complex interfaces dealing with different physical domains, and typically requiring extensive automatic control capabilities. Prototype testing for a system such as a WHLS is likely to be prohibitively expensive and enhances the attractiveness of a reliable multi-domain system model.

6. CONCLUDING REMARKS

A Model Based Design philosophy was used by Babcock in the development of a novel countermeasure launching system and this enabled the design to be progressed from concept to production units in a very short timeframe. The countermeasure system is covert, inherently safe and scalable for different store sizes.

Submarine systems are often highly coupled and influenced by interactions with different physical domains. Model Based Design has significant potential for cost savings by reducing duplication of design effort across the different domains and significant reduction – or even, elimination – of prototype testing.

The Development of an Improved Submarine Escape System

M Ratcliffe[1], T Peacock[2] & R Manion[2]
1 Babcock Pty Ltd., Adelaide, SA, Australia
2 Babcock, Bristol, UK.

ABSTRACT

The modern submarine comes in many sizes and configurations and performs different missions. Despite the differences in size and role, all submarines share a number of attributes; the fundamental quality is that they operate at (often significant) depth. This prime capability means that escape from a stricken submarine is inherently perilous.

The 'tower escape method' is adopted by a number of navies worldwide. These escape towers operate according to a simple design philosophy that involves flooding an enclosed chamber with seawater until the air pressure is equalised with the depth pressure. Once equalised pressures are achieved, an exterior hatch is opened and the escaping submariner may float to the surface.

Ideally, a system would allow escape from shallow depths all the way to the limits of human endurance. It also needs to be reliable, minimise risk to escapees and use air efficiently to permit a large crew to escape.

The features required from a modern submarine escape system collectively form a set of stakeholder requirements. This paper discusses these requirements and describes the work that is being conducted by Babcock to design, develop and test an improved submarine escape system that is safer, more reliable and efficient than current systems and that may – crucially – be retrofitted to existing classes of submarine.

1. INTRODUCTION

The extreme environment in which a submarine operates presents a number of risks. Any serious incident has the potential to sink the submarine, and these incidents can occur for a number of reasons, including:

- Flooding;
- Fire;
- Collision;
- Grounding;
- Explosion; and
- Propulsion failure.

The history of the submarine is littered with catastrophes. Although it is well known that the US has lost only one submarine since the adoption of SUBSAFE following the sinking of the USS Thresher, high-profile incidents have continued to occur regularly.

The most extreme recent example was the loss with all hands of the Russian Kursk, but recent history includes a number of serious incidents, including near catastrophes due to flooding on HMAS Dechaineux and the collision of USS San Francisco with an underwater mountain.

It is apparent that some accidents are inevitable, despite extensive safeguards attempting to reduce the risk to As Low As Reasonably Practicable (ALARP).

2. EVACUATING A STRICKEN SUBMARINE

There are three main considerations for crew attempting to evacuate a stricken submarine. These are discussed briefly below.

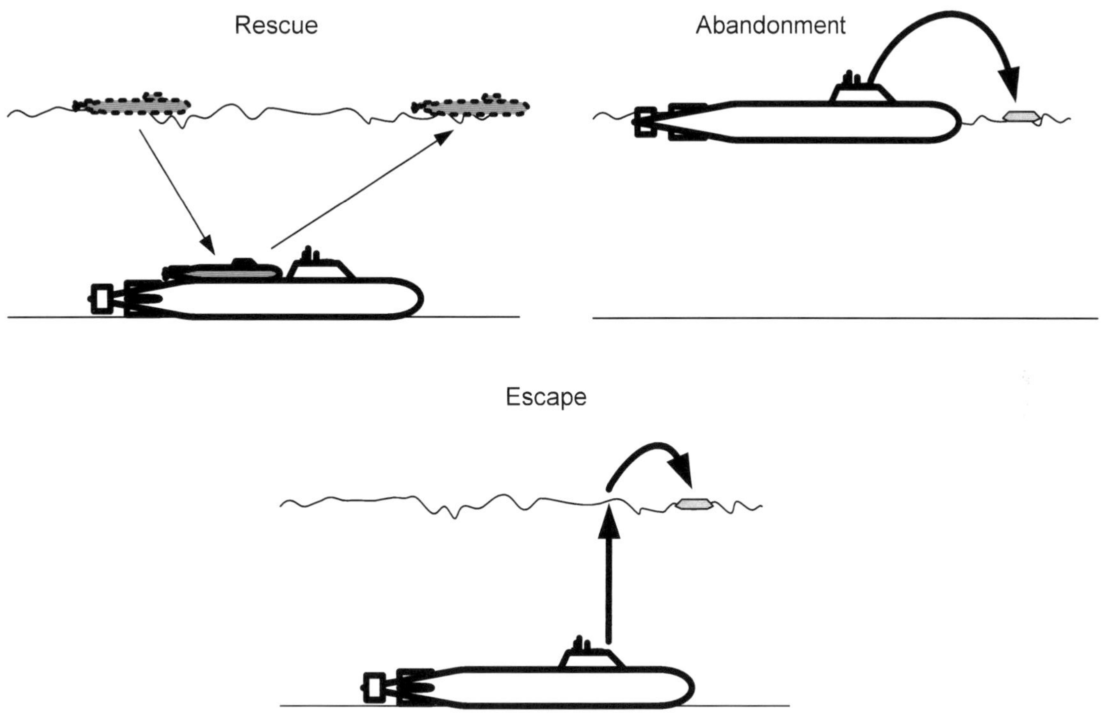

Figure 1 – Rescue, Abandonment and Escape from a Stricken Submarine

2.1 Rescue

Rescue is the preferred method and involves a rescue vehicle that mates to the damaged submarine in the vicinity of the escape hatch. There are a number of rescue vehicles available worldwide, such as the British LR5 and the US Submarine Rescue Diving Recompression System (SRDRS). Many submarines worldwide have adopted standard rescue seats around rescue hatches to facilitate the use of international rescue vehicles.

Rescue vehicles fall into two categories, namely tethered diving bells (or Submarine Rescue Chambers, SRCs) or free-swimming submersibles (Submarine Rescue Vehicles, SRVs). An SRC, the McCann Rescue Chamber, was used in the only successful extraction of personnel from a stricken submarine – the USS Squalus in 1939.

The NATO Submarine Rescue System is a state-of-the-art rescue vehicle that is fully transportable by air or road to the nearest port where it can use an appropriate ship of opportunity as a mothership (MOSHIP).

2.2 Abandonment

If a stricken submarine is on the surface, the immediate threat may appear to be reduced. However, it may not be possible to remain within the submarine. If the submarine has had to surface as a response to flooding, for example, then the submarine is likely to sink again, resulting in a limited time available on the surface. Similarly, fire is likely to result in the evolution of toxic gases.

In either case, rapid evacuation of the submarine is likely to be required. Even at the surface, the environment remains hostile to human life due to extreme cold and the risk of drowning as well ongoing hazards presented by the damaged submarine.

A level of protection is provided by immersion suits, but life rafts provide a far more robust environment and greatly increase escapees' chances of survival.

2.2.1 Babcock Life Raft

Babcock has designed and successfully trialled an undercasing system for the Royal Netherlands Navy. This is an encapsulated system stored within a GRP pressure vessel that may be operated either on the surface or just below. When operated underwater, the capsule floats to the surface before ejecting and inflating a standard marine life raft.

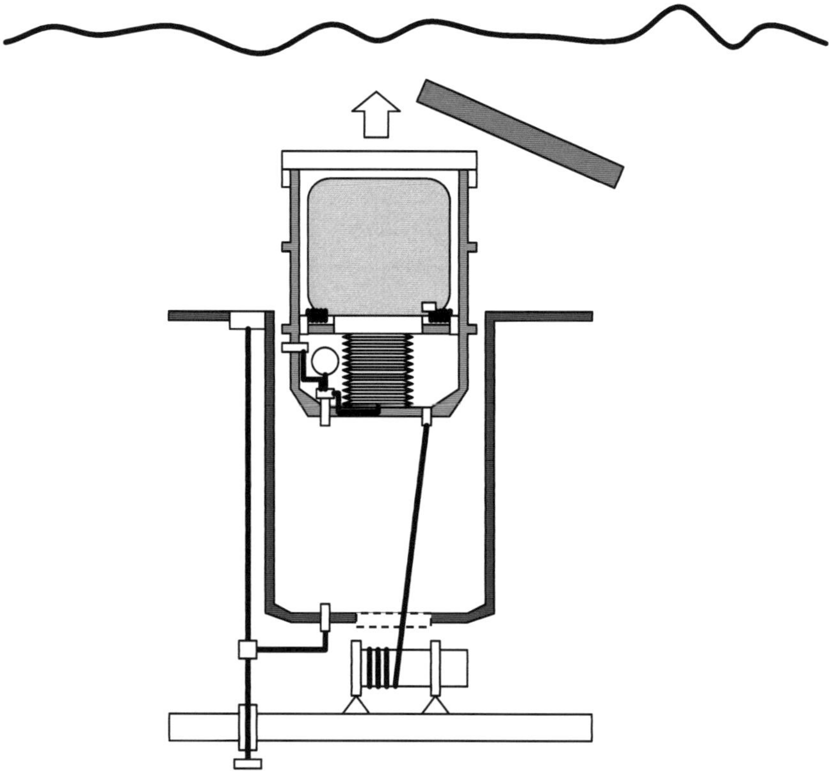

Figure 2 – Schematic of Babcock Undercasing Lift Raft

2.2.2 Inflatable Freeboard Extenders

A further key safety device used for abandonment is the Inflatable Freeboard Extender. These can be deployed to increase the freeboard of a submarine on the surface to reduce the risk of water entering the submarine when hatches need to be opened to allow rapid exit from a surfaced submarine.

2.3 Escape

Two methods of escape are available, namely Rush (or Compartment) escape and Tower escape.

2.3.1 Rush escape

For rush escape, the inside of the submarine is essentially opened to the underwater environment. The escape compartment is flooded and pressurised, and escapees (who have access to a breathable air supply) make their way out through the exit hatch or tower sequentially. This has a major disadvantage due to the amount of time that escapees spend at depth pressure, for reasons discussed below in Section 4.

2.3.2 Tower escape

Tower escape is the preferred method of escape, especially for larger submarines. Three components are required, namely an immersion suit, an escape tower and an air system. Tower escape forms the main topic of the remainder of this paper.

3. SUBMARINE ESCAPE IMMERSION EQUIPMENT (SEIE)

Once the decision to escape has been made, all surviving crew don a Submarine Escape Immersion Equipment (SEIE), which has replaced the Steinke Hood as a wearable rescue device.

The SEIE is made up of several key components, including:

- Immersion suit and hood;
- Thermal Liner for temperature insulation; and
- Single Seat Life raft.

Breathable air is used to inflate the hood and stole, allowing the escapee to breathe. The pressure of the air is maintained above tower pressure to ensure that the hood remains inflated.

The air within the hood and stole acts as a buoyancy aid and helps the escapee float to the surface. On reaching the surface, the single person life raft is activated. The SEIE therefore provides a significant measure of protection from the elements, although it is still considered as the last resort.

4. ISSUES/HAZARDS

Decompression sickness (DCS) is ultimately a result of dissolved gases coming out of solution in the body and forming gas bubbles. If decompression occurs slowly enough, then the human body can deal with the gas coming out of solution by gaseous exchange in the lungs. However, where decompression is too rapid to allow this to take place, DCS can occur with bubbles forming in the blood or other tissues. DCS can result in a range of symptoms, varying from minor joint pain, through lung damage to brain and spinal cord injury, and ultimately, death.

The risk of DCS can be significantly reduced by reducing the time spent under pressure. So where a scuba diver will spend a significant time at depth, followed by a slow ascent including decompression stops, an escaping submariner may ascend quickly provided that the time spent under pressure is minimised.

In addition to DCS, other conditions can occur under high pressures, including nitrogen narcosis and oxygen toxicity.

Despite these hazards, the human body is able to survive for a short period at enormous depths, as indicated by the accomplishments of competitors in the sport of free-diving. Participants within the different variants of the sport have achieved astounding feats including breath-holding underwater for more than ten minutes and no-limits dives to greater than 200m.

5. ESCAPE FUNCTIONAL REQUIREMENTS

As stated above, the human body is capable of surviving at astonishing depth. However, most submarine escape systems have been designed for depths of up to approximately 180m. Conversely, rescue vehicles may not be able to operate at shallow depths due to the low hydrostatic pressures and strong wave-induced currents that make the process of mating to the submarine difficult.

Escape from submarines is – almost by definition – a very hazardous activity that can never be made risk-free. Nevertheless, risk can be suitably minimised to ALARP by considering the physiological effects of depth on the human body. The key is to manage the conflicting requirements of minimising the amount of time spent at increased pressure, consistent with achieving rapid evacuation but to keep the rates of change of pressure low enough to avoid barotraumas.

At high pressure, some of the gases that comprise atmospheric air can become powerful narcotics. In order to avoid these effects, diver breathable air is provided to escapees via dedicated air systems. As the amount of air available is limited on a submarine, the time required for individuals to escape should be minimised to allow for all survivors to escape.

The considerations described above result in three primary requirements, namely:

- Operate at all depths down to the limits of human survivability;
- Minimise the risk of harm to escapees by managing exposure to pressure and pressure gradients; and
- Make best use of air to ensure that all survivors can escape.

6. PRINCIPLES OF OPERATION

Submarine escape towers operate according to a relatively simple procedure, as follows:

- Enter escape tower;
- Flood tower (resulting in pressurisation to depth pressure);
- Escape from tower; and
- Drain tower (into submarine).

6.1 Enter escape tower

Entry into the escape tower is influenced by a number of criteria. The size of the access hatch is clearly pertinent but its exact configuration is also critical. A hatch that opens into the escape tower (i.e. outwards from the main body of the submarine) will be harder to open, especially if there is some deadweight obstruction inside the escape tower.

Such a hatch will also require more internal volume inside the tower, but will be easier to seal against the hull. As the "last man" must be able to close the hatch unaided, this may be a critical design driver.

Following a successful embarkation into the tower, the escapee will attach the escape suit into the air supply and will be ready to receive the air that is needed for breathing. The air also serves to inflate the life jacket and the hood of the escape suit, as well as contributing to the pressurisation of the tower.

6.2 Pressurise

The rate of change of pressure needs to be managed to avoid barotraumas. The escapee cannot be exposed to depth pressure instantaneously; rather, the pressure increase must be gradual. Studies have shown that doubling the pressure every four seconds is the optimum, representing a compromise between risking barotraumas and causing the escape process to be too slow.

Typically, a fixed size orifice valve is used to control the flood and pressurisation phases. This causes a number of problems, as the flow rate through the fixed orifice is a function of differential pressure, itself a simple function of depth.

With a fixed orifice, the pressurisation will be too rapid at significant depths (resulting in the system being unsafe), but too slow nearer the surface (leading to increased risk of DCS and inefficient use of air, as well as increasing risk for the remaining crew in the main body of the submarine).

The exact method of pressurisation may vary, and vented and unvented systems may be used to optimise the use of air. An unvented system requires the entire tower to be pressurised, which can permit quite rapid escapes, at the expense of raised temperatures. With a vented system, air is permitted to escape into the submarine during the flood phase until the height of water reaches a particular level. At that point, the vent closes, and the pressure is increased, either by allowing additional inflow of water, or by pumping in additional air.

Although it is clearly preferable to use water, in practice, the pressure inside the tower is increased by the addition of both water and air for breathing purposes and hood inflation.

6.3 Escape from tower

In order to reduce the amount of time spent at maximum pressure, the top hatch needs to open quickly, allowing the escapees to exit as quickly as possible. The top hatch can be sprung loaded to open automatically once the pressures are equalised. The top hatch is then closed via a linkage from inside the submarine.

6.4 Drain tower (into submarine)

Once the escapees are on their way to the surface, the tower must be drained into the submarine as quickly as possible. Where the access hatch opens into the escape tower, almost all the water needs to be removed via a drain valve before the hatch can be opened. For a hatch that opens into the main body of the submarine, the drain time can be significantly reduced by simply opening the hatch. However, a number of safeguards will be required to avoid hazards from the volume of water within the tower.

7. BABCOCK SYSTEM DEVELOPMENT

Babcock has been developing a new tower escape system which has considered the functional requirements listed above.

This section describes theoretical work that has been conducted to investigate and optimise the pressurisation and flood cycles in the light of the physiological requirements discussed above.

A review of ideal design characteristics is also described below.

7.1 Pressurisation cycle

A system model was developed from first principles to examine the development of an escape tower's pressurisation characteristics. The theoretical model was compared against test data from an in-service escape tower, which showed excellent agreement. The validated theoretical model was therefore able to be used with confidence for sensitivity studies into the pressurisation performance.

When a fixed orifice optimised for 180m depth was used, the theoretical model was used to demonstrate that the pressure increases too rapidly for safety for escapes below the design depth. It also showed slow pressurisation when above the design depth of 180m. These results agreed with the expected behaviour based on engineering judgement. The results added further confidence in the performance of the theoretical model, and formed an important baseline for sensitivity studies.

A profile controlled pressure valve was added into the system model to investigate the potential for optimising the pressure increase rate (i.e. slow enough not to cause barotraumas, but fast enough for rapid escape and efficient use of air). The valve model was able to accommodate the pressure within the tower and the depth pressure, and achieved the desired pressure doubling time across the entire desired range of operation.

Valves with prescribed pressure profiles have significant pedigree in submarine equipment, and are used within the Weapon Discharge System of the Collins Class. However, the profile controlled valve requires electrical power to perform correctly, and it is by no means certain that this would remain available on a stricken submarine.

A variable orifice valve could also be used for simple depth compensation. This would not be able to compensate for changes in pressurisation rate as the pressure in the tower increases, but remains operational for the entire depth range. This would therefore represent a major improvement over a fixed orifice design, where the pressurisation rate below 180m is too rapid for safe use.

7.2 Flood

Ideally, the flood phase is performed as quickly as possible, since no physiological stress occurs to the escapee. With a fixed orifice size, the flooding phase is slow at shallow depths, and faster at significant depth. A Depth Compensating Flood Control Valve could be used to control the rate of flooding and achieve a rate at all depths equal to that achieved at the design point of 180m. This could be achieved by an automatic mechanically operated system, which requires no operator input or external power source.

As the flood phase should be performed quickly, with no prescribed minimum time to flood, simply increasing the size of a fixed orifice flood valve is a valid proposition. This would effectively move the design point determining the orifice size from 180m to a shallower depth (perhaps the surface).

In addition to the theoretical work investigating the flood and pressurisation cycles, a number of design activities are ongoing at Babcock; these are described below.

7.3 Tower Capacity

There are a number of design requirements that have been derived for the tower itself. These concern tower capacity, ergonomics and overall geometry.

Many submarine escape towers are designed for an individual escapee, while two man towers are also used on larger submarines. Work conducted by Babcock has shown that most two man escape towers have unnecessarily large diameters, which results in overly long escape times coupled with inefficient use of air. Moving to a three man escape tower would require only a modest increase in tower size and would result in a much more efficient use of resources allowing three survivors to escape in much the same time as two escapees from a two man tower.

Mounting escape equipment outside the tower would provide a number of benefits. It would enable the inside of the tower to be streamlined, removing a number of potential snagging points that could tear an escape suit. It would also allow the overall tower volume to be decreased which would reduce the flood and pressurisation time, as well as allowing escape equipment to be more easily inspected and maintained.

In addition to the considerations discussed above, designing the tower so that escapees are able to stand upright provides important benefits versus one that requires escapees to crouch, as strain on the lungs is reduced, resulting in lower amounts of air consumption.

7.4 Lower hatch design

Babcock has considered a number of different concepts for a redesign of the lower hatch of escape towers. The current design of lower hatch requires the escape tower to be nearly completely empty before it can be opened. The requirements derived for the lower hatch design were:

- to allow escapees to access the tower;
- enable last man operation;
- allow the hatch to be opened against a dead weight; and
- achieve a rapid draining function.

Analysis conducted by Babcock into a number of concepts – with and without rapid drain capabilities – revealed that those concepts that did incorporate a rapid drain feature offered a clear performance advantage.

8. ESCAPE SYSTEM ACCEPTANCE

As well as the critical functional requirements discussed throughout the current paper, there are a number of additional requirements that are crucial to the acquisition of the appropriate escape capability. The escape towers need to be reliable and available as well as enabling an appropriately de-risked acquisition programme. For retrofit to existing classes of submarine, the escape system also needs to be able to be installed with minimal disruption to the submarine programme.

Integration of the system can circumvent several of the problems associated with current designs, where a number of different OEMs provide individual hardware items. Adopting an integrated approach where a single company has overall system ownership can provide significant benefits in reliability and system performance. A key reason for this is that the technology can readily be tested at a system level by a design authority.

To achieve the appropriate technology readiness level, Babcock is developing a whole-system test facility that would enable the system to be tested for the entire escape cycle at depths down to 600m. A schematic of the test facility is presented in Figure 3, below which shows the test facility concept. It consists of an external pressure vessel that develops depth pressure acting on the external portion of the escape tower under test.

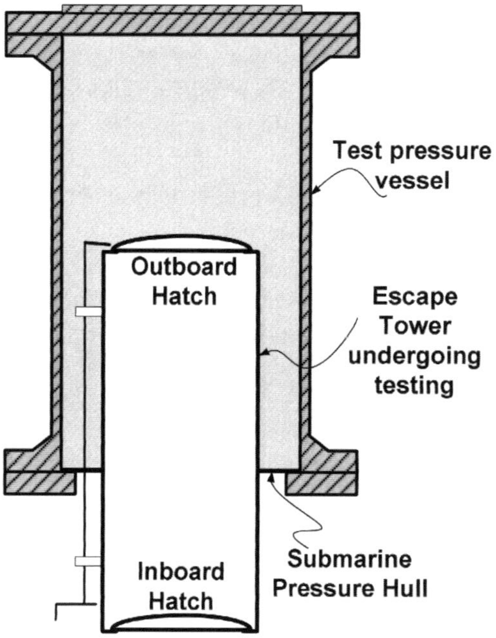

Figure 3 – Schematic of Babcock Escape Tower System Test Facility

The fact that the test facility allows the system to be tested as a whole provides a significant benefit in that interactions between components and sub-systems are automatically included within the testing environment. This allows the design to progress along the Technology Readiness Level (TRL) scale and be adopted with significantly more confidence than could otherwise be achieved.

9. CONCLUDING REMARKS

Mature options for escape from stricken submarines are essential considerations for submarines both at the design phase and for through-life capability enhancement.

This paper has discussed the work being undertaken by Babcock to develop a reliable technology for tower escape. It has discussed the primary requirements for tower escape in the context of the physiological demands made upon the body by depth pressure.

A number of findings have been reported in this paper including:

- The consumption of air can be reduced by considering tower geometry and allowing for three man escape configurations; and
- The use of profile controlled or depth compensating pressurisation valves can make submarine escape possible at depths far beyond the current design point for existing systems.

Finally, the paper has briefly discussed the test facility that Babcock is developing which enables escape systems to be tested as complete systems, significantly de-risking development and acquisition of the escape capability.

Boat Launch and Recovery – A Key Enabling Technology For Flexible Warships

Andy Kimber,
BMT Defence Services Limited, Bath UK
Info@bmtdsl.co.uk

ABSTRACT

The ability to launch and recover both manned and unmanned boats is a common feature of many future warship concepts; many also feature flexible mission bays to allow a range of vehicles to be stowed and deployed. This reflects the expectation that unmanned vehicles may offer the key to greater flexibility and effect from an individual surface vessel, mitigating the reducing numbers available to many navies. However, the transition from concept to operational capability will require the development of existing launch, recovery, stowage and movement systems; they will be required to interface with a range of different vehicles rather than be designed for one specific boat, the operational envelope will need to be maximised and issues regarding safe operation will have to be addressed. This paper seeks to review the progress made and identify the issues to be addressed in the immediate future if the concepts are to be fully released. It addresses the difficulty of incorporating operable and affordable systems from a ship designer's perspective and takes a view on how future ship / boat interfaces may develop.

BACKGROUND

The trend toward greater offboard capabilities

It has become a noticeable trend in warship design to allow for the carriage of greater numbers of offboard vehicles and boats. This appears to be the result of a combination of the well documented shifting operational focus (particularly western navies) from blue water, cold war to global policing together with a maturing of unmanned vehicles for use on the next generation of vessels. With current operations having a focus on constabulary type missions, the size of boats to be carried are increasing beyond commercial Fast Rescue Craft (FRC). Currently 7.5m RHIB (e.g. the RN Pacific 24) are common with aspirations for 11m or even 14m RHIB's in the future. Also, the number of boats is increasing e.g. two or even four boats can be used in a given operation, whilst still requiring at least one ships boat to remain available with the host ship to support helicopter operations and Man Over Board (MOB).

With Unmanned Underwater Vehicles (UUV) now entering service for MCM reconnaissance operations there is an aspiration to deploy these from a range of surface vessels and Unmanned Surface Vehicle (USV) technologies are also developing for both MCM and surveillance missions. These vehicles vary greatly in size, shape and speed, which adds further complication to launch and recovery requirements if an extensive range of vehicles are to be operated.

Hence, the traditional port / starboard davit arrangement for two dedicated ship boats will no longer be suitable for many of the next generation of warship designs. Many designs now seek to incorporate a "mission bay" where a range of vehicles, boats and support equipment can be stored and deployed. Alongside this, the need to launch and recover this range of assets in high sea states to maximise the envelope of operations is complicated by the need to handle these with a minimum of deck hands whilst achieving high safety standards. As some assets are likely to be unmanned, the launch and particularly recovery is further complicated.

Commercial Comparisons

Davits, stern ramps and unmanned vehicle launch and recovery systems are all currently utilised in commercial vessels. There are many similarities between naval boat davits and commercial FRC / MOB davits, although it is noted that naval systems have to be designed for more frequent use and potentially larger, heavier boats. Stern ramps are used in paramilitary and particularly coast guard vessels due to the fast response time. Some offshore vessels also utilise stern ramps particularly for workboats.

Of particular interest to navies are the offshore support vessels designed to operate Remotely Operated Vehicles (ROV), because of the parallel with naval vessels designed to conduct MCM utilising UUV's. As with ships boats, there are a variety of equipments used in launch and recovery of ROV's in the offshore environment, including cranes, A frames, telescopic davits and moonpools. Fitted with heave compensation, these systems are capable of operation to upper Sea State 6 (Ref. 1). Another interesting solution is the containerised launch and recovery system for UUV's used in underwater surveying, which allows the use of ships of opportunity rather than operating dedicated vessels.

Figure 1 ROV LARS system (left, image courtesy of Cargotec) and UUV in containerised LARS (right, image courtesy of Kongsberg)

LAUNCH AND RECOVERY TECHNOLOGIES FOR WARSHIPS

For the purposes of illustrating the relative merits, some analysis has been conducted. This analysis is part of ongoing work and at this stage should be taken as more illustrative than definitive in terms of performance numbers. The seakeeping analysis was performed on a 90m length concept (BMT Venator) representing a small multirole naval platform (refer to Ref. 2).

Davits and Cranes

The current generation of davits are typically designed for fast rescue or man overboard duties as required by SOLAS; for naval ships the davits are also designed for more regular use as the rescue craft are often also the ships boats.

Typically installed port and starboard on naval vessels to allow flexibility in the side on which to deploy, there are a variety of types available but typical types used in naval vessels include:

- Slewing davits, with a pedestal forward of the boat cradle and the arm stowed over, swinging in a horizontal plane to deploy / recover the boat;
- C type frames, with integrated boat stowage, where the arm swings in a vertical plane;
- Telescopic deck head mounted davits, used where the boats are stowed in bays.

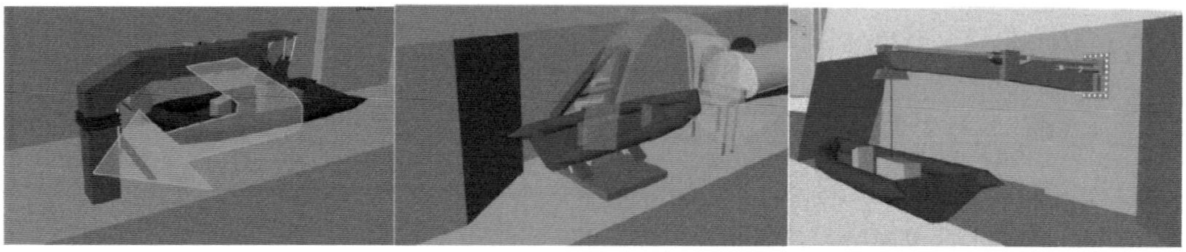

Figure 2 Types of davit – Slewing, C frame, Telescopic

The appropriate type of davit will depend on the arrangement to be adopted and space available (C type davits have a smaller footprint than slewing, but a slewing arm has a lower profile for example). In most applications the davit is associated with a single cradle and boat.

SOLAS requires fast rescue craft to be stowed "in a state of continuous readiness in not more than 5 min" and makes further reference to requirements within the LSA Code (Ref 6). Although no specific sea state requirements are defined for davits on commercial vessels, this latter document includes the following requirements for rescue craft launching appliances:

- "Shall be fitted with a device to dampen the forces due to interaction with the waves when the fast rescue craft is launched or recovered";
- "The winch shall be fitted with high-speed tensioning device which prevents the wire from going slack in all sea state conditions…";
- "Lowering speed for a fast rescue craft with its full complement of persons and equipment shall not exceed 1m/s";
- "A fast rescue boat launching appliance shall be capable of hoisting the fast rescue boat with 6 persons and its full complement of equipment at a speed not less than 0.8m/s."

For a naval vessel, the launch and recovery sea state limit of ships boats is often driven by the need to use them for fast rescue craft and specifically helicopter guard duties, hence the boats need to be recovered in sea states equivalent and often beyond the operating limits of the helicopter. A compensated davit may be expected to perform in Sea State 6, which is adequate for this purpose. The factors governing safe operation of the davit in high sea states include:

- Pendulum effects causing the boat to collide with the recovering vessel;
- Hoisting speed, too slow and there is a risk of snatching or the boat overturning if caught by a wave; too fast and the acceleration is not safe for the crew in the boat;
- The effect of ship motions on personnel on deck and resultant safety for the deck crew.

The davits are typically single point lift with heave compensation incorporated to control the distance between the boat and the sea surface as the boat is lowered / raised. However, it is also necessary to control the boat to prevent adverse swinging and, therefore, normal operations will involve a bow line being used to constrain the swinging motion of the boat as it is lifted or lowered. This is manually held at the deck of the recovering vessel, typically by deck crew a little distance from the davit location to provide sufficient length and angle in the line to control the boat. Depending on the sea state, this can require a number of deck crew to control the davit and hold the lines and the ability of the deck crew to safety execute the operation can often become the limiting factor, with operations becoming interrupted, taking longer to perform or ultimately becoming unsafe altogether.

To investigate the pendulum effect, a tool has been developed which provides a simple model of a weight on a line. The model does not include passive or active compensation but it allows the line to be shortened or lengthened to reflect deployment or recovery. The motions of the vessel are modelled using data generated from a seakeeping programme and these are used to excite the pendulum by moving the point at which the line is "held". Figure 3 illustrates how the diameter of the pendulum swing varies with height above the water surface as a simulation is conducted lifting a weight (for various ship headings in Sea States 5 and 6). The load being lifted is 2.2 tonnes (i.e. a typical RHIB) and the motions represent a davit at approximately midships on a 90m vessel. Additionally, lines representing the horizontal distance of the hull from the vertical origin of the pendulum are illustrated, taking into to account the movement of the hull due to roll. These two motions, the hull and the pendulum, are coupled as the movement of the hull provides the excitation to the pendulum as shown in the sketch. Where the pendulum diameter ("A") exceeds the horizontal distance to the hull ("B") then this would represent the potential for the swinging boat to "impact" the ships side (i.e. where the lines cross, and noting that the distances account the width of the RHIB as 1.5 metres).

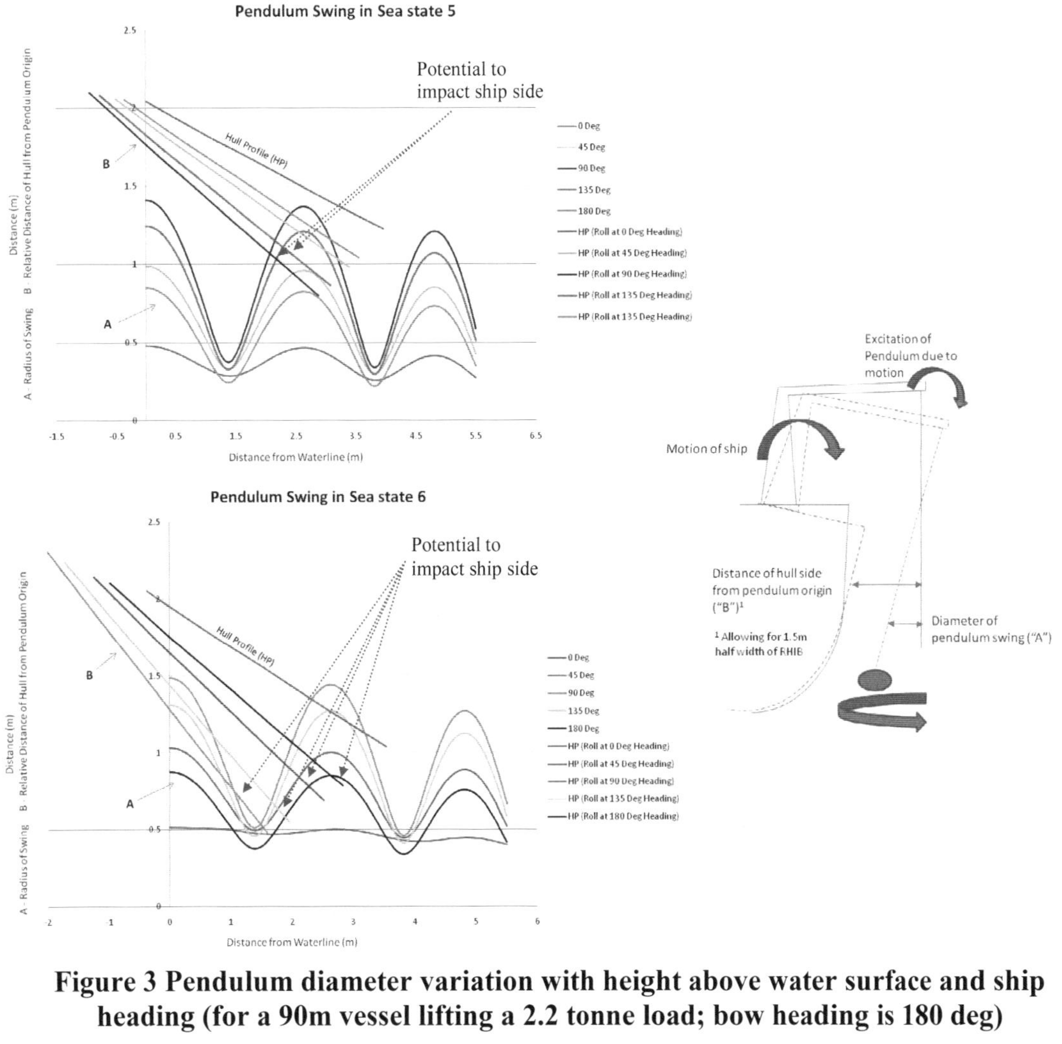

Figure 3 Pendulum diameter variation with height above water surface and ship heading (for a 90m vessel lifting a 2.2 tonne load; bow heading is 180 deg)

These curves illustrate:

- The maximum pendulum diameter occurs as expected for beam headings;
- The least diameter occurs in stern seas;
- For bow headings (as it is more common to recover in headings around the bow as better vessel directional control is possible), the pendulum diameter is greater than stern seas but still provides adequate hull clearance in Sea State 5 and has limited contact in Sea State 6.

As previously indicated, the LSA Code provides both minimum and maximum speeds for the hoisting of fast rescue craft. In most cases, the LSA Code requirement for FRC appliances on Ro-Ro vessels that hoist speed should exceed 0.8m/s should be sufficient for a warship in Sea State 6, although if sea states are required in excess of commercial norms then it should be checked that the boat can be hoisted sufficiently fast to remain clear of the profile of the wave surface.

The analysis also considered the impact of ship motions on personnel. The limits applied were: roll < 4 deg RMS, pitch < 1.25deg RMS, lateral and vertical accelerations < 0.1g. This analysis indicated potential limitations from Sea State 5 for the 90m length vessel considered, and demonstrates that depending on the size of the vessel these may become the limiting factor rather than the lifting appliances.

Stern Ramps

The incorporation of stern ramps in combatant designs became common practise a few years ago and a number of examples exist, most notably the USCG National Security Cutter which also incorporates a gantry overhead to lift RHIB's into the ramp and more recently the DCNS Gowind OPV L'Adroit.

Stern ramps have a variety of geometric configurations and these are explained to good effect in Reference 3 which reported a survey of different arrangements in service. A ramp:

- Will have an angle to the horizontal typically 7 to 12 degrees – a higher ramp aids self launching under gravity but can be harder to incorporate due to the height required;
- The ramp may be shaped to the boat hull or use rails and rubbing strips or rollers to provide protection to the boat and ramp whilst offering a suitable amount of friction;
- The end of the ramp (typically at the transom) will be sufficiently below the design waterline such that the boat will contact the ramp ("ground") at a point beyond the end. The end of the ramp is the sill and the sill depth is the depth of water above the sill to the design water line.

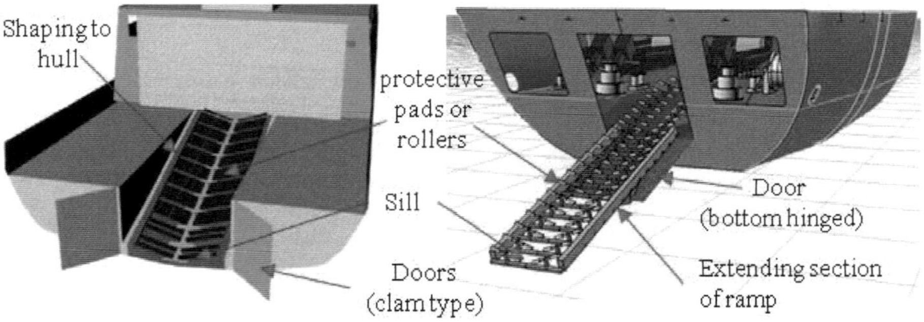

Figure 4 Typical Stern Ramp Geometry (internal left, extending right)

A principle advantage to the stern ramp is the rapid launch and recovery times possible and the reduced deck crew required for operations. From Reference 3 typical launch times that can be achieved are 10 seconds or less and if gravity launched then minimal manpower is required. It is recognised that there are some variations in the methods and procedures used but generally the following applies. To launch, the boat is normally partially lowered but held on a quick release hook such that the propulsors are submerged and the engines can be started (to prevent overload of the engines). The hook is released and the boat will self launch under gravity provided the ramp angle is sufficiently steep. Recovery times are typically 10 - 20 seconds and involve the boats coxswain lining up behind the vessel, then at the appropriate moment driving the boat partially up the ramp under its own momentum. The hook is then passed to the boat to capture it and attached to pull it fully up the ramp and secure it. There may also be doors to close the ramp to the sea. For both launch and recovery, there is no manual handling of the boat by line, other than capture or release by a quick release system. The latter system is an area for development as more automated and safer systems are sought.

The primary limitation in using a stern ramp is the sill water depth. In order to prevent damage to the embarking boat, it should contact the ramp beyond the end such that the boat is "grounded" on the shallow ramp angle. To achieve this, the sill depth must exceed the draught of the embarking boat as it crosses the sill (Figure 5).

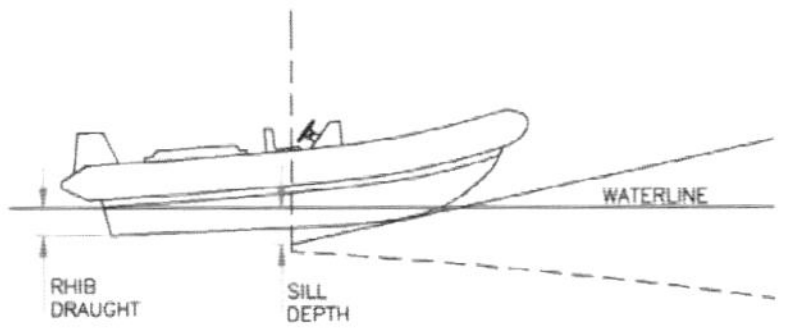

Figure 5 Sill Water Depth

Firstly, as a static geometric problem, the size and therefore draught of the embarking boat defines the sill depth required below the design waterline; the greater the sill depth that can be achieved, the larger draught boat that can be embarked. However, as the depth of the sill increases, this requires a greater transom immersion for the host vessel. This has adverse affects on the performance of the hullform and constrains the allowable propeller diameter.

Secondly, as the dynamic effect of waves is included, the profile of the wave surface combined with the pitching of the host vessel will cause the sill depth to change and the sill may even become exposed. Hence, for a given sea state there is a (reasonably cyclic) period of time during which the ramp is not available as there is insufficient clearance over the sill for the draught of the boat. As the boat is recovered typically under its own power by the coxswain selecting the right moment to drive into the ramp, there is a required window of time defined by the reaction time of the coxswain, time for the boat to accelerate on to the ramp, the actual time to recover on to the ramp plus suitable margins. The availability of the stern ramp becomes a factor of comparing the required window of time to recover against the predicted window for which the sill depth is sufficient in a given sea state (Reference 4 offers much further detail).

An initial analysis conducted on a 90m length concept vessel indicates that the ramp has good availability in Sea State 3 conditions (Figure 6). The ramp availability is illustrated in Figure 6 for a speed of 6 knots for the host vessel, and Figure 7 illustrates that this does not appear to be a significant factor as little difference is observed for 4, 6 and 8 knots speeds. A minimum and preferable time window has been selected based on interpretations of References 3 and 4, although this would be subject to further consideration and testing. For Sea States 4 and 5, the availability window has reduced and becomes less than preferred and in bow headings is close to minimal. Compared to Reference 1, which indicates similar results for smaller vessels, this shows that the availability of the stern ramp is not (significantly) scaleable with vessel size. In fact, as the host vessel increases in size its pitch motion becomes increasingly different to the recovering boat (the latter will tend to follow the wave profile, whilst a larger warship will be close or exceed the wave length and hence the stern motions will become out of phase with the waves).

One aspect that does appear to affect the availability is that as sea state increases, the associated wave period is also assumed to increase resulting in the sill remaining immersed for a longer period. This results in the plots for Sea State 4 and 5 in Figure 6 being much closer than that for Sea State 3. However, whilst the results for Sea State 4 and 5 have similar availability times, the larger amplitude of the waves in Sea State 5 will result in a faster change in the water depth and the wave travelling further up the ramp. In Figure 7 it can also be observed that the plot is shifted aft, an effect of the increased wave encounter frequency for bow headings due to the forward speed of the vessel. These results are clearly sensitive to the environment modelled and the behaviour of the ship, neither of which has been explored in detail here.

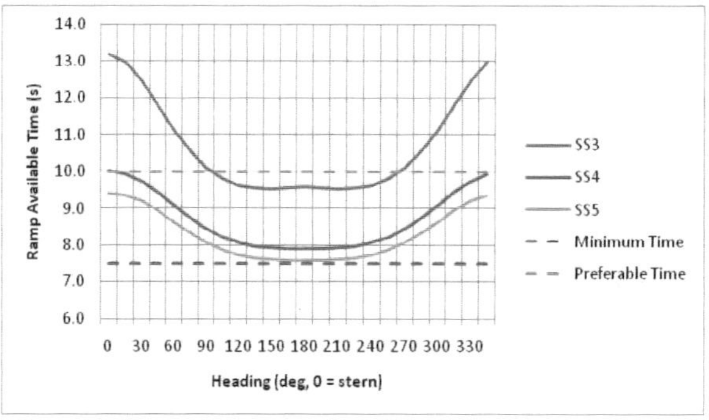

Figure 6 Stern Ramp Availability (90m concept design)

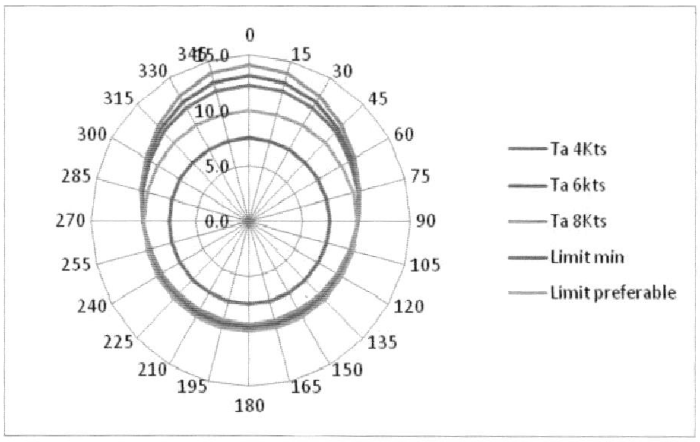

Figure 7 Ramp Availability Time (s) versus heading (degs) and ship speed (kts)

A further complication is the embarkation of unmanned vehicles. If stern ramp recovery is achieved by the coxswain selecting the correct moment to embark based on judgement of the sill depth variation, then doing so remotely will be more challenging. A more autonomous approach using motion prediction combined with accurate relative position systems would offer a solution without the human in the loop judgement. However, such systems require analysis of the highly complex interactions of two different vessels on the sea surface ideally in Sea State 4 and beyond. Such systems will require significant research and testing and are currently beyond the Technology Readiness Level (TRL) to be suitable for the vessels currently being developed.

In the case of UUV recovery, as these vehicles are slow in comparison to a RHIB (circa. 4 kts typically), the recovering vessel will either need to stop or proceed at very low speed and would hence suffer from poor controllability. This can in part be overcome if a drogue system is used. In this case the UUV is captured by a drogue which is deployed and towed behind the recovering vessel. Once captured by the drogue, the UUV can be recovered at higher speed by winch rather than under its own propulsion. Whist the UUV still has to catch the drogue, this can be achieved by paying out the drogue cable to reduce the relative speed of the drogue, whilst allowing the ship to proceed at a suitable speed.

Mission Bay Areas

In order to fully realise the potential of offboard vehicles, manned or unmanned, the host vessel needs to carry a number of vehicles, potentially different types and to be able to select and launch / recover the appropriate one without significant material handling.

Without entering the operational requirements domain in any detail, it can be considered that a future reconfigable combatant could require:

- Two or four mission boats to conduct constabulary tasks or USV's for MCM;
- At least one ship's boat to allow helicopter operations to continue when mission boats are deployed;
- Two large UUV's to conduct underwater reconnaissance;
- Potentially tens of single shot type UUV's for MCM disposal.

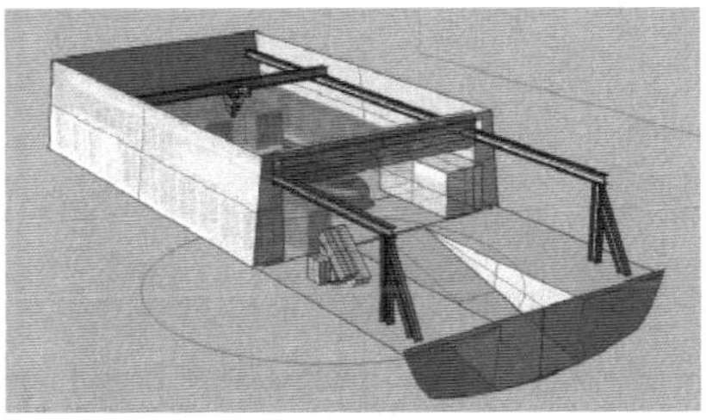

Figure 8 Mission Bay Arrangement (Reference 2)

The sensible conclusion is that a "garage" or mission bay area is required in which to store, configure and maintain all these vehicles in a protected environment (Figure 8 illustrates an example arrangement). However, with RHIB's now expected in the range 8 to 14 metres for some roles, the mission bay has to be of significant size and have a required height of approximately 6 metres or 2 deck heights to both stow and move items. The method of movement of vehicles from stowage to deployment and back also needs to be considered. The ability to handle vehicles and boats needs to match the aspiration for operating limits, e.g. at least Sea State 4 but ideally 5.

In Reference 2 a range of options were considered and in summary:

- Gantries are the most flexible and could move loads throughout the garage area; however, they may require greater deck head clearance and as the load is subject to pendulum forces, higher sea states will be more challenging;

- Using rollers or tracks on and in the deck offers the potential for safer movement in higher sea states, but are less flexible as the handling routes are defined and will incur a maintenance burden;
- Skidding also offers greater control in higher sea states, but obviously will require management of the surfaces due to wear;
- A combination – for example rollers transversely to move large items like boats, combined with a small gantry for lighter items such as UUV's.

A significant issue for the ship designer is the incorporation of a large mission area into the design. The structural integration into the main strength planes of the ship, the self supportability of the enclosing structure of a potentially large open area and also fore and aft access all need to be considered. Figure 9 illustrates the potential problems to be considered, where the garage in this case is two decks high and results in a structural arrangement where the fore and aft deck integration cannot be easily achieved and careful consideration is required around the discontinuities.

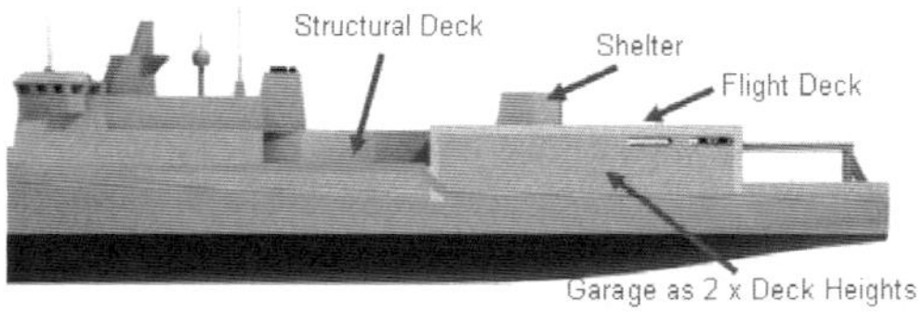

Figure 9 Garage and Flight Deck details

POTENTIAL SHIP SOLUTIONS

Initial concepts for host type vessels, such as in the USN LCS classes or as illustrated in the BMT Venator concept in Figure 10, utilised a stern ramp and mission bay. However, as described the stern ramp operability does not scale with size and the resulting sea state limits are below the aspirations. The further complications of recovering unmanned vehicles make the choice of a stern ramp much less attractive.

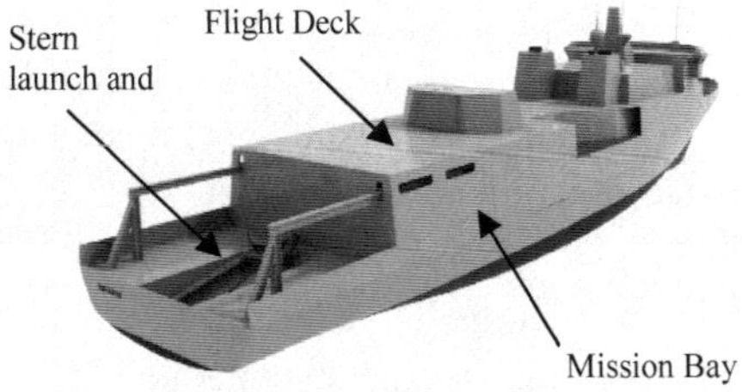

Figure 10 Concept With stern Ramp and Aft Garage

An alternative is to adopt a midships mission area utilising davit launching to either side. In the illustration in Figure 11, the mission bay area is located further forward in the same concept in order to utilise davit or crane systems amidships. As a result, the flight deck is now aft and lower from the original concept. However, the mission bay area is now in the prime midships region of the ship, requiring accommodation and operational spaces in the superstructure to be pushed forward. Access between the forward superstructure and aft of the mission bay is also difficult to arrange. This solution would have similarities to those adopted in the offshore industry where ROV and research ships adopt a similar arrangement. However, it is notable that these ship tend to have larger superstructures to accommodate the living areas of the ship forward of the working deck and as a result they are larger beamed and heavier displacement than warships and OPV's of similar length.

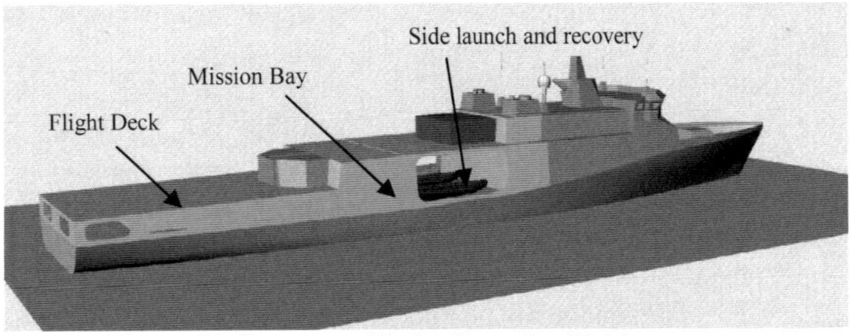

Figure 11 Concept with davits and midships garage

A further alternative is a mix of launch and recovery solutions, as proposed in the "Securitor" solution (Figure 12, Ref. 5). Here, the boats or vehicles are carried in side openings and launched and recovered via davits. However, the design also featured a small "dock" (although a stern ramp is equally applicable) allowing one boat or vehicle to be ready for more rapid deployment. The key reasoning in this concept was the expectation that boats or unmanned vehicles may be expected to spend much more time deployed as endurance is improved. Hence, once carried into theatre the host vessel becomes more of a support vessel. In many instances, more modest sea states would prevail and a floodable dock would allow refuelling, crew transfer and some maintenance operations to be conducted without lifting the vehicle out of the water.

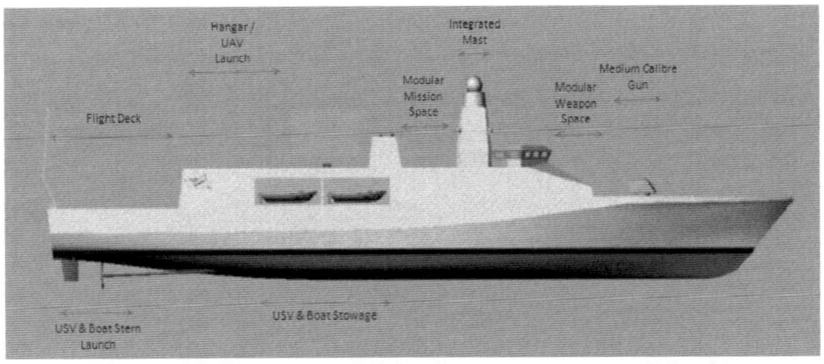

Figure 12 Concept with both stern ramp and davits

CONCLUSIONS

The inclusion of a stern ramp has advantages in terms of deployment and recovery speeds, but ultimately a stern ramp lacks availability in the higher sea states in which a davit is still capable of operation. The stern ramp availability is driven by sill depth and pitch motions, whilst a davit needs to consider a range of constraints associated with the environment, including pendulum effects, hoisting speed and the safety of deck crew.

Hence, a vessel designed to operate in benign to moderate sea conditions and requiring rapid boat response times may be appropriate for a stern ramp, as illustrated by their use mainly in coast guard and smaller patrol vessels. A vessel operating in higher sea states such as offshore environments, including conducting MCM and environmental assessment may select davit arrangements, accepting the increase in handling procedures to gain a greater environmental envelope. A flexible, multi-role naval vessel is therefore likely to err towards the later, because of its greater flexibility and sea state envelope. A mixed solution may offer a compromise between these but suffers from the need to find space for both systems and requires training to cover several deployment methods, although the rapid response would be beneficial when operating in constabulary missions and may be an advantage to consider. In this case, the stern ramp is the secondary method of deployment and the mission space would be located midships with the davit system(s).

ACKNOWLEDGEMENTS

The opinions expressed in this paper are those of the author and not those of BMT Defence Services Limited, other companies, the UK MoD or any of its agencies. The author wishes to acknowledge the co-operation and able assistance of the BMT staff in the writing of this paper, particularly Andrew Jones and Rob Armstrong.

REFERENCES

1 "MacGREGOR innovations ensure safe, efficient and effective offshore operations and service", 2009 – Macgregor website, http://www.macgregor-group.com
2 Kimber A., Giles W. (2006) "Minor Warship Roles - How technology is leading to a new vessel type", Pacific International Maritime Conference 2006.
3 Sheinberg R; Minnick P V; Beukema T G; Kauczynski W; Silver A L; Cleary C (2003) "Stern Boat Deployment Systems and Operability", SNAME Transitions 2003.
4 Dalzell JF (2003) "A Proposed Criterion for Launch Ramp Availability", Naval surface Warfare Center.
5 Kimber A., Booth J. (2010), "Novel Technologies for Enhancing the Performance of Constabulary and Security Orientated Ships" RINA Warship Conference 2010.
6 LSA Code, 2010 Edition.

Military Threat Analysis: The Challenge for Class Societies

Thomas Grafton
Det Norske Veritas AS, Norway

ABSTRACT

Combat effectiveness is the defining characteristic for any naval combatant in the same way the oil cargo capacity is for an Oil Tanker. However, unlike an Oil Tanker where the cargo capacity is easy to set and measure, combat effectiveness is not simple to define, quantify or measure. The biggest challenge is perhaps deciding what level of combat effectiveness is acceptable and then to ensure that this level is maintained through life.

Increasingly, navies are asking Class Societies to help verify combat effectiveness, in particular the response of the ship to military threats, for example air blast, internal blast, underwater shock, fragmentation and whipping. This may be limited to the response of the structure only, or a more complete analysis of the ship and associated systems.

In this paper the services a Class Society can offer are explored, based on practical experience where available, highlighting the critical phases needed to support verification, the roles and responsibilities of all parties, possible conflicts and conflict resolution.

To aid the debate a comparison is made between the war-fighting capabilities of a naval ship and the revenue winning capabilities of a large cruise ship, exploring the roles and responsibilities of the Class Society and the owner in maximising the ships potential in each case.

SURVIVABILITY – IN THE NAVAL CONTEXT

Survivability is defined as the capability of a naval ship to avoid and/or withstand a man-made hostile environment while performing its mission (1). This is considered to be made up of three main elements: 1) susceptibility, the ability to avoid detection; 2) vulnerability, the ability to withstand a hit; and 3) recoverability, the ability to reconfigure and restore the damaged systems after a hit. This can be described pictorially with time in figure 1, taken from figure 8 in reference (2).

In the last one hundred years nearly all the damaging attacks on naval ships have taken place in two world wars, skewing the number of attacks to the first twenty years of the range. In this period a wealth of data on survivability and combat effectiveness was gained (through hard earned experience) and qualitative knowledge was accumulated and put into naval design standards. This was followed by a period of full scale testing, often using the large number of spare hulls freed up by peacetime. These tests were then used to make the first quantitative assessments of survivability, supported in later years by the development of the computer and more advanced analysis.

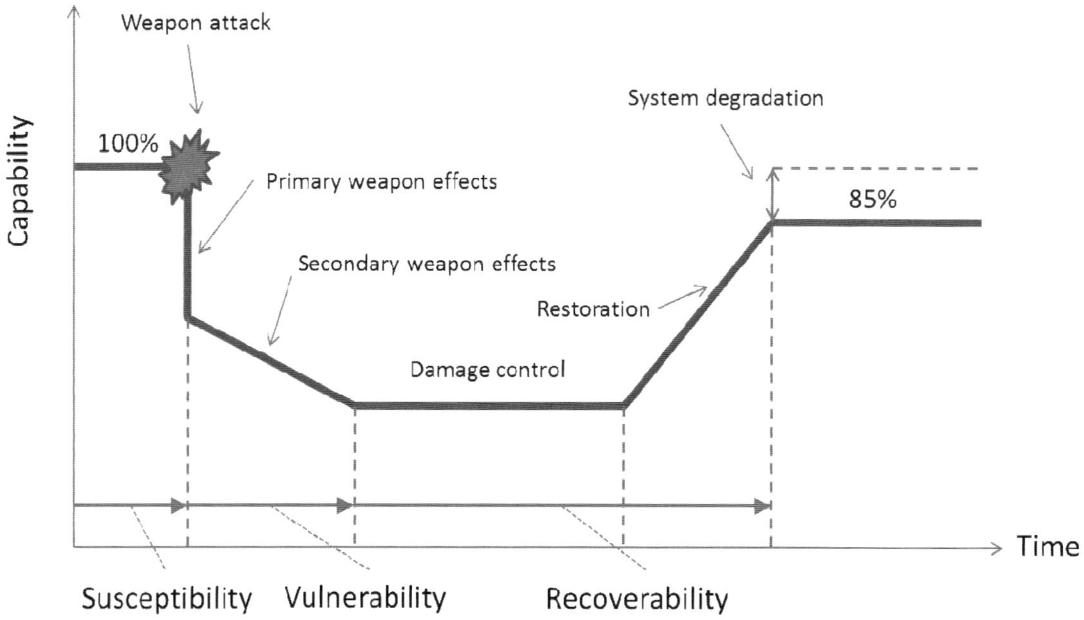

Figure 1: Survivability, the loss and recovery in capability over time, as presented by MacDonald and Horstmann, (2)

In the last twenty to thirty years the number of combat attacks reduced to only a handful and the state of the art was furthered through increasingly complex survivability assessments, supported by rapidly increasing computer processing power. Nowadays, full scale testing is rare (although the results are always helpful for validating methods) and quantitative methods have largely replaced the qualitative methods of the past. For the first time it was possible to assess the survivability in the design stage, enabling trade off studies reducing the need for prescriptive rule requirements. Today, computer codes are available commercially for survivability assessment, for example PREVENT, (3), and SURVIVE, (4).

The focus in survivability assessment has thus shifted from good design based on testing and experience to good design through analysis and iteration.

Here begins the first challenge for Class Societies. Although Class Society involvement in warship design is well documented, (1), with many IACS Class Societies publishing rules for naval ships, most rule requirements are prescriptive and must be met (and continue to be met in service) to maintain the class notation. Thus, class rules tend to focus on good design practice, such as general arrangement design (separation of machinery, subdivision, blast resistant structure, structural fire protection, etc.), rather than more complex whole ship assessments. Many Class Societies publish optional survivability notations covering topics such as above water weapons effects, external and internal blast, underwater weapons effects and whipping.

CONTRASTING WARSHIP SURVIVABILITY WITH THE CRUISE SHIP BUSINESS

It is interesting to contrast Class Society involvement in a complex commercial ship project with a naval project. The link between a cruise ship and a warship is not at first glance obvious. However, we can consider that a cruise ship owner 'fights' a war with other cruise ship operators to 'win' passengers for a cruise and needs to do this both profitably and safely. Thus a cruise ship owner must consider 'survivability' elements during design and throughout operation to provide a winning service whilst remaining profitable.

Cruise Ship Survivability:

The ship must 'survive' in the marketplace, the critical elements necessary for this are:

- Make money throughout the design life
- Operate safely and be perceived by passengers (and regulators) to be safe
- Contribute to building and maintaining the owners brand image

To do this the ship needs a 'wow' factor, interesting holiday destinations and an appropriate level of luxury and facilities for the market (e.g. 1 star verses 5 start rated cruising).

The ship will be 'attacked' by other cruise operators on price, quality, special features and destinations. In addition, the ship may be 'attacked' by manmade disasters or criminal acts such as terrorism, fire, and fast spreading viruses requiring specialist medical attention, and so on. The measure of survivability will be partly the profitability and partly that the ship has a long, safe life, without incident or accident (or to have survived incidents and accidents that have occurred). Finally, both the public and international regulators must be convinced that the ship can survive threats that are reasonably foreseeable (fire, flood, etc.).

The 'threat scenarios' for a cruise ship are shown pictorially in figure 2 overlaid with possible survivability features that can counter these threats. The list is not exhaustive and other threats could also be included, such as upcoming environmental regulations or falling behind in marketing and innovation. Shore side systems that play an important part in the 'battle' to survive in the market place could also be included, for example internet and booking systems.

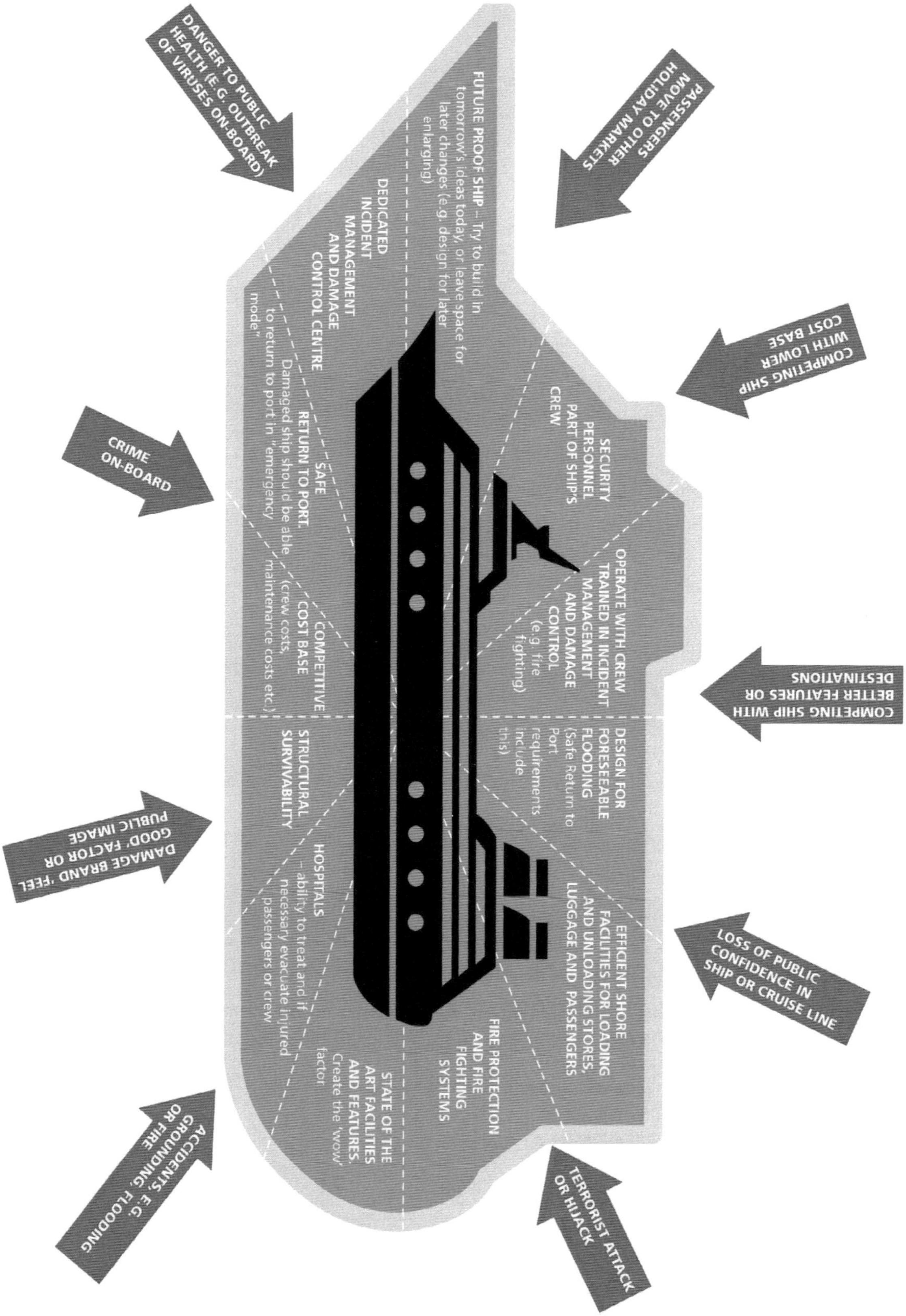

Figure 2: Survivability elements and threats for a cruise ship operator

How is a Cruise Ship Owner Organised to do this?

A cruise ship operator must set up an organisation, both ashore and at sea, that allows the cruise line to at least survive and preferably flourish in the market place. Management functions for a typical cruise line are shown in Figure 2 and contrasted with the equivalent for an operator of naval ships. In the figure a cross-reference between equivalent departments in attempted. The picture quickly becomes complex as typically the Navy and the Department of Defence are separately managed organisations both contributing to particular management functions.

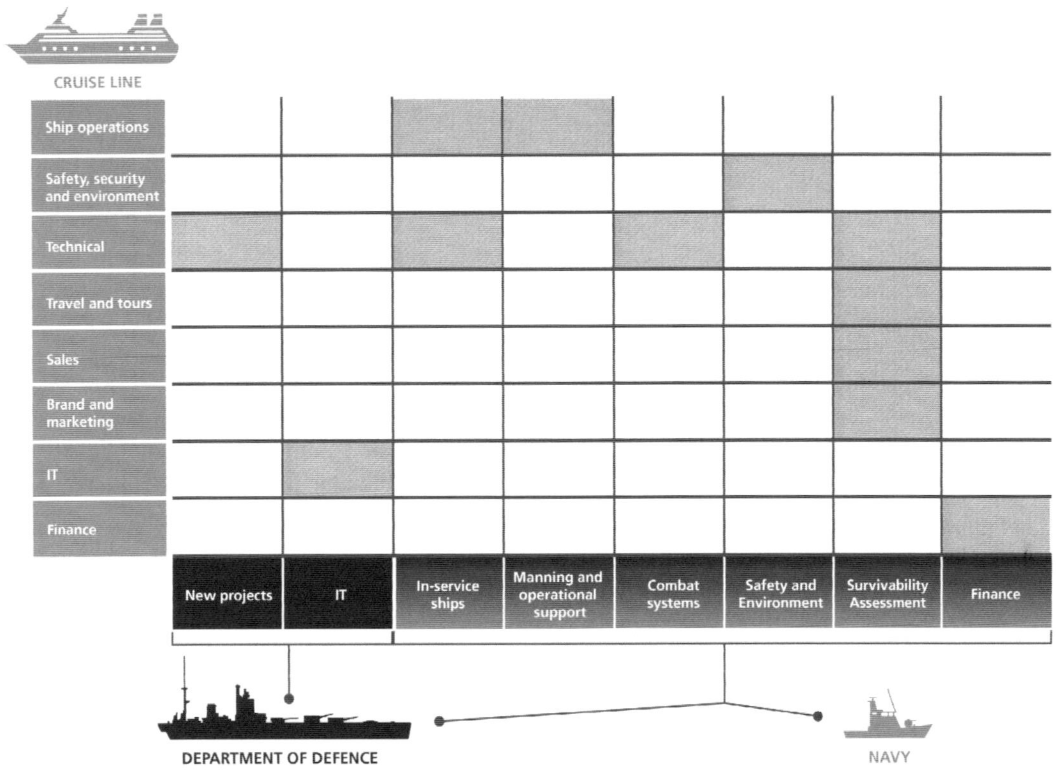

Figure 3: Management functions for a cruise ship operator and a naval ship operator

The 'survivability' elements for a cruise line are well defined and separated into clearly differentiated functions. A separate overarching survivability function is not necessary as a suitable assessment of survivability is financial performance and brand value. A healthy brand is an indication that the cruise line is performing well now and will continue to do so into the future.

For the naval ship the assessment of survivability is much more complicated, as described in the previous sections, and requires a degree of subjectivity. The assessment requires input from designers, operators, subject experts (e.g. missile designers), and intelligence experts (to define enemy threats) and can take some time.

Affordability and financial performance are not easy to calculate for a naval ship operator, primarily because the ships are rarely used for their intended purpose (in combat). A cruise ship is operated 'in combat' every day and so the financial performance is easy to measure.

DIFFERENCES BETWEEN CLASS SOCIETY INVOLVEMENT IN THE NAVAL AND CRUISE BUSINESSES

For a cruise ship operator Class Societies are used to comply with international law and where their involvement demonstrably adds value. Typical focus areas for the Class Society are technical advisory, risk reduction and ship Classification (a mandatory requirement). This support may involve both maritime and business assurance experts. Class usually has no involvement in the Travel and Tours; Sales; Brand and Marketing; and Finance functions. Because these functions are so crucial for the company's survival they have large in-house teams of experts working in these areas. The cruise company has relatively fewer people working in the other functions and some activities are contracted out to smaller companies or consultants. The cruise line focuses its best resources in areas critical to its survivability: that is those areas which maximise revenue and maintain and develop the brand.

The naval ship owner on the other hand has a broadly equal number of people in all functions, perhaps skewed away from technical were consultants can be used for non-military features. Small specialist consultants are increasingly being used for military functions (e.g. combat systems) as it is no longer practical to keep large in-house resources when there are relatively few projects with large time gaps in-between. The naval ship operator is not required to use Class Societies and uses them only when they can add value, or when internal policies require it. The naval owner has to demonstrate fairness and value for money in every decision it makes to demonstrate that public money is being spent wisely. This often leads to a relatively larger organisation than for a cruise line.

For a naval ship owner, technical resources are often limited and spread over a wider range of functional areas. Therefore, Class Societies often work together with a larger number of departments in a naval organisation compared with a cruise operator. Typical focus areas for the Class Society are much the same as for a cruise ship: technical advisory, risk reduction, ship Classification and management support, for example helping create a naval flag administration or implementing equivalent merchant ship regulatory systems into the naval environment. However, the scope for Class tends to be broader and the partnership between Class and the naval operator tends to begin relatively earlier in the project life-cycle. The first steps are usually as advisors or consultants, with ship Classification coming later.

For a cruise ship owner *Class focuses on technical risk reduction and leaves market competitiveness and 'survivability' issues to the owner*. The cruise owner relies on Class to a greater extent for technical and statutory support, focusing its main resources elsewhere.

CAPABILITY CLOUD

Returning to naval ship survivability, it is interesting to focus on the beginning of the process: The definition of the problem and subsequent quantification to give design parameters with associated acceptance levels. Before any survivability assessments can be completed the following information is needed:

- Threat weapons to be considered
- Damaging effects of these weapons
- Critical systems need to be defined (that should survive the attack)
- Acceptance criteria, e.g. for structural strength, stability, system operability, etc.

The eventual survivability is often assessed by considering the ship's ability to Float, Move and Fight after the attack. Very quickly a complex web of threat weapons, critical systems and associated analyses is created, with the ships survivability being affected to a differing degree by each attack. The ship design must then be modified to improve the overall survivability to ensure that there is a balance between the vulnerability of the ship to each attack and the likelihood of the attack occurring. This can be simulated through 'wargaming', adding another complex process to the assessment. This fast becomes a cloud of action, involving iterative loops and intuitive leaps as described in reference (5). Out of this cloud comes a requirement against which the ship can be assessed.

The assessment typically requires a detailed model of the ship and systems, more complex perhaps than other models used for the design of the ship. When the analysis is complete it is quite possible that the ship will not meet the requirement, and either the requirement must be re-assessed or the ship re-designed. We then re-enter the cloud.....

The whole process is made more complex by the need to periodically re-enter the cloud to consider new threat weapons and threat scenarios through the life the ship. Theoretically, the ship can be designed to meet the requirement however it may not be practical or cost effective to do so. Eventually, a degree of subjectivity is needed to enable a decision to be made to stop the process and accept that the best compromise has been achieved.

It should be apparent that this complicated process is not suitable for a simple class notation which needs a clear requirement, assessment method and acceptance criteria that must be met and continue to be met throughout the life of the ship.

CLASS SERVICES FOR NAVAL SURVIVABILITY

Class Notations

The most well-known output of classification of naval ships is the class notation, for example:

✠1A1 Naval ICE-C E0 HELDK NBC-2

This is a typical notation taken from a recent newbuilding to DNV class. The characters of the notation are explained in the table below:

✠	The symbol ✠ will be given to craft built under the supervision of DNV
1A1	The notation **1A1** will be assigned to ships with hull, machinery, systems and equipment found to be in compliance with applicable rule requirements.
Naval	A ship flying a naval flag and administered by a national naval administration with additional naval requirements appropriate for a combatant (for example general arrangement design, fire safety, escape and evacuation, protection from radiation hazards and storage of ammunition).
HELDK	Helicopter landing area covering basic strength requirements (can also include additional naval requirements for helicopter decks by separate agreement)
E0	Instrumentation and automation to allow an unattended machinery spaces.
ICE-C	Ships with basic ice strengthening in the bow region (ice thickness approximately 0.4 metres).
NBC-2	Nuclear, Biological and Chemical protection measures have been applied to the ship to Level 2 (the highest level), requiring a pressurised citadel and other special features to facilitate operations of long duration in a contaminated area.

Table 1: Characters of a naval class notation

There are survivability elements to this notation. The combination of the **Naval** notation and **NBC-2** leads to requirements for the ship to be split into self-contained zones for fire control, damage control and gastight division to contain damage and to avoid the ingress of Nuclear, Biological or Chemical (NBC) agents. The ship is to be subdivided by watertight bulkheads as necessary to fulfil the stability criteria that shall be watertight to the damage control deck. There are rule requirements for intact and damaged stability suitable for a naval ship which has been subjected to attack. No openings are permitted in the bulkheads below the damage control deck and any valves to operate piping systems shall be placed on or above the damage control deck. Machinery systems are to be located in separate compartments so that damage to one compartment does not cause a loss of the propulsion or electrical system.

In addition, there are many other naval requirements that will be familiar to experienced naval ship designers. The **Naval** notation is intended to cover the whole ship, offering survivability to a basic level appropriate for a combat ship.

The survivability measures described above are prescriptive rules based on experience. DNV offer an additional optional notation for Combat Survivability to cover the weapons effects on hull structure and/or critical systems. This **CBT** notation (found in Pt.6 Ch.18 of the DNV rules, reference (6)) can be given on request and uses a 'contractual damage extent' to assess the loss of structural or systems capability. The contractual damage is defined by a damage circle radius (or minimum damage zone size), a damaged survival seas state and speed in the survival sea state. At the present time the contractual damage must be specified by the owner, if the damage extent is not specified the damage radius is set to 40% of breadth.

Verification against Other Military Standards

A class notation contains requirements that must be met, and must continue to be met through the operational life of the ship. Thus, a class notation guarantees compliance with the rule requirements through independent verification throughout the life of the ship.

However, some navies have their own standards that they prefer over DNVs Rules. These standards vary between nations and so it has not been possible to cover every navy's expectations in one Rule set. To allow for these national variations the **Naval** class notation can be appended with **(navdist)** to show that the requirements have been modified (or tailored) to suit the navy in question. Any deviations from the rules, alternative methods or standards are documented in the appendix to the Classification Certificate.

In some instances the standards the navy wishes to use may be difficult to comply with fully. This typically occurs when the standard is written with a certain ship type or arrangement in mind which is different to the solution chosen for the new ship. If the class scope had been modified to include the standard then it would not be possible to offer the class notation if the requirements are not met (class notations follow, in general, an all or nothing approach). Therefore, it is better for Class to offer independent verification services to show the level of compliance with the standard, and to help find remedies to enable compliance as far as practical.

Typical standards which Class may be asked to work with include: Allied Naval Engineering Publications, ANEPS, e.g. ANEP43, Edition 2, on Combat Survivability (Classified); NATO Standardization Agreements, STANAG's, e.g. STANAG 4142, Edition 2, on Shock Resistance Analysis of Surface Ships; or other national such as Warship Stability Standards (covering survivability implicitly through survivable damage scenarios).

The most common example of this for DNV is for shock analysis. On recent newbuildings, when shock analysis is requested, it is offered as a verification service separate from Class of the ship using standards requested by the Navy. A separate Classification Note covers the testing of components to demonstrate that the required shock level had been achieved, (7).

Survivability Assessments

In the defence industry the knowledge base and tools needed for warship survivability analysis is limited, and the necessary expertise is thinly spread between a small number of navies and their preferred specialist partner companies. Examples include BMT Defence Services PREVENT software, (3), and QinetiQ's SURVIVE software, (4).

For an owner, completing a survivability assessment can be difficult due to the small market of suppliers and because the expense in gaining expertise is high. These difficulties have resulted in Class Societies receiving increasing numbers of requests for assistance.

For this reason few, if any, Class Societies have been involved in whole ship survivability assessments (the author can find no publically available references). An additional contributing factor may be that the market for these services is limited and the expense in gaining expertise is high. But, it is precisely because of these difficulties that Class Societies are receiving increasing numbers of requests for assistance.

Class Societies have the skills necessary to complete an assessment of the survivability of a ship using complex computational analysis. DNV undertakes similar investigations for analysis of explosions on Offshore Structures to help with the risk analysis and mitigation during design, as well as assessments of the survivability of the structure afterwards. However, at present, Class Societies have limited experience in *warship* survivability, in particular using the specialist tools necessary for a full and complete ship survivability assessment.

Present Class Position

It is ***easy*** for Class Societies to write class notations covering good design features for survivability. All Class Societies offering naval rules do this to varying extents.

It is ***relatively easy*** for Class Societies to write class notations covering structural and system vulnerability ***if the initial size and type of damage can be specified by the navy***. This is covered by DNV class notations such as **CBT** and **NBC**.

It is ***moderately difficult*** for Class Societies to verify designs against other military standards, often due to the challenge of finding persons with the necessary subject matter knowledge and experience (a common problem shared by all Class Societies and providers of specialist engineering services).

It is ***difficult*** for Class Societies to complete whole ship survivability assessments due to the limited number of available assessment tools, each requiring specialist knowledge and experience to operate, coupled with the need to define the threat weapons and their lethality as inputs to the analysis. DNV could leverage off experience with offshore structures to assist.

SO, WHAT SHOULD CLASS DO?

Class Societies should continue to produce Rules covering good design features that reduce vulnerability and ease recoverability after attack. However, this is only a small part of the overall picture, see figure 4, and other services can be offered in addition.

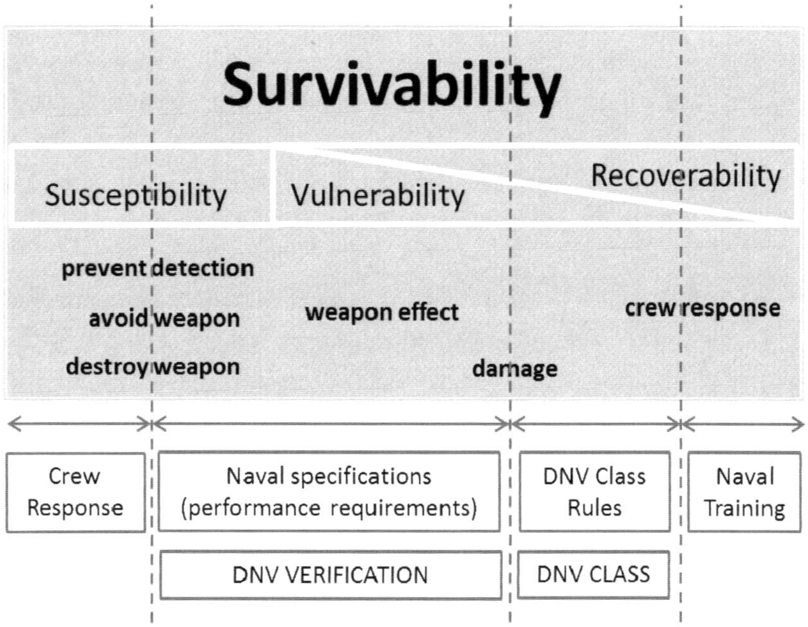

Figure 4: DNV Services contributing to survivability

Class Societies can work together with navies and warship survivability specialists on susceptibility assessment and threat modelling. The most value will be extracted from the Class Society when they are used as an independent verifier of the modelling and subsequent analysis.

Whole ship survivability assessment currently needs specialist analysis software and experienced operators to do the modelling and interpret the results. Class Societies could offer verification services to support this activity, but this will be complicated as the software code is owned by other companies and the methods and experience needed for independent verification may not be readily available.

Recoverability assessment is subjective and depends on human actions in many cases. It requires deep experience of naval operations and human factors. Recoverability is often tested by running training scenarios on ships usually in support of certification activities to demonstrate that the ship is combat ready. Experience gained from these tests should be collected and where sensible converted to simple rule requirements for design, or as areas where analysis and verification are needed.

CONCLUSIONS

Today the DNV class notations **Naval** (and **Naval Support** when used with appropriate additional sub-notations) include basic design features that contribute to survivability.

Class Societies can and should improve on these design rules in consultation with navies. It is important to separate essential elements that should be included in the core rules from desirable elements that can be covered by optional notations.

Navies want combat survivability notations but their survivability assessment processes are not aligned with classification processes and timescales. This creates a problem as class requirements must be met if the class notation is to be given. Often the navies threat assessment process and survivability assessment processes are on-going, and a realist design 'threat' cannot be specified in time to allow assessment during classification.

Cruise ship designers do not use class services for aspects of their business that can be viewed as equivalent to survivability features in warship design. We are seeing that Class Societies often work together with a larger number of departments in a naval organisation compared with a cruise ship operator and that class is being used to assist in filling capability gaps.

In the future, Class Societies could offer more services to navies to assist with survivability assessment. To do this they need to focus on learning more about this topic and working together with navies to better understand the survivability assessment process as well as how navies specify design features that improve survivability for newbuildings. More class notations may not be the answer.

REFERENCES

(1) Frieze, P.A. and Shenoi, R.A. (editors), 2006, "Committee V.5 Naval Ship Design", Proceedings of the 16th Ship and Offshore Structures Congress, 20-25 August 2006, Southampton, UK. Volume 2 pages 213 – 263.

(2) MacDonald, G. and Horstmann, P., 2008, "Cost Benefit Evaluation Methodology in Warship Survivability Design", Pacific 2008 International Maritime Conference, Sydney, Australia.

(3) Heywood, M. and Lear, T., 2006, "PREVENT – A Tool to Reduce Vulnerability Early in the Design", Warship 2006: Future Surface Warships. London, UK, June 2006, 9 pages.

(4) Schofield, J., 2009, "SURVIVE and SURVIVE Lite – Survivability Assessment from Concept to Operational Support", QinetiQ Publication Number QINETIQ/09/00063, presented at the American Society of Naval Engineers ASNE Day 2009, 8th – 9th August 2009, Gaylord National Resort & Convention Centre in National Harbor, MD.

(5) Brown, D.K., 1986, "Defining a Warship", Naval Engineers Journal, Volume 98, Issue 2, Pages 31-40, March 1986.

(6) Det Norske Veritas, 2012, "Rules for High Speed, Light Craft and Naval Service Craft", January 2012 edition, Høvik, Norway.

(7) Det Norske Veritas, 2002, "Classification Note No.57.1, Shock Testing of Equipment and Systems – Naval Applications", January 2002, Høvik Norway

Safety Assurance of Small Craft Operated by Governments and Navies

Thomas Grafton
Det Norske Veritas AS, Norway

ABSTRACT

Various government departments and navies operate a wide variety of small craft with many different uses, for example routine harbour patrol; boarding operations or high speed insertion of Special Forces. These craft may carry a wide range of personnel, from highly trained crew to personnel with basic sea survival training or even passengers. To provide necessary safety assurance there is a need to certify these craft to demonstrate that:

- they have been designed with an adequate level of safety for intended operation;
- they have adequate safety equipment for the embarked persons;
- the craft has sufficient survivability.

The above can be achieved partially by applying design standards, safety codes and certification schemes for small commercial craft or leisure craft. However the concept of operation for some government craft can be quite different and the use of these standards may not give the required level of safety. In this paper DNV discusses the approach adopted and standard developed by DNV to address possible shortfalls in safe operation of small government vessels.

TYPES OF CRAFT OPERATED

Craft based on Commercial Designs

Many governments and navies operate a range of small craft for activities such as coastal patrol and policing, customs and excise duties, training or work boat duties. These craft may be crewed, operated and maintained in a similar way to commercial craft or recreational craft and so are often based on an existing design from the commercial or recreational markets. The craft may however have modified fire fighting, lifesaving and rescue arrangements to suit their intended operation.

These craft are often designed and built to standards for small commercial craft or recreational craft as no bespoke standards exist. The owner must then specify additional features or design requirements to ensure that they are fit for their intended purpose. A typical example is given in Figure 1.

Figure 1: A 16 metre Police boat for the New South Wales Water Police, built at Austal Ships (http://www.austal.com/en/products-and-services/sub-category-listing.aspx?cat=defence-products&subcat=auxiliary-vessels)

Complex Military Craft

Navies often need specialist craft for military operations such as boarding operations or to insert troops to hostile places at high speed. These craft are expected to operate in combat and may carry weapons or troops with their own weapons. The craft may operate for extended periods away from base support, often in hostile places. These craft tend to be purpose built using either bespoke military standards or a comprehensive military specification which may also specify commercial craft standards. A typical example is given in figure 2.

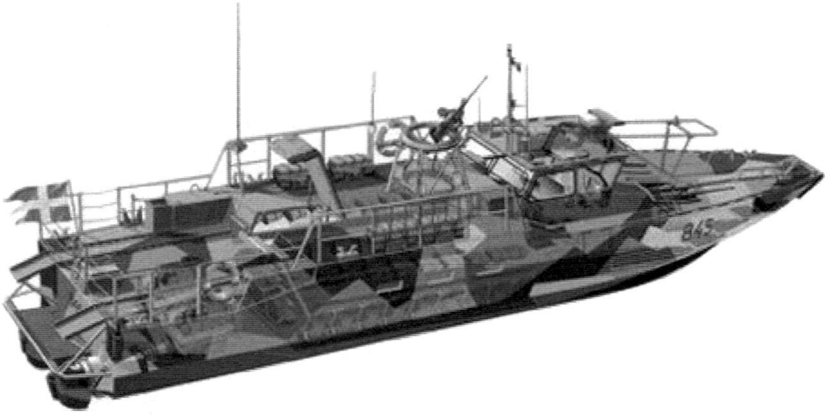

Figure 2: A Dockstavarvet Combat Boat 90 (http://www.dockstavarvet.se/Products/Combat_patrol_boats/CB_90_H.aspx)

Quasi-Combat Craft

Some small craft need to be armed with guns or other weapons due to their specialist role but can otherwise be based on a commercial, patrol or recreational craft design (as no additional military features are necessary). A common example is a rigid hull inflatable boat fitted with a permanent machine gun mounting. These craft are normally adapted from commercial designs with additional structure and analysis to support the mounting and use of the gun system. Additional safety features may also be incorporated (for example to cover the storage of ammunition).

Boats Operated From Larger Ships

Nearly all large naval ships embark one or more small craft that can be used as tenders or for naval operations such as boarding other vessels or deploying troops. Some of these craft may be commercially designed rescue craft and approved under SOLAS (and the International Life Saving Appliance Code, Chapter V, Section 5.1) for this purpose. However, as these craft can be operated in harsh conditions and are regularly launched and recovered many manufacturers offer special craft for military uses; see for example Zodiac Milpro (www.zodiacmilpro.com) shown in figure 4.

Figure 3: A rigid hull inflatable boat operated from a naval ship

EXISTING REGULATIONS

Small recreational craft and small commercial vessels that operate within the territorial seas of a nation have to comply with the regulations of that state. Example national standards include (1) for Australia or (2) and (3) in the United Kingdom. Standards assume implicitly that craft remain within the territorial sea area and operate close to the shoreline (within approximately 150 nautical miles). Less restricted operation can be given, however this effectively brings the craft under the regulatory systems in place for larger commercial ships.

The rules and regulations applicable to small commercial vessels and recreational craft implicitly assume that assistance from shore based rescue services or other vessels is available at all times and requirements have been developed on that basis. Further, they assume that the craft can seek refuge in an emergency, for example to avoid exposure to extreme sea conditions or to make it easier for rescue services to give assistance.

Small naval craft may operate in remote areas, where assistance from shore based rescue services or other vessels is unlikely (due to remoteness) or will not be given due to the military nature of the craft. The operational range for any small craft is limited and so small naval craft will always be within range of an operational base, however, this operation base could be another ship or other military forces located in a hostile place. Therefore, the concept of rescue through local assistance is not always applicable. Similarly, the concept of the craft seeking a nearby place of refuge to avoid prolonged exposure to harsh seas or to aid rescue is not applicable for the same reasons.

Other government craft, for example police launches or search and rescue craft, cannot always expect assistance in an emergency, often because their purpose is to provide emergency assistance to other vessels. It is thus sensible that these craft should have a higher level of safety in their design and operation to account for the fact that rescue or assistance in an emergency is less likely.

Thus, we see that existing rules for commercial craft or recreational craft are unsuited to small naval or government operated craft.

SAFETY GOALS

It is sensible that any standard which aims to assure the safety of small craft operated by governments and navies should cover the same goals as standards for small commercial craft and recreational craft; however the method of achieving the goals may be different. A review of the existing standards, e.g. (1) to (6), gives five top level goals that must be fulfilled:

1. Operate within the safe limits placed on the craft during design
2. In the event of an accident, to stay afloat long enough to allow safe evacuation from the craft into the sea or to another rescue vessel
3. To enable safe evacuation of personnel into the sea or another rescue vessel offering assistance (this may include other boats, ships or rescue helicopters)
4. Be designed to prevent the ignition of flammable substances as far as reasonably practical. In the event of a fire then reasonable means of extinguishing to be provided, along with means for limiting the spread of fire for long enough for rescue to be effected or to allow for safe abandonment
5. To enable passengers or untrained persons to safely evacuate the craft with minimum assistance

In the following series of tables the methods of achieving each goal for small government and naval craft are compared with small commercial and recreational craft.

Goal 1: Operate within the safe limits placed on the craft during design		
Achieved by:		***Differences***
Small Commercial Vessels and Recreational Craft	***Small Naval Craft and Government Craft***	
Rules for material and manufacturing suitable for the construction methods and materials in common use	Rules for material and manufacturing suitable for the intended materials, manufacturing methods and operational environment(s).	Naval craft may be manufactured from advanced materials and be operated in extreme climatic conditions
Construction rules defining loads and strength suited to coastal operation	Construction rules defining loads and strength appropriate for the intended operation of the craft	Loads will be different for naval craft (beaching, lifting and/or docking on land or on other ships, coming alongside in heavy seaway, etc.) Additional 'robustness' and design margins are needed as it may be harder to stay within safe limits
Provision of operator guidance (for example design sea state and significant wave height)	Provision of operator guidance including information on ultimate limit states	For naval craft, operational guidance should give an understanding of the ultimate capacity of the ship as it may not be possible to stay within safe limits
Operational restrictions to stay within design limits (either stated explicitly or an implicit assumption of the rules and regulations)	-	Naval craft should not have operational restrictions as these will be difficult to comply with (e.g. not always possible to seek shelter)
Quality System Production Certificate (to certify construction works)	Quality System Production Certificate (to certify construction works)	Same requirement
Type approval certificate (for series production)	Type approval certificate (for series production)	Same requirement
Product Verification and Certification to confirm adherence to design rules	Product Verification and Certification to confirm adherence to design rules	Same requirement
-	In-service verification of the material state and of repairs and changes (as changes more likely)	For naval and government craft it should be possible for the crew to inspect the craft after damage (access to structure needed)

Table 1: Achieving the safety goals (1 of 5)

Goal 2: In the event of an accident, to stay afloat long enough to allow safe evacuation from the craft into the sea or to another rescue vessel		
Achieved by:		***Differences***
Small Commercial Vessels and Recreational Craft	***Small Naval Craft and Government Craft***	
Rules for stability, subdivision and watertight integrity suitable for coastal operation	Rules for stability, subdivision and watertight integrity appropriate for the intended operation of the craft	Naval and government craft may need to stay afloat longer as time to rescue could be long. The damaged craft may also need to function as a 'safe haven' for the crew
Construction rules (damaged strength for bulkheads etc.)	Construction rules (damaged strength for bulkheads etc.)	Same requirement
-	Damage control and repair equipment kept on-board	Damage control by crew rather than abandon after damage

Table 2: Achieving the safety goals (2 of 5)

Goal 3: To enable safe evacuation of personnel into the sea or another rescue vessel offering assistance (this may include other boats, ships or rescue helicopters)		
Achieved by:		***Differences***
Small Commercial Vessels and Recreational Craft	***Small Naval Craft and Government Craft***	
Adequate number of appropriate lifesaving appliances	Adequate number of appropriate lifesaving appliances	Same requirement but possibly a different solution
Rules for safety equipment and systems	Rules for safety equipment and systems	Same requirement but possibly a different solution
Training in the use of lifesaving appliances	Training in the use of lifesaving appliances	Same requirement but possibly a different solution
	Survival equipment to allow an extend time in the water or extended time on-board the damaged craft	Concept of evacuation into the sea for rapid rescue by other vessels may not be relevant for naval or government craft.

Table 3: Achieving the safety goals (3 of 5)

Goal 4: Be designed to prevent the ignition of flammable substances as far as reasonably practical. In the event of a fire then reasonable means of extinguishing to be provided, along with means for limiting the spread of fire for long enough for rescue to be effected or to allow for safe abandonment		
Achieved by:		***Differences***
Small Commercial Vessels and Recreational Craft	***Small Naval Craft and Government Craft***	
Adequate structural fire protection	Adequate structural fire protection that allows for extended rescue time	Longer protection time after a fire as rapid rescue after damage cannot be expected for some types of naval and government craft
Adequate fire detection systems	Training in operation of fire fighting systems	Same requirement but possibly a different solution
Adequate fire fighting equipment to extinguish or fight fires for long enough to enable safe evacuation or rescue	Adequate fire fighting equipment to extinguish or fight fires with limited or no external assistance	Improved fire safety equipment may be needed as the consequence of a fire on crew safety is greater (as external assistance is unlikely for some naval craft)
Training in operation of fire fighting systems	Training in operation of fire fighting systems	Same requirement

Table 4: Achieving the safety goals (4 of 5)

Goal 5: To enable passengers or untrained persons to safely evacuate the craft with minimum assistance		
Achieved by:		***Differences***
Small Commercial Vessels and Recreational Craft	***Small Naval Craft and Government Craft***	
Clearly visible and marked emergency exits	Clearly visible and marked emergency exits	Passenger carrying small government and naval craft tend to operate as harbour tenders or for crew transfer. Therefore, no reason to have different requirements to equivalent commercial craft
Additional stability requirements to give longer evacuation time (due to untrained passengers)	Additional stability requirements to give longer evacuation time (due to untrained passengers)	
Extra fire safety provisions to give longer evacuation time in case of a fire	Extra fire safety provisions to give longer evacuation time in case of a fire	
Additional lifesaving appliances suited for safe evacuation of untrained persons	Additional lifesaving appliances suited for safe evacuation of untrained persons	

Table 5: Achieving the safety goals (5 of 5)

WRITING A NEW STANDARD

Parameters to Define Small Government Craft in a New Standard

To enable rules and standards to be developed some categorisation is necessary so that a wide range of craft can be covered by one standard. The parameters and sub-categories are chosen based on the goals discussed previously, and after consultation with navies through the bi-annual DNV Naval Craft Committee meetings, see table 6.

Parameter	***Sub-categories***	***Notes***
Length	< 24 metres	
Speed	< 45 knots	Vessels with very high speeds will require special consideration and may need different design rules due to high design accelerations
Type of persons on-board	Crew / Special Persons / Passengers	**Special persons** are persons who are not crew or passengers that have received basic sea survival training and familiarisation training for the craft and its safety equipment
Availability of assistance after an accident	Rapid / Extended / None	**Rapid** means assistance is available from land based professional rescue services or other professional rescue vessels within 3 hours. **Extended** means it is possible to get limited assistance to the craft within 24 hours **None** means the damaged craft or the embarked persons will need to escape to a place of safety without assistance
Night time operation	Yes / No. With or without overnight accommodation	For navigation equipment, accommodation standards and whether night time rescue is expected
Type of use	Heavy / Medium / Light	Related to the crafts ability to withstand loads, survivability after damage and distance from assistance and availability of assistance during normal operational scenarios. **Heavy** duty shall have some built in damage tolerance to allow for misuse.

Table 6: Categorisation of small government craft

Topics to Be Included In a New Standard

Given the wide range of craft and operational concepts the first activity should be a review of a document describing the concept of operation of the craft. This will allow the correct rules to be selected and ensure that the standard is appropriate for the craft and the operational environment.

The remainder of the document should use the same statutory scope set by the national administrations for small commercial vessels, recreational craft and High Speed Light Craft. This allows future standards and safety management systems to re-use parts of existing documentation, limiting the amount of new documentation necessary. The following topics are covered as the core scope:

- Hull structure, machinery and systems
- Stability and watertight integrity
- Fire safety
- Lifesaving appliances
- Navigation and radio equipment
- Pollution prevention

For military craft it may be sensible to include some additional items:

- Carriage of dangerous goods (e.g. ammunitions or hand weapons)
- Lifting craft with people on-board in a seaway
- Military load cases (for example additional loads on the hull structure due to "non-standard" military uses, normal operation should be covered in the core scope). Examples include weapons recoil effects on hull structure and beaching loads for landing craft.

The content of the core items needs some further debate. For example, loads on the structure of a harbour tender may be very different from those on a military raiding craft which is expected to operate at high speeds in rough seas and be grounded on a beach for offloading of personnel. Some load cases may need to be considered for all craft and others as optional for more specialised craft.

CERTIFICATION OPTIONS

Small commercial vessels and leisure craft are usually *certified* by a Class Society rather than being subjected to the more holistic *classification* process. The term *Certification* is used to describe a verification process where the craft has been shown to meet the requirements written in a standard at one point in time (usually when delivered). For small commercial craft and leisure craft DNV certify the ship using Standard for Certification No. 2.21 – Craft, April 2010, (5). This standard covers the design and build of craft from 6 to 24 metres and with speeds up to 45 knots. This process is shown on the top of figure 4.

Classification includes periodic inspection and survey to ensure that the requirements of the applicable standards are met throughout the service life and includes a system for managing the re-certification as well as logging any deviations or conditions upon which the classification is based. The craft will be issued with a Class Certificate and this will remain valid so long as the craft operates within the limits given in the certificate. The craft will need to be periodically re-certified and the class certificate typically lasts for around 5 years. DNV can class craft with a length greater than 18-20 metres (experience shows that the rule requirements can be difficult to meet for craft smaller than this size) These craft are usually classified using the current edition of the DNV Rules for High Speed, Light Craft and Naval Service Craft, (6). This process is shown on the bottom of figure 4.

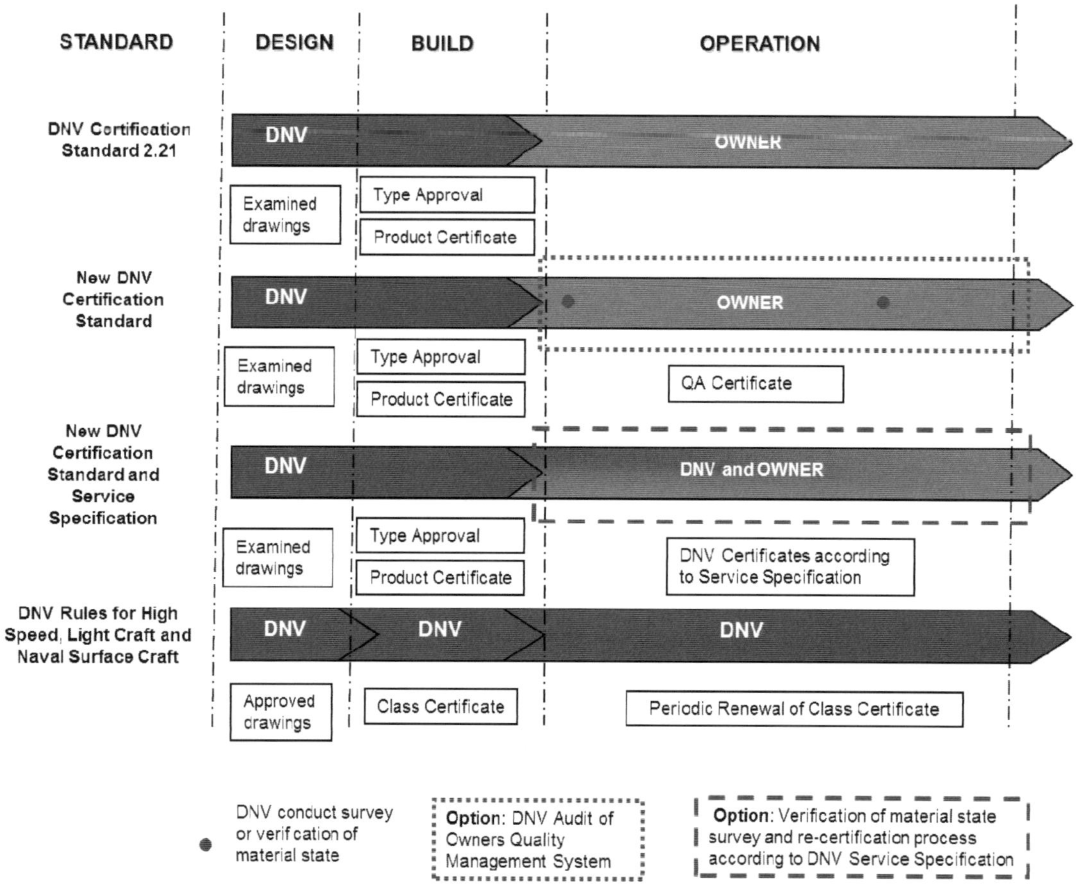

Figure 4: Options for the certification of small naval or government operated craft

To cover the widest range of craft, with very different concepts of operations and technical complexity, a degree of flexibility is needed in the certification/classification process. Two intermediate options have been proposed, these are shown in the middle of figure 4. For the first option a new DNV Certification Standard would be written to cover the design and build of small naval or government operated craft. Certification in the operational phase would be covered by the owner. To support this DNV could conduct surveys or verification of the material state when instructed by the owner. Further, the owner's existing management system for certifying the vessels in the operational phase could be audited and a quality assurance certificate issued.

For the second option the new DNV Certification Standard is used for design and build along with a DNV Service Specification describing the in-service phase (mirroring the DNV approach to certification of offshore structures). The DNV Service Specification would describe the process for verification of the material state of the ship (e.g. by survey) and re-certification. This would be more flexible than full Class and could involve the owner's surveyors or management processes.

CONCLUSIONS

DNV is developing rules and services for the safety assurance of small government operated craft and naval craft. The work completed so far has shown that existing standards are not suitable for these craft as the operational philosophy is different, requiring different approaches and solutions.

Any new standard for these small craft should at least match the safety goals for small commercial craft or leisure craft to enable the maximum re-use of existing standards, only developing new rules and regulations where necessary. This will enable the new standard to be accepted both by builders of small craft and the government departments and navies that operate them.

The standard will become part of a service covering the design, build and operational phases. This service is flexible and tailored to meet the needs and risks faced by naval and government departments.

Work on both the DNV Certification Standard and the DNV Service Specification is on-going and interested navies or government departments are encouraged to participate.

REFERENCES

(1) Combined USL/NSV 2010: Uniform Shipping Laws and National Standard for Commercial Vessels, consolidated 2010 edition, National Marine Safety Committee (www.nmsc.gov.au), Australia.

(2) Merchant Shipping Notice, MSN 1827 (M), Categorisation of Waters. Maritime and Coastguard Agency (MCGA), Southampton UK, February 2011.

(3) Merchant Shipping Notice, MSN 1823 (M), Safety Code for Passenger Ships Operating Soley in UK Categorised Waters, Maritime and Coastguard Agency (MCGA), Southampton UK, April 2010.

(4) Directive 2009/45/EC of the European Parliament and the Council of 6 May 2009 on Safety Rules and Standards for Passenger Ships (Recast), as amended on 1 June 2010.

(5) Det Norske Veritas Standard for Certification No. 2.21 – Craft, April 2010, Høvik, Norway.

(6) Det Norske Veritas Rules for High Speed, Light Craft and Naval Service Craft, January 2012 edition, Høvik, Norway.

Residual stress measurements through the hull section of submarines

X. Ficquet[1], M. Körner[2], E.J Kingston[1]

[1]VEQTER Ltd, Park Row, Bristol BS1 5UB, UK
[2] DCNS, Rue Choiseul, 56311 Lorient, France
xavier.ficquet@veqter.co.uk

ABSTRACT

The fatigue lifetime of a submarine hull depends on the loads generated by hull contraction under the effect of hydrostatic pressure and the residual stresses existing in the absence of external loading. The use of numerical simulation allows a straightforward calculation of the stresses induced by the hydrostatic pressure. However it is more intricate to determine the residual stresses resulting from the sheet bending process combined with the sheet assembly using a multipass welding process.

This paper presents a review of work carried out to ascertain the residual stresses present within three submarine hull sections. An appraisal of available measurement methods is first carried out, and then the calculation of the residual stresses generated by the bending process is presented with a measurement technique comparison on a bent beam. A compendium of measurements of residual stress through T-welds show that the general stress profiles are similar but there is scatter between measurements. Finally a comparison of the residual stress distributions within the submarine hull structures is shown. The comparison shows a very good agreement in profile and magnitude for all submarines compared.

INTRODUCTION

Efforts have been made to establish and improve ultimate submarine design life limits. The residual stresses induced by manufactures, like rolling and welding, are variables required for analysis of various degradation mechanisms associated with the aggressive conditions experienced by a submersible (buckling, fatigue, stress corrosion cracking). In a submarine structure, the hull is essentially composed of rolled plates and T-frames welded together. The residual stresses produced by the bending process can be calculated. The resulting through thickness distribution has a typical "Z" shape with tension at the inner surface of the hull and compression at the outer surface. The residual stresses induced in T-butt welds are highly tensile in and around the weld which can have important effects on the mechanisms of crack growth. Consequently, it is very important to predict residual stress and its effect on structural failure in the analysis of fracture mechanics for the structure.

The objective of this paper is to present through thickness residual stress measurements carried out on submarine hulls from the toe of the T-butt weld to the outer surface of the hull. A review of residual stress measurement techniques capable of measuring thick mechanical components is first presented. Then the calculation of the residual stresses from the three rolls bending process is explained. In order to show the capabilities of commonly available residual stress measurement techniques an elastic-plastic bent beam was manufactured and measured using four types of measurement technique. A compendium of measurements of residual stress of a T-weld is presented and compared with a "standards" assessment procedure. Finally a comparison of the residual stress distributions within thc submarinc hulls is shown. The Deep Hole Drilling (DHD) measurement technique was employed for the two submarines investigated. In total six measurements were carried out at different angular positions to detect variability in manufacture on a Rubis class submarine and two measurements were performed on an IZAR submarine. Comparisons were made with the Neutron Diffraction (ND) residual stress measurements available in the published literature.

RESIDUAL STRESS MEASUREMENT TECHNIQUES

The most commonly available residual stress measurement techniques that can be applied to cast irons and steels are shown in Figure 1. Descriptions of these techniques have been summarised in by Wither et al. [1] and Kandill et al. [2]. The techniques can be categorised in terms of penetration depth and the degree of material removal required. As it can be seen in Figure 1 only a few techniques are available to measure components thicker than 10mm.

Non-invasive methods are generally based on measuring dimensional changes to the crystal lattice or some physical parameter affected by the crystal lattice dimensions. High energy synchrotron radiation provides high spatial resolution and three-dimensional maps of strain distributions. The neutron diffraction method has the ability to obtain bulk measurements up to about 40mm through the thickness of steel samples. For the latter two methods, it is necessary to take the sample to the radiation source and therefore the methods have practical limitations. Often the complex geometry or the size, as in this case, precludes direct non-invasive measurement.

Fundamentally, invasive and semi-invasive methods rely on strain relaxation procedures by removing a volume of material resulting in stress relaxation and hence a measured deformation. The degree of invasion depends on the extent of material removed from the specimen. Semi-invasive and invasive methods arc commonly used to perform measurements in sub-surface regions or the bulk of a specimen.

The DHD technique is semi-invasive and able to measure residual stresses deep inside a component [3, 4]. Semi-invasive methods allow several measurements to be made on the specimen but not at the same location. To increase the measurement range through the thickness a combination of methods could be used. For example, from 0.1mm to 2mm the Centre-Hole Drilling technique could be employed and then the DHD technique could be used to perform the measurement through the rest of the component thickness.

Finally, invasive techniques such as the Block Removal Splitting and Layering (BRSL), slotting and contour methods allow through-thickness measurements of residual stress to be made. However they destroy the specimen completely so that no further measurements can be obtained [5].

RESIDUAL STRESS FROM THE THREE ROLLS BENDING PROCESS

Analytical solution for the three rolls bending process

There is an analytical solution to determine the hoop residual stress field produced by the three rolls bending process, see Figure 2, used to form the submarine hull structure. The solution is described by De Aguiar et al. [6] and Queener et al [7], which uses the constitutive elastoplastic bending, bending moment and elastic springback relationships to calculate the residual stresses. Finally, using those relationships, the hoop residual stress $\sigma_{r\theta}$ inside the rolled sheet after unloading can be determined by adding the elastic stress $\sigma_{u\theta}$ caused by unloading at every position y of the sheet from the stress already there σ_{θ}, see Figure 2, can be calculated using the following equations:

$$\sigma_{r\theta} = \sigma_{\theta} + \sigma_{u\theta} \tag{1}$$

with:

$$\sigma_{u\theta} = E' \cdot \frac{y \cdot r_b}{y + r_b} \cdot \left(\frac{1}{r_c} - \frac{1}{r_b} \right) \tag{2}$$

If $|y| < y_0$, then:

$$\sigma_{\theta} = E' \cdot \frac{y}{r_b} \tag{3}$$

If $y_0 \le |y| \le t/2$, then:

$$\sigma_{\theta} = sign(y) \cdot k \cdot \left(\frac{2}{\sqrt{3}} \right)^{n+1} \cdot \frac{\left(\frac{|y|}{r_b} \right)^{n}}{1 + k \cdot \left(\frac{2}{\sqrt{3}} \right)^{n+1} \cdot \left(\frac{|y|}{r_b} \right)^{n-1} \cdot \frac{n}{E'}} \tag{4}$$

With:

y_0 is the transition point, $E' = \frac{E}{1 - \nu^2}$, k is the strength coefficient

E the elastic modulus, ν the Poisson's ratio and n the work hardening index

Measurement comparisons of a four point elastic-plastic bent beam

In order to compare different measurement techniques, Goudar [10] used a 50mm square bent beam. The beam was plastically deformed on a servo-hydraulic testing machine in a four-point bend fixture. Different measurement techniques were then employed to measure the residual stresses within the uni-axial stress field with a constant bending moment, i.e. between the two inner rollers.

The Incremental Centre-Hole Drilling (ICHD), Deep-Hole Drilling, Neutron Diffraction and Ring Core techniques were carried out in the central region of the beam. The analytical, Fe and measurement results are plotted in Figure 3 for the near surface residual stresses and Figure 4 for the full through thickness residual stresses.

At the surface, the ICHD technique provides results from about 0.2mm to 1mm depth. The ring core technique provides results from about 0.8mm to 3.8mm from the surface. The DHD technique provides results through the thickness of the component from the first millimetre of the component surfaces. The neutron diffraction technique used a gauge volume of 3x3x3 mm^3 and provided results from about 3mm of each component surface. Overall the comparisons of all techniques are in very good agreement. Compressive stresses are found on one surface of the beam, then increasing to tension at about 1/3 of the beam thickness, then decreasing to compression at about 2/3 of the thickness and finally reaching tensile stress at the other surface of beam. This example shows that using a combination of measurement technique can be beneficial to gain the entire residual stress distribution and also to increase the confidence given to any measurement. In most cases a comparison between the diffraction and strain relieving measurement technique is ideal. Diffraction techniques are sensitive to the material texture and the grain size while strain relieving techniques are affected by plastic relaxation when the residual stress is highly triaxial.

RESIDUAL STRESS FROM A T-BUTT WELD

Assessment procedure

The treatment of residual stresses in welded components is considered in defect assessment procedures such as BS7910 [9 and R6 [10]. The R6 and BS7910 equations (Eq.5, 6 and 7) used to predict residual stresses for T-butt and fillet welds are based on the electrical energy of the largest weld run at the weld toe. These equations give the hoop and axial residual stresses for ferritic steel.

For $r \leq r_0$:

$$\sigma_R^{H,A} = \sigma_y \{1 - (r / r_0)\} \qquad (5)$$

For $r > r_0$:

$$\sigma_R^{H,A} = 0 \qquad (6)$$

$$r_0 = \sqrt{\frac{122 E_1}{\sigma_y}} \qquad (7)$$

With: (r) the distance from the weld toe into the component, (E_1) the electrical energy per unit length of the largest weld run at the toe and (σ_y) the yield strength.

Measurements comparison of a T-butt weld

Figure 5 shows a comparison of longitudinal residual stresses within T-butt welds for a range of materials, T-section geometries and measurement methods. The normalised results are shown from the weld toe through the thickness of the component and stresses in relation to the yield strength. The measured residual stress data in the literature [3, 11] shows a wide range of variability in the magnitude with values from 0% to 120% of the yield strength near the weld toe. The trends however are very similar, the longitudinal stress starts at its highest magnitude at the weld toe then decreases to a compressive stress at about 30% of the thickness of the component. Finally the longitudinal stress increases toward a negligible stress value near the opposite surface. Figure 5 shows the mean line for all data, as well as the mean plus and minus two standard deviations. It can be observed that the standard deviation start at the weld toe with a highest value of about 0.4 and decreases progressively to a minimum of 0.1 at 90% of the thickness.

The differences in the stress level can be attributed directly to the difference from weld to weld. The 100mm T-section from Wimpory et al [11] was measured using two different techniques, ND and DHD, the results show a very good agreement through the thickness of the component with the exception of the compressive zone where the DHD peaks at -0.5 while the ND only shows -0.2. However in this component, the neutron diffraction measurements were carried out on a 25mm section cut from the original T-section which may explain this variation due to longitudinal stress relaxation.

RESIDUAL STRESSES IN SUBMARINE HULL SECTIONS

The submarine hull section measured was constructed from bent plate welded together to create the hull with added T-frames welded to the inner surface. The expected residual stress is therefore created from the two sources of manufacturing processes seen in the previous sections, i.e. the residual stresses from bending and the residual stresses from a T-butt weld. To measure the residual stresses in the submarines the DHD technique was employed.

The first DHD measurements were carried out on a full scale mock-up of an IZAR submarine section at the T-butt weld toe through the thickness of the hull [3]. The second set of DHD results were taken from a Rubis class submarine with a hull section of 3800mm in radius [12]. Within the Rubis class submarine three measurements were carried out on the port side and three measurements were carried out on the starboard side, as defined in Figure 6. All six measurements were carried out along a radial line through the thickness of the hull section. The axial location of the measurements was chosen so that the exit of the reference hole was approximately 12mm from the T-frame face and at the weld toe. The IZAR and Rubis class submarine results are compared to those carried out by Pearce et al [13] on a similar submarine structure using ND technique. Pearce's measurements were carried out on a curved plate, SAW, T-butt sample removed from a full scale Collins class submarine section and measured 1200mm × 1200 mm. The ND measurement was carried out directly beneath the T-butt weld with a gauge volume of $4\times4\times20mm^3$. The thickness of the curved hull plate was greater than the maximum penetration depth of the ISIS ENGIN-X ND instrument. Therefore results from the weld toe to about a third of the thickness were not measured.

The measured normalised hoop stresses showed the same trends for all three mock-ups, high tensile stresses close to the weld root decreasing to compressive stresses with radial depth at about 0.35, then increasing to tensile stress between 0.6 and 0.7 depth before decreasing to compression close to the outer hull surface. The first compression zone at about 0.35 is important as it will have an impact on the propagation of any crack from the weld toe and therefore a key step in determining the continuing integrity of a structure in service. Figure 6 also shows the bending stress calculated for the Rubis class submarine. It can be observed that all three sets of results show a similar "Z" shape profile typical of a bending induced stress.

CONCLUSION

An appraisal of available measurement methods was conducted to assess their ability to obtain through thickness residual stress distributions in a submarine hull section. This paper then showed the calculation of the residual stresses introduced during the three rolls bending process and provided a comparison of residual stress measurement technique results for on a similar four-point bent beam process. The results showed an excellent agreement between all the techniques, at the surface with ICHD and Ring Core, and the sub-surface with DHD and ND. A compendium of residual stress measurements and finite element analyses was then provided for a T-butt weld component. The comparison showed a similar trend for all results and agreed with the predictions of the "standards" assessment procedures. Important variations in the amplitude were found near the weld root, ranging from about 40% to 100% of the yield strength for the measurements only. The variation in magnitude decreased towards the opposite surface to that weld.

Finally, residual stress measurements within two the submarine structures showed a combination of residual stresses from both the bending and welding processes. With high tensile stresses near the weld root, followed by a, bending "Z" shape residual stress profile, with peak stress reaching about half of the yield stress at 1/3 and 2/3 of the thickness.

REFERENCES

1. Withers, P.J. and Bhadeshia, H.K.D., (2001), Material Science and Technology,.17: p. 355-365(11).
2. Kandill F.A., Lord J.D., Fry A.T., and Grant P.V., (2001), NPL Report.: MATC(A).
3. Kingston E.K., Stefanescu D., Mahmoudi A.H., Truman C.E. and Smith D.J., (2006), "Novel applications of the deep-hole drilling technique for measuring through-thickness residual stress distributions", Journal of ASTM International, Vol. 3, No. 4, Paper ID JAI12568
4. Leggatt R.H., Smith D.J., smith S.D., and Faure F., (1996), "Development and experimental validation of the deep-hole method for residual stress measurements" Journal of Strain Analysis. 31: p. 177-189.
5. George D., (2000), "Determination of residual stresses in large section stainless steel welds", University of Bristol, PhD thesis.
6. De Aguiar J.B., Barbosa G.M., Batalha. G.F., (2001), "Sheet bending theory applied to a three roll process". 1° Congresso Brasileiro de Engenharia de Fabricação..
7. Queener C.A., De Angelis. R.J. (1968), "Elastic Springback and Residual Stresses in Sheet Metal Formed by Bending". Transactions of the ASM, Vol. 61.
8. Goudar D. (2011), "Quantifying uncertainty in residual stress measurement using hole drilling technique", University of Bristol, PhD thesis.
9. BS7910, (2005), "Guide on methods for assessing the acceptability of flaws in metallic structures", British Standard Institution

10. British Energy, (2001), "Assessment of the Integrity of Structures Containing Defects, R/H/6-Revision 2", British Energy Generation Limited

11. Wimpory R.C., May P.S., O'Dowd N.P., Webster G.A., Smith D.J., Kingston E., (2003), "Measurement of residual stresses in T-plate weldments", Journal of Strain Analysis for Engineering Design, 38 (4), pp. 349-365.

12. Ficquet X., Körner M., Kingston E.J, (2011), "Measurement of residual stresses in a hull section of a Rubis class submarine", MAST Conference Proceeding, Marseille

13. Pearce S.V. et al., (2008) , Materials Science and Engineering A 480 411–418

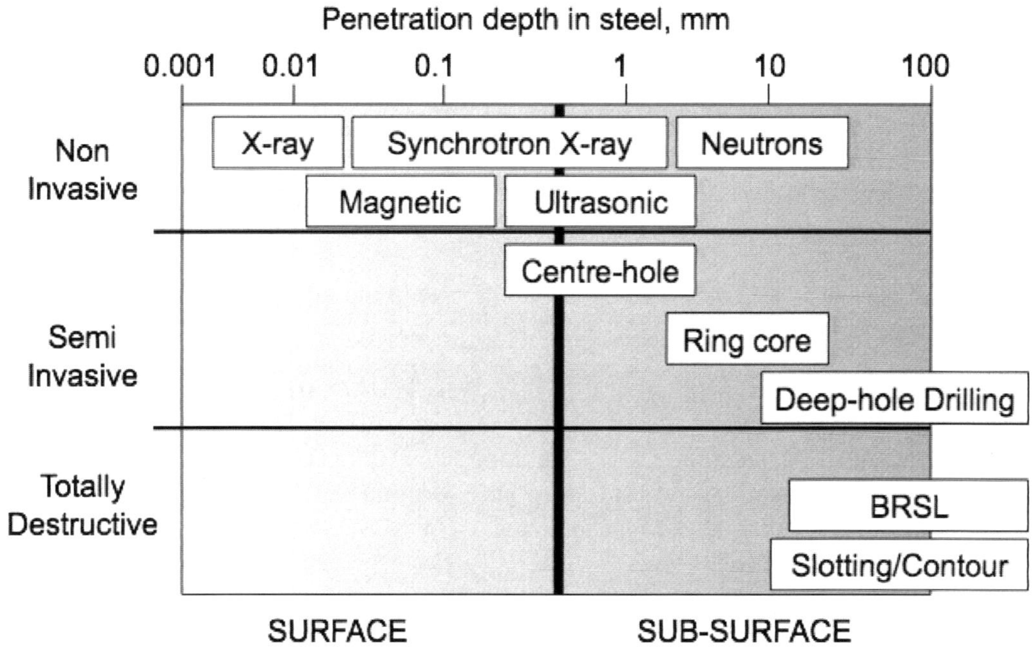

Figure 1: Residual stress measurement methods in steel.

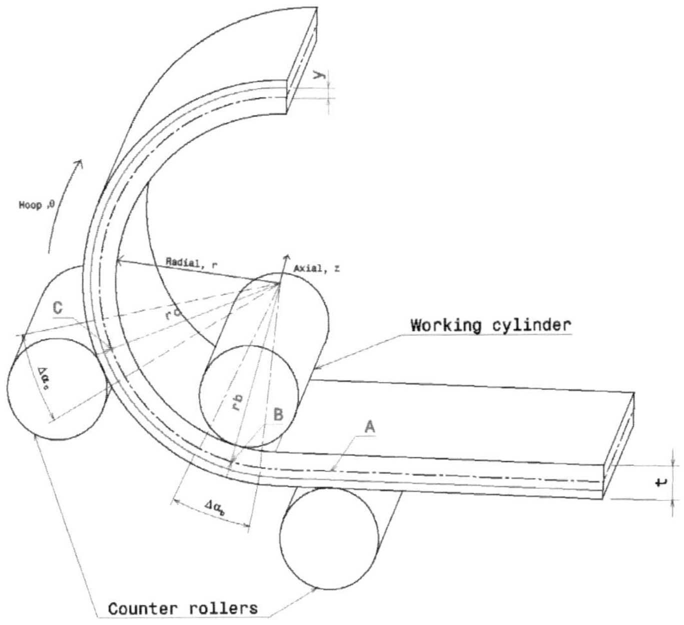

Figure 2: Three rolls bending process

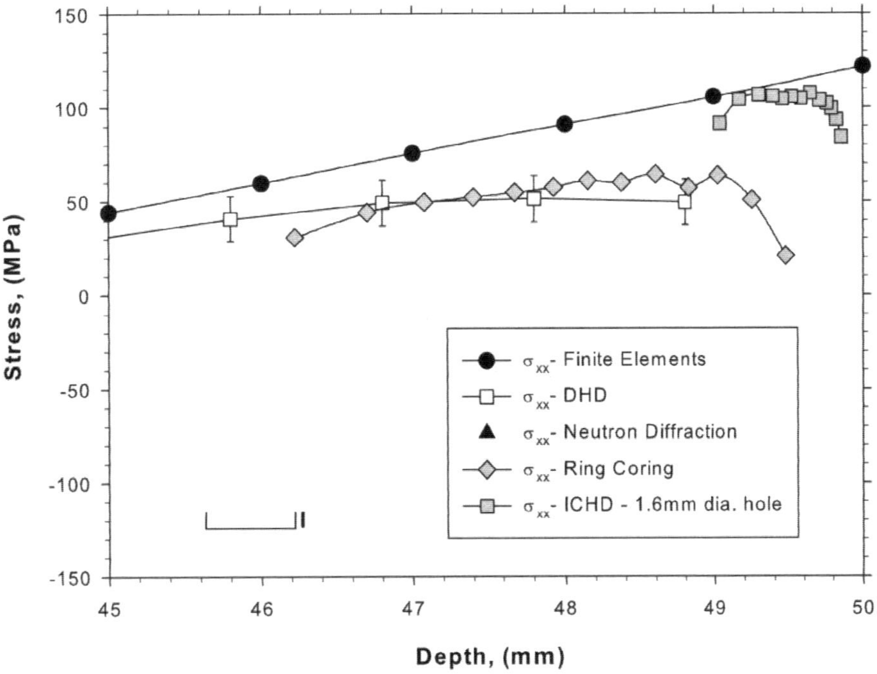

Figure 3: Comparison of near surface, longitudinal residual stresses in a four point plastically bent beam

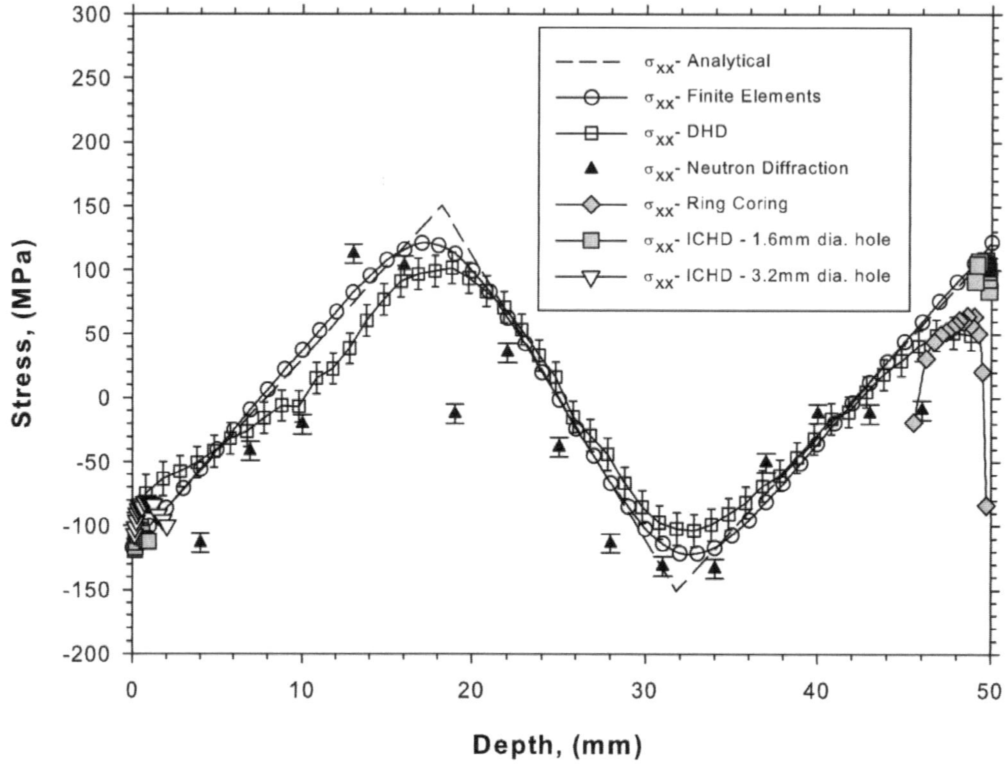

Figure 4: Comparison of longitudinal residual stresses in a four-point plastically bent beam

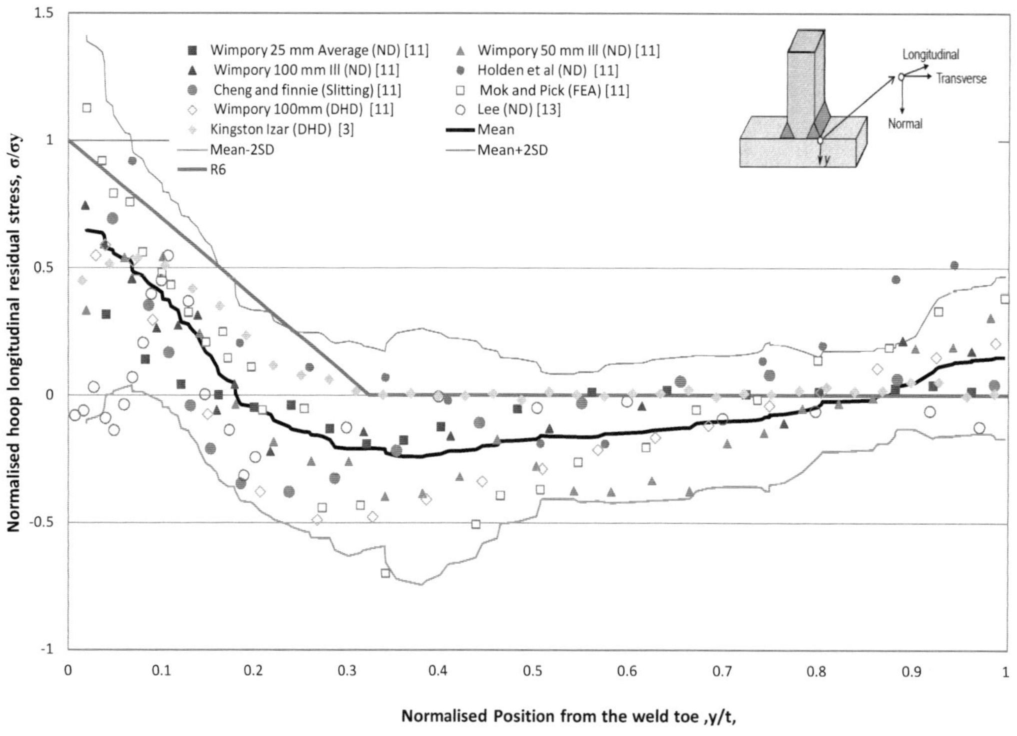

Figure 5: Longitudinal residual stress distributions for a range of welded T-joints

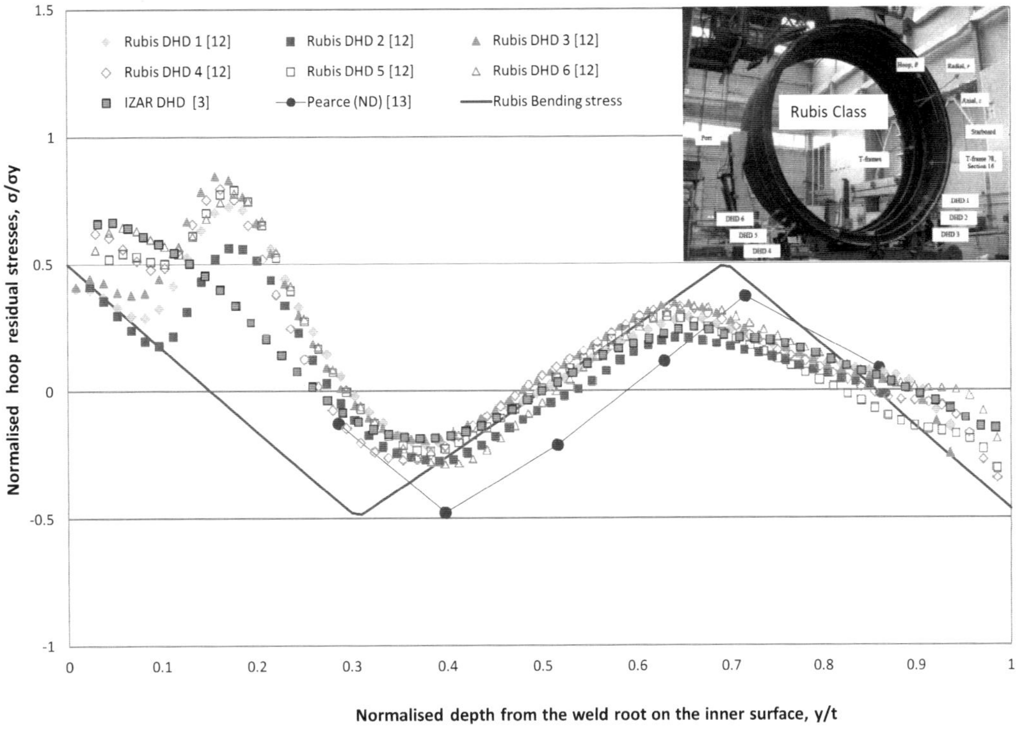

Figure 6: Comparison of hoop residual stresses from three types of submarine

Understanding the Manoeuvring Performance of an X-Plane Submarine in Deep Water and Near the Free Surface

Authors Name: P Crossland, N Kimber and N Thompson
Affiliations: QinetiQ Ltd, UK
Email contact: pcrossland@QinetiQ.com

1. INTRODUCTION

In the UK, the current submarine capability is heavily targeted towards providing support to a submarine fleet with traditional cruciform stern arrangements; however, it is thought that performance improvements may be achieved with the use of alternative control surface arrangements, such as X-planes. Operationally, the advantages of X-planes, over the more traditional cruciform arrangements typical of UK Royal Navy submarines, are that they may provide greater safety; if an X-plane submarine was to experience a single plane jam then the remaining three planes would be able to not just counteract the effects, but also enable the submarine to remain operational. The disadvantages are that when sizing the X-planes to provide a desired horizontal plane performance this effectively sets the performance in the vertical plane too. This trait can be offset to some extent by having an X-plane design with the planes not set at a 90 degree angle to each other, but at an angle that creates a greater component of lift in the horizontal plane than the vertical plane.

However, before such X-plane concepts can be considered as design options for potential future UK submarines, the capability to understand the performance of such configurations during the design process must be developed and validated beforehand. A validated method of predicting the performance of a manoeuvring submarine will provide the ability to understand the merits of alternative stern configurations; and reduce risk during the design and procurement of such submarines. As part of a four year programme of work, a validated means to enable the investigation of the relative merits of a range of aft appendage configurations, including X-plane stern arrangements, was developed by QinetiQ Ltd. This paper describes the experiment techniques used to derive the coefficients from captive model tests and the results from free running model tests in deep water and in close proximity to the free surface and, where applicable, compares the model tests with numerical predictions.

2. X-PLANE DESIGN

The aim of the study was to develop an X-plane stern arrangement for an existing design of submarine to allow testing in the Ship Tank and Rotating Arm at QinetiQ Haslar and allow tests using a free running model. Due to the requirement to make early design decisions to enable a complete experimental programme to be undertaken in the desired timescales, it was recognised that the X-plane design would not be optimal for that particular hull form but would represent an arrangement that was suitable for generating validation data. The "design space" is represented in Figure 1. In addition to the time constraints, there were also design constraints; principally that any X-plane stern should easily interface with the existing physical captive model and had to be configurable to be able to cost-effectively implement the design on to the Submarine Research Model (SRM) developed by QinetiQ Ltd and covered later in this paper. So, as a result of the design constraints, the forward and aft radii of the cone section were fixed to interface with the existing model, the cone angle and stock locations were fixed to ensure configurability with SRM and the X-plane configuration had to be orthogonal. Furthermore, the span of the planes is effectively fixed since it should be contained in a box defined by the maximum beam of the hull. Once the span had been fixed the decision was made to maintain the same aspect ratio of the upper rudder of the existing submarine design which, in effect, fixed the root and tip chords. Thus, each of the X-plane appendages were geosims of the upper rudder from the existing submarine; the result was an increase of around 33% in total lifting surface in the horizontal plane compared with the original upper and lower rudder combined.

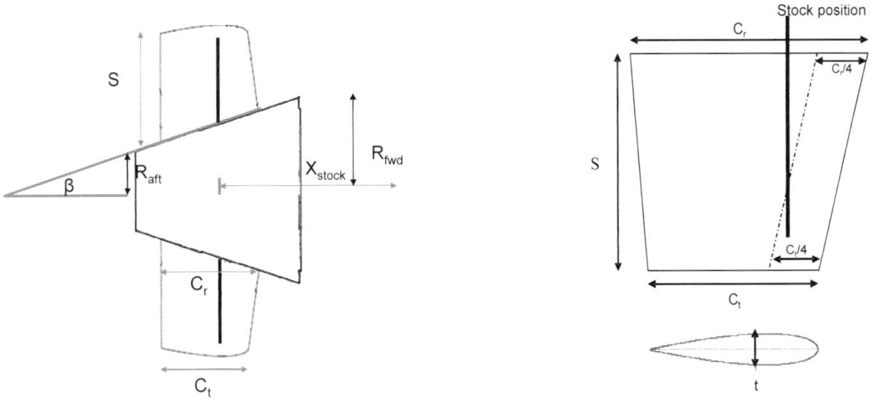

(a) Stern configuration (b) Appendage design

Figure 1: Design variables for an X-plane

3. COEFFICIENT BASED TECHNIQUES

There are a number of methods used for predicting the manoeuvring performance of a deeply submerged submarine, see (ITTC, 2008) for approaches for surface ships. Generally, in order to predict manoeuvring parameters such as advance, transfer overshoots etc, the fundamental trajectories of the submarine undertaking specific manoeuvres are required. The exceptions to this are the stability and control indices, (Spencer, 1968), given as:

$$G_V = 1 - \frac{M'_{uw}(m' + Z'_{uq})}{M'_{uq} Z'_{uw}}, \quad G_H = 1 + \frac{N'_{uv}(m' - Y'_{ur})}{N'_{ur} Y'_{uv}}$$

The traditional system-based methods require a physical model to be manufactured and tested in a series of captive model experiments to measure directly the hydrodynamic forces and moments acting on the appended submarine as a function of the motion state variables , see (Lewis, 1989) for example. This is based on the assumption that the problem can be regarded as quasi-steady state whereby each component of the force and moment is a function of the instantaneous values of the velocity and acceleration components of the body, and the deflection of the control surfaces. One mathematical approach to determining the quasi steady state forces and moments on a manoeuvring submarine is described by (Gertler and Hagen, 1967). For example the equation for pitch moment is given as:

$$M' = \left(M'_{uu} + M'_{uw}w' + M'_{vw}v'^2 + M'_{uu\delta b}\delta b + M'_{uu\delta s}\delta s\right) + \left(M'_{wv}w'v' + M'_{u|w|}|w'| + M'_{|wv|}|w'v'|\right) + \left(M'_{\dot{w}}\dot{w}' + M'_{uq}q' + M'_{vp}v'p' + M'_{vr}v'r'\right)$$
$$+ \left(M'_{qv}q'v' + M'_{u|q|\delta s}|q'|\delta s\right) + \left(M'_{\dot{q}}\dot{q}' + M'_{pp}p'^2 + M'_{rr}r'^2 + M'_{rp}r'p' + M'_{q|q|}q'|q'|\right) - \left(m'g'x'_G - B'x'_B\right)\cos\phi\cos\theta - \left(m'g'z'_G - B'z'_B\right)\sin\theta + M'_{wave}$$

where the ' represents the non-dimensionalised pitch moment given as: $M' = \frac{M}{\frac{1}{2}\rho U^2 L^3}$

For an X-plane configuration, the terms involving the movement of the stern plane, δs, are more complex since the use of independently actuated planes means there are potentially individual contributions from each appendage. Hence, for an X-plane arrangement there will be a separate set of hydroplane coefficients derived for each individual control surface (X_i), as shown in Table 1. Furthermore, the experimental design is fundamentally a function of the coefficients that are to be derived; that is, the experiment must be designed so that the state variables can be controlled as independent variables.

Cruciform			X-plane
Rudder	Sternplane		
$X_{uu\delta r\delta r}'$	$X_{uu\delta s\delta s}'$		$X_{uu\delta X\delta Xi}'$
$Y_{uu\delta r}'$	-		$Y_{uu\delta Xi}'$
-	$Z_{uu\delta s}'$	→	$Z_{uu\delta Xi}'$
$K_{uu\delta r}'$	-		$K_{uu\delta Xi}'$
-	$M_{uu\delta s}'$		$M_{uu\delta Xi}'$
$N_{uu\delta r}'$	-		$N_{uu\delta Xi}'$

Table 1: Comparison of cruciform and X-plane appendage coefficients

4. CAPTIVE MODEL TESTS

A 5m Glass Reinforced Plastic (GRP) model of the existing submarine design was modified such that the standard cruciform appendage configuration was replaced with the orthogonal X-plane configuration. For the experiment, the model had a balance fitted that measured the forces and moments acting on the model during both steady state and dynamic load conditions.

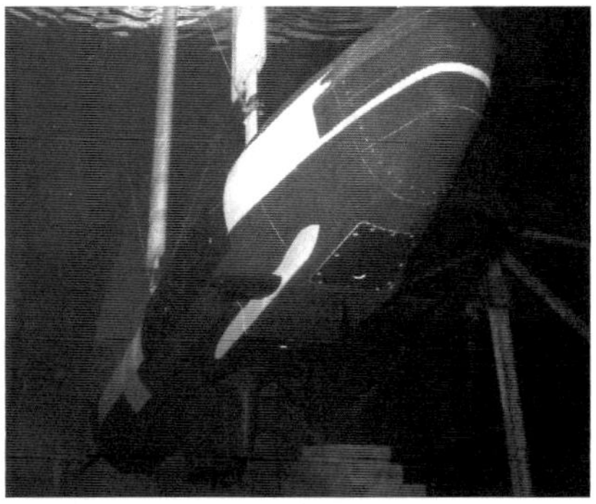

(a) Ship Tank

(b) Rotating Arm

Figure 2: Captive model experiments

The model was towed in the Ship Tank to provide the relationship between the forces, moments and velocities and the various state variables such as body angles of attack and plane angles of attack. During these runs the four X-plane appendages were moved individually, as adjacent dual planes, opposing dual planes, and all four planes together, over their operational range to help understand any interference effects. The experimental runs on the Rotating Arm were designed to measure relationships between the forces, moments and velocities and the various state variables for non-zero pitch and yaw rate. Figure 2 shows the captive model mounted on the Ship Tank carriage and the Rotating Arm; in total approximately 2,700 runs were undertaken in order to capture the full effects of the X-plane design. One particular area of extensive investigation was whether there were any significant interference effects due to the all moving X-planes. That is, can the effect of each X-plane be simply added together to get the effect of any combination of planes moving together? Figure 3 shows the resulting heave force and pitch moment associated with moving all four planes (as stern planes) and comparing with the summation of the heave forces and pitch moments associated with moving the planes individually. This appears to show that except at the largest angles of attack the interference effects are negligible.

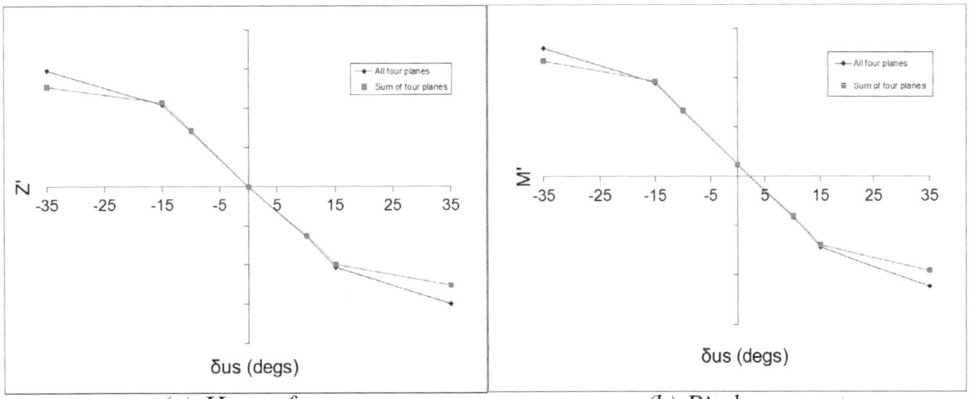

(a) Heave force *(b) Pitch moment*

Figure 3: Comparison between forces due to moving all planes and moving planes individually

5. CFD TECHNIQUES

Determining the unsteady forces and moments on a manoeuvring submarine using computational methods represents a challenging problem and perhaps these methods are not sufficiently mature for use as a design tool; the alternative to determining these forces and moments directly is to assume that the hydrodynamic forces and moments acting on the appended submarine can be represented as a function of the motion state variables in the same way as that assumed for experiments. QinetiQ Ltd has used the commercially available CFD code ANSYS CFX5 with some success in capturing the steady state forces and moments on a submarine at angles of attack. In parallel to the extensive experimental programme, a series of CFD calculations were undertaken using the same X-plane geometry (propulsor not shown), Figure 4.

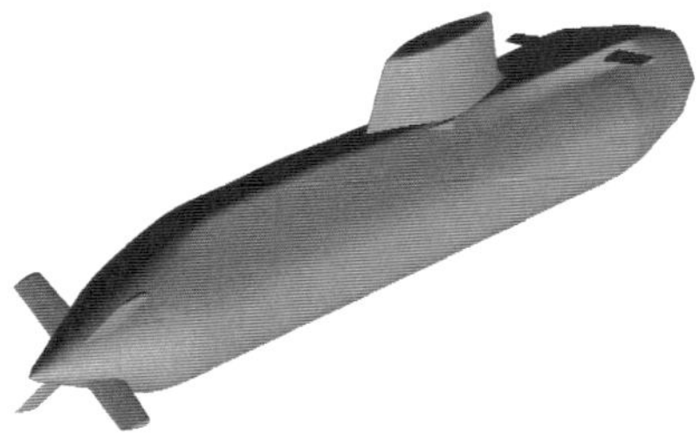

Figure 4: Hull with X-plane

A (Baseline) Reynolds Stress model of turbulence was used in all the calculations which included an "Automatic Wall Function" capable of resolving the turbulent boundary layer near solid boundaries down to the viscous sub-layer. Transition effects were not modelled and the flow was assumed to be fully turbulent over the whole domain. Propulsor effects were modelled by applying a uniform momentum source distribution over the volume occupied by the propulsor. In cases of non-zero pitch and yaw rate, the calculations were carried out in a rotating frame of reference. The entire computational domain was assumed to be rotating at a constant angular velocity about a fixed centre of rotation, giving a quasi-steady flow solution. Figure 5 shows an example of the quality of the predictions when compared with the experiments. In this particular case, CFD predictions were undertaken and compared with the conditions from the experiments. These data were taken from the Ship Tank experiments but similar quality of predictions was observed for the Rotating Arm experiments for non-zero pitch and yaw rate.

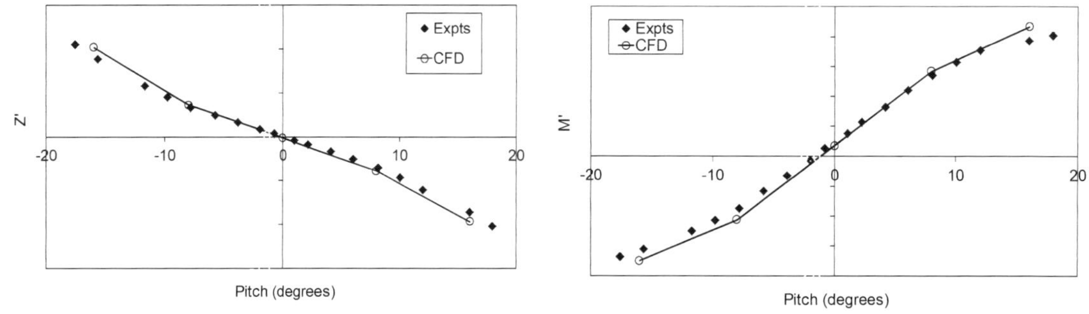

(a) Heave force *(b)Pitch moment*

Figure 5: Predicted and measured forces and moments

The coefficients relating to the acceleration terms in the force and moments equation were derived using the Submarine Component Added Mass program (SCAM) which is based on the approach developed by (Watt, 1988) and is used to estimate the added mass coefficients. In SCAM, each component of the submarine is approximated by an equivalent ellipsoid. SCAM, thus, takes account of the added mass contributions from the appendages as well as the hull.

6. IMPLEMENTATION OF MATHEMATICAL MODEL

As mentioned earlier, the basis for the hydrodynamic coefficient set which populates the simulation model comes from the constrained model experiments. The next step in the design evaluation process was to incorporate this coefficient set into a format suitable for use in simulation. Until free-running model experiments have been conducted, there are no validation data available for simulations using this mathematical model. However, the initial simulation model can be used in the analysis of submarine stability and design of control algorithms that are usually based on simplified linear models of the submarine dynamics, or so called linear derivatives. Having obtained the values for the derivatives for the X-plane design, the vertical and horizontal stability indices were calculated as $G_v = -0.11$ and $G_h = -0.04$. The accepted stability criteria are that G_v and G_h should be positive to indicate vertical and horizontal stability respectively, suggesting that this particular X-plane would not meet the design criteria for both the horizontal and vertical plane. Experience shows that submarines with negative G_v and G_h are not typical; indeed, this was not the case for the original cruciform design which has flapped stern planes with stabilisers. So, the result of adding an X-plane, with all four appendages moving, has imparted some dynamic stability issues on this design. The key question is how this instability would manifest itself in any free running model with this configuration, which will be demonstrated later in this paper.

7. AUTOPILOT DESIGN

One key aspect of the design using X-planes is the design of the control algorithms that accompany them; the design is different to that for a cruciform arrangement. A conventional cruciform dive command, which would just be applied to the stern planes, would be applied to all four X-plane control surfaces, as shown in Figure 6(a). (The view is from aft looking forward.) Similarly, when a conventional cruciform initiates a turn, this would just be applied to the rudder, which for the X-plane would be applied to all the control surfaces as shown in Figure 6(b).

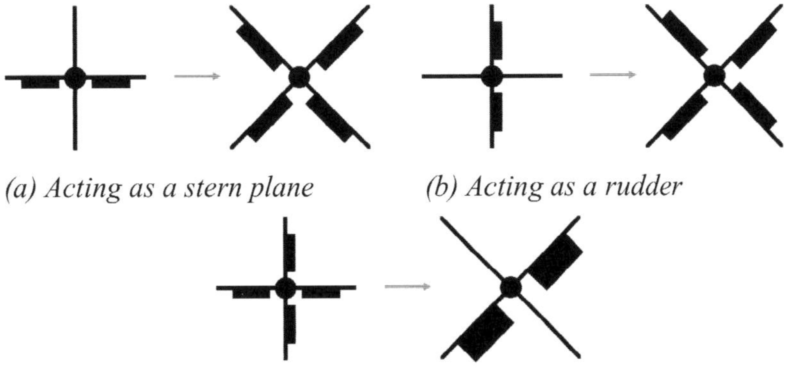

(a) Acting as a stern plane *(b) Acting as a rudder*

(c) Combination of depth and heading control

Figure 6: Cruciform to X-plane transformation

When simultaneous depth and heading demands are made, the demanded control surfaces are simply summed. Figure 6(c) shows an example of a combined dive on the planes and a starboard rudder. The upper starboard and lower port control surfaces sum their deflections, while the upper port and lower starboard demands cancel out. During the development of the coefficient set and implementation in simulation, it was found that the lower pair of X-plane control surfaces was slightly more effective than the upper pair, despite being geometrically identical, see Table 2. This is reflected in the pertinent coefficients, where the lower pair values are around 8% greater than the upper pair values. This is possibly due to the presence of the bridge fin creating vortices that are transported downstream and impact on the upper pair of control surfaces. The table also shows the relative power (as a magnitude of the X-plane coefficient divided by the pertinent coefficient from the cruciform) of the X-plane relative to the cruciform; this table demonstrates the increase in control authority that is attained from the four all moving appendages. Indeed there is around 90% increase in control authority in the vertical plane which is not particularly the plane in which agility is a desired feature.

Coefficient	Cruciform (baseline)	X-plane upper pair	X-plane lower pair	X-plane combined
Zds′	1	0.91	0.98	1.89
Mds′	1	0.92	0.98	1.90
Ydr ′	1	0.67	0.71	1.38
Ndr ′	1	0.69	0.76	1.45

Table 2: Relative power of the X-plane

The approach to the design of the course and depth autopilots was to treat them separately, with the demands summed at the output to the control surfaces themselves as illustrated in the above example. The course control (and similarly depth control) was divided into two modes, course-keeping and course-changing (or depth-keeping and depth-changing for the vertical plane). The controllers themselves take the form of a high order state-space system; embedded in the autopilot is a procedure for switching between the different modes in the horizontal and vertical plane. All the controllers were run in parallel to ensure instabilities between plane angle demands were avoided during the switching process.

8. FREE RUNNING MODEL TESTS

Once the autopilot algorithm had been developed it was implemented into a real time operating system for use in the free running model, the QinetiQ Submarine Research Model (SRM). The model is capable of replicating most of the manoeuvres that would be pertinent on a full scale submarine and in some areas more; one primary area is that of exploring extreme manoeuvres such as hydroplane jams and the subsequent recovery strategies. The SRM was configured as a geosim, Figure 7, of the model used in the constrained experiments.

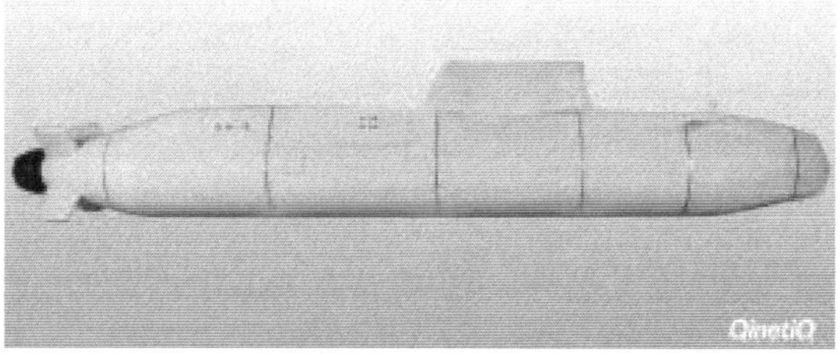

Figure 7: Profile view of SRM clad as an X-plane

The experiments fall into two distinct parts: manoeuvres in the Ocean Basin and manoeuvres at the deep-water reservoir. In general, the Ocean Basin manoeuvres were limited to the slow and moderate speed runs and some limited depth changes to investigate turning circles, zig-zags and autopilot performance. The bulk of the programme was conducted at the reservoir where the available space allowed for the higher speed manoeuvres to be conducted as well as those which required larger depth changes, such as jam scenarios.

On directional instability, during a turn at 8 knots where a pull-out manoeuvre was initiated, the yaw rate actually persisted beyond setting the rudder amidships. The yaw rate does decay, but settles on a non-zero value, i.e. the submarine keeps turning. Further experiments demonstrated that whilst in the vertical plane there was some transient instability, the design can be controlled by the planes without incurring excessive plane activity. Figure 8 shows an example of the data from the free running model tests; Figure 8(a) shows how the tactical diameter varies with rudder angle for a range of speeds tested. These turning circles were conducted using all four control surfaces as rudders, with additive depth control applied as required. Because of the requirement for depth control, some of the 30° rudder turns did not quite achieve this angle. The data show a consistent variation in the tactical diameter with rudder angle and are largely independent of speed. There is a distinct "flattening off" at the higher rudder angles with no improvement in diameter at 30° rudder over 25° rudder, possibly as a result of the planes stalling.

A series of tests included the application of a single hydroplane jam following a steady period of straight and level running. All other control surfaces remained under autopilot authority, and the initial response was to "do nothing", i.e. allow the autopilot to simply carry on with the current ordered depth and heading. An example of an 8 knot jam to rise is shown in Figure 8(b). Generally, for modest jam angles of say 10° and 20°, any pitch, depth and yaw excursions were minimal. However, for the higher 30° jam to rise on the upper port plane, the excursions were more considerable, and yaw became uncontrolled. The initial response of the course keeping and depth keeping controllers is to control yaw and pitch equally; however, both failed in this scenario. When the heading error reached

10°, the autopilot switched to the course-changing controller which had no integral action. As such the demands of the heading control then become swamped by the depth control so the submarine is no longer controlled in yaw. As a result the control of depth and pitch are regained at the expense of increasing the yaw rate.

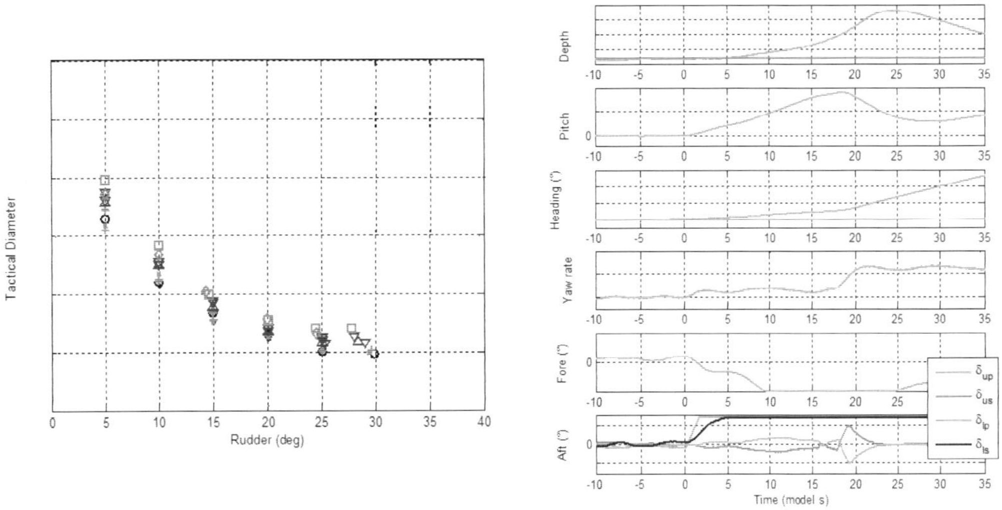

(a) Tactical diameter *(b) Single stern plane jam*

Figure 8: Free running model data

In general, single hydroplane jams at moderate angles were comfortably dealt with by the autopilot without any further action, with depth and heading successfully maintained. However, single hydroplane jams at the maximum deflection were not successfully controlled by the "do nothing" strategy. An alternative strategy of slowing down and allowing heading changes meant that depth and pitch were better maintained.

9. SUBMARINES NEAR THE FREE SURFACE

There are times when a submarine is required to operate close to the sea surface when using the periscope or, in the case of a diesel-electric boat, when snorting during the period of recharging the batteries, the submarine can experience complex oscillatory motion [Musker *et al*, 1988] that can present problems in that the submarine may broach the surface, increasing the possibility of detection, or in the event of a temporary loss of the suction force the submarine may sink below depths that periscope and snorting operations can be undertaken. This issue creates a design challenge associated with the development of submarine autopilots controlling depth, (Burcher and Rydill, 1995), which manifests itself as two technical problems – one of size and actuation of lifting surfaces such as hydroplanes and the other of capacity of any onboard trim and ballast system. The main concern to the designer is how to accommodate the effects of the second order suction force, which could mean that if the planes and trim systems are poorly designed they could become saturated in certain wave conditions.

The non-linear behaviour of the submarine means that it is susceptible to the effects of wave grouping. Groups of large waves can be passing over the submarine for a period of time, whereby the suction force is temporarily increased, which must be counteracted in order to prevent the submarine from broaching (Burcher and Rydill, 1995). One method of quantifying wave groups is by the Smoothed Instantaneous Wave Energy (SIWEH) function, introduced by (Funke and Mansard,1980), which is computed by squaring the wave surface elevation and applying a low pass filter (based on the spectrum peak period).

The logic behind this approach is to enable a method of predicting the dynamic response of a floating structure which responds only to low frequency drift excitation; this is not dissimilar to the phenomenon creating the second order effects on a submarine at periscope depth. For a surface elevation η with spectral peak period of T_p the SIWEH is:

$$E(t) = \frac{1}{T_p} \int_{-\infty}^{\infty} \eta^2 (t+\tau).Q(\tau) d\tau$$

where Q(τ) is a smoothing function given as

$$Q(\tau) = \begin{cases} 1 - \dfrac{\tau}{T_p} & for \; -T_p \le \tau \le T_p \\ 0 & elsewhere \end{cases}$$

Figure 9 gives an example of the spectral responses in a scaled sea state 3 (with an ITTC spectrum) in head seas with the speed of the model equivalent to around 8 knots. There are clear first order responses at frequencies contained in the wave spectrum; there are also higher order responses at frequencies much lower than any contained in the wave spectrum.

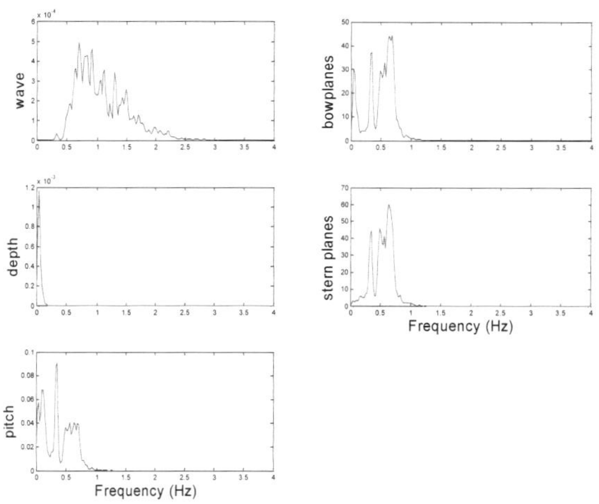

Figure 9: Response spectra in a sea state 3

The first order depth response (heave) has been practically removed, which is as a result of the autopilot in the model responding to depth error; the autopilot is aimed at maintaining depth and thus induces small angles of pitch in the model to achieve this. Again, as a result of the autopilot, the planes are responding to both the first and higher order responses; with the bow planes (or fore planes) correcting for most of the low frequency components.

Figure 10 shows the spectral analysis of the SIWEH function (of the wave) compared with the pitch and depth response, for the same run conditions illustrated in Figure 9. The results show that the frequency content of the SIWEH function is very similar to that of the depth and pitch response, in contrast to the encounter wave spectrum. The spectral ordinates have been normalised by the maximum spectral ordinate for each response to enable the plots to be compared more easily. For the pitch response, there are frequency components that are contained in the wave spectrum but clearly lower frequency components that are not present in the waves. There are frequencies that coincide with the frequency content of the SIWEH function suggesting that pitch response is related to the group structure; there are however, intermediate frequencies where there is no clear forcing function. For the depth response, there is very little response at the wave frequencies (likely due to the effect of the autopilot) but there is a significant response at frequencies close to those contained in the SIWEH function, again suggesting the possible relationship between depth response and group structure.

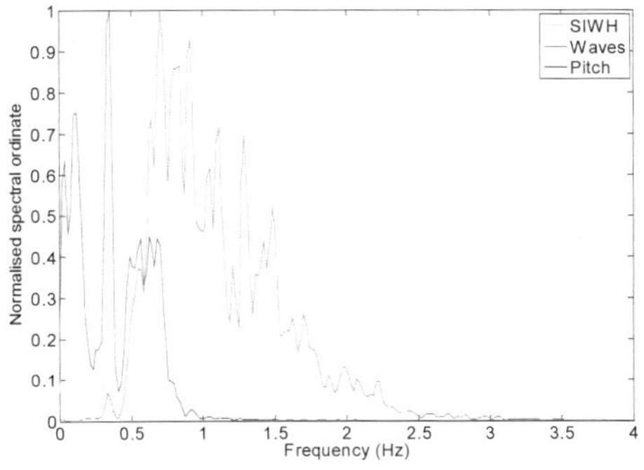

(a) Pitch

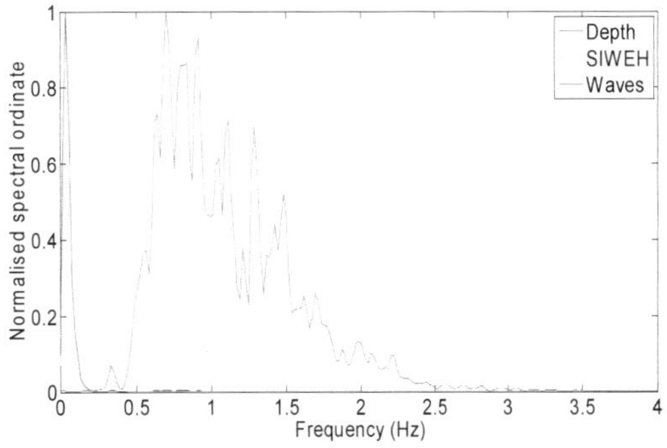

(b) Depth

Figure 10: Normalised response spectra for pitch and depth

10. CONCLUSIONS

This paper has described the principal elements of a four year research programme undertaken by QinetiQ Ltd to develop the numerical and experimental capability to assess the performance of an X-plane submarine design. The paper has brought together the development of a design toolset and an extensive experimental programme of captive and free running model tests to provide the capability to evaluate novel stern plane configurations and provide safe operator guidance.

11. ACKNOWLEDGEMENTS

The authors acknowledge the UK MoD for supporting this research programme and acknowledge the efforts of the QinetiQ Hydrodynamics Team for their support in undertaking the work described here.

12. REFERENCES

25th International Towing Tank Conference (ITTC), The Manoeuvring Committee report. Proceedings, Vol I, Fukuoka, Japan, 2008.

Burcher, RK., Rydill, L. Concepts in submarine design, Cambridge University Press, 1994. ISBN 0 521 41681 7.

Funke ER., Mansard EPD., On the synthesis of realistic sea states. Proceedings of the 17th International conference on Coastal Engineering, 1980, pp167-168.

Gertler, M., Hagen, GR., 'Standard equations of motion for submarine simulation' Naval Ship Research and Development Centre, Report No. 2510, USA, 1967.

Lewis, EV., Principles of Naval Architecture, Vol III, Motions in waves and controllability, SNAME, 1989.

Musker, AJ., Loader, PR., Butcher, MC., Simulation of a submarine under waves. International Shipbuilding Progress, 35, No. 404, 1998, pp 389-410.

Spencer, JB., Stability and control of submarines – Parts I-IV, Journal of the Royal Navy Scientific Service. Vol 23, No. 3, May 1968.

Watt, G.D., 'Estimates for the Added Mass of a Multi-Component, Deeply Submerged Vehicle', RINA WARSHIP Conventional Naval Submarines, UK, 1988.

Damage Control Tools and Technologies for Naval Platforms

Jeffery Robinson, Grant I. Gamble, Brigitta Suendermann, Michelle R. Grech and Warren Roberts

Defence Science and Technology Organisation, Maritime Platforms Division, Melbourne, Victoria, Australia
Grant.Gamble@dsto.defence.gov.au
Jeffrey.Robinson@dsto.defence.gov.au
Brigitta.Suendermann@dsto.defence.gov.au
Michelle.Grech@dsto.defence.gov.au
Warren.Roberts@dsto.defence.gov.au

ABSTRACT
Damage control involves identifying, managing and suppressing damage sustained by a naval ship. Current damage control activities involve complex, time critical and resource intensive tasks. Such labour intensive operations become a key concern for future ship acquisition programs particularly since maritime platforms are now being designed to operate with fewer crew. Hence, the drive is to develop more effective damage control tools and technologies, not only in terms of improving the task at hand, but also offer the added advantage of improving the overall effectiveness of the response and repair phase to the platform system as a whole.

In this regard, DSTO is conducting a research and development program looking at improving current damage control tools and technologies on naval ships. Research has incorporated the evaluation of commercial off-the-shelf and in-house damage control tools that can be effectively applied for the containment of damage (e.g. quick acting pipe repair clamp). A critical aspect that is also being considered in the research and development of such tools is human systems integration. Human systems integration analysis ensures that the full potential of any new technology in the area can be successfully integrated for optimum crew safety and capability of the platform.

This paper will discuss why the need for research and development in this area is critical followed by a description of damage control tools and technologies being developed within DSTO including the importance of human systems integration input in the development of such tools. This paper will conclude by providing an indication of research and development being pursued by DSTO in this area.

INTRODUCTION
The survivability of a naval platform can be defined as the ability to complete a mission successfully in the face of a hostile environment with minimal losses [1]. Survivability of a platform comprises susceptibility, vulnerability, and recoverability, which are defined as:

- Susceptibility: the likelihood of the platform being hit by a threat weapon;
- Vulnerability: the likelihood of the platform being disabled or damaged after being hit;
- Recoverability: the likelihood of recovering the capability after having been hit and encompasses the damage control and repair/restoration of systems on the platform.

Figure 1 provides a graphical representation of the effect of a damage event on the mission capability of a platform. Recoverability aims to return the platforms' lost capability (through damage) either fully or partially, to enable the mission to continue. If the mission cannot be continued, then recoverability will improve the likelihood of safe retreat from the hostile environment. Recoverability is also applicable to peace time damage incidents (as a result of grounding, collision, fire, etc).

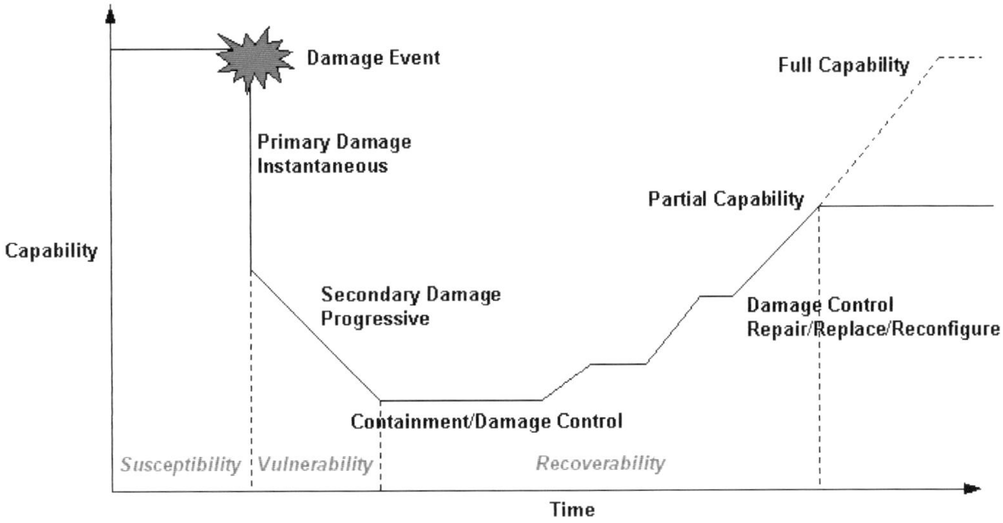

Figure 1: Temporal representation of the damage process from being detected (through the damage event) until final capability recovery [1]

Recoverability involves a number of activities. The initial phase of recoverability is damage control (DC), carried out by a distributed team of trained personnel with a range of tools who act to limit and repair damage caused by, for example, fire and smoke, and flooding.

Modern warships use a combination of automated and manual techniques and equipment for damage control processes. Automated systems can play an important role and, at least partially, influence crewing numbers on platforms. These include systems to detect damage locations and prevent spread (for example, sensors and fire suppression systems). To date however, the majority of damage control activities are labour intensive. If an automatic system fails due to the damage incident, then traditional manual techniques are applied.

An important aspect to consider in the development of DC tools and technologies is human systems integration, which has not always been effectively considered in ship design. Damage events often put the personnel operating these systems into very demanding, confusing, and stressful situations which can push human capabilities to their limits. An example of human systems integration shortfalls involved tests to firemain systems by the U.S. Damage Control Automation for Reduced Manning (DC-ARM) and Integrated Fleet Survivability Programs, on board the Ex-USS Shadwell [2]. The ex-USS Shadwell is a full scale DC research and development experimental platform owned by the US Navy for conducting ship survivability investigations and analyses such as testing of fire fighting agents, DC systems and technologies, and predictive models. Under the 1998 doctrines and practices, it took over 20 minutes for DC teams to control damage to firemain systems. The time line was attributed to the lack of information to quickly identify and locate the rupture; effort and time that could have been deployed more effectively elsewhere. A subsequent DC-ARM program [2], again based on the firemain system, focused on the evaluation of supervisory control systems (SCS) and enabled greater situation awareness for the damage controller to direct human actions, control ship systems remotely and to integrate ship functions with human requirements and roles. The outcome was that effective damage control was achieved in a shorter time with 60% fewer personnel than in the previous evaluation.

The drive is therefore to develop more effective damage control tools and technologies to improve repair processes and situation awareness. In this regard, DSTO is conducting a research and development program aimed at improving current damage control tools and technologies on naval ships, and incorporating human systems integration principles. The research ranges from basic high labour intensive technologies through to automation. Some of these technologies will be discussed below.

DC TOOLS AND TECHNOLOGIES

This paper will focus on pipe damage recovery including pipe patching techniques from smaller leaks to totally breached pipes and automated isolation of ruptures in fluid piping systems using smart valves. We will also discuss the DSTO involvement with fire suppression technologies.

Pipe Patching Tools

Piping systems are critical for the operation of naval platforms. This includes the supply of sea water, fresh water and chilled water through piping systems for cooling machinery, surveillance and weapon systems, and for fire fighting purposes. Damage to these piping systems from weapon fragments can cause a reduction in pressure, and subsequent leakage of water which can cause damage to platform systems and flooding. The current method used by the Royal Australian Navy (RAN) and other navies around the world to train crews in preventing leakages or major flooding involves the use of banding clamp tools or similar devices for temporary repair of damaged or split piping.

The US Naval Research Laboratory (NRL) has conducted research and reported the results of several sets of experiments involving the implementation of naval damage control equipment [3]. Of interest to DSTO was the application of temporary pipe patching repair techniques employed on pressurised firemain pipes (that is, where water flow in the pipe could not be stopped during the repair) with diameters ranging from 10 to 15 cm. The experiments undertaken by the NRL involved the use of the Jubilee clamp and the pipe wrench with banding kit.

Results from NRL pipe patching experiments are shown in figure 2 as a plot of leak rate reduction versus implementation time for the Jubilee and the banding kit pipe patching clamps. Both of these pipe patching technologies exhibited variability and limitations in their effectiveness, implementation time and leakage rate, following attachment as is evident in figure 2. The results indicate that the Jubilee clamp performed better than the banding clamp in reducing the leak rate and had a shorter implementation time. Ideally, results should be in the upper left of the graph (large leak reduction and quicker implementation times).

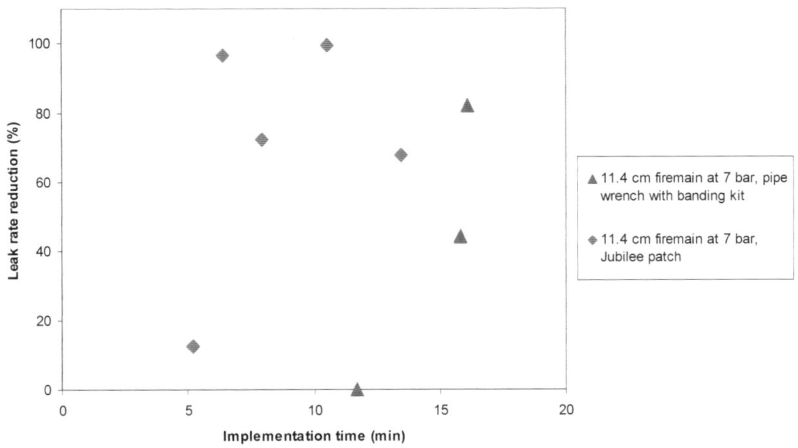

Figure 2: Results for pipe patching experiments by NRL comparing the Jubilee clamp with the banding kit [3]

Pipe patching tools developed by DSTO

A novel pipe patching clamp, known as the 'lever clamp' has been developed by DSTO. The initial aim was to develop a pipe patching clamp which would perform better than current pipe patching tools and meet the following criteria:

- Requires a minimal number of separate parts to implement;
- Does not require additional tools to attach;
- Reduces implementation time when compared to existing clamping techniques and;
- Maintains or improves effectiveness when compared to existing clamping techniques.

The resultant lever clamp, shown in figure 3, has a shaped spring steel plate which covers half to three quarters of a 10-15 cm diameter pipe. The inner surface has a rubber sheet which acts as a gasket over the rupture. On the open side of the steel, nylon strapping with a buckle is looped across a hook for loose attachment to the pipe. The lever can be adjusted, by wing nuts, and used to tighten the clamp firmly to reduce leakage. The lever clamp can be produced in varying lengths, with the option to increase or decrease the number of levers to suit the length. Single lever versions can be used with flexible sealing material such as rubber which can wrap around damaged elbow pipe sections.

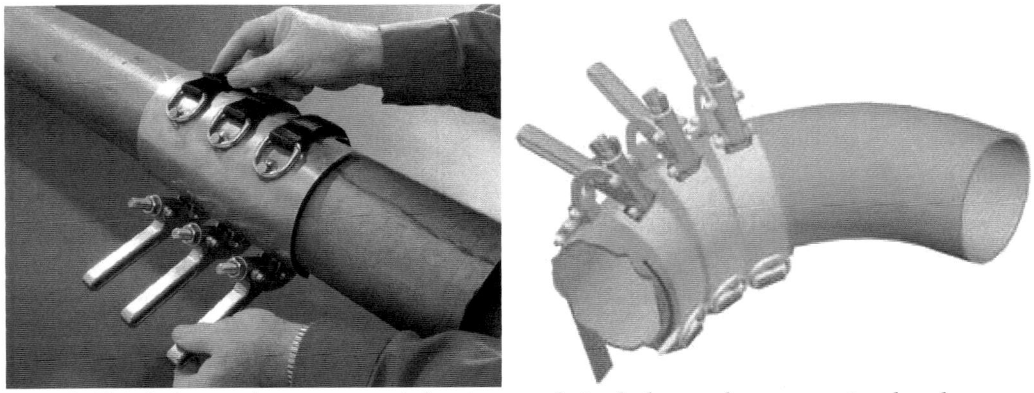

Figure 3: Triple lever clamp on straight pipe, and single lever clamps on pipe bend

Preliminary trials carried out at the RAN School of Survivability and Ship Safety at HMAS Cerberus in Victoria in 2008 showed the 'lever clamp' to be 80% quicker to fit than the banding clamp. Another trial, on an enhanced clamp, was carried out in 2010 with the evaluation incorporating some human factors aspects. This trial aimed to evaluate the modified design against the banding clamp currently in use by the RAN. The three clamping methods evaluated included:

- Triple lever clamp,
- Banding clamp and accessories, which is the current method used by the RAN, and
- 3-single lever clamps – 3 separate single lever clamps employed, over a piece of rubber and a copper sheet.

In this case, the application of the banding clamp included a piece of rubber, overlaid by copper sheet loosely attached using cable ties to the pipe next to the rupture. The patch and cable ties were slid over the leak and the metal bands applied to seal/reduce the leak.

Participants in the evaluation study included four RAN personnel at the Able Seaman rank with damage control experience. They were instructed in the use of the lever clamp and given practical experience prior to testing. Measures of effectiveness employed in the study included:

- Time taken to fit clamps,
- Ease of use through subjective evaluation by participants using post trial questionnaires, and
- Clamp effectiveness based on water leakage following clamp implementation, evaluated by subject matter experts and rated as high, medium or low.

As shown in figure 4, each of the three clamping methods were tested on pipes of 10 cm diameter, positioned at waist height (approximately 110 cm) level and overhead (approximately 190 cm). The waist height pipe had three cuts of 10 cm by 0.5 cm and was pressurised by a pump with a rating of 4.8 bar. The overhead pipe contained a 20 cm by 0.5 cm cut along the pipe and was pressurised by a pump with a rating of 5.5 bar. The participants were requested to repeat each clamping a minimum of two times on each pipe, to ensure repeatability and reliability of data. For each test, two personnel were used to fit the banding clamp and single lever clamps. Only one person was used to fit the triple lever clamp. Each person rotated between clamp methods following each test. For the banding clamps, three bands were applied for the waist height pipe, and four bands for the overhead pipe.

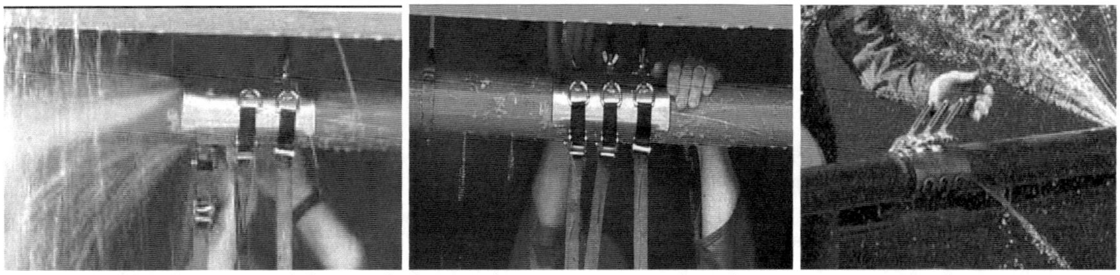

Figure 4: Application of the triple lever clamp to an overhead pipe rupture, and a waist height pipe rupture.

Figure 5 shows the time and subjective effectiveness (at stopping leaks) for the waist height (110 cm) and overhead pipe (190 cm). The findings indicate that the triple-lever clamp performed better than the banding clamp and the 3-single lever clamp for its implementation time and effectiveness at stopping leaks for both the 110 cm and 190 cm pipes tested. The 3-single lever clamp appeared to improve the implementation time over the banding clamp particularly at the pipe height of 190 cm but not to the same extent as the triple-lever clamp. Furthermore, although the 3-single level clamp had faster fitting times than the banding clamp at the 190 cm pipe height, it still generally had only medium subjective effectiveness at leak stopping. Variation in the results for the 3-single lever clamp, particularly at the 110 cm height, was attributed to different material properties in the rubber pieces over which the clamps were applied (as indicated by arrow). Further testing is required to optimise this combination.

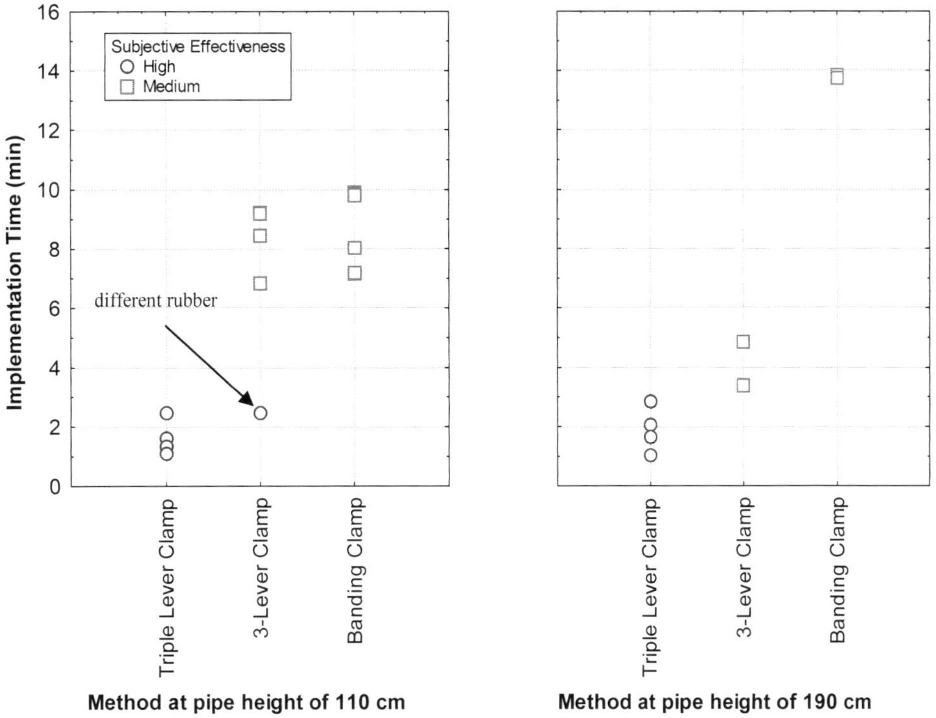

Figure 5: Time and subjective effectiveness (at stopping leakage) results

The results of the study clearly show that the triple lever clamp performed better than the banding clamp and the 3-single lever clamp in terms of reducing the subjective leak rate evaluation and implementation time.

During the trial the four subjects were requested to provide a subjective rating on ‘ease of use’ for each clamp, at each height. The findings indicate that subjects rated ‘ease of use’ higher (i.e. better) for the triple-lever clamp (median = 9.5) and the 3-single lever clamp (median = 8.5) when compared to the banding clamp (median = 5). As expected these trends replicate the observed implementation times discussed above.

The development of the lever clamp and evaluation of clamping methods provided evidence that there is potential to improve the efficiency of damage control tools which then may contribute to optimising the recoverability processes.

The Pipe Jumper

In the event that firemain piping is severely damaged by, for example blast, water supply would be turned off at the nearest valves to avoid compartment flooding. Most navies, including the RAN no longer stow sections of pipe onboard to repair such failures, and rely on redundancy. However, redundancy is unlikely to be available for all piping including sprinkler system supply lines for example. In this regard DSTO have been prototyping a pipe jumper device to re-supply water across a breach and ensure that critical systems are able to continue providing a level of functionality during the recovery process. Figure 6 shows an example of a pressurised pipe jumper undergoing testing.

To use a pipe jumper to breach a damaged pipe, the section of damaged pipe is cut out, leaving two near square cut ends. The pipe jumper has internal seals that fit firmly over each cut end of the firemain (rolled steel pipe) and it is then locked into position by its four tapered locking bolts. A temporary hose section is created from commercial off-the-shelf parts consisting of a piece of fire hose cut to the required length with each end fitted over a hose barb and secured by bolt clamps, as shown in figure 6. The temporary hose section then clips onto the pipe jumpers, and the water supply to the firemain loop can be re-established.

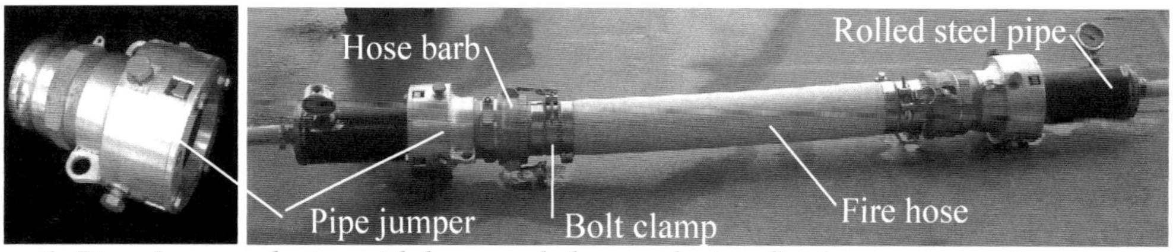

Figure 6: Pipe Jumper device with hose and clamps during fluid pressure testing

The pipe jumper connection devices were individually tested under a static pressure of 10 bars over six hours. A small leak was observed on one device over that time. A full flow test was then conducted using a flow circuit consisting of two jumper devices and fire hose configured as in figure 6, a water tank, pump, restriction gate and inline pressure gauge. The gate was closed to produce 10 bar within the device. Again, a slight seeping occurred around the seals, which is acceptable in a damage control situation.

This device was only tested on four inch black rolled steel pipe, with a wall thickness of 1/4". Further testing of this device should be conducted with other pipe dimensions and materials. It is anticipated that future versions will include the hose barb as part of the design with the outer body made of polypropylene to make it lighter.

The pipe jumper has the potential to restore liquid flow to damaged piping systems to restore their function and therefore improve overall ship capability.

Valve Systems

Piping systems on modern naval platforms, such as firemains, are designed with redundancy in the event of damage. If damage to a piping system occurs, redundancy is achieved using loop systems with cross connections and valves to isolate damaged sections, which are often remotely controlled. A disadvantage associated with this current set up is that systems that enable the control of these valves can be vulnerable to the incident which damaged the piping system in the first place. A solution to this is to use automatic valves, with local control and power supply systems. These 'smart' valves are able to close automatically to isolate damaged sections of piping to maintain flow and pressure, and limit flooding from damaged piping.

Development and evaluation of smart valves have been undertaken by the NRL [4, 5, 6], including an assessment of different methods for detecting and isolating piping system ruptures, which has led to proposals for fitting smart valves to new and existing ships [7, 8]. Hydraulic Resistance Logic appears to be a reliable method for controlling smart valves; it uses valves which operate independently based on pressure, flow rate, flow direction, and distance from a supply pump. The flow direction and rate are determined from the differential between pressure sensors on either side of each valve. Essentially, if high flow and low pressure are detected a predetermined delay begins. If the conditions remain at the end of the delay, then the valve will close. The delay allows valves furthest from the system supply to close first, minimising the number of isolated sections. Different delays may be set for each flow direction at each valve.

A simple example is provided to demonstrate smart valve behaviour, figure 7. It is a looped piping system model built in the Pipe Flow Expert fluid flow and pressure loss calculations software [9]. A rupture is enabled by closing the valve between the two outlets, and opening the outlets to atmosphere. The Pipe Flow Expert software was used to calculate the pressure and flow in the system, to which the smart valve logic was applied. For each rupture, the smart valve logic closed the closest two valves, correctly isolating the ruptures. For example, for pipe breach 1 smart valves 1 and 2 closed to isolate the rupture.

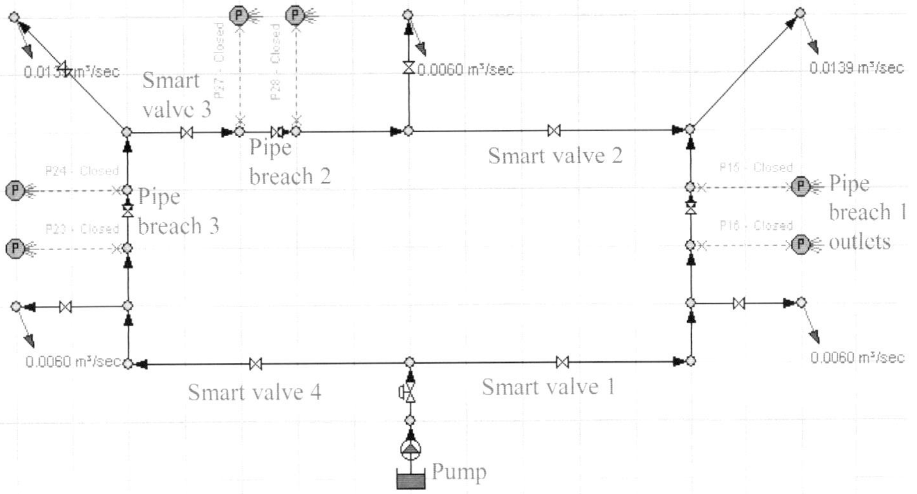

Figure 7: Piping system simulation example with three ruptures and four smart valves

A bench top piping system is being constructed by DSTO to further investigate smart valve systems and practical issues surrounding their use in the RAN. It will make use of 2.5 cm piping and commercially available control valves and sensors, combined with computer based logic control.

Further work will involve the analysis of the restoration of piping system capability using technologies such as the pipe clamp and pipe jumper described previously. This will be based on damage control performance analysis including the modelling of personnel movement for the collection of tools and transit to damage location, and repair implementation.

Fire Fighting and Suppressant Systems

Over the last decade much research has gone into investigating fire fighting and control systems [10]. This is mainly due to the adverse effects associated with the use of halon and carbon dioxide based fire suppressant agents. These dangerous substances are mainly used in situations where flammable spray and liquid fires are most likely to occur (e.g. engine rooms and machinery spaces). Research has led to development and implementation of alternative systems which are as effective, more environmentally friendly and safer to use. One of the most researched is the water mist system. This type of technology has been actively pursued since the 1970's with significant progress being made over the past decade [10].

DSTO are currently evaluating the use of water mist technologies on RAN ships. A series of tests with water mist in ISO 9705 rooms [11] with combustible boundaries has been undertaken, with both wooden cribs and diesel (figure 8) as fire fuel [12]. These tests indicated that water mist is capable of rapidly cooling compartment atmospheres and suppressing fires, and while not always extinguishing the fires, the water mist systems reduced the fire sizes so that they would be manageable by manual techniques.

Figure 8: Flashover and water mist suppression of ISO 9705 compartment fire, and rapid cooling of an ISO 9705 compartment following water mist aplication.

A graph of the rapid temperature drop that water mist can achieve is shown in figure 9 [12]. Ten temperature profiles are shown, each at different heights, with the upper temperatures reaching 600°C prior to water mist activation, and those temperatures dropping sharply following activation.

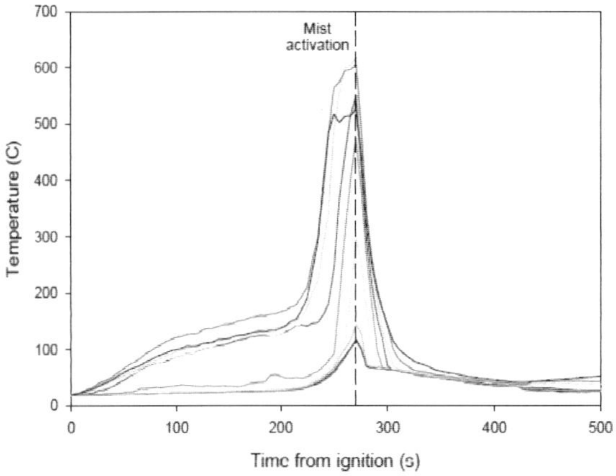

Figure 9: Rapid cooling of an ISO 9705 compartment following water mist application [12]

It is critical to note that any advances in water mist or other fire suppressant technologies will need to be incorporated within the current DC doctrine. This will require changes in DC training, manuals and procedures and the upgrading of current training facilities to reflect the use of water mist or other technologies and automated remote control systems (e.g. automated boundary cooling in the secondary zone, the integration of sensor data, etc). The damage control actions required for water mist use, including personnel movement into the compartment and extinguishment of residual fire, will be analysed and modelled using techniques currently being developed to optimise procedures [13].

DISCUSSION AND FUTURE WORK PROGRAM

This paper has described some of the work that DSTO has conducted for improving DC repair tools and describes some of the research conducted on the evaluation of commercial off-the-shelf (COTS) and in-house developed DC tools and technologies. From the evaluations contained within this report these methods were effectively applied and may result in better containment of damage and fire than conventional techniques.

DSTO aims to gain a thorough understanding of DC tools and technologies and the knowledge developed will enable advice on these systems to be provided to the RAN for procurement and fitment to an existing platform or as part of a new platform design. The implications of different tools and technologies on damage control processes will also be examined. The main issue is to ensure that all these systems offer a whole of platform improvement in the recoverability process. Currently, evaluations of each system have been carried out separately. However, DSTO will be focussing on improving the evaluation process by modelling the damage control activities and processes, including personnel movement and damage control tool implementation. The aim is to develop a capability that will allow an analysis of how new DC technologies will improve the recoverability process for the whole of the ship.

REFERENCES

1. Deyong, L., (2008), *Integrated Ship Survivability for the Royal Australian Navy*, Proceedings Pacific 2008 International Maritime Conference, Sydney, Australia.
2. Williams, F. W., Farley, J. P., Tatem, P. A., Durkin, A., Nguyen, X., Luers, A. C., et al. (2003), *DC-ARM final demonstration report,* NRL/FR/6180--03-10,056, Naval Research Laboratory, USA.
3. Williams, F.W., Hill, S.A., Campbell, C. and Kuzma, H., (1997), *Results of Damage Control Tactics and Equipment Tests - Phase III: Cold Water Tests*, NRL/MR/6180-97-7919, Naval Research Laboratory, USA.
4. Lestina T., Runnerstrom E., Davis K., Durkin A., and Williams F.W., (1999), *Evaluation of Firemain Architectures and Supporting Reflexive Technology*, NRL/MR/6180-99-8346, Naval Research Laboratory, USA.
5. Durkin A., Williams F.W., Lestina T., Bradley M., Downs R., and Runnerstrom E., (2000), *Evaluation of Reflexive Valve Logic for a Shipboard Firemain,* NRL/MR/6180--00-8425, Naval Research Laboratory, USA.
6. Lestina T., Bradley M., Downs R., Runnerstrom E., Durkin A., Williams F.W., and Farley J., (2001), *Evaluation of DC-ARM Reflexive Smart Valve,* NRL/MR/6180--01-8552, Naval Research Laboratory, USA.
7. O'Rourke, R., (2009), *Navy Aegis Cruiser and Destroyer Modernization: Background and Issues for Congress*, Congressional Research Service, Library of Congress, Washington, DC, USA.
8. Naval Surface Warfare Center - Carderock Division, (2008), SeaFrame, Volume 4, Issue 1, West Bethesda, MD, USA.
9. Daxesoft Ltd., (2010), *Pipe Flow Expert User Guide Version 5.12*, PipeFlow.co.uk, United Kingdom.
10. Darwin, R. L. and Williams, F. W. (2000) *The development of water mist fire protection systems for U.S. navy ships,* Journal of American Society of Naval Engineers 112 (49).
11. International Organisation for Standardisation, (1993), *Fire Tests - Full-scale Room Test for Surface Products*, ISO 9705:1993(E).
12. Griffin, G.J., Xu, Q., Bradbury, G.P., Bicknell, A.D., Preston, C.M.L., White, N., Apte, V., Burch I. and Mathys, Z., (2008), *Investigation of the Fire Behaviour and Mist Suppression of an ISO 9705 Room Lined with Glass Reinforced Polymer (GRP) Panels,* CSIRO Technical Report CMSE-2008-036.
13. Grech, M.R., Drohan, D., Roberts, W. and Gamble, G.I., (2012), *Ship Recoverability: How Important is the Human Factors Input?*, Proceedings Pacific 2012 International Maritime Conference, Sydney, Australia.

Advanced Nondestructive Inspection for Submarine Structural Integrity Management

Dr Cayt Harding and Dr Geoff Hugo
Defence Science and Technology Organisation
cayt.harding@dsto.defence.gov.au

Abstract

Nondestructive inspections (NDI) are relied upon to assure submarine structural integrity and to enable effective and efficient through-life maintenance, thereby underpinning the safety, availability and capability of the platform. Advances in NDI technologies provide opportunities for substantial enhancements to maintenance practices, such as the use of automated or remote scanning to inspect otherwise inaccessible structure and to reduce damage inflicted through disassembly for maintenance. This paper explores the role of advanced NDI in submarine structural integrity management as well as in modern maintenance approaches such as condition-based and reliability-centred maintenance.

1. INTRODUCTION

Modern diesel submarines are among the most complex of military systems, incorporating sophisticated control and combat systems at the cutting edge of technology. However, by contrast, the structural components of the submarine are often designed and managed based on relatively simple structural analyses and maintenance strategies that make little use of the advanced analytical tools currently employed for other advanced engineering systems, such as aircraft and nuclear power generation. The design and maintenance approaches applied to submarines have their heritage in the ship-building industry and, whilst practices have evolved substantially through the SUBSAFE program for management of safety-critical structure, many of the current practices for maintenance and structural integrity management fall short of best practice as applied in the aerospace industry.

In a climate of diminishing defence budgets, but with escalating community expectations for best engineering practice in due diligence and risk management, there is substantial motivation to make use of better tools to nondestructively characterise the structural integrity of submarine structures. However, improved nondestructive inspections are only of benefit if they interface to a structural integrity management framework that can effectively use the additional information to better manage the structure. This paper examines modern nondestructive inspection methods and discusses how these might be used to provide more cost-effective and reliable assurance of submarine safety, availability and capability.

2. NONDESTRUCTIVE INSPECTION OF SUBMARINES

Nondestructive inspections (NDI) are used extensively during both the fabrication and sustainment of submarines to detect structural and material defects and to verify the health and integrity of structural materials. Different NDI techniques are used depending on the material, the component geometry to be examined, and the sizes and types of defects that must be detected. Common NDI techniques include: eddy current, ultrasound, radiography, magnetic testing methods, fluorescent penetrant and detailed visual inspections, including the use of boroscopes to view inside pipes, pressure vessels and other hidden structures.

Inspections performed during fabrication of submarines and submarine systems will include many NDI techniques used routinely in high-end manufacturing, ranging from ultrasonic testing of the steel plate during production to radiographic examination of electrical components to confirm correct assembly. Of particular interest to this discussion are the inspections that underpin the structural integrity of the submarine, the most prominent of which are the inspection of welds in the hull structure. Safety-critical welds are inspected during fabrication using ultrasonics, penetrant, magnetic particle, radiography and visual inspections to assure the weld quality [1]. These weld inspections serve a dual purpose: firstly, a quality assurance function to provide confidence that the welding process is operating consistently within specification and, secondly, to detect significant defects which must be repaired such as cracking, lack of fusion, inclusions or porosity [2].

Throughout their service life, submarine structures degrade through processes of corrosion, fatigue, stress corrosion cracking, breakdown of protective coatings, impact* and other processes of material degradation†. The detection, assessment and treatment of corrosion consume considerable resources in the sustainment of submarines. Critical structures that may experience corrosion include seawater tanks, heat exchanger tubing and other piping. Access to these structures to perform inspections will generally be very difficult. Detection of fatigue cracking and stress corrosion cracking comprise the most critical through-life inspections for ensuring the safety of a submarine. Some of the most challenging submarine NDI problems involve inspection of hidden structure, either through an energy-absorbing coating such as anechoic tiles, or within multi-layer structures or enclosed spaces. Frequently, the structural disassembly required to gain access for inspection causes additional damage and exacerbates general issues with reliability of the platform.

3. ADVANCED NONDESTRUCTIVE INSPECTION TECHNOLOGY

The demarcation between "traditional" and "advanced" NDI is imprecise. The most useful distinction is to draw the line between those traditional methods that are well-established in manufacturing and maintenance, and those newer more advanced methods that are not yet widely used but have potential to significantly enhance maintenance practices. In some cases, advanced NDI techniques share the same underlying physics as the traditional methods but offer improvements in automation of data acquisition and analysis, enabled by increasing computing power.

3.1 Automated Scanning NDI

Traditional NDI methods such as ultrasonic or eddy current inspection involve a skilled operator either taking point-by-point measurements or manually scanning over an inspection area using a single hand-held probe. Reliable interpretation of the response from the inspection instrument requires knowledge of the local structure and a good understanding by the operator of the interaction between the interrogating signal (ultrasound, x-ray, eddy current etc) with potential defect types. The NDI technician must mentally correlate the position of the probe on the structure with the instrument response to interpret whether the response signal is likely to be due to a defect.

In contrast, newer automated scanning inspection methods use a mechanical scanner to raster scan the probe over the inspection area to build up a three-dimensional data set, comprising the NDI response measured at an array of points throughout the inspection area. Analysis software can post-process these data to generate a variety of images such as cross-sections

* Impacts may include large scale impacts with the hull, such as impact with test torpedoes, or smaller-scale impacts on internal structure due to dropped tools or other minor impacts.

† Other processes of material degradation may include time- and/or temperature-induced microstructural changes in metallic materials which can degrade their properties, degradation of composite material through matrix cracking or fibre damage, and ageing of polymeric materials such as rubbers or plastics.

through the structure showing defects or the extent of damage. Interpretation of these images is more intuitive, accurate and reliable than manual inspection. In a fully automated analysis, the data are automatically processed against predefined criteria to determine whether or not defects are present. However, fully automated analysis is generally only feasible for high-throughput inspections of relatively simple geometries. Whilst mechanised NDI scanning has been available for multiple decades, advances in computer power and analysis algorithms have enabled substantial improvements in both the quality of data which can be collected and the sophistication of the analysis and interpretation tools for identification, location and sizing of defects.

3.2 New Inspection Methods

This paper examines how advanced NDI could be used for cost effective assurance of submarine safety, availability and capability. Consequently, this discussion is limited to mature technologies that already have demonstrated acceptance in other industries – specifically phased array ultrasonics, digital and computed radiography and thermography.

Analogous to phased-array radar and sonar systems, phased array ultrasonic testing (PAUT) utilises an array of ultrasonic transducer elements, each of which can be pulsed with different relative delays in transmitting and receiving the ultrasonic beam. The superposition of the individual beams generates an ultrasonic beam that can be electronically manipulated to vary the focus and angle of the beam, and to rapidly scan the generating and receiving apertures over a 1- or 2-D array. This manipulation may allow inspection of the region of interest without physically moving the array probe. PAUT instruments are now commercial off-the-shelf (COTS) equipment and PAUT is gradually becoming a mainstream NDI technique. Its deployment is currently limited only by the cost of the equipment, training of personnel, and the need to incorporate references to PAUT into relevant structural design codes.[‡]

There are several new COTS radiographic systems available which provide substantial enhancements compared to traditional wet-chemical film-based industrial radiography. These include:

- Film digitisation for improved data storage and image manipulation of traditional film-based inspection.
- Computed radiography using flexible, reusable phosphor plates instead of traditional film, which require shorter exposures and have wider dynamic range. The images on the phosphor are read into a computer using a laser. Subsequent digital processing of the radiographic images can optimise defect detection.
- Digital radiography using solid state amorphous silicon detector panels, which can be used to provide near real-time images.
- Computed x-ray tomography, which can reconstruct the three-dimensional structure, enabling cross-sectional images to be produced without interference from overlying and underlying areas of the structure.

All of these variations lower consumable costs and reduce the occupational health and safety compliance burden when compared to traditional film radiography. They also offer faster inspection time, increased sensitivity and a permanent record of the inspection.

Thermographic NDI detects subsurface defects by sensing spatial variations in the surface temperature of a component or structure using an infrared camera. Typically, the component

[‡] Whilst PAUT is not specifically excluded as a means of compliance, many structural design codes specify NDI requirements and defect accept/reject criteria explicitly in terms of traditional single-element ultrasonic instruments. This restricts the exploitation of new PAUT capabilities until the codes and standards are updated to include references to PAUT as an acceptable means of inspection.

is thermally excited using a pulsed or periodic heat source (such as a flash lamp). The incident heat diffuses into the component and any subsurface defects which disrupt the flow of heat through the component are revealed by a change of temperature on the surface immediately above them. Thermography is capable of very rapid inspections of large areas of composite materials for defects such as impact damage, disbands and delaminations.

3.3 Obstacles to Widespread use of Advanced NDI

A common theme for advanced NDI methods is that they offer multiple options for analysis and presentation of data, and provide considerably more detailed information about the condition of the structure. Advanced NDI methods also have better capabilities to size defects accurately and to show the exact extent of damage, enabling the assessment and disposition of detected defects to be based on more sophisticated engineering analyses. However, advanced NDI often reveals other deviations and anomalies compared to the design blueprint that may have previously been hidden. Subsequent engineering assessment of that information may be time-consuming and expensive and existing engineering systems may struggle to deal with the uncertain serviceability that may be derived from a detailed assessment of the structure.

The skill sets required to conduct advanced NDI may be very different to those required for traditional NDI using hand-held instruments. NDI technicians tend to be trade-qualified and, without understating their skill with traditional NDI techniques, their skills in the use of computer-based equipment vary and thus NDI technicians vary in their natural aptitude for advanced NDI. In addition, a thorough interpretation of the inspection data to assess the implications for the structure may require deep understanding of the physics of the inspection as well as the engineering context, including consideration of load paths and potential failure mechanisms. This may require professional rather than trade level competencies. The engineering and NDI functions must be well integrated in order to capitalise on the benefits of advanced NDI.

A further challenge to the use of advanced NDI lies in the lag in extending governing standards and syllabi for personnel training and certification to cover the new NDI methods, as well as in revising structural design codes and standards to explicitly recognise the new NDI capabilities as accepted means of compliance. Faced with these challenges and often under time pressure, engineering authorities frequently fall back on the less-capable but better established traditional NDI methods, rather than striving to exploit advanced NDI for the long-term benefits of a better inspection.

4. RELIABILITY OF NONDESTRUCTIVE INSPECTION

NDI is not performed for its own sake. Every inspection has a purpose and inspections are only performed in order to learn something about the health of the structure. Consequently, every inspection has a reliability specification which is either explicit or implied by the context. Identification of exactly how the inspection contributes to the integrity of the platform will reveal the reliability or other performance characteristics required for the NDI.

A key criteria for appropriate application of an NDI technique is the size and type of defect the technique is capable of detecting with high reliability, relative to the sizes and types of defect that could be structurally significant [3]. Detection of defects by NDI is best described as a probabilistic process with a probability of detection (POD) that normally increases with defect size but which is also influenced by defect type, orientation, morphology and location. A typical POD curve as a function of defect size is given in Figure 1. A defect size having a defined minimum POD is often identified as the "reliably detectable defect size" and is considered to be the largest defect that might remain undetected in the structure following an inspection, as anything larger will have been detected and either removed or managed explicitly through engineering analysis. For in-service inspections, the reliably detectable defect size is used to set the inspection interval, whilst for production inspections the detectable defect size can drive design parameters, such as the peak design stress allowed for a component.

Traditionally, probability of detection has been measured via empirical POD trials, in which a large number of inspections are performed under field (or simulated field) conditions on a set of representative specimens containing representative defects. To account for the sampling variability inherent in an empirical statistical trial, it is normal to determine a lower confidence limit on the estimated POD, which provides a lower bound on the possible POD consistent with the observed data [4]. Measuring POD through empirical trials is very expensive and may be infeasible for many applications due to the impracticality of preparing genuinely representative specimens containing representative defects. Alternative approaches to POD assessment are being developed to use physics-based or empirical models to predict POD. Such approaches are known as model-assisted probability of detection (MAPOD) [5].

It is essential to distinguish between the sensitivity and reliability of an inspection. NDI reliability is primarily concerned about the largest defect that might possibly be missed by an inspection and therefore remain undetected in the structure. The ability of an inspection system to detect very small defects – the sensitivity of the inspection – is not a good indicator of reliability. The sensitivity of NDI is usually driven by the physics of the inspection and describes performance under optimal conditions. The reliability must take into account human factors in correctly performing the inspection, including all possible real-world causes for a defect to be missed.

Probability of detection is not the only metric for characterising NDI reliability and an appropriate choice of metric depends on the purpose of the inspection. For example, in process control the false call rate is as important as probability of detection. When supporting rectification of in-service problems, the ability to size the defect may be more important than either POD or the false call rate. However, NDI reliability literature tends to focus on POD as it is a key input to analysis of safety-critical structure [3 , 6].

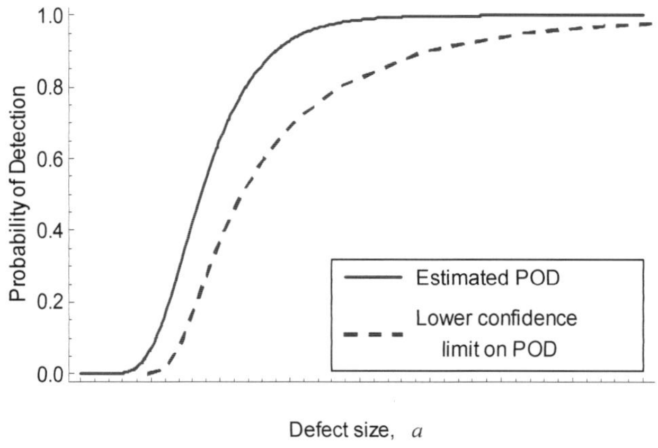

Figure 1 Probability of detection and lower confidence limit on POD plotted against defect size, a.

5. INTERFACE OF NDI TO STRUCTURAL INTEGRITY MANAGEMENT

Structural integrity is the subset of technical integrity that relates to physical structures. In the case of submarines, structural integrity includes both the watertight integrity of the hull and the ability of the internal and external structures to withstand the operational loads. Of key interest to this paper is the way in which nondestructive inspection feeds into structural integrity management of submarines. A key question to be explored is: *What assumptions are made within the engineering analyses to justify the structural integrity of the submarine that could not be made if the inspections were not carried out?*

Some applicable standards and design analyses for submarines are examined in section 5.2. However, it is valuable to first review the use of NDI reliability information in aircraft structural integrity management, where the airworthiness standards explicitly set out the NDI performance required.

5.1 The Perspective of Aircraft Structural Integrity Management

For aircraft structural integrity management, there is a clearly articulated link between the capability and reliability of NDI and the remaining life of the structure. Assuming adequate control of corrosion, the dominant process that limits the service life of aircraft is metal fatigue and consequently the airworthiness of all modern aircraft structure relies heavily on a fracture mechanics based analysis of fatigue crack growth and failure.

The damage-tolerant design philosophy, also known as safety-by-inspection, uses a damage tolerance analysis (DTA) to assess the ability of the structure to withstand service usage and loads in the presence of damage such as a fatigue crack. A DTA is typically a deterministic analysis in which the growth of an assumed defect is predicted, based on in-service loading and experimentally determined crack growth rates. A 'safe' inspection interval is determined as a prescribed fraction (typically half) of the time it takes for the defect to grow from the minimum reliably detectable defect size, a_{NDI}, to the critical defect size, a_{crit}, at which the structure could fail under service loads, Figure 2. A DTA requires extensive engineering analysis to determine the loads, stresses, crack growth rate and critical defect size, often necessitating detailed finite element analyses and load models applicable for the local structural geometry. The US Defence Joint Service Specification Guide for Aircraft Structures [7] specifies conservative flaw sizes (a_{NDI}) that shall be assumed to exist initially in a structure as a result of the manufacturing process, normal usage and maintenance, and following an in-service inspection. If a designer or maintainer wishes to use an a_{NDI} smaller

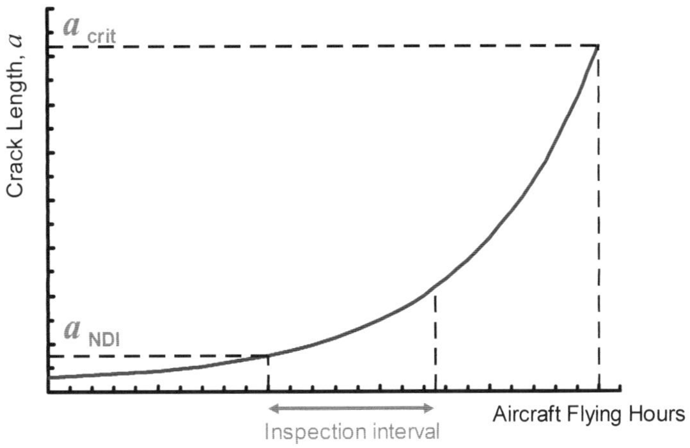

Figure 2 Inspection interval determined from a_{NDI} and crack growth curve (schematic)

than those specified, the detectable defect size should be shown to have at least 90% probability of detection demonstrated with at least 95% statistical confidence: this defect size is known as $a_{90/95}$.

An alternative approach to the deterministic DTA is to perform a probabilistic risk analysis to evaluate the overall risk of structural failure for an individual component, or the entire aircraft. It should incorporate probability distributions for each of: the initial defect size, the defect growth rate, the material properties (such as fracture toughness) and the load-spectrum describing the probability distributions of in-service loads. Integration over these distributions, projecting the growth of defects with flight cycles, gives an overall risk of structural failure. Periodic NDI reduces the risk of failure by detecting and allowing the removal of a proportion of the larger defects from the component or structure. The POD curve is a key input, as it describes the risk that defects will remain in the structure after the inspection. Advances in computing resources are making risk analysis approaches more practical and airworthiness authorities are keen to capitalise on the benefits of this approach. However, its use in practice is presently limited by the scarcity of data for the distributions of the input parameters.

5.2 Submarine Structural Integrity Management

The role of NDI in submarine structural integrity management is not so readily identified. RAN Technical Regulatory System sets out requirements for regulation of technical integrity of ADF maritime materiel in the Materiel Requirements Set (MRS). The MRS functional and performance requirements for submarine structures [8] states that the initial imperfections and residual stresses expected during construction are to be taken into account in all relevant calculations (paragraph 5.2.4), whilst paragraph 5.10.1 requires that nondestructive examination of welds provide assurance that no surface or subsurface defects exist in any plane (e.g. transverse or longitudinal with respect to the weld). However, this is at odds with the engineering reality that defects are present in every structure.

The RAN Materiel Requirement Set also requires that a hull survey be conducted periodically to ensure adequate strength is maintained throughout the service life. The survey is primarily visual inspection, supplemented by thickness measurements or nondestructive examinations as required [9]. The hull survey is essential for through-life detection of corrosion to ensure submarine structural integrity, although it does not search for cracks growing in fatigue or stress corrosion cracking.

The SUBSAFE program is a key tool in the management of structural integrity for submarines, which captures all aspects of submarine safety including occupational health and safety and hazard management, submarine operations, materiel certification, licensing and maintenance management [10]. The SUBSAFE documentation sets out very clearly the components of the submarine that are safety-critical, known as the SUBSAFE certification boundary. The technical requirements within the SUBSAFE program are centred around Objective Quality Evidence, being the documentary evidence required to confirm an item meets all the design, construction and installation requirements [11]. A fundamental tenet of SUBSAFE is to maintain the integrity of the pressure envelope. However, the implementation focus within SUBSAFE is on materiel management requirements and the details regarding the structural design and maintenance philosophies are left to the designer. Specific requirements for NDI of welds are mandated by reference to TS-020 [1], which specifies the minimum size of *indications* that must be reported, thus implicitly setting limits on the defect sizes that are captured in the MRS requirement that "no surface or subsurface defects exist". However, this is not an explicit specification for the reliability of the NDI because it does not consider (i) the inherent statistical variability in NDI response versus actual defect size, and (ii) the possibility that a large defect might be missed due to reliability issues such as human factors in the inspection.

In the absence of an explicit specification for NDI reliability and POD, there are a variety of structural integrity engineering analyses that directly or indirectly incorporate expectations for NDI performance and reliability:

- the structural design includes implicit assumptions about the of size defects present at manufacture,
- the through-life management of problem locations may require regular inspections at defined intervals based on the reliably detectable defect size,
- disposition of detected defects will incorporate explicit or implicit assumptions about the accuracy of defect sizing by NDI, and
- condition surveys should account for the reliability of NDI to give an indication of what size defects may not have been detected by the survey.

5.3 NDI Reliability Required for Submarine Weld Inspections

This section looks in detail at structural design assumptions regarding the size of defects present in welds following production NDI under one available design standard.[§] The UK Sea Systems Publication SSP74 on Design of Submarine Structures [12] presents a design philosophy centred around design for interframe collapse or overall pressure collapse, with the hull structure treated as pressure vessel subject to a static external pressure. This failure mode is a static strength problem and not sensitive to small defects in the structure. The reliability limitations of NDI are unlikely to be a concern for static strength. Hence, fatigue is still a primary driver for the performance of NDI for crack detection. The fatigue life design philosophy applied in SSP74 is for a safe-life structure designed in accordance with BS 7608:1993 *Code of practice for fatigue design and assessment of steel structures* [13]. Although the hull primarily experiences compressive stresses, some regions of the structure and shell plating are subject to bending which may cause local tensile stresses. In addition, residual stresses from welding or fit-up can give rise to high tensile stress and therefore induce low-cycle fatigue loading when combined with the dive cycle of the submarine [14]. Under fatigue loading, small defects may grow undetected and, if not detected, significant fatigue cracks could result in a leak or fast fracture of the hull.

The fatigue safe life determined under BS 7608 is based on empirically determined safe S-N curves for different class welds.[**] The weld classes incorporate maximum defect sizes that are assumed to remain in the welds following fabrication, specified as upper limits of flaw areas for different class welds. Typical welds used in submarine construction, such as plain butt welds and tee butt welds, would be assigned classes D, E or F, where class D is the highest class with the most stringent process control and inspection requirements. The upper limits of flaw area depend on the thickness and an extract from BS 7608 is given in Table 1 for class E and F welds. For class D welds, the upper limits of flaw size are to be determined in accordance with BS 7910:2005 *Guide to Methods for Assessing the Acceptability of Flaws in Metallic Structures* [15]. Application of BS 7910 requires detailed knowledge of the local geometry and fatigue loading, and could generally be expected to have more stringent limits than those for Class E welds.

[§] This review is based on the limited submarine design standards available in the open literature. The UK MoD mandatory requirements for submarine design are contained in DEF STAN's 08-163 and 09-50, which were not available for this review.

[**] BS 7608 divides welded joints into various classes for fatigue calculation purposes, based on the joint geometry, weld detail and preparation, stress orientation, locations of possible crack initiation, and methods of inspection. The weld classifications in BS 7608 are different to those defined in TS-020 [1].

Table 1 Upper limits of flaw area of planar flaws and slag inclusions in butt welds extract from BS 7608 [13]

	20 mm plate thickness		50 mm plate thickness	
Class	Near surface	Away from surface	Near surface	Away from surface
E	17 mm^2	40 mm^2	38 mm^2	110 mm^2
F	45 mm^2	80 mm^2	116 mm^2	260 mm^2

There is a key difference between the NDI requirements for welded structures for submarines and those for damage tolerant aircraft design. For BS 7608, whilst assumed maximum defect sizes are provided in the standard, the actual weld fatigue life is determined empirically from testing of welds manufactured with the same process. In contrast, aircraft damage tolerance analysis explicitly incorporates the crack growth rate from an initial defect size so that the fatigue life is critically dependent on a_{NDI} and requires adequate probability of detection to be demonstrated.

Thus, we have arrived at an assumption that is made about the health of the structure following nondestructive inspection. For 20 mm thick plate, it is assumed that the fabrication process, including NDI, will ensure that there are no defects larger than 17 mm^2 (equivalent to a 4.6 mm diameter circular flaw) near either surface or 40 mm^2 (7 mm diameter) away from the surface.[††] This provides a starting point for assessing whether the reliability of the inspection is adequate, although no attempt will be made here to assess the probability of detection for weld inspections against these defect sizes.

Section 5.3 looked at the NDI reliability assumptions embedded in the process to assure weld integrity for submarines, which is valuable to illustrate that the reliability and performance requirements are found, not in the NDI standards and procedures, but in the engineering analyses that rely on assumptions about the defects present in the structure.

6. RELIABILITY-CENTRED MAINTENANCE

The reliability of weld inspections during fabrication discussed above is just one aspect of the roles of NDI in submarine safety and sustainment, albeit one that is essential to get right. The broader impact of advanced NDI technology will be seen through more effective inspections that provide better information about the health of the structure throughout its service life.

Reliability-centred maintenance (RCM) and condition-based maintenance (CBM) focus on identifying underlying causes of failure and performing maintenance only as it is required. The RCM framework defines what failure management strategies should be employed to ensure a system achieves the desired levels of safety and reliability, including the use of CBM and condition monitoring processes and technologies [16 , 17]. Condition monitoring processes include vibration monitoring, oil analysis, nondestructive inspections, structural health monitoring and prognostics, all of which can be employed to identify signs of impending failure.

Reliability-centred maintenance, condition-based maintenance, condition monitoring, prognostic health monitoring, advanced prognosis, risk-based inspection and risk and reliability analysis are all strategies for cost-effective assurance of safety, availability and capability of engineering assets. Each of these strategies is dependent on an ability to

[††] A more sophisticated analysis would be required to draw these conclusions for any specific given weld, including consideration of which surface experiences high loading –near or far from the inspected surface.

adequately characterise the current condition of particular components and, as such, these maintenance strategies can only be fully implemented using advanced inspection technology to provide a better picture of the health of the structure.

7. CONCLUSION

Advances in nondestructive inspection provide new possibilities for cost-effective sustainment of military submarines through remote inspection of previously inaccessible structure and reduced damage inflicted through disassembly for maintenance. Modern condition-based and reliability-centred maintenance approaches rely on advanced NDI, combined with structural and machine health monitoring to provide detailed assessment of materiel condition. However, where NDI is performed on components within safety-critical systems, the reliability of the technique and the reliably detectable defect size become an essential part of the overall risk management strategy and the application of fracture-mechanics based structural analysis tools require quantitative knowledge of the inspection reliability being achieved in practice.

8. REFERENCES

1. *Quality Control Regulations for Inspection and Examination of Welds and Rules for Evaluation and Approval.* (2004) TS-020 Revision 14, ASC Collins Class Submarines Engineering Standard.
2. Nondestructive Inspection of Weldments, Brazed Assemblies and Soldered Joints. (2002) In: *ASM Handbook Volume 17 Nondestructive Evaluation and Quality Control.* Vol. 17, ASM International.
3. Harding, C. A. and Hugo, G. R. (2011) *Guidelines for Interpretation of Published Literature on Probability of Detection for Nondestructive Testing.* DSTO-TR-2622, Defence Science and Technology Organisation, Melbourne, Australia.
4. Harding, C. A. (2008) *Methods for Assessment of Probability of Detection for Nondestructive Inspections.* PhD Thesis, The University of Melbourne, Victoria, Australia.
5. Harding, C. A., et al. (2009) *Model-Based Methods for Reducing Costs Associated with POD Studies.* TTCP/MAT-TP-5/7/2008, The Technical Cooperation Program.
6. Nickisch, S., et al. (1997) Alternative Approaches to the Validation of Nondestructive Testing Methods. In: Thompson, D. O. and Chimenti, D. E. (eds.) *Review of Progress in Quantitative Nondestructive Evaluation.* Vol. 16, New York, Plenum Press, pp. 2037-2044.
7. *Joint Service Specification Guide Aircraft Structures.* (1998) JSSG-2006, Department of Defense, USA.
8. *Submarine System Requirements - Submarine Structures.* (2009) DEF(AUST) 5000 Volume 09 Part 01 Issue 01, Australian Defence Force Maritime Materiel Requirements Set.
9. *Submarine System Requirements - Submarine Hull Survey.* (2001) DEF(AUST) 5000 Volume 09 Part 02 Issue 01, Australian Defence Force Maritime Materiel Requirements Set.
10. *RAN Submarine Safety (SUBSAFE) Manual Volume 1 Subsafe Program General Requirements.* (2010) ABR 6013 (SUBSAFEMAN 1), Royal Australian Navy.
11. *RAN Submarine Safety (SUBSAFE) Manual Volume 1 Material Requirements.* (2010) ABR 6091, Royal Australian Navy.
12. *Design of Submarine Structures.* (2001) Sea Systems Publication SSP74, Defence Procurement Agency, Sea Technology Group, United Kingdom.
13. *Code of Practice for Fatigue Design and Assessment of Steel Structures.* (1993) BS 7608, British Standards, United Kingdom.
14. Burcher, R. and Rydill, L. (1998) *Concepts in Submarine Design.* Ocean Technology Series 2, Cambridge, UK, Cambridge University Press
15. *Guide to methods for assessing the acceptability of flaws in metallic structures.* (2005) BS 7910, British Standards, United Kingdom.
16. *Condition Based Maintenance Plus (CBM+) for Materiel Maintenance.* (2007) DoDI 4151.22, United States Department of Defence Instruction.
17. *Reliability Centered Maintenance (RCM).* (2011) DoDM 4151.22-M, United States Department of Defence Manual.

Developing Australia's Future Submarine

Hans Wicklander
© Deep Blue Tech Pty Ltd
hans.wicklander@deepbluetech.com.au

ABSTRACT

Developing Australia's future submarine will be a significant undertaking. Deep Blue Tech has started the development of the capability and capacity required to take on this challenge. It involves building the team, the processes, the plans and the tools constituting the core of an enduring submarine development enterprise, which underpins a level of sovereign independence for Australia's future submarine capability.

Deep Blue Tech (a wholly owned subsidiary of ASC) has since its inception in 2007 developed the capability to create conceptual submarine solutions from high level needs, applying a life-cycle perspective and system engineering principles.

It is an integration challenge activity requiring broad domain knowledge across the subsystems while taking into account the many interdisciplinary speciality engineering domains. A significant part of the challenge is the fact that the submarine, including its crew, is a complex system of systems.

The approach, while requirement driven, is fundamentally evolutionary and iterative making use of best practice and lessons learned in all phases of the life cycle of previous submarines. It is based on the unique Australian circumstances including the fact that development has to start from the current state.

The paper outlines the key elements of the challenge and the steps taken to get ready.

1. INTRODUCTION

This paper outlines some important aspects of the development of Australia's Future Submarine and the steps Deep Blue Tech has taken to prepare for this work. It is based on the work performed by Deep Blue Tech and the author's more than 20 years of study and practical experience of submarine development, its processes and constraints, and what works and what does not. The current emphasis is on the early phases of the programme and the design element of the submarine enterprise. The views expressed in this paper are those of the author, and do not necessarily reflect those of ASC or the Commonwealth of Australia.

2. AUSTRALIA'S STRATEGIC CIRCUMSTANCES AND THE NEED FOR SUBMARINES

Australia needs Submarines

Much has been written about Australia's need for the Future Submarine. This paper does not intend to deal with this interesting subject. The Australian Government published the Defence White Paper Force 2030 in 2009 [1] which clearly articulates the strategic need for the Future Submarines. Also, allied submarine deployment in the Asia Pacific region will be limited due to the declining size of the US submarine fleet over the coming decades, making a contribution from Australia increasingly important.

The need for submarines and their capabilities will evolve over time

The need for a submarine capability will remain, but change over time. The evolution of technology will push new solutions that may offer improved capability or reduced costs. The pull of military requirements based on evolved tasks for the submarines or the increased capability of potential adversaries will drive new requirements.

The Defence White Paper is updated periodically to reflect the evolution of Australia's strategic circumstances. Some elements, like geography are less likely to change (Figure 1). It is more or less certain that technology will continue to evolve. The geopolitical development is harder to predict, even if it is clear that the period after the Cold War has brought significant change. The rate of change has increased. Globalisation and the rise of the Asia Pacific region makes Australia even more dependent on seaborne trade. Australia's wealth of natural resources is underpinning the wellbeing of the nation and is worth defending in any reasonable contingency [2]. Submarines are becoming more capable and prolific in the Asia Pacific region [3]. It is clear that Australia will need submarines with a commensurate capability edge for at least the next few generations and that the capability edge will need to evolve over this time span.

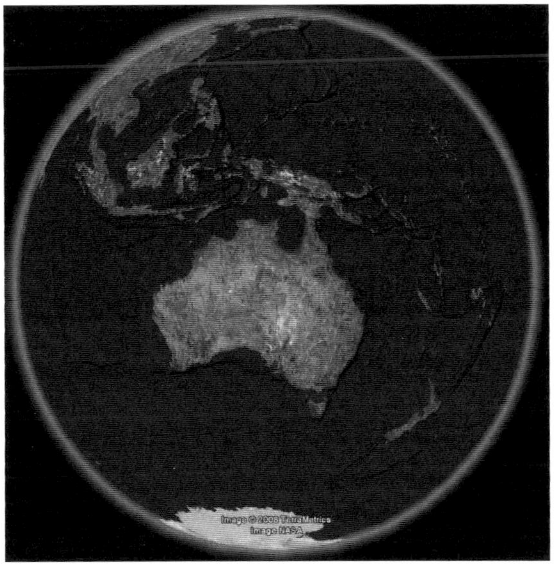

Figure 1: Australia – A Remote Island Continent (Google Earth)

The Future Submarine will be non nuclear

The Australian Government has decided not to pursue nuclear submarines. This option has been debated from time to time. Even if there was a willingness to pursue nuclear submarines and the necessary funding could be made available, it is reasonable to consider this option only when Australia has introduced a civil nuclear power industry and the underpinning foundations including elements like training, regulatory bodies, special facilities and waste management are in place. Developing a new submarine is a large and complex task without having to also concurrently develop a new industrial sector. Pursuing nuclear propulsion may also jeopardise the necessary political and public support for the Future Submarine Programme.

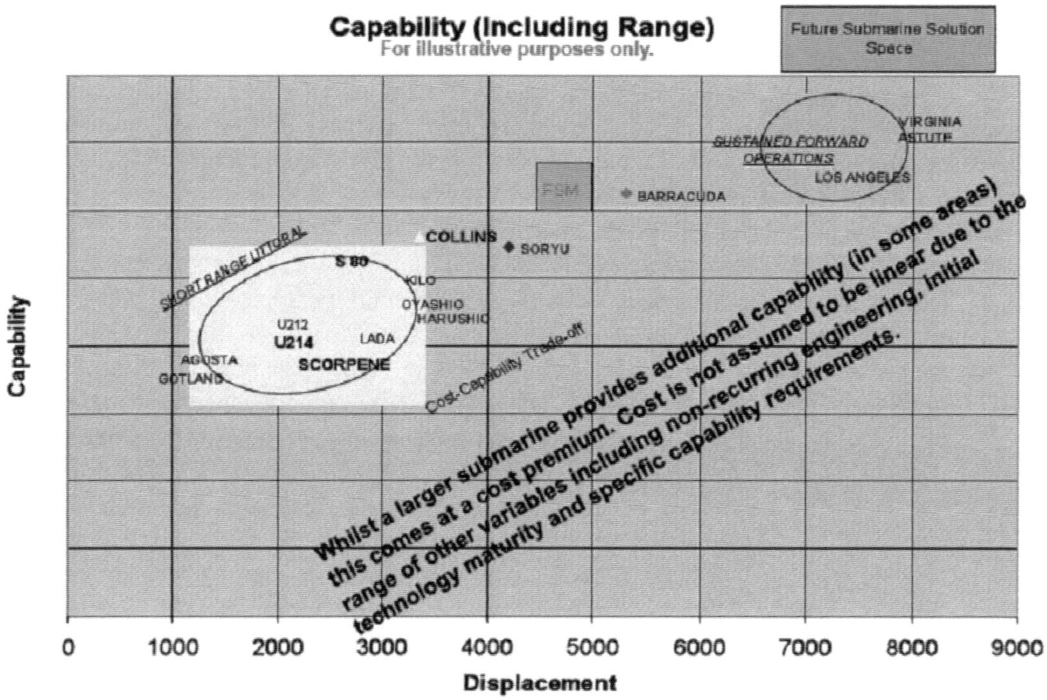

Figure 2: MOTS, Collins, FSM and SSNs presented by the Chief of Navy [4]

There is no MOTS solution - The Future Submarine will be a developmental programme

The Defence White Paper [1] describes a unique set of needs based on Australia's strategic circumstances. No other nation has developed a submarine capability to meet Australia needs. Figure 2 indicates that the capability needs are greater than offered by existing conventional submarines, but less than what nuclear attack submarines offer. The capabilities of the Collins class are arguably closer to Australia's future requirements than anything on the market. The Japanese Soryu class is also close, but Japan does not export defence materiel, including submarines. None of these Military off the shelf (MOTS) submarines satisfy Australia's current or future requirements. This has been much debated in the media, often by advocates of certain procurement models or salesmen with a product to sell. The reality is clear, the solution for Australia is not available off the shelf, it has to be developed.

Australia will be the Parent Nation for the Future Submarine

A unique submarine developed to satisfy the Defence White Paper will give Australia a capability she requires. The benefits of being a sovereign nation with a level of sovereign independence come with associated responsibilities. No one else will have these responsibilities. The responsibilities can not be contracted out or outsourced offshore. The Parent Nation has the sole responsibility for its child.

3. EVOLUTION

Evolution is any change across successive generations of populations. Charles Darwin was the first to formulate a compelling scientific argument for the theory of evolution by means of natural selection [5]. Evolution by natural selection is a process inferred from three facts:

- more offspring are produced than can possibly survive,
- traits vary among individuals, leading to differential rates of survival and reproduction, and
- trait differences are heritable.

Thus, when members of a population die, they are replaced by individuals that are not born from random parents. Natural selection is the only known cause of adaptation.

Adopt an evolutionary development process

Submarine development follows a similar evolutionary process. Evolution requires a starting point. No nation develops a new submarine from nothing. Evolution is the most common method employed by nations successfully developing submarines [6][7][8][9][10]. Evolution will reduce project cost and duration; reduce risks by drawing on previous experience and investments. It applies not only to the initial acquisition process. The evolutionary approach is equally applicable to maintain the capability edge through life, while capitalising on the evolution of technology and managing obsolescence.

Australia's starting point is the Collins class

Evolution requires a starting point. Australia has evolved to where it is, not just in submarine capability and operations but also in its industrial base. This is where Australia is today, and evolution has to start from there. The Collins class is the only current submarine of which Australia has intimate and first hand knowledge. Figure 2 shows that it is also closest to Australia's future requirements.

Evolving the Future Submarine from Collins is the natural way ahead. Retain what is good, keep good principles and solutions. Some completely new systems will be developed where there are changed or new requirements, as a consequence of new emerging technology, new threats or new operational imperatives. When making significant changes or introducing new systems, it is essential to consider the original design principles, the pedigree. Few, if any, drawings or components will be the same, however the underlying principles remain. This approach will reduce the risks and the work required. It will allow focus on areas requiring development. Revolutionary design should only be used by exception in a few critical areas.

Australia is a young nation. Her evolution from British colony to a first tier nation is still ongoing. Australia's submarine capability shows a similar pattern. Australia's first submarines, the *AE1* and the *AE2*, were commissioned in 1914, achieved fame for operations, but were lost in 1914 and 1915 respectively. Australia acquired Oberon class submarines in the sixties. The submarines were built in the UK and the UK acted as Parent Nation. Australia operated and maintained the submarines. Towards the end of the Oberon era, the UK gradually withdrew from its Parent Nation role and Australia undertook a number of independent capability upgrades, mostly concerning weapons and sensor systems. The next step in the evolution was the Collins class in the eighties. The submarine was evolved from the Swedish Västergötland class to meet Australia's requirements. The submarine was designed by Kockums and built in Australia by the Australian Submarine Corporation. The Collins project was large, ambitions and generally successful. Much has been written about it [11].

Some elements of the original Collins programme, including the challenge of local production of subsystems, the combat system and sufficient consideration of the natural environment, were probably not optimal. The build programme was successful, short in duration and met very ambitious targets for Australian industry participation. The Collins programme also caused second order effects like more efficient industrial relations and modern quality systems to the benefit of Australia's industry as a whole. The transition into service, the initial arrangements for through life support and the long term planning to end of life could have been more successful.

The Parent Nation role requires attention

Many of the well publicised Collins problems are related to not embracing the Parent Nation role. The significance of this role seems to have been underestimated, and this may well be the root cause to many of the problems experienced. Priority and resources applied may have been insufficient. There are signs that Australia now is recognizing and addressing the problems. A long term whole of Government view is required.

Apply the lessons from all phases of the Collins program

The Collins lessons represent a significant investment that will enable Australia to evolve the successful parts of design, build, transition into service, maintenance and upgrade while avoiding the problems that were experienced.

Learn from other nations and apply best practice

It is human nature to learn only from your own mistakes. It is harder, but still possible to learn from others. The RAND Corporation has studied the submarine sectors in the US and the UK and completed a large number of studies dealing with optimum approaches considering cost, long term demand and necessary skills for design and build [12][13]. Most recently Australia has also been assessed [14].

RAND has investigated the many challenges a successful Submarine Enterprise has to manage. Much has also been written on the highly successful US nuclear submarine enterprise [15]. The UK conducted a brief but costly procurement experiment with the initial Astute programme. Competition on the first tier and shifting responsibility to industry did not work. The UK has adopted a collaborative approach for the new Successor ballistic submarine programme [16].

Common features for successful Submarine Enterprises are:

- Clear objectives,
- A national mandate,
- Clear structure including clear long term roles for key participants,
- A life cycle perspective, and
- Stable long term plans with limited fluctuations in workload.

These features are necessary conditions for success. A new approach will be required to achieve a successful Australian Submarine Enterprise. Mechanisms for demonstrating value for the taxpayer's money other than competition between legal entities will be required to select participants of an enduring enterprise [17]. All successful Parent Nations of submarines have come to this conclusion. The necessary preconditions for open market competition do not exist [18]. There is an emerging recognition of the need for a Submarine Enterprise in the Australian debate [19].

Australia needs selected help from allies and friends

Being behind the leading edge in the evolution process makes it possible to evolve faster. Transfer of knowledge from friends and allies or recruitment of overseas staff with relevant experience and knowledge can facilitate this process. Best practice should be considered when building a Submarine Enterprise charged with delivering the capability with a life cycle perspective. Submarine technology is generally closely guarded. The special solutions and knowledge that make a real capability difference are only available though government to government arrangements. Typically, it is necessary to have a level of knowledge to be able to ask for the right help and to be trusted to receive it. The US technology transfer to the UK is a good example [20]. The US AN/BYG-1 Combat Control System for Collins is another example. Special conditions for the use and protection will generally apply to such technology. For example, Australia has to protect Swedish and US technology used on Collins. These conditions heavily constrain which parties can be included and their level of participation in the Submarine Enterprise.

Australia needs an enduring Australian Submarine Enterprise

The next step in the evolution for Australia is to accept the responsibility and become the Parent Nation for the Future Submarine. Design capability is the difference between first and second tier nations and will be the last capability Australia has to acquire to possess all the enabling elements necessary to become a successful Parent Nation.

Some would argue that only larger nations with nuclear submarines like the US, Russia, France and the UK are capable of being Parent Nations. This is not the case, Japan, Sweden and Germany have longstanding and successful submarine enterprises. Japan and Sweden have developed submarines to meet national needs. More recently, South Korea and Spain seem to be on the same path. Germany has taken a different route focusing on export. It is worth noting that Sweden, with a smaller population and smaller defence budget than Australia, considers submarines as a key required national capability and maintains a high level of sovereign independence.

An enduring Australian Submarine Enterprise will enable Australia to successfully conduct the Parent Nation role.

4. DEVELOPING A SUBMARINE

To develop is to bring into being and to grow into a more advanced state by deliberate effort over time. Engineering is the discipline, art, skill and profession of acquiring and applying scientific, mathematical, economic, social, and appropriate experience or tacit knowledge, in order to design and build systems and processes that realise a need. One who practices engineering is called an engineer. They are needed to conduct a development programme.

The development of the Future Submarine is a multidiscipline engineering programme

Creating an appropriate mathematical model of a problem allows engineers to analyse it, and to test potential solutions. Usually, multiple reasonable solutions exist, so engineers must evaluate the different design choices on their merits and choose the solution that best meets their requirements. Engineering is different from science. Scientists try to understand nature. Engineers try to make things that do not exist in nature, they solve problems.

Standards like ISO15288 [21] establish a common framework for describing the life cycle of systems created by humans. It defines a set of processes and associated terminology. These processes can be applied at any level in the hierarchy of a system's structure. Selected sets of these processes can be applied throughout the life cycle for managing and performing the stages of a system's life cycle. This is accomplished through the involvement of all interested parties, with the ultimate goal of achieving customer satisfaction.

Deep Blue Tech's Concept Formulation process is based on a top down systems approach aligned with modern systems thinking, supported by state of the art tools, combined with bottom up domain knowledge, experience and best practice [22][23]. Submarine design is an unforgiving trade that still requires art and science based on experience. In particular, the challenge is to design a system that can conduct submarine missions. The system is an autonomous system of technical systems and crew. The crew is within the system boundary and has to be considered as an integral part. The submarine is like a ship, but operates in three dimensions, like an aircraft, but its missions last for months not hours and like a spacecraft, but in a substantially harsher environment and with adversaries to deal with.

The design process needs to consider the whole submarine system. In order to understand the performance on the whole submarine system it has to be decomposed into subsystems. The characteristics of the subsystems need to be modelled and supported by assessment of available technical solutions, their performance and state of evolution. Deep Blue Tech has studied state of the art for subsystems that are available or under development on the world market. This is done though literature studies, attending conferences and exhibitions and by establishing relationships with providers. Being independent of companies selling products makes Deep Blue Tech able to take an impartial, solution independent, view that builds the trust of others.

The first step of the development process is to transform the need to requirements. As discussed in section 1, the set of requirements for the Australia's next submarine is unique and include:

- Enormous area of interest, from cold to tropical waters,
- Very long ocean transits and long mission duration,
- High levels of mobility and endurance,
- Low signatures across all spectrums, including at higher speed,
- Capability edge over potential adversaries, and
- Interoperable with the US.

Deep Blue Tech has identified areas of particular interest based on the unique Australian requirements and lessons learned from Collins. Examples of these are energy and propulsion, hull shape and hydrodynamics and human factors and habitability.

Models have been developed for energy and propulsion systems to interpret operational requirements, assess performance and to understand the impact at the whole of submarine level [24][25][26][27].

Deep Blue Tech has developed a Computational Fluid Dynamics capability to address significant improvement required in hull shape and hydrodynamic performance. A project in cooperation with the University of Adelaide and the Defence Science and Technology Organisation supported by the Australia Research Council aims to develop an experimentally validated model to predict submarine noise for the purposes of submarine design, operation and procurement [27].

Australia has experienced problems in attracting and retaining submariners. While there are multiple reasons that contribute to this issue, one factor is the submarine habitation design. Submariners are volunteers and the Designer should offer a submarine environment that is not only safe and functional, but also aligned to community exceptions. Deep Blue Tech has studied how this may be achieved [31][32][33]. Interior architects and industrial designers from academia have been engaged to support the work.

While some aspects of Australia's needs are unique, there are other aspects that are common to submarine design in general [6][7][8][9][10][29][30]. The following set of characteristics makes submarine design an almost unique art and science compared to other human endeavours:

- Safety criticality,
- Extremely large and complex system of systems,
- Significant costs,
- Long life cycles, 50+ years from identification of threat to disposal,
- Requirements will evolve significantly over the life span - An approach for dealing with change is required,
- Of national interest and significance,

- Secret and closely protected technology,
- No whole of submarine prototypes - We have to get it right on the whole of submarine level, and
- Some aspects of performance are critical - If the submarine is engaged in a duel scenario, then it is no good to be second best.

Integration is the core of whole of submarine design

A characteristic for whole of submarine design is that it is all about integration. It is about integration of technologies and systems using components developed for other purposes. The defence market is generally not driving technology, as it did during the Cold War. The challenge is to be an early adopter and integrator of emerging technologies developed for other purposes. Open systems based on recognised standards make this less challenging than in the past. A clear structure, robust processes and good information management is required to maintain a consistent set of information to be used by all, the single point of truth. The US Virginia program successfully introduced integrated process and product development, with a focus on cost reduction [34]. It was a major improvement compared to the earlier Seawolf program.

Invest in early phases - It is a cost saving

Design has a major impact on all aspects of the programme. It is therefore essential to invest in early phases, to get it right early. More than half of the life cycle cost can be committed during the design phase [35]. Design does not cost much in relation to the whole programme, but it can provide great risk reduction and ensure that balanced customer requirements are developed using informed trade-offs. An investment in the early phases can give great cost savings through life. It is not just to avoid risk. Equally important is to explore uncertainty and seize opportunities. Often only the negative side of uncertainty, risk, is given appropriate attention.

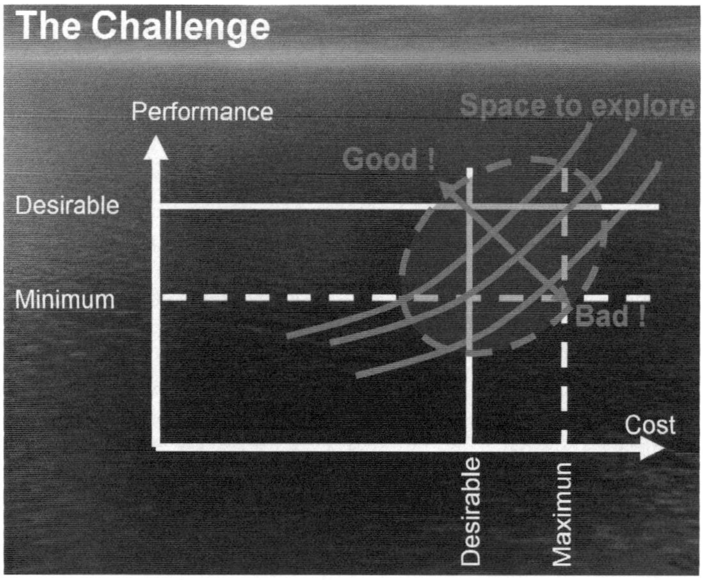

Figure 3: The Solution Space

The challenge to find a balanced solution in the solution space is illustrated in Figure 3. Often, but not always, increasing performance results in increased cost. On the whole of submarine level, requirements and solutions are interdependent, often through complex and indirect relationships. Deep Blue Tech has developed a Concept Formulation process supported by a number of tools and methods to address this challenge. By evolving many solutions, the solutions can be compared and contrasted. Using an iterative process the fittest options survive and generate offspring in the following iterations. Thereby the solution space can be explored and a balanced solution identified, offering best value for money. The same process can be used to find the minimum cost for a defined level of capability, or the maximum capability for a given cost.

A technical solution that can not be costed is only of academic interest. There are various methods for addressing this problem. Deep Blue Tech has developed cost models to support the design work. One example is a parametric cost model based on available data on a large number of submarine programmes. This model can be used for top down cost assessments. Bottom up models are used when the focus is on comparison of system options.

A high level representation of Deep Blue Tech's Concept Formulation process is shown in Figure 4. The process translates the needs expressed in the Defence White Paper [1] to requirements. The requirements are developed through mission analysis and requirements engineering and captured in a requirements management tool. The next step is architectural design where functions that can satisfy the requirements are developed, grouped and linked to the high level system architecture. The system architecture is the whole of submarine level solution. It is an abstract representation describing a system architecture in terms of behaviour and structure. The next step, synthesis, moves from the abstract to the physical world. As the arrows indicate, the process is iterative. Feedback and adjustment is required until a balanced solution is achieved. "Balanced" will have different definitions, depending on the purpose of the work undertaken and the stage in the lifecycle where the solution is determined. In the early stages, balance may require a certain level of consistency, using only a low level of fidelity.

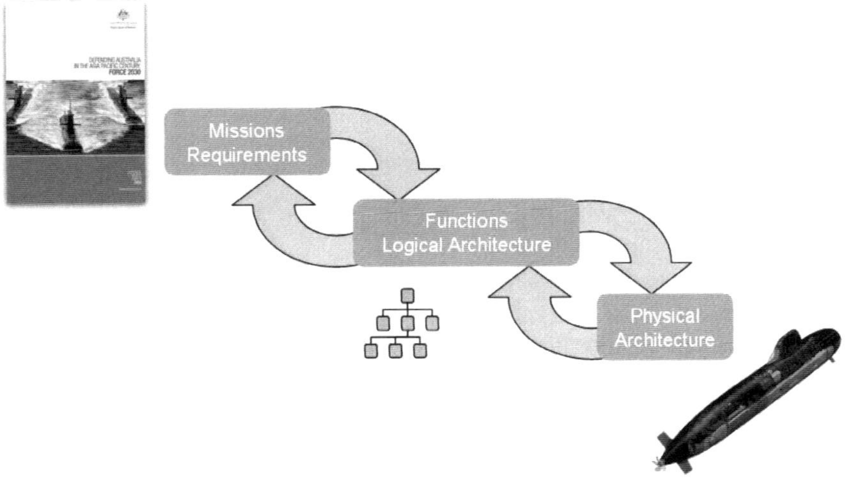

Figure 4: Concept Formulation Process – Levels of abstraction

Deep Blue Tech has adopted a Model Based Systems Engineering approach using the System Modelling Language (SysML) and modern computer based tools. SysML is an international standard allowing exchange of data or information across tools and organisations. All the work shown in Figure 4 is captured in the tool. The tool offers a single point of truth, traceability and multiple viewpoints allowing different parts of the organisation a view that best meet their needs, while maintaining consistency.

Figure 5 provides a different view of the Concept Formulation Process. It shows the iterative nature of the process. Iterations are used to refine or vary the solution.

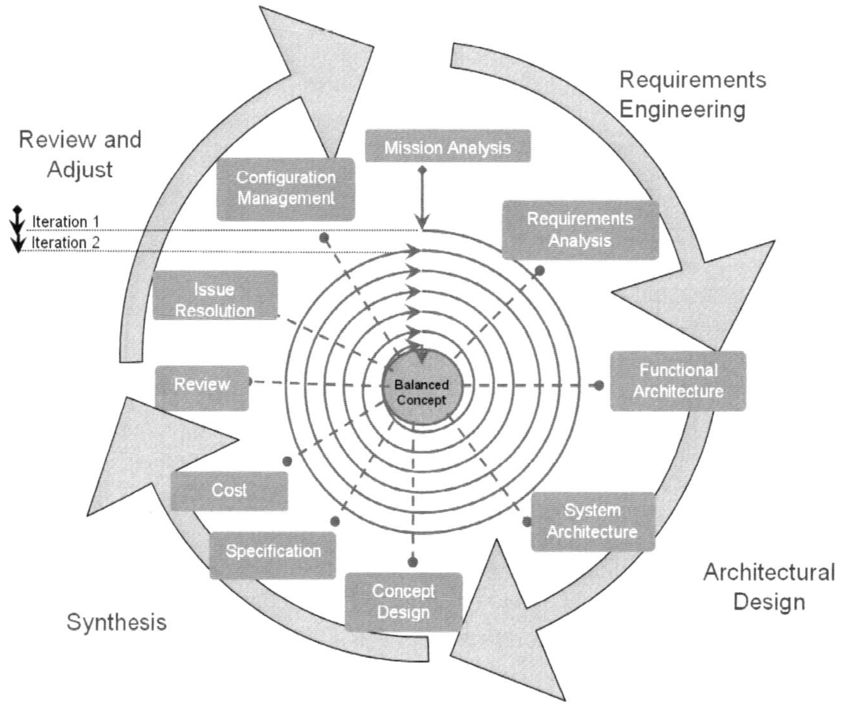

Figure 5: Concept Formulation – A converging Iterative Process

While Figures 4 and 5 show the high level Concept Formulation process on the whole of submarine level it is necessary to define the submarine and its subsystems in order to assess feasibility and cost. Figure 6 shows the next level of the Concept Formulation process. In this concept design process the subsystems are defined and space, weight, power consumption, integration requirements and performance determined. A hull including a spatial arrangement is created in three dimensional parametric CAD and subsystems allocated space and arranged. The traditional naval architecture activities coordinate the balancing of weight and volume, hydrostatics and hydrodynamics and powering. The concept design process is also iterative.

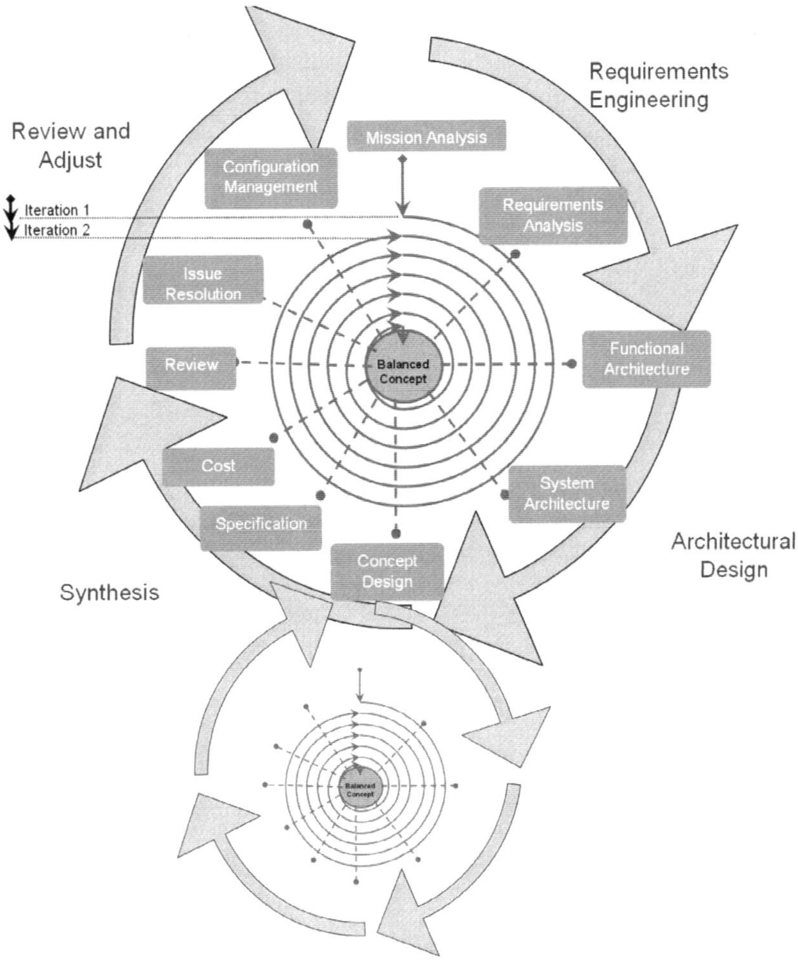

Figure 6: Concept Design – A subset of Concept Formulation

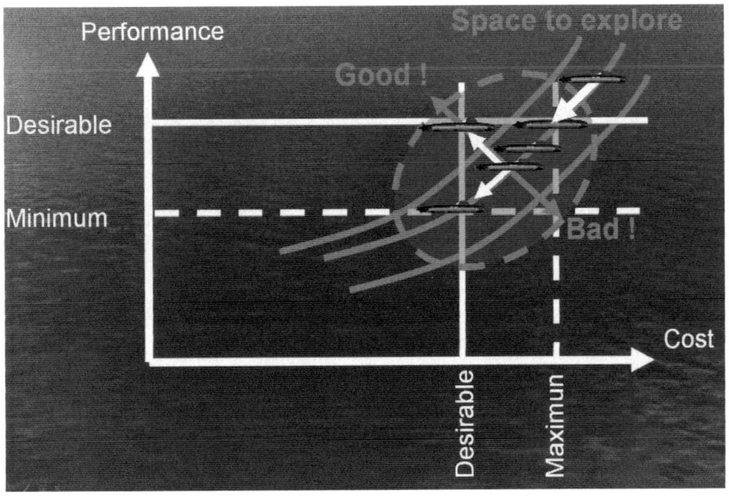

Figure 7: Exploring the Solution Space

The best submarine Australia can afford can be evolved though a Concept Formulation process as discussed above. The results on the highest level could be shown as in Figure 7. The Concept Formulation process can present results in a way that allows the customer to ask "what if" questions and make informed trade off decisions based on exploration of the impact of cost and performance drivers. A competent customer can interpret the relative merits of various cost-capability options and thereby make the iterative process converge to a balanced solution. Deep Blue Tech has developed tools and methods to support rapid exploration of the solution space [36]. The size and duration of the effort will depend on the maturity of the requirements and the range of options to explore.

Process is a means - Not an objective

The discussion above shows that developing a submarine for Australia is a complex undertaking with many unique features and constraints. Deep Blue Tech has developed a Concept Formulation process supported by tools for this purpose. The process has been developed to manage the complexity while maintaining simplicity. It is essential that all processes are developed or tailored to the need.

Develop in Batches

Twelve or more Future Submarines will create a solid foundation for a sustainable Submarine Enterprise. Dividing the programme into batches allows a continuous process of development to maintain the capability advantage. Parallel streams of research and subsystem development allows new capability to be introduced when proven ready. The total cost of ownership should be considered from the very start. Designing for build, sustainment and upgrades are essential elements.

A spiral Capability Maintenance process will maintain the capability edge and the submarine design capability

The submarine system is a system of systems. On the highest level we have a class of 12 submarines. That can be divided in three batches of four as indicated in Figure 8. Each batch becomes a submarine programme in its own right. A continuous process of development allows up to date systems to be introduced into a submarine when first delivered, and later upgraded. Each subsystem has a lifecycle of its own. Obsolete systems can be replaced at regular points in time. Some parts, like the hull, may survive the whole lifecycle. Others elements, like software, may be upgraded every second year. A costed plan to end of life is essential to manage the capability. The plan has to integrate maintenance and upgrades. The focus should be to maintain the capability to minimum life cycle cost.

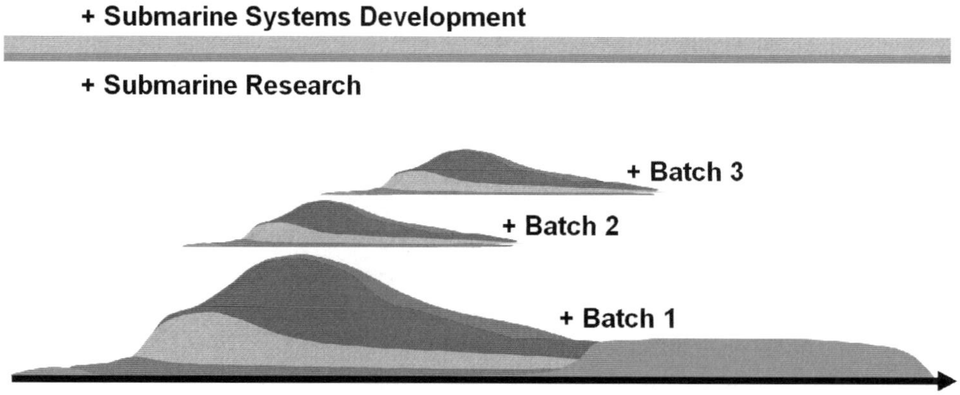

Figure 8: A Lifecycle Approach

Modelling and simulations are important activities to get things right early and avoid expensive corrections at later stages. Deep Blue Tech is developing an integrated simulation environment to support whole-of-submarine design and engineering throughout the entire submarine life cycle [37][27] that includes the environment the submarine system operates in [38].

5. MEGAPROJECTS AND THE FUTURE SUBMARINE PROGRAMME

The successful delivery of complex mega-projects requires the effective management of numerous interrelated variables – often on a global scale. Megaprojects generally undergo a protracted gestation process. Their national significance calls for serious preparation and rigorous scrutiny prior to a national commitment. A megaproject should be hard to start, but equally difficult to change or stop.

Large engineering development programmes have positive spin-off and spill-over effects on the wider economy

The Sydney Harbour Bridge exemplifies both the gestation difficulties experienced by iconic megaprojects, but also highlights that great projects can be successfully executed in Australia [39]. The project had been discussed for many decades when John Bradfield was appointed Chief Engineer in 1912. The construction started in 1923 and the bridge opened 1932. It stands proudly and is recognised as an Australian icon.

Probably the most relevant Australian example, other than the Collins programme, is the Snowy Mountains Scheme [40]. It includes all the elements of a nation building programme.

In 1958, then Australian Prime Minister, Sir Robert Menzies, spoke of the triumph of the Snowy Mountains scheme:

"In a period in which we in Australia are still, I think, handicapped by parochialism, by a slight distrust of big ideas and of big people or of big enterprises ... this scheme is teaching us and everybody in Australia to think in a big way, to be thankful for big things, to be proud of big enterprises ... to be thankful for big men."

Menzies' words are still relevant today.

The development of the Future Submarine will be no different. Much could be said about the nation building aspects of megaprojects. It has been shown that very large engineering development projects can have positive spin-off and spill-over effects. The Swedish tax payers enjoyed a 0.43% of Gross National Product (GNP) return of 0.17% of GNP investment for the Gripen fighter jet programme during the period 1982-2007 [41].

6. BUILDING THE DESIGN CAPABILITY

The design capability is based on competent people working together in an organisation using suitable processes and tools evolving a sufficient knowledge base. The knowledge base has to be broad enough to cover a wide range of disciplines and deep enough to understand critical details of each discipline. Deep Blue Tech is building all these elements and developing a capable integrated team. These elements are necessary parts of the capability and can not be addressed in isolation.

Many areas are strategically important for submarine design including:

- Understanding submarine operations,
- Safety,
- Whole of Submarine Design including general arrangement, system integration, modelling, simulation, performance analysis,
- Signatures,
- Sensors,
- Structural design,
- Materials and surface treatments,
- Energy and propulsion systems,
- Weapons and other payloads,
- Subsystem specification,
- Test and evaluation (including test facilities and ranges),
- Verification and validation,
- Cost modelling, and
- Modular design and construction.

There are some areas where significant improvements from Collins are required including:

- Upgrade paths,
- Design for maintainability, reliability and availability,
- Flexible integration of sensors, weapons and other payloads,
- Information systems, human-machine interaction,
- Habitability,
- Life cycle planning, and
- Long term sustainable development.

Most of the areas requiring improvement are related to Capability Maintenance. Deep Blue Tech has adopted an asset management approach which in combination with the systems engineering approach discussed earlier provides a proactive life cycle perspective. Maintenance, obsolescence management and capability upgrades are considered holistically during the design phase and planned to end of life. This is a significant improvement compared to the traditional model where upgrades are seen as surprising disturbances to the maintenance programme. The overall objective is to deliver the required capability to minimum through life cost for the nation, while creating a solid foundation for the next submarine programme.

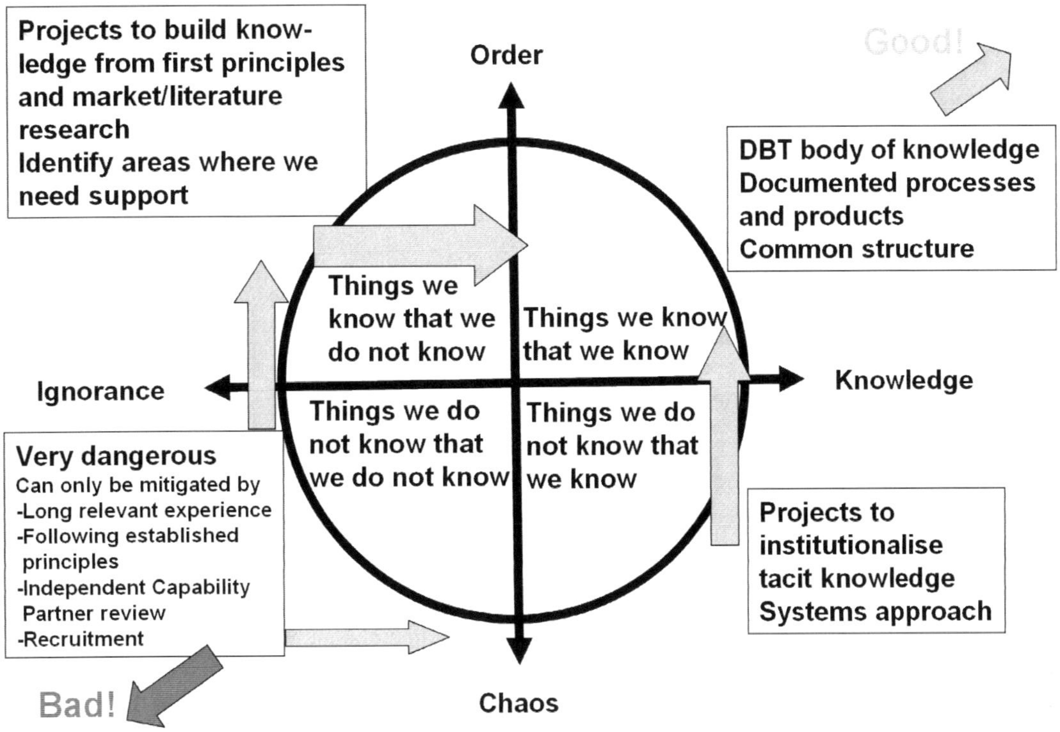

Figure 9: Managing Knowledge

Deep Blue Tech has developed a framework for managing the knowledge development (Figure 9). A body of knowledge can be developed using a scientific approach and first principles. This can be done by intelligent and well educated staff. By recruiting experienced staff with tacit knowledge, projects can be devised to institutionalise such knowledge, thereby significantly accelerating the knowledge growth. Having staff with extensive domain experience allows Deep Blue Tech to identify knowledge gaps that need to be filled. For some identified areas it is acknowledged that external assistance is needed. There is still a risk for unknown knowledge gaps. This risk is mitigated by engaging eminent experts for external review who also foster intellectual development.

The most obvious sources of help are the US, Australia's alliance partner, and Sweden where Collins has its origin. The US may offer critical elements that can not be sourced elsewhere. Sweden offers design and non nuclear submarine expertise not present in the US. A global network of suppliers for subsystems, equipment and services is also needed.

It is necessary to understand all aspects of the capability, at least to the level of detail that allows long term planning and assessment of risks and performance at the highest level. Capability is therefore more important than capacity. It also includes being a competent customer of goods or services sourced from others, being able to specify the goods or service, assess its acceptability when delivered and then to be able to integrate the goods or service in the larger system.

The designer has a central role in the lifecycle of a submarine. The Customer, the Designer, the Builder, the Maintainer and the User are key elements of the Submarine Enterprise. The science and technology community in Defence and academia is also an important element with close links to the Designer. Deep Blue Tech's Mission is to be the

Designer for the entire lifecycle of Australia's Future Submarines. The Designer can be explained by defining its responsibility, its scope of work and the activities conducted through the lifecycle.

The Designer's overall responsibility is that the submarine system fulfils it specification, if it is built to the design.

The Designer's scope includes the following activities:

- Design the Submarine System,
- Manage the systems engineering process,
- Develop technical procurement specifications,
- Develop material specifications,
- Integrate the Submarine System,
- Develop and deliver information that enables the Builder to build the submarine system,
- Define verification activities and assess verification results,
- Interact with Customer, User, Builder and Maintainer and subsystem providers,
- Design for the full life cycle – sustainment, total cost of ownership,
- Manage issues and changes, and
- Develop and deliver information that enables the Customer and Maintainer to manage, maintain and further develop the Design through life.

The Designer participates in all parts of the lifecycle. The life cycle can be divided in any number of phases with various names, but essentially includes a number of steps:

- Feasibility Studies. Early stage studies of operational needs, potential solutions, emerging technology. Major inputs from the Science and Technology community.
- Concept Design. Focus on whole of submarine requirements, performance and whole of submarine integration. Evaluation of options.
- Preliminary Design. Focus on major systems.
- Detailed Design. Focus on producing input to build, specifications for build and procurement.
- Support build. Focus on supporting build, change management and issue resolution
- Support verification. Focus on verifying whole of submarine performance.
- Support while in service. Focus on maintaining capability edge, upgrades, changes and obsolescence management.
- Support disposal.

Australia produces the engineering graduates necessary to sustain an enduring submarine design capability.

The Design element of the Submarine Enterprise is relatively small. RAND has studied required workforce in the US [12] and the UK [13]. The US enterprise is self sufficient, independent of other nations, and has ambitions to be at the forefront of development. The US enterprise therefore requires a large number of people. The UK accepts some help from others and requires less manpower. Both the US and the UK develop nuclear submarines that require a significant overhead. Sweden requires significantly less, but still maintains an advanced submarine design capability, more focused on a best of breed philosophy. A reasonable assumption for the size of the team Australia requires would be that it is smaller than in the UK, but larger than in Sweden. Some question if such an ambition is achievable for Australia.

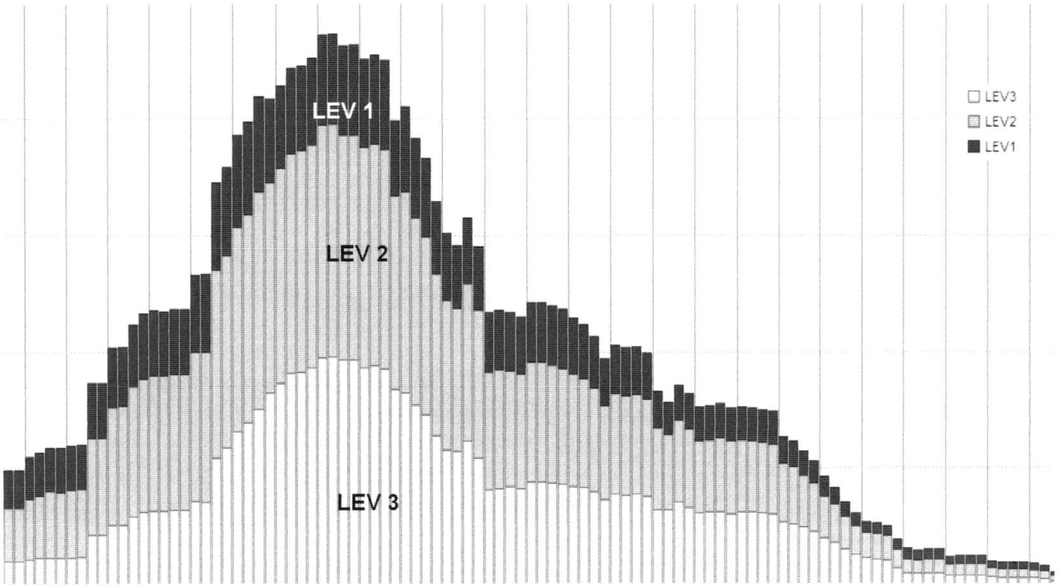

Figure 10: Design Workforce to Develop a Submarine

Figure 10 shows an example of the workforce required to design a submarine. The number of staff needed in various disciplines over time needs to be determined. It is equally important to manage the mix of staff with different levels of experience. Level 1 staff are the most experienced and level 3 staff are the least experienced.

The main challenge is to grow to the first peak. The Deep Blue Tech solution to this challenge is to import competency to get started. Experienced staff have been sourced from the UK and Sweden. A Workforce Model has been developed, taking into account the number of staff over time for various disciplines and skills levels. Gap analysis is used to guide recruitment and professional development. The model assumes that it takes 20 years to become fully proficient in a submarine development domain.

A serious constraint applies to the Work Force Model. The Collins enterprise needs to continue supporting the Collins class while the first new submarine is developed. For obvious reasons, the bulk of the submarine knowledge currently resident in Australia is engaged in Collins activities. Only a very limited flow of personnel from the Collins enterprise can be accepted, focussing on a limited number of experienced staff that have been subject to the Collins technology transfer programme. A truly new capability is under development in Deep Blue Tech.

Once the peak is achieved, we can maintain a steady state where an enduring capability will only require a steady inflow of graduates. Deep Blue Tech has been successful in attracting excellent young engineers from around Australia.

To test this proposition, let us assume that the design element needs 500 people. Work Force Modelling undertaken by Deep Blue Tech indicated that the number is smaller.

If engineers are recruited when they graduate and continue for an average of 20 years, 25 graduates are needed every year. It will be a challenge to attract and retain competent staff. Interesting work is the key. Australia produces more than 5000 engineering graduates per year. Top quality people are required. If the intake is selected from the top 10%, only 25 of 500 or 5% of the available pool is needed. Few, if any other programmes will offer a more challenging engineering task of national significance.

Developing and maintaining the required workforce is achievable

The best method to develop engineers is to work on a challenging programme under professional guidance of more experienced peers. It takes about 20 years to become a true professional in a submarine engineering domain. A good basic science based academic education, like an engineering degree, is an essential precondition before entering the workforce.

The engineering workforce needs continuous professional development. An important aspect is to attain higher education including masters and doctoral degrees. ASC has supported the development of three education programmes to address the need: the Masters of Marine Engineering, Military Systems Integration and System Support Engineering [42]. The programmes have been developed in cooperation between industry and academia.

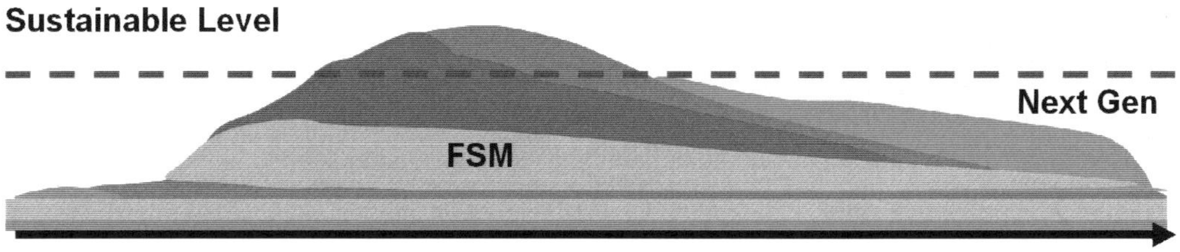

Figure 11: An Enduring Design Capability – The First 60 Years

An Australian submarine development workforce can be developed within 20 years as indicated in Figure 11. Deep Blue Tech has already created the core of the capability.

Planned properly, the Design capability will be a sustainable system with the following attributes:

- All competencies always needed,
- Continuous development,
- Limited fluctuations in demand,
- Creates and maintains the Capability Edge, and
- Nation Building.

7. CONCLUSIONS

Developing Australia's future submarine will be a significant undertaking. The following observations have been made above:

1. Australia needs submarines.
2. The need for submarines will evolve over time.
3. The Future Submarine will be non nuclear.
4: There is no MOTS solution - The Future Submarine will be a developmental programme.
6. Australia will be the Parent Nation for the Future Submarine.
7. Adopt an evolutionary development process.
8. Australia's starting point is the Collins class.
9. The Parent Nation role requires attention.
10. Apply the lessons from all phases of the Collins programme.
11. Learn form other nations and apply best practice.
12. Australia needs selected help from allies and friends.
13. Australia needs an enduring Australian Submarine Enterprise.
14. The development of Future Submarine is a multidiscipline engineering programme.
15. Integration is the core of whole of submarine design.
16. Invest in early phases - It is a cost saving.
17. Process is a means - Not an objective.
18. A spiral Capability Maintenance process will maintain the capability edge and the submarine design capability.
19. Large engineering development programmes have positive spin-off and spill-over effects on the wider economy
20. Australia produces the engineering graduates necessary to sustain an enduring submarine design capability.
21. Developing and maintaining the required workforce is achievable.

Australia can develop the Future Submarine. We have a good starting point in the Collins experience.

It is not necessary to know everything about everything to get started. It may be the other way around. Sufficient knowledge exists today to commence the journey. Deep Blue Tech stands ready to make a significant contribution to Australia's Submarine Enterprise.

REFERENCES

[1] Department of Defence, Defending Australia in the Asia Pacific Century : Force 2030, Defence White Paper 2009

[2] Babbage, R., Australia's Future Underwater Operations and System Requirements, Kokoda Paper No. 4 April 2007

[3] Pacey, B., Under the Air Sea Gap: Australia's Anti-Submarine Warfare Challenge, Kokoda Paper No. 16 May 2011

[4] Crane, R., Submarines in Australia 's Maritime Strategy, Submarine Institute of Australia Conference 2010

[5] Darwin, C., On the Origin of Species, John Murray London 1859

[6] Kormilitsin Y. N. and Khalizev, O. A., Theory of Submarine Design, Saint Petersburg State Maritime Technical University, 2000

[7] Gabler, U., Unterseebootbau, Bernard & Graefe 1997

[8] Allmendinger, E. E., Submersible Vehicle Systems Design, SNAME 1990

[9] R. Burcher & L. Rydill, Concepts in Submarine Design. Cambridge University Press, 1994

[10] Nordin, M., *A systems theoretical methodology for Naval Integrated Complex Systems*, Chalmers University of Technology, RINA Naval Submarines, Bath 2011

[11] Yule, Peter; Woolner, Derek (2008). *The Collins Class Submarine Story: Steel, Spies and Spin*. Cambridge University Press 2008

[12] John Birkler et al, *The U.S. Submarine Production Base: An Analysis of Cost, Schedule, and Risk for Selected Force Structures*, RAND Corporation 1993

[13] John F. Schank et al, *The United Kingdom's Nuclear Submarine Industrial Base, Volume 1 Sustaining Design and Production Resources*, RAND Corporation 2005

[14] John F. Schank et al, *Learning from Experience: Lessons from the Submarine Programs of the United States, United Kingdom, and Australia*, RAND Corporation 2011

[15] Rockwell, T., *The Rickover Effect*, Naval Institute Press 1992

[16] Lloyd, S. *Early Lessons from the Successor SSBN Programme*, Submarine Institute of Australia Technology Conference 2011

[17] Birkler, J., et al, *From Marginal Adjustments to Meaningful Change - Rethinking Weapon System Acquisition*, RAND Corporation 2011

[18] Smith, A, *An Inquiry into the Nature and Causes of the Wealth of Nations*, W. Strahan and T. Cadell, London 1776

[19] Davies, A., *Subaqueana australis - the future evolution of Australia's submarines*, ASPI 2011

[20] Mackby, J and Cornish, P., *U.S.-UK Nuclear Cooperation after 50 years*, Chatham House London 2008

[21] ISO/IEC 15288:2008 Systems and software engineering - System life cycle processes, ISO 2008

[22] Pearce, P., *An Integrated Approach To Submarine Concept Design Formulation,* Systems Engineering Test and Evaluation Conference, SETE 2010, Adelaide, Australia

[23] Pearce, P., *Model-Based Systems Engineering and its Application to Submarine Design*, Submarine Institute of Australia Technology Conference 2011

[24] Williams D., *Submarine Energy & Propulsion Considerations*, Submarine Institute of Australia Technology Conference 2011

[25] Williams, D., and Folbert A., *Energy considerations for a long range diesel electric submarine*, Electric Ship Technologies Symposium (ESTS), 2011 IEEE
[26] Williams, D. and Andreasson, T., *The Australian Air Independent Propulsion (AIP) Dilemma*, UDT Pacific 2008
[27] Wardle, C. and Pearce P., *Simulating Energy Management within the Virtual Submarine*, SimTecT 2011, Melbourne, Australia
[28] http://www.arc.gov.au/pdf/LP11_R1/ARC_Linkage_Projects_2011_Round_1-Application_Detail_Summary_The_University_of_Adelaide.pdf
[29] Rickover, H. G., *Doing a job*. Speech at the Columbia University School of Engineering November 5 1981.
[30] Duncan, F., *Rickover and the Nuclear Navy : The Discipline of Technology*, Naval Institute Press 1990
[31] Folbert A., *Australia's Future Generation Submarine - Crew Requirements*, MAST Europe, Stockholm 2009
[32] Folbert, A, *A New Approach to Submarine Habitability*, DHSS 2010
[33] Folbert, A., *Human Factors and Habitability*, Submarine Institute of Australia Technology Conference 2011
[34] General Dynamics Electric Boat, *The Virginia Class Submarine Program : A Case Study*, Groton, 2002
[35] Blanchard, B. S. & Fabrycky, W. J., *Systems Engineering and Analysis*, Prentice-Hall 1998
[36] Plant, P., *Early Stage Tools for Submarine Design : Boat-by-Numbers*, Submarine Institute of Australia Technology Conference 2011
[37] Pearce, P., *A Roadmap for Simulation-Based Acquisition*, SimTecT Adelaide 2009
[38] Harrison, R., *Ocean Environment Modelling - Interpreting a Large Data Structure in Real Time*, SimTecT 2010, Brisbane, Australia.
[39] Lalor, P, *The bridge: the epic story an Australian Icon : the Sydney Harbour Brigde*, Allen & Unwin 2005
[40] Collins, B., Snowy: *The Making of Modern Australia*, Hodder & Stoughton Ltd 1991
[41] Gunnar Eliasson, *Advanced Public Procurement as Industrial Policy*, Springer 2010
[42] Law, M., *Engineering Masters Programs to Support Submarine Design and Maintenance*, Submarine Institute of Australia Technology Conference 2011

The use of modern high strength steel grades in the Australian Air Warfare Destroyer (AWD)

J.P.W. Verdaasdonk [1,2], R Spyker [1,3], Z. Sterjovski [4]
1 Air Warfare Destroyer Alliance
2 Raytheon Australia Pty Ltd
3 ASC Pty Ltd
4 Defence Science and Technology Organisation (DSTO)

Jan.Verdaasdonk@ausawd.com

INTRODUCTION

The Australian Defence Force (ADF) is seeking to markedly improve the overall capability of the surface combatant force by the acquisition of a new Air Warfare Destroyer (AWD) through Project SEA 4000. At least three ships are to be built in Australia.

The AWDs must have a capability to operate in extreme cold temperatures which could be experienced in the Southern Ocean. In such extreme conditions, steel can become susceptible to brittle fracture, particularly if subjected to combat damage above the water line. As a consequence, the AWDs must be constructed of a steel grade that has adequate low-temperature toughness to provide acceptable resistance to cold-brittle fracture.

The steel grade originally used in the baseline Spanish F100-class did not meet Australian requirements. Consequently, the steel for the AWD design was changed from AH36 to DH36 for the hull, and DH55 was retained for the crack-arresting strakes.

In collaboration with DSTO, the AWD Alliance characterised the low-temperature-toughness properties of DH36, and DH55. The notch toughness of these steels was investigated at a range of cold temperatures by Charpy V-notch (CVN) impact testing and explosive-load testing (ELT). The ductile-to-brittle-transition temperatures of these steels were determined to assess the risk of brittle fracture in cold environments.

This paper presents the process of validation and testing performed for the DH55 and DH36 grade steels used in the construction of the AWD to ensure they support the required cold weather capability and satisfy ADF requirements.

BACKGROUND

Following the evaluation of various acquisition options, in June 2007 the Commonwealth selected the F104 (as in-service with the Spanish Armada with certain modifications) as the preferred design for the platform system for the Air Warfare Destroyers.

The AWD Program is executed by the AWD Alliance consisting of the Commonwealth of Australia, and the Industry Participants, namely ASC and Raytheon Australia. The AWD Alliance is empowered and is responsible for the conduct of the AWD Program.

The Commonwealth has engaged Navantia as Platform System Designer (PSD). Construction is taking place at the ASC shipyard adjacent to the SA government's

Common User Facility at Techport, Osborne with blocks being built at sites in Adelaide (ASC), Ferrol (Navantia), Melbourne (BAES) and Newcastle (FORGACS).

AWD HULL DESIGN

The AWD is a destroyer class ship with an overall length of 146.7m, beam of 18.6m and a maximum displacement of 7000t, carrying a crew of 234 at a top speed of 28+ knots.

The F104 is a typical example of warship design with longitudinal stiffening of the hull panels, supported by web frames. The longitudinal structure of the AWD is a combination of T-profiles and Holland "Bulb" Profiles, with the first being installed in the double bottom and the latter on the vertical side shell.

In accordance with the USN Standard [11], High Yield Steel shear strakes are included in the shell plating at the bilge and deck edge to act as crack arrestors. The shell plating of the Spanish F104 is constructed in AH36 for the majority of the shell plating and DH55 for the crack arrestor strakes (Figure 1). The combination of all these design features creates a high strength light weight hull structure.

The AWD hull required an improved brittle fracture performance in low temperature environments to allow for deployment into Australia's southern waters. This was achieved through the use of DH36 grade steel in lieu of AH36 grade steel. This material change is implemented throughout the whole ship structure and the suitability was agreed to be confirmed by an investigation into the material selected for the AWD.

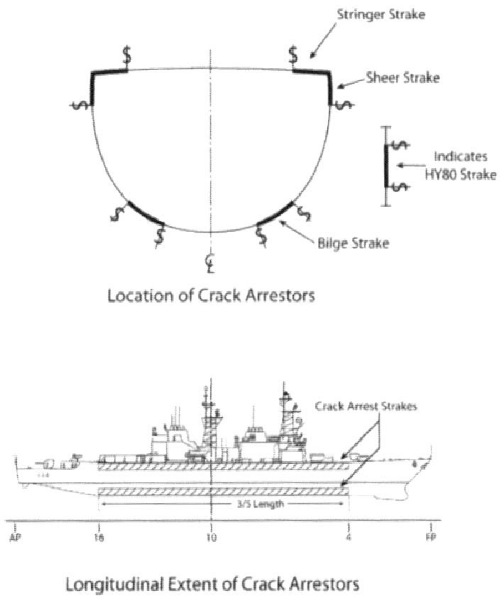

Figure 1: General crack arrest strake locations [11]

STEEL SELECTION CRITERIA

The AWD hull design is based on the Spanish F104 hull design, which follows the USN Design Data Sheets as detailed in the NAVSEA Structural Design Manual for Naval Surface Ships [6] or structural design of the hull.

In the past it was usual for the USN to build combatant ships with High Tensile Steel (HTS) and HY80 steel. The HY80 was normally employed as a crack arrestor at the critical high stress areas such as the stringer, sheer and bilge strakes and the HTS steel was employed in the remainder of the ship's hull. In accordance with NAVSEA Design Manual

[6], the specification for HTS plate is MIL-S-16113. However, through time, this specification has been superseded by MIL-S-22698C [13]. In Section 6.4 of MIL-S-22698C, a table is provided which incorporates data for the HTS and lists DH36 steel grade as an equivalent steel to HTS grade-type 1. This information is also incorporated in the American Bureau of Shipping (ABS) Guide for Building and Classing Naval Vessels Rules Part 2, Chapter 11, Section 4 [10].

The steel grading is based on the tensile strength of the steel, where the Prefix of A, D or E is based on the temperature where the CVN impact tests are performed (at resp. 0°C for A grade, -20°C for D grade and -40°C for E grade) The XX36 steel has a minimum yield stress of 355 N/mm^2 and XX55 grade steel a minimum yield stress of 550 N/mm^2. [7]

The use of DH36 steel grade for the main part of the AWDs hull and superstructure is therefore in accordance with the intent of the USN structural standard and has been accepted as equivalent to HTS as indicated above. Nonetheless, the selection of DH36 was evaluated for Australian operating conditions (see "Materials Testing" section). It should be noted that the ABS DH36 is equivalent to the Lloyd's DH36 steel grade.

SHEAR STRAKE MATERIAL

Hull design in line with the NAVSEA Design Manual [6] requires the implementation of sheer and bilge strakes of a tougher material than the rest of the hull (to inhibit crack propagation). The toughness of these arrestor strakes would prevent brittle fracture, paragraph 8.5.1of NAVSEA Design Manual [6] states that a suitable sheer strake material would be grade HY80.

The design of the AWD incorporates DH55 for the shear strakes instead of HY80. Although the tensile strength of both materials is equal, DH55 is not equivalent to HY80 with respect to specified toughness.

COLD WATER OPERATIONS

The AWD may be required at times to operate in cold climates and therefore the toughness of the hull steel at very cold temperatures is an important consideration. The temperatures the ship will be exposed to in these waters will be as low as -2 ºC in seawater and -20 ºC in air. There is an increased risk to the onset of brittle fracture if the steel toughness at cold temperatures is low or unspecified, such as A-Grade steel which has no toughness requirement below 0°C. For this reason, Lloyd's Rules [5] advises that the hulls of ships intended for operation at temperatures below 0°C should not be built of steel of A-Grade steel. Also, Lloyd's Rules for the Manufacture, Testing and Certification of Materials [7] have a section dedicated to ferritic steels for low temperature service. In this section, the minimum design temperature for the various low temperature grades of steel are specified.

The selection of a suitable hull steel will reduce the risk of brittle failure since the ductile-brittle transition temperature is highly sensitive to metallurgical factors such as composition, grain size, and precipitate morphology [8]. The AWD has been designed against the onset of brittle fracture to improve its overall operational capability.

BRITTLE FRACTURE

As described in the literature [2] Brittle fracture is the sudden, extensive cleavage fracture with negligible plastic deformation under a tensile stress at stress concentrations in ferrous steels. It manifests itself in a sudden rapid propagation of cracks through a stressed material. The applied stress level can be significantly lower than the catalogued allowable stress of the material.

The three main contributing factors to brittle fracture are:

- Low environmental and operating temperatures,
- A tri-axial state of stress,
- A high rate of loading. There is an increased risk with rapid or shock loading or strain rate > 0.1/sec (e.g. ballistic attack on tanks or explosive attack on ships and submarines).

All three of these factors do not have to be present at the same time to produce brittle fracture. A tri-axial state of stress, such as exists at a notch, and low temperature are responsible for most service failures of the brittle type. However, since these effects are accentuated at high rates of loading, many types of impact tests have been used to determine the susceptibility of materials to brittle behaviour. Steels that have identical properties when tested in tension or torsion at slow strain rates can show pronounced differences in their tendency for brittle fracture when tested in a notched-impact test and other test methods.

MATERIALS TESTING – METHODS, RESULTS AND DISCUSSION

Explosive-loading tests (ELTs) were undertaken to assess the fracture performance of 8 mm thick DH36 and 10 mm thick DH55 steel plates. The sample plates were sourced from the steel purchased to build the AWD. The ELT, which was developed by DSTO as a modification of MIL-STD-2149A [12], was selected because it provides a more realistic representation of high-strain-rate military-loading events that the AWD may be exposed to in-service (as compared with the conventional CVN impact test).

In the ELT (Figures 2- 3), a 50 mm long brittle weld is deposited onto the centre of a steel test plate. The plate is cooled in a freezer, then placed over a circular die and dynamically loaded with an explosive charge once the plate has reached the designated test temperature. The brittle weld, often referred to as a crack-starter weld, introduces a fine crack (with high-notch acuity).

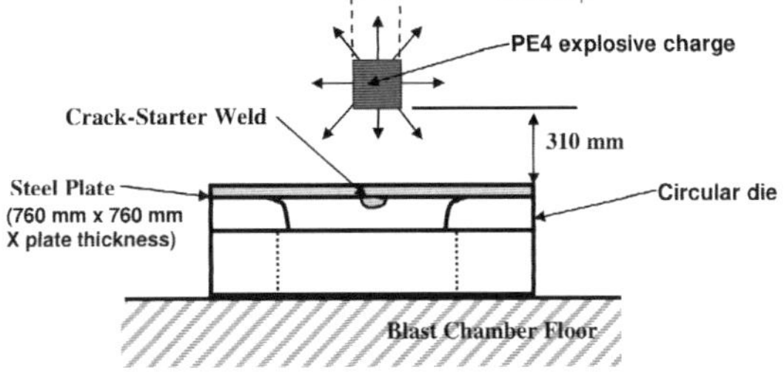

Figure 2 Schematic diagram of test set-up for the explosive-loading test (not to scale).

As the plate is forced through the circular die, the fine crack in the brittle weld usually propagates into the steel plate. The appearance of the surface of the crack in the steel plate is then evaluated with a view to ultimately determine the ductile-brittle transition temperature ($DBTT_{ELT}$) for particular steels.

For both 8 mm DH36 and 10 mm DH55, ELTs were carried out over a range of temperatures to converge on the ductile-brittle transition temperature ($DBTT_{ELT}$). Steels typically exhibit a shift from ductile to brittle fracture as temperature decreases, and the transition from predominantly ductile to brittle fracture usually occurs over a range of temperatures. The crack/fracture appearance from each ELT is characterised according to its percentage of ductile fracture, and this information is then used to establish the $DBTT_{ELT}$. The $DBTT_{ELT}$ for the current work was defined as the temperature at which the fracture appearance is 50% fibrosity (or ductile shear) and 50% brittle cleavage.

Figure 3: Photographs illustrating the key elements of the ELT procedure.(a) Removal of the test plate from the freezer, (b) the circular die, (c) placement of the test plate on the circular die, (d) setting up the stand-off distance between the PE4 charge and the plate using the cardboard box or by suspension with nylon netting, (e) monitoring plate temperature into the control room in preparation for the blast, and (f) the plate and die block after the ELT.

Ductile fractures (Figure 4(a)) absorb high levels of energy and are characterised by appreciable plastic deformation prior to and during the propagation of the crack. On the other hand, brittle fracture (Figure 4(b)) in steels is a comparatively low energy event and is characterised by an extremely rapid rate of crack propagation and minimal plastic deformation at the fracture surface. Ductile shear lips (Figure 4(b)) are evident in predominantly brittle fractures. These shear lips are indicative of a plane-stress state that is common in the fracture of relatively thin steels and they contribute to the percentage fibrosity on the fracture surface.

The rules pertaining to the classification of naval surface ships only consider CVN impact toughness [7]. In view of that, the CVN ductile-brittle transition temperature ($DBTT_{CVN}$) and impact energy for DH36 and DH55 were also determined. These tests were carried out in accordance with AS1544.2-2003 [9]. However, the CVN impact test has limited usefulness in that it does not truly reflect the strain rates that the AWD may be exposed to in-service. Additionally, the acuity of the machined notch in the CVN impact specimen does not truly represent the acuity of planar crack-like defects that typically exist in real-life structures.

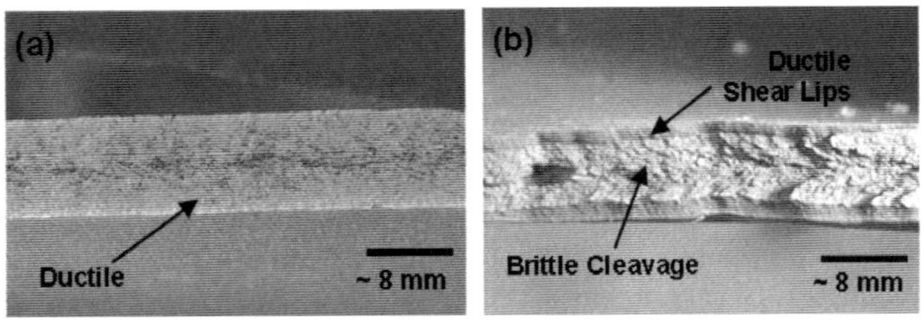

Figure 4 Typical fracture surface appearance for a predominantly (a) ductile fracture, and (b) brittle fracture, in test plates subjected to ELTs.

The CVN impact test is comprised of a small-scale machined-notch specimen that is subjected to impact loading by a swinging pendulum which results in the fracture of the specimen (Figure 5). The nominal test temperature for each CVN impact test is achieved by immersing the test specimen in a low temperature liquid bath.

As was the case for the ELT, the fracture appearance of each specimen was evaluated in order to determine the percentage of fibrosity (or ductile shear fracture) on the fracture surface. The current paper only presents results for the $DBTT_{CVN}$ of 8 mm DH36 and 10 mm DH55 for specimens oriented transverse to the plate rolling direction. However, it should be noted that DH36 and DH55 both met all the impact energy requirements as specified by Lloyd's Rules for the Manufacture, Testing and Certification of Materials [7].

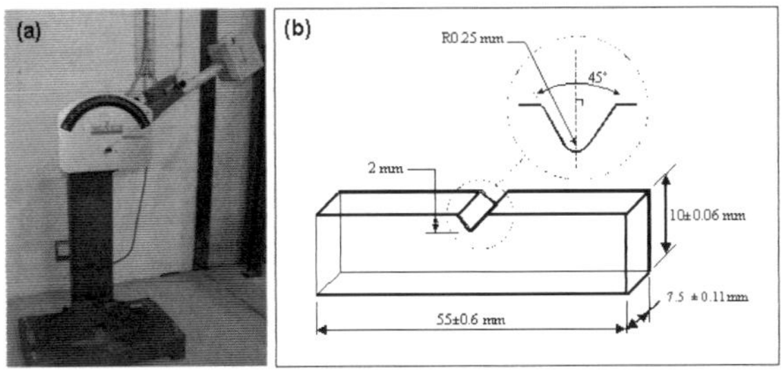

Figure 5(a) Photograph of the Charpy impact machine used in the current work, and (b) a schematic diagram of a 7.5 mm sub-size CVN specimen (Note: the cross-section for a standard specimen is 10 mm x 10 mm).

The experimental test program for the explosive-loading and CVN-impact tests of DH36 and DH55 are given in Tables 1 and 2, respectively. Tests were carried out at a variety of temperatures to converge on the $DBTT_{ELT}$ and determine the $DBTT_{CVN}$.

Table 1: ELT program for 8 mm DH36 and 10 mm DH55 steel plates.

Specimen	Thickness	Specimen size	Test Temperatures (°C)
DH36	8 mm	760 mm x 760 mm	-37, -32, -27.5, -27, -24,-22.5, -20, -17.5, -17, -13, -5, 22, 32
DH55	10 mm	760 mm x 760 mm	-28, -27, -26, -24.5, -23.5, -18.5, -14.5

Table 2: CVN Test program for 8 mm DH36 and 10 mm DH55 steel plates. The sample was oriented transverse to the direction of plate rolling. The specimen sizes used are in accordance with AS1544.2-2003 [10].

Steel	Specimen size	Test Temperatures (°C)
8 mm DH36	Sub-size (7.5 mm)	22, 0, -20,-40,-60,-80,-100,-110
10 mm DH55	Standard (10 mm)	-20,-40,-60,-80,-100,-106,-112

The $DBTT_{ELT}$ and $DBTT_{CVN}$ curves for 8 mm DH36 and 10 mm DH55 are shown in Figure 6. From this figure, it is evident that there is a significant difference between the $DBTT_{ELT}$ and $DBTT_{CVN}$. For 8 mm DH36, the $DBTT_{ELT}$ was determined to be -22.5°C, whereas the $DBTT_{CVN}$ was -66°C. For 10 mm DH55, the $DBTT_{ELT}$ was determined to be -26°C, whereas the $DBTT_{CVN}$ was -109°C. This significant difference in the transition temperature between the two test types is due to differences in notch acuity and strain rate. The transition temperature for both DH36 and DH55 in the high-strain rate ELT is below the minimum design temperature (-20°C) for the AWD, which suggests that the use of these steels for the hull of the AWD should satisfactorily reduce the risk of brittle fracture in cold environments.

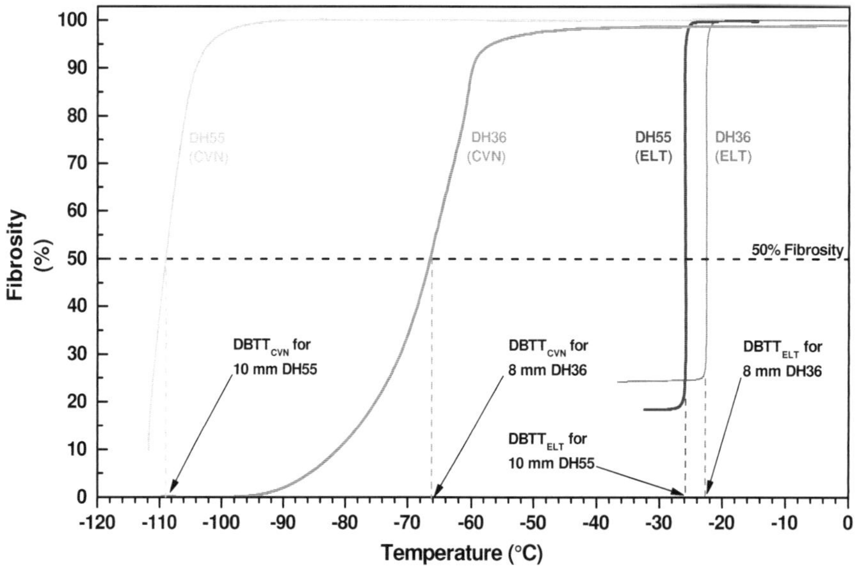

Figure 6 Fibrosity (%) versus test temperature for DH36 and DH55 test plates subjected to explosive-loading and CVN impact tests. The ductile-to-brittle transition temperature is defined as the temperature at 50% fibrosity.

The difference in the ELT transition temperatures between DH36 and DH55 was only 3.5°C. This small difference between these two steels also tends to suggest that the DH36 steel used for the AWD is of high quality and should perform as required for Australian conditions. The original ship steel specification called for Grade A steel, which was replaced by Grade D steel and as a result of this collaborative experimental program the project confirmed the change to DH36 has significantly reduced the risk of brittle fracture in cold environments.

The superior resistance of DH55 to both plastic deformation and cracking is evident when explosively loaded DH55 and DH36 steel plates are compared (compare photographs in Figure 7 to Figure 8). The increased resistance to deformation of 10 mm DH55 relative to 8 mm DH36 is evidenced by the smaller bulge depths in DH55, which may be attributed to the higher strength and greater plate thickness of DH55. The DH55 steel also outperformed DH36 in resisting high strain-rate fracture, illustrated by the absence of significant cracking at temperatures above the $DBTT_{ELT}$ in DH55 (compare Figure 7(a-e) to Figure 8(a-f)). Additionally, at temperatures below the $DBTT_{ELT}$, the severity of cracking in DH36 was much greater than in DH55 (compare Figure 7(f-i) to Figure 8(g-i)).

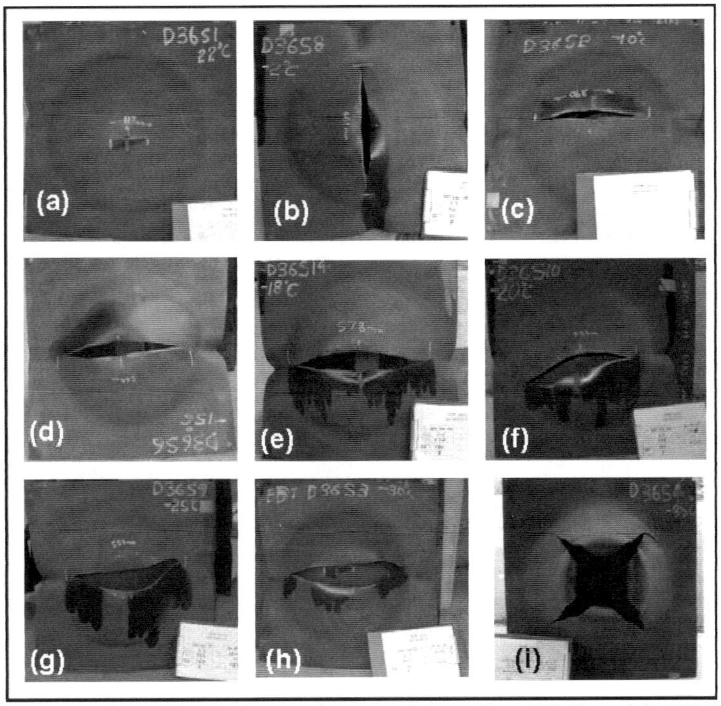

Figure 7 Photographs of DH36 test plates subjected to ELT at (a) 22°C, (b) -5°C, (c) -13°C, (d) -17.5°C, (e) -20°C, (f) -22.5°C, (g) -27.5°C, (h) -32°C and (i) -37°C. Note: The temperature scribed on the test plates are the thermocouple temperature (and not the temperature of the centre of the plate).

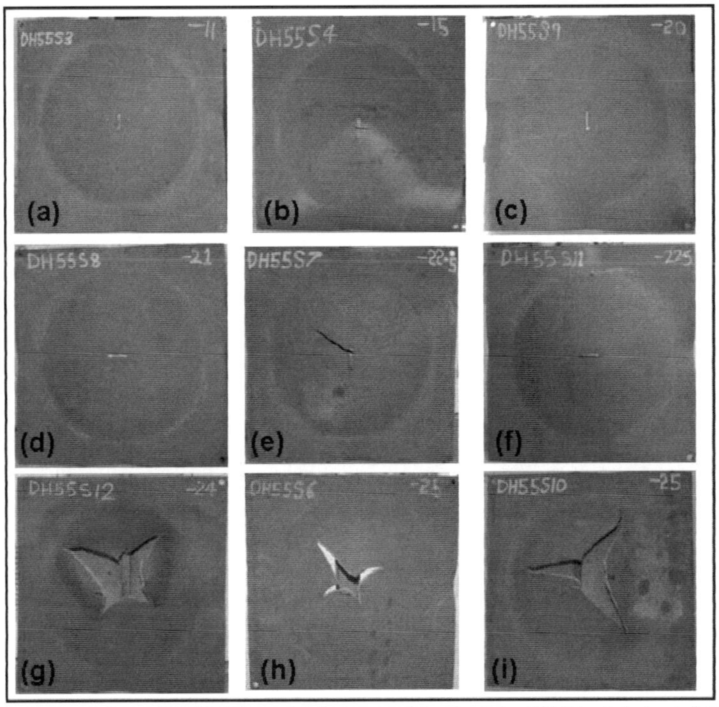

Figure 8 Photographs of DH55 test plates subjected to ELT at (a) -14.5°C, (b) -18.5°C, (c) -23.5°C, (d) -24.5°C, (e) -26°C, (f) -26°C, (g) -27°C, (h) -28°C and (i) -28°C. Note: The temperature scribed on the test plates are the thermocouple temperature (and not the temperature of the centre of the plate).

Additionally, Figure 5.2 in the NAVSEA Structural Design Manual shows a graph of the CVN temperature transition curve for typical production of HY80 plate. When the results of the CVN tests for DH55 steel grade are superimposed onto this plot it is evident that the supplied AWD DH55 steel exceeds the theoretical upper range of notch toughness of typical HY80 grade steel.

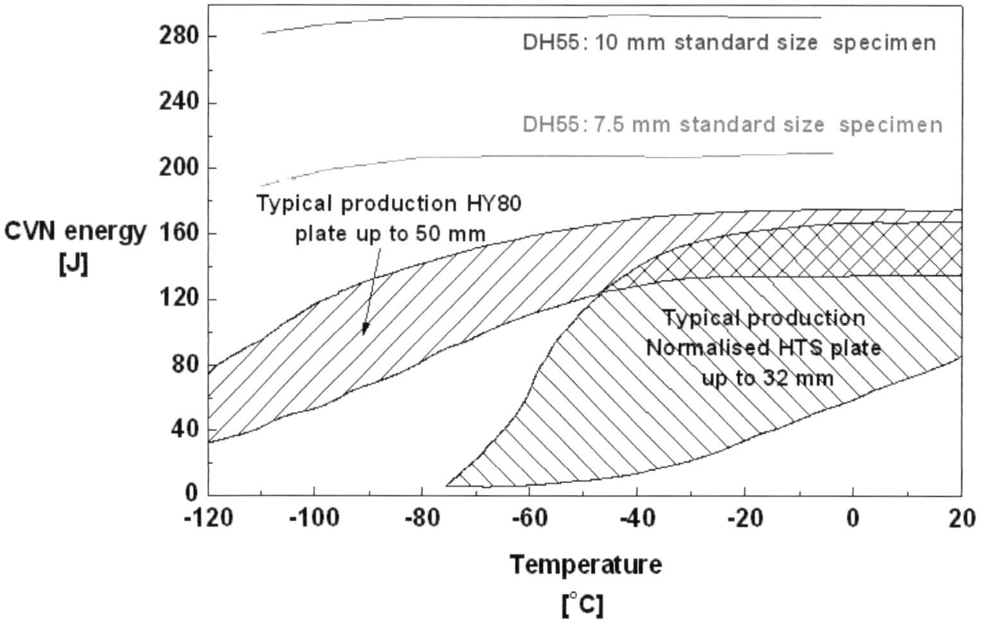

Figure 9 NAVSEA [6] temperature transition curves for typical HY80 based on standard size CVN specimens. (Modified). DH55 temperature transition curve data for standard and 7.5 mm sub-size specimens are superimposed to show that their CVN toughness values are greater than those typically determined for HY80.

The results and discussions in this section show that the steel currently selected for the AWD is suitable for the operations defined for the platform. In order to ensure the structural integrity of the AWDs, the notch toughness of each batch of steel is verified by CVN impact testing before being released into production.

CONCLUSIONS

Original selection of steel grades

The original selection of steel for the F100 Class with AH36 steel grade for all plating (except the stringer, sheer, and bilge strakes, where DH55 was selected) was considered unsuitable for a front line naval ship when operating in cold conditions. The NAVSEA Design Manual used in the design of the F100 does not include special requirements for cold weather operation. It was noted that for cold water operations a minimum Grade of DH would be required for the AWD to comply with Classification Society Rules for surface combatants operating in cold weather climates.

Final selection of steel grades

Based on considerations noted above, the AH36 steel grade was replaced with DH36 for the main hull, while the DH55 steel remained for the high stress areas at the stringer, shear and bilge strakes.

The results of the experimental testing program, as described in this paper, have shown that the notch toughness and ductile to brittle transition temperature of the DH36 and DH55 steel used in the construction of the AWDs is appropriate for the expected operating environment of the AWDs.

The suitability of replacing AH36 with DH36 was validated by the results from the DBTT transition temperature testing for DH36. Further, the results confirmed that the notch toughness of DH55 exceeds the minimum requirements for HY80, which indicates the DH55 should be an effective crack arrestor material.

The AWD Alliance, in collaboration with DSTO, has modified the AWD materiel specification to meet the demands of a cold weather operating environment and in so doing has potentially enhanced the ADF's surface combatant operational capability.

REFERENCES

1. T9074-BD-GIB-010/0300 Appendix B Steel Plate, Alloy, Structural, High Yield Strength (HY-80 and HY-100)
2. Engineering Materials 1, M.F. Ashby & D.R.H. Jones, University of Cambridge, 1998
3. DSTO-CR-2507-2009, DSTO Report Explosive Load Testing of the Grade DH-36 and DH-55 Steel Plates Selected for the Air Warfare Destroyer, 2009
4. AWD-0002776 Selection Of Hull Steel For The HOBART Class Air Warfare Destroyers (Internal Report)
5. Lloyd's Register of Shipping Rules and Regulations for the Classification of Naval Ships, Vol 1, Part 6
6. NAVSEA 0900-LP-097-4010, Structural Design Manual for Naval Surface Ships, 15 Dec 1976
7. Lloyd's Rules, Rules for the Manufacture, Testing and Certification of Materials, July 2007
8. G.E Dieter, Mechanical Metallurgy – SI Metric Edition, McGraw-Hill Book Company, 1988.
9. Standards Australia, AS1544.2-2003 Method for impact tests on metals Part 2: Charpy V-notch
10. American Bureau of Shipping (ABS) Guide for Building and Classing Naval Vessels Rules Part 2, Chapter 11, Section 4
11. MIL-HDBK-519 Handbook For The Design And Assessment Of Naval Surface Ship Structure
12. MIL-STD-2149A, TEST METHOD STANDARD,: STANDARD PROCEDURES FOR EXPLOSION TESTING FERROUS AND NON-FERROUS METALLIC MATERIALS AND WELDMENTS, 1990 US Department of Defense
13. MIL-S-22698C. MILITARY SPECIFICATION: STEEL PLATE, SHAPES AND BARS, WELDABLE ORDINARY STRENGTH AND HIGHER STRENGTH: STRUCTURAL (29 JUN 1988)

Employing System Architecting to Achieve Acquisition Risk Mitigation and Sustainment Focus on the Warfare Destroyer Program

Steven J. Saunders, FIEAust CPEng,
Raytheon Australia Pty Ltd
Steve.saunders@ausawd.com

Major defence acquisition programs, like the SEA 4000 Royal Australian Navy Air Warfare Destroyer (AWD) Program involve complex integration and hence risks to on-budget and on-time delivery. These risks manifest from the high level of complexity of organisational, programmatic and technical inter-relationships. Focusing the engineering effort on near term acquisition considerations may achieve delivery of the required capability, however these acquisition considerations must be balanced with the longer term sustainment considerations for the capability in order to ensure total ownership costs are minimised.

This paper describes how the AWD Program has employed system architecting early in the program to establish a robust architecture in order to minimise risk to acquisition. This architecture has established flexible evolution and technology refresh paths to ensure the capability is able to be sustained for the 30 year life of the capability. Lessons from the AWD Program are presented in order to provide strategies that may be successfully leveraged into future Mission System Integration programs undertaken by the Australian Defence Force as well as around the world.

INTRODUCTION

Organisations providing Mission System Integration (MSI) services are faced with Systems and Systems-of-Systems (SoS) involving high levels of complexity. The complexity is manifested in a number of dimensions such as; system scope, multiple viewpoints from numerous stakeholders, temporal scope covering many technology evolutions and system capability change. This complexity has long since surpassed the ability to conceive and design a balanced solution that best addresses all the concerns from all of the stakeholders over the full life of the system using simple design tools and within the cognitive capacity of the human mind. Previous strategies, using traditional engineering tools and drawing on "expert" engineers have since been augmented with the employment of Enterprise Architecture practices.

The RAN SEA4000 Air Warfare Destroyer Program has adopted a strategy to collaboratively work with RAN Capability Development group and industry prior to the construction phase to ensure the best capability can be successfully delivered within the constraints established for the program. Instrumental during this early phase was the adoption of System Architecting to ensure the combat system could be designed, procured and integrated within cost and schedule, whilst ensuring the system is sustainable for the life of the system.

Techniques used to ensure the best capability could be delivered on time focused on aligning the integration strategy with the system architecture and available integration facilities from around the world. The program is currently seeing the return from this strategy with significant progress seen on combat system integration being conducted in six countries concurrently. The integration program has focused on mitigating the higher risk interfaces first and is working to an integration program with 12 months of float in the schedule.

A significant portion of the architecting effort has been to map out the combat system sustainment for the 30 years the combat system will remain in service. Techniques have provided flexibility for future technology refreshes by enabling different areas of the combat system, with differing technology drivers, to be updated independently. Similar features allow new capabilities to be added over time without requiring entire combat system re-certification activities. These features will ensure the HOBART Class Combat System remains sustainable and continues to provide the highest possible capability to the RAN for the first half of this century.

SYSTEM COMPLEXITY

System complexity takes on a number of dimensions. First reactions generally associates complexity with a high number of system components and interfaces (technical complexity). However, system complexity takes many forms and may include;

- Complexity associated with component and interface count,
- Complexity in dealing with the multiple, and often conflicting needs from stakeholders,
- Complexity in dealing with technology change possible over a system lifecycle; measured in decades,
- Complexity in dealing with system capability insertion that may be required of the system over the system lifecycle,
- Complexity in the magnitude and partitioning of information processed by the system,
- Complexity with respect to the effect that the system environment and the broader enterprise has on the development and use of the system,
- Complexity in the managed transition from current to new systems in the context of the ongoing enterprise, and
- Complexity with respect to how the environment (external system) in which the system is to operate within evolves over time.

Each of these dimensions on their own provides unique challenges to an organisation intending to deliver systems meeting all stakeholder needs. However, when all dimensions are processed simultaneously, an acceptably balanced solution becomes increasingly difficult to define. When systems cover longer and longer time spans, complexity increases due to the increasing uncertainty in the multitude of variables within the system under definition, as well as the changing expectations of the stakeholders.

Use of Architecture to Manage Complexity

The ability to manage complexity is related to the underlying limitation in the human's ability to handle multiple sensory stimuli (Miller 1956). Miller describes experiments where limitations are observed and linked to limitations in the human short term memory, or how much information can be processed at the one block of time. The ability to manage more information increases by organizing information in a number of dimensions, re-coding and "chunking" information. This allows more information to be processed whilst not exceeding the fundamental cognitive limits of the human mind for absolute judgment, span of attention or span of immediate memory.

Architecture development involves building abstract models to represent the system solution in a way that allows the architect to make informed decisions and communicate system architectural features that support the various stakeholders' views. This architecture definition phase is aimed at balancing the often conflicting needs of the stakeholders in order to develop a system architecture that amplifies features to deliver a higher valued system to the various stakeholders. During the architecture trade-off phase, high levels of detail/complexity become a hindrance, swamping the system architect with information overload and masking potentially important features and relationships in the system.

To manage this enormous complexity, various architecture tools, processes, models and frameworks are available to the architect. These are employed to address the unique problems presented by the system being defined. At the beginning of a program, architecture planning should be utilized to scope the architecture products that best suit the system to be defined (TOGAF 2007). Not all architecture products or processes available in the literature or industry are required, nor is one model or framework necessarily complete. A minimum but comprehensive selection of architecture products is generally all that is required.

Architecture definition involves grouping and modeling the system architecture, as well as providing multiple views into the architectural model as depicted in Figure 1. These viewpoints should not be generated ad-hoc rather they need to be aligned to the key stakeholders identified in the initial steps of architecture definition (Saunders 2007). There will normally be many views generated, each aimed at providing information that allows (human) reasoning into how the architecture supports various stakeholder needs. For example, a security view of the system architecture will highlight security features included in the architecture. A system evolution view highlights how the system architecture will support managed evolution over time. These views only provide details relevant to the specific stakeholder interests in order not to unduly overload the intended audience with non-relevant information

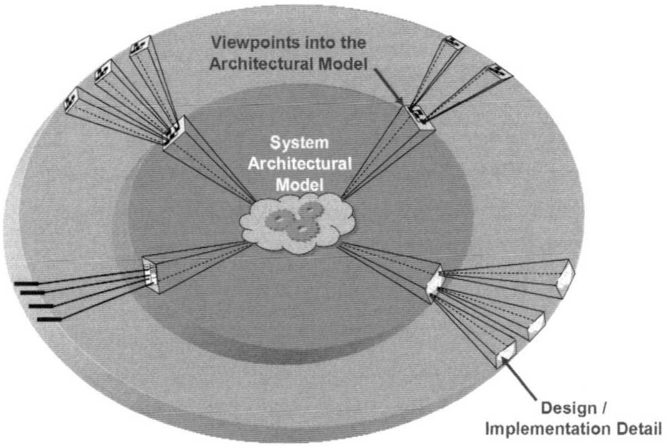

Figure 1 Abstract Architectural Model with various Viewpoints into the Model

When comparing to research performed by Miller (1956) and Pollack (1952), it becomes apparent that architectural models provide;

- Ordering of information into varying dimensions (viewpoints), and
- Re-coding and "chunking" of information into lager information blocks (nodes, components, systems).

By hiding complexity within the models the architectural models enable decisions to be made on the underlying structure and relationships within the model to highlight and develop those features valued by the stakeholders.

SYSTEM ARCHITECTING vs SYSTEM DESIGN

The terms architecture and design are often used interchangeably, however this is improper. The terms architecture and design should be used to refer to distinct phases of system development. "Architecture" is defined as *"the structure of components, their relationships, and the principles and guidelines governing their design and evolution over time"* (DoDAF 2004). Architecture products consist of abstract models and views that describe underlying structure and relationships between system components.

This single system architecture can be translated into a number of separate design realisations. Each design is real, tangible and employs specific technologies in the solution. The design is captured in engineering drawings and specifications which translate to real products. The architecture is "instantiated" by a selected technology set and the abstract models of the architecture are translated into a real instance of that architecture.

Evidence that a single architecture can be realised with a number of designs is seen in marketplaces that are required to change products rapidly to track a fast evolving marketplace. Excellent examples of this are mobile phone companies who employ a stable "reference architecture", within a number of mobile phones in a product line (McGregor 2004). Each mobile phone can be rapidly designed as the underlying architecture has been purposely developed to support flexibility and change.

Maximising Delivered Value Through Architecting

It has been shown from statistical analysis of projects run by the US DoD, the majority of a program's cost is committed in the initial phases of a program with a relatively small expenditure of resources (Defense Acquisition University 1993). This is illustrated in Figure 2, where it is seen that up to 70% of the final program costs will have been committed by the decisions made in the early phase of the program.

Significantly, a large portion of the final system's cost is committed prior to the system design commencing. This is due to the impact of the structure (architecture) upon the acquisition and through life costs. Of note is this structure (architecture) is put in place either by overt action (the act of architecting) or by default; if no architectural effort is applied. If not overtly architected, the default architecture will be used which (logically) is expected to impact the through life cost in a random manner, and generally in a detrimental manner.

Extrapolating from this, it can be asserted that developing a good system architecture at the start of the program will lead to a more effective design that meets the stakeholder's needs with a higher probability of achieving lower through life costs.

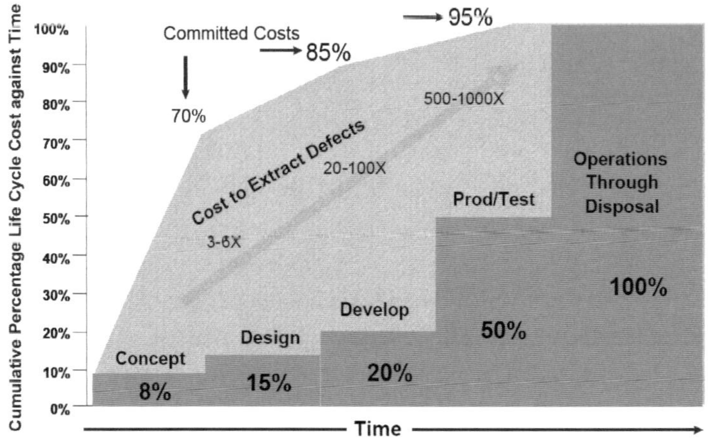

Figure 2 Committed / Expended Program Cost vs. Program Phase

AWD PROGRAM

Overview

The Royal Australian Navy (RAN) SEA4000 Air Warfare Destroyer (AWD) Program is the largest Naval Program undertaken in Australia. It will deliver in a timely and efficient manner an affordable, effective, flexible and sustainable Air Warfare Destroyer capability for the security of Australia. The AWD program must develop an Australian combat system, integrating components mandated by the Australian Government, with commercial components and systems onto the Spanish F100 class platform. Three ships are to be developed and delivered along with land based support systems.

History

SEA 4000 was established in 2002, after which the AWD capability was defined and the Aegis Weapon System was selected as a mandated component for the combat system. In May 2005, Raytheon Australia was selected as the Combat System Systems Engineer (CSSE) to develop and deliver the HOBART Class Combat System for the AWD. During SEA4000 Phase 2, and in the period between May 2005 and October 2007 a HOBART Class Combat System architecture was developed along with two combat system designs based on this architecture. In accordance with Government direction, the two design solutions provided options to the Australian Government to choose a solution that best meets capability within the cost/schedule/risk profile acceptable to the Government. The two designs included;

- "Evolved Solution", being a solution that is based on a parent Navy ship but modified to best meet the Capability Requirements defined in the Function and Performance Specification and Operational Concept Document, and
- "Existing Solution", being a solution that is based on an Off-The-Shelf ship with minimal platform modification as required to meet mandatory requirements.

In June 2007, the "Existing Solution" was selected by the Australian Government and contracts were signed in October 2007 for the AWD alliance partners to enter the build phase (phase 3). The existing solution involved the design of an "Australianised" combat system to be integrated into the Navantia F104 reference platform.

Initiating the AWD Architecture Effort

On the AWD Program, program risk was significantly reduced in phase 2 by adopting an integrated approach to architectural definition, industry surveys, employment of Cost-As-An-Independent-Variable (CAIV) principles, capability modeling and feedback to Capability Development on expected delivered capability. Architectural definition included modeling operational views of the system to describe and analyse fundamental inputs to capability (FIC) and the ways the AWD may be employed in the Australian Defence Force. Additionally, architectural features have been included to reduce the impact of system design changes that may occur during acquisition and through the life of the ship. Architecture definition has been supported by a series of market surveys to validate the required technologies exist and their associated costs and risks fit within the system and program constraints. Programmatically, a rigorous adoption of CAIV ensured the budgetary constraints were adhered to as the system definition continued to be evolved.

Figure 3 illustrates the high level architectural process adopted in the system definition phase. Working with the capability development group, system architecture tools were used to capture and translate a number of operational scenarios into operational activity models. The operational scenarios were referred to as vignettes, with each vignette representing a snapshot in time of an important operational situation. The set of these vignettes used in the operational analysis covered the significant scenarios for which the AWD would be required to operate. In DoDAF terminology, the operational activities represent the Operational Activity view, or OV-5, for the AWD. In deriving the system requirements, these operational activities were analysed to separate those activities which would be performed by the crew from those which would be automated by the AWD system. The completion of this translation resulted in a set of functions that the system must perform (the SV-4 view in DoDAF) along with indicators of required performance (the Measures of Performance; MOPs).

Planning and Problem Definition

Initial planning of the architecture products was undertaken in the initial stages of phase 2 in order to ensure sufficient and suitable architecture products would be produced to achieve the required system definition and to ensure no nugatory or low value effort is expended. On the AWD program, a number of DoDAF products were selected to be developed, supplemented with a number of concepts from TOGAF and the Zachman framework used to assess if any deficiencies may exist in the planned architecture products.

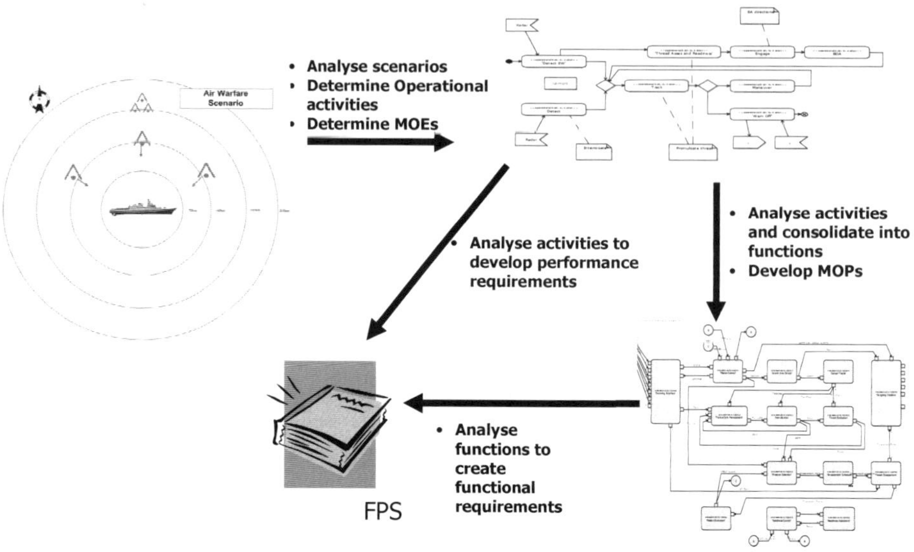

Figure 3 System Definition: Operational Analysis / Requirements Definition

In order to understand the stakeholders and their needs, early effort was applied to identify and list the key AWD stakeholders. These stakeholders were documented and regularly checked to ensure the architectural considerations covered the key interests of each stakeholder. As the problem space was being defined, high level architecture principles and an architecture vision were captured. These principles and vision were used to ensure architecture effort remained focused on the key features required to meet stakeholder concerns. The high level HOBART class combat system architectural vision was defined as;

> *"Develop a robust architecture for the HOBART Class Combat System that enables a low risk First of Class deployment, that is stable over the life of the system, and that supports multiple design variants to enable managed evolution in capability and technology".*

At the conclusion of phase 2, the RAN Capability Development Group and industry had refined the system concept to a mature set of artifacts covering capability requirements, system architecture, cost estimates and error bounds (acquisition cost and total ownership cost), program schedule milestones, risks and opportunities. By including the end users in the upfront collaboration effort, the achievable capability was better understood helping with eventual final acceptance when the ships enter service in the latter half of this decade.

Mitigating AWD Acquisition Risk

In order to lower the risk to the on-time, on budget delivery of the combat system, system architecting was employed to take consideration of the factors that place program delivery at risk. Considerations to this architecture effort include;

- Ensuring the architecture can accommodate the existing functional boundaries present in MOTS/COTS equipment or equipment classes to avoid new developments,
- Understanding and aligning the integration strategy for the program with the tightly coupled groups of equipment/subsystems ("federations of equipment"),

- ➢ Understanding and retaining existing certification boundaries that may exist in products/sub-systems (e.g. weapon systems)
- ➢ Minimising complexity and criticality of interfaces between groups of equipment within the architecture,
- ➢ Understanding and factoring in security boundaries within the architecture,
- ➢ Considering data types and data separation concerns.

On the SEA 4000 AWD Program, all above considerations, as well as many more second order considerations have been factored into the architecture. Multiple architectural alternatives had been considered, rejected or refined before a final stable architecture was arrived at. This architecture formed the basis of the structure of the design that would follow, and has remained stable since 2006 until the present time. Figure 7 illustrates the AWD HOBART Class Combat System architecture, "pulled apart" to show the key architecture federations.

During the design phase, equipment selections can be made based on requirements that are concordant with the element within the architecture that the equipment fulfils. Product selections can therefore focus on the capability strengths and costs of the products. By careful consideration of available equipment categories, it was found that there was a natural match to the architecture, and hence minimised the need to perform significant new (risky) development work during the design phase.

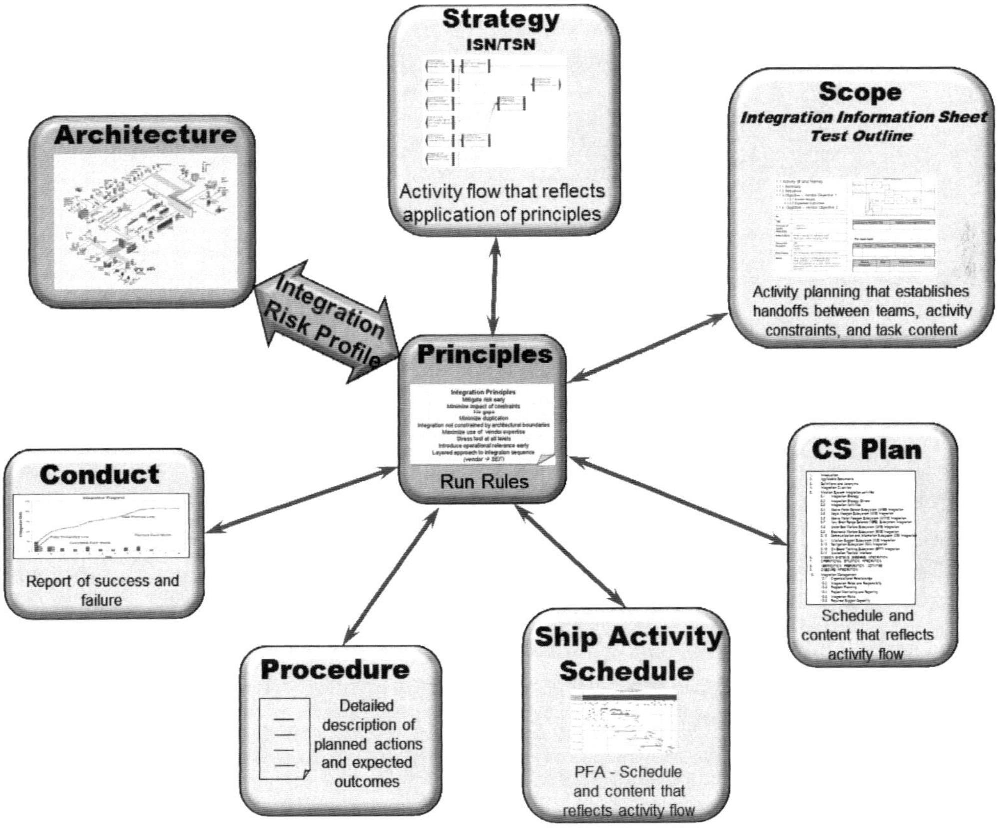

Figure 4 HOBART Class Combat System Integration Planning

In order to help minimise the risk during the integration phase, the integration strategy was developed to closely align with features built into the architecture to allow higher risk interfaces to be integrated earlier in the program as well as leveraging off existing integration facilities that exist in Australia as well as around the world in vendor facilities. Figure 4 illustrates the high level strategy used in the integration planning phase. What is not evident is the level of consultation between the integration group and the architecting group during the planning phase, to collaboratively arrive at an integration strategy and architecture that are complementary. This effort characterised integration risks for every interface and by aggregating these, was able to provide an integration risk profile that was continually managed as equipment selections were made during the tendering phase of the program. The graph shows a low number of high risk interfaces that were managed into the architecture as well as the historical changes in integration risk profile as alternative architectures were refined (up to 2nd pass approval), and then during the equipment selection and design maturation out through PDR and CDR and up to today.

Figure 5 illustrates the interface risk profile as tracked through the early architecting and design phases of the program. The high "churn" prior to second pass is the result of testing alternate architectures before the final architecture was selected.

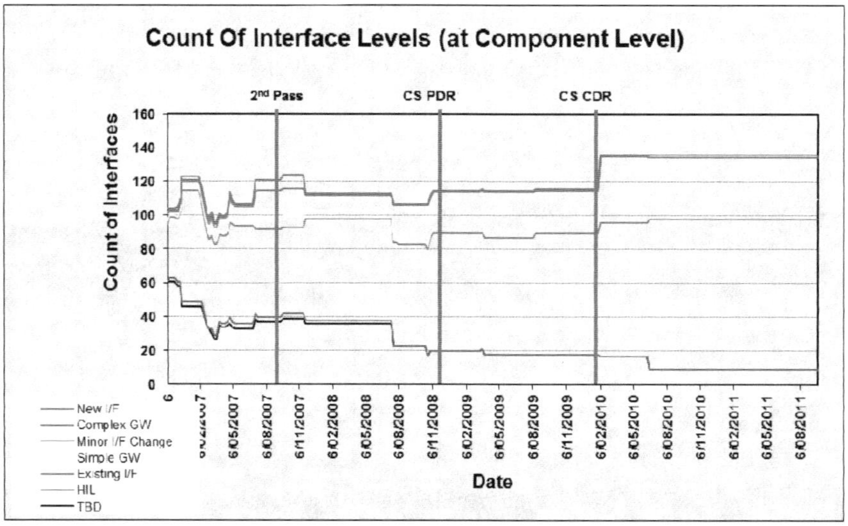

Figure 5 Tracking / Managing Interface Risk Profile

The result of aligning the integration strategy with the architecture has been an integration program that has minimised high risk interfaces and has intentionally integrated the higher risk interfaces earlier in the program. Figure 6 illustrates the initial "designed in" low integration risk approach and the time phased reduction in integration risk since the start of combat system integration in 2010. The integration is able to be conducted concurrently in multiple locations around the world, and building in float to the integration program to ensure the "traditional" integration risk does not place the Program in jeopardy.

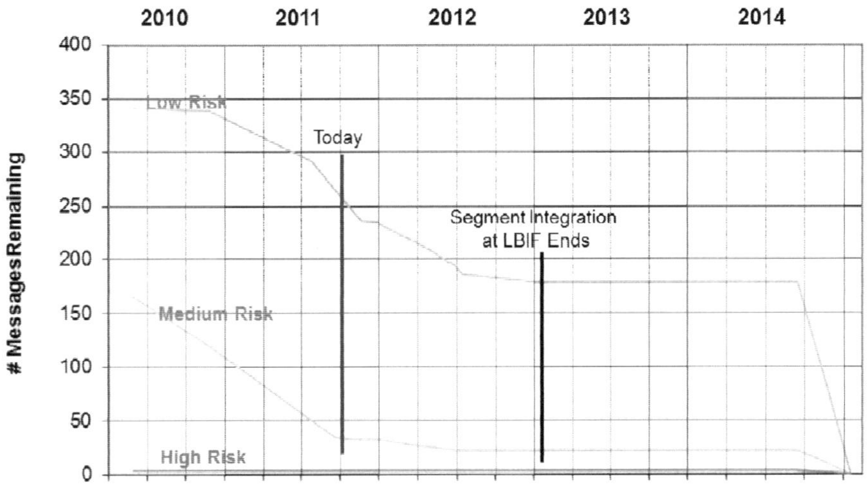

Figure 6 HOBART Class Combat System Integration Risk Burn Down

Mitigating AWD Sustainment Risk

Modern Combat Systems have an expected life that spans many decades. On the AWD Program, the time horizon set for the architecture effort was 40 years (10 years in development and delivery plus 30 years life of type). Understanding the impacts of new capability, changing environment and technology change may have on the system over this 40 years is critical in order to develop an architecture that supports the system as it evolves over this time span.

It is unrealistic to be able to predict all capability growth needs, system environment changes or all technology changes that may occur over the life of the system. However architectural principles can be applied that minimise the downstream impact of these changes. During the architecture definition phase, considerations are able to be factored into the architecture to;

- Separate tightly coupled integration "federations of equipment",
- Align integration and system support strategies with these "federations",
- Retaining equipment certification boundaries fully within tight "federations",
- Aligning like technology components to the "federations of equipment".

Architectural features adopted to architect the system for low risk during acquisition also have a high likelihood of also supporting the goal to minimise the impact of change to the architecture over time. Techniques adopted from TOGAF have been used to assist in understanding the impacts of technology change on the system. This has resulted in the grouping of technology areas centered around;

- USN supplied military equipment, leveraging off USN technology refresh cycles, rates and roadmaps,
- Commercially procured military equipment which are closer aligned to the ADF capability roadmap areas, and

➢ Computing and Infrastructure components which provide flexibility to track the rapidly evolving commercial Information, Computing Technology (ICT) sector for technology trends.

In order to assist the analysis of future capability change, a derivative of Architecture Tradeoff Analysis Method (ATAM) (Bass 2003) was employed to assess the robustness of the HOBART Class Combat System architecture. Numerous "change scenarios" (or hypothetical changes) have been run over the combat system architecture and included numerous independent reviews. The goal was to ensure the architecture is robust and will remain stable over the life of the program. Figure 7 shows the resultant HOBART Class Combat System architecture with the fundamental federations pulled apart to illustrate the decoupling provided. Each of the federations of equipment enables;

➢ Independent technology refresh to occur,

➢ Independent capability insertion to be provided,

➢ Support of re-certification and integration of localized changes rather than requiring complete system re-test.

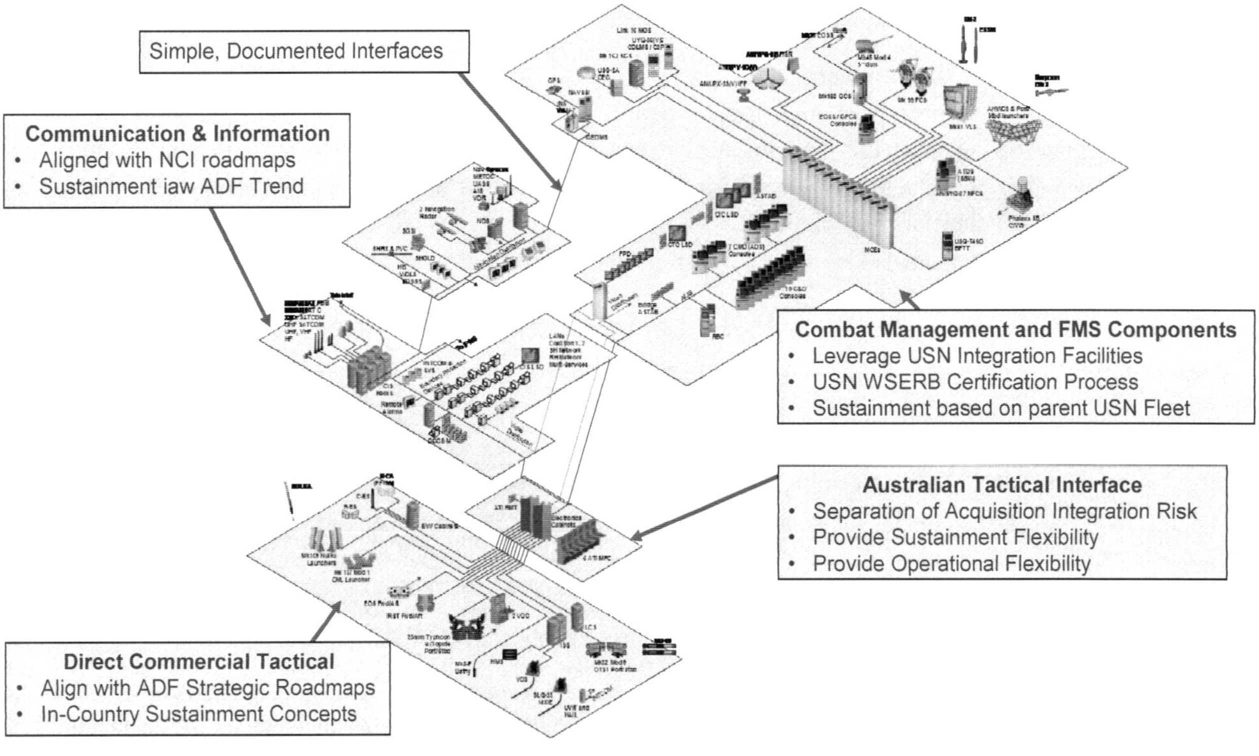

Figure 7 HOBART Class Combat System Physical Architecture

Status of the HOBART Class Combat System

The SEA 4000 Air Warfare Destroyer Program is four years into the design and production phase and has been progressively integrating components of the combat system since 2010. Integration is occurring in Australia and five other countries at various vendor sites, including the United States Navy's Combat System Engineering Development Site (CSEDS) site. Ultimately, the underlying architecture has enabled this incremental / piecewise integration between systems to be conducted as systems get developed using worldwide facilities that best suits that integration.

To date, 30% of the Combat System interfaces have had an initial level of functional integration conducted. A direct outcome of the intentional alignment of interface risk to the integration strategy is interfaces with higher risk levels are integrated earlier in the program. After all combat system components are procured and integrated to the extent supported by the land based facilities, a significant float in the program will remain before ship board testing commences as is illustrated in Figure 8.

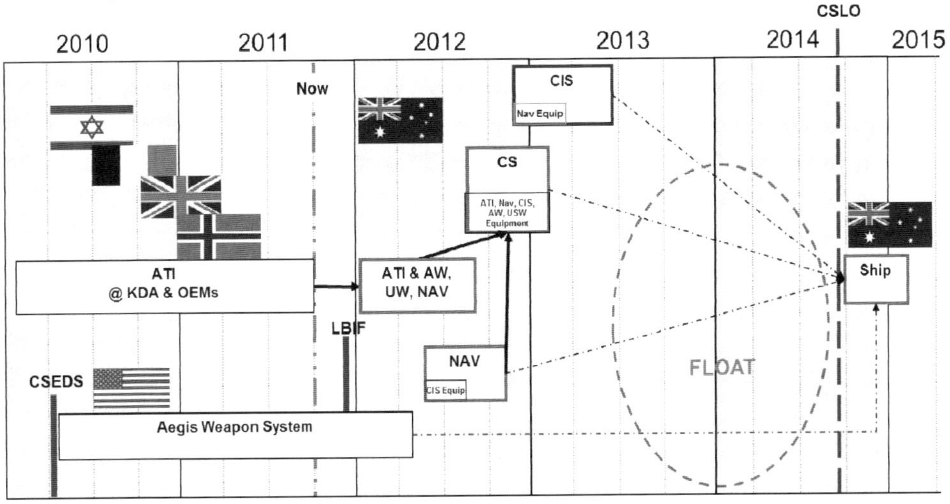

Figure 8 HOBART Class Combat System Integration Strategy

CONCLUSION

Current and future systems and systems-of-systems exhibit complexity that challenges Mission Systems Integrators in making informed decisions at a holistic level, on the system features that best add system value to the business or enterprise. Limitations of the human short term memory and cognitive capacity now require the use of enterprise architecture practices that allow simplification of representative abstract models of the system so informed reasoning and decisions on architecture can be made. Earlier "contemporary design centric" approaches to system definition often capture significant detail but it is this detail that often impedes balanced decisions to the benefit of the whole system over the full lifecycle of the system.

Mission Systems Integrators are now employing enterprise architecture practices in order to deliver higher valued systems and system-of-systems to the enterprise. The use of enterprise architecture practices not only assists in managing complexity, but it also supports the ability to manage change. Higher valued systems can be defined with the full system time horizon in mind. The Air Warfare Destroyer Program has established a stable architecture prior to the start of the design effort. This has enabled the product selection and detailed design phase to focus on selecting the best capability for the allocated budgets to deliver the most capable combat system possible within the program constraints. The underlying architecture has ensured the fundamental features that allow the system to be built, integrated and delivered are preserved.

Additionally, this architecture has been aligned with the integration strategy to ensure the system is able to be integrated and delivered with the lowest risk to the program. Throughout the architectural effort, a significant focus has been retained on the sustainability of the system. Features have been included to provide flexibility to the RAN to select capability upgrade options to be included in future evolutions of the combat system as well as features that provide flexibility in how often and in which order to adopt technology refresh for the system. By considering near term and the long term needs, the final Combat System design has been able to balance the competing needs and ensured the overall total ownership cost is minimized, covering the 30 year life of the capability.

REFERENCES

Air Warfare Destroyer Internet Home Page, www.ausawd.com

Bass, Len & Clements, Paul & Kazman, Rick, "*Software Architecture in Practice*", Addison-Wesley, 2003

Defense Acquisition University, 1993

DoDAF, "*Department of Defense Architecture Framework (DoDAF)" Version 1*, U.S. Department of Defense & C4ISR Architecture Working Groups, 9 Feb 2004

McGregor, John D. "Software Architecture", in *Journal of Object Technology*, vol. 3, no. 5, May-June 2004, pp. 65-77. http://www.jot.fm/issues/issue_2004_05/column7

Miller, George, *"The Magical Number Seven, Plus or Minus Two: Some Limits on Our Capacity for Processing Information"*, The Psychological Review, 1956, vol. 63, pp. 81-97

Pollack, I. *"The information in elementary audi-tory displays*". Journal of the Auditory Society of America 24: 745-749. 1952

Saunders, Steven *"Promoting the Real Value of Systems Engineering using an Extended SCARIT Process Model"*, Proceedings of the INCOSE 2007 Symposium, 2007

TOGAF, *"The Open Group Architecture Framework (TOGAF 8.1.1)"*, The Open Group, 2007

BIOGRAPHY

Steve Saunders is an Engineering Fellow for Raytheon Australia. He received his Bachelor of Electrical Engineering, from the University of Technology Sydney (UTS) with first class Honors in 1990. He has worked with Rockwell International, Boeing Australia and now Raytheon Australia on Australian Defence projects in various Systems Engineering Management, Requirements Development, Architecture, Design and Test roles.

Steve has been involved in the Royal Australian Navy's Air Warfare Destroyer Program since 2005 as the Combat System Chief Architect working in phase 2 of the Program to establish the Combat System architecture and has recently transitioned to the Combat System Chief Engineer role. Steve has written numerous articles on Systems Engineering and has a strong interest in improving System Engineering and System Architecting maturity and the agility of Systems Engineering to support the rapidly evolving technology environment and complexity within the defence industry.

Design to Production for the Air Warfare Destroyer

J.A Carr
Air Warfare Destroyer Alliance
jim.carr@ausawd.com

INTRODUCTION

The Air Warfare Destroyer (AWD) Program is executed by the AWD Alliance consisting of the Commonwealth of Australia and the Industry Participants, namely ASC and Raytheon Australia. The AWD Alliance is empowered and is responsible for the conduct of the AWD Program, including the platform system design.

The Commonwealth has engaged Navantia as Platform System Designer (PSD). The AWD is constructed at ASC in Adelaide with blocks being built at sites in Adelaide (ASC), Newcastle (FORGACS), Ferrol (Navantia), and Melbourne (BAE Systems).

The existing F104 ship baseline was upgraded to accommodate design modifications introduced in F105 ship which is in construction concurrently with the AWD. It also includes modifications resulting from obsolescence, Australian legislative compliance, and the introduction of an Australian Combat System.

BACKGROUND

The construction of the AWD involves a widely dispersed manufacturing capability both nationally and internationally. The build strategy for AWD has remained the same since the selection of ASC as primary shipbuilder in May of 2005 to construct the ship in more than one building location. This strategy was implemented to both mitigate labour resource limitations within any one major industrial location of Australia, as well as increased flexibility to select the best capability for specific manufacturing tasks.

The selection of the F100 existing platform in June of 2007 introduced a new challenge for the Alliance in maintaining the build execution in multiple locations, as the F100 design was structured to a single location build strategy. Additionally the PSD Contract made provision for the design to be released under a technical data package that is predominantly a 2D paper construction drawing based with limited provision for the release of Computer Aided Design (CAD) data.

The limitations in the technical data package provisioning arrangement have resulted in increased production engineering responsibility for the Alliance in supporting manufacturing in multiple locations. If left untreated, these limitations would have also significantly constrained the use of computer-aided manufacturing techniques leading to inefficiency in the production design and production support activities.

The Alliance engineering team has worked with the PSD to maintain the original AWD dispersed build strategy and to exploit opportunities across the AWD manufacturing streams to allow for efficient use of modern CAD/Computer Aided Manufacture (CAM) production design tools. This paper will outline some of those areas exploited and adopted from initial structural fabrication through to waterfront technical problem solving.

Structural NC Cutting Post Processing

The AWD structural design released from the PSD began in early 2009. The design was released as a combination of structural drawing products and a set of cut files as Numerical Code (NC). The Alliance shipbuilding engineering team, working with the PSD, conditioned the NC data for plasma/oxy cutting machines used to cut steel for AWD. The steel fabrication occurs in locations in Australia using different burning technology.

The PSD structural application (FORAN) includes the provision of automatic plate nesting and cutting of ship's hull plate and structural sections. The FORAN output CAD/CAM data in its native format was not suitable for all the NC plate cutting machines employed at the AWD manufacturing sites. To achieve full use of the data, a post-processing step was developed. The post processing for each fabrication site was achieved by building a series of test plates for each fabrication area to validate that cutting actions required were executable. The resultant CAM data output for both ASC and Block Subcontractor sites as depicted in Figure 1.

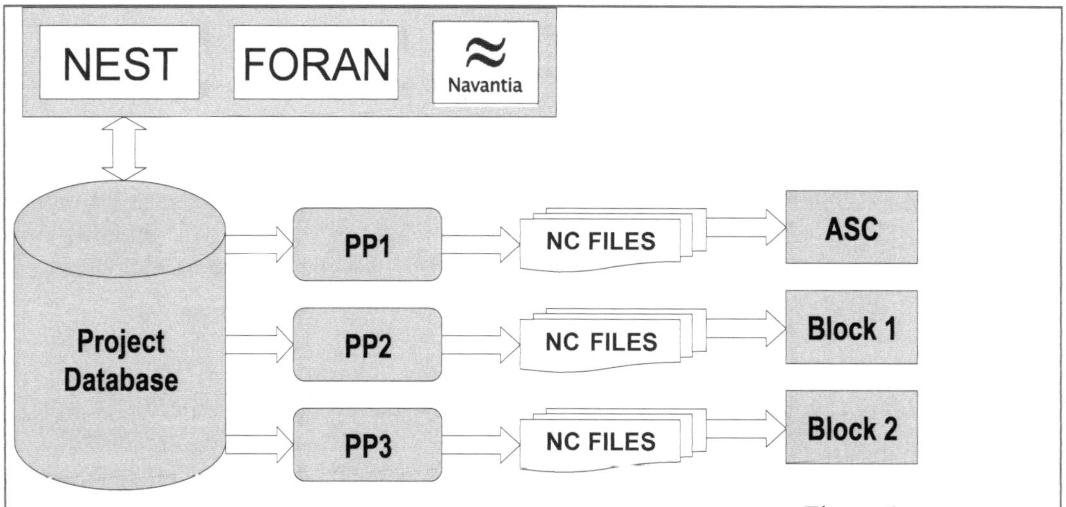

Figure 1

The production output file can currently be deployed in any of three locations in Australia or in Spain without having to modify the baseline design product in the native CAD file.

Common Welding Processes and Procedures

The welding requirements for the AWD at the time of contract in 2007 were those set by the F100 baseline specifications and were conflicting in direction. The requirements specified that welding was to be in accordance with the PSD production procedures which were supplied as reference data and were developed to AH grade material. The requirements also mandated material selection for welding and inspection to be in accordance with Lloyds Register rules. The shipbuilding team within the Alliance chose at the time of contract to develop and qualify weld procedures for AWD that were in accordance with Lloyds Naval ship rules from January 2006 as that was the most current set of rules invoked at the time of contract award. The decision, although resulting in some added engineering work for the Alliance during start-up phase of the contract, has benefitted the program in common weld processes, procedures and qualifications across all structural welding sites that have been approved by the Class Society. The common procedures also resulted in common weld consumables at all sites and the sharing of lessons learned by each of the block builders.

Lifting Handling and Transportation

The AWD build strategy has block construction in multiple locations with different facility constraints. Blocks are then transported to ASC via a sea-going barge. Each lift and transport must be engineered to insure safe working loads. The shipbuilding engineering team employs a group of computer engineering tools to conduct the design and to analyse each element. Concept designs are developed initially in 2D CAD and used for liaising with production and planning groups regarding requirements and applicability of desired mocks, jigs and fixtures. Final designs including calculations, reports and CAD detail drawings are analysed in a Strand 7 finite model tool. The below figures 2 and 3 depict completed assemblies. Figure 2 is the design for lifting the aft deckhouse block. The modelling identifies back up structure requirements and picks optimum pad eye location to support safe lift.

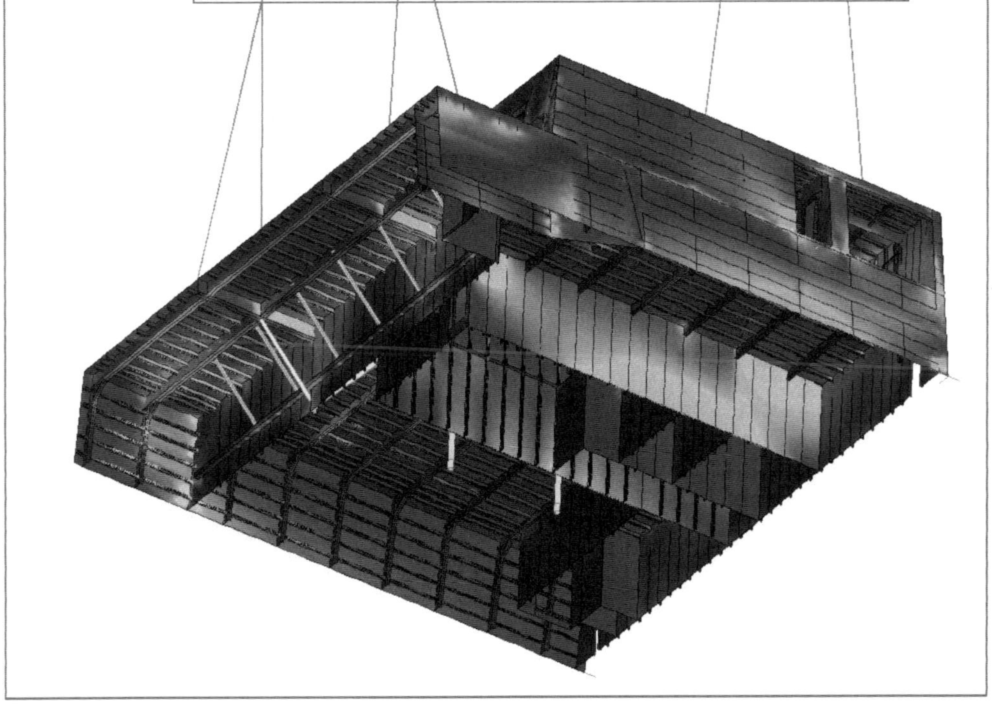

Figure 2 Block 703 Block Straight Lifting Analysis

Figure 3 is the completed analysis for rotation of a sub-block assembly. Each sub-block is analysed to allow for use of a common rotating beam and lugs at ASC. This approach allowed for the rotations to be set up as a repeating work station under the as large gantry crane. Figure 4 shows the completed assembly in the yard being rotated.

Figure 3 Block 411 Sub Block Rotation Analysis

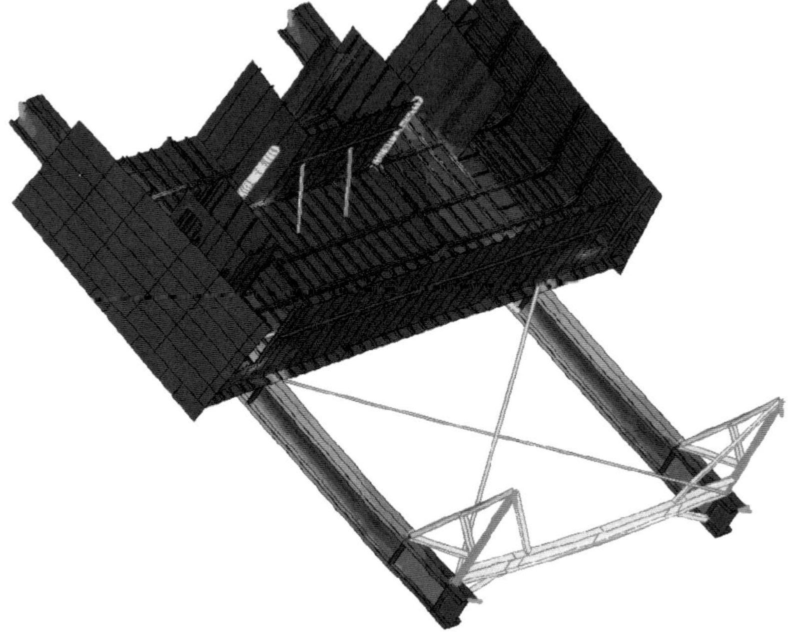

Figure 4 Block 413 Sub-block rotation

Piping Support System (PSS) & Electrical Support System (ESS)

The Alliance shipbuilding engineering team develops and maintains a Piping Support System (PSS) and an Electrical Support System (ESS) throughout the Production and Test phases of the project. PSS and ESS allow management and analysis of data related to piping and electrical cable and component installations on AWD. The AWD piping systems contains more than 22,000 unique pipe assemblies to be fabricated tested and installed in the ship. The AWD ship contains more 25,000 unique cable runs and 14,000 items of electrical equipment to be installed. The majority of data to support this activity is supplied directly by the PSD, whereas other data is developed by shipbuilding engineering, trade planning and/or captured during production. Both systems allow production to track piping or electrical systems at any phase of the AWD build and input data for analysis and optimisation for each subsequent AWD.

PSS and ESS both have the capability to generate a broad list of reports tailored to the end user. The reports are customisable by means of applicable selection criteria. PSS and ESS reports generated are designed to correspond to specific production and testing tasks. Typical excerpt of a report generated is shown in the figure 5.

Figure 5

Smart Plant Model

The PSD supplies the AWD project with a set of reference 3D models with the release of drawing products. The reference models are a visualisation tool that enables the shipbuilding team within the Alliance to interrogate design areas of the ship for all disciplines.

The 3D model is used first in the build planning phase prior to the start of construction of each major block or area. The model allows the shipbuilding planner an opportunity to define the work sequencing and create snapshots of the represented areas to assist the mechanic during construction. Figure 6 depicts a typical build strategy excerpt whereby a piping module which has been planned for installation in the inverted position of the block can be depicted in the build plan document.

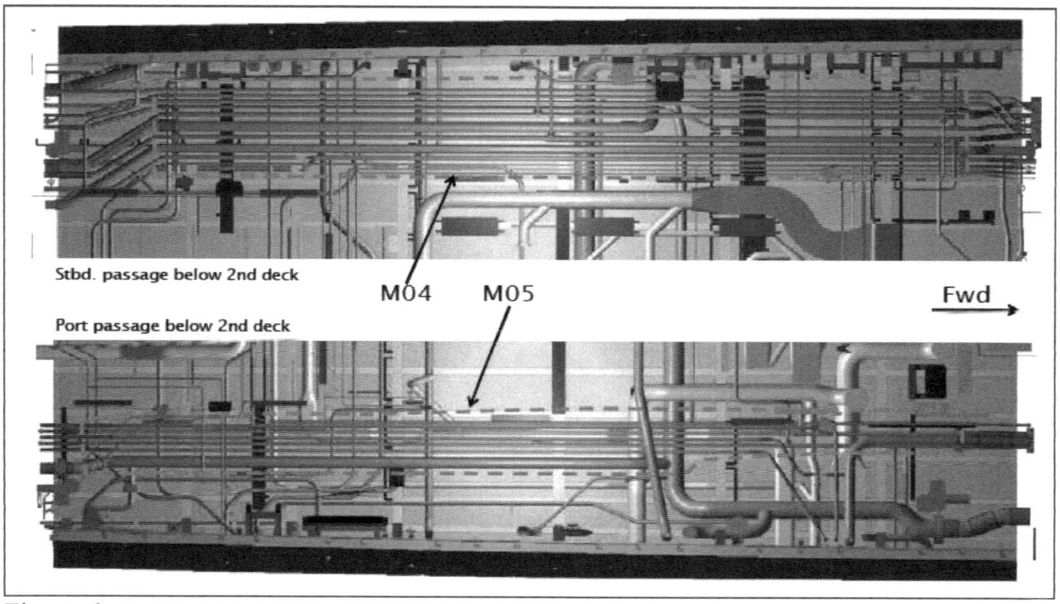

Figure 6

The 3D model is also used for resolution of production issues whereby the design cannot be accommodated during construction. Alternatively, if the design has an error which is found during construction, the field engineer can quickly give direction to the mechanic on solving the issue and a drawing update can follow. Figure 7 is an example of a design clash found in construction that was interrogated in the model to resolve a production installation issue. The result was the light was relocated at the waterfront using the model information as guidance and the correction was forwarded back to the PSD to have the drawing corrected to match the outcome.

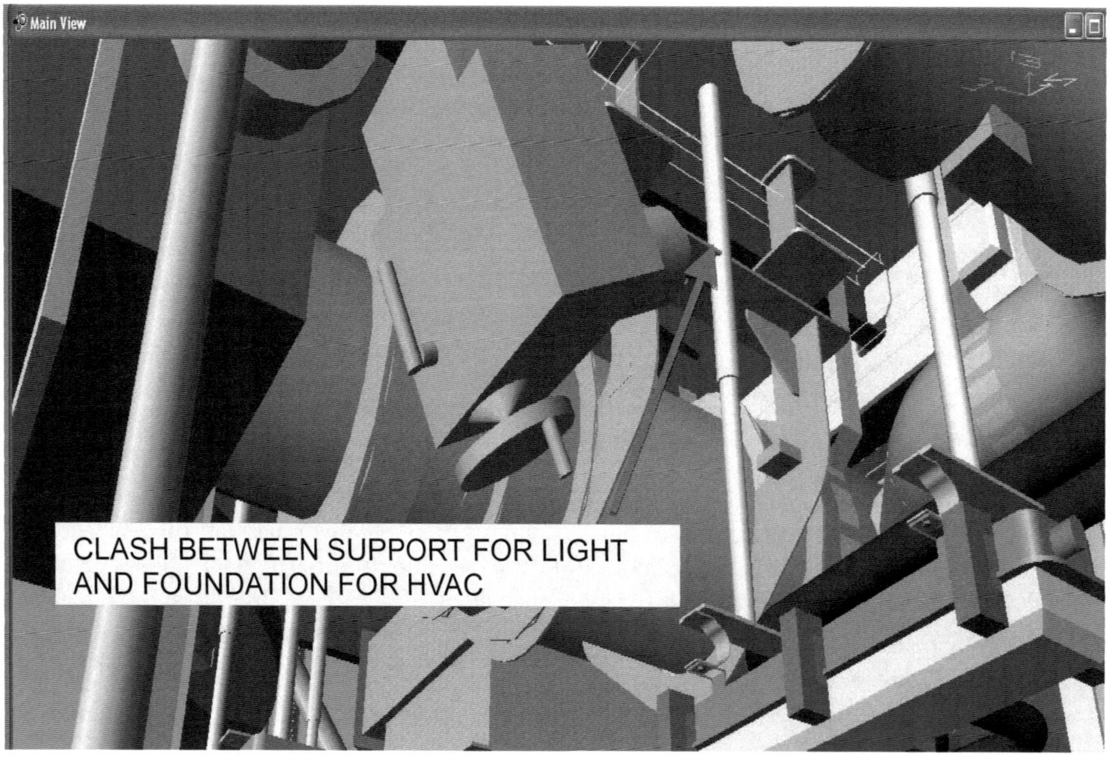

CONCLUSION

For modern warship design the use of Computer Aided Design (CAD) and Computer Aided Engineering (CAE) is the most accurate and practical way to implement design for manufacturing. Limitations in contracting decisions that prevent full use of CAD/CAE applications can be treated, but should be avoided as they introduce inefficiency in the production design and production support activities.

"The Build Sequence' – the art and science behind the consolidation process

Simon Ridgway, Peter Davie
ASC Shipbuilding
simon.ridgway@ausawd.com; peter.davie@ausawd.com

INTRODUCTION The AWD Program is executed by the AWD Alliance consisting of the Commonwealth of Australia, and the Industry Participants, namely ASC and Raytheon Australia. The AWD Alliance is empowered and is responsible for the conduct of the AWD Program, including the platform system design.

The Commonwealth has engaged Navantia as Platform System Designer (PSD). The AWD is constructed at ASC in Adelaide with blocks being built at sites in Adelaide (ASC), Newcastle (FORGACS), Ferrol (Navantia), and Melbourne (BAES).

The existing F104 ship baseline was upgraded to accommodate design modification introduced in the F105 ship which is in construction concurrently with the AWD as well as modifications resulting from obsolescence, Australian legislative compliance, and the introduction of an Australian Combat System.

F100 AND AWD BLOCK BREAKDOWN

For construction purposes the AWD and F105 retain the same block definition as the first F100 flight. The block breaks correspond naturally to the key structural elements of the ship and are defined to ensure a maximum consolidation weight of 200 tonnes to conform with the crane capacity over Navantia's Ferrol slipway. The AWD block breakout and construction zone definitions are illustrated in figure 1.

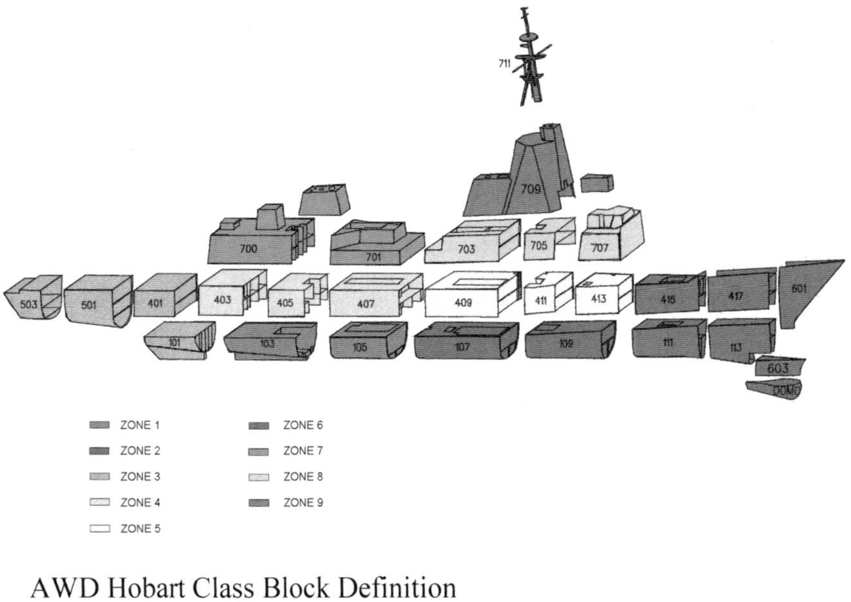

Figure 1. AWD Hobart Class Block Definition

The block and zone based design of the ship is driven by Navantia's "Integrated Construction Methodology" (ICM) which is an application of the "Product Work Breakdown Structure" derived from IHI methods of ship construction (ref 1). The ICM focuses on maximizing fitting of welded outfit items at the sub-assembly or block assembly stages and the use of modular outfitting assemblies. This approach limits the amount of work that needs to happen "on-block" and is a key enabler for ASC's philosophy of maximizing outsourcing both at the block level and, for in-yard work, at the sub-assembly level.

Figure 2. Block 413 during Inverted Outfitting Stage

NAVANTIA FACILITY

Navantia has consolidated the five F-100 ships on the inclined slip in its Ferrol Shipyard in North-Western Spain. Access, stability and clearance considerations dictate the ship configuration at launch. In the case of the F105 the bow sonar and sonar dome (block 603) is consolidated following launch whilst the ship is in the dry dock to avoid damage. The A-Brackets, propeller shafts and propellers are also installed post launch due to the difficulties of installation on the slipway. The need to achieve hull completion in order to launch and dry dock the ship is a key driver of Navantia's consolidation sequence.

Figure 3. Navantia inclined ways

Figure 4 Navantia dry dock

AWD FACILITY

Blocks for the Air Warfare Destroyer program are being constructed in multiple yards around Australia and some at Navantia's Ferrol facility. ASC's role will be the consolidation and systems integrations of the ship. It will also build the Aegis Tower area that will be subject to the greatest levels of change to incorporate the "Australianised" combat system configuration.

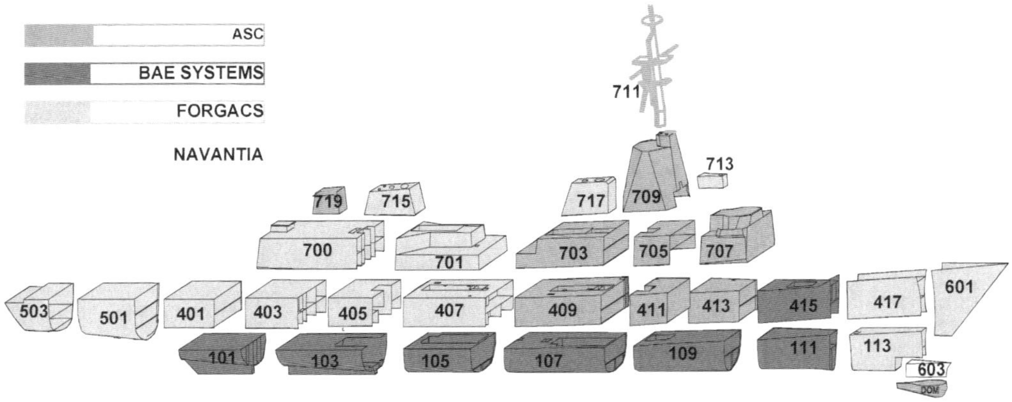

Fig 5. Block Allocations to Subcontractors (Ship 1)

Consolidation of AWD Blocks will occur on the South Australian Government Techport Common User Facility (CUF) commencing in October 2012. This newly constructed facility offers the benefit of a level construction site coupled with a side-transfer facility and a Syncrolift® shiplift with a capacity in excess of 9,400 tons and 160m in length. This system allows the option to dock and undock a fully outfitted AWD if required underpinning a build philosophy and schedule designed to maximise outfit levels prior to launch in preference to driving to a relatively earlier launch with significant work requiring completion in the dry-dock.

Fig 6. South Australian Government Techport Common User Facility and ASC shipyard

In addition to the land-level site, ASC has invested in a 900 tonne crane with a 102m boom crane capacity to enable the lifting of larger more heavily outfitted consolidation units and four 200 tonne mobile transporters providing the ability to drive machinery blocks onto the

CUF with major equipments (reduction gears, gas turbines and propulsion diesel engines) incorporated rather than waiting to include these items on the building berth.

Figure 7. ASC Self Propelled Modular Transporters

A final feature of the ASC facility designed to further exploit the efficiencies available from level construction is the placement of two "outfit support towers" (OST) beside the ships in construction. These towers allow for a permanent facility including tool cribs, workshops and amenities available to workers in the ships avoiding the need to leave the work site for extended periods during the day. By placing suitable access cuts in the hull it is possible to significantly improve the time spent "on the job" and alleviate some of the significant movement penalties associated with working in a completed ship. A schematic of the AWD personnel access plan and the outfit support towers in figure 8.

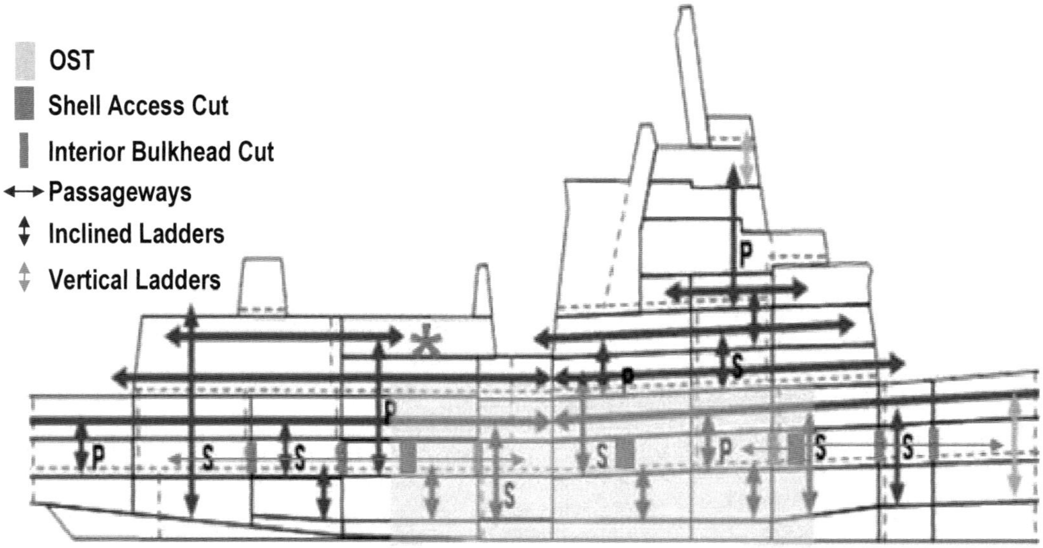

Figure 8. Use of side access cuts and OSTs to support efficient on-board work

COST OF DELAYING WORK

It is generally accepted that there are significant cost penalties for executing work later in the production cycle. Whilst the quantum of the relation is a product of many factors one generally accepted metric is the 1:3:5 ratio for hours in pre-outfit: hours on-board: hours in the water (ref 2). As ASC is still only in the pre-outfit stage the actual ratio for the yard isn't known, but as the 1:3:5 reflects a mature shipbuilding operation it is probably conservative for a new yard providing a strong impetus to drive work early.

INCREASING PRE-OUTFIT COMPLETION

The focus of Navantia's ICM is to exploit the efficiencies inherent in maximizing pre-outfit levels. In ASC's case further efficiencies should be possible due to the facility differences described above as larger and more complete units can be consolidated.

One method of increasing the total outfit executed at the pre-outfit stage is to increase the size of the consolidation units through combining multiple units into "grand blocks". This approach maximizes enables the completion of the scope of work across block breaks as well as allowing the alignment and joining of blocks in the outfitting hall reducing the overall consolidation duration.

For the AWD build ASC has taken advantage of the facility capability to plan joining of the machinery spaces along the keel into two grand-blocks, as well as grand-blocking the blocks containing the key combat systems spaces. Additional grand blocks covering the accommodation spaces in zones 4 and 5 are also under investigation as current plans do not use all the "grand blocks" identified by Navantia for the F105 build and the land-level facility also allows for "driving out" rather than lifting keel-line units into place.

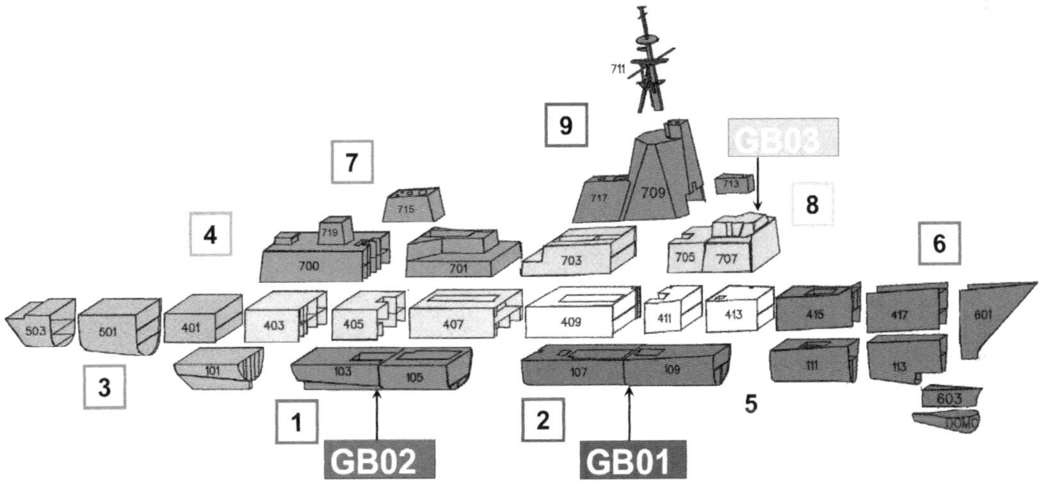

Figure 9 Proposed grand blocks for AWD 1 consolidation

BALANCING PRE-OUTFIT AND ON-BOARD CRITICAL PATHS

The consolidation sequence of the ship is a key attribute in the planning and control of the overall program schedule. Defining the consolidation sequence sets the demand profile for the blocks, which in turn allows the "back scheduling" of work through the pre-outfit facilities to support that plan. The order and placement of the blocks during consolidation also dictates how early the key work-fronts needed to integrate the ship system can be opened up.

In the case of an AEGIS ship, there are two key critical paths defining the duration between the initial keel laying and the final completion of the ship. The primary consideration is the commencement of the cable pulling and compartment load-outs necessary to support the combat system integration and set to work. The next critical path is the completion of the machinery installation and alignment. Unlike a slipway construction, the final completion of the hull to allow launching is not critical. The selection of the launch point is driven primarily by the scope of work that cannot be completed on the land level. In the case of the first Hobart Class ship, this work includes the need to perform final alignments prior to both main engine and combat system light-offs.

SETTING THE CONSOLIDATION SEQUENCE

As the pivot point for the ship construction schedule, the consolidation schedule provides a critical tool for managing changes to the ship program. The setting of the sequence is a simple "stacking" exercise. It is controlled by a number of heuristics including, ensuring all necessary supporting blocks are in place; avoiding the creation of "key-holes" in the sequence; and avoiding the consolidation of the blocks "closed ends" in a manner that requires staging to complete joints. The complexity arises when this conceptually simple exercise needs to accommodate situational factors, such as construction delays in critical blocks and design restrictions that affect block support requirements and outfitting access due to temporary bracing.

EVOLUTION OF THE AWD CONSOLIDATION SEQUENCE

When the initial consolidation sequence was developed following the award of the AWD contract, it was planned to use an optimal consolidation sequence driving up the combat tower to support main cable pulls with a secondary emphasis on building the stern to support machinery alignment in order to maximize the value from the ASC land-level facility (figure 10). This also allowed the establishment of a repeating work stations to sequentially outfit grand blocks 1 and 2 at a constant pace throughout the program maximizing learning.

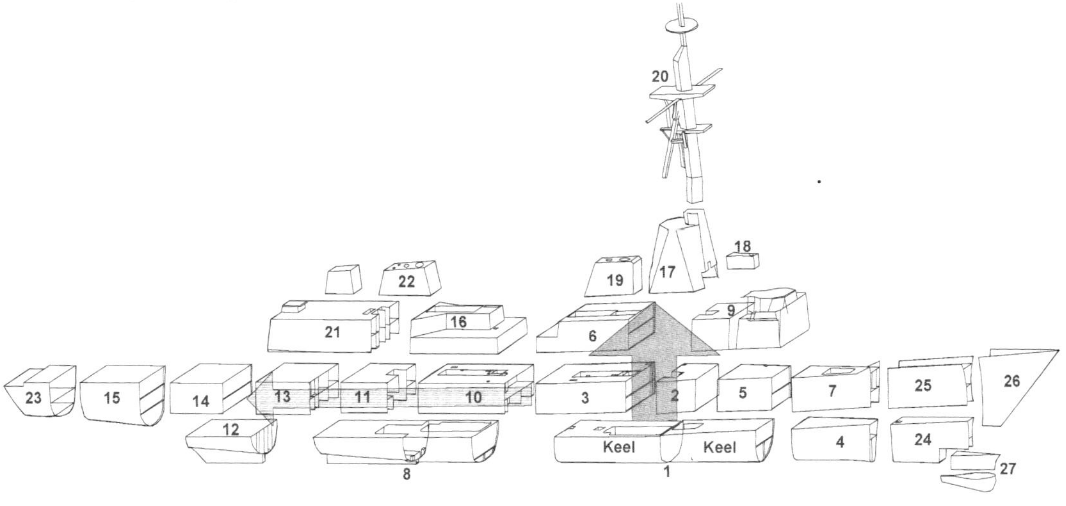

Figure 10 Ideal CUF consolidation sequence optimizing cable pull as the primary critical path

This philosophy has been challenged in a number of ways as the program has progressed, knowledge of the ship design has increased and real world events have intervened.
One example of the types of factors that may drive a change in the consolidation sequence can be seen as the result of difficulties in the construction of the ship 1 machinery blocks. The resolution of these issues has significantly reduced the value of the repeating work station concept for follow ships and lead to an erosion of the time available for serial outfit

of the ship 1 machinery spaces. To minimize the impact on the program, the consolidation sequence has been adjusted from the "ideal" to a more stern focused approach shown in figure 11. This revised sequence allows earlier completion of the keel line to support machinery alignment whilst maintaining the relative completion dates of the combat systems tower to the initial keel laying.

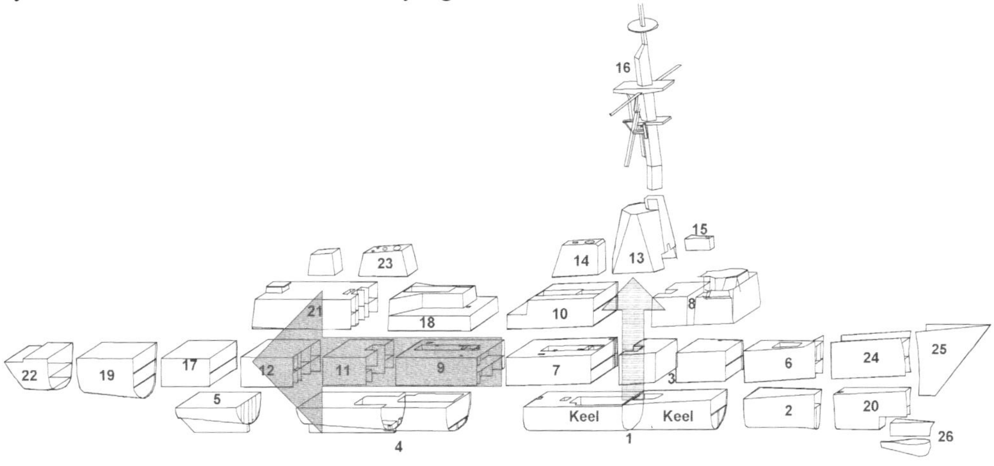

Figure 11 Revised consolidation sequence to accommodate keel and machinery issues but maintaining relative starts for cable pulls

FUTURE CHALLENGES

With the keel laying for HOBART less than 12 months away the most immediate challenge around the consolidation sequence will be achieving the necessary rate of hull consolidation to support the plan. With a unit consolidation pace of one every two to three weeks in a new shipyard, it will be necessary to operate the consolidation operations over multiple shifts to ensure completion of the makeup activity of early blocks prior to commencing consolidation of the blocks abutting them. Achieving the required rate will also demand high accuracy in the blocks presented for consolidation to prevent rework. ASC is investing heavily in total station technology and the training of a significant Accuracy Control team in order to minimize the cost and time associated with the scribing and cutting necessary to support final fit of the blocks.

The short overlap (12 months) between the delivery dates for the Hobart class ships presents a challenge in the incorporation of lessons learned from ship to ship. In particular, the ability to change the consolidation sequence as a result of learning may be challenging as the keel laying of the follow ships occurs shortly before the completion of the preceding hull integration. Given the lead-times through pre-outfit, adjusting the consolidation sequence of Ship 2 will prove difficult but with the knowledge gained from both the ship 1 launch and consolidation of ship 1 and most of ship 2, opportunities to gain improvements through the consolidation of ship 3 are expected to be substantial.

The AWD project remains on a tight schedule, with many significant challenges to overcome. However, the AWD Alliance is committed to seeking further improvements in productivity. As more experience is gained with the consolidation and set-to-work of the ships, opportunities to challenge accepted practice and safely increase productivity will be applied at every opportunity in order to deliver the capability promised by the AWDs.

REFERENCES

1. Y Okayama, (1980) "Product Work Breakdown Structure", i
2. Kvaerner Masa Marine Inc, (1996) "Shipyard Cost Model Using Activity-Based Costing Methods,

Marine propulsion technologies: transforming grey ships to green

by Christoph Fenske.
Senior Manager, Application Center Marine, Naval Surface Craft
MTU Friedrichshafen GmbH, Germany
christoph.fenske@mtu-online.com

ABSTRACT

Exhaust gas emissions of ships have been and will be further restricted by the International Maritime Organisation (IMO) as well as by regional and national legislation for civil shipping. Despite the fact that Navies are not legally forced to fulfil these requirements, many Navies and Coast Guards already comply or intend to comply with the latest emission regulations. With IMO Tier 2 and Tier 3 emission limits in force or on the horizon, the requirement specifications for many new ship procurement programs demands compliance with this legislation. Public and political pressure to spend taxpayers' money on green technology is evident in both industrialised and emerging countries.

This paper describes the current status of IMO's diesel engine emission limits and discusses the boundary conditions of IMO Tier 2 and IMO Tier 3 NOx limits. It highlights the challenges that engine manufacturers face when developing engines for the new requirements and gives insight into the technological road map of MTU.

Eventually the paper outlines the impact on ship designs such as superstructures and supply systems of the technologies under discussion, namely engine-internal technologies and exhausts gas after-treatment systems. The paper concludes with an outlook on alternative fuel concepts.

CURRENT AND FUTURE EMISSION LEGISLATION

The MARPOL 74/78 conference on emissions from ships defined restrictions in Annex VI, Regulation 13, on all nitrogen oxide (NOx) emission of civil ships with diesel engines of more than 130 kW. The latest revision defines Tier 2 requirements for all international waters, already in force for ships with keel laying in 2011, and Tier 3 requirements for so-called Emission Controlled Areas (ECAs), enforced for ships with keel laying not earlier than 2016.

Other than the United States, which also specifies reduction of emissions of unburned hydro carbons (HC), carbon monoxides (CO) and particulate matters, the current emission regulations put up by MARPOL comprise only NOx reduction.

While the step from Tier 1 to Tier 2 limits was a slight reduction of 20% only, the Tier 3 requirements call for a massive reduction of NOx emissions by 80%. Besides a couple of ECAs being under discussion, the only confirmed ECA so far is the North American Coast, as displayed in **Figure 1**. Australia, too is considering confirming its coastlines as ECAs.

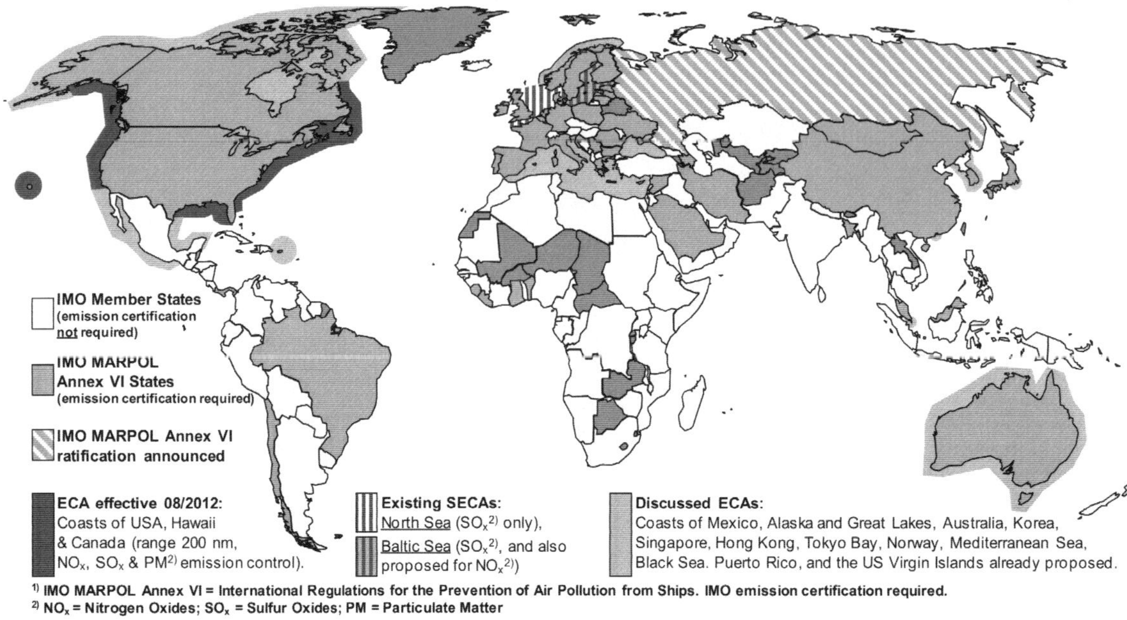

Figure 1: Emission Controlled Areas (ECAs)

What does all this have to do with Naval ships, especially as the regulations are only binding for commercially operated ships? Under Article 3 "Application", it is also clearly addressed that: "The present convention shall not apply to any warship, naval auxiliary or other ship owned or operated by a State and used for the time being, only on government non-commercial service". However, Article 3 also states that it should be applied if the operational capabilities are not impaired. Whatever the conclusion may be, it will be interesting to see who will eventually judge the impact of the additionally required technologies on operational capabilities.

But what drives Navies to demand fulfilment of more stringent emission requirements for their new building projects, if they obviously are not obliged to comply?

Principally, motivations heard from our customers are:

- Public opinion: "If we spend taxpayers' money, let us do it for green technology"
- Political will: "We are at the technological forefront", also "We belong to the First World"
- Green image / Ecological awareness: "Let us be as green as possible, to reduce the ecological footprint"
- Technical / Economical feasibility: "Let us do it, because we can."

Of course, it is not up to industry to argue about the motivations to comply with the latest exhaust emission requirements, but it is our role to demonstrate and make visible the impact on ship design and operation. This should eventually allow decision makers to judge for themselves, on the interference of this new technology with operational capabilities.

ENGINE TECHNOLOGIES TO COMPLY WITH IMO REQUIREMENTS

To understand how the currently applied technologies work it is important to understand how nitrogen oxides are generated during the combustion process.

Nitrogen oxides are generated during the combustion process when the nitrogen from the intake air reacts with the oxygen from the intake air at temperatures above 1,000 degrees celsius. The generation increases exponentially with temperature, so the higher the temperature and the longer the reaction time at these high temperatures, the higher the nitrogen oxide content of the exhaust gas. Since slow and medium speed diesel engines[1] have more time for their combustion process, the formation of NOx is specifically higher and this is reflected in the MARPOL regulation, as shown in **Figure 2**.

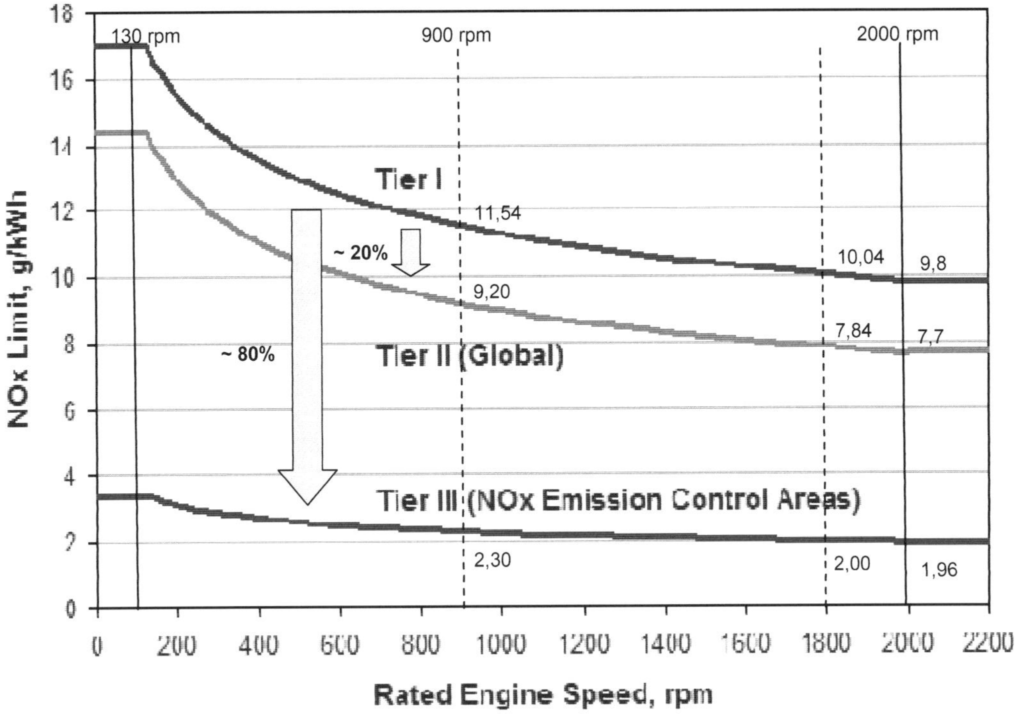

Figure 2: Emission limits as specified in MARPOL 74/78 Annex VI, Regulation 13

Thus, all engine internal technologies aim at reducing the combustion temperature.

---

[1] Slow speed diesel engines: nominal speed less or equal 250 rpm
Medium speed diesel engines: nominal speed more than 250 rpm, but less or equal 1,000 rpm
High speed diesel engines: nominal speed more than 1,000 rpm

Technologies to meet IMO Tier 2: Retarded Injection Timing

The most simplistic means to adapt an engine to lower NOx emissions is through the retardation of the injection point, resulting in lower combustion peak temperatures. But since the combustion is retarded in total, the exhaust gas temperatures are increased. Older engine designs have limitations in retarding the combustion, due to the violation of the admissible temperature ranges for the materials in the exhaust gas system. Aside from being an interesting alternative to the significant re-design of an engine, the fuel consumption is increased by typically 2-5% of the consumption of the IMO Tier 1 complying engine. This makes it increasingly difficult for Sales people to actually sell; as only an engine that has very special characteristics – for instance an exceptional power-to-weight ratio – might still be sellable.

Typically, engine designers do not throw away mature designs light heartedly, but emissions are the major driving force behind the obvious reassessment of product portfolios or of applications served.

Technologies to meet IMO Tier 2: Miller Combustion Cycle

The two principal paths to improve emission behaviour of internal combustion engines are internal technologies or external (after-treatment) technologies. Conventional technologies such as modifying or modernizing the turbo charging system, the injection system or combustion forming of the classical Diesel process, have reached their limits with IMO Tier 2. See **Figure 3**.

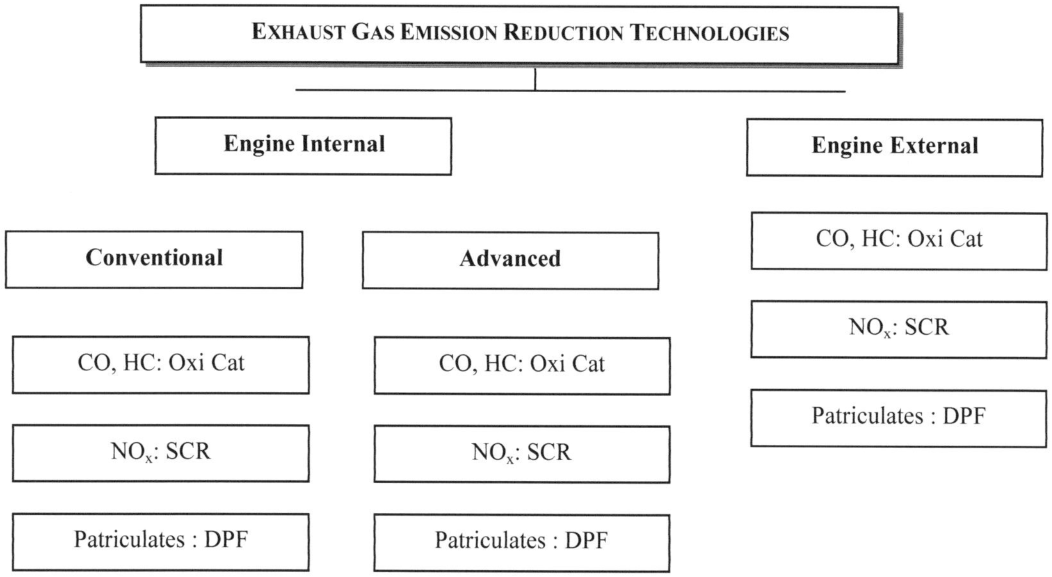

Figure 3: Emission reduction technologies

In the 1990s, the injection of fuel-water emulsions was considered an easy way to lower NOx emissions, but they have proven to be unreliable and sensitive to insufficient maintenance, as typical in commercial applications. Also in applications with a good share of transient operations, corrosion in the fuel system, the combustion chamber and related components, were evident; so much so that engine manufacturers investigated in other technologies.

Eventually, the Miller Cycle was introduced in a range of engine designs. The process was established in the 1940s but only implemented in the last decade on series produced industrial engines. The Miller Cycle uses the inherent mechanical processes of an engine without the significant redesign of the core engine, utilisers the four-stroke principle. With the inlet valve closing earlier than in the classical Diesel process, the intake air is de-compressed after the inlet valve has closed. The air temperature at the start of the compression stroke, and thus the peak temperature, is lower. While reducing the NOx emissions, the fuel consumption is kept on a similar level; therefore there is no fuel consumption penalty.

Still, redesigning an engine to implement the Miller Cycle is not an easy task, since the entire combustion process, the turbo charging and the camshaft need to be modified. The turbo charging is designed to produce higher charge air pressures to counteract the higher choking effect of the inlet valve.

A good example of the application of these technologies is the next generation of the MTU Series 1163 engine, currently under development and ready for delivery in mid-2013. Above and beyond the mentioned modifications, the 1163M04 will benefit from an electronically controlled, high pressure fuel injection system – also known as "Common Rail" fuel system. The two-stage turbo charging system will be retained but with a more modern compressor and turbine-style design, to increase the charge air pressure. To control the engine, the latest-generation electronic governor and engine control system is implemented, providing more diagnostic functions to the operator.

While the original engine series features a conventional injection system, the summary of new technologies applied allowed for a significant reduction of fuel consumption of up to 7%, while power-to-weight ratio, acceleration behaviour, and maximum continuous power remain virtually or actually unchanged.

The described technologies are allowing for full use of standard Naval consumables such as standard marine fuel NATO F76 even with a sulphur content as high as 500 ppm, despite more effort needing to be put into fuel conditioning to minimize particles and free water from the bunker fuel.

Technologies to achieve Emissions lower than IMO Tier 2: Exhaust Gas Recirculation (EGR)

Current emissions for other non-road applications, such as the rail application in Europe, are already more stringent than marine emissions today and will become even more stringent in 2012, with limits to EU Stage IIIB coming into force. Since MTU's Series 4000 is present in both the marine and rail market, the engine is a good example to demonstrate the different emission scenarios to be met. While the IMO Tier 2 requirements define a maximum NOx cycle value of 7.8 g/kWh for 1800 rpm nominal speed, the imminent EU Stage IIIB regulation demands a combined NOx and HC value of 4 g/kWh. MTU's Series 4000R04 is the first certified engine that complies with the requirements, keeping the fuel consumption and power-to-weight ratio down to levels of its predecessors. The complexity of the engine however has increased significantly, with a two-stage turbo charging system featuring six instead of two turbo chargers and the exhaust gas cooler and related flap system to adjust the recirculated gas rate. Both engines however are set up with the Miller combustion cycle.

Exhaust Gas Recirculation (EGR) is currently one of the most promising technologies to achieve lower NOx emissions (see **Figure 4**). As the name suggests, a part of the exhaust gas is taken and recirculated to take part in the combustion process. In order to achieve a comparable power-to-weight ratio, the recirculated exhaust gas is cooled by means of an air-to-water heat exchanger and then mixed to the fresh air. The low-oxygen exhaust gas reduces the combustion temperature significantly and thus lowers the formation of NOx by at least 40% of the non-EGR engine.

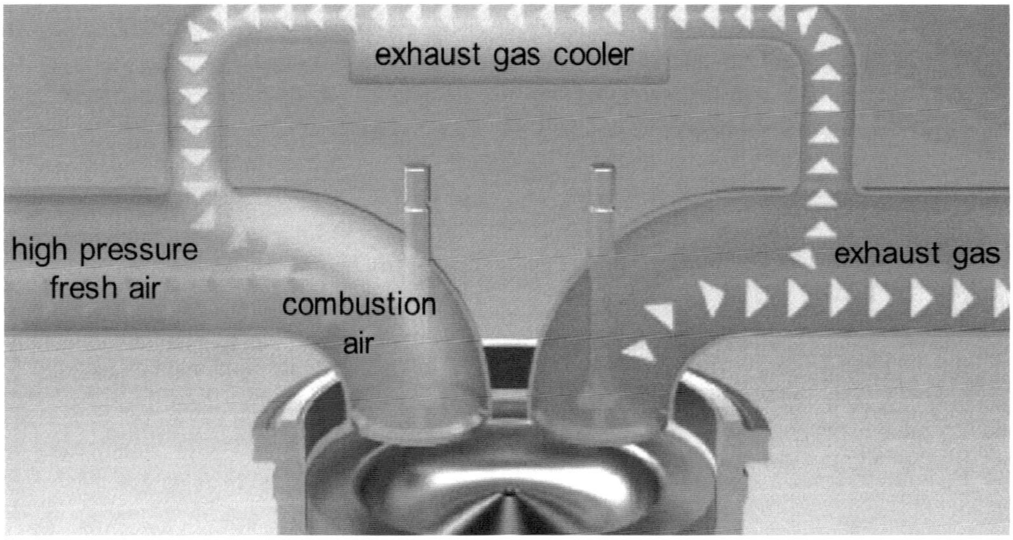

Figure 4: Exhaust Gas Recirculation (EGR)

One of the major obstacles for EGR is the cooling of the hot exhaust gas that contains water and sulphur oxides as products of the combustion. In the EGR cooler, the dew point temperature is under-run and sulphuric acid could result in corrosion in the cooler. Thus, today's typical marine fuel qualities are not feasible for EGR-engines. Global availability of ultra-low sulphur fuels would overcome this disadvantage of EGR engines, but it is assumed that this is still some years away from now. Additionally, EGR engines require a higher charge air pressure, so more adjustments need to be done on the turbo charger system (e.g. sequential turbo charging) to achieve comparable acceleration behaviour, i.e. performance map.

It must not be underestimated that these engines are more mechanically complex than their predecessors, with more intensive maintenance and more complex tasks driving up life cycle costs. With low manning concepts to counteract the challenges of finding skilled seamen, this may become a real challenge for Navies and Coast Guards.

EGR has the highest potential to meet IMO Tier 3 limit engines internally, but the gap between the currently achieved values and IMO Tier 3 is still significant.

Technologies that meet IMO Tier 3: Selective Catalytic Reduction (SCR)

Leaving the field of engine internal technologies, we will focus on the engine external technology or "Selective Catalytic Reduction" (SCR) in the following section. SCR is not a new technology, but fielded for many years if not decades. The concept reduces exhaust gas emissions off-engine and thus can be combined with existing or older engine designs. The core component is the reactor, containing ceramic hollow blocks, i.e. the actual catalyst. A typical arrangement is shown in **Figure 5**.

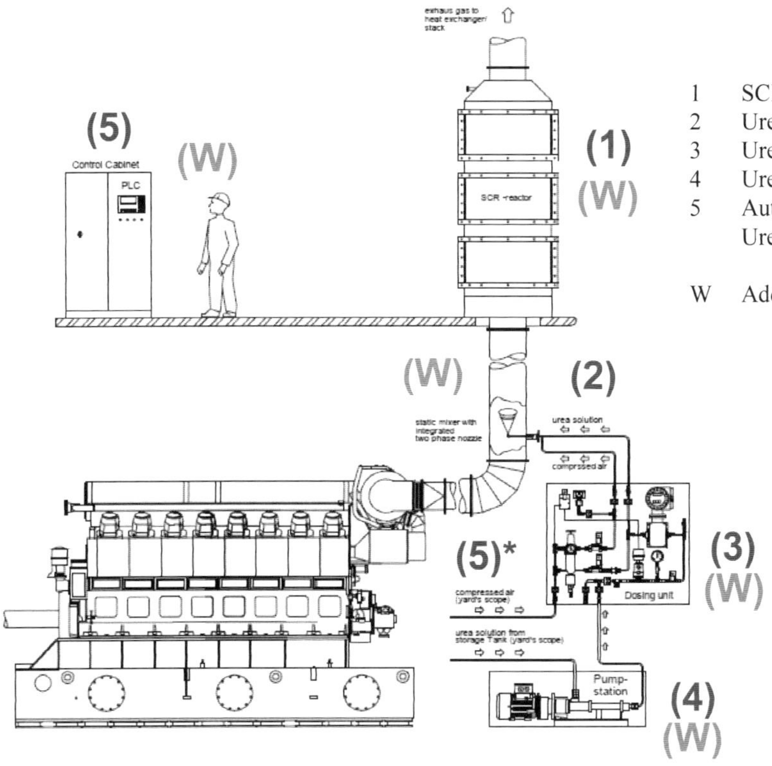

1 SCR Reactor
2 Urea Injection Pipe (stainless steel)
3 Urea Dosing Unit
4 Urea Pump
5 Automation
Urea Tank (not shown)

W Additional Space / Maintenance required

Figure 5: Typical Selective Catalytic Reduction (SCR) arrangement

In the reactor, on the surface of the catalyst, the nitrogen oxides (NO, NO_2) of the exhaust gas are reduced to non-toxic water and nitrogen. Also taking part in the reaction is ammonia (NH_3), provided by an agent (urea solution or "AdBlue") injected in a mixing pipe or chamber to achieve a homogeneous mix of gas and ammonia. The injection needs to be controlled closely in order to:

a) Reduce the NOx as required; and
b) Not allow for ammonia slipping, since ammonia is a toxic gas.

In order to ensure continuous reaction, a minimum gas temperature is mandatory. In case the temperature dips below the minimum temperature of 400°C, injected ammonia may slip out of the reactor and NOx levels in the exhaust gas may exceed the admissible limits. Hence, engines need to be compliant with emission requirements even if the SCR is not working and injection needs to be stopped, i.e. with engine internal technologies.

MTU performed installation studies together with some naval shipbuilders in order to understand the impact on existing ship designs. A typical SCR reactor for a 20V 8000 M71L (developing 9100 kW at 1150 rpm) is shown in **Table 1**.

Feature	Value
Reactor volume [m^3], appr.	3,5 – 4,0
Dimensions [L x B x H] [mm], appr.	2500 x 2500 x 4000
Mass EAT – System [kg], appr.	6000
Length of injection pipe [mm], appr.	3000
Diameter of injection pipe [mm], appr.	DN 1300
Required power [kW], appr.	2,3
Urea consumption [% of B], appr.	7 – 10
Compressed air supply required [V_n in m^3], appr.	80 (7 bar)

Table 1: Typical features of an SCR reactor (IMO Tier 3) for a Series 8000 engine

It is obvious why shipbuilders call for an engine internal solution, since the unit is heavy, bulky and requires additional supply and control systems to be installed in the superstructure, which might already be packed with mission relevant – i.e. C4ISR – equipment and weapons. Additionally, the urea dosing system and the mixing pipe upstream of the reactor is required to be made of stainless steel and will need to comply with Naval safety standards. What makes the accommodation of the system even more arduous is the need to completely overhaul the reactor after 12,000 operating hours or five years, whatever occurs first, so that the reactors would have to be located close to the ship's hull for unobstructed removal and re-installation.

On the other hand, reactors feature a certain degree of exhaust gas noise attenuation, making the installation of exhaust gas silencers potentially obsolete. A careful balancing of required noise attenuation, actual damping of the reactor and the given space is needed on a case-by-case basis.

Navies have a different approach to the shock capability of the system. In any case, the system might not inadvertently effect the war fighting capabilities of the ship, so the ceramic elements in the reactor must not block the exhaust gas line. By-pass solutions would additionally consume valuable space, and the last word on this system seems not to be spoken.

However, the impact of the SCR technology on NOx emissions is the highest of all discussed technologies. It allows for reduction of more than 90% of the raw nitrogen oxide emissions, so that even an engine optimized for IMO Tier 1 easily remains within the Tier 3 limits.

Technologies to meet IMO Tier 3: LNG as Fuel

The commercial market currently sees a boom in the acceptance of Liquefied Natural Gas (LNG) as an alternative to standard commercial fuels such as heavy fuel oils. Two principal paths are followed by the industry, either dual-fuel engines, allowing for the alternating use of both LNG and standard diesel, or a gas-only engine concept. LNG already allows the reduction of NOx emissions down to IMO Tier 3 levels without after-treatment of the exhaust gas and a significant reduction of CO_2 emissions.

Even for commercial shipping, the insufficient supply base for LNG is still critical, however, regionally as in Europe, the supply base is developing whenever sufficient demand is present. The much lower energy density of LNG consumes almost double the space for the same sea endurance as standard diesel fuel. On the other hand, heavy and bulky after-treatment systems can be avoided.

It remains to be seen whether or not Navies and Coast Guards will accept LNG as an alternative. It needs to be noted that:

- LNG poses a higher fire hazard to the ship than standard diesel fuels.
- The operational behaviour of duel fuel engines is significantly different to a modern high speed diesel engine relative to load acceptance behaviour, and low load capability. However, gas-only engines – namely the Rolls-Royce Bergen range – are superior to duel fuel engines.
- Power-to-weight ratio will be significantly lower than a typical high speed diesel engine consuming more space and adding more weight to the platforms, which might be an issue especially for semi-planing and planing hulls.
- First demonstrator engines are entering the field with long-term feasibility studies underway for proving the LNG technology for high speed diesel engines.
- Naval logistics would need to be set up to provide global supply.

Nevertheless, LNG will be one of the prime emission reduction paths for commercial shipping and even for Coast Guards the above questions might satisfyingly be answered.

CONCLUSION

The new emission legislations for marine applications pose a challenge for both the engine manufacturers and the ship designers. It appears that there is no golden path, many paths seem to eventually lead to the fulfilment of the requirements.

For high speed diesel engines, a combination of engine-internal and Exhaust Gas After-treatment (EGA) measures to achieve the optimum in operational and economic performance appears to be needed in order to meet the IMO Tier 3 emission limits.

Duel fuel engines seem to be a viable solution for commercial shipping whereas gas-only engines fuelled by LNG are superior especially when it comes to operational behaviour and ship design impact. Further development and field testing is needed to adapt high speed diesel engines to gas-only operation. For planing and semi-planing hulls the only viable solution so far are diesel fuelled high speed diesel engines.

Navies and Coast Guards have principally the freedom of choice to comply with the latest emission requirements, but there are good reasons for them to do so anyhow. In any case, they should be aware of the engine and ship design and operational impact of the IMO Tier 3 compliance.

Design of Smart Valves for Damage Control and Survivability

James Rudolph, Piers Fraser
Australian Marine Technologies
james.rudolph@ausmarinetech.com.au, piers.fraser@ausmarinetech.com.au

ABSTRACT

Action damage to fluid systems aboard naval vessels is a significant risk to the function of the undamaged parts of the system, to ship stability, and the ability of the ship to continue its mission. One approach to limit the loss of fluid in the event of damage to the pipes is to install smart valves that detect burst pipes and close automatically. This paper outlines the project background, design theory, and preliminary design undertaken by Australian Marine Technologies in the design of an integrated smart valve solution.

The most common method for detecting burst pipes has been measuring system pressure, often in combination with flow rate. Until recent developments, within both the USN and water utilities, there has been little demand for alternative approaches.

Initial design efforts favoured using pressure sensors to detect a pipe rupture. Testing onboard HMAS BALLARAT showed similarities between regular operation and pressure changes due to a simulated burst. AMT is well placed to continue development of a pressure sensing solution to control valve actuation, however it has become clear through the course of the design process and testing that there is significant unique design effort required to ensure that a fluid pressure based system can accurately identify pipe failure events. Further, using pressure measurements to determine if a pipe has failed relies on secondary data to infer the failure.

The future development focus will be to investigate methods that directly determine if the pipe has failed, followed by scale tests of a prototype pipe system to determine the effectiveness of both pressure and direct measurement techniques and correlate modelling data. With further refinements resulting from analysis of test results, the final detail design will be completed including prototyping, shock qualification, packaging and design for production.

INTRODUCTION

Closed fluid systems contain a finite fluid volume. In systems such as chilled water circulation mains, the loss of water affects the operation of equipment being cooled. Furthermore, the heat load of the equipment cooled by the chilled water system cannot be supported until the system is re-filled, any air removed, and the water chilled to its operating temperature. It is therefore advantageous to minimise the loss of working fluid from these types of systems.

Damage to open systems such as the firemain is also a concern, as this may cause loss of fire fighting capability in other parts of the ship, and result in rapid flooding of the compartment where the damage occurred.

In order to limit the loss of fluid in the event of damage to the pipes, strategically placed valves can be triggered to rapidly close and isolate the damaged pipework. This paper outlines the project background, design theory, and preliminary design undertaken by Australian Marine Technologies in order to solve the issues outlined above.

BACKGROUND

The Royal New Zealand Navy (RNZN) is currently upgrading the platform systems aboard their two ANZAC Class Frigates. A key element of the upgrade is to increase the Chilled Water (CW) system capacity to support elements of the Combat System (CS). In the event of action damage to the CW system, the loss of fluid could potentially diminish the war fighting capability of the vessel. The major equipment and fittings used within the Chilled Water System meet military shock standards and while the piping system is quite robust it stretches from stem to stern making it the most vulnerable component of the system. To further complicate matters, by its very nature chilled water is a closed loop system with limited reserve water capacity making the reliable supply of chilled water during a Damage Control (DC) incident difficult.

The traditional approach, and indeed the one currently being implemented by the RNZN, is to provide duplication and separation coupled with remote controlled isolating valves to allow the operator to configure the chilled water system as circumstances require. The ability to break the system into sections aligning with the major DC zones and to separate the CS consumers from the rest of the system provides an acceptable level of survivability. However, in the event of damage to pipework there is a limited amount of time to react to a leak and isolate it from the system. This occurs during the period of peak stress and information overload for the DC Repair Bases, immediately after the ship has suffered action damage. Realistically any damage to the chilled water pipework within a DC zone will result in loss of supply for the entire zone.

To address this AMT investigated the feasibility of installing Smart Valves within the CW distribution main. These valves were to automatically close if a pipe burst was detected, and communicate with the new Integrated Platform Management System (IPMS) also installed as part of the RNZN's Platform System Upgrade. Communication with IPMS was deemed essential so that in the event of action damage, the CW system could be reconfigured to maintain supply to consumers affected by the damage. Using Smart Valves the process of isolating and reconfiguration can be fully automated with the operator workload reduced to a monitoring function. The short reaction times that are possible with Smart Valves provide the real possibility of minimising fluid loss and restoring chilled water supply to the remaining intact equipment within the impacted zone.

Damage Control Automation

For some time navies around the world have been grappling with how to reduce the manning levels needed to support DC incidents and to reduce the workload placed on those personnel. A number of directions are being actively pursued within contemporary IPMS installations including Incident Report and Recording, Fire Fighting Systems, Real-time Flooding and Stability Monitoring and Automated Incident Responses etc.

In the latter case the isolation of damaged pipework to ensure continuity of supply of seawater for fire fighting or cooling water for machinery and combat systems is time consuming and heavily reliant on the system knowledge of the sailors involved. The benefits associated with being able to automate and remotely monitor this process include: freeing skilled personnel to focus on other tasks; limiting secondary flooding caused by pipe breakage; and increased availability of the Combat and DC Systems.

The Royal Australian Navy (RAN) through the Defence Science and Technology Organisation (DSTO) (Clark, 2001) have been monitoring global developments with the view to exploiting automation to achieve manning reductions without degrading a ship's ability to respond to DC incidents and experimenting with ways to automate DC isolations particularly in relation to the ship's firemain. Similarly the Royal Navy (RN) and the Royal Canadian Navy (RCN) are looking to automation as a means for providing quick and reliable alternatives to traditional DC systems (Hiltz, 2005). It is believed that the RN has installed this capability onboard the Type 45 DARING Class although the exact form of this implementation is not known.

Ex-USS SHADWELL DC-ARM Trials

The United States Navy (USN) has similarly pursued these goals through their DC-Automation for Reduced Manning (DC-ARM) program but on a far grander scale. The decommissioned Landing Ship (LSD-15) SHADWELL, previously transferred to the Naval Research Laboratory (NRL) in 1988, has recently been the scene of a number of trials evaluating various automated DC systems including Smart Valves. Termed Reflexive Valves by the NRL, a series of developmental trials were conducted in the late 1990's (Lestina et al, 1999) with a successful prototype being demonstrated in 2001.

Figure 1: USS SHADWELL in her heyday

Further trials simulating action damage tested the operation of Reflexive Valves within both the Firemain and Chilled Water systems (Lestina et al, 2001). The valves where linked to an IPMS which integrated the normal range of systems sensors and remote controlled valves. This combination proved effective in identifying pipe ruptures and successfully reconfiguring the pipework to restore supply where possible and to maintain supply to unaffected areas. As a result of this success the technology is being installed onboard the DDG-51 ARLEIGH BURKE Class.

EXISTING TECHNOLOGY AND APPROACHES

Several approaches to this problem have been developed both within Defence and the wider community. Hydraulic fuses are available within the nuclear and the gas/fuel industries to protect against the environmental and health hazards of accidental discharges due to pipe failure. These systems utilize the pressure differential across the valve to distinguish between normal and excessive flows (Alcaro, 2011). They are rugged and simple with the sole goal of cutting flow as soon as practicable, with some valves designed to require fluid pressure to be restored on both sides before re-opening (Yokota, 2011).

More interesting lines of research and development can be found within the water utilities sector where leaks can take a long time to discover and can cause considerable loss and damage both to the utility and to property adjacent to the leak. The extensive and complicated nature of a water distribution system has resulted in a search for measurable signals, either active or passive, that can be used to identify the occurrence of a leak and then to localise its position.

Active detection methods have included: injecting sound signals into the fluid and measuring the return echo (Hunaidi & Wang, 2000); or injecting high frequency electrical signals into the skin of the pipework and measuring return echoes.

Passive methods include: measuring system pressure and/or flow and looking for values that exceed the normal range of variation; detecting naturally occurring pressure waves to discriminate the likely location of the pipe failure (Misiunas et al, 2005, p.2); monitoring the natural frequencies within the pipe system looking for step changes; or mass balance auditing where total flows in and out are balanced and discrepancies assessed more closely.

While it is apparent that there are a number of avenues for development, to date measuring system pressure in combination with flow rate has been the most widespread approach. There is an extensive history of development using this method, dating back to the early days of the nuclear and gas/fuel industries. The focus of efforts by the USN and water utilities has renewed interest in pipe failure detection, creating opportunities for investigation into alternative solutions.

DESIGN BRIEF

Functions

The functions of the smart valve package are defined in Figure 2, with the representation indicating the dependency of each function. The functions are a guideline as to what the final design will actually do. As with the requirements below, the functions are likely to be refined, with new or more specific steps added as the project evolves. This is part of the iterative approach taken to the project, and it was considered essential to define the basic functions to form a basis for the comparison of different approaches to the problem as the solutions arose.

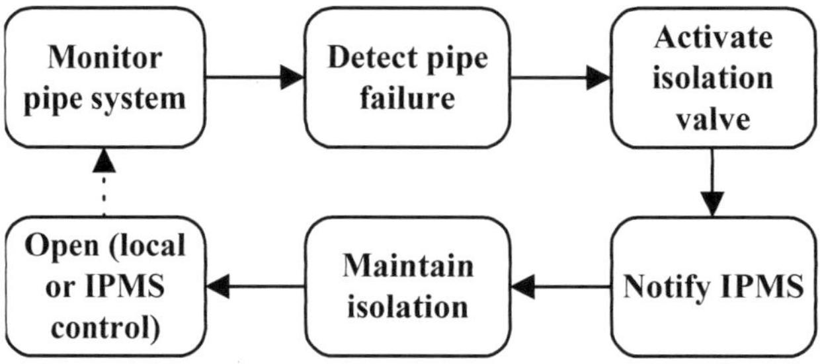

Figure 2: Functions to be performed by smart valve package

Requirements

In order to guide development of concept solutions, a basic set of requirements and performance specifications were created. The design requirements are clarified as the design process continues, but at this stage of the project the requirements have been kept simple and open to ensure alternative solutions are not ruled out prematurely.

To assess the strength of various concepts, the requirements have been used to create criteria that allow each proposed design to be compared directly. The requirements are shown in Table 1, which includes the criteria and relative importance weighting, used to determine the overall score for each concept. Where performance specifications are known, they have been listed against the relevant requirement.

Table 1: Design concept selection criteria

Requirements	Criteria	Relative Weight	Specification
Detect pipe failure	-	-	Shall comply
Prevent false positives	False positives/duty period	4	TBC*
Autonomous isolation	-	-	Shall comply
Communicate with ship management system	Level of integration	3.5	ie. fully integrated, signal only etc
Time to close valve - closed system	Volume of fluid through valve	5	<volume (m^3) of system surge tank
Time to close valve - open system	Volume of fluid through valve	5	$<1\ m^3$
Impact on ship systems	Minimise re-work/changes	2	-
Retrofit capability	Minimise re-work/changes	2	-
New design integration	Flexibility of application	1	-

*Specifications to be defined as performance details of various approaches become available.

DESIGN CONCEPTS

The existing approaches outlined above provided the initial direction to the development of design concepts. These approaches can be broadly grouped into three categories: fluid sensing, electrically sensing and measuring physical characteristics. The following sections outline the concepts developed in each of these categories.

Fluid Sensing

Using pressure sensors to detect pipe failure is an established method, and was the focus of our efforts in the early stage of this project. Using sensors located either side of each isolation valve in the system; a micro-controller monitors the pressure fluctuations in the working fluid. Figure 3 shows this arrangement.

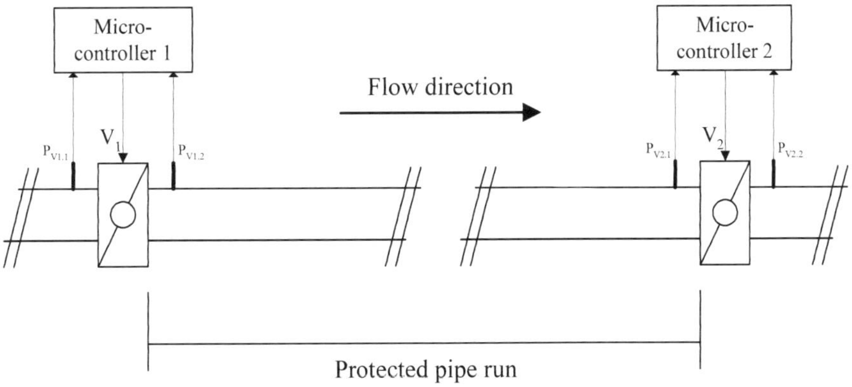

Figure 3: Pressure sensor and valve arrangement

The pressure data may be monitored directly using a micro-controller to analyse the data. Activities such as opening/closing valves, and starting/stopping pumps can produce large pressure fluctuations (Wood et al, 2005, pp1.3-1.4). These fluctuations may be similar to those measured when a pipe bursts, typically a large negative change in pressure with little or no recovery. Pipe size and material, fluid temperature and flow velocity also contribute to the magnitude and transmission of pressure pulses. It is therefore necessary to carefully consider the amplitude and operating conditions to avoid false positives.

The data from either side of the valve can also be converted into fluid velocity. If the valve loss factor is known, the difference in fluid velocity can be predicted for normal operating conditions. A rapid increase in fluid velocity will occur if a pipe downstream of the valve fails, and this can then be used to shut the valve. This approach also needs to be treated with caution, demand changes in other parts of the pipe system may also alter the flow velocity and generate false positives.

Electrically Sensing

Measuring the electrical characteristics of the pipe is an approach that was given some consideration for the simplicity of the approach, and because it directly measures the condition of the pipe itself. The method with the most potential utilises reflectometry. This involves passing a high frequency electrical signal along the pipe material. Where there is a discontinuity part of the wave is reflected; its magnitude and the time taken for the reflection to reach the reflectometer can be measured (Lo, 2005, p.97). The isolation valves would be triggered when the magnitude of the reflected wave exceeds a pre-set limit value. One of the advantages of this system is that it directly measures the integrity of the pipe.

Physical Characteristics

The physical characteristics of the pipe system can be measured using strain and vibration sensors. Directly measuring the shock from a weapon strike has also been considered as a way of determining if pipes have failed.

The use of piezoelectric strips to measure vibration has been established as a cost effective monitoring method. The sensor is inexpensive, and as such can be placed in more locations around the pipe system. Vibration of any pipe spool is driven by excitation inputs, such as the water flow and mechanical vibration from propulsion and auxiliary machinery. Piezoelectric strips are used to provide vibration data to a micro-controller, which then performs averaging calculations over a set time period to establish a baseline measurement of the pipe harmonic. If the pipe is damaged during an engagement, the vibration of the pipe will change, triggering the isolation valves. The key benefits of this system are the distributed nature of the input sensors, low cost and compatibility with non-conductive pipe materials increasingly being used for ship applications.

Combination Measurements

In addition to the approaches outlined above, a combination of any of the above may yield greater fidelity and control, lowering the likelihood of false positives. The key to successfully combining any of the concepts is to ensure that the cost of additional sensors and processing complexity does not outweigh the benefits gained in reliability and accuracy.

FIELD TESTING

As stated above, the initial effort of this project favoured using pressure sensors to detect a pipe rupture. As the technical aspects of this approach were being investigated, the opportunity arose to conduct a test onboard HMAS BALLARAT, a Royal Australian Navy ANZAC Class frigate. The test procedure was established to gain insight into the effect of valve actuation on pressure in pipes, the travel time of pressure waves, and ideally to produce pressure/time graphs of representative pipe burst events. The following paragraphs outline the test and results.

Test Procedure

Two pressure sensors were connected to a data logger which recorded samples at 2000Hz, or one sample every 0.0005 seconds. Figure 4 shows the pressure sensors threaded into a small test section which is attached to the forecastle hydrant onboard HMAS BALLARAT.

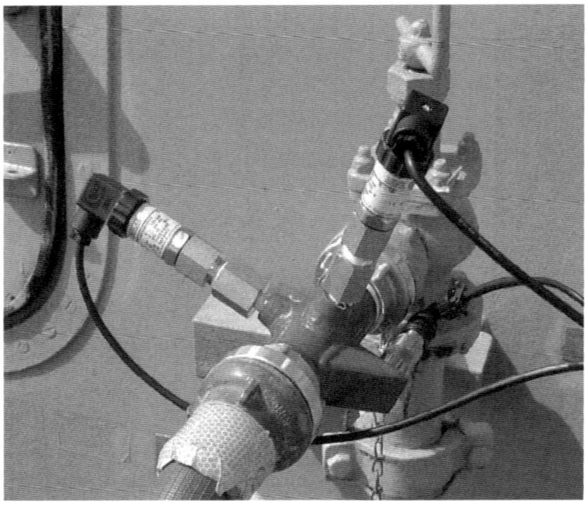

Figure 4: Test section showing pressure sensors

A 15m fire hose is attached to the downstream end of the test section. In the first test the hose was fully charged with the nozzle closed. The nozzle valve was opened rapidly and run for several seconds before the valve was closed again. When a valve is opened the mean pressure should theoretically drop as the flow rate increases. A pressure pulse due to the change in pressure will also occur, and be reflected in the pipe with diminishing magnitude over time. This profile is similar to that expected on the supply side of a burst in a pipe with an established steady flow. A comparable pressure change occurs when the nozzle valve is shut, however the mean pressure rises, with larger magnitude pressure pulses.

The second test aimed to model a burst pipe from the downstream side of the failure. The hose was run continuously fully open, and the valve at the hydrant shut quickly, with the hose allowed to drain down naturally.

Results

The results from these tests are shown in Figure 5 and Figure 6 respectively. Figure 5 indicates that there is significant pressure recovery following both the valve opening (at ~0.5s) and when the valve is shut (at ~3.6s). The fluctuations after the valve is opened are less symmetrical, as the fluid velocity affects the time in which the reflected pressure pulse can travel along the pipe. If a pipe burst downstream of a pressure sensor, it is expected that the pressure profile would reflect that measured between 0.5 and 3 seconds in Figure 5. As the flow accelerates due to the leak, the pressure drops and fluctuates, however the pump is able to continue supplying water and therefore the pressure will not fall away to zero as seen in Figure 6.

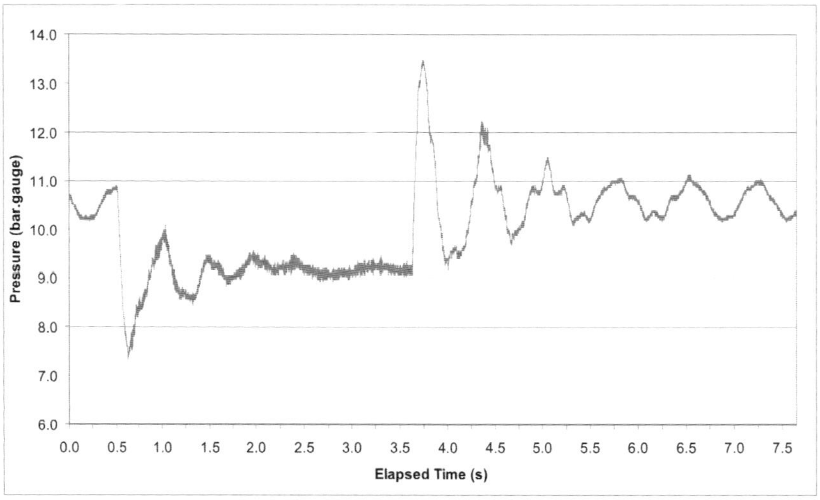

Figure 5: Pressure/Time graph of valve open/close test results

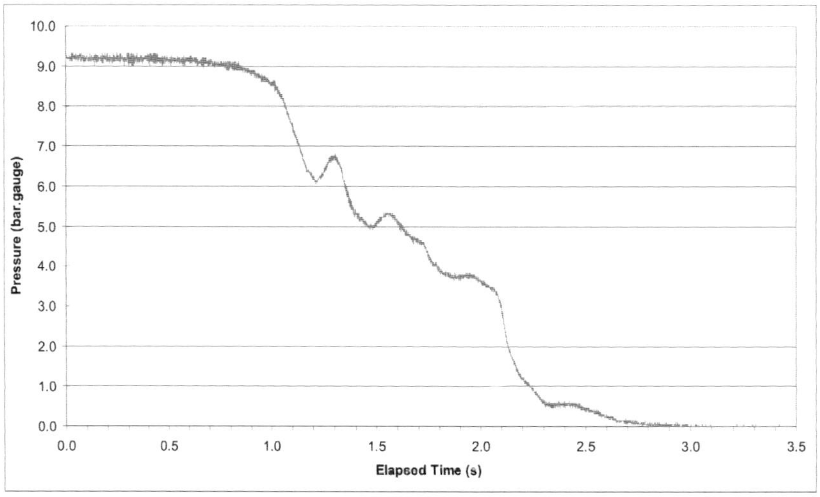

Figure 6: Pressure time graph showing pipe burst test results

The results of the second test (Figure 6) show that the fluctuations, caused by pressure pulses are still present, however the rapidly diminishing pressure in the hose alters the wave form. This data shows that if a pipe bursts there should be a significant and rapid negative pressure gradient detected downstream of the burst, with no appreciable pressure recovery towards the normal operating pressure.

The steep decline in pressure seen immediately after the "burst" event (at ~1s, Figure 6) occurs for almost two thirds of a second prior to any recovery. Contrast this with data in Figure 5, where after the initial drop in pressure, the pressure gradient becomes positive in around 0.25s. This implies that downstream of the pipe failure a clear indicator of a burst may be detected, however on the supply side of the burst the pressure profile may be confused with a normal operation, for instance the closing of a valve.

To reliably use pressure data to trigger an isolation valve, it is important that any analytical software run by the valve micro-controller can discern the difference between event pressure profiles, and reliably identify changes in pressure due to routine system operation as opposed to a pipe failure. Further testing is proposed to measure the pressure fluctuations either side of a simulated pipe failure concurrently, to gain further understanding of the pressure profiles at different locations in a pipe system.

AUTONOMOUS CONTROL

To satisfy the need for autonomous valve control, a processing and control module is required. The use of open source hardware for this purpose will contain the cost of the final product and allow maximum flexibility both in the development phase and when implemented.

In recent years a rapid rise in the use and application of open source hardware in the hobbyist and enthusiast spheres has occurred. The wide uptake in the domestic community has seen the emergence of several alternative open-source manufactures and has reduced the cost of Programmable Logic Controllers (PLCs). This has lead to rapid adaption by academic and research organisations as a viable alternative to expensive proprietary PLC products. Further, the large community of users provides an extensive knowledge base for programming and hardware setup.

The frontrunner in the open source PLC market is produced by Arduino. As an open source platform, several alternative models and brands exist, and employ the same or similarly specified processing chips and hardware, with the same input/output pins and general arrangement as the Arduino branded versions. The various Arduino-type prototype boards use a 16MHz, 8-bit processor to drive several digital and analogue input/output pins. The chip is programmable using the C+ language, and can store up to 32Kb. The advantage of this is that the user can select a model that meets quality and cost guidelines, without compromising compatibility and function.

All the concept designs developed rely on a microprocessor to process the signals from various sensors, and actuate the isolation valve if a pipe rupture is detected. The Arduino-type open source board offers a customisable platform which accepts multiple inputs, can provide control outputs, and which can be programmed to process data as required for different parameters being detected. During development this will aid in prototyping and allow changes to both the program and hardware configuration to be made as the need arises.

ONGOING DEVELOPMENT

From the tests conducted to date, and the theoretical knowledge developed, there is sufficient understanding of pressure based systems to continue design development in this direction.

It has become clear through the course of this project that there is a significant design effort required to ensure that a pressure based system will work in different locations on a ship, and in different ships. This is because the pipe system layout, pipe size, working fluid and valve type are integral in determining the various pressure pulse magnitudes to be expected in a system during normal operation. With poorly defined "normal operating" parameters, it is likely the system will fail to work as required, and may result in a large number of false positives. Additionally, a greater understanding of the pressure fluctuations at points upstream and downstream of a pipe burst is required before a reliable system can be implemented.

Using pressure measurements to determine if a pipe has failed relies on secondary data to infer the failure. While it is common practice in control system design to measure indirect parameters, it is more desirable for both fidelity and response time, to measure a direct quantity relating to the control. The ongoing development focus will investigate the feasibility of, and theory behind, methods which can directly determine if the pipe has failed. This will occur concurrently with continued exploration of a pressure based approach. Once the theoretical constraints of the various approaches are well understood, the intention is to develop real world representative scale tests and a prototype pipe run to determine the effectiveness of the techniques and correlate any theoretical predictions with actual results.

The test results will allow refinements to be made to the design requirements and specifications, to ensure that the functions are fully satisfied by the final design. Once the design requirements have been refined and are fully defined, the design concepts will be compared and a preferred design selected. The detailed design phase will focus on developing a marketable solution, and involve further prototyping, shock qualification, packaging and mounting solutions, and design for production.

REFERENCES

Alcaro Hydraulic Services, 2011, *Velocity Fuses*, viewed 5/11/11, <www.alcarohydraulics.com.au/velocity-fuses>.

Clark G.A., Birch I.A., 2001, *Emergent Technologies for the Royal Australian Navy's Future Afloat Support Force*, Maritime Platforms Division, Aeronautical and Maritime Research Laboratory.

Hiltz J.A, 2005, *Damage Control and Crew Optimisation*, Defence R&D Canada – Atlantic, Technical Memorandum DRDC Atlantic TM 2005-010.

Hunaidi O & Wang A, 2000, *LeakFinder – New Pipeline Leak Detection System*, National Research Council Canada, Institute for Research in Construction, Ottawa, Canada, October

Lestina T, Runnerstrom E & Davis K, 1999, *Evaluation of Firemain Architectures and Supporting Reflexive Technology*, Naval Technology Center for Safety and Survivability Chemistry Division.

Lestina T, Bradley M, Downs R & Runnerstrom E, 2001, *Development of DC-ARM Reflexive Smart Valve*, Naval Technology Center for Safety and Survivability Chemistry Division.

Lo, C, 2005, *IEEE Transactions in electromagnetic compatibility,* Vol 47, No 1, pp97-104.

Misiunas D, Lambert M.F., Simpson A.R & Olsson G, 2005, "Burst detection and location in water distribution systems", *Water Science and Technology: Water Supply*, Vol 5, No 3-4, pp.71-80.

Wood D, Lingireddy S & Boulos P, 2005, *Pressure Wave Analysis of Transient Flow in Pipe Distribution Systems,* MWH Soft, Pasadena, CA, USA.

Yokota Manufacturing, August 2011, *Emergency Shutoff Valve*, viewed 5/11/11 <www.aquadevice.com/>.

Status and way forward for LNG as a maritime fuel

Lars Petter Blikom
DNV, Norway
E-mail: lars.petter.blikom@dnv.com

ABSTRACT AND INTRODUCTION

Over the past 10 years, LNG has been proven as a technically viable fuel for commercial ships. Currently, fuel prices and environmental legislation are driving forces for the first steps of an exponential growth in the number of ships fuelled by LNG. This presentation will include a review of current LNG fuelled ships in operation, the order book for LNG fuelled ships under construction, and key lessons learnt so far. The presentation will also focus on the major challenges to future development: bunkering operations and availability of LNG. DNV has also recently finalized a study for expected market penetration of LNG versus other maritime fuels under various scenarios for future development; the results of this study will be presented.

THE PAST TEN YEARS OF LNG FUELLED SHIPS

The history of the application of LNG as a maritime fuel is short, but is now set to take off. DNV developed the first classification rules for gas fuelled ships in 2000 along with the construction of the world's first LNG fuelled ferry. This very first LNG fuelled ferry project was launched because of a direct requirement from the Norwegian Government to run this specific ferry route on natural gas – they were motivated by two things: Increased domestic use of natural gas and close to elimination of local emissions from the ferry operation.

Today the fleet of gas fuelled ships has grown to 22 ships covering ferries, offshore supply vessels and coast guard vessels. They are all operating in Norwegian waters, and a major reason for their very existence is a continued push by the Norwegian Government. However, have now passed ten years of operation, the shipowners and their crews seems fairly unanimous in their positive experience with this type of fuel. The complete list of LNG fuelled ships in operation is shown in Table 1.

Table 1: Total global fleet of LNG fuelled ships in operation

Year	Type of vessel	Owner	Class
2000	Car/passenger ferry	Fjord1	DNV
2003	PSV	Simon Møkster	DNV
2003	PSV	Eidesvik	DNV
2006	Car/passenger ferry	Fjord1	DNV
2007	Car/passenger ferry	Fjord1	DNV
2007	Car/passenger ferry	Fjord1	DNV
2007	Car/passenger ferry	Fjord1	DNV
2007	Car/passenger ferry	Fjord1	DNV
2008	PSV	Eidesvik Shipping	DNV
2009	PSV	Eidesvik Shipping	DNV

2009	Car/passenger ferry	Tide Sjø	DNV
2009	Car/passenger ferry	Tide Sjø	DNV
2009	Car/passenger ferry	Tide Sjø	DNV
2009	Patrol vessel	REM	DNV
2009	Car/passenger ferry	Fjord1	DNV
2010	Patrol vessel	REM	DNV
2010	Car/passenger ferry	Fjord1	DNV
2010	Patrol vessel	REM	DNV
2010	Car/passenger ferry	Fjord1	DNV
2010	Car/passenger ferry	Fjord1	DNV
2010	Car/passenger ferry	Fosen Namsos Sjø	DNV
2011	PSV	DOF	DNV

Table 2 contains all the ships for which a yard order has been signed. It is worth noticing that the list now expands outside of Norway: The four Ro-Ro vessels operated by Sea-Cargo and Norlines will operated on routes within Northern Europe and the Baltic Sea. The RoPax operated by Viking Line will serve a route between Sweden and Finland. The High speed RoPax operated by Buquebus will be built in Incat, Tasmania for operations between Argentina and Uruguay. The two most recently ordered platform supply vessels will be operated in the Gulf of Mexico by Harvey Gulf.

In addition, it is promising to see a list of four vessels already decided for conversion to LNG propulsion. This indicates the beginning of a potentially huge market for conversion projects.

Table 2: Total global fleet of LNG fuelled ships under construction

Year	Type of vessel	Owner	Class
2011	Car/passenger ferry	Fjord1	DNV
2011	PSV	Solstad Rederi	DNV
2011	General Cargo	Nordnorsk Shipping	DNV
2012	PSV	Olympic Shipping	DNV
2012	PSV	Eidesvik	DNV
2012	PSV	Eidesvik	DNV
2012	Ro-Ro	Sea-Cargo	DNV
2012	Ro-Ro	Sea-Cargo	DNV
2012	High speed RoPax	Buquebus	DNV
2012	PSV	Island Offshore	DNV
2012	PSV	Island Offshore	DNV
2012	PSV	REM	DNV
2013	Ro-Ro	Norlines	DNV
2013	Ro-Ro	Norlines	DNV
2013	Car/passenger ferry	Torghatten Nord	DNV
2013	Car/passenger ferry	Torghatten Nord	DNV
2013	Car/passenger ferry	Torghatten Nord	DNV
2013	Car/passenger ferry	Torghatten Nord	DNV
2013	RoPax	Viking Line	LR
2013	PSV	Harvey Gulf Int. Marine	ABS
2013	PSV	Harvey Gulf Int. Marine	ABS

Planned conversion

Year	Type of vessel	Owner	
2011	Car/passenger ferry	Fjord1	DNV
2011	Chemical tanker	Tarbit Shipping	GL
2013	RoPax	Fjordline	DNV
2013	RoPax	Fjordline	DNV

CURRENT DRIVING FORCES

There are at least two very important driving forces for the move towards LNG as a maritime fuel:

- Emissions regulation leave LNG as one of very few options for compliance
- Expected future price development for oil versus natural gas indicates a significant difference in favour of natural gas, and hence LNG.

The International Maritime Organization's (IMO) plan for combating local emissions from shipping is two-fold; first, SOx emissions will have to be reduced from both existing and newbuild ships operating in so-called emission control areas (ECA). Secondly, NOx emissions will have to be reduced for all newbuild ships operating in international waters. The requirements are being implemented gradually and will have full force from 2015 and 2016, for SOx and NOx, respectively. For their existing fleets, ship owners are left with only three options if they wish to continue trading in ECAs from 2015: switch to low-sulfur fuel, install an exhaust gas purification system (scrubber), or switch to LNG. For newbuild ships delivered after 2016 the choices are even fewer; either switch to LNG fuel or install selective catalyst reduction (SCR) system for cleaning the exhaust gas of NOx. This system would come on top of the scrubbers for cleaning of SOx.

Figure 1 shows the already decided ECA areas, but it should be expected that these areas will be extended and decided also for other areas.

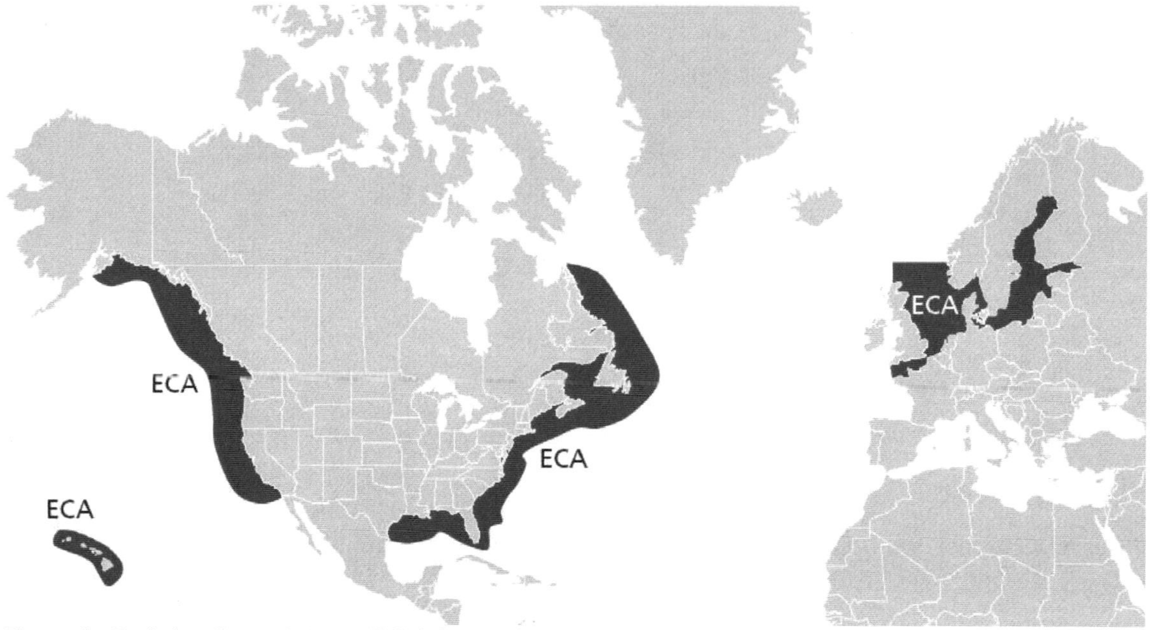

Figure 1 - Emission Control Areas (ECA) for special emissions requirements

The driver related to the price difference between the various fuel options is much more difficult to quantify. Analysts are consistent in always failing to foresee the both oil price and the natural gas price, separately. So we should not expect them to foresee the relative difference between these two fuel sources over the entire lifecycle of the ship, which is what we need in order to make a qualified decision about which fuel offers the better total economic performance.

Regardless of the difficulties in foreseeing future prices, there is a general perception in the market that natural gas will be cheaper than oil on an average per energy content basis on the long term. The reason is simply that there is more natural gas reserves left to be developed than there is oil, which should also translate into lower production costs for natural gas than oil as reserves becomes more and more scarce and difficult to extract.

In any case, it is worth noticing that this very uncertain difference in future fuel prices is more or less the only thing that matters to the economics of future shipping. With reference to Figure 2 it is very clear that fuel represents the lion's share of daily costs for operating almost any type of ship.

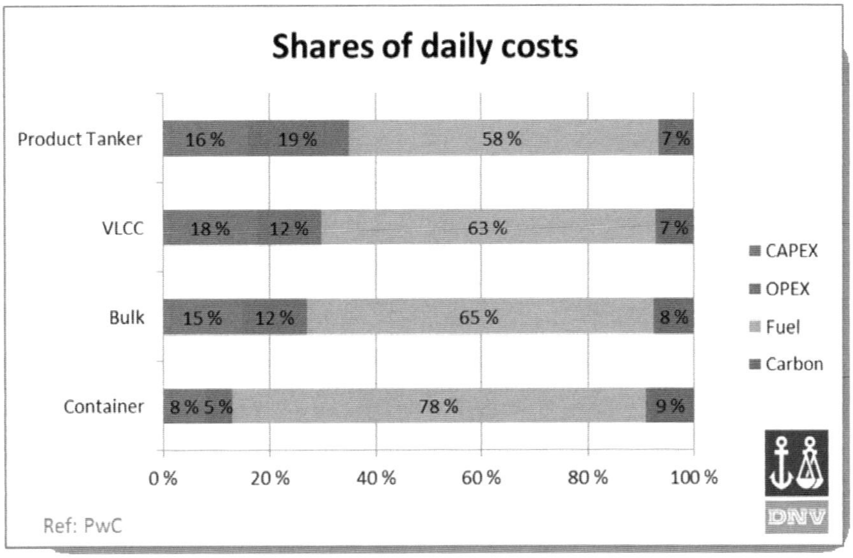

Figure 2 - The main cost contributors for some typical ships

OUTLOOK

In order to try to quantify the impact of these two drivers described above, and others, we decided to apply a scenario planning methodology. This is a methodology that fits well when there is a high degree of uncertainty and you have acknowledged that any single point forecasts would be merely naive. An overview of the method is provided in Figure 3.

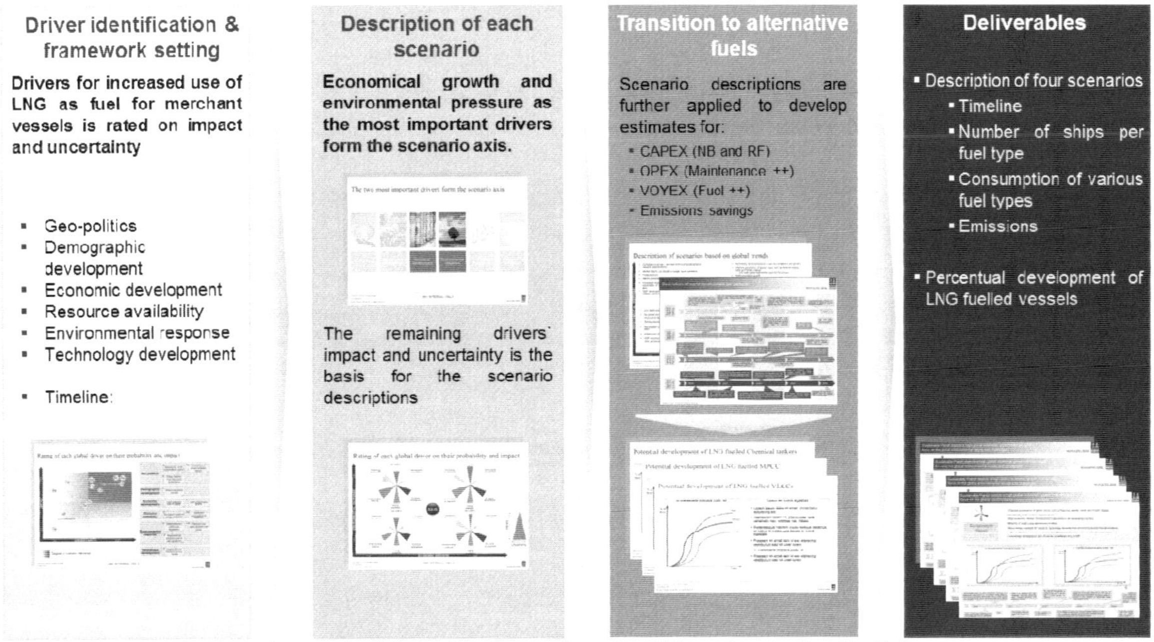

Figure 3 – Scenario planning methodology for developing plausible estimates for the future

After deciding on the most important drivers, these were used to define four different scenarios. Inside these four scenarios various other drivers are included in order to describe a plausible future regulatory and business environment. Then estimates for the implementation of each of the available options compliant with regulations are developed taking all parameters of the scenario into account. Figure 4 summarises the result.

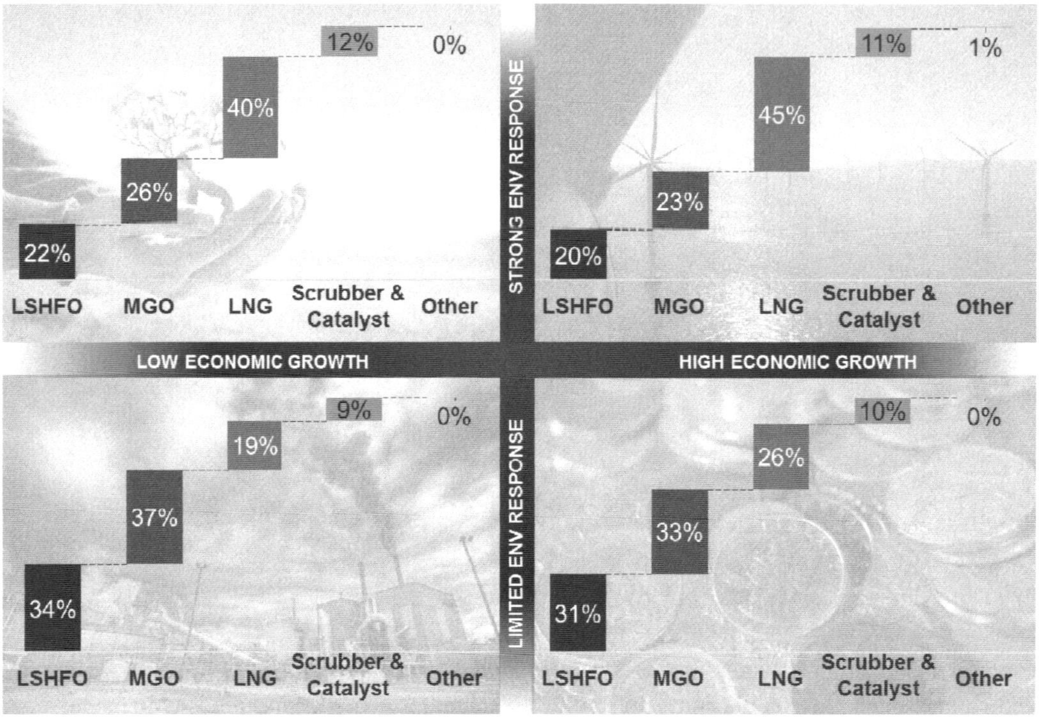

Figure 4 – The results of the scenario planning process

CONCLUSION

As Figure 4 clearly shows, regardless of which scenario develops, the shipping industry will be divided in their choice of fuels. This leads to several key observations: Firstly, there will be a significant demand for LNG to the maritime industry, so infrastructure for supply to main ports and bunkering locations should be developed as soon as possible in order to ensure that this option is a reality by the time ship owners are forced to choose. Secondly, it is clear that the cost picture for ships in the future will be highly dependent on their choice in fuels, which means the choice of fuel will have a huge impact on the competitive situation between different ship owners. Preparing for this new framework for shipping should be in the biggest interest of all actors involved.

Economical Multi-Role Solutions

Authors:
Tim Speer (Austal) and Simon Carroll (CEA Technologies) Email
contact: tims@austal.com and simon.carroll@cea.com.au

INTRODUCTION

It has long been recognised that there is a significant cost and schedule risk associated with the acquisition of a new class of naval vessels. A recognised strategy for reducing this risk is to standardise platforms across multiple missions. Standardised platforms result in a reduction of non-recurring design costs, improved supportability in-service and a reduction in the crew training burden. With the increasing adoption of modular mission system capabilities, the standardised platform approach is being considered by a number of navies including the Royal Navy with their Type 26 frigates, designed to be easily adapted to change roles and capabilities depending on the strategic circumstances. The Littoral Combat Ship Program in the US is defined as a multi-mission combatant, which can be reconfigured with mine warfare, submarine warfare or a surface warfare mission packages. The Australian Defence Capability Plan (DCP) describes the SEA 1180 program as a rationalisation of the Navy's patrol boat, mine countermeasures, hydrographic and oceanographic forces, potentially into a single modular multi-role class.

In this paper we provide an overview of some of the likely mission capabilities and sub-systems required to deliver the SEA 1180 Offshore Combatant Vessel (OCV). We also provide guidance on the relative cost and complexity associated with each of these sub-systems. The key challenge associated with each sub-system is to limit the level of complexity associated with the system integration. The use of a modular approach needs to be carefully managed to ensure that the ship platform is not configured to support every possible module configuration and that every module does not require seamless integration into a combat management system.

THE ADVANTAGES OF THE TRIMARAN HULL

The Austal 80m Multi Role Vessel (MRV) has been developed over several years based on the lessons learned from three generations of trimaran hull design. The first generation 127m vehicle-passenger ferry, *Benchijigua Express* has now been in successful commercial service for six years, while the second generation Littoral Combat Ship, *USS Independence* was handed over to the USN in December 2009. The third generation 102m vehicle-passenger ferry constructed by Austal completed sea trials in 2010.

The MRV hull provides:

- Improved seakeeping performance when compared with similar sized monohull and catamaran platforms, to improve the operability including for aircraft and small craft, launch and recovery
- An extremely efficient hull form offering lower resistance, lower fuel consumption and less required propulsion power than monohull designs
- Significantly improved resistance due to the lightweight aluminium hull as demonstrated on the Armidale Class Patrol Boats
- Substantial deck area to support a large flight deck
- A large internal mission bay to support a variety of vehicles and mission equipment

Austal has recently completed a towed resistance test program on the MRV hull form to verify the predicted hull resistance compared with numerical predictions (correlated against full-scale trials results). Figure 1 illustrates the model operating at speeds of 10, 20 and 29 knots. It is difficult to visually determine the operating speed due to the lack of any apparent wake at the bow of the model. It is only apparent at the stern of the model where transom separation is occurring at higher vessel speeds. The hull design of the MRV utilises a plum stem configuration to increase the design waterline and reduce the half angle of entrance of the hull. This hull configuration was also used on the 102m vehicle-passenger ferry discussed above.

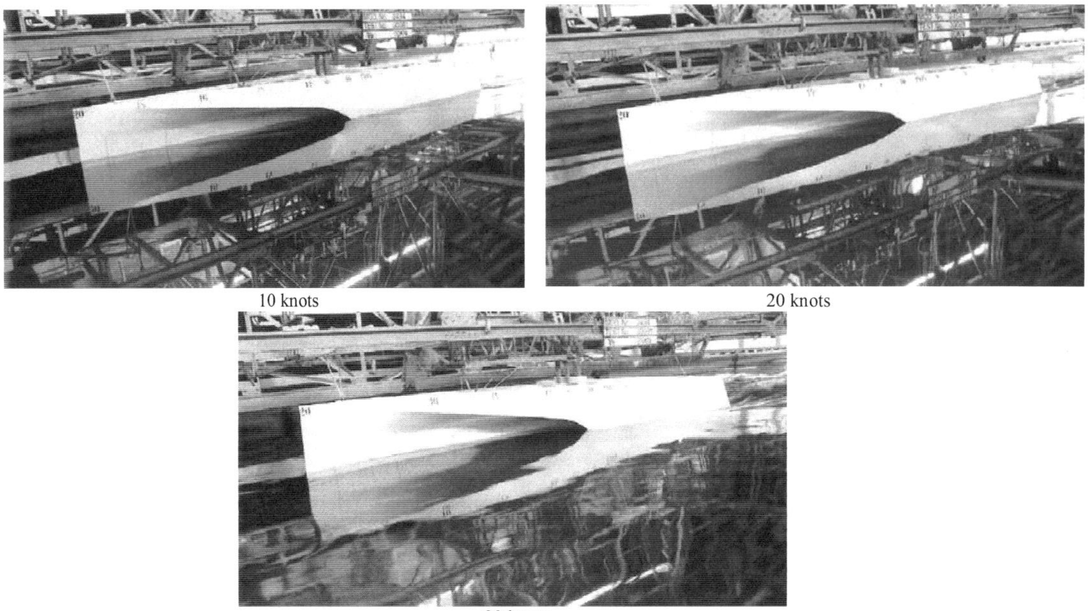

10 knots 20 knots

29 knots

Figure 1 The towed resistance model testing of the 80 MRV hull form demonstrated excellent performance with extremely low wave making.

The Austal MRV basis design provides a significant capability in support of the stated OCV missions:

A. Maritime Constabulary
B. Maritime Littoral Warfare
C. MCM
D. Maritime Geospatial

This basis design has significant organic capability associated with its flight deck and its large mission bay. Austal's experience with the LCS program in the US has demonstrated the utility of these two design features.

The flight deck not only multiplies the surveillance capability of the platform; it also provides additional capability options in support of Maritime Littoral Warfare, MCM and Maritime Geospatial missions. The LCS program MCM package includes systems such as the Northrop Grumman AN/AES-1 (Airborne Laser Mine Detection System), the Raytheon AN/ASQ-235 Airborne Mine Neutralization System (AMNS) and the EDO Marine and Aircraft Systems ALQ-220 Organic Airborne Surface Influence Sweep (OASIS). The LCS MCM capability relies heavily on the organic aviation capability of the platform to support these systems. The Fugro LADS Mk3 ALB System is a key tool in the Australian context, in support of Maritime Geospatial collection of bathymetry data.

The mission bay provides substantial working area in support of a variety of missions. Various surface and underwater vehicles may be stored and launched from this area. The large ramp allows the loading of the vehicles and modules. Refrigerated storage, accommodation modules and medical facilities may all be embarked in this area depending on mission demands.

In addition to the basis design characteristics, a number of additional optional system capabilities are investigated in this paper, to analyse cost versus benefit and trade-offs associated with the MRV platform and its integrated systems. The discussion is organised into the following areas:

- Hull Structure
- Machinery and Electrical Systems
- Communications

- Navigation
- Weapons and Sensors (and their integration)
- Offboard Vehicles (and their integration)
- Modular Systems (and their integration)

The DCP states that the SEA 1180 Program for a single modular multi-role class or family of around 20 OCVs has an anticipated acquisition cost \$3B–\$5B. This acquisition cost would result in a cost of \$150M to \$250M per vessel. Austal has reviewed the cost associated with the MRV basis design and would be able to deliver the platform well within this budget. The following cost estimates are calculated in relation to the predicted acquisition cost of the basis design.

HULL STRUCTURE

The proposed aluminium hull structure is designed around the requirements of the DNV HSLC Rules and the global hull structure can withstand the dynamic loads associated with hollow landing and crest landing in severe sea conditions. In addition to these loads the vessel may also be designed in accordance with the additional classification notation ELT (extended life time). This notation allows the fatigue life of the platform to be verified for 30 years or more as is the case for the LCS design. Austal's experience indicates that if the hull girder is rigid enough to withstand the class specified dynamic global bending loads, it will also have adequate stiffness to withstand underwater explosion (UNDEX) whipping loads. The shock qualification of the entire design will, however, add significantly to overall platform cost, based on the qualification requirements for all the installed equipment and the associated equipment foundations. Even if only selected essential systems such as the propulsion system are shock qualified, then the anticipated increase to the platform cost would be more than 10%.

Any requirement for a Collective Protection System on the platform would also result in some significant design changes associated with the Citadel zone, including the ventilation system, air locks, decontamination stations, counter-measure wash down systems and chemical biological radiological (CBR) detection systems. The best way to limit these impacts is to limit the size of the Citadel. A compact ventilation system to serve 200m^2 and 30 crew may be incorporated for a modest cost (<1% of the platform cost).

MACHINERY AND ELECTRICAL SYSTEMS

One of the key requirements of the maritime constabulary and maritime littoral warfare missions is the ability to operate economically at low speed for extended periods of time. A typical operating profile for a vessel of this type is illustrated in Figure 2. This operating profile clearly illustrates that operation at speeds between 5 and 19 knots are a key system capability. In order to dramatically improve the propulsion system performance in this operating range, Austal in cooperation with MTU are evaluating the utility of a hybrid propulsion system.

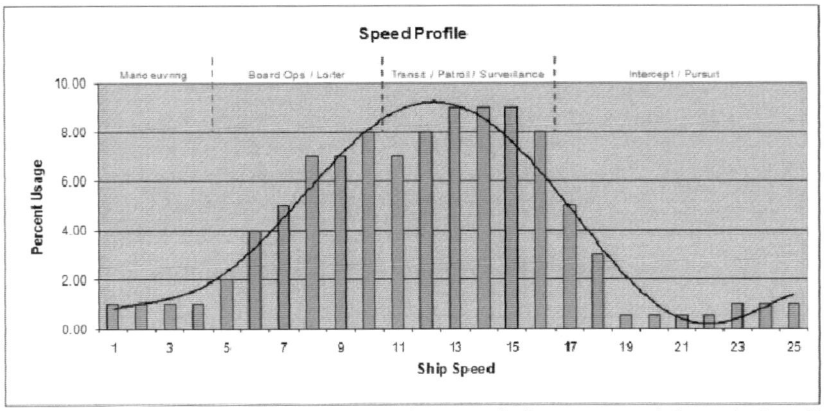

Figure 2 A typical operating profile may require the OPC platform to spend the majority of its time between 5 and 19 knots, which enhances the importance of efficient operation at low power

Traditional diesel-electric propulsion systems add significantly to the weight and cost of the propulsion machinery due to the need for substantial generators and electric motors to support the full power of the propulsion system. MTU have developed a novel hybrid propulsion system which allows the vessel to operate at low speed and transit speeds using electrical power and only relies on the main propulsion engines for higher speeds. The arrangement of diesel engine and crank shaft generator is illustrated in Figure 3.

Figure 3 The MTU hybrid propulsion system enables the selective use of the main propulsion diesel engines to optimise fuel consumption, reduce engine running hours and reduced maintenance costs

The different potential operating modes of the system are illustrated in Figure 4.

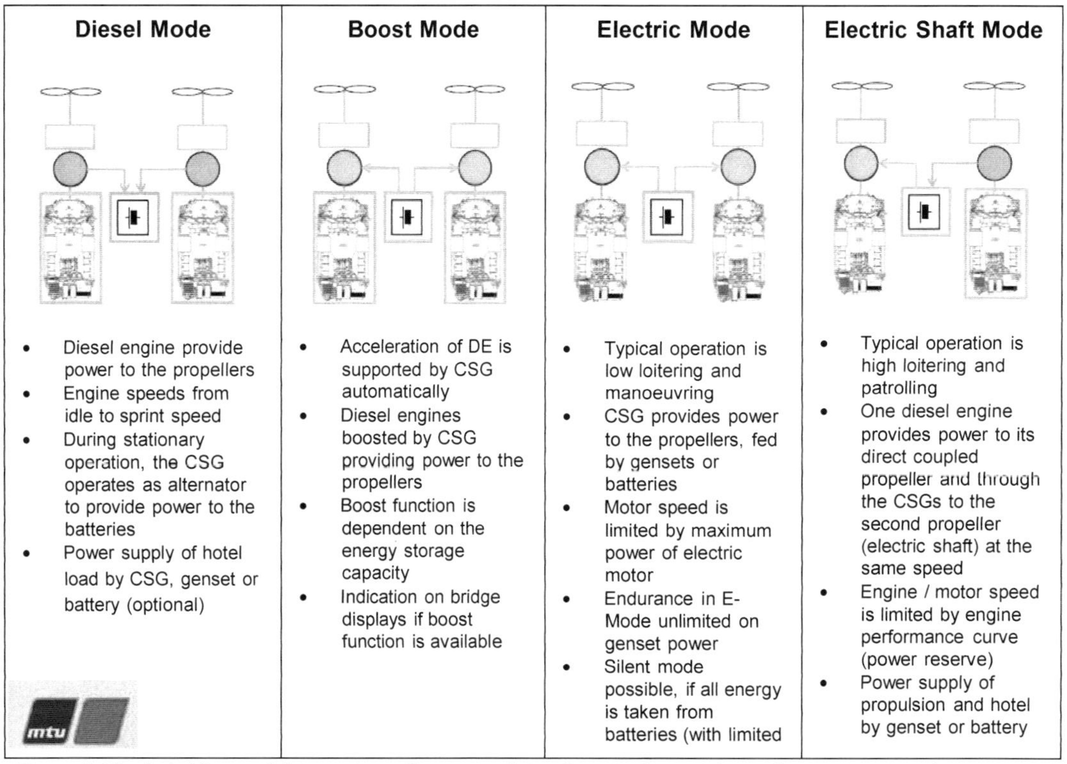

Figure 4 The hybrid propulsion system allows a variety of operating modes to better support the range of operating speeds required of the MRV platform

The key advantage of the hybrid propulsion system is that the installed ship's generators may provide propulsion power at low ship speed, with much more efficient specific fuel consumption. The generators may operate at 80-90% of their MCR, compared with less than 20% of the MCR of the main propulsion diesel engines. As demonstrated in the above operating profile, this may equate to more than 80% of the operating hours of the platform. The potential fuel saving resulting from this hybrid arrangement is as much as 15% over the entire operating profile, while reducing the operating hours of the main propulsion diesels and reducing the system maintenance cost. The other operational capability of the system which may be useful during Maritime Littoral Warfare missions is the ability to run in silent mode from battery power for limited periods.

The cost associated with this hybrid propulsion system has been estimated at an additional 3% of the platform cost.

COMMUNICATIONS

The range of available communication system options is wide and the final selection depends on the requirements of the operator. The Austal MRV design includes a commercially compliant GMDSS A3 communication suite and an allowance of more than 3% of the total platform cost for a secure military communications system.

Austal has contracted CEA Technologies to provide this level of communications capability for both domestic and international programs, including the recent Cape Class Patrol Boat for the Australian Customs and Border Protection Service.

The CEA Integrated Ship Communications System (CEA-ISCS) provides a fully integrated internal and external communications suite that is designed to meet the constraints of a minimum manned vessel, while providing the functionality to achieve modern day network centric communications requirements. The CEA-ISCS is traditionally included in the overall topside design of the platform and provides the following capabilities:

- Internal communications networks – point to point; conference circuits;
- External communications (secure and unsecure) – HF, VHF, UHF and SATCOM voice and data networks;
- Communications management terminal; message handling terminal
- Digital voice recording system;
- Tactical intercom; public address system; general alarms; and
- Integrated PABX.

An Electro Optical Surveillance System (EOSS) positioned on the main mast provides maximum visibility for surveillance, man overboard and crew operations. The thermal imaging camera provides high performance passive infra-red imaging in day, night and poor visibility. The camera utilises full TV Medium Wave IR detectors coupled with the latest generation of advance image processing electronics to deliver superior imaging performance. There are many of these systems available commercial off the shelf, and the system cost varies dramatically depending on the required system capability (stabilisation, thermal capabilities and cooling requirements, optical and digital magnification capabilities, etc.). If the system is not required to be integrated with any of the weapons systems then the integration cost may be avoided.

A number of ship's boats situational awareness systems are available in the market today. These systems dramatically improve the networked capability of the deployed ship's boats and the associated boarding parties. The Austal Mobile Asset Command and Control (MAC^2) system allows the secure exchange of voice, text and imagery via VHF, GSM or satellite networks. The total cost of this system is quite modest but provides a significant increase in the capability of one of the key platform assets when engaged in wide range of mission activities.

Another capability which may be fitted to the platform at modest cost is a long range acoustic device (LRAD) which may be used for both, communication and counter terrorism / force protection requirements. An appropriate LRAD for a vessel of the size of the MRV could be installed on the bridge wings.

NAVIGATION

The installed navigation system is configured to address the requirements of a minimum manned vessel with inter-switchable X and S band radars and ECDIS displays. The bridge layout complies with the nautical safety requirements of a high speed craft, providing 360 degree visibility and allowing safe navigation with a minimum of two personnel at the bridge navigating console. In addition to satisfying the standard IMO navigation radar requirements, an option exists to provide a more advanced navigation and situational awareness radar. When considering the operating roles of the proposed SEA 1180 platform a radar such as the newly developed SeaGuard X-Band Coherent Radar (XBCR) provides a total solution for these capabilities.

Developed by CEA Technologies and Northrop Grumman the SeaGuard radar is an advanced Doppler radar solution that:

- Provides target discrimination in a wide range of conditions, from calm to heavy sea states and littoral environments;
- Detects small targets in a cluttered environment;
- Interfaces to Automated Identification Systems (AIS);
- Tracks up to a total of 250 air and surface targets;
- Provides complete Automatic Radar Plotting Aid (ARPA) capability plus standard navigational radar functionality; and
- Incorporates helicopter approach and control

The SeaGuard XBCR uses the industry standard VisionMaster (VM) FT display providing a naval display and numerous other functions for the operator. The COTS based display controls the SeaGuard XBCR, as well as providing interswitching to other radar sources. This flexibility ensures that the anticipated operational requirements of all four roles detailed for the SEA 1180 platform are met with a common radar for whole of class service and support. The cost of the SeaGuard XBCR is less than 1% of the platform cost.

WEAPONS AND SENSORS

The weapons and sensors fit for the MRV is heavily dependent on the role chosen for the platform and as such may be part of the core equipment fit or may be part of a modular capability. Systems such as towed decoys, modular missile systems and mobile close-in weapons systems are systems that can be either fixed or mobile, while others such as larger calibre guns and rotating radars are more suited as a fixed and/or permanent capability.

Surface Search and Air Search Radar Options

The inclusion of a dedicated surface search and/or air search radar capability on-board the MRV platform is an option that has previously required a permanent radar capability. With the advent of multi-function radars a modular radar arrangement has become more viable for vessels such as Austal's MRV, as this type of radar can perform a range of tasks through a single interface while only occupying a single permanent radar mount on-board the platform.

Through the use of their scalable active phased array technology, CEA Technologies is able to incorporate the functions and features of a multi-function radar into a mobile and modular capability suitable for use with the MRV platform. This concept provides the user with the ability to transfer the radar module between platforms in order to suit the operational requirement of each vessel. The installation of the radar module takes a matter of hours and has minimal impact on the remaining sensors and systems on-board the platform requiring only power and fibre connections for effective operation. In addition, due to the modular nature of the proposed system it is easily freighted by air to meet the operational requirements of the ship system should a module need to be replaced or fitted at short notice.

A variant of the radar module is shown in Figure 5 and includes the S-Band CEAFAR Active Phased Array Radar (2 faces) and the X-Band CEAMOUNT Active Phased Array Illuminator (1 face). The CEAFAR radar is a fully digital, multi-function active phased array radar that provides three dimensional air and volume surveillance and surface search capabilities. The S-Band CEAFAR radar has been designed to meet the challenges of the littoral environment and provides superior clutter rejection capabilities affording the operator the ability to detect and track targets in all conditions. The CEAMOUNT illuminator is included as an option should a semi-active missile be incorporated into the platform as it integrates seamlessly with CEAFAR without requiring additional interfaces to the combat management system.

Figure 5 CEAFAR and CEAMOUNT Modular Active Phased Array System provides a scalable, mobile and transportable option for a 3D Air and Surface Search Capability

Due to the scalable nature of the radar and illuminator architecture the number and size of radar and illuminator faces within the CEA modules can be designed to meet the mission requirements. For example, if there were no semi-active missile capability for the platform then the CEAMOUNT illuminator would be removed and the module would only contain CEAFAR faces. Each module could contain two or three CEAFAR faces to achieve 180° of air and surface surveillance, with two modules required to achieve all round 360° surveillance. This provides further flexibility to the platform as the modules are not required to be co-located, allowing them to be positioned at the most convenient topside location.

Figure 6 MRV variant with fixed face CEAFAR Phased Array System provides a potential hull option for those vessels dedicated to the Maritime Littoral Warfare Role

Should the requirement exist for a permanent air and surface search capability, on some or all of the platforms, Austal and CEA have also worked together to combine the capabilities of the MRV platform with a dedicated phased array mast. As seen in Figure 6, the MRV is well suited to be fitted with a single mast that can house the CEAFAR radar and CEAMOUNT illuminator should this sensor specification be required. Such a concept provides the benefit of the MRV hull technology along with 4th generation active phased array radar technology for platforms that may be dedicated to the Maritime Littoral Warfare role where the "combat" focus of these vessels is a key requirement.

Combat Management System (single consoles versus fully integrated CMS)

The inclusion of a Combat Management System (CMS) into the MRV platform is also highly dependent on the role of the platform. While a fully integrated CMS as a core capability in a dedicated operations room is clearly more suited to the combat and warfare related roles of the MRV, options exist to reduce the size of the equipment required for this type of CMS while also maintaining a high level of CMS functionality.

Systems such as Saab's 9LV Situational Awareness Terminal (9LV SAT) provide a single operator terminal that can be situated on the bridge and provide a coordinated and integrated tactical picture from a range of on-board systems and sensors. This type of capability removes the need for a dedicated operations room while still providing a level of combat system functionality that can be configured to meet the role of the host platform. Systems such as this are also highly compatible with the communications and situational awareness systems that may be fitted to ship's boats providing a higher level of integration and coordination with these organic platforms.

A range of different combat system solutions are available depending on the required combat capability of the platform. The level of integration of this system with the on-board weapons and sensors will have a significant impact on the overall ship system cost. The SAAB 9LV system cost equates to less than 1% of the platform cost, however a more comprehensive system with multiple consoles and integrated with all weapons and sensors may cost more than 15% of the platform cost. For this reason, the required level of CMS integration needs to be carefully evaluated early in the requirements definition phase.

Decoys

The Terma SKWS Decoy Launching System can support the control of either six or twelve firing tubes (Figure 7) per side of the vessel to provide 360 degree coverage against attacking anti-ship missiles. A range of different decoys can be supported including Chaff, Infrared rounds, Dual mode RF and IR rounds and Torpedo decoys. While the decoy system cost only equates to around 2.5% of the platform cost, there is a requirement for the system to be integrated with the installed combat management system to manage ship situational data and to detect and engage incoming threats. As discussed above, this level of integration may substantially increases the complexity of the platform and the associated cost of the integrated combat system.

Figure 7 The Terma decoy launching system provides 360 degree coverage against a variety of incoming anti-ship missiles

Hull Mounted Sonar

There are a variety of different hull mounted sonar options to support both MCM and Maritime Geospatial missions. Suppliers such as ITT, Ultra and L-3 offer a range of different capabilities to perform differing tasks. The forward looking sonar can operate as an obstacle avoidance sonar, and as a forward looking detection and classification sonar. There are also side scan and other hull mounted options which improve the capability to search, detect, track and classify multiple targets. The complexity, level of integration and cost of the installed sonar system depends on the required system capability. A simple forward looking sonar equates to less than 1% of the platform cost while a more capable hull mounted sonar may equate to 10% of the platform cost. These systems may be operated using dedicated consoles, however as the level of integration into the ship combat management system increases so too does the cost.

Towed Arrays

A towed array may easily be deployed from the aft end of the Austal MRV design, with the wide transom and available mission bay volume. Towed arrays offer more capability than a hull mounted sonar as the array is able to operate away from the interference of the hull. The towed array supports long range detection and classification in littoral or deep water and detection of torpedos. The integration cost associated with a towed array includes foundations and hydraulics for the large winch. This system also may be operated using dedicated consoles, however as the level of integration into the ship combat management system increases so too does the cost. The system cost of a towed array may be as much as 15% of the platform system cost.

A number of integrated sonar systems exist in the market to support integrated hydrographic survey or surface ship torpedo defence (including decoys). These integrated systems vary in cost depending on the system capability and range between 5% and 12% of the platform cost.

Primary Gun Options

A vessel of the size of the Austal MRV can accommodate a variety of primary weapon options up to the Bofors 57. The level of integration required for a 20mm gun is limited, only requiring electrical power and a structural attachment location. As the gun size increases so too does the integration cost. Larger weapons require a dedicated magazine and ammunition hoist, power, ventilation air, and control system integration. The weapon cost associated with a 57mm weapon, excluding the associated integration cost, would be in the range of 10% of the platform cost. For a smaller weapon in the 20mm range the cost would be less 3% of the platform cost without any of the significant integration cost associated with a larger weapon.

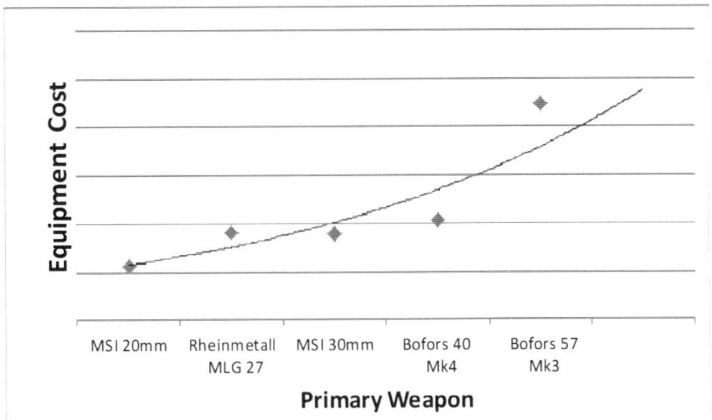

Figure 8 While the equipment cost grows exponentially with the calibre of the weapon, so to do the system integration costs associated with the foundations, the magazine, power, cooling and the CMS.

Close in Weapons Systems and Missile Systems

There is a wide variety of close in weapons systems available in the market to provide platform self defence capability and an even wider variety of missile systems to provide self-defence or offensive capability. These systems include the Phalanx, SeaRAM (Raytheon) and the MARTE and CAMM missile systems (MBDA). The CIWS systems provide a last line of defence for the platform against incoming anti-ship missiles. The advantage offered by Phalanx and SeaRAM (Figure 9) are that they offer a self-contained automated fire control system, enabling it to automatically search for, detect, track, engage and confirm kills using its computer-controlled radar system. This capability is of interest on a vessel without an integrated combat management system and significantly reduces any associated integration cost.

Figure 9 The SeaRAM system is installed on the Austal built Littoral Combat Ship and provides an automated self defence capability with an 11 missile tubes.

MBDA MARTE missile system provides a lightweight system with four MARTE Mk2/N missiles (Figure 10), each capable of sinking an enemy vessel up to 600 tons and at a range in excess of 30km. It is ideal for smaller combatant vessels without requiring dedicated sensors for its operation. The Missile Control Unit takes inputs from the ship navigation system and targeting data from the CMS, however the system can operate in a stand-alone configuration. The missile has fire and forget capability once it is launched, with no in-flight communication or re-vectoring required. The MARTE missile system cost is in the range of 8% of the platform cost and may be installed without a significant integration burden on the platform.

Figure 10 The MBDA MARTE missile system provides a significant SSM capability for the Maritime Littoral Warfare Mission.

OFFBOARD VEHICLES

Key capabilities of the Maritime Constabulary mission are the ability to conduct high speed interception, interdiction or pursuit activities and to perform boarding operations. The existing Armidale Class Patrol Boat has demonstrated the importance of the ship's boats and their launch and recovery arrangements to support these operations. Austal has extensive experience with the heave compensated Vestdavit side launch arrangement. This arrangement provides the ability to launch and recover the manned ship's boats at vessel speeds of up to 10 knots. This system will also be utilised on the eight Cape Class Patrol Boats currently under contract at Austal for the Australian Customs and Border Protection Service. The Austal MRV design includes a port and starboard davit and ship's boat to support rapid deployment. The cost of an additional ship's boat and launching davit is less than 1% of the platform cost.

In addition to this side launch arrangement, the MRV also includes a stern notch for the launch and recovery of a variety of different manned and unmanned vehicles. A gantry cane is arranged over the notch to support the lifting of vehicles into the mission bay. There are a variety of different unmanned underwater vehicles available in support of Maritime Littoral Warfare, MCM and Maritime Geospatial missions. These include the SAAB (Double Eagle, AUV62), Atlas Electronic (Sea Otter Mk II, Seafox and Seawolf) and Bluefin Robotics (21, 12D, 12S) vehicles.

The Bluefin Robotics 12S is capable of operating to depths of 200m with and endurance of up to 26 hours at a speed of 3 knots (with a side scan sonar). RF, Iridium, acoustic communication is available as well as Ethernet and flash drive data connection when recovered. The larger 12D offers operation to a depth of 1500m and an endurance of 30 hours, while the even larger Bluefin 21 offers operation to a depth of 4500m and an endurance of 25 hours. These systems may carry a variety of payloads including side scan sonar, digital camera and strobe and a multi-beam echo sounder. Each vehicle may be launched and recovered using a single point lift arrangement. An A-frame launching arrangement, which is available as part of the Bluefin scope of supply, is illustrated in Figure 11. The Bluefin 21 and its associated launching arrangement would equate to approximately 5% of the platform cost, however the smaller 12S would equate to less than 2% of the platform cost.

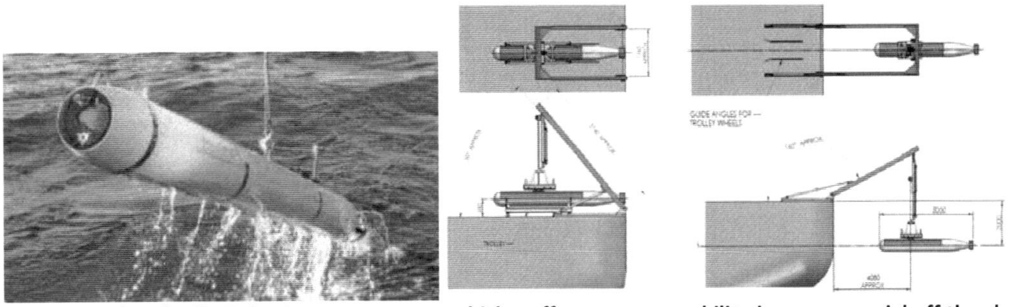

Figure 11 The Bluefin Robotics autonomous vehicles offer proven capability in a commercial off the shelf product including a modular A-frame launching arrangement.

There are many un-manned aerial vehicle (UAV) options available in the market depending on mission payload and endurance requirements. A small UAV like the Insitu Integrator or Scan Eagle offer an excellent surveillance capability at a relatively low cost. The Integrator can operate with a maximum payload of 17kg and an endurance of up to 24 hours. The Integrator's baseline sensor package includes inertially stabilized electro-optic (EO), long-wave infrared (LWIR), and mid-wave infrared (MWIR) cameras, with IR marker and laser rangefinder on a single, long-endurance vehicle. The line-of-sight communications range is over 55 nautical miles (nm) with extended beyond line-of-sight mission radius of up to 550 nm. The vehicle can be launched with a pneumatic catapult and recovered with a SkyHook wingtip capture mechanism. The cost of this UAV option would be in the range of 2.5% of the platform cost.

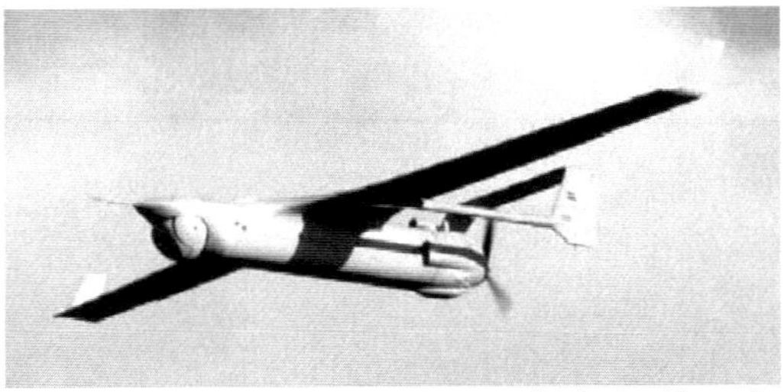

Figure 12 The Insitu Integrator offers a surveillance system which significantly increases the capability of the ship system.

The more capable MQ-8B Fire Scout UAV manufactured by Northrop Grumman performs the following missions:

- Long Dwell Surveillance
- SAR Search
- Force Protection Multiplier
- Mine Warfare (compliment MH-60S)

This vehicle is in use for the Littoral Combat Ship program in the United States. This rotary wing aircraft requires a more significant level of integration into the platform with a Ground Control Station incorporating the Tactical Control System (TCS) (utilizing Tactical Common DataLink (TCDL)) and the Unmanned Common Automatic Recovery System (UCARS).

MODULAR SYSTEMS

The large mission bay available on the Austal MRV allows the integration of a variety of different mission packages including :

- Modular hospital facilities (including surgical facilities and four hospital beds)
- Modular personnel accommodation (including berthing for 20 personnel in five TEU sized modules)
- Containerised refrigerated and frozen storage

Each of these modular capabilities may be fitted to the MRV for less than 1% of the platform cost.

CONCLUSION

The MRV concept is extremely adaptable in order to meet the various specified mission requirements of the Offshore Combatant Vessel. One of the key challenges associated with the OCV requirements definition is to establish the level of required combatant capability. Many countries have recognised the advantages of multi-role platforms to reduce the cost and risk associated with the design, construction, operation and support of many different platform types.

The flexibility and adaptability of the Austal MRV is a key discriminator in this regard. Through Austal's experience in programs such as the USN LCS program, we have been able to develop a platform that is capable of delivering the necessary OCV capabilities in a cost effective manner. Figure 13 illustrates the base MRV cost and the relative costs associated with configurable and/or additional capability options. This approach helps the end user to shape not only their requirements but also fit their requirements within budget and capability constraints.

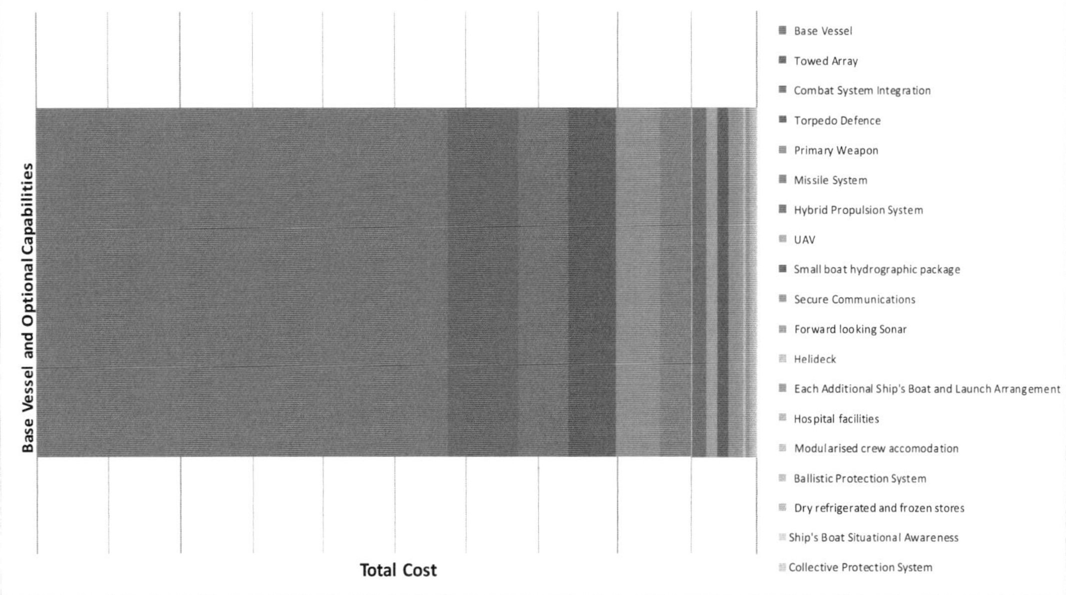

Figure 13 Estimated Breakdown of MRV Base Vessel and Optional Capabilities Cost

Through working closely with technology leading industry partners such as CEA Technologies, Austal is able to continue to offer innovative and flexible system solutions. We continue to advocate regular dialogue between industry and the customer to ensure that both sides are as well informed as possible and best equipped to respond to each other's needs.

REFERENCES

1. "Project SEA 1180, Offshore Combatant Vessel (OCV), Initial Capability Description", Version 1.0, 31 August 2011, Approved by CCDG, AIRMSHL Harvey and CN, VADM Griggs.
2. "Defence Capability Plan", 2011, Public Version.
3. MTU, "Hybrid Propulsion for Patrol Boats", Friedrichshafen, September 2011.
4. Australian Customs and Border Protection Service, RFT Number 102039, Conditions of Contract, Attachment A, Annex A, Part A, Function and Performance Specification, Version 1.0
5. Speer T and Armstrong T, "The Development of an 80m Multi-Role Trimaran Vessel, and the evolution of the LCS-2 design", PACFIC 2010, Australia, January 2010.

Investigation into the Roll Behaviour of Landing Craft in Deep Water

J. Hill[1], T. Turner[1], E. Dawson[1], G Macfarlane[2]
1 Defence Science and Technology Organisation, Melbourne, Australia
2 Australian Maritime College, University of Tasmania, Launceston, Australia
Jenny.Hill2@dsto.defence.gov.au

ABSTRACT

Understanding and predicting the seakeeping and stability behaviour of a landing craft in both deep and shallow water is critical to assuring their safe and effective operation. The Australian Defence Science and Technology Organisation (DSTO) and the Australian Maritime College (AMC) are undertaking a research program to develop a numerical based capability to predict the motions of landing craft during operation. A series of model scale experiments have generated data for landing craft roll decay and motion response in calm water and in beam seas for both deep and shallow water conditions.

This paper presents the outcomes of an investigation into the roll behaviour of a geometric series of landing craft hull forms in deep water. Numerical simulations were performed using FREDYN, a non-linear, six degree of freedom time-domain ship motion prediction code. The results are validated using experimental data. An empirical method proposed by Ewers and Freathy [1] for calculating roll damping coefficients of barge like hull forms, in addition to the Ikeda-Himeno-Tanaka (IHT) and Fast Displacement Ship (FDS) methods, are evaluated against experimental data using the existing roll damping functionality within FREDYN [2].

The significance of predicting the seakeeping and stability performance of conventional landing craft, without the need for scale model testing, is discussed as well as the potential to use the tool to develop stability criterion, enhance platform operability and safety and facilitate simulator based operator training.

INTRODUCTION

The Australian Defence Organisation is currently undertaking an acquisition program (JP2048) to upgrade its existing amphibious capability. This acquisition program is split into several Phases with Phase 3 being the replacement of its amphibious watercraft. These watercraft are required to deliver personnel and equipment from ship to shore through a variety of wave environments. The environments include that experienced within the well dock of the parent ship to deep water offshore and the inshore surf zone. Understanding the seakeeping and stability behaviour of the landing craft in both deep and shallow water is critical to their safe operation. The Australian Defence Science and Technology Organisation (DSTO) and the Australian Maritime College (AMC) are undertaking a research program to develop a numerical based capability to predict the motions of landing craft in a variety of operational environments.

Part of this capability includes the modification and validation of an existing ship motion prediction software tool (FREDYN). Once validated for landing craft hull forms, this numerical tool will have a variety of applications and provide significant benefits to many of the world's Navies. The application of this tool will support the development of more appropriate intact and damage dynamic stability criteria for landing craft. The stability criterion that is currently applied to landing craft by many Navies is based on an empirically derived criteria developed during the 1960's for Naval combatants operating in a deep water environment [3]. This criterion is not necessarily applicable to the barge like hull forms of conventional landing craft or to operations in a shallow water environment.

FREDYN can also be integrated into ship handling training simulators and utilised by landing craft operators to ensure that amphibious operations are performed with the highest level of preparedness and competency. Some of the potential advantages of classroom based simulation training, as opposed to "hands on" training, are a reduction in training costs, an increase in training capacity, an increase in the proficiency and awareness of operators and an inherently safe training environment. Studies undertaken by the United States Coast Guard have shown that these benefits can be achieved through their use of a small boat simulator as a training tool [4].

Another application of this capability is the ability to define and fully understand the operational boundaries of both an intact and damaged landing craft. The tool might also form the basis of cargo loading software to assist the management of the landing craft's load condition and support the safe and effective completion of amphibious operations.

If this tool is to be utilized by Navy in the applications described above it is imperative that the accuracy of the predictions is verified. DSTO's research program is focused on developing and validating components within the tool that significantly influence the prediction of the motions of the landing craft. These include: viscous roll damping formulations, ship motions, manoeuvring, resistance and propulsion.

SCOPE

The focus of this paper is the development and implementation of a viscous roll damping method for the accurate prediction of landing craft roll motion. The preliminary results of both an experimental and numerical study into the roll behaviour of a geometric series of landing craft hull forms in deep water are presented. For an accurate prediction of landing craft roll motion, an appropriate roll damping method is required. Ewers and Freathy [1] have proposed an empirical method for the prediction of the viscous roll damping method coefficients for offshore barge type hull forms. Numerical predictions using this method and other roll damping formulations, including the Ikeda-Himeno-Tanaka (IHT) and Fast Displacement Ship (FDS) methods were completed using FREDYN. The results of these predictions are evaluated against experimental data.

EXPERIMENTAL STUDIES

A series of model tests were undertaken at the Model Test Basin at the AMC in Launceston, Tasmania. Roll decay tests with initial heel angles of 5, 10 and 15 degrees were conducted for three different landing craft models with varying length to beam (L/B) ratios. The L/B = 3.12 and 4.37 models are geometrically similar and represent the typical size relationships of landing craft. The L/B = 3.77 model is representative of a conventional naval landing craft. The particulars of the three models are shown in Table 1. Figure 1 shows a photograph of the L/B = 3.12 model during testing.

L/B	Draught AP (m)	Draught FP (m)	Displacement (kg)	Roll Gyradius k_{xx} (m)	VCG (m)	Beam (m)	Scale Factor
3.12	0.113	0.113	59.99	0.144	0.131	0.481	16.067
3.77	0.100	0.100	100.59	0.230	0.210	0.640	10.000
4.37	0.123	0.123	54.51	0.125	0.135	0.372	14.831

Table 1 Model particulars

Figure 1 Model (L/B = 3.12) during roll decay test

The motions of the landing craft models were measured in all six degrees of freedom using a QUALISYS three-dimensional digital video motion tracking system. This system is calibrated using a series of 16 permanent reference markers which have been surveyed into position.

The model setup and test procedures adopted were consistent with the standard Model Test Basin experimental procedures and are based on the recommended procedures of the International Towing Tank Conference (ITTC) [5].

The estimated level of uncertainty associated with the model's measured parameters is presented in Table 2. Both the L/B = 3.12 and 4.37 models' vertical centre of gravity (VCG) was determined through the conduct of an inclining experiment while the VCG of the L/B = 3.77 model was determined using the tilt frame. The roll radius of gyration (k_{xx}) was measured using a tilt frame for all of the models.

Estimated levels of uncertainty (to 95% confidence level)	
Model beam	± 1.00 mm
Model length	± 1.50 mm
Model displacement	± 100 grams
Model VCG	± 1.00 mm
Measurement of roll	± 0.10 degrees

Table 2 Estimated measurement uncertainty for the landing craft models

NUMERICAL STUDIES

The Cooperative Research Navies (CRNav) currently comprises the Navies of Australia, Canada, France, The Netherlands, the United Kingdom and the United States Coast Guard. The CRNav Dynamic Stability group was established in 1989 to initiate a research program focussed on increasing the understanding of the dynamic stability of naval vessels. This was to be done from a physics based approach rather than an empirically derived approach such as that developed by Sarchin and Goldberg in the 1960s. The work of CRNav has led to the development of FREDYN, a non-linear time domain numerical tool capable of predicting the motions of a ship in moderate to extreme seaways.

Roll damping is an important component of any ship motion simulation. FREDYN currently has three roll damping methods to choose from [2]. The Fast Displacement Ship (FDS) method is recommended for frigate type hull forms operating at moderate to high speeds. The Ikeda-Himeno-Tanaka (IHT) method is for conventional merchant type hull forms operating at low to moderate speeds. FREDYN has a third option for the roll damping formulation referred to as MANDAMP. This method allows the user to input appropriate damping coefficients obtained from either model tests or other formulations. The coefficients required are:

B_0^{lin} the linear speed independent roll damping coefficient;
B_u^{lin} the linear speed dependent roll damping coefficient;
B_0^{qua} the quadratic speed independent roll damping coefficient; and
B_u^{qua} the quadratic speed dependent roll damping coefficient.

Collectively, these coefficients are used to calculate a roll moment as a reaction to the ship's rolling rate.

Ikeda, Himeno and Tanaka have developed an alternative approach to that described above to calculate the roll damping coefficients. This approach gives consideration to the pressure distribution around the hull that results in the formation and shedding of vortices [6]. Predictions using this method showed improvements over earlier formulations but the solution is sensitive to the pressure distribution at the bilges. Ewers and Freathy [1] extended this work and developed an empirical formula for the quadratic term (B_0^{qua}) of the roll damping method for barge like hull forms, see Equation 1. In this formulation B_0^{lin}, B_u^{lin} and B_u^{qua} are considered to be equal to zero, however, care is required when completing numerical simulations to ensure that these parameters are only accounted for once. Therefore, in the application of FREDYN it is necessary to set B_0^{lin} = -40 kN-m-sec as the linear component is pre-calculated as part of the hull's non-viscous added mass and damping coefficients. Numerical predictions were completed using the MANDAMP method in conjunction with the Ewers and Freathy formulation for B_0^{qua}.

$$B_0^{qua} = \frac{4L\rho R_r^2 \omega \cdot \left(3.9 - 51.4\frac{R_b}{B}\right) \cdot \left[\left(0.5B - R_b\right)^2 + \left(D - R_b\right) \cdot \left(D - R_b + 1.5H_h\right)\right]}{9\pi} \quad \text{Equation 1}$$

Where:

L	=	Length between perpendiculars (m)
ρ	=	Density of water (kg/m^3)
R_r	=	Distance from bilge to centre of gravity (m)
R_b	=	Bilge radius (m)
B	=	Beam (m)
H_h	=	KG – D (m)
D	=	Draft (m)
KG	=	Vertical location of centre of gravity above the keel (m)
π	=	Pi [3.1416] (constant)
ω	=	Natural roll frequency (rad/sec)

RESULTS

Numerical simulations of roll decay were completed for three L/B ratio landing craft models and at three different initial heel angles using the viscous roll damping methods previously described. In all cases the roll response (as a function of time) predicted using the FDS method was the same as that predicted by the IHT method. For this reason, only the results of the IHT simulations will be shown in this paper. Figure 2 to Figure 10 show the comparison between the experimental and numerically predicted roll decay time traces for the three landing craft models at the three different initial heel angles.

It is evident that the predictions made using the IHT method and the Ewers (B_0^{qua}) based method (Equation 1) do not match the experimental results in either period or amplitude. The roll period is a function of the model's mass and its added mass and can be modified by changing the k_{xx} value. Within FREDYN, the added mass is calculated using a strip theory approach. This approach is valid only for slender hull forms (L/B > 4). Therefore, the numerically predicted added mass may need tuning for low L/B hull forms such as those considered in this study.

It was decided to investigate the effect of changing the initial k_{xx} for each of the models with the aim of tuning the numerical prediction to the experimental results. Consequently, additional simulations were completed to obtain a match in the roll response period by modifying the roll radius of gyration (k_{xx}) input value. The vertical centre of gravity of the model was assumed to be fixed and therefore GM_T remained unchanged.

Once the predicted roll response period was matched to the experiment, and an appropriate k_{xx} value was obtained, a new B_0^{qua} based on the revised k_{xx} was calculated using Equation 1. A comparison between the numerical predictions with this revised B_0^{qua} value and the experimental results are also shown in Figure 2 to Figure 10. The percentage change in k_{xx} required to tune the numerically predicted roll period for each of the models is noted in each figure. These results show that when using the revised B_0^{qua} value, a significant improvement in both the roll period and amplitude can be achieved for the MANDAMP with Ewers method for all of the conditions when compared to the experimental data.

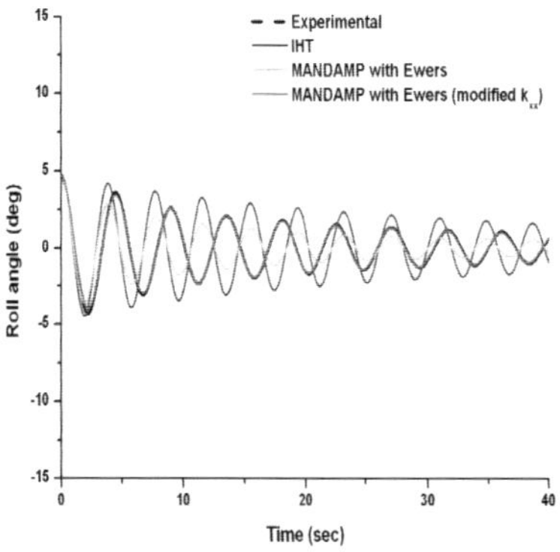

Figure 2 Roll decay response; Model L/B = 3.12; Initial angle of heel = 5 degrees; k_{xx} increased by 21%.

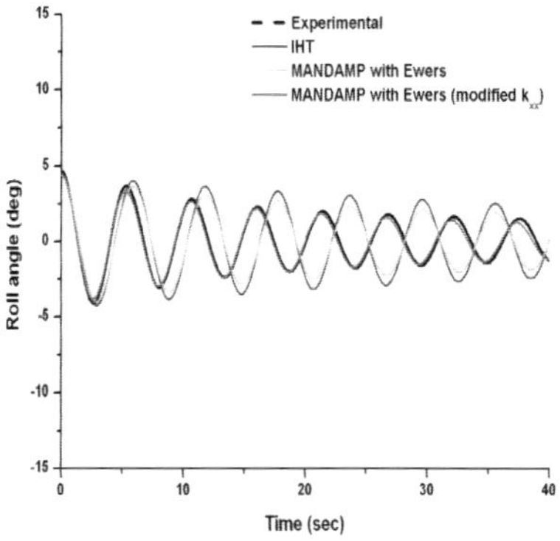

Figure 4 Roll decay response; Model L/B = 4.37; Initial angle of heel = 5 degrees; k_{xx} reduced by 15%.

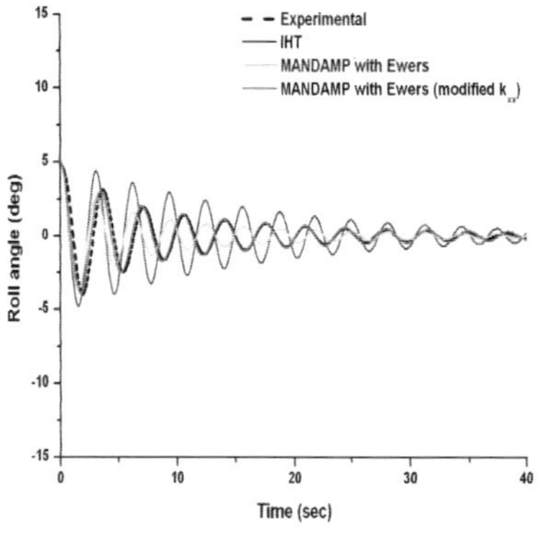

Figure 3 Roll decay response; Model L/B = 3.77; Initial angle of heel = 5 degrees; k_{xx} increased by 15%.

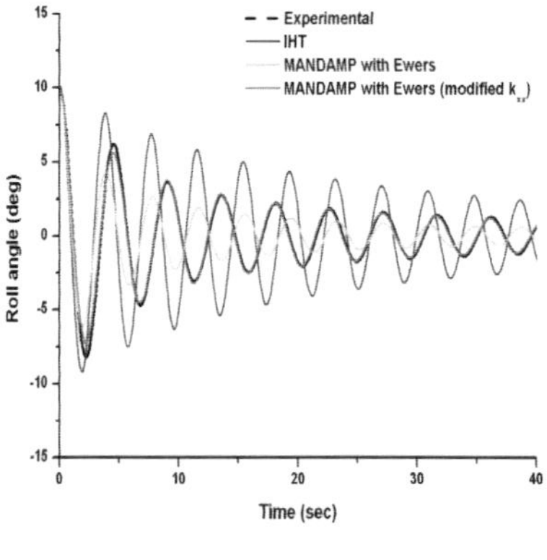

Figure 5 Roll decay response; Model L/B = 3.12; Initial angle of heel = 10 degrees; k_{xx} increased by 21%.

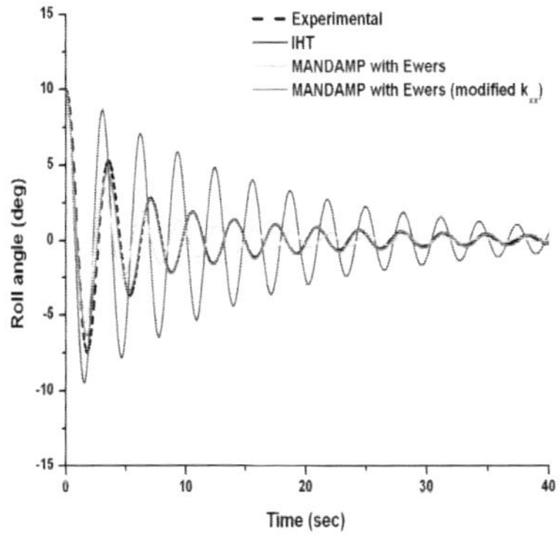

Figure 6 Roll decay response; Model L/B = 3.77; Initial angle of heel = 10 degrees; k_{xx} increased by 15%.

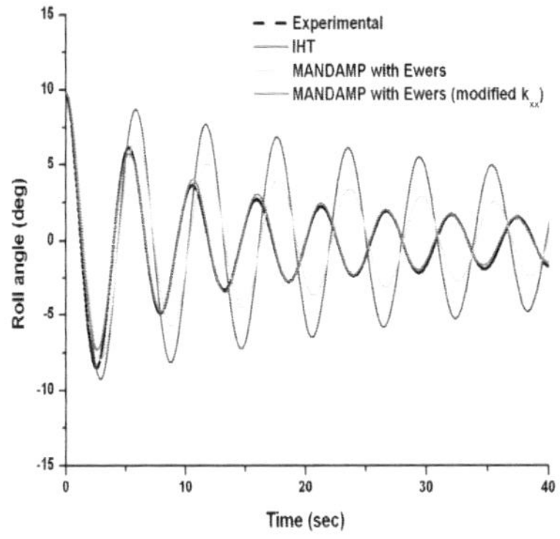

Figure 7 Roll decay response; Model L/B = 4.37; Initial angle of heel = 10 degrees; k_{xx} reduced by 15%.

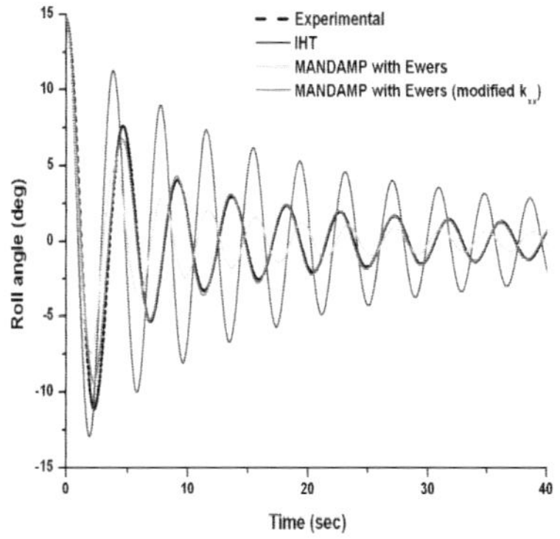

Figure 8 Roll decay response; Model L/B = 3.12; Initial angle of heel = 15 degrees; k_{xx} increased by 21%.

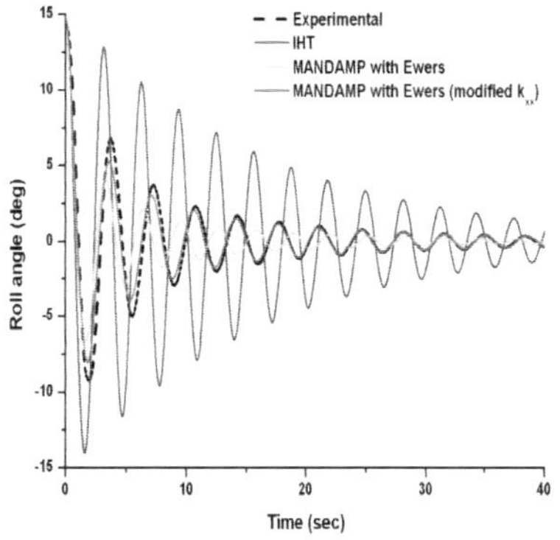

Figure 9 Roll decay response; Model L/B = 3.77; Initial angle of heel = 15 degrees; k_{xx} increased by 15%.

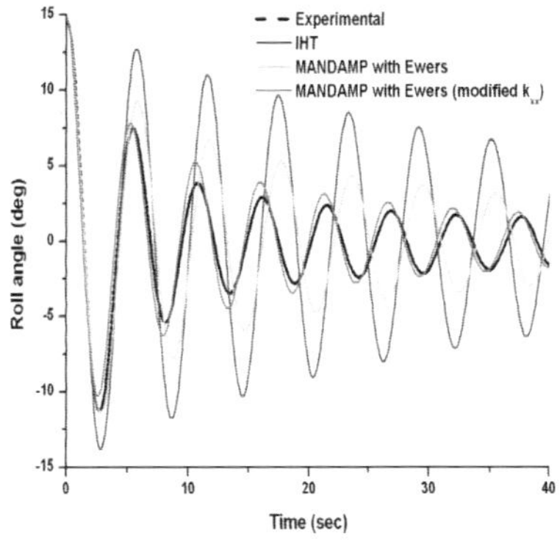

Figure 10 Roll decay response; Model L/B = 4.37; Initial angle of heel = 15 degrees; k_{xx} reduced by 15%.

Figure 2 to Figure 4 show the predicted roll decay for the 5 degree initial heel angle condition. Excellent agreement with the experimental results for all models was observed for the predictions using the MANDAMP with Ewers method with the revised k_{xx} and B_0^{qua} values appropriate for each landing craft.

Figure 5, Figure 6 and Figure 7 show a comparison between the numerical predictions and experimental results for the 10 degree initial heel angle for the L/B = 3.12, L/B = 3.77 and L/B = 4.37 models respectively. The results show that in comparison with the experimental data, both the period and the amplitude are predicted with a good level accuracy for all models when initially heeled to an angle of 10 degree.

At the larger initial heel angle of 15 degrees the MANDAMP with revised Ewers predictions show good agreement with the experiment in both period and amplitude for the L/B = 3.12 and L/B = 4.37 models, see Figure 8 and Figure 10 respectively. The results for the L/B = 3.77 landing craft show that the MANDAMP with revised Ewers method slightly under predicts the amplitude but is in excellent agreement with the period, see Figure 9. It was noted that the L/B = 3.77 landing craft had a shallower draft than the other two models. It is surmised that at the larger initial angle of 15 degrees there may be additional interaction between the viscous flow around the bilge and the free surface resulting in an increase in roll damping. This is consistent with the observations made by Ewers and Freathy [1].

DISCUSSION

For ship motion prediction it is important that motion damping is accurately modelled. The majority of ship motion damping arises from the oscillating hull radiating energy in the form of waves. In roll motion the amount of energy dissipated in the form of the radiated waves is small relative to other forms of damping which include the energy lost due to skin friction, vortex shedding and appendage forces [7]. Hull forms with relatively sharp bilges, for example landing craft and barges, will shed vortices as they roll. This vortex shedding phenomenon absorbs a significant amount of energy and is therefore a significant contributor to the roll damping of these types of hull forms. Numerical seakeeping codes account for this loss of energy by applying a roll moment based on viscous roll damping coefficients.

As previously discussed, FREDYN currently offers three methods to predict the viscous roll damping coefficients for ships. Of these methods, the FDS formulation was obtained from analysing systematic model tests on fast displacement type vessel (frigates) and is applicable for frigate type vessels at forward speeds of Froude number 0.15 and greater. The IHT approach is based on model tests of conventional merchant type ships at low to moderate speeds [2]. The results presented in this paper show that neither the FDS or IHT methods are suitable for barge like hull forms such as conventional landing craft.

The Ewers and Freathy formulation based predictions using the MANDAMP method show significant improvement in the accuracy for both the amplitude and period for all landing craft models considered. As the L/B ratio increased, the difference between the experimentally obtained and the numerically predicted amplitudes of the roll decay increases slightly. This may be due to the slenderness of the hull form tending towards the limits of the roll damping method's range of application.

A factor that may need to be considered is the relationship between roll damping and roll velocity when vortex shedding is the dominant source of energy loss. Viscous roll damping is generally proportional to the square of the roll velocity but in the case of barge type hull forms this relationship may no longer be valid, that is, the roll damping may vary to some other power.

Overall, the Ewers and Freathy formulation for the calculation of the viscous roll damping coefficient has been shown to be acceptable for the simulation of the roll behaviour of a landing craft in deep water. Further studies are now required to determine if this formulation is still applicable when modelling the response of these low L/B type vessels in shallow water and in waves.

CONCLUSION

To enable an accurate numerical simulation of landing craft hull form roll behaviour to be achieved, an appropriate viscous roll damping method is required. This paper has shown some of the preliminary outcomes of a study into the identification of such a method. Experimental studies were undertaken to validate the numerical simulations of the roll decay behaviour of several landing craft models with varying L/B ratios. This data was used to evaluate the performance of the three different viscous roll damping methods available within the CRNav time domain seakeeping tool, FREDYN. The FDS and the IHT methods proved to be unsuitable when modelling landing craft type hull forms. The FREDYN simulations which used the MANDAMP method, with roll damping coefficients based on the formulation proposed by Ewers and Freathy [1], showed reasonable accuracy when predicting the roll decay of the three hull forms considered. The accuracy of these simulations was found to be strongly dependent on the value of roll radius of gyration used.

Further work is required to develop a process to determine the appropriate roll radius of gyration to be used as the basis of the viscous roll damping component for any landing craft motion simulation. This process should eliminate the need to undertake any model testing and the associated resource requirements. Additional investigations into the applicability of the Ewers and Freathy formulation to model the landing craft roll behaviour for different loading conditions, in shallow water and in regular and irregular waves should also be undertaken.

Once the method is developed and implemented, FREDYN can be used for various applications such as ship handling simulators to ensure that amphibious operations are performed with the highest level of safety and mission readiness.

ACKNOWLEDGEMENTS

The authors acknowledge the valuable input from the staff at the Australian Maritime College whilst undertaking the experimental component of this work.

REFERENCES

1. Ewers, W. A. and Freathy, P. E. (1981) *Barge Motions Research Project - Phase 2 Report, Part 3 Further Work on Viscous Roll Damping*. Atkins Research and Development
2. MARIN (2011) *FREDYN Version 10.3 Computer Program for the Simulation of a Steered Ship in Extreme Seas and Wind.* CRNAV, Maritime Research Institute Netherlands.
3. Sarchin, T. H. and Goldberg, L. L. (1962) Stability and Buoyancy for U.S. Naval Surface Ships. *SNAME Transactions, Society of Naval Architects and Marine Engineers* 70
4. Miller, E. R., Jakobsen, B. K. and Fox, B. W. (2011) Development of a Prototype Small Boat Operator Trainer for the USCG. *SNAME Chesapeake Section* March 16
5. International Towing Tank Conference. *Recommended Procedures*. (2008) [Accessed September 2011]; Available from: http://ittc.sname.org/documents.htm.
6. Ikeda, Y., Himeno, Y. and Tanaka, N. (1977) On Eddy Making Component of roll Damping Force on Naked Hull. *Journal of the Society of Naval Architects of Japan* 142
7. Lloyd, A. R. J. M. (1989) *Seakeeping: Ship Behaviour in Rough Waves*. Sussex, England, Ellis Horwood

The Quest for Optimum High Speed Planing Craft Design

A. F. Mohamad Ayob, T. Ray and W. F. Smith
University of New South Wales @ Australian Defence Force Academy
w.smith@ adfa.edu.au

High speed planing craft play key commercial roles in niche passenger ferrying and high value cargo transport in addition to supporting several critical maritime activities such as coastal surveillance, reconnaissance, and life-saving operations. Despite their significant use across a range of domains, formal optimization frameworks have rarely been proposed to deal with the design challenges for high speed planing craft. Highlighted in this paper is an optimization framework for the conceptual and preliminary design of high speed planing craft. A summary of the components of the optimization framework is presented, followed by several optimization examples, namely; calm water resistance minimization and resistance minimization in a seaway operation. The concept of *scenario-based* hydrodynamic design optimization is also introduced using an example of a small rescue craft operating in a predefined sea-state. Finally, a multi-objective optimization case study considering total resistance, steady turning diameter and vertical impact acceleration is presented to demonstrate the capability to explore trade-offs while at the same time providing an understanding of the design intent of a basis ship.

NOMENCLATURE

B	Beam (m)	*STD*	Steady turning diameter
L	Length (m)	*SD*	Standard deviation
T	Draft (m)	COM	Component Object Model
η	Vertical impact acceleration	NSGA-II	Non-dominated Sorting Genetic Algorithm II
C_v	Speed coefficient	SA-EA	Surrogate Assisted Evolutionary Algorithm
R_T	Total resistance (kN)	IDEA	Infeasibility Driven Evolutionary Algorithm
R_C	Calm water resistance (kN)	USCG	United States Coast Guard
FC	Fuel consumption	WPB	Notional name for USCG patrol boat
t	Operational tempo		

1 INTRODUCTION

High speed planing craft play key commercial roles in niche passenger ferrying and high value cargo transport in addition to supporting several critical maritime activities such as coastal surveillance, reconnaissance, and life-saving operations. Since the 1930's there has been significant interest in building planing craft, in particular Patrol Torpedo boats (Rottman, 2008). It followed that some rigorous research of planing craft design was conducted in the 1970s (Clark *et al.*, 2004). Although the motivation for the authors' work has been aimed at improving seakeeping while retaining as much speed as possible, it is interesting to note that an inquest into an incident off Christmas Island, Australia, that resulted in the deaths of 30 asylum seekers in December 2010, concluded that 'life-saving rescue vessels (that have been) designed are not capable of being operated in adverse weather' (Jones, 2011). Such rescue vessels are predominantly planning craft.

Over the course of history, the development of vessels that can operate well in their intended environment has evolved. Traditionally, such improvement was based on the knowledge gained through incremental change and trial and error. In the mid 1900's a more systematic analysis of the process gave rise to the ship design spiral as originally proposed by (Evans, 1959). Further advances flowed from the use of design optimization, which provides the opportunity to promote the identification of *superior* solutions, rather than simply *satisfactory* designs has been promoted. There are a significant number of optimization frameworks for the design of displacement vessels. Examples include the works of (Ray, 1995) and (Ganesan, 1999) on container ships, (Mistree *et al.*, 1990) and (Smith, 1992) on destroyer and frigate forms, Peacock (Peacock, 1998) on offshore patrol vessels, and (Gammon, 2004) and (Majumder *et al.*, 2002) on fishing vessels. Other recent studies of other displacement craft optimization are: (Neu *et al.*, 2000), (Zaraphonitis *et al.*, 2003), (Grigoropoulos, 2004), (Wolf *et al.*, 2005), (Neti, 2005), (Koh *et al.*, 2005), (Mason and Thomas, 2007) and (Grigoropoulos and Chalkias, 2010).

The development of ship design optimization frameworks has generally been targeted towards displacement craft rather than planing craft. Factors contributing to this include the wide application of displacement forms, the availability of standard naval architectural tools and the demand for producing designs that minimize the use of resources in competitive environments (e.g. pressure to reduce fuel and building costs). In contrast, the development of optimization frameworks for high speed planing craft has been minimal. Reasons for this are likely the limited sources of published experimental data, the specialized use of the hull form in restricted commercial, recreational and military domains and the lack of tailored analysis tools. However, several articles dedicated to planing craft optimization are (Jons *et al.*, 1985), (Almeter, 1995) and (Herrington and Latorre, 1998).

In order to support design optimization of planing craft, an underlying framework should:

1. Allow easy incorporation of different scenarios, design criteria etc. with analysis modules providing different levels of fidelity.

2. Allow shape representation and shape manipulation that is able to generate different variants of hull forms within the required fairness and defined chines.

3. Include an optimization method that is capable of dealing with single and multi-objective optimization problems with constraints. Furthermore, since the performance evaluations may be computationally expensive, the optimization algorithms employed should be efficient.

Presented in this paper is a summary of the accumulated knowledge gained during the development of a state-of-the-art high speed planing craft optimization framework by the authors at the University of New South Wales at the Australian Defence Force Academy. Discussed are the proposed methods for dealing with typical coast guard vessel optimization problems having single and multiple objectives, leading to improved and innovative designs that can operate in calm water and in a seaway with the consideration of various operational profiles. Case study problem statements are outlined and followed by highlights of the results to date. The intent is to demonstrate a capability that could be leveraged to explore future designs more thoroughly.

2 DESIGN OPTIMIZATION FRAMEWORK

A planing craft design optimization framework as described in (Mohamad Ayob *et al.*, 2010b) and (Mohamad Ayob *et al.*, 2011c) has been developed over the last 4 years. The framework incorporates a surface information retrieval module, an optimizer module and several accepted naval architectural performance estimation methods. The framework utilizes a *basis ship* (or parent ship) approach and performs automatic surface information retrieval from a standard offset data table using B-splines. It interprets the resulting ship geometry as the basis hull form. The naval architectural tools that are incorporated in the framework include: hydrostatics calculations, calm-water resistance (Savitsky, 1964), seakeeping estimation (resistance due to waves and vertical impact acceleration) (Savitsky and Brown, 1976) (Savitsky *et al.*, 1993), and manoeuvring performance estimation (Denny and Hubble, 1991) (Lewandowski, 2004). The optimizer module included in this framework consists of a set of population-based, stochastic algorithms, namely the Non-dominated Sorting Genetic Algorithm (NSGA-II) (Deb *et al.*, 2002), the Infeasibility Driven Evolutionary Algorithm (IDEA) (Singh *et al.*, 2008) (Ray *et al.*, 2009), and the Surrogate Assisted Evolutionary Algorithm (SA-EA) (Isaacs *et al.*, 2007). For the sake of brevity, the specific details of the optimization algorithms are not included herein but can be found in the respective references. However, as they are stochastic, the general approach is to run a number of cases and present results including statistical characteristics.

A flowchart of the planing craft optimization framework is shown in Figure 1. The planing craft optimization process starts with a set of offset data that is fed into the surface information retrieval module, where the resulting surface representation of the basis hull form is generated. The chosen optimizer (NSGA-II, IDEA or SA-EA) generates candidate designs with new values of length (L), beam (B) and draft (T) which are subsequently used to generate new hull form geometries. The performance of the candidate designs (e.g. hydrostatics, calm-water resistance, seakeeping and manoeuvring) are evaluated and used by the optimizer. The framework is built using the Microsoft Component Object Model (COM$^{©}$) framework that allows for future expansion of the analytical modules.

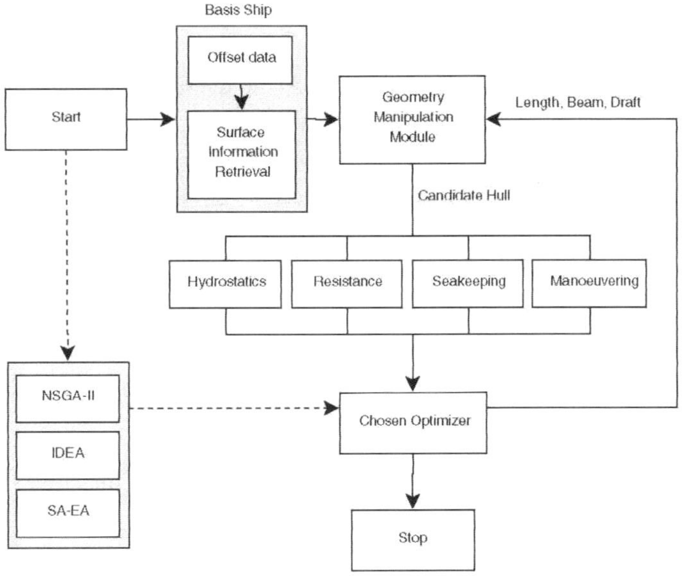

Figure 1: Planing craft optimization framework

3 RESISTANCE MINIMIZATION OF HIGH SPEED PLANING CRAFT

Hull resistance minimization has been a popular subject in ship design optimization studies. Several works (Day and Doctors, 1997), (Gammon, 2004), (Koh *et al.*, 2005), (Mason and Thomas, 2007) and (Grigoropoulos and Chalkias, 2010) also tackle resistance minimization in calm water and in a seaway using genetic algorithms. With respect to high speed planing craft design optimization, minimization of calm water resistance has been demonstrated using a well known patrol craft, the USCG WPB-110ft, where the published data can be found in the work of (White and Savitsky, 1988). The profile view of the patrol craft is shown in Figure 2. Shown in Figure 3(a) and 3(b) are the body plans of the actual vessel and that of the mathematical model of the WPB-110ft craft generated using the proposed framework. Prior to optimization, the 3-dimensional mathematical model of the patrol craft was validated through comparison with the published numerical experimental data. As discussed in (Mohamad Ayob *et al.*, 2011a), between the measured hull and the mathematical model, the geometric values (of length, beam and draft) are matched within 1.3% and displacement is matched with a 2.23% of error. A small error in the speed coefficient, C_v, and the calm water resistance, R_C, which is less than 5% is argued to demonstrate the match between the actual vessel and the model is acceptable. Following the basis ship approach method, the numerically optimized candidate designs are later compared with the modelled craft, hereafter referred to as the 'basis ship'.

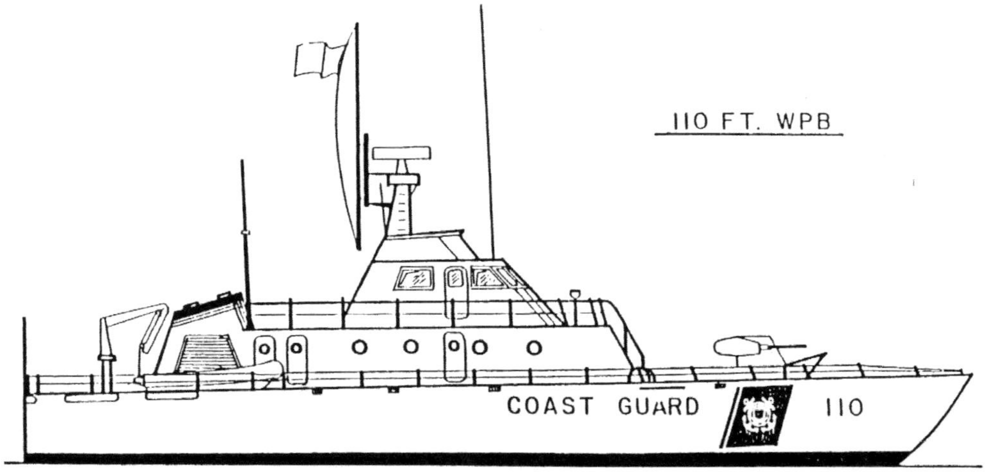

Figure 2: Outboard Profile View of the USCG WPB-110ft
(reproduced from the work of (White and Savitsky, 1988))

Calm water resistance minimization of the basis ship has been conducted using the proposed framework resulting in a significant reduction of R_C as shown in Table 1, which is reproduced from (Mohamad Ayob *et al.*, 2011a). It was reported that the proposed framework was capable of identifying a candidate design which posses lower calm water resistance, up to a 10% reduction, when compared to the basis ship. The best, mean, and median designs and the standard deviation of the designs across 30 individual runs are reported to indicate consistency in the results obtained using the proposed algorithms.

Shown in Table 2, reproduced from (Mohamad Ayob *et al.*, 2010b), are the results of the minimization of resistance of the basis WPB-110ft ship operating in a seaway. The experiment was aimed to identify a superior candidate design to operate in conditions modelling Northern Australian coastal waters. The results indicate significant reduction of total resistance, R_T, is possible – up to 7.74%, 6.11% and 4.19% (indicated by bold typeface) for operations in sea-states 3, 4 and 5 respectively. When compared to the results for calm water resistance in Table 1 and across different sea-states in Table 2, it is observed that the percentage of resistance reduction was reduced due to the demand to maintain speed at higher seas.

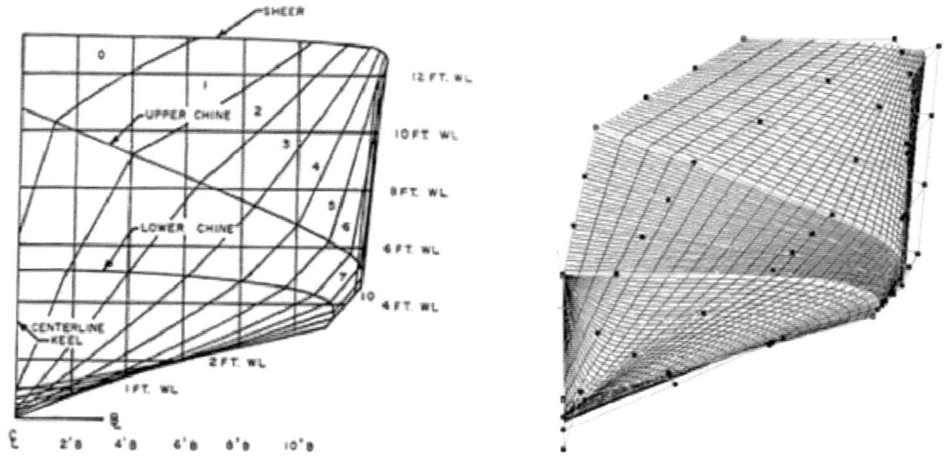

(a) Ship lines of the WPB-110ft (White and Savitsky, 1988)

(b) B-spline surfaces defining WPB-110ft

Figure 3: Body plan of the USCG WPB-110ft (actual and basis ship)

Table 1: Summary for the best candidate solutions (calm water resistance minimization) obtained using NSGA-II, IDEA and SA-EA

	NSGA-II	IDEA	SA-EA
Best (kN)	122.38	122.43	122.40
Mean (kN)	123.14	122.95	122.84
Median (kN)	122.87	122.68	122.64
SD	0.693	0.778	0.523
% R_C reductions	10	9.9	10

Table 2: Summary for the best candidate solutions (minimization of resistance in a seaway) obtained using NSGA-II, IDEA and SA-EA – Function Evaluations and Improvement

		NSGA-II	SA-EA	IDEA
Sea State 3	Avg. Num. of Fun.Eval.	720	688.4	720
	% of Minimized Resistance	7.35	7.71	**7.74**
Sea State 4	Avg. Num. of Fun.Eval.	880	856.6	880
	% of Minimized Resistance	5.69	5.94	**6.11**
Sea State 5	Avg. Num. of Fun.Eval.	760	744.3	760
	% of Minimized Resistance	**4.19**	4.02	3.99

4 SCENARIO-BASED HYDRODYNAMIC DESIGN OPTIMIZATION

It is preferable to design a ship based on a defined operational profile. Several operational scenarios that include seaway conditions, survivability, energy consumptions and emergency events can be simulated in the early design stage. The operating scenarios can be derived from the required operational profile combined with performance measures of the ship such as survivability (Boulougouris and Papanikolaou, 2004) (Akhil *et al.*, 2008), fuel consumption (Radan *et al.*, 2006) and electrical generation (NAVSEA, 2007). Scenario consideration based on an operational profile is referred to in this work as *scenario-based optimization*. Although increasing interest in including operational profiles in ship design optimization has been observed, these studies have been limited to minimizing fuel consumption and increasing the probability of survival. Furthermore, limited studies on the investigation of scenario-based design optimization of high speed planing craft are reported in the literature.

The numerical experiments conducted at the UNSW@ADFA contributed to the study of scenario-based hydrodynamic design optimization through the introduction of several possible scenarios. These include; (i) a multiple speed operation in a single sea-state (Mohamad Ayob *et al.*, 2011b), (ii) a single speed operation in multiple sea-states (Mohamad Ayob *et al.*, 2010b), and (iii) the investigation of the effect of vertical impact acceleration, η, to small craft performance (Mohamad Ayob *et al.*, 2010a).

As an example, shown in Table 3 is the comparison of fuel consumption between the candidate and basis ship designs for an operational time of t = 3000 hours/year as indicated in (USCG, 2010) and (Karafiath *et al.*, 2001). The optimization formulation aims to identify the superior candidate designs given three operational speeds of patrol (16 knots); transit (25 knots); and intercept (30 knots). Also, given a predefined operational time, t, the division of the operational tempo is set as 60% at 16 knots, 20% at 25 knots and 20% at 30 knots. Four optimization problems were executed which cater for minimization of resistance at 16 knots, 25 knots, 30 knots and lifetime profile operation which leads to candidate designs being labelled D^{16}, D^{25}, D^{30} and D^{L}, respectively. The details of the implementation and results can be found in (Mohamad Ayob *et al.*, 2011b). It can be observed in Table 3 that a significant fuel saving of 10.2% can be achieved if the basis ship is optimized to operate at 16 (D^{16}) and 30 knots (D^{30}). Fundamentally and conceptually, the proposed framework enables a ship designer to identify candidate designs for operations across different speeds with a predefined lifetime operational profile.

Table 3: Fuel consumption comparisons among candidate and basis designs (t=3000 hours/year)

	FC [kg/hr]	Fuel saving (%)
Basis ship	175.39	-
D^{16}	157.48	10.21
D^{25}	160.33	8.58
D^{30}	157.5	10.2
D^{L}	157.66	10.11

5 MULTI-OBJECTIVE OPTIMIZATION CASE STUDY

Summarized in this section is the capability of the proposed framework to conduct a multi-objective optimization exercise, again using the example of the WPB-110ft planing craft. The multi-objective optimization problem posed is described as the identification of planing craft hull form geometries with minimum total resistance, R_T, vertical impact acceleration, η, and steady turning diameter, *STD*, subject to constraints on displacement and stability (transverse metacentric height), operating with a speed of 30 knots. The seakeeping scenario for sea-state 3 was modelled using a Pierson-Moskovitz energy spectrum with a significant wave height of 1.34m. The details of the implementation can be found in (Mohamad Ayob *et al.*, 2011a). All the solver algorithms (NSGA-II, IDEA and SA-EA) were able to obtain non-dominated solutions. The non-dominated solutions from all algorithms were combined to form a pool of non-dominated solutions which consists of a well-spread solution surface with good diversity. The results indicate a range of designs with R_T varying between 162.58kN and 235.21kN, *STD* between 245.37m and 505.52m and η between 0.549g and 0.987g. The basis hull has a R_T of 176.65kN, *STD* of 373m and η of 0.75. It can be observed in Figure 3 that the design with minimum R_T has maximum *STD*. Similarly the design with minimum *STD* has maximum R_T. This finding is not surprising. The third axis not shown reflects the seakeeping performance.

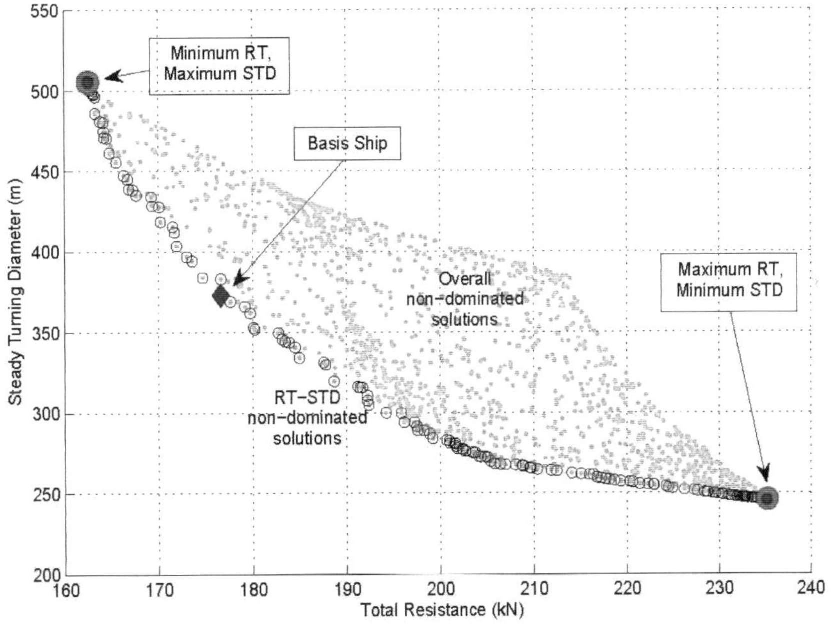

Figure 3: Non-dominated set for R_T and *STD*

As shown in Figure 3, the basis ship is positioned on the non-dominated front of two objectives, R_T and *STD*. This is an interesting revelation that implies that the basis ship is among the superior candidate designs. From the figure, what is evident is that the original designer chose a good compromise between resistance and steady turning diameter and did not select a design representative of a single-objective analysis (either extreme). Using the proposed framework, demonstrated here is the capability to capture the design decision trade-off while at the same time providing an understanding of the design of the USCG WPB-110ft basis ship.

6 CONCLUSION

The aim of the work summarised in this paper was to develop a framework for optimum design of high speed planing craft in conceptual and preliminary design stages. To achieve this, a flexible geometry modeller and several experimentally derived naval architectural analysis tools were combined with state-of-the-art optimization algorithms. This optimization framework utilized a full geometric representation of hard chine planing craft, where superior designs were identified based on stated operational requirements.

Through the use of the proposed planing craft geometry representation method, combined with experimentally derived naval architectural tools, a validated 3D mathematical model to represent the complete geometry of a high speed planing craft was identified and shown to compare favourably with experimental data published in the literature. Such a capability is valuable for ship designers for the visualization. Furthermore, it provides for valid model generation capability within an optimization framework.

Indicative case studies in the optimization for high speed planing craft have been thoroughly explored. The investigation started with the minimization of calm water resistance and resistance in a seaway, followed by a multi-objective optimization case study of total resistance, vertical impact acceleration and steady turning diameter. Later, optimization problems based on operational scenarios were explored. The capability offered by the proposed framework provides unlimited possibilities for creating designs for specialized missions of high speed planing craft.

7 REFERENCES

Akhil, K.R., Ghose, D., Koteswara, R.S., 2008. Optimization Deployment of Multiple Decoys to Enhance Ship Survivability. In Proceedings of the 2008 American Control Conference.

Almeter, J.M., 1995. Resistance Prediction and Optimization of Low Deadrise, Hard Chine, Stepless Planing Hulls. Ship Technology and Research Symposium (STAR). Society of Naval Architectures and Marine Engineers.

Boulougouris, E., Papanikolaou, A., 2004. Optimisation of the Survivability of Naval Ships by Genetic Algorithm. In Proceedings of the 3rd Conference on Computer Applications and Information Technologies in the Maritime Industries (COMPIT'04), Siguenza, Spain.

Clark, D.J., Ellsworth, W.M., R., M.J., 2004. The Quest for Speed at Sea. Naval Surface Warfare Center, Carderock Division.

Day, A.H., Doctors, L.J., 1997. Resistance Optimization of Displacement Vessels on the Basis of Principal Parameters. Journal of Ship Research 41, 249–259.

Deb, K., Pratap, A., Agarwal, S., Meyarivan, T., 2002. A Fast and Elitist Multiobjective Genetic Algorithm: NSGA-II. IEEE Transactions on Evolutionary Computation 6, 182-197.

Denny, S.B., Hubble, N.E., 1991. Prediction of Craft Turning Characteristics. Marine Technology 28, 1-13.

Evans, J.H., 1959. Basic Design Concepts. Naval Engineers Journal 71, 671–678.

Gammon, M.A., 2004. Ship Hull Form Optimization by Evolutionary Algorithm. Istanbul.

Ganesan, V., 1999. A Model for Multidisciplinary Design Optimization of Containership. Virginia Polytechnic Institute and State University, Blacksburg, Virginia.

Grigoropoulos, G.J., 2004. Hull Form Optimization for Hydrodynamic Performance. Marine Technology 41, 167-182.

Grigoropoulos, G.J., Chalkias, D.S., 2010. Hull-Form Optimization in Calm and Rough Water. Computer-Aided Design 42, 977 – 984.

Herrington, P.D., Latorre, R.G., 1998. Development of an Aluminum Hull Panel for High Speed Craft. Marine Structures 11, 47–71.

Isaacs, A., Ray, T., Smith, W., 2007. An Evolutionary Algorithm with Spatially Distributed Surrogates for Multiobjective Optimization. Progress in Artificial Life. Springer Berlin / Heidelberg, pp. 257-268.

Jones, L., 2011. Emergency Vessels not 'Equipped' for High Seas. Canberra Times, Canberra.

Jons, O.P., Koelbel, J., Sheldon, R., 1985. A New Generation of High-Performance Planing Craft. Naval Engineers Journal, 234-247.

Karafiath, G., Cusanelli, D.S., Jessup, S.D., Barry, C., 2001. Hydrodynamic Efficiency Improvements to the USCG 110 ft WPB ISLAND Class Patrol Boats. Transaction of Society of Naval Architects and Marine Engineers (SNAME) 109, 197–220.

Koh, L., Atlar, M., Mesbahi, E., Janson, C.-E., Larsson, L., Abt, C., 2005. Novel Design and Hydrodynamic Optimisation of a High-Speed Hull Form. 5th International Conference on High Performance Marine Vessels, Shanghai, China.

Lewandowski, E.M., 2004. The Dynamics of Marine Craft: Maneuvering and Seakeeping. World Scientific, Computer Sciences Corporation.

Majumder, M., Akinturk, A., Calisal, S.M., 2002. Optimized Design of Small Craft. Marine Technology 39, 67-76.

Mason, A., Thomas, G., 2007. Stochastic Optimisation of IACC Yachts. 6th International Conference on Computer and IT Application in the Maritime Industries (Compit 2007), Cortona, pp. 122-141.

Mistree, F., Smith, W.F., Bras, B.A., Allen, J.K., Muster, D., 1990. Decision-Based Design: A Contemporary Paradigm in Ship Design. Transactions of the Society of Naval Architects and Marine Engineers (SNAME) 98, 565-597.

Mohamad Ayob, A.F., Ray, T., Smith, W., 2010a. A Framework for Scenario-Based Hydrodynamic Design Optimization of Hard Chine Planing Craft. In: Bertram, V. (Ed.), 9th International Conference in Computer and IT Applications in the Maritime Industries (COMPIT'10). Technische Universitat Hamburg-Harburg, Hamburg, Gubbio, Italy, pp. 7-19.

Mohamad Ayob, A.F., Ray, T., Smith, W., 2010b. Hydrodynamic Design Optimization of a Hard Chine Planing Craft for Coastal Surveillance. Pacific 2010 International Maritime Conference (Pacific 2010), Sydney, Australia, p. 10.

Mohamad Ayob, A.F., Ray, T., Smith, W.F., 2011a. Beyond Hydrodynamic Design Optimization of Planing Craft. Journal of Ship Production and Design 27, 1–13.

Mohamad Ayob, A.F., Ray, T., Smith, W.F., 2011b. Scenario-based Hydrodynamic Design Optimization of High Speed Planing Craft for Coastal Surveillance. 2011 IEEE Congress on Evolutionary Computation (CEC), New Orleans, LA.

Mohamad Ayob, A.F., Ray, T., Smith, W.F., 2011c. Uncovering Secrets Behind Low-resistance Planing Craft Hull Forms Through Optimization. Engineering Optimization (in press).

NAVSEA, 2007. Report to Congress on Alternative Propulsion Methods for Surface Combatants and Amphibious Warfare Ships. . In: Naval Sea Systems Command, U.S.N. (Ed.).

Neti, S.N., 2005. Ship Design Optimization using Asset. Virginia Polytechnic Institute and State University, Blacksburg, Virginia.

Neu, W.L., Hughes, O., Mason, W.H., Ni, S., Chen, Y., Ganesan, V., Lin, Z., Tumma, S., 2000. A Prototype Tool for Multidisciplinary Design Optimization of Ships. Ninth Congress of the International Maritime Association of the Mediterranean, Naples, Italy.

Peacock, D.A., 1998. Decision-Based Hydrodynamic Design of Displacement Monohulls. School of Mechanical and Manufacturing Engineering. University of New South Wales, Sydney.

Radan, D., Sorensen, T.A., Johansen, A.J., Adnanes, A.K., 2006. Probability Based Generator Commitment Optimization in Ship Power System Design. WSEAS Transactions on Systems.

Ray, T., 1995. Preliminary Ship Design in the Multiple Criteria Decision Making Framework. Indian Institute of Technology, Kharagpur.

Ray, T., Singh, H.K., Isaacs, A., Smith, W., 2009. Infeasibility Driven Evolutionary Algorithm for Constrained Optimization. Constraint-Handling in Evolutionary Optimization. Springer Berlin / Heidelberg, pp. 145-165.

Rottman, G., 2008. US Patrol Torpedo Boats, World War II. . Osprey Publishing.

Savitsky, D., 1964. Hydrodynamic Design of Planing Hull. Marine Technology 1, 71-95.

Savitsky, D., Brown, P., 1976. Procedures for Hydrodynamic Evaluation of Planing Hulls in Smooth and Rough Water. Marine Technology 13, 381-400.

Savitsky, D., Koelbel, J., Joseph G., 1993. Seakeeping of Hard Chine Planing Hull.

Singh, H.K., Isaacs, A., Ray, T., Smith, W.F., 2008. Infeasibility Driven Evolutionary Algorithm (IDEA) for Engineering Design Optimization. 21st Australasian Joint Conference on Artificial Intelligence: Advances in Artificial Intelligence. Springer-Verlag Berlin, Heidelberg, Auckland, New Zealand.

Smith, W.F., 1992. The Modeling and Exploration of Ship Systems in the Early Stages of Decision-Based Design. University of Houston.

USCG, 2010. United States Coast Guard Acquisition Directorate.

White, J.A., Savitsky, D., 1988. Seakeeping Predictions for U.S.C.G Hard Chine Patrol Boats. N.Y. Metropolitan Section of SNAME.

Wolf, R., Dickmann, J., Boas, R., 2005. Ship Design using Heuristic Optimization Methods. 46th AIAA/ASME/ASCE/AHS/ASC Structures, Structural Dynamics and Materials Conference. Massachusetts Institute of Technology, Austin, Texas.

Zaraphonitis, G., Papanikolaou, A., Mourkoyannis, D., 2003. Hull-Form Optimization of High Speed Vessel with respect to Wash and Powering. Proceeding of the 8th International Marine Design Conference, Athens, pp. 43-54.

Development of New RAN Personal Lifesaving Equipment

D. Drohan[1], B. McNeice[1], P. Luckin[2],
WF. Smith[3] and S. Burdekin[3]
[1] Navy Engineering Division, Department of Defence, Australia
[2]Navy Health Service, Royal Australian Navy
[3]University of New South Wales @ Australian Defence Force Academy
david.drohan1@defence.gov.au

ABSTRACT

The Evacuation and Lifesaving unit of the RAN Navy Engineering Division has conducted a literature review of personal lifesaving equipment. The findings have influenced further research and experimental trials to update the requirements for personal lifesaving equipment for both Navy and Army.

Trials have indicated the current RAN abandon ship lifejacket (essentially a 1950s design) and the current abandon ship suit (a 1980s design), had never had their integration assessed. Further to this, testing showed discrepancies against contemporary requirements. This has led to the development of a new Personal Abandonment Lifejacket & Suit (PALS) concept. The revised requirements consider aspects, including airway clearance, body thermal protection and a waist belt pouch to protect the PALS when worn at Action stations. Modern packaging will allow servicing intervals to be increased to 5 years, thus reducing onboard maintenance. Existing commercial standards are modified where necessary to include the additional military requirements.

In addition to PALS, there are currently several lifejackets that both Navy and Army use for very similar operations. The Multi Purpose Lifejacket (MPL) will rationalise these. The operational requirements for both Navy and Army have been considered. Buoyancy trials have shown the need to focus on improved lifejacket fit and additional buoyancy for heavily encumbered personnel. Further research with a marine manikin in a wave tank will address the issue of maintaining an airway when encumbered with equipment.

INTRODUCTION

The philosophy behind naval lifesaving equipment is considerably different to the commercial passenger ship counterpart. The naval crew's priority is to save the ship. They will spend considerable effort in damage control and fire fighting to save a stricken ship and only abandon when it has become obvious that the battle has been lost. As such, there may be no time for dry shod evacuation. They will usually use wet shod evacuation (jumping into the water), to abandon ship. This is in stark contrast to a passenger ship where the first priority is to evacuate the passengers in an orderly fashion using dry shod means, either by the use of entering lifeboats prior to lowering them into the water or the use of a Marine Evacuation System to gain direct access to liferafts. In addition many military operations require personnel to be encumbered when operating over water. This can include operations with Army's standard webbing and associated equipment or Navy's boarding party equipment. These military requirements have influenced the development of specialised personal lifesaving equipment for military use. The purpose here is to explain the development work currently being undertaken by the RAN's Navy Engineering Division (NED) on requirements for the next generation of personal lifesaving equipment and represents a work in progress.

DEVELOPMENT OF PERSONAL LIFESAVING EQUIPMENT

Little thought was given to naval ship's lifesaving equipment during WWII with a subsequent high loss of life. Since WWII there have been considerable studies into understanding the issues of cold water and protecting a person from drowning, with the results being the introduction of new lifejackets to protect the airway and suits to protect the body from cold water.

Lifejackets

Post war research identified the preferred location of the lifejacket buoyancy. The lifejacket must have buoyancy in front of the chest to self right and behind the head to maintain an acceptable freeboard. The posture of the body to ensure the face is kept clear of the water is depicted in Figure 1. An inclination of between 30 and 60 degrees is the preferred body floating angle [1] (Lee 1965[1]) as cited by Brooks [2].

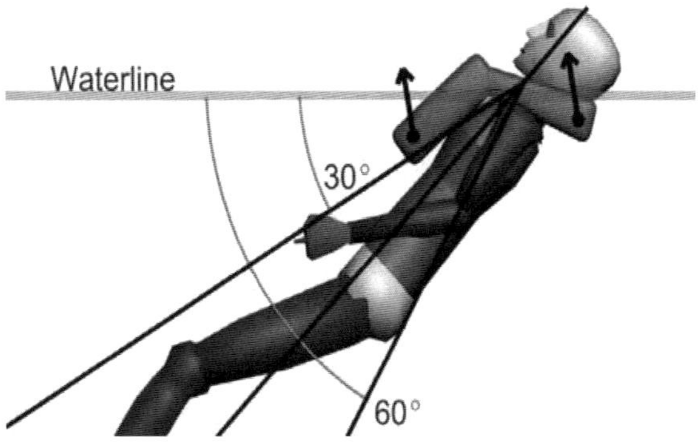

Figure 1 - Preferred lifejacket floating angles

Higenbottam and Slater [3] used marine manikins fitted with lifejackets to measure the volume of water ingress in the oral cavity in the open ocean. They identified that it does not take many splashes into the mouth cavity to cause inhalation of 200ml of water, which may be enough to cause drowning[2]. The invention of the spray hood has greatly improved the protection of the airway by greatly reducing mouth cavity water ingress [4].

The International Marine Organisation (IMO) Maritime Safety Committee (MSC) subcommittee has recently considered the issue of head angles for an unconscious person[3]. The recommended angles are shown in Figure 2. Although not identified in the MSC submission, a head tilt angle (to the body) of up to 20 degrees can be measured[4].

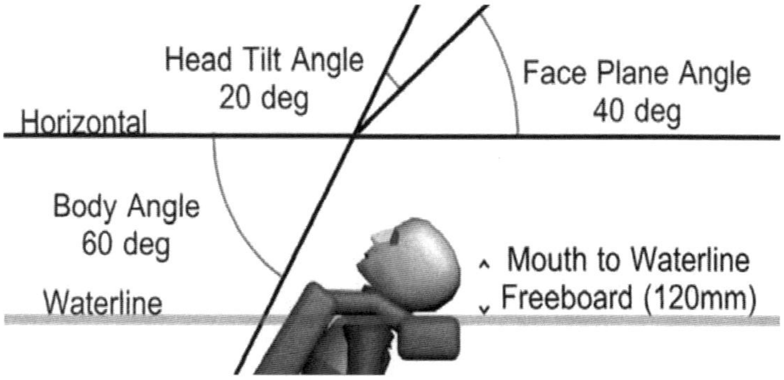

Herrmann and Stormer's research [5] into oral freeboard identified that in general as the freeboard increased the oral flooding decreased. In some cases the results suggest that life-jackets with freeboard heights of around 85 mm had approximately the same or even lower frequency of flooding than some of the jackets with a freeboard of

[1] Lee (1965). This paper can not be sourced, but the information was cited by Brooks

[2] Note: the statement that 200 ml of water may drown a person is disputed by Dr Luckin, (a medical specialist with extensive experience of survival in cold water and drowning), as the figure is based on a study that drowned sheep.

[3] Revision of testing requirements for lifejackets – IMO 55th session Agenda item 18 IAW resolutions MSC.81(70) and MSC.200(80) respectively, submitted by Norway 14 Jan 2011.

[4] The 40° Face plane angle subtracted from the 60° Body angle provides the 20° Head tilt angle

120-140 mm.

Figure 2 - Ideal face and body plane angles in the water

This implies that a target minimum freeboard required may be approximately 85mm. In addition, the fit of a lifejacket is most important as any lifejacket slippage away from the body will result in lower mouth freeboard [5]. The Life Saving Appliance (LSA) code [6] states the minimum oral cavity freeboard shall be at least 120mm[5]. This requirement appears to be first introduced in 1974[6]. Unfortunately the rationale on how this requirement was determined is not explained.

A review of current survival equipment that is available for encumbered personnel indicates these types of lifejackets may be exceedingly bulky or can only act as a buoyancy aid (good mouth freeboard clearance is provided, but may not provide adequate support for exhausted personnel). As Hermann and Stormer suggest the mouth to water freeboard could be lower, further research into this issue is warranted to identify the balance between body angle, head tilt and oral freeboard.

Abandonment Suits

The full effect of immersion in cold water (considered by some to be less than 15° C) has only been recently fully understood. During WWII between 30,000 and 40,000 men in the Royal Navy lost their lives due to drowning or the effects of cold water and/or hypothermia once they had abandoned ship [7]. It is now known there are four distinct ways in which a human immersed in cold water may become incapacitated and die [8] These are as follows:

- Cold shock (within seconds to minutes) – Uncontrollable gasping for air with increased heart and breathing rate.
- Functional disability - Decreased central coordination and cold muscles affecting manual dexterity.
- Progressive hypothermia – Reduction in body core temperature below 35° C.
- Post rescue collapse – Rescue in the vertical position can lead to death, due to hypovolemia[7] and hypothermia.

If the person's head goes underwater after jumping overboard and the cold shock reflex causes a gasp and inhale (and simultaneously decreases breath holding ability) the person may breathe in water and drown before they resurface.

Abandonment suits have now been developed to help protect personnel from the effects of cold water immersion. It is most important that any abandonment suit keep the person dry inside. Studies have identified that water leakage into abandonment suits can significantly reduce their insulative properties. A 500ml leak over the limbs produced a response similar to no leakage, whereas 500ml water ingress on the torso reduces body core temperature [9]. A recent

[5] Note: The current version of the LSA code was published in 2010. The previous prescriptive requirements (2003 edition) have been replaced by performance requirements that are acquired from a Reference Test Device (RTD) as detailed in Annex 1 of the current code. Lifejacket testing uses the RTD criteria to provide average results for each series of tests. The author has chosen to apply the 2003 LSA prescriptive requirements for this paper and the marine manikin study, rather than construct a RTD to determine the criteria.

[6] Brooks (1995) Designed for life – Lifejackets through the ages.

[7] Hypovolemia: Low blood volume.

presentation on the development of the immersion suit reinforced that survival time in 13° C water with one litre of water ingress in the suit is significantly less than 1 hour, in Beaufort winds 5 and higher [10]. Further, a person with pooled water inside their suit may not have the strength or hand dexterity to pull themselves into a liferaft.

Brooks noted that persons floating in an immersion suit and lifejacket do not assume the LSA code prescribed floatation angle, nor did they self right in the required time [11] [12]. In the 1960s, Lee's[8] observations on posture noted that a vertical posture offers less resistance to wave action and places the person at great risk of periodic immersion of the airway. A supine[9] position (that many abandonment suits assume) places the person at a greater risk of airway obstruction, especially if unconscious. Whilst this is less of a problem for a conscious person, it becomes a greater concern as a person becomes more exhausted, hypothermic and lapses into unconsciousness. When unconscious, the tongue can fall back inside their pharynx[10], thus blocking the airway with fatal results.

It is a standard first aid procedure to place an unconscious person in the recovery position, on their side to maintain an open airway[11]. This is impossible to achieve in the water. The best alternative is to attempt to maintain an open airway, by a positive head tilt angle, in an effort to stop the tongue blocking the airway once it becomes relaxed. The left diagram shown in Figure 3 depicts how the tongue blocks the airway with no head tilt. The right diagram depicts how the tongue might be prevented from blocking the airway by a simple head tilt. This simple technique can be applied to a person wearing a lifejacket in an attempt to keep their airway open, even if they lapse into unconsciousness.

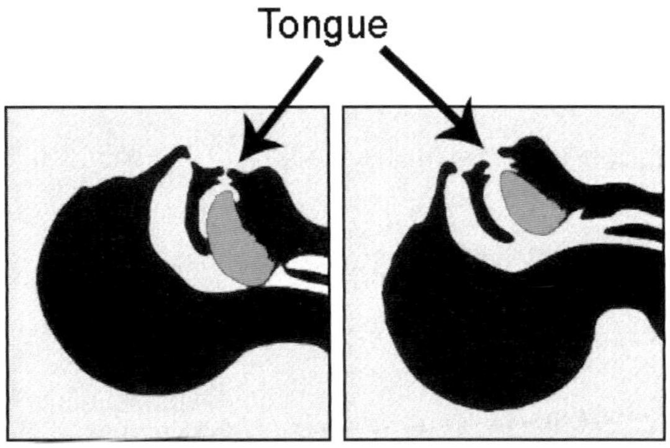

Figure 3 - Head tilt angle affect on the tongue

(Modified from V. Atanassova)

The head tilt angle issue is complicated when a lifejacket is worn in conjunction with an abandonment suit. Most abandonment suits create a shallow body angle due to air trapped within the suit that provides additional buoyancy to the lower torso and legs. The LSA code [6] identifies that a lifejacket must maintain a freeboard of 120mm from the mouth to the water. This should correspond to a face plane angle of 40 degrees to the water line.

Thus when a compliant SOLAS lifejacket is worn with an abandonment suit, the resulting head tilt angle could be as much as negative 30 degrees, (ie; the head is tilted forward) as shown in Figure 4. The lifejacket is omitted for clarity in this and subsequent figures.

[8] Lee 1965, as cited by Brooks (2003).
[9] Supine: Lying on the back or having the face upward.
[10] Pharynx: The passage that leads from the mouth, nose cavity to the lungs and stomach.
[11] Australian First Aid, St John Australia, 4th Edition - 2006: Page 73 Recovery position

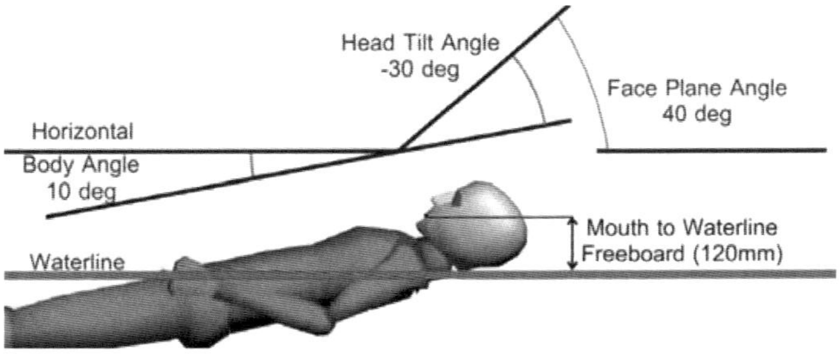

Figure 4 - Maintaining mouth freeboard at the expense of head tilt

This angle may not cause problems for a fully conscious person who has not been in the water for long. However, if the person becomes fatigued, develops hypothermia and progressively declines into unconsciousness, the tongue may relax and block the airway. Attempting to maintain an adequate airway angle with a shallow body angle results in a reduced mouth freeboard clearance as shown in Figure 5. The development of any new abandonment suit should consider this issue. A head tilted back might clear the airway, but it is more difficult to keep the airway clear of water, if the body is in a near horizontal position on the water.

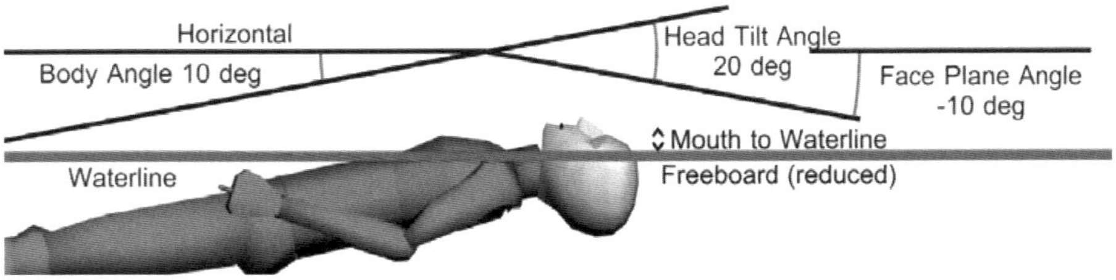

Figure 5 - Maintaining head tilt at the expense of mouth freeboard

DEVELOPMENT OF NEW RAN PERSONAL LIFESAVING EQUIPMENT

With an understanding of the development history of lifejackets and abandonment suits, new work has been undertaken within the RAN's Navy Engineering Division (NED).

Current RAN Personal Lifesaving Equipment

The RAN has adopted the Pattern 50N lifejacket (an evolved 1950s design) as their current abandon ship lifejacket. It is gas filled, manually operated, double chambered and is fitted with a spray hood. For limited thermal protection in the water, the RAN adopted the Thermal Protection Suit (TPS) that uses a draw string around the neck for the sealing method. Small and large crew members wearing the TPS and uninflated Pattern50N lifejacket are shown in Figure 6.

The Directorate of Navy Platform Systems (DNPS) within NED undertook a trial and determined that the TPS tends to permit water ingress. Functional[12] and survival times predicted by AVMED[13]

[12] Functional time - The time available before a human can not perform basic tasks, lapses into unconsciousness.

[13] AVMED – Aviation Medicine at HMAS Albatross.

using the computer software CESM[14] have confirmed the thermal protection of the TPS could be improved. In addition the LSA [6] freeboard and self righting requirements are not satisfied with the Pattern50N when worn in conjunction with the TPS. The DNPS trial report provided the evidence that an improved abandonment lifejacket and suit could be developed for the RAN.

Figure 6 - Current personal lifesaving items

RAN Personal Abandonment Lifejacket & Suit (PALS) Requirements

Based on the findings from the DNPS lifejacket and TPS trial, a set of performance requirements for a Personal Abandonment Lifejacket & Suit (PALS)[15] have been developed. These have been based on commercial standards, but modified for military requirements where required. It is intended PALS could be worn together (both the lifejacket and suit) when in cold water. When in warmer tropical waters only the lifejacket may need be worn, as it is unlikely cold water shock or hypothermia will be an issue prior to boarding a liferaft. Their primary requirement is to provide airway and thermal protection from the time of abandoning ship through to self boarding a liferaft, when external rescue may be many hours or days away. The packed lifejacket and suit are stowed inside pouches that are worn on a waist belt that will be a similar size to the existing pouches as shown in Figure 7.

The pouch requirements call for toughened material to allow PALS to be worn at Action Stations and provide a level of protection against tearing from hot jagged metal similar to which a human could survive. To reduce through life costs, the intention is to increase the service life interval up to five years. This may be achieved through vacuum sealed foil packaging stowed in lockers inside the ship. The issue of airway protection for an unconscious person wearing an abandonment suit is so complex and difficult to resolve, that it may have to be mitigated by a Standard Operating Procedure such that all crew are aware of the issue and therefore help to keep the airway clear for any unconscious persons in the water.

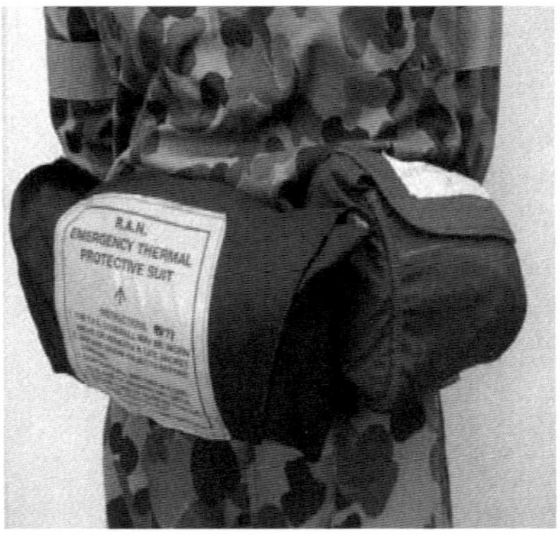

Figure 7 - Current RAN TPS and Lifejacket pouches

Proposed Multi Purpose Lifejacket

The RAN wishes to rationalise the number of types of army and navy lifejackets to be stored onboard their ships. Therefore, DNPS is undertaking development work to determine suitable requirements for a Multi Purpose Lifejacket (MPL) that should replace a number of existing lifejackets that are used for day to day operations such as boat operations, where there is greater risk of falling overboard. Preliminary trials have suggested that a yoke style, inflatable lifejacket with a buoyancy of approximately 275N could fulfil the requirements of these activities. The MPL is intended to be worn by encumbered military personnel. External rescue from other fleet units is most likely available to any personnel who have fallen in the water within one hour. The minimum winter water temperature is unlikely to be below 10° C, because operations should not occur in the Southern ocean without special survival suits. The CESM predication of survival

[14] Cold Exposure Survival Model (CESM) Version 3.02 developed from Tikuisis's research in 1997.
[15] Personal Abandonment Lifejacket & Suit (PALS) Requirements: Version 5.1 dated Sept 2011.

time is over two hours without a survival suit in this context. This issue does highlight the conflicting priorities of adequate protection from a sudden fall into cold water whilst maintaining operational abilities.

A recent DNPS trial has demonstrated the buoyancy of the MPL prototype MOTS 275N lifejackets could not meet the LSA Code of 120mm freeboard from mouth to water when personnel are encumbered with up to 40 kgs of equipment. In some cases, the airway was compromised. In addition the body angle when encumbered was also compromised. This lack of lifejacket freeboard occurs because of problems with lifejacket fit and as well as the weight of additional kit.

Lifejacket fit issues

There are three components of the prototype MPL restraint system, being a waist strap, back strap and crotch straps. The waist strap holds the lobes of the bladder to the lower torso, the back strap connects the neck section of the bladder to the back and the crotch straps hold the back strap and waist strap in their desired location. All subjects had these straps tightened prior to entering the water, yet a number of fit issues were observed.

1. The ability for the MPL to fit correctly relies on the tension of the crotch straps to hold the other straps securely in place. This could not be achieved as they were unable to be fastened securely causing the entire jacket to ride up the torso. Note how the crotch straps (highlighted in the figure) ride up the posterior of the subject in Figure 8. This is a safety issue rather than just comfort, as mouth freeboard has been reduced by the crotch straps failing to hold the encumbered body weight from settling further into the water.

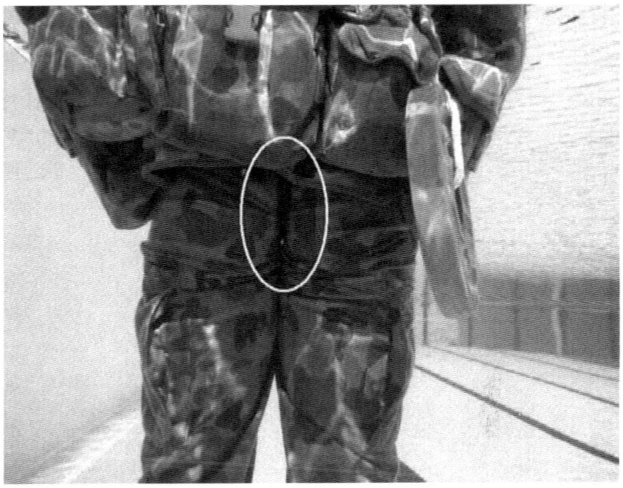

Figure 8 - Effect of crotch straps

   (Note: gym weights were used to simulate the weapon and ammunition. It is acknowledged the simulated weights may have different densities to the actual items.)

2. The addition of the ballistic vest can alter the shape of the torso, from a relatively box like shape to more of a pyramid like shape. Webbing was often attached in the same position as the lifejacket waist strap, making the possibility of the waist strap sliding up even greater. These issues can be seen in Figure 9.

Figure 9 - Ballistic vest affecting lifejacket fit

Figure 10 Lifejacket separation

3. The back strap appeared to be partly ineffective as there was considerable separation between the bladder and the subject's torso. This can be clearly seen in Figure 10. (Note; the near vertical position of the torso).

4. The size and shape of the bladder around the head caused a number of problems, such as incorrect head tilt and additional pressure when wearing helmets. The inflated bladder puts pressure around the neck and forces the head forward. Helmet pressure and a kink in the bladder can be seen in Figure 11.

Helmet head pressure   Bladder kink

Figure 11 - Pressure on head caused by bladder

Encumbered personnel study

Providing adequate freeboard for encumbered personnel is a significant challenge. DNPS has recently purchased and modified a marine manikin to investigate these issues. The purpose of this research is to determine if a reduced freeboard that has been identified by the study [5] could be applied to encumbered personnel.

A joint study between DNPS, the University of NSW at ADFA and the Australian Maritime Collage (AMC) is currently underway to investigate measuring water ingress at reduced lifejacket freeboards with the marine manikin in regular waves up to 0.25m. Wind affects can not be included, but will be considered by generating short steep waves to provide a chop effect.

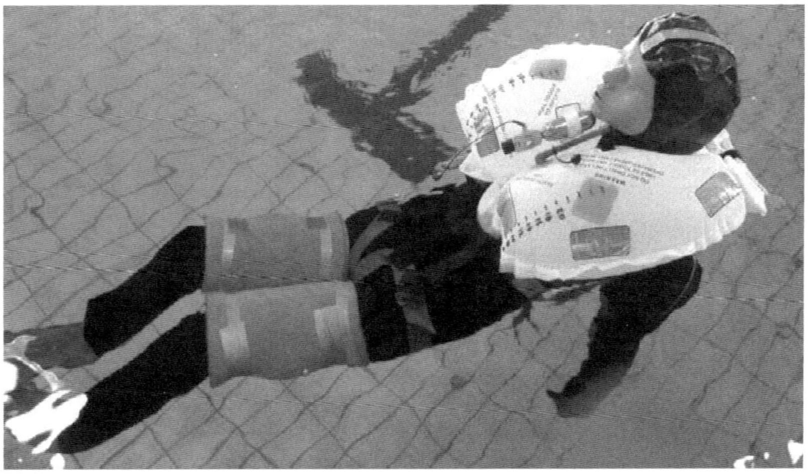

Figure 12 - The manikin undergoing flotation angle tests

Figure 12 depicts the marine manikin undergoing flotation tests in a swimming pool. Note: the use of close cell foam to adjust the floatation angle and prototype thigh straps to improve lifejacket fit. The manikin has had its mass, volume, centre of gravity and flotation angles measured. Good validation against a human's relevant measurements has been achieved [13]. It is acknowledged the manikin can not simulate a conscious survivor, as it can not replicate a person's mouth closure, that would be used to protect their airway from water ingress.

DNPS and ADFA are currently investigating the issue of lifejacket fit for encumbered personnel that could be improved with the use of thigh straps similar to those used on a climbing harness. The thigh harness fits below the pouches on a ballistic vest and is connected to the chest harness of the MPL. Initial experiments with the manikin indicate this system prevents the lifejacket from riding up an encumbered body. It should also be more comfortable for personnel during operations. In addition another option under investigation is the addition of small inflatable bladders attached to the thigh straps to provide extra buoyancy. Not only does this system provide some additional buoyancy, the leverage effect provided by the thigh straps improves the body angle for the encumbered person. This study includes calculating the density of most items used by encumbered personnel.

Modular MPL requirements

There are a number of military requirements that the MPL will have to meet and the intention is to provide a modular system to meet these requirements. Some of the major requirements are:

- Protect navy personnel who find themselves in the water when undertaking special duties, boarding party (encumbered) and flight deck fire fighting operations. One new concept being investigated with support form DSTO, is developing two types of a second valise to fit over the lifejacket to either provide limited stab protection to the lifejacket bladder (for boarding party operations) or fire insulation protection to the lifejacket bladder for flight deck fire fighting operators.
- Protect heavily encumbered army personal who find themselves in the water whilst undertaking ship/air to shore operations. The testing of the MPL will also involve drops from

20m and surfacing from 20m underwater to simulate a fall from a high sided vessel or escape from a sinking vessel or helicopter.

- Ensure the MPL when used by army in an assault role, is capable of being doffed quickly once the troops have landed is planned.

ADDITIONAL RESEARCH

It is anticipated the joint marine manikin testing program should discover the limits of potential MPL requirements with water ingress performance measured against reduced freeboards. This will be useful knowledge for developing requirements for personnel who are heavily encumbered and can not meet the LSA code freeboard requirement.

One interesting challenge currently being investigated is providing real time data logging of water ingress into the manikin's oral cavities in conjunction with recording the manikin's body roll and pitch movements in the wave tank. The intention is to identify the amount water ingress over a period of time. Testing the manikin at AMC's wave tank facility is scheduled for March 2012. A number of wave tank runs with the manikins mouth freeboard set a various freeboards ranging from 120mm down to 80mm with and without the lifejacket spray hood deployed are planned. The aim is to measure wave splashes and duration into the mouth for each run in the wave tank. Once the water ingress data has been obtained, it is envisaged a computer model of the manikin can be generated in order to conduct further studies with interpolated data. It is anticipated a maximum load that the lifejacket can safely support will be determined.

In summary, the outcome of this research and future trials is to achieve an effective lifejacket for heavily encumbered military personnel, reducing the number of lifejackets that need to be stored and maintained on RAN ships.

REFERENCES

1. Schiff and Hafen. *The man in the aqueous element.* 1959 [cited 2011, 10/01/2011]; Available from: http://www.secumar.com/secumar/c/download/the_man_in_the_aqueous_element.pdf.
2. Brooks, C. *Survival in cold waters: Staying alive.* 2003 [cited 2011, 17-3-2011]; Available from: www.tc.gc.ca/eng/marinesafety/tp-tp13822-menu-610.htm.
3. Higenbottam, C. and P. Slater, *The dymanic behaviour of the floating survivor in sea waves and its effect on airway protection*, in *International Conference on Environmental Ergonomics.* 1990. p. 200-201.
4. Herrmann, R. *Life jacket spray-hood function and improvements.* 1998 [cited 2011, 15/12/2010]; Available from: http:www.lboro.ac.uk/departments/ids/groups/EEC/ICEE/textsearch/1998/herrmann-1998.pdf.
5. Herrmann, R. and A. Stormer, *Results of recent research on life-jackets.* Journal of the Royal Naval Medical Service, 1985. 71: p. 161-168.
6. IMO, *International Life-saving Appliance Code.* 2003, London: International Maritime Organization.
7. Talbot, A., G, *Report of the Talbot committee on naval lifesaving*, Admiralty, Editor. 1946, Ministry of Defence: London.
8. Steinman, A., RADM and G. Giesbrecht, *The four stages of cold-water immersion.* On Scene - The Journal of the US Coast Guard Search and Rescue, 2006. 2006(Fall): p. 13.

9. Tipton, M., J and P. Balmi, J, *Assessment of immersion suit performance: Human compared to immersion thermal manikin tests.* European Journal of Applied Psychology, 1996. 72: p. 394-400.

10. Coleshaw, S.R.K. *Immersion suit and lifejacket developments.* 2008 [cited 2011, 12/3/2011]; A powerpoint presentation for the UK oil and gas industry]. Available from: http://www.oilandgasuk.co.uk/downloadabledocs/271/8.%20Sue_Coleshaw_SRK_Coleshaw.pdf.

11. Brooks, C., et al. *Survival at sea for Mariners, Aviators and Search and Rescue personnel - Lecture notes HFM-152.* 2008 [cited 2011, 14/2/2011]; Available from: www.rta.nato.int/Pubs/RDP.asp?RDP=RTO-AG-HFM-152.

12. Brooks, M., *Designed for life, Lifejackets through the ages.* 1995, Richmond: Mustang Engineered Technical Apparel Corp.

13. Kreighbaum, E. and K.M. Barthels, *Biomechanics, A qualitative approach for studying human movement.* Third ed. R. Miller. 1990, New York: Macmillan Publishing Company.

Future Role of Major Lighthouses: The Sarawak Perspective

Ahmad Faizal AHMAD FUAD[1], Saharuddin ABDUL HAMID[1]
University Malaysia
afaf@umt.edu.my

Abstract

Lighthouse falls under category of visual aids to navigation and serves as important structure for marine navigation. With the introduction of Global Navigation Satellite System (GNSS) to ships, the role of traditional lighthouse for navigational reference had reduced to secondary. Presently in Sarawak, commercial ships that fitted with GNSS and other electronic navigational equipments do not rely on lighthouse for navigation reference. On the other hand, fishing vessels highly rely to lighthouse for their navigational needs. This is due to most of the fishing vessels are operating within 12 nautical miles from the coast and not carrying GNSS and nautical chart. In order to recover the role of lighthouse to become relevant again for commercial shipping, lighthouses should be equipped with system that link them with shipboard navigational equipment in order to improve safety of navigation and justify the light dues. Such devices are AIS and DGPS. This paper discuss on the setup of all major lighthouses in Sarawak with AIS and recommends measures to recover the role.

Key Words: Lighthouse, Aids to Navigation, Safety of Navigation.

1. Background

As party to SOLAS Convention 1974, Malaysia is obliged to provide aids to navigation service to shipping community in order to ensure the safety of navigation (International Maritime Organization, 2004, 2011). One of the systems that fall under aids to navigation is lighthouse. Lighthouse is an important structure for marine navigation and mainly use for positional reference. According to IALA[2], lighthouse is a conspicuous structure on land, close to shoreline, or in water that acts as the daymark and provides a marine signaling light with a range of up to 25 nautical miles (IALA, 2001).

Sarawak is one of the states in Malaysia and located in island of Borneo. The first lighthouse was introduced to this state in 1872 (Percetakan Nasional Malaysia Berhad, 2011). Currently, there are seven lighthouses operational along the coast of Sarawak of nearly 1000 kilometers long. With the introduction of Global Satellite Navigation System (GNSS) in the 90's that provides a reasonably accurate, reliable, and affordable navigation system to commercial shipping; lighthouse roles for navigation had been

[1] Faculty of Maritime Studies and Marine Science, University Malaysia Terengganu, 21030 Kuala Terengganu, Terengganu, Malaysia
[2] International Association of Marine Aids to Navigation and Lighthouse Authorities

reduced to secondary. This had raised question whether the lighthouse services should be continue in the future. This study tries to answer this question.

2. The Present Situation

2.1 Physical and System Information

The current information of the seven lighthouses shows by table in Annex I. The information comprise of physical structure and lighting system for each lighthouse.

2.2 Light's Distance

In order to determine distance of light produce by each lighthouse, the manufacturer measured the light intensity and use the formula below published by IALA to calculate the effective intensity (Pharos Marine, 1997; Pharos Marine & Automatic Power, 2002, 2006).

$I_e = I_0 \text{ x } t / a + t$

I_e = effective intensity in candelas
I_0 = peak intensity during the flash in candelas
t = duration of flash in seconds
a = value obtained by a table which refers to flashes produce by blanking or rotating optical apparatus

The results (intensity in candelas) then refer to "Table for Darkness Luminous Range in Nautical Miles at T-factor 0.74[3]" in order to determine the luminous range of each lighthouse according to its given light's character. The distance of lights radius for each lighthouse shows by figure in Annex II.

2.3 The Navigational Equipments on Board Merchant Ships

In order to determine the present usage level of lighthouse, the current navigational practice should be study. Currently, the navigational practice is depends on navigational equipment carry by ships which in accordance with carriage requirements stipulate in Chapter V SOLAS 1974 (Regulations 1 and 19) and the Merchant Shipping Ordinance 1952 (national shipping law). According to the Regulation 19, all ships irrespective of size shall carry magnetic compass, compass bearing device, nautical charts and publication, and receiver of GNSS or radio terrestrial radionavigation system (International Maritime Organization, 2004). In addition to the aforementioned, ships of 300 gross tonnage and upwards shall carry echo sounder, 9 GHz RADAR, speed and distance measuring device and Automatic Identification System (AIS). For carriage of

[3] Atmospheric Transmission Factor of 74% (0.74) which is the light's intensity remain when a light beam has passed a distance of one nautical mile

nautical charts, it can be replaced with Electronic Chart and Display Information System (ECDIS) (International Maritime Organization, 2004).

In order to know the carriage of navigation equipment by vessels operating in waters of Sarawak, samples of 25 vessels which comprise of commercial and government vessels of various types are use (Annex III). All of these vessels are complying with the local shipping law and SOLAS 1974. The samples show that the vessels are carrying GNSS receiver as the principle means for positional fix. For preventing collision, all vessels are carrying RADAR[4] and most are carrying AIS. Few vessels are carrying ECDIS, which apply the integration of multiple navigation equipments in order to display it on a single screen of ECDIS. The samples show the local vessels are complies with the carriage requirements stipulate in SOLAS 74 and the local shipping law. Therefore, with the carriage requirement of navigational equipment institute by the relevant law had reduced the role of lighthouses to secondary and redundant for current navigational requirement.

As the responsible agency that provides aids to navigation service and collect light dues from commercial ships, Marine Department Malaysia should look into approach to recover the role of lighthouses for current navigational requirement. The system of choice should complement the shipborne navigational equipment by reducing workload during watch keeping and justify the light dues.

2.4 The Navigational Equipments on Board Fishing Vessels

According to Department of Fisheries Malaysia, number of registered fishing vessel of Sarawak in 2010 is 5681 (Department of Fisheries Malaysia, 2010; Marine Fisheries Department Sarawak, 2011). This number is comprises of fishing vessel of zones A, B, C and C2. Fishing vessels for zones A and B referred as coastal fishing vessel while C and C2 are deep sea fishing vessel. Zones A and B are cater for small fishing vessel with fishing ground of not more than five and twelve nautical miles consecutively from coast. While zones C and C2 are for large fishing vessels of at least 40 and 70 gross tonnage respectively, which the fishing ground starts 12 nautical miles from the coast to the boundary of Exclusive Economic Zone (EEZ). Among all of these categories, fishing vessel of zone A has the highest percentage of 86% while the combine percentage of zone C and C2 is 10%.

The minimum carriage requirement for fishing vessels in Malaysia according to zones shows in Table 1. With respect to Sarawak, 86% of the fishing vessels are Zone A and only require carrying magnetic compass. These vessels are operational within the visual range from the coast. Therefore, these vessels rely on prominent landmark ashore for their navigational reference such as hill or telecommunication tower. In locations where lighthouse is available, it serves as the all day navigational reference. Therefore, lighthouse is very important for Zone A fishing vessels as these vessels carry very basic navigation equipment and rely on visual and local knowledge for navigation. For fishing vessels of zone C and C2, these vessels are required to carry Magnetic Compass, GNSS receiver and Echo Sounder. However, the carriage of nautical chart is not compulsory by

[4] Radio Detection and Ranging

the fisheries department. This is possibly due to most of skipper of fishing vessel are not train to use nautical charts and nautical publications. Nevertheless, from observation, some of the vessels are carrying GNSS receiver with built-in electronic chart plotter. Although the plotter automatically fixes the position, the sources of charts use by the plotters are not approved by any maritime authority. Thus, this type of electronic chart plotter does not replace nautical paper chart as ECDIS. Therefore, to ensure safety of navigation, fishing vessels of C and C2 are still relying on lighthouses to verify their position as given by the electronic chart plotter.

Table 1: Carriage Requirement of Navigation and Communication Equipment for Fishing Vessel According to Fishing Zone (Source: Fisheries Department Malaysia)

Fishing Zone	Area of Operation from Coast	Carriage Requirement of Navigation Equipments
A	Within 5nm	Magnetic Compass
B	5nm to 12nm	Magnetic Compass, GNSS Receiver, Echo Sounder, and Wireless Radio
C	12nm to 30nm	Magnetic Compass, GNSS Receiver, Echo Sounder, Wireless Radio, and Satellite Phone
C2	30nm to Boundary of EEZ	Magnetic Compass, GNSS Receiver, Echo Sounder, Wireless Radio, and Satellite Phone

3. The Measures for Relevant ness

3.1 The Linkage between Lighthouses and Onboard Navigational Equipments

In the era of electronic navigation, the automatic linkage between lighthouses and the shipborne navigational equipment should use the existing technology or equipment available. The example of such technology is data sharing or interchange by using radio link. The available system or device to perform such function is AIS.

3.2 AtoN-AIS Project

Marine Department Malaysia had setup the shore based AIS monitoring system along the Sarawak coast in 2007. The information of the system shows in Annex IV. The main objectives of the system are to monitor the shipping traffic, enable coastal state to gain information on ships and its cargoes, and serve as tool to improve safety of navigation as recommended by IALA (IALA, 2002). The system also combines Marine Aids to Navigation and Automatic Identification System (AtoN-AIS). The combination of these two systems are improving the relevant ness of visual aids to navigation especially lighthouse to commercial ships by linking aids to navigation ashore with navigation equipments onboard. In order to gain the full extent of the system, Automatic Weather Station (AWS) is integrates with the AIS and enable ships to receive weather updates from the AIS along with information of the lighthouses. Another advantage gain from the AIS system is the authority able to monitor the operational status of the lighthouses from the office and complement the existing SCADA system.

4. Discussion on the Future of Lighthouse

For commercial ships that increase in size and speed, the linkage of the lighthouse with onboard navigational equipment by AIS or other radio-based equipment in the future would be useful in verify their position provided by GNSS receiver. The combination of AtoN-AIS may also complement GNSS in the event of GNSS failure. Information provides by the AtoN-AIS are status of the lighthouse, distance and bearing from the lighthouse, and local weather updates. The information can be used to fix the ship's position on paper chart and complement the usage of visual bearing for positional fix. Another application of electronic navigation to lighthouse years before the integration with AIS is Differential Global Positioning System (DGPS). In certain countries, lighthouses are use as platform for DGPS Broadcasting Station. The station uses known position of the lighthouse as correction for GPS position. The correction data is transmitted by MF radio link to ship's DGPS unit in order to correct current ship's position (Ahmad Fuad, 2002). However, between these two systems, AIS is more feasible fitted to the lighthouse as it provides navigational reference within VHF radio range and operates independently. Unlike AIS, failure of GPS satellites would cause failure to terrestrial DGPS broadcasting stations. Furthermore, the carriage of DGPS is not a requirement by any national or international convention. Even ships in the samples are not carrying DGPS receiver. Therefore, AIS is a capable system to recover the role of lighthouses.

Compare with the commercial ships, most of fishing vessels only carry magnetic compass for navigation. The majority of these fishing vessel (90%) are operating within 12 nautical miles (Zone A and B) from the coast of Sarawak. Therefore, they rely heavily on lighthouse for navigational reference especially during night or low visibility. Therefore, lighthouse is important for fishing vessel.

By looking into the requirements of commercial ships and fishing vessels, the traditional lighthouses should continue to operate with certain improvements. For the requirement of commercial ships, the lighthouse authority should spend more resources in linking their lighthouses with the navigation equipments onboard by using AIS or other radio-based equipments. Although this would incur more cost to the authority in addition to existing cost of maintaining the lighting systems and lighthouse's structure, it would recover the role of traditional lighthouse as navigational reference to commercial ships. Annex V shows the light dues collection and expenses in Sarawak from 2001 to 2003. The expenses cover annual operational[5] cost and development[6] projects. The amount of saving annually gain from 2001 to 2003 was small as the differences between the collection and expenses were not significant. The saving is important to finance development projects related to AtoN and hydrography as conferred by Merchant Shipping Ordinance Sarawak 1960. However, the AIS and AtoN-AIS projects in 2006 were financed by the government under the 8th Malaysia Plan instead of the light dues. The savings from the light dues collection during the time may not sufficient to finance the AIS project along with existing AtoN development projects.

---

[5] Example of operational cost are procurement of AtoN spare parts and fuel for maintenance vessel

[6] Example of development projects are upgrading lighting system and new AtoN tower construction

In tandem with requirement from the fishing vessels that require the continuation of traditional lighthouse operation, the light's distance of more than 25 nautical miles should be reduce between 12 to 20 nautical miles as majority of the fishing vessels (90%) are operating within 12 nautical miles from the coast. This measure would reduce the operational cost of the lighting system and the existing power supply system (solar panels and batteries), while the saving can be used to support the operation of AIS.

Therefore, the installation of AtoN-AIS to lighthouses and review the range of light can fulfill both the requirements of commercial ships and fishing vessels.

5. Conclusion

The usage of traditional lighthouse for navigation is depends on the current navigational requirements of commercial ships. The requirements are determine by the carriage requirements of navigational equipment which subject to the relevant international conventions and the national law. The carriage of GNSS receiver has reduced the role of traditional lighthouse for navigation and become redundant for current navigation needs. On the other hand, fishing vessels in Sarawak highly rely on lighthouse for navigation. This is due to their carriage requirement of navigation equipment is less strict than commercial ships and in according to fishing zones as stipulated in the national law. Furthermore, the operators of fishing vessels are not carrying navigation chart and not train for its usage. These factors are causing fishing vessels highly depend on lighthouse for their navigational reference as the majority of the fishing vessels are operating within 12 nautical miles from the coast.

In order to recover the role of the lighthouses in Sarawak, AIS is used automatically to link the lighthouses and navigational equipment on board ships. Along with the installation of AtoN-AIS to the lighthouses, the present range of light should be review to range of between 12 to 20 nautical miles as majority of fishing vessels are operating within 12 nautical miles from the coast. Therefore, the changes to lighthouses are required in order to cater for current navigational requirements and the continuance of the lighthouse service.

6. Recommendations to Recover the Role of Lighthouse for Navigation

In order to ensure the relevant ness of lighthouse to cater for future navigation needs, the changes should be conducted as the following:

i. Reduce the range of light between 12 to 20 nautical miles. The reason for this recommendation is commercial shipping currently do not require lighthouse for positional fix due to availability of GNSS receiver. Other reason is most of fishing vessels (90%) in Sarawak are operating within 12 nautical miles from the coast.

ii. Lighting system of the lighthouse should be less complicated and using low power than the traditional system. The example is such application is using the blanking light instead of rotating light for light's flash character (Atkins and Drennan Marine Consultancy, 2010)

iii. Lighthouse should be fitted with equipment that would link between shipboard navigational equipment and the lighthouses such as AIS and RACON[7].

[7] Radar Beacon. Equipment which use the shipboard RADAR frequency and display the designated modulation on RADAR display

References

Ahmad Fuad, A. F. (2002). *The Sarawak Marine Department Differential Global Positioning System and Its Contribution.* Paper presented at the Ninth Emergency Response Planning (ERP) Seminar on Disaster Management, Bintulu, Sarawak, Malaysia.

Ahmad Fuad, A. F. (2010). Sarawak Marine Department Automatic Identification System (AIS) Shore Facilities *Sarawak Nautical Almanac.* Kuching: Sarawak Marine Department.

Atkins and Drennan Marine Consultancy. (2010). *Assessment of the Provision of Marine Aids to Navigation around UK and Ireland: Final Report.*

Department of Fisheries Malaysia. (2010). *Kedudukan Sektor Perikanan di Malaysia 2009*: Department of Fisheries Malaysia.

IALA. (2001). IALA Aids to Navigation Guide (4th ed.): IALA.

IALA. (2002). Recommendation on the Provision of the Shore Based Automatic Identification System (AIS).

International Maritime Organization. (2004). *International Convention for the Safety of Life at Sea 1974*: International Maritime Organization.

International Maritime Organization. (2011). Status of Conventions. Retrieved 12 march, 2011

Marine Fisheries Department Sarawak. (2011). *The Number of Fishing License Release in 2009 and 2010 According to Fishing Zone.*

Percetakan Nasional Malaysia Berhad. (2011). *Sarawak Government Almanac 2011.*

Pharos Marine. (1997). Refurbishment of Tg. Sirik Lighthouse.

Pharos Marine & Automatic Power. (2002). Tanjung Jerijih Lighthouse: Installation, Operation and Maintenance Manual.

Pharos Marine & Automatic Power. (2006). Refurbishment of Tanjung Po Lighthouse.

Sarawak Marine Department (Cartographer). (2003). *Tanjung Datu to Labuan.*

Annex I

Information of Physical Structure and Lighting Equipments of Lighthouses in Sarawak

No	Name of Station	Information
1	Tanjung Datu	**Physical Structure :** • Cylinder shape concrete structure with Tideland aluminium lantern house on top. • Height of structure is 8 meter from ground level and 171 meter elevation from mean high water level. • Location : 02^0 04' 48'' N 109^0 38' 18'' E • The cylinder structure was built in 1990 was painted with bright red in 2006.
		Lighting System : • Pharos Marine Rotating Beacon model APRB – 252 D.C Standard Wattage • Lens size is 250 mm height with total number of six lens panels. • Flasher lampchanger used is APCL-5 (60 watts max) the 6-bulb automatic lampchanger • The current bulb used is 12V/50Watt single filament type. • Distance of light at T = 0.74 is 16.9 nautical miles and at T = 0.85 is 25.9 nautical miles. (Range in SAR1 is 18nm at T=0.74) • Flashing character is one flash every 15 seconds
	Tanjung Po	**Physical Structure :** • Rectangular shape triple storey pagoda shape house with steel and glass lantern house on top. • Height is 15 meter from ground level and elevation is 131 meter from mean high water level. • Location 01^0 43' 30.8''N 110 ^{0} 31' 12.6''E • The lighthouse structure was built in 1872 while the current lantern house was installed in 1936 for the housing of the Stone Chance lens.
		Lighting System : • Rotating Beacon model PRB-20 series from Pharos Marine • Lens size is 500mm Stone Chance Lens • The flasher/lampchanger is UVLA 2 bulb automatic lampchanger that using 400 watt metal arc as main and 100 watt as standby. • Distance of light at T = 0.74 is more than 20 nautical miles. (Range in SAR 1 is 24nm at T=0.74) • Flashing character is two flashes every 10 seconds (3 RPM).
	Tanjung Jerijih	**Physical Structure :** • Steel tower structure with aluminium lantern house on top. • Height of tower is 36 meter from ground level and elevation is 38 meter from mean high water level. • Location 02^0 09' 22.68''N 110 ^{0} 09' 48.02''E • The lighthouse daymark is easy to spot from distance of 5 nautical miles due to its large size and the usage of bright colour of red and yellow.
		Lighting System : • Main lighting system is using major rotating beacon model PRB-24 Mark 2 from Pharos Marine. • The main light is consist of six panels of light where each panel consist of two sealed beam light align in vertical formation • Three panel active and three panel standby. • Bulb used is 12volt 40 watts tungsten halogen bulb

		• The rotating beacon is rotate four rounds per minute (4 RPM) in order to achieve 3 flashes every 15s. • Fixed intensity is 1,500,000 candelas. • Tanjung Jerijih Lighthouse is the most powerful lighthouse in Sarawak in term of light intensity. • Distance of the main light is 22.2 nautical miles with effective intensity of 280,000 candelas. (Range in SAR 1 is 23nm at T=0.74).
	Tanjung Sirik	**Physical Structure :** • Steel tower structure with aluminium lantern house on top. • Height of tower is 36 meter from ground level and elevation is 38 meter from mean high water level. • Location 01^{0} 43’ 30.8’’N 110^{0} 31’ 12.6’’E
		Lighting System : • Main lighting system is using major rotating beacon model PRB-20 from Pharos Marine incorporate the former lighthouse cut and polish glass fresnel lens from Stone Chance. • The optic system is 500mm diameter cut and polish fresnel lens with two lens panels. • Bulb used is 100 watts metal arc (UVLA Lampchanger) and 100 watt for standby. • The rotating speed of the beacon is three revolution per minute which giving the flashing character of flash one every 10 seconds. • Fixed intensity of light is 597,839.9 candelas. • Distance of main light is 22.5 nautical miles at T = 0.74 with 280,106.7 candelas. For T=0.85 is 35.8nm. (Range in SAR 1 is 23nm at T=0.74) • The emergency light used is FA-250 flashing beacon from Pharos Marine. • Distance of the emergency light is only 8 nautical miles. This is due to flashing type beacon of shorter range compare to rotating beacon.
	Tanjung Kidurong	**Physical Structure :** • Steel tower structure with aluminium lantern house on top. • Height of tower is 21 meter from ground level and elevation is 90 meter from mean high water level. • Location 03^{0} 16’ 30’’N 113^{0} 03’ 49.2’’E • The lighthouse is not fitted with daymark.
		Lighting System : • Main lighting system is using major rotating beacon model ZL-HP3A from Zeni Lite of Japan. • The optical system is 300mm diameter concave reflector • The main light is consist of two tier Halogen Reflector Array • Bulb used is 12volt 40 watts tungsten halogen bulb • Fixed intensity is 725,000 candela. • Distance or main light is 22.2 nautical miles with effective intensity of 280,000 candela. (Range in SAR 1 is 23nm at T=0.74) • Light character is flash 2 every 15 seconds.
	Tanjung Lobang	**Physical Structure :** • Steel tower structure with stage on top to mounted the light. • Height of tower is 26 meter from ground level and elevation is 57 meter from mean high water level. • Location 04^{0} 21’ 44.13’’N 113^{0} 57’ 43.59’’E • The lighthouse is fitted with a daymark. • The new structure was built in 2006.

		Lighting System : • Rotating Beacon model TRB-300 from Tideland Signal • Lens size is 300 mm height with total of six lens panels. • Flasher lampchanger used is APCL-5 (60 watts max) the 6-bulb automatic lampchanger • The current bulbs used are 12v 36watt single filament type. • Flashing character is one flash every 10 seconds. • The current bulb used is 12V/36 Watt single filament type. • Distance of light with the flashing character is 4,304 candela at T = 0.74 is 12.2 nautical miles and at T = 0.85 is 17.9 nautical miles. (Range in SAR 1 is 16nm at T=0.74) • Distance of light with fixed intensity of 21333 candela is 15.9nm at T=0.74 and 24.1 at T=0.85
	Kuala Baram	**Physical Structure :** • Steel tower structure with aluminium lantern house on top. • Height of tower is 21 meter from ground level and elevation is 90 meter from mean high water level. • Location 03^0 16' 30''N 113 ^{0} 03' 49.2''E • The lighthouse is not fitted with daymark.
		Lighting System : • Main lighting system is using major rotating beacon model ZL-HP3A from Zeni Lite of Japan. • The optical system is 300mm diameter concave reflector • The main light is consist of two tier Halogen Reflector Array • Bulb used is 12volt 40 watts tungsten halogen bulb • Fixed intensity is 725,000 candela. • Distance or main light is 22.7 nautical miles with effective intensity of 329,000 candela (3 RPM). Range in SAR 1 is 23nm at T=0.74. • Light character is 3 flash every 20 seconds.

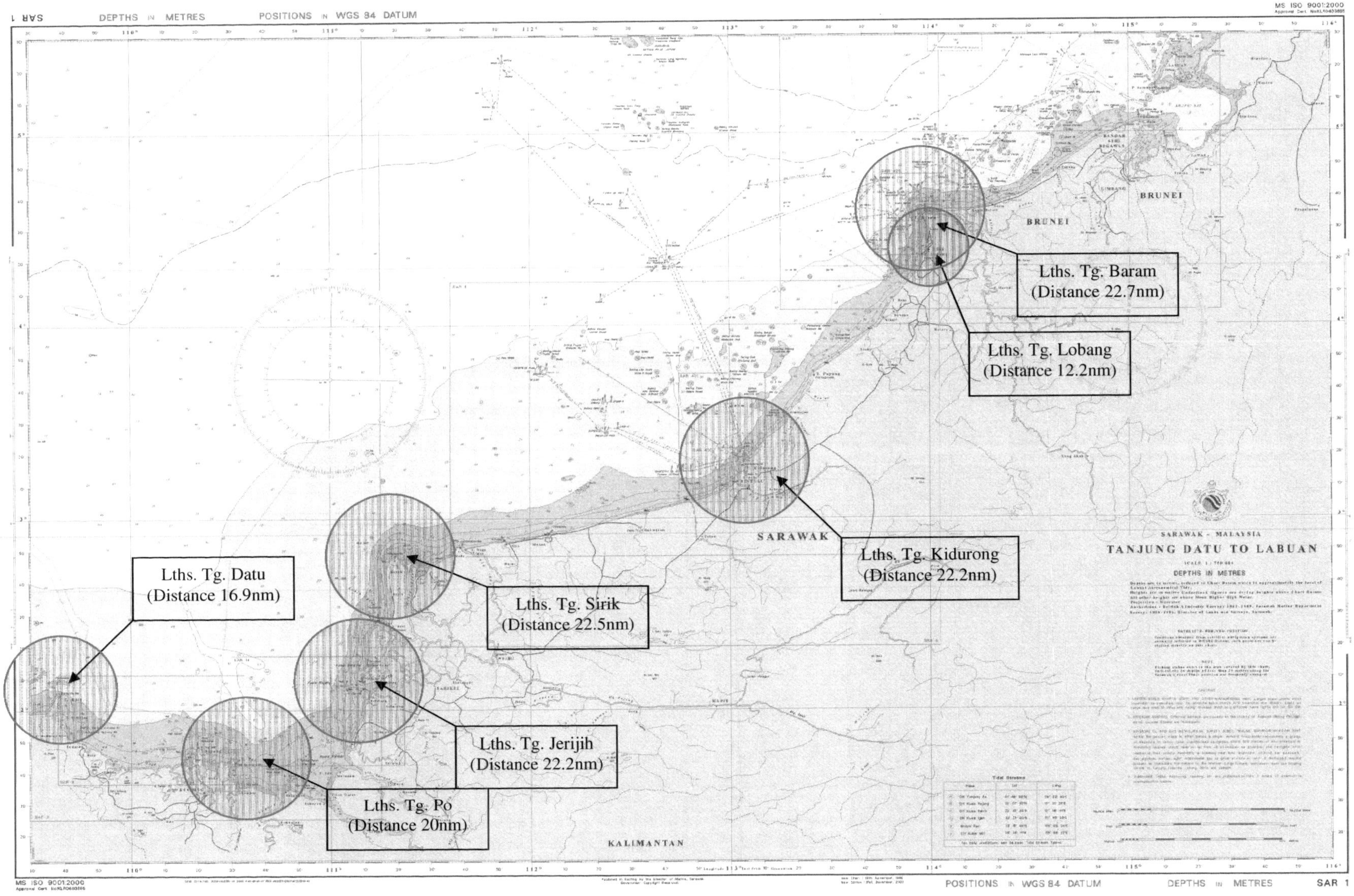

The Coverage of Lighthouses of Sarawak (Source: (Sarawak Marine Department, 2003)

The Sample of Commercial and Government Ships that Operate along the Coast or Ports of Sarawak

No.	Ship's Name	Port of Registry	Gross Tonnage	Type	Navigational Aids
1.	MV Sinaran Laut	Kuching	1,073 GT	Buoy Tender Vessel (Government Ship)	1. GPS 2. RADAR 3. Echo Sounder 4. Gyro Compass 5. AIS 6. Electronic Chart Plotter
2.	MV Bahtera Laut	Kuching	Approx. 700 GT	Oil Spill Response Ship (Government Ship)	1. GPS 2. RADAR 3. Echo Sounder 4. Gyro Compass 5. AIS
3	MV Permata Hidro	Kuching	Approx. 300 GT	Hydrographic Survey Vessel (Government Ship)	1. GPS 2. RADAR 3. Echo Sounder 4. Gyro Compass 5. AIS 6. ECDIS
4	MT Budi Mesra	Port Klang	6,921 DWT	Product Tanker	1. GPS 2. RADAR 3. Echo Sounder 4. Gyro Compass 5. AIS
5.	MT Budi Ikhlas	Port Klang	10,225 DWT	Product Tanker	1. GPS 2. RADAR 3. Echo Sounder 4. Gyro Compass 5. AIS
6	MV Borcos Fateh	Kota Kinabalu	225 GT	Utility /Fast Crew Boat	1. GPS 2. RADAR 3. Echo Sounder 4. Gyro Compass
7	MV Borcos Fadel	Kota Kinabalu	225 GT	Utility	1. GPS 2. RADAR 3. Echo Sounder 4. Gyro Compass
8	MV Borcos Firdaus 2	Kuching	238 GT	Utility	1. GPS 2. RADAR 3. Echo Sounder 4. Gyro Compass
9	MV Borcos Tasneem 7	Kuching	Approx. 1700GT	Offshore Support Vessel	1. GPS 2. RADAR 3. Echo Sounder 4. Gyro Compass 5. AIS
10	Puteri Class (5 ships) and Puteri Satu Class (6 ships)	Port Klang	88, 634 GT and 93,000 GT respectively	LNG Carrier	1. GPS 2. RADAR 3. Echo Sounder 4. Gyro Compass 5. AIS 6. ECDIS
11	Tenaga Class (5 ships)	Port Klang	80,346 GT	LNG Carrier	1. GPS 2. RADAR 3. Echo Sounder 4. Gyro Compass 5. AIS

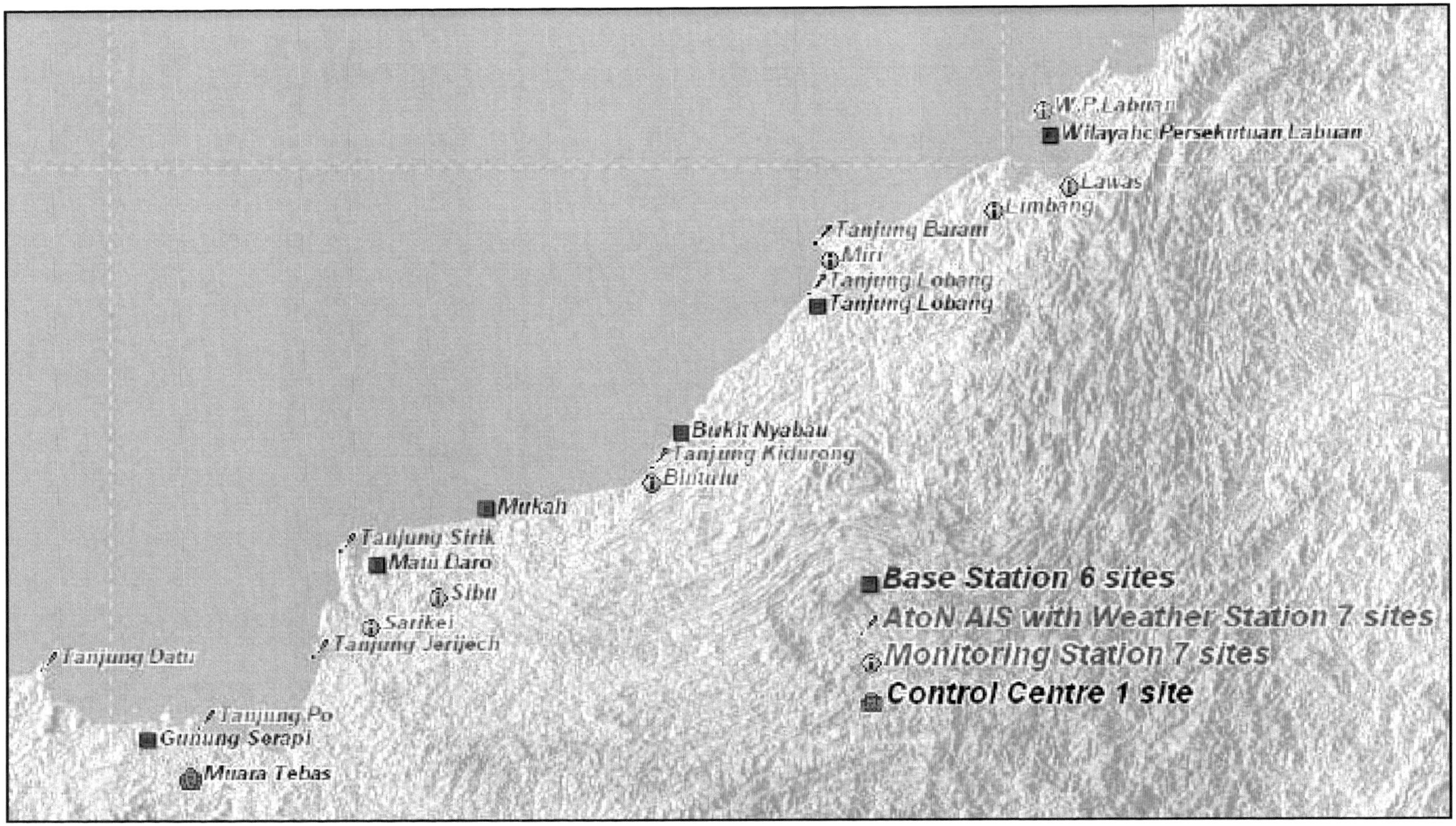

The Information of AIS System of Sarawak (Source: (Ahmad Fuad, 2010)

Annex V

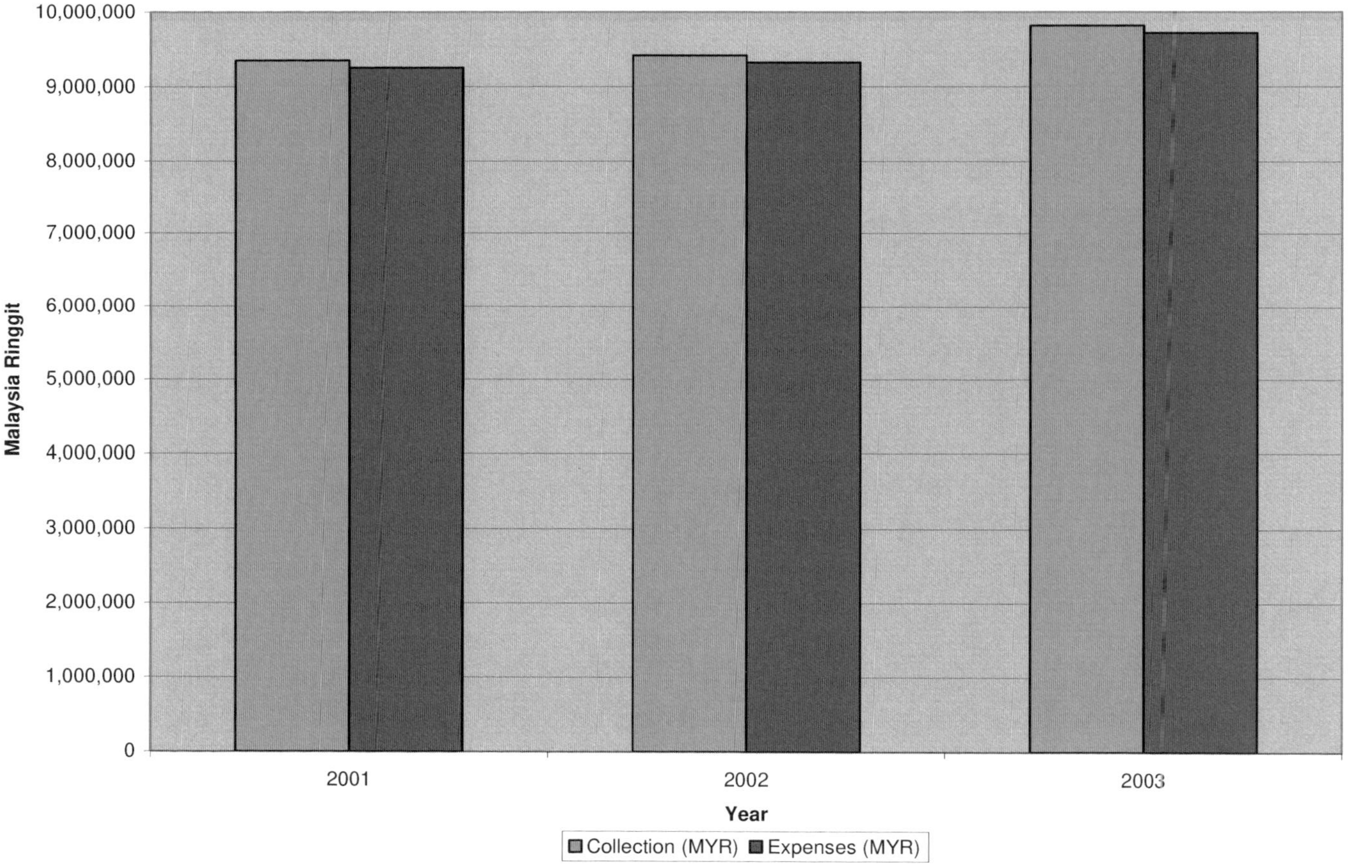

Collection and Expenses of Sarawak Light Dues (generated using data from Marine Department Malaysia)

Naval Landing Craft Stability

Peter Hayes[1], Warren Smith[2], Martin Renilson[3] and Stuart Cannon[4]
[1]DNPS, Department of Defence, peter.hayes5@defence.gov.au
[2]UNSW@ADFA, w.smith@adfa.edu.au
[3]AMC, University of Tasmania, mrenilson@amc.edu.au
[4]DSTO, Department of Defence, stuart.cannon@dsto.defence.gov.au

ABSTRACT

Landing craft have been operated by various navies around the world, including the Royal Australian Navy (RAN), for many decades to transport vehicles and personnel, often from ship to shore. The RAN assesses the stability of landing craft using contemporary naval stability criteria, perhaps unnecessarily penalising their design and operation. Over the years, multiple landing craft stability issues have arisen, with varying degrees of comfort felt for the solutions developed. The stability criteria that need to be applied to this type of craft are currently being investigated through various research programs sponsored by the Defence Science and Technology Organisation and the Directorate of Navy Platform Systems.

One of these research programs involves extensive captive model tests in beam waves together with a parallel ship motions and stability simulation analysis. Other factors, such as wind heeling and water ingress over the bulwark, are also being examined. The ultimate aim of this research program is to derive intact stability criteria that are based on physical principles while being suitable for application by the mainstream naval architectural community.

INTRODUCTION

The RAN assesses the stability of landing craft using contemporary naval stability criteria. Current stability requirements are provided in DEF(AUST)5000 Vol3 Pt2 (2003) which is presently being revised. Over the years, multiple landing craft stability issues have arisen, with varying degrees of comfort felt for the solutions developed. There is a push that commercial stability standards should suffice and thus avoid naval "gold plating". The stability criteria that need to be applied to this type of craft are currently being investigated through various research programs by the Defence Science and Technology Organisation (DSTO), the Directorate of Navy Platform Systems (DNPS), the University of New South Wales college at the Australian Defence Force Academy (UNSW@ADFA) and the Australian Maritime College (AMC), with the outcomes now contributing to a collaborative research project between the UK and Australia.

This paper presents a discussion of some intact stability issues that can apply to landing craft in extreme environments. The paper also outlines a research program that aims to determine the effects of landing craft parameters on the capsize behaviour of these craft, presenting results to date. The ultimate aim is to derive stability criteria for landing craft that are based on physical principles while being suitable for application by the mainstream naval architectural community.

STABILITY RESEARCH

Existing stability criteria are based on the still water characteristics of the vessel, incorporating factors to account for operation in severe environments. Some, such as the basic International Maritime Organisation (IMO) criteria, require nominated characteristics of the righting arm curve, including minimum areas under the GZ curve and minimum GM values. These were based on earlier work, such as that by Rahola (1939). Because these types of criteria have been derived empirically they are strictly only applicable to the data set and the environments used in their derivation. Yet these criteria have been extended to many vessel types and sizes not in the original data set, and to environments markedly different and more severe than those original environments.

Quasi-static criteria have been introduced in more recent decades and these attempt to include the effects of wind and waves as overturning forces to be resisted. In these criteria, wave effects are usually introduced to the still-water righting arm curve by a 'roll-back' angle. Wind effects are introduced by determining the upright wind heeling lever. International Maritime Organisation (IMO) criteria apply a constant wind heeling lever whereas naval criteria, first presented by Sarchin and Goldberg (1962), vary the upright wind-heeling lever by $\cos^2\Phi$ where Φ is the heel angle. Few, if any, criteria allow for the effects on stability of the entry of green seas into wells.

The wind/wave criteria do not address other forms of instability in waves such as may result from a broaching event. There is now considerable research being conducted to include more aspects of dynamic stability. The Cooperative Research Navies (CRN) project, of which Australia is a member, is developing the ship motion and dynamic stability simulation tool, FREDYN. Use of this type of tool, in conjunction with the results of model and full scale trials as benchmarking tests, will ultimately allow more comprehensive stability criteria to be derived than those currently available.

DEFINITIONS

Any discussion of quasi-static stability needs to deal with definitions of areas under and above the GZ or righting arm curve. These are illustrated in Figure 1.

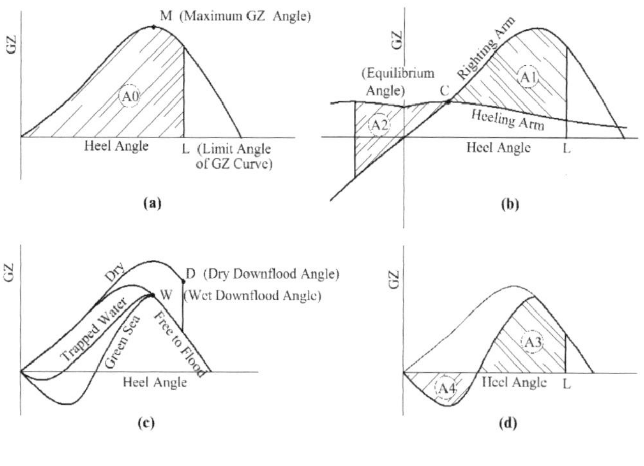

Figure 1: GZ Curve Area Relationships

The limit angle of the GZ curve, L, could be a downflooding angle or some other limiting criterion that may be applied.

The GZ curves of Figure 1(c) consist of:

- Dry — There is no water in the well.
- Free to Flood — Water is assumed where the deck is below the external waterline.
- Trapped Water — Water trapped in the well as the bulwark emerges after a large heel.

- Green Sea — The well is completely filled by a green sea at upright and sheds water as the landing craft heels.

The downflooding angles of Figure 1(c) are:

- Dry, D — The angle at which the bulwark inner edge touches the water, assuming that up till then the well is dry.
- Wet, W — The angle at which the bulwark inner edge touches the water, assuming that the well is free to flood. The wet downflooding angle will be less than the dry downflooding angle as the vessel is effectively heavier and therefore has less freeboard.

The areas that have been defined in Figure 1 consist of:

- A0 — Area under the righting arm curve and above the heel axis, extending from the initial heel axis intercept to the limit angle, L.
- A1 — Area under the righting arm curve and above the heeling arm curve, extending from the equilibrium angle, C, to the second intersection or limit angle, L, whichever is less. Approximately proportional to the energy restoring the vessel to equilibrium.
- A2 — Area under the heeling arm curve and above the righting arm curve, extending back from the equilibrium angle a nominated rollback. Approximately proportional to the energy disturbing the vessel from equilibrium.
- A3 — Area under the righting arm curve and above the heel axis, extending from the angle of loll to the limit angle. Conceptually the same as A0 but uniquely identified to avoid confusion. Approximately proportional to the energy restoring the vessel to equilibrium.
- A4 — Area above the righting arm curve and below the heel axis, extending from zero to the angle of loll. Approximately proportional to the energy gained in rolling from upright to the angle of loll.

STABILITY OF VESSELS HAVING WELLS

Landing craft typically have a large well area formed by the bulwarks and ramp(s). What are the effects if a large green sea were to be taken onboard in rough sea conditions? Typically, there will be a number of baffled air intakes/exhausts situated on the inside of the bulwark. Should the vent sill be used to define the downflooding point for intact stability analysis? Alternatively, can the downflooding point be taken as the top of a full-length bulwark — this is the approach seen in some contemporary stability criteria.

These situations have been investigated for a simple rectangular barge shape with bulwarks. A spreadsheet was developed that allowed key dimensions to be varied and determine the static GZ curves defined in Figure 1(c). The green-sea case sheds water as it heels from the upright. An example of the output, analogous to an existing landing craft, is presented in Figure 2.

It can be seen that the 'dry' downflooding angle exceeds the 'wet' downflooding angle, due to the greater weight of the craft with trapped water and the correspondingly reduced freeboard. The volume trapped on return to upright is not that large, about 17% of the total well volume in this case, and results in an angle of loll of about 8 degrees. If a green sea were to completely fill the well, Area A3 far exceeds Area A4. This vessel would very likely survive a green-sea event.

The spreadsheet allows the effects of changing parameters to be examined. For example, what would happen were the bulwarks only plate with no buoyant volume as shown in Figure 3? The situation is drastically worse. The angle of loll from the trapped water is excessive at about 17 degrees and Area A4 exceeds Area A3. If a green sea were to fill the well, the vessel would immediately capsize. If the 'dry' GZ curve to downflooding was being used to assess stability and flooding occurred, the vessel is very close to switching to a negative GZ value.

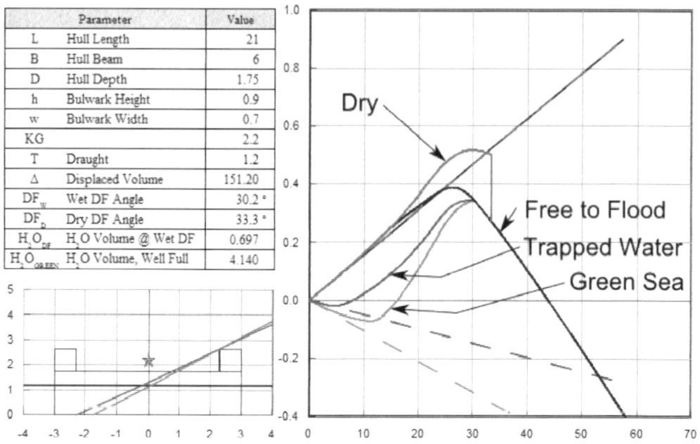

Figure 2: Landing Craft with Buoyant Bulwarks

These results suggest some stability criteria:

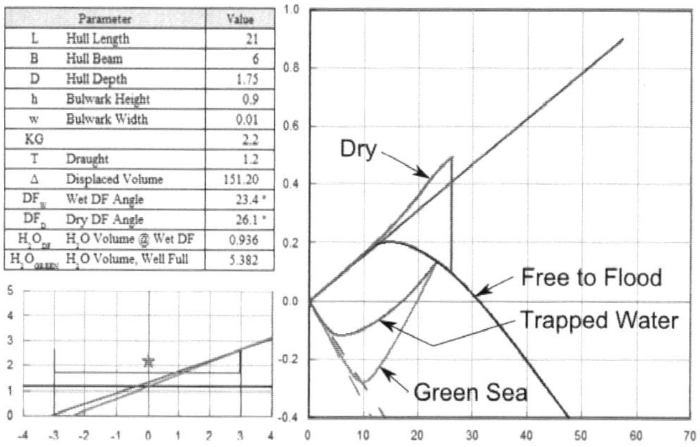

Figure 3: Landing Craft with Plate Bulwarks

- If the dry GZ curve is to be used for a stability assessment, then it must be terminated at a suitable angle that assures a measure of safety. At the very least, this can be no more than the angle of vanishing stability of the free-to-flood GZ curve. It may be more sensible to limit the dry GZ curve to the wet downflooding angle.
- If the vessel is to be operated in an environment where a green-sea event is possible, Area A4 must be less than Area A3 by a suitable margin to allow for dynamic effects. This should be determined by some future study but, noting that it is easier to subsequently relax a criterion, in the interim a value of 50% is suggested.

INTACT STABILITY EXPERIMENTAL PROGRAM

Figure 4: Generic Models

Hayes (2010) presented an intact stability experimental program based on a series of three models with different freeboards, fitted at times with a bulwark making an enclosed well. The models are shown in Figure 4. The intention was to experimentally determine roll responses to waves at different values of freeboard and KG. The models were 810 mm long, 250 mm wide and had a draught of 42 mm. The low depth model (50 mm depth) had a freeboard to the deck of 8 mm. The other two models had depths of 70 mm and 90 mm. Three KGs were tested, ranging from 76 mm to 116 mm. In an experimental series, conducted in the towing tank at AMC, Launceston, in 2009, the models were tested in steep regular beam waves at the low KG. A second experimental series was conducted in 2010, also in steep regular waves. This utilized the same models but at the two higher KGs. A report of this second experimental series is in preparation.

Note that these models were intended to be generic rather than representative of existing designs. At a scale of 1:26, the 70 mm depth model approximates a contemporary RAN design. Where full scale or full size behaviour and results are discussed in this paper, a scale of 1:26 applies.

Significant data has been collected over the two experimental programs. Almost all of it is roll and heave response of models in regular waves. Wave heights were large for the model, corresponding to 1.3 m to 3.9 m wave heights for a full scale 21 m vessel. The models were tethered by a constant force mooring to limit sway, surge and yaw.

An examination of maximum roll angles versus wave heights has revealed some tentative trends:

- Maximum roll angles tend to be similar between the various model configurations, provided that the freeboard (to top of bulwark or to the deck depending on configuration) was about equal to or larger than the wave amplitude.
- Maximum roll angles are reduced if freeboard is less than the wave amplitude, usually coinciding with some water on deck.
- The maximum roll angle is a function of KG, reducing for higher values. This was most noticeable at the larger wave amplitudes, where the double roll amplitude ranged from 38° to 34° for the highest freeboard case.

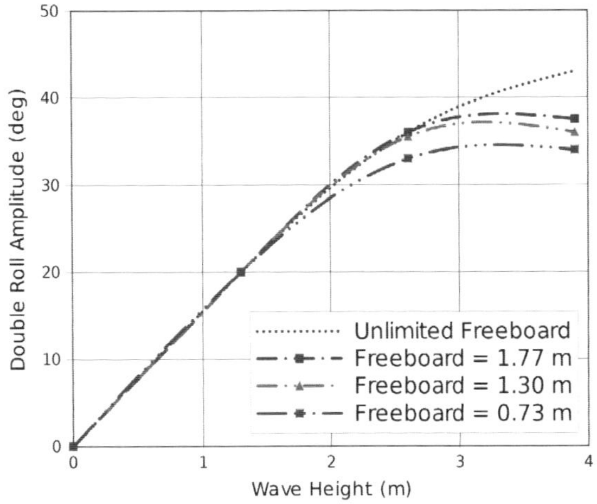

Figure 5: Maximum Roll Angle as a Function of Wave Height, Low KG Case

Figure 6: Arduino Based Motion Logging System

The approximate relationship for the low KG case is shown in Figure 5. The 'Unlimited' curve is imaginary, the other curves are labelled by freeboard value. There was not enough data collected to provide a definitive relationship, but this certainly suggests an avenue for future research.

DNPS conducted a model experimental test program in 2011 using an accurate scale model of an RAN landing craft. The experiments were performed in much the same way as the earlier experiments — the model was tethered. Significant water passed over the bulwark at times. An Arduino based motion logging system, shown in Figure 6, was being developed by DNPS at the time and was trialled during this experimental series. Figure 7 shows a comparison between towing tank data and the Arduino based data-logger results for a roll decay test. The results are almost identical except for a short period while the data logger wrote to the memory card. This problem is currently being addressed.

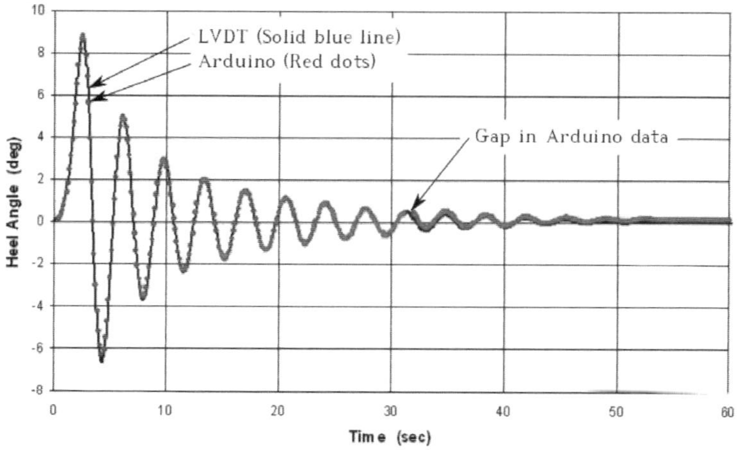

Figure 7: Roll Decay Data Comparison

A tethered run that exhibited significant water on deck was repeated with the model free to drift. There was now no water on deck and motions were easier. No doubt this was partly due to changing the encounter frequency due to sway (drift), but it was felt that removing the tethering was also a factor. This will be explored in an experimental test program, programmed for February 2012.

The full size GZ curves for all the model configurations are presented in Figure 8. The various model configurations are identified as follows:

- 50/70/90 = Actual model hull depth in mm (corresponding to hull depths of 1.30 m, 1.82 m and 2.34 m respectively at 1:26 scale)
- H = Hull only
- BC = Bulwark with closed freeing ports fitted (bulwark was 45 mm (1.17 m) high by 25 mm (0.65 m) wide each side, closed at each end).

The GZ curves with a shaded background capsized or it was considered that they would capsize (the 50H model at the highest KG was not tested as it capsized at the lower KGs). Of note is that if the free-to-flood GZ curve exceeded 0.2 m — a basic IMO criteria — the vessels did not capsize. This was despite being exposed to relatively large seas, resulting in large roll angles and significant water on deck and into the well. This may be a coincidence — the models were tethered and only tested in regular beam waves, which are not necessarily a good representation of realistic wave conditions. Also to be noted is that the only model with a bulwark forming a well that failed the proposed green-seas criterion capsized.

Every model, excluding the 90H with bulwark which was not tested, experienced water on deck and into the well for some combinations of wave height and frequency. If the freeboard was less than about the wave amplitude, there was significant water on deck at higher frequencies. If the freeboard was greater than about twice the wave amplitude, then there was no water on deck. The low KG models experienced more water on deck than the high KG models. This is possibly a function of the natural roll period — lower KG resulted in shorter natural periods which would, in turn, mean steeper waves at the vessel's natural roll period.

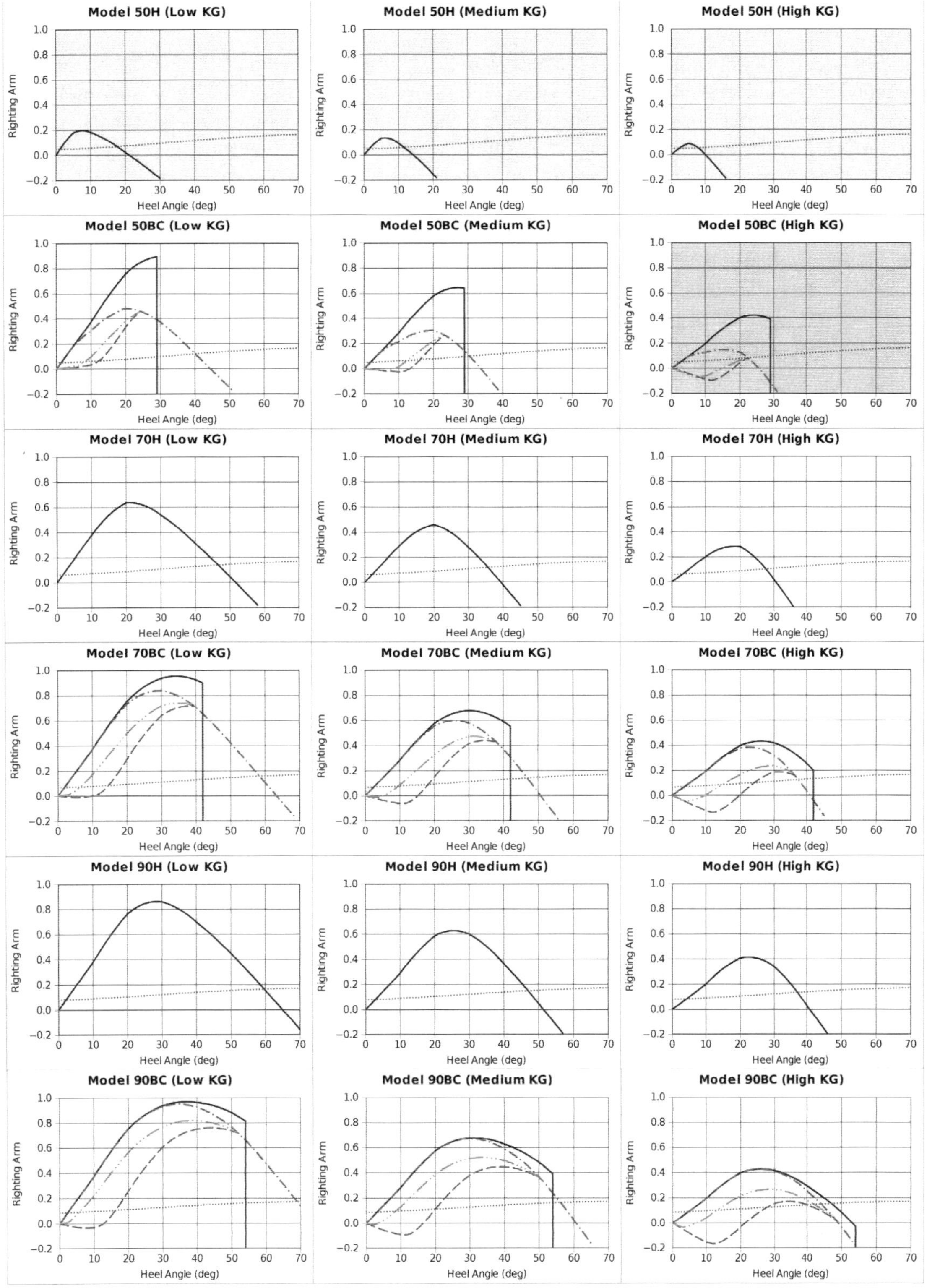

Figure 8: GZ Curves at Full Scale (26:1) of all Model Configurations

FREDYN SIMULATION OF THE MODEL RESULTS

Two weeks of model experiments provided data for only a limited subset of all possible freeboard/KG combinations, let alone other length/beam ratios and so on. The intention was to conduct roll simulation studies using FREDYN and compare to the model data. If adequate correlation could be obtained, FREDYN could be used to fill in the data set.

The hulls were modelled in Rhino at full size. The Rhino definition was then used to develop the FREDYN data. In the case of models fitted with a bulwark, the hull was modelled solid to the top of the bulwark and the well modelled as a series of tanks. The well was subdivide into four equal sections longitudinally, and two transversely. Openings to the sea were defined on the top surface of all tanks and each common tank boundary was defined as an opening. This scheme was adopted to provide for some measure of progressive flow in the FREDYN flooding module.

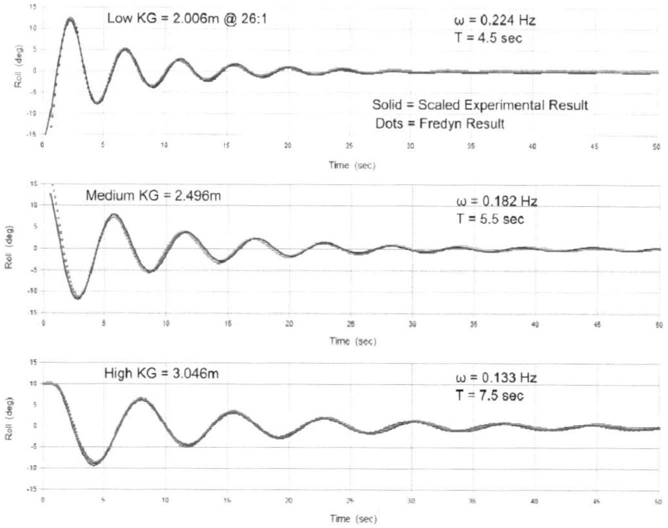

Figure 9: FREDYN/Model Roll Decay Comparisons

The roll damping coefficients were determined by trial, achieving a very good correlation between FREDYN and the roll damping experiments, as shown in Figure 9.

The tethering arrangement was modelled in FREDYN by two methods:

- Sway and yaw fixed to simulate that the model was largely restrained in these degrees of freedom.
- Mooring lines from the model to fixed positions corresponding to the tank sides, with the spring constant of the mooring lines selected by trial to match observed deflections as best as possible. Note that this is different to the tank arrangement, where lines passed through pulleys to suspended weights, creating fixed force mooring lines of variable length.

FREDYN was run at the scaled wave heights and periods corresponding to the experimental data. This was done for the sway/yaw and mooring line cases, and also with yaw fixed to simulate a model free to drift but kept beam on to the waves.

The results obtained so far are mixed. In many cases, especially those involving little or no water on deck, FREDYN roll motions are very close to the observed model motions. In some cases the FREDYN results are dissimilar. Generally, FREDYN seems to be doing a reasonable job of simulating the roll motions of a landing craft in regular waves but tends to be conservative — larger roll angles and more capsizes are predicted. However, further analysis and checking of the data needs to be completed before any definitive conclusions can be made.

WIND HEELING

Wind heeling is included as a capsizing force in contemporary wind/wave criteria. Where this is a significant factor, the wind heeling lever versus heel-angle function applied must be appropriate. The original naval stability analysis reported by Sarchin and Goldberg (1962) assumed a $\cos^2\Phi$ function, as if the ship were a thin plate. The results of the analysis of the (mainly) WW II data led to an A1/A2 ratio of 1.4 as being appropriate — refer to Brown and Deybach (1998). The IMO criterion assumes a simple straight-line function and requires an A1/A2 ratio of 1.0. Neither function is necessarily correct — it is clear that ships are not thin plates and that they will float on their side with an exposed windage. Brown and Deybach (1998) proposed that the wind-heeling arm decrease from the upright value in a cosine function to a minimum with the ship lying on its side with half the hull submerged. However, ships typically float higher than half the beam. If the ship is considered as a simple rectangle, it can be shown that the heeling arm at 90° becomes:

$$WHA_{90} = \frac{\frac{1}{4} C_D \rho V^2 L B^2 \left(1 - \frac{d}{D}\right)}{1000\, g\, \Delta} \qquad (1)$$

where:
- WHA = Wind Heeling arm, m (subscript is the heel angle)
- C_D = Drag Coefficient
- g = acceleration due to gravity, m/s^2
- ρ = Air density, kg/m^3
- V = Wind speed, m/s
- Δ = Ship displacement, t
- L = Ship length, m
- B = Ship beam, m
- D = Ship depth, m
- d = Ship draught upright, m

A number of existing ships and a model used to experimentally determine the wind heeling moment in a wind tunnel (Khee 2010) were analysed by projecting the silhouette at various angles onto the vertical plane and developing the wind heeling lever. The waterline was determined to give the correct displaced volume and trim was held constant. The results are presented in Figure 10.

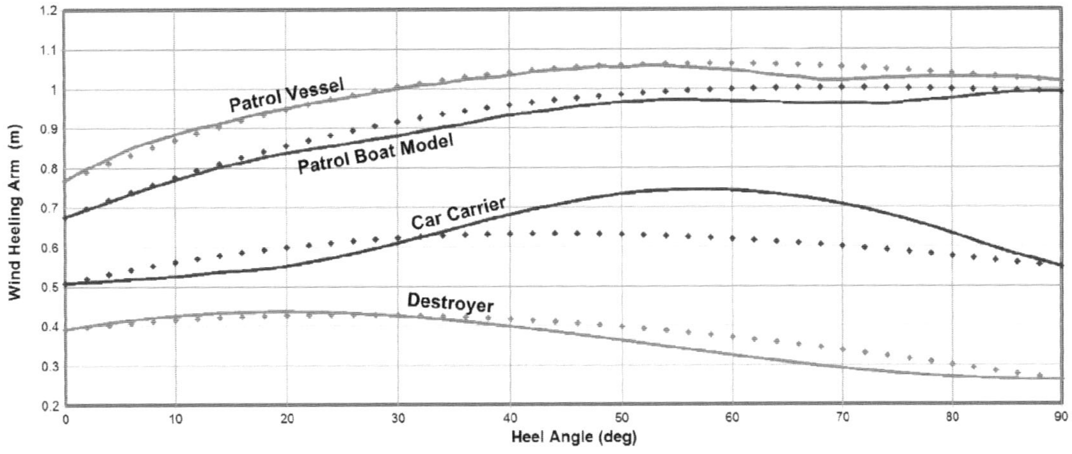

Figure 10: Wind Heeling Relationships

The best fit equation was found by trial to be:

$$WHA_{\phi} = WHA_0 \cos(\phi) + \frac{WHA_{90}\,\phi}{90} \qquad (2)$$

where: WHA = Wind Heeling arm, m (subscript is the heel angle)
□ = Heel Angle (deg)

In Figure 10, the solid line is the analysed wind heeling lever and the dots are the equation fit.

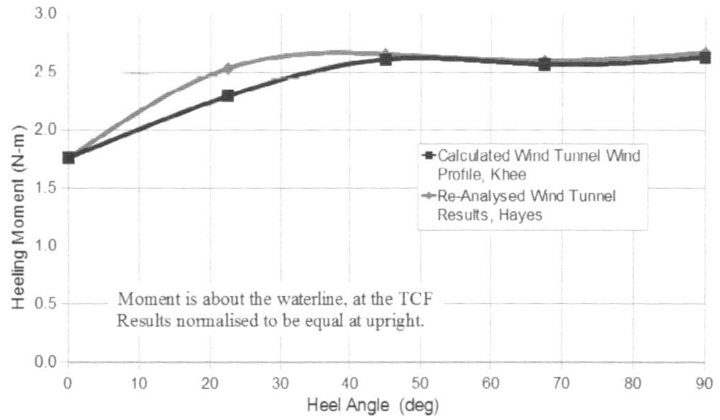

Figure 11: Wind Tunnel Correlation

Khee (2010) presented wind tunnel results of a patrol-boat-shaped model heeled from 0° to 90°. The data was reanalysed separately and a good correlation to the basic silhouette analysis method obtained. This is shown in Figure 11 for wind heeling moments about the waterline. The basic trends of Figure 10 are shown by the wind tunnel experiments.

The work to date suggests that the wind heeling lever verses heel angle relationship should be developed by wind tunnel experimentation. Failing this, an approximate relationship based on the 0° and 90° heel levers has been developed. Further work is required to determine how this revised wind heeling function would be used in wind/wave-type criteria. Ideally, the original data used in developing the naval criteria of Sarchin and Goldberg (1962) should be re-examined.

CONCLUDING COMMENTS

Dynamic effects can lead to a vessel surviving waves when, statically, the vessel has very limited stability. This happened with the low-freeboard models. At the low KG, the vessel withstood a number of wave heights and frequencies, even though at all times water flowed freely across the deck. However, depending on dynamic effects, especially if based on the limited subset examined in the experimental studies, would seem risky. There are also static situations, such as passenger crowding and off-centre cargo, that need to be considered.

The model experiments suggest that:

- the IMO criterion of the minimum value of the GZ curve of 0.2 m should apply to the free-to-flood GZ curve;
- the green-seas criterion needs to be applied if freeboard is less than the wave height (double wave amplitude), at least for the length/beam ratio examined; and
- model experiments, when determining behaviour in more extreme waves, perhaps should be performed free-to-drift rather than tethered.

The maximum roll angles experienced suggest that a roll-back angle of less than the naval value of 25° could be applied. There was a definite relationship of maximum roll angle versus wave height from the experimental data. This needs to be checked against the IMO roll-back formulation, as well as against the variations to the IMO roll-back formulation more recently proposed. A preliminary examination suggests that the basic IMO formulation incorrectly determines the natural roll frequency, especially for higher KG cases.

The FREDYN modelling showed promise, but further work is needed before conclusions can be made. This could become a valuable tool for extending the currently limited model-derived data set.

The Australian Department of Defence, along with ADFA and AMC, is actively pursuing research into landing craft stability. This paper presents results from a postgraduate research program that have highlighted some interesting issues that need to be researched further. The ultimate aim of the program is to significantly advance the science of ship stability in general and of landing craft in particular, and to develop more-rational stability criteria applicable, in the first instance, to landing craft.

REFERENCES

1. DEF(AUST)5000 ADF Maritime Material Requirements Set, Vol 3 Pt 2, *Stability of Surface Ships and Boats*, Issue 2, May 2003.
2. Hayes, P. 2010, *Landing Craft Stability Model Testing Report AMC February 2009*, Report A017159, Directorate of Navy Platform Systems, Department of Defence, Canberra.
3. *FREDYN: A Computer Program for the Simulation of a Steered Ship in Extreme Seas and Wind*, Version 10.2, 2010, CRNAV, Maritime Research Institute Netherlands.
4. Rahola, J. 1939, *The Judging of the Stability of Ships and the Determination of the Minimum Amount of Stability*, doctoral thesis, The University of Finland, Helsinki, Finland, May.
5. Sarchin, T.H. & Goldberg, L.L. 1962, 'Stability and Buoyancy Criteria for U.S. Naval Surface Ships', *SNAME Transactions*, vol. 70.
6. Khee, C. H. 2010, *Variation of Wind Heeling Moment with Angle of Heel of Ships*, undergraduate thesis, The university of New South Wales, Sydney, Australia, October.
7. Brown, A. J. & Deybach, F. 1998, 'Towards a rational intact Stability Criteria for Naval Ships', *Naval Engineers Journal*, January, pp. 65-77..

Developing Australia's Indigenous Marine Vehicle Manoeuvring Analysis and Evaluation Capability

E. Dawson[1], B. Morris[1], J. Duffy[2]
1 Defence Science and Technology Organisation, Melbourne, Australia
2 National Centre for Maritime Engineering and Hydrodynamics, Launceston, Australia
Edward.Dawson@dsto.defence.gov.au

ABSTRACT

The manoeuvrability and control of marine vehicles, whether commercial or naval platforms, is critical to their safe, efficient and effective operation. It is a regulatory requirement of the International Maritime Organization that specific craft meet prescribed manoeuvring performance criteria to mitigate the risk of collision and catastrophic failure. In-harbour operational efficiency in the marine transport sector is a significant financial driver, while Naval war-fighting doctrine demand that surface combatants can out-manoeuvre and out-perform the myriad of present day and future threats to enable them to maintain their dominance as a naval deterrent.

This poses the question: are we as a nation in a position to assess and understand the manoeuvring performance of marine vehicles to the level required by the statutory, commercial and defence stakeholders? If not, then what would it take to bridge that capability gap and is an investment into the development of an indigenous capability the best solution?

To date, science and industry have developed methods to analyse and evaluate the manoeuvring and control characteristics of marine vehicles. These range from physical experimentation to systems identification and more recently, advanced numerical simulation. The abilities and limitations of these methods are generally well understood such that they can be appropriately implemented.

In this paper, systems thinking is utilised to explore the complex issue of how to develop an indigenous marine vehicle manoeuvring analysis and evaluation capability. The strategic and technical aspects surrounding this capability, along with some of the risks involved, are assessed and discussed. Concept solutions to fulfil the capability's requirements are proposed with the aim to provide the best outcome based on the needs of the maritime industry and its customers.

INTRODUCTION

The manoeuvring and control of marine vehicles, whether commercial or naval platforms, is critical to their safe, efficient and effective operation. It is a regulatory requirement of the International Maritime Organization (IMO) that specific craft meet prescribed manoeuvring performance criteria to mitigate the risk of collision and catastrophic failure with the aim to safeguard human life, maritime infrastructure and the marine environment. In-harbour operational efficiency in the marine transport sector is a significant financial driver, while Naval war-fighting doctrine demand that surface combatants can out-manoeuvre and out-perform the myriad of present day and future threats to enable them to maintain their dominance as a naval deterrent.

The prediction and evaluation of marine vehicle manoeuvrability and control resides as a discrete component within the traditional design cycle, such as that presented by Rawson & Tupper [1]. However, this is a component of the design process that has only recently been established as an area for calculation and analysis. Similarly, it is only in relatively recent times that regulatory bodies began to mandate that a large portion of the world's fleet meet specific manoeuvring performance criteria. Traditionally, the design of a ship's steering and control system was left to the shipbuilder who would use empirical methods incorporating simple ship parameters. In many instances this resulted in ships with poor handling characteristics and course keeping ability [2].

Today, mathematical based manoeuvring models of varying complexity are employed to simulate the performance of marine vehicles for the purpose of research, design, certification and operator training. The development of these models has been realised through the advancement of experimental and numerical techniques in combination with full-scale trials and systems identification methods. The application of these models has provided significant benefits to the commercial and defence maritime sectors. Perhaps their greatest return has been to contribute to the reduction of the inherent risk and cost of designing and operating marine vehicles.

Given the current state of manoeuvring analysis technologies and the importance of fit for purpose design and safe operation, the question begs: does a comprehensive manoeuvring analysis and evaluation capability exist within Australia to enable analysis at various stages of a platform's lifecycle and does it successfully deliver to the needs of the in-country commercial and defence maritime industries? At first glance it appears that such a capability does not exist and while some of the industry's needs are being met, these are oftentimes done so through the use of overseas service providers and at significant cost. However, while we do not currently have a mature capability we do have the capacity within Australia to develop one. Now the question becomes: is there a need to develop an indigenous capability and what would it look like?

To answer this question a systems thinking methodology is employed to facilitate structured problem identification. Fundamental areas of Australia's maritime industry are reviewed at a high level to identify their existing and future needs in the context of marine vehicle manoeuvring analysis and evaluation. Concept solutions are developed based on the inherent dysfunctions identified during the study. These solutions are assessed using a comparative risk assessment approach.

PROBLEM IDENTIFICATION METHODOLOGY

Systems thinking is said to have emerged in the 1940s due to the inability of mechanistic, or reductionist, thinking to adequately explain the emergent properties of biological organisms which couldn't be derived from the sum of their parts [3]. Systems thinking formed the basis of the classical systems engineering methodology, which has been applied successfully to real-world technical problems such as a production system. These applications came to be known as "hard" systems. Later, several authors (see for example [3], [4] and [5]) adapted systems thinking to management science and in particular, the solution of complex or disorganised problems. These problems can include social, economic and technological aspects and have come to be known as "soft" systems.

In the present work, the development of an indigenous marine vehicle manoeuvring analysis and evaluation capability is seen to be a soft system due to the social and technological aspects involved. The social aspects include the differing concerns of the various stakeholders, such as the marine industry, regulatory bodies, Defence and researchers, all of whom have an interest in the capability's development. This has led to the selection of Hitchins' Rigorous Soft Method (RSM) approach to addressing complex problems. RSM is not domain specific and can be used by both individuals and groups [4].

RSM utilises a process of inquiry that is similar to the approach a General Practitioner may use to diagnose a patient's illness [4]. In essence, RSM can be seen as a diagnostic procedure that begins with the identification of a problem, or soft system, and then interrogates the system to identify its inherent deficiencies. Remedial concepts or solutions can then be developed in response to the identified deficiencies. Finally, the proposed remedies are assessed to verify whether they can successfully resolve the deficiencies without adversely affecting the system. The RSM diagnostic procedure is shown in Figure 1. The RSM diagnostic procedure presented by Hitchins [4] concludes with the identification of remedies or solutions that address the implicit deficiencies. Equipped with this structured body of information the solution development and implementation phase can be initiated.

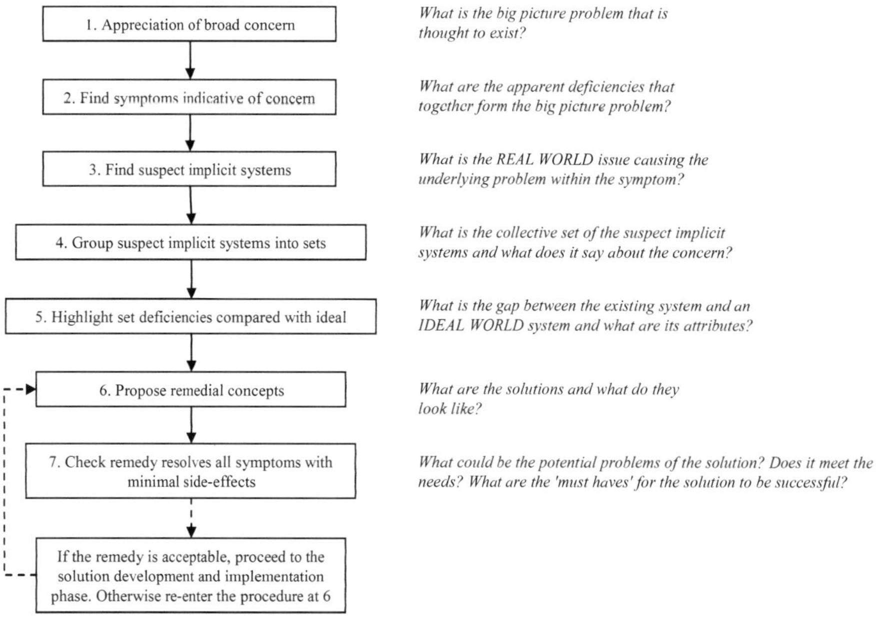

Figure 1: RSM diagnostic procedure and questions posed by the Authors of this paper for each stage [4]

Appreciation of Broad Concern

The broad area of concern for our soft system of interest is that Australia does not have a comprehensive indigenous marine vehicle manoeuvring analysis and evaluation capability. Such a capability would be of significant benefit to the Australian maritime sector, both commercial and defence, as it would promise the opportunity to furnish the industry's data needs at a reasonable cost, with a low level of risk and to the level of quality required.

Symptom Identification

The existing Australian maritime sector was reviewed to identify symptoms that exemplify the area of concern. The following technology areas within the sector were found to present symptoms that reflect the area of concern and merit further investigation and analysis:

- Ship Handling Training Simulators;
- Commercial Fast Ferry Industry;
- Australian Maritime Defence Acquisition;
- Maritime Science and Research; and
- Navy Materiel Certification.

Although the scope of the investigation is not exhaustive, it does consider a significant portion of the industries and organisations within this sector that engage in marine vehicle manoeuvring evaluation and simulation. The symptoms that were identified indicate that there is a significant dependence on accurate manoeuvring analysis and consequently the prediction or measurement of platform specific hydrodynamic forces and coefficients. The symptoms are presented in Table 1.

Technology Area	Symptoms Indicative of Concern
Ship Handling Training Simulators	A significant amount of resources is needed to develop simulation models to the level required for the provision of operator training. This includes platform specific hydrodynamic coefficient data. This data is often produced by conducting full scale trials. For some cases there can be issues related to the availability of data required to populate ship handling simulation models, in particular the cost, time and effort required to establish a complete and accurate data set. Ship handling simulators are also used to conduct feasibility studies of a specific ship operating in a particular port. The data requirements for such studies are very intensive. These studies may be a pre-cursor to actual trials, and in cases where the vessel is not yet built, the hydrodynamic data must be predicted. This data may support the claim of compliance with regulatory criteria and proposed design requirements, however, it is difficult to verify or validate the claim.
Commercial Fast Ferry Industry	Analysis and evaluation of fast mono-hull and multi-hull marine vehicle manoeuvring and control is important for determining the performance of safe and effective in-harbour handling and operation at sea. There are also regulatory implications invoked by the prevailing High Speed Craft Code [6] that require a vehicle's performance to be quantitatively verified prior to in-service operation. These evaluation techniques require hydrodynamic coefficient data that are specific to each platform. A complicating factor is that high-speed multi-hull hydrodynamics may diverge from the traditional empirical methods used to predict displacement monohulls. Also, accurate prediction of low-speed manoeuvres such as berthing are important to minimise turn around times.

Technology Area	Symptoms Indicative of Concern
Australian Maritime Defence Acquisition	A manoeuvring capability could be (but is not currently) used to support Defence's role in developing platform/system operational performance requirements and evaluation of platform concepts. The RAN operates a diverse range of platforms from landing craft to multi-hulls, frigates to fleet auxiliaries and submarines.
Maritime Science and Research	Numerical simulation of marine vehicle hydrodynamic performance is an invaluable research tool. However, simulation of novel hull forms and operation in extreme conditions require the mathematical models and their output to be verified and validated before any quantitative outcomes can be realised. Simulation of marine vehicles within a multi-system simulation, such as multi-platform or platform-weapon interaction, requires the manoeuvring model to be populated with accurate and reliable hydrodynamic coefficients.
Navy Materiel Certification	The Royal Australian Navy (RAN) publishes the DEF(AUST) 5000 Materiel Requirements Set which contains requirements for the manoeuvrability of its platforms. In some of the requirements, such as course keeping, empirical equations have been used as the basis for assessing whether the platform can meet the requirement [7]. Deliverables for platforms assessed against these requirements include numerical simulations as a precursor to a full scale trial.

Table 1: Symptoms indicative of concern

Suspect Implicit Systems

Further investigation of the identified symptoms of concern leads to their root cause or suspect implicit systems. Each of the symptoms representing the five technology areas was interrogated to elicit the underlying cause or causes. The suspect systems are presented in Table 2. In this context, the term 'data' represents hydrodynamic coefficients along with the associated information required to furnish a complete manoeuvring model.

Technology Area	Suspect Implicit Systems
Ship Handling Training Simulators	1. Platform specific hydrodynamic coefficient data used to populate a ship handling simulator may not be available as a result of the design process or the data set may not be complete. 2. Simulation models populated with full scale trials data are subject to the limitations of the trials data: influences of bathymetry, currents, waves and the above water environment. Operator practices and procedures used during the trials are also a significant influence on the results. 3. The cost of conducting full scale trials is prohibitive and can result in limited data being generated.
Commercial Fast Ferry Industry	1. Simulation models developed using full scale trials data may not be accurate for complex low-speed manoeuvres conducted with the assistance of secondary propulsion devices, such as berthing. 2. The parameters and operational speeds of high-speed hull forms may lie outside the limits of applicability of existing hydrodynamic coefficient prediction techniques.
Australian Maritime Defence Acquisition	1. Tactical manoeuvring performance requirements are largely undefined for platform acquisition programs. 2. Contractual implications can lead to limited or no technical information being supplied by the designer/provider on a platform's manoeuvring performance.
Maritime Science and Research	1. The platforms under analysis are often dissimilar in configuration and operation to those established in existing manoeuvring models and consequently no relevant data is available. 2. Existing empirical methods to determine hydrodynamic data are also not readily applicable to unique hull forms.

Technology Area	Suspect Implicit Systems
Navy Materiel Certification	1. The empirical equations that form the basis of several manoeuvrability requirements may be generated from hull forms that are not representative of current and future platform designs. 2. Inaccurate numerical prediction of a platform's manoeuvrability, submitted as part of the platform's certification, has the potential to render the certification assessment invalid until a full scale trial can be performed.

Table 2: Suspect implicit systems

Based on the suspect implicit systems identified in this study, it is apparent that deficiencies in the following areas exist:

- Unavailability of data
- Unacceptable or poor quality of data
- Data is not relevant
- Data is incomplete
- Relatively time consuming and expensive to produce data

In line with Hitchins' RSM, the outcomes of the procedure so far provide a clearer understanding of the problem and the inherent dysfunctions within the implicit systems.

Comparison of Current System with Ideal System

In order to develop a remedy or remedies it is necessary to identify the ideal world system attributes. These attributes represent what is considered to be required to enable the technology areas to meet their objectives effectively and efficiently, as opposed to the current state of being where there are unresolved deficiencies. The ideal system (or capability) attributes are presented in Table 3.

Technology Area	Ideal System Attributes
Ship Handling Training Simulators	The system would: 1. Provide a complete set of hydrodynamic data for a specific platform or class of platforms and for multiple load conditions; 2. Allow manoeuvring data to be generated for an extensive range of waterway configurations; 3. Allow data to be generated quickly and at reasonable cost when compared to full scale trials.
Commercial Fast Ferry Industry	The system would: 1. Enable the prediction of hydrodynamic data for a wide range of manoeuvres, including berthing with the assistance of secondary propulsion systems, such as bow thrusters; 2. Provide techniques to predict hydrodynamic coefficients for a wide range of hull forms that may currently lie outside the limits of applicability of readily available empirical prediction techniques.
Australian Maritime Defence Acquisition	The system would: 1. Provide data and models, if necessary, that permit Defence users to establish platform tactical manoeuvring performance requirements in response to the prevailing operational concepts and Navy doctrine; 2. Provide data and models, if necessary, that allow Defence users to evaluate the manoeuvring performance of a specific platform; 3. Accommodate the range of current and future navy platform types; 4. Allow data to be generated quickly and to the required level of accuracy.

Technology Area	Ideal System Attributes
Maritime Science and Research	The system would: 1. Enable hydrodynamic data to be generated for an extensive range of maritime platform types; 2. Accommodate a broad range of operational environments and up to six degrees of freedom of motion.
Navy Materiel Certification	The system would: 1. Facilitate the generation of up to date manoeuvrability requirements for RAN platforms; 2. Provide numerical predictions of platform manoeuvrability that are sufficiently accurate to reduce the reliance on full scale trials for platform certification.

Table 3: Ideal system attributes

Remedial Concepts

It is apparent from the investigation so far that the gap between the existing manoeuvring capability and a new capability, comprising the existing and ideal system attributes, is now more tangible. With a detailed understanding of the functional requirements of each of the technology areas it would be possible to develop tailored solutions that successfully deliver to each area's requirements. However, at this stage of the RSM it is only necessary to form conceptualised remedies, which are presented in Table 4.

While remedies could be developed using novel techniques, existing overseas service providers and yet to be developed infrastructure, an objective of this study is to identify opportunities where Australia's existing technology and intelligence resources can be used to bridge the gap.

Technology Area	Remedial Concepts
Ship Handling Training Simulators	Development and implementation of an independent platform specific hydrodynamic manoeuvring data generation service. The provider would be capable of providing the data needed by the ship handling simulator custodians on an as required basis and to their specific requirements. This data is likely to be orientated around mercantile ships, naval ships and harbour service fleets.
Commercial Fast Ferry Industry	Development and implementation of an independent platform specific hydrodynamic manoeuvring data generation service. The provider would be capable of providing the data needed by platform designers in the fast ferry and high-speed ship industry on an as required basis and to their specific requirements.
Australian Maritime Defence Acquisition	Development and implementation of a simulation tool that enables the evaluation of a platform's tactical manoeuvring performance. The tool would be populated with platform specific data on an as required basis. This data would allow the tool to accurately predict platform manoeuvring performance in extreme or unconventional operating scenarios. This required data would be generated within Australia using an integrated system comprising existing infrastructure, technologies and skill sets.
Maritime Science and Research	Development and implementation of an independent platform specific hydrodynamic manoeuvring data generation capability. This capability would provide the data needed by the maritime science, research and development sector on an as required basis and to their specific requirements. This data is likely to be required for any type of platform, from conventional to non-conventional, naval to commercial. This data would be generated within Australia using an integrated system comprising existing infrastructure, technologies and skill sets.

Technology Area	Remedial Concepts
Navy Materiel Certification	Implementation of a validated simulation tool that can generate a set of RAN manoeuvring requirements aligned with the platform's concept of operations (CONOPS) and based on specific hull form characteristics. Ideally, this simulation tool would enable the accurate evaluation of a platform's manoeuvring performance against these requirements. The tool would be populated with platform specific data on an as required basis.

Table 4: Proposed remedial concepts

The remedial concepts proposed in Table 4 share the central theme of establishing relevant, accurate and validated manoeuvring simulation and assessment tools. This implies that their current method of development, with respect to the underlying hydrodynamic manoeuvring data, lacks some of the features (quality, timeliness, etc.) that would remedy the symptoms identified in Table 1. The remedies, or solutions, that are likely to facilitate a suitable method or methods for generating manoeuvring data, can now be grouped into options according to the means available to develop them here in Australia. These options are presented in Table 5.

Option	Type	Description
1	Experiment based development	In this approach, results from model scale experiments form the basis of the data utilised in the development of the simulation tool.
2	Numerical based development	This approach to tool development uses numerical techniques, such as slender body theory, low aspect ratio theory or computational fluid dynamics as its basis of generating manoeuvring data.
3	Full scale trials based development	Full scale ship characteristics and trial data is used in the development of the simulation tool.
4	Combined experiment-numerical based development	In this approach a combination of experimental and numerical methods is used to generate data to support the simulation tool's development.
5	No Change (Existing)	This option embodies the current state of the capability used by the Australian maritime industry. It comprises full scale trials, numerical analysis and overseas combined methods.

Table 5: Indigenous remedy options

The final step in the RSM procedure (Figure 1) is to check that these remedies resolve the symptoms identified with minimal side effects. The adopted approach to assess the different development options is to utilise the Comparative Risk Assessment (CRA) methodology.

Comparative Risk Assessment

Comparative Risk Assessment is a decision making aid that can be used in situations where there are multiple needs and various technological options available for fulfilling those needs [8]. The key component of a CRA is the comparative analysis, where the benefits, costs, risks, adoptive constraints and uncertainties associated with each development option are assessed. However, first these attributes need to be identified, ranked and weighted for each solution option.

In order to baseline the CRA, the existing method of obtaining data is also included. Consequently, there are five options included in the CRA. This means that the top ranked option receives a score of five times the weighting, the second ranked option scores four times the weighting, and so on. The scores are then summed to give an overall total for each development option. In order to complete the CRA the following assumptions were made:

- Each development option will be based on existing indigenous technologies and processes for generating data;
- Each development option will be capable of producing a complete set of data points based on the needs of the simulation and evaluation tools.

The weightings, rankings and scores for each of the options that comprise the CRA are presented in Table 6. The weightings and rankings assigned to each option were done so on the basis of the Authors' experience in and knowledge of the Australian maritime industry's needs and the technology used in the proposed options. The assigned weightings are the average of the values nominated by the authors. The individual rankings were given according to the author group's consensus. Ideally, industry stakeholders would be engaged to provide the weightings (or level of importance) for each of the criteria, while rankings would be determined based on the normalised levels of performance offered by prospective service providers for each option. However, for the purpose of developing concept solutions, the present approach is seen as a valid starting point.

	Weight	DEGREE OF EFFECT									
		OPTION 1 Experimental		OPTION 2 Numerical		OPTION 3 Full Scale Trials		OPTION 4 Combined (Op1 & Op2)		OPTION 5 Existing (No Change)	
BENEFITS (Most beneficial option ranks highest)		Rank	Score	Rank	Score	Rank	Score	Rank	Score	Rank	Score
Option is able to provide accurate data	1.00	2	4.0	3	3.0	4	2.0	2	4.0	3	3.0
Option is able to provide results in a timely manner	0.90	3	2.7	3	2.7	3	2.7	3	2.7	3	2.7
Option is able to provide data relevant to stakeholder needs	1.00	1	5.0	2	4.0	3	3.0	1	5.0	4	2.0
Option is able to provide a complete set of data	1.00	2	4.0	2	4.0	4	2.0	2	4.0	3	3.0
SUM	19.50		15.7		13.7		9.7		15.7		10.7
COSTS (Lowest cost option ranks highest)		Rank	Score	Rank	Score	Rank	Score	Rank	Score	Rank	Score
Option development costs	0.80	4	1.6	2	3.2	5	0.8	3	2.4	1	4.0
Option sustainment costs	0.90	3	2.7	2	3.6	5	0.9	3	2.7	1	4.5
Cost to stakeholders (Option pay per use costs)	1.00	4	2.0	3	3.0	5	1.0	3	3.0	4	2.0
SUM	13.50		6.3		9.8		2.7		8.1		10.5
RISKS (Lowest risk option ranks highest)		Rank	Score	Rank	Score	Rank	Score	Rank	Score	Rank	Score
Option will not be sustained for required lifecycle	0.85	1	4.5	1	4.5	1	4.5	1	4.5	4	1.8
Option may be unavailable at times (workload > resources)	1.00	3	3.0	2	4.0	4	2.0	2	4.0	5	1.0
Simulation environment changes leads to option redundancy	0.65	3	1.8	3	1.8	4	1.2	2	2.4	2	2.4
No requirement for individual stakeholders to 'own' the option	0.70	1	3.5	3	2.1	4	1.4	2	2.8	3	2.1
SUM	16.00		12.8		12.4		9.1		13.7		7.3
ADOPTIVE CONSTRAINTS (Least constrained ranks highest)		Rank	Score	Rank	Score	Rank	Score	Rank	Score	Rank	Score
Perceived inaccuracy	1.00	2	4.0	5	1.0	3	3.0	3	3.0	4	2.0
Regulators reliance on full scale trials	0.85	2	3.2	3	2.4	1	4.0	2	3.2	3	2.4
SUM	9.25		7.2		3.4		7.0		6.2		4.4
UNCERTAINTIES (Least uncertain option ranks highest)		Rank	Score	Rank	Score	Rank	Score	Rank	Score	Rank	Score
Option performance	0.90	2	3.6	3	2.7	3	2.7	2	3.6	3	2.7
Option reliability	1.00	2	4.0	3	3.0	4	2.0	2	4.0	3	3.0
Ability to develop and implement option	0.80	3	2.4	3	2.4	2	3.2	2	3.2	1	4.0
SUM	13.50		10.0		8.1		7.9		10.8		9.7
Total	71.75		52.0		47.4		36.4		*54.5*		42.6
%			72.7		66.3		50.9		*76.2*		59.6
Overall Ranking			2		3		5		*1*		4

Table 6: Comparative risk assessment results

From the CRA it can be seen that the option of a combined experimental and numerical prediction based solution (Option 4) ranks as the most beneficial solution. It is interesting to note that Option 4 only ranks marginally better than Option 1 (the experimental based solution) and that both Option 4 and Option 1 are relatively superior to Option 2 (the numerical prediction solution). The implicit systems are directly related to the option Benefits presented in Table 6.

The result of this CRA implies that an indigenous marine vehicle manoeuvring analysis and evaluation capability has the potential to be beneficial for each of the technology areas assessed in this study. Of course, each of the three leading options (Option 4, 1 and 2) are able to be tailored to meet the needs of the stakeholder. Any tailoring of the proposed options could alter their position within the CRA. As with any development process, the development of each of the remedial concept options will be bound by implicit stakeholder constraints of cost, schedule and risk level.

DEVELOPMENT OF EXISTING INDIGENOUS CAPACITY

An example of the level of readiness of the existing capacity to be developed into a capability is the work currently being conducted by the Australian Defence Science and Technology Organisation (DSTO) Surface Platform Systems (SPS) branch and the Australian Maritime College (AMC). Located in Launceston Tasmania, the AMC operates Australia's core hydrodynamic testing facilities.

Since 2010 the DSTO-SPS branch has been collaborating with the AMC to develop an experimental surface platform manoeuvring analysis capability. In 2011 preliminary manoeuvring experiments were conducted for a generic landing craft in the AMC's towing tank using the existing DSTO/AMC Horizontal Planar Motion Mechanism (HPMM) and the DSTO surface platform manoeuvring dynamometer. As this development work continues the project will deliver a system that will provide multi-degree of freedom force measurement and accommodate a broad range of platform types. Verification and validation (benchmarking) studies will be completed as part of this work program to establish the operational limits of the capability. Previously, such a system was not available, yet it has taken only a modest investment of time and resources to advance the existing infrastructure to enable experimental based manoeuvring analyses. The data produced from these experiments will be used to populate the manoeuvring model within an advanced time domain ship motion analysis software tool used by DSTO.

While the AMC's facilities are unique within Australia, they are only one example of the relevant experimental facilities maintained within the country. For example, there are several wind tunnels currently operated by research institutes and commercial organisations that can be, and in some instances are, used for control surface design.

Unlike the aforementioned experimental facilities, numerical prediction tools such as CFD and other codes are readily accessible with respect to cost and useability to most practitioners and organisations within Australia. Consequently, there are a substantial number of Australian based engineering businesses, academic institutions and advanced research groups that possess the fundamental experience and tools that could be developed to support a numerical based manoeuvring analysis and evaluation capability. With the continuing development of open source codes, the cost to operate these tools and the available capacity to produce results becomes more favourable.

While these examples only represent a portion of what is likely to be required to form a comprehensive and useful capability, it does indicate that an indigenous manoeuvring analysis and evaluation capability could be realised.

EVALUATION

A systems thinking approach was used to conduct a preliminary investigation into the need to develop an indigenous marine vehicle manoeuvring analysis and evaluation capability. A structured method of inquiry based on Hitchin's Rigorous Soft Methodology was used to complete the study. The results of the study show that this approach has been successful in: identifying the inherent dysfunctions of a suspect system, identifying the gap between the existing and ideal system attributes and developing solution options. In addition, this methodology and procedure provides a high level of traceability between the stakeholders and the system under analysis.

A comparative risk assessment approach was used to evaluate the remedies or solutions identified as part of the RSM investigations. This approach has qualitatively and quantitatively evaluated the ability of each of the solution options to resolve the inherent system dysfunctions. It is worth noting that the data utilised in the CRA is somewhat subjective due to the closed group (i.e. the authors) that developed the information. However, this type of assessment is useful during concept development as it provides a structured approach to comparing options based on tacit knowledge. While the CRA has been successful for its level of application in this study, it would be of greater benefit to include this approach in future investigations where representatives of the Australian maritime industry are engaged and surveyed. In doing so the CRA would be based on the distinct needs and priorities of the technology area stakeholders and the ability of the service providers to meet these needs.

CONCLUDING REMARKS

This study is considered to be suitable as a starting point for a more detailed investigation into the current and future manoeuvring analysis and evaluation needs of the Australian maritime sector. It is evident, even after a cursory assessment, that there are symptoms indicating that the manoeuvring data needs of some of Australian maritime sector's stakeholders are not being met to a satisfactory standard. To better understand the requirements of these stakeholders it would be prudent to undertake a more detailed study and involve representatives from the operational, regulatory, design and research areas of industry. The objective of such a study would be to clearly define a unified set of requirements and their significance.

While an outcome of this study has been to identify a combined experimental and numerical remedy to overcome the existing deficiencies, an equally important outcome has been to identify that the capacity to develop this solution already exists within Australia.

REFERENCES

1. Rawson, K. J. and Tupper, E. C. (2001) *Basic Ship Theory*. Oxford, England, Butterworth Heinemann
2. Watson, D. G. M. (1998) *Practical Ship Design*. Oxford, England, Elsevier Science Ltd
3. Flood, R. L. and Jackson, M. C. (1991) *Creative Problem Solving: Total Systems Intervention*. Chichester, UK, John Wiley & Sons
4. Hitchins, D. K. (2007) *Systems Engineering: A 21st Century Systems Methodology*. Wiley Series in Systems Engineering and Management, Sage, A. P. ed. Chichester, UK, John Wiley & Sons
5. Checkland, P. (1999) *System Thinking, Systems Practice: A 30 Year Retrospective*. Chichester, UK, John Wiley & Sons
6. 'International Maritime Organisation (2000) *HSC 2000 Code - International Code of Safety for High-Speed Craft, 2000.*
7. Royal Australian Navy (1999) *DEF(AUST) 5000 ADF Materiel Requirement Set*. In *Volume 3: Hull System Requirements, Part 7: Ship Manoeuvrability Requirements and Analysis*. Canberra, Australia.

8. Lawless, E. W., Jones, M. V. and Jones, R. M. (1984) *Final Report On: Comparative Risk Assessment - Toward and Analytical Framework*. Midwest Research Institute, National Science Foundation of USA

SOVEREIGN AND CROWN IMMUNITY FOR WARSHIPS? – OR ACTING IN A MANNER CONSISTENT WITH IMO ENVIRONMENTAL CONVENTIONS

J Mansell[1] & M Tsamenyi[2]
University of Wollongong
jnkmansell@gmail.com & Tsamenyi@uow.edu.au

ABSTRACT

Merchant ships engaged on international voyages are required to comply with the standards for protection of the marine environment developed by the member States of the International Maritime Organization (IMO). Measures are in place in international law to ensure that these standards are implemented, monitored and enforced. Warships interact with these merchant ships on all oceans, face the same perils of the sea, and share many of the same issues of prevention of pollution. In principle warships, troopships, naval auxiliaries and government owned ships on non-commercial service have sovereign immunity from compliance with IMO environmental protection standards. However, to manage public expectations, reputational risks and liability, many States are increasingly noting the exhortation in IMO environmental Conventions to adopt appropriate measures to ensure that warships act in a manner consistent with those Conventions, where reasonable and practical. At the national level, some States have extended the application of national policies and legislation on environmental protection to their own warships and other vessels entitled to sovereign immunity under international law. This paper examines the international regulatory regime for merchant ships, analyses the international law aspects of sovereign immunity for warships, and discusses the international standards for environmental protection, and the implications of these standards, for warships. The paper also examines some State practice. It is concluded that IMO Conventions provide an appropriate standard for all ships, including warships, and that it is reasonable for Governments to apply these standards to warships where practicable.

INTRODUCTION

Merchant ships and warships share the world's oceans and seaways, face the same age-old perils of the sea, and pose similar threats to the marine environment. And yet, these two classes of ship have different obligations placed upon them in international law for prevention of marine pollution. Merchant ships are universally required to meet well established international environmental protection standards whilst warships, in theory, have sovereign immunity from almost all of them. Governments, as contracting parties to IMO Conventions, have the discretion to grant or not grant immunity to their own vessels from many of these international regulatory standards. The warships of some navies are required to meet international standards for prevention of marine pollution but this is by no means universal. Increasingly high standards are being developed for merchant ships to prevent or minimize all types of pollution of the marine environment; both oceanic and atmospheric. There is a clear expectation in international law that warships should meet

[1] John Mansell PhD, Visiting Senior Fellow, Australian National Centre for Ocean Resources and Security, University of Wollongong, NSW, Australia
[2] Martin Tsamenyi PhD, Professor of Law, Director, Australian National Centre for Ocean Resources and Security, University of Wollongong, NSW, Australia

these standards. This paper examines the practical and operational implications of application of IMO environmental protection standards to warships.[3] The starting point is to summarise international law for regulation of ships.

INTERNATIONAL LAW FOR REGULATION OF SHIPS

In April 1958 a suite of four Conventions were adopted in Geneva at the first United Nations Conference on the Law of the Sea (UNCLOS I) The most relevant of these Conventions to regulation of ships is the High Seas Convention (HSC),[4] the preamble to which reflected the desire of the States Parties to "codify the rules of international law relating to the high seas", including the customary primacy of the flag State in all matters to do with jurisdiction and control of ships.[5] A further United Nations Conference on the Law of the Sea (UNCLOS III) was convened to settle many outstanding and developing issues, particularly a legal regime for deep seabed mining[6] and growing concerns over pollution of the world's oceans. Over a prolonged period of negotiation at UNCLOS III between 1973 and 1982, the seminal Law of the Sea Convention (LOSC) was developed. This Convention is expressly stated to '...prevail over the Geneva Conventions on the Law of the Sea of 29 April 1958'[7] – as between the parties to the LOSC. The LOSC is the umbrella framework of international law of the sea; brought into effect for merchant ships through IMO Conventions which are, in turn, brought into effect for individual ships flying the flag of a State through the national law of the flag State. States with ships registered in their territory - flag States, have duties under the LOSC to effectively exercise jurisdiction and control in administrative, technical and social matters over ships flying their flags.[8] The LOSC defers to the IMO to set the international standards for the regulation of merchant ships. This is done through conferring in the 'competent international organization'; generally accepted as a reference to the IMO in matters of safety of ships, security and pollution prevention. Similarly, it is universally accepted that reference in the LOSC to 'generally accepted standards' is recognized in international law to mean the Conventions developed by the 170 member States of the IMO.

SOVEREIGN AND CROWN IMMUNITY

Warships and merchant ships are both subject to the jurisdiction of the flag State. The 1958 High Seas Convention grants warships 'complete immunity from the jurisdiction of any State other than the flag State'.[9] The LOSC reaffirms the exclusive jurisdiction of the flag State for warships, stating that 'Warships on the high seas have complete immunity from

---

[3] United Nations Convention on the Law of the Sea (LOSC), Montego Bay, 10 December 1982, in force 16 November 1994. 21 ILM 1245 (1982), Article 29, *Definition of warships,* For the purpose of this Convention, "warship" means a ship belonging to the armed forces of a State bearing the external marks distinguishing such ships of its nationality, under the command of an officer duly commissioned by the government of the State and whose name appears in the appropriate service list or its equivalent, and manned by a crew which is under regular armed forces discipline.

[4] Convention on the High Seas (HSC), Geneva, 29 April 1958. In force 30 September 1962. 450 *UNTS* 133

[5] The preamble to the HSC gives all States the right to sail ships under that State's flag on the high seas (HSC Preamble and Article 4). Every State is under a duty to fix conditions for the grant of its nationality to ships, for the registration of ships in its territory, and for the right to fly its flag (HSC Article 5). Ships are to have the nationality of the State whose flag they are entitled to fly. A second United Nations Conference (UNCLOS II) in 1960 narrowly failed to settle the outstanding issues of the breadth of the territorial sea and fishery limits.

[6] Churchill RR and Lowe AV, *The Law of the Sea*, 3rd Edition, Manchester University Press, 1999, pp. 15/16

[7] LOSC, Article 311, *Relation to other conventions and international agreements*

[8] Ibid, Article 94, *Duties of the flag State* This is referred to as flag State control

[9] HSC, Article 8

the jurisdiction of any State other than the flag State'.[10] Merchant ships are subject to the exclusive jurisdiction of the flag State on the high seas in the exercise of legislative and enforcement jurisdiction[11], 'save in exceptional circumstances expressly provided for in international treaties or in [the LOSC]'[12]. Sovereign immunity, with regard to warships, means immunity from investigative or coercive jurisdiction by States other than the flag State. The derivation of this rule in customary international law is the principle of reciprocity of respect between monarchs from the era when most States were headed by a sovereign – 'an equal does not have power over an equal'[13]. The resultant immunity is extended to representatives of the State, the government, and to vessels owned or operated by the government on the premise that such vessels are under the direct control of the government.[14] Warships are granted immunity from almost all IMO Conventions through being excepted from the requirements of those instruments, for example as 'ships of war and troopships'[15] or non-application of the Convention 'to any warship, naval auxiliary or other ship owned or operated by a State and used, for the time being, only on government non-commercial service'.[16] The ability of States to continue this sovereign immunity for warships at the domestic level in national legislation is commonly referred to as crown immunity. The sovereign immunity of a warship in the territorial sea of another State is confirmed in LOSC Article 30 which provides that; 'If a warship does not comply with the laws and regulations of the coastal State concerning passage through the territorial sea and disregards any request for compliance therewith which is made to it, the coastal State may require it to leave the territorial sea immediately' It is made clear that the flag State of a warship 'shall bear international responsibility for any loss of damage to the coastal State resulting from the non-compliance by a warship with the laws or regulations of the coastal State concerning passage through the territorial sea or with the provisions of this Convention or other rules of international law'.[17] Similar provisions and liabilities apply to warships in transit through straits used for international navigation.[18] Under the heading in the LOSC of 'Sovereign Immunity'[19], governments are required to adopt appropriate measures for warships for protection and preservation of the marine environment;

> the provisions of this Convention regarding the protection and preservation of the marine environment do not apply to any warship, naval auxiliary or

[10] LOSC, Part VII, High Seas, Section 1. General provisions. Article 95, *Immunity of warships on the high seas*. The LOSC, Article 96, *Immunity of ships used only on government non-commercial service*, similarly provides immunity for ships used only on government non-commercial service

[11] Churchill and Lowe, p.208

[12] LOSC, Article 92, *Status of ships*.

[13] Opeskin B and Tsamenyi M, The Law of the Sea, *Public International Law*, Eds. Blay S, Piotrowicz and Tsamenyi M, II Edition, Oxford University Press, Melbourne, 2005, pp 176/177

[14] Accordingly, if the crew of a warship were to mutiny or turn to piracy the ship would not be deemed to be under the control of the government and not entitled to immunity but would be assimilated to a pirate vessel in which case the provisions of Article 105 of the LOSC relating to *Seizure of a pirate ship or aircraft* would apply.

[15] SOLAS Chapter 1, regulation 3, *Exceptions*

[16] MARPOL, Article 3, *Application*

[17]LOSC, Article 32, *Immunities of warships and other government ships operated for non-commercial purposes*. This liability could extend to the flag State of a warship, for example, from pollution damage resulting from non-compliance of a warship with the traffic routeing measures put in place by a coastal State within its territorial waters, a grounding or an operational discharge of oil..

[18] Ibid, Article 42 (5) *Laws and regulations of States bordering straits relating to transit passage*. 'The flag State of a ship or the State of registry of an aircraft entitled to sovereign immunity which acts in a manner contrary to such laws and regulations or other provisions of this Part shall bear international responsibility for and loss or damage which results to States bordering straits'.

[19] Ibid, Article 236, *Sovereign immunity*

> other vessels or aircraft owned by or operated by the State and used, for the time being, only on government non-commercial service. However, each State shall ensure, by the adoption of appropriate measures not impairing operations or operational capabilities of such vessels or aircraft owned or operated by it, that such vessels or aircraft act in a manner consistent, so far as is practicable, with this Convention.[20]

The LOSC takes pains to clarify that, with the abovementioned exceptions, and the restrictions placed upon the activities of warships regarding innocent passage through a territorial sea,[21] 'nothing in this Convention affects the immunities of warships and other government ships operated for non-commercial purposes'.[22]

IMO ENVIRONMENTAL PROTECTION CONVENTIONS

There are a total of 51 IMO Conventions and Protocols falling into three main groups; namely; safety, prevention of pollution, and liability and compensation for damage caused by pollution. The great majority of these Conventions are not applicable to warships.[23] The three principal IMO instruments are those dealing with safety of life at sea (SOLAS), prevention of marine pollution (MARPOL 73/78), and standards of training, certification and watch keeping for seafarers (STCW). Sovereign immunity is granted to warships in respect of these three Conventions. The IMO Conventions of most significance to warships, and those where there are expectations in international law and policy that Governments will not grant crown immunity, are those dealing with prevention of pollution of the marine environment, namely;

- International Convention for the Prevention of Pollution from Ships, (MARPOL), 73/78
- International Convention on Civil Liability for Oil Pollution Damage (CLC), 1992
- International Convention on Civil Liability for Bunker Oil Pollution Damage (BUNKER), 2001
- International Convention on the Establishment of an International Fund for Compensation for Oil Pollution Damage (FUND), 1992
- International Convention on the Control of Harmful Anti-fouling Systems on Ships (AFS), 2001
- International Convention for the Control and Management of Ship's Ballast Water and Sediments (BWM), 2004

These instruments will be analysed in turn to examine the implications for warships if immunity is waived by the contracting party to the Convention.

MARPOL 73/78

The most significant IMO Convention regarding appropriate equal standards for merchant

---

[20] This wording is clearly derived from that of MARPOL 73/78; adopted when negotiations were commencing on development of the LOSC.

[21] LOSC, Article 19, *Meaning of innocent passage*

[22] Ibid, Article 32, *Immunities of warships and other government ships operated for non-commercial purposes*

[23] A clear exception, which applies to all ships, is the Convention on the International Regulations for Preventing Collisions at Sea, London, 20 October 1972. In force 15 October 1977. *UKTS* 77; *ND IV*

ships and warships, is MARPOL 73/78[24]; the principal IMO Convention dealing with prevention of pollution of the marine environment by ships from both operational[25] and accidental[26] causes. Like the majority of IMO Conventions MARPOL grants sovereign immunity to 'warships, naval auxiliaries and other Government owned vessels used for non-commercial purposes'[27]. However, unlike many other non-environmental IMO Conventions, MARPOL requires each Party to ensure that such ships, 'by the adoption of appropriate measures not impairing the operations or operational capabilities of such ships owned or operated by it, [to] act in a manner consistent, so far as is reasonable and practicable, with the present Convention'.[28]

MARPOL, which entered into force in October 1983, comprises six annexes; the first two of which are mandatory with the remainder being optional;

Annex I	Regulations for the Prevention of Pollution by Oil (1983)
Annex II	Regulations for the Control of Pollution by Noxious Liquid Substances in Bulk (1983)
Annex III	Regulations for the Prevention of Pollution by Harmful Substances Carried by Sea in Packaged Form (1992)
Annex IV	Regulations for the Prevention of Pollution by Sewage from Ships (2003)
Annex V	Regulations for the Prevention of Pollution by Garbage from Ships (1988)
Annex VI	Prevention of Air Pollution from Ships (2005)

150 States representing 99.14% of world tonnage have ratified Annexes I and II of MARPOL as at 31 July 2011. Other annexes of MARPOL have received varying degrees of generally high support, particularly in percentage of world tonnage.[29]

Annex I

Annex I of MARPOL applies to all ships with oil tankers of 150 gross tonnage and above and all ships of 400 gross tonnage and above being required to undergo surveys and carry certification. The practical effect of Annex I for warships is that discharge of oil into the sea is tightly controlled[30] through the use of oily-water separators, sludge or holding tanks and oil record books. No ballast is to be carried in fuel tanks, and no oil forward of the collision bulkhead, so far as is reasonable and practical.[31] A shipboard oil pollution

---

[24] International Convention for the Prevention of Pollution from Ships, London, 2 November 1973, as amended by the protocol, London, 1 June 1978 (MARPOL 73/78) 1340 *UNTS* 61, 62 (E)

[25] 'operational' causes would include oil spills resulting from operational errors such as during bunkering, or through pumping of oily bilge water overboard in excess of allowable ratios of oil to water.

[26] 'accidental' causes include oil spills arising from accidents such as collision, fire and grounding

[27] MARPOL, Article 3(3), *Application.* The present Convention shall not apply to any warship, naval auxiliary or other ship owned or operated by a State and used, for the time being, only on government non-commercial service".

[28] Ibid

[29] MARPOL Annex III; 135 States and 96.45% of world tonnage, Annex IV; 128 and 86.62%, Annex V, 141 and 97.36% and Annex VI; 63 and 88.90%. *Lloyd's Register/Fairplay World Fleet Statistics,* Summary of Status of Conventions, 31 December 2009. http://www.imo.org/About/Conventions/StatusOfConventions/Documents/status-x.xls accessed 25 April 2011

[30] Discharge of oily-water is strictly controlled within certain special areas; Mediterranean Sea, Baltic Sea, Black Sea, Red Sea, the "Gulfs area", the Gulf of Aden, North West European waters, the Oman area of the Arabian Sea and Southern South African waters, and completely banned within Antarctic waters (south of 60 degrees South). Discharge of oily-water is prohibited in Particularly Sensitive Sea Areas (PSSA), e.g. the Great Barrier Reef. There are a total of 12 PSSAs

[31] MARPOL, Annex I, Regulations 14, 15 and 24

emergency plan (SOPEP) is required to be carried.[32] Surveys of construction and equipment are required to verify compliance with Annex I and the ability is provided for the Administration (flag State) to delegate this function to a recognized organization[33], usually a classification society. The surveyor would inspect the warship to the same empirical standards as a merchant ship and issue an International Oil Pollution Prevention Certificate on behalf of the flag State to attest to compliance.[34] An amendment to Annex I, which took full effect from 2010,[35] required an accelerated phase out of single hulled tankers to minimize the risk of an oil spill in the event of a collision or grounding. This amendment has already had a significant impact upon the operational capabilities of single-hulled naval tankers with some being restricted to carriage of oil only in centre tanks in order to comply with the requirement for a double hull. Under another amendment to Annex 1[36] single-hulled tankers are required to restrict their total carrying capacity of heavy grade oil to less than 5000 ton and obtain the permission of the Governments of other countries before entering their ports. Annex I of MARPOL was amended in 2007 to require newly built or converted ships with total oil capacity greater than 600 cubic metres to have structural features (double hulls, protective location and restrictions on the size of individual tanks) to limit the outflow of oil in the event of a collision or grounding.[37] Another amendment to Annex I, which could have implications for certain warships, is a ban on the carriage or use of heavy fuel oil in Antarctic waters[38] from 01 August 2011.[39] An exception is provided for vessels involved in emergency rescues.

Annex II

The second annex of MARPOL is that dealing with 'the control of pollution by noxious liquid substances in bulk'; more simply expressed as 'chemicals', and aimed primarily at the prevention of pollution of the sea by chemicals carried in bulk. This annex has very limited relevance to warships as it largely requires chemical tankers to comply with the International Bulk Chemical Code (IBC Code) which contains international standards for the safe transport of cargoes of bulk chemicals. As Annexes I and II of MARPOL are mandatory a State must apply them both through its national law if it does not wish to grant crown immunity to its warships. The national law of such a State would typically provide, with regard to Annex II of MARPOL, that the law applied to 'all ships of the State, including warships and other ships of the Defence Force, and to all foreign ships under the State's jurisdiction, that carry noxious liquid substances in bulk'.[40]

Annex III

Annex III of MARPOL, which deals with the prevention of pollution by harmful

---

[32] Ibid, Regulation 26 *Shipboard oil pollution emergency plan*

[33] Ibid, Regulation 4(3)(a), *Surveys and inspections*. "...the Administration may, however, entrust the surveys either to surveyors nominated for the purpose or to organizations recognized by it"

[34] Ibid, Regulations 5, 7 and 8

[35] Ibid, Regulation 20, *Phase out for single hull tankers*

[36] Ibid, Regulation 13H, *Carriage of heavy grade oil*

[37] Ibid, Regulation 13, *Fuel oil tank protection*

[38] Antarctic waters are deemed to be all those waters south of Latitude 60 degrees South

[39] MARPOL, Annex I (*Regulations for the prevention of pollution by oil*) on *Special requirements for the use or carriage of oils in the Antarctic area,* a new chapter 9 with a new regulation 43, which would prohibit the carriage, in bulk as cargo, or carriage and use as fuel, of: crude oils having a density, at 15°C, higher than 900 kg/m3; oils, other than crude oils, having a density, at 15°C, higher than 900 kg/m3 or a kinematic viscosity, at 50°C, higher than 180 mm2/s; or bitumen, tar and their emulsions. http://www.imo.org/about/conventions/pages/action-dates.aspx accessed 06 May 2011

[40] Wording based upon that contained in New Zealand Marine Environment Protection Rule 142B – *Documents (Record Books and Manuals) – Noxious Substances*

substances carried by sea in packaged form, became mandatory in 2004 through an amendment to the SOLAS Convention[41] which made the International Maritime Dangerous Good (IMDG) Code mandatory.[42] The IMDG Code is not applicable to warships as the provisions of the code 'are applicable to all ships to which [SOLAS] applies…'[43] However, if a State waives immunity for warships the elements of the IMDG Code that cover marine pollutants would be applicable to those ships.

Annex IV

Annex IV; prevention of pollution by discharge of sewage from ships, could well be included in application of MARPOL to warships through lack of crown immunity. The annex requires ships to undergo surveys, hold certification and maintain a sewage record book. Under the original annex, in force 27 September 2003, ships are prohibited from discharging sewage into the sea within specified distances from the shore unless they have a sewage treatment plant. An amendment to Annex IV, dated 01 August 2005, requires new ships to be equipped with a sewage treatment plant, or a sewage comminuting and disinfecting system, or a sewage tank. These provisions will be extended to existing ships from 27 September 2013. The discharge of sewage will be permitted more than 3 nautical miles from the nearest land if the ship is equipped in accordance with the 2005 amendments. If the sewage is not comminuted or disinfected it must be discharged more than 12 miles from the nearest land.

Annex V

Annex V of MARPOL deals with prevention of pollution by garbage[44] from ships. Similarly to oil and sewage, ships are required to undergo surveys, carry certification and maintain a garbage record book. The garbage book must record all instances of disposal or incineration of garbage. A garbage management plan is also required. Amendments arising from a review of the effectiveness of Annex V were considered and adopted by IMO in July 2011 and are expected to enter into force on 01 January 2013.[45]

Annex VI

The newest annex to MARPOL, brought into force through a Protocol adopted in 1997 and entering into force in 2005, deals with the vexed and growing issue of air pollution from

[41] International Maritime Organization, International Maritime Dangerous Goods (IMDG) Code, http://www.imo.org/Safety/mainframe.asp?topic_id=158 accessed 07 May 2011

[42] SOLAS ,Chapter VII, Part A, Regulation 1

[43] IMDG Code, *Application of the Code*, para. 4.5

[44] MARPOL Annex V, *Prevention of pollution by garbage from ships*. 'Garbage includes all kinds of food, domestic and operational waste, excluding fresh fish generated during the normal operation of the ship and liable to be disposed of continuously or periodically. Annex V totally prohibits disposal of plastic anywhere into the sea and severely restricts discharge of garbage from ships in to coastal waters and "Special Areas"' http://www.imo.org/OurWork/Environment/PollutionPrevention/Garbage/Pages/Default.aspx accessed 07 May 2011

[45] The main changes include the updating of definitions; the inclusion of a new requirement specifying that discharge of all garbage into the sea is prohibited, except as expressly provided otherwise (the discharges permitted in certain circumstances include food wastes, cargo residues and water used for washing deck and external surfaces containing cleaning agents or additives which are not harmful to the marine environment); expansion of the requirements for placards and garbage management plans to fixed and floating platforms engaged in exploration and exploitation of the sea-bed; and the addition of discharge requirements covering animal carcasses. http://www.imo.org/MediaCentre/PressBriefings/Pages/43%20MEPC62ENDS.aspx accessed 11 August 2011

ships.[46] Annex VI prohibits deliberate emissions of ozone depleting substances such as halons and chloroflurocarbons (CFCs), with new installations containing these substances prohibited.[47] It is also prohibited under Annex VI to incinerate contaminated packaging products and polychlorinated biphenyls (PCBs). Strict limits are set on airborne emissions from ship exhausts of sulphur oxide and nitrogen oxide with the maximum allowable limit for sulphur oxide decreasing from 4.5% to 0.5% by 2020. A limit of 1.5% already applies in some Emission Control Areas (ECA) introduced under an amendment to Annex VI which entered force in July 2010.[48] If ships cannot meet this emission standard in these areas they must fit an exhaust gas cleaning system or use other technological methods to limit sulphur oxide and/or nitrogen oxide emissions. The practical effect of the limits on sulphur oxide and nitrogen oxide will be to preclude the use of heavy fuel oil aboard ships by 2020. IMO is to review the availability of fuel oil with a sulphur content lower than 0.5% in 2018. Some ships already voluntarily changeover their generators from heavy fuel oil to diesel oil in port to minimise pollution of the atmosphere. Amendments to Annex VI in 2010 also introduced progressive reductions in nitrogen oxide emissions with the most stringent controls coming into force on 1 January 2016 for so-called "Tier III" engines; those installed in ships constructed on or after 1 January 2016 and operating in ECAs. IMO approved a further amendment to Annex VI in July 2011 which establishes mandatory measures to reduce emissions of greenhouse gases from ships. New ships will be required to be built to an Energy Efficiency Design Index (EEDI) and all ships are required to operate in accordance with a Ship Energy Efficiency Management Plan (SEEMP).[49] These amendments are expected to come into force on 1 January 2013 and apply to all ships of 400GT and above.[50]

LIABILITY FOR OIL POLLUTION DAMAGE

The IMO has three instruments that deal with liability and compensation for damage caused by oil pollution from ships, namely;

- International Convention on Civil Liability for Oil Pollution Damage (CLC), 1992
- International Convention on Civil Liability for Bunker Oil Pollution Damage (BUNKER), 2001
- International Convention on the Establishment of an International Fund for Compensation for Oil Pollution Damage (FUND), 1992

---

[46] MARPOL, Annex VI, *Air pollution and greenhouse gas emissions*, http://www.imo.org/OurWork/Environment/PollutionPrevention/AirPollution/Pages/Default.aspx accessed 07 May 2011

[47] New installations containing hydro-chloroflurocarbons (HCHCs) are permitted until 1 January 2020.

[48] The North Sea and Baltic Sea ECAs are already in effect with an ECA covering sulphur, particulate matter and nitrogen extending 200 nautical miles from the coats of the USA and Canada due to come into effect on 01 August 2012. ECAs can be approved for sulphur dioxide and particulate matter or nitrogen oxide or all three. Limits for sulphur oxide in ECAs are reduced to 1% from 1 July 2010 (from the previous 1.5%) and further to 0.1% by 1 January 2015.

[49] IMO News, *Proposed GHG amendments to MARPOL convention circulated*, Issue 1, 2011, p.7. 'The EEDI is a non-prescriptive, performance-based mechanism that leaves the choice of technologies to use in a specific ship design to industry. As long as the required energy-efficiency level is attained, ship designers and builders would be free to use the most cost-efficient solutions for the ship to comply with the regulations...the SEEMP establishes a mechanism for a shipping company and/or a ship to improve the efficiency of ship operations'.

[50] There is an exception provision allowing an extended transition period under which Administrations can waive the requirement for new ships from complying with the EEDI where the contract is placed up to four years after entry into force of the amendment, the keel is laid up to four years and six months after entry into force, and delivery is later that six years and six months after entry into force. *Lloyds List* 19 July 2011

The first two Conventions deal with liability in cases of oil pollution from cargo or bunkers with the third establishing an international fund, accrued though levies on the importers of oil[51], to compensate the victims of oil pollution from ships. The Bunkers Convention is modeled upon the CLC, both of which require the ship to have insurance cover for oil pollution damage, and to carry evidence of such cover. The Conventions apply to pollution damage on the territory of a State, in its territorial waters or in its exclusive economic zone. Warships have sovereign immunity from liability for pollution of the marine environment[52] and do not have any obligation to pay compensation for the results of pollution damage[53]. There is no expectation in the application of these three Conventions that sovereign immunity and crown immunity should be waived by a contracting Government, as is the case in MARPOL. Regardless of whether a State waives immunity from these liabilities, the flag State would be liable for compensation arising from pollution of the marine environment from warships or Government owned ships on non-commercial service.[54]

ANTI-FOULING SYSTEMS

The International Convention on the Control of Harmful Anti-fouling Systems on Ships (AFS) was adopted in 2001 and entered into force in September 2008. The wording on the application of the AFS Convention mirrors that of MARPOL in that it does not apply to warships but asks contracting Parties to ensure '...by the adoption of appropriate measures not impairing operations or operational capabilities of such ships owned or operated by it, that such ships act in a manner consistent, so far as is reasonable and practicable, with this Convention'.[55] The AFS Convention, which applies to ships greater than 24 metres in length on international voyages, prohibits the use of harmful organotins in anti-fouling paint and ensures that the use of harmful substances in anti-fouling paints would not happen in the future. Ships of above 400 gross tonnage must undergo an initial survey of their anti-fouling system before entering into service; before the required International Anti-fouling System Certificate is first issued, and if the anti-fouling system is changed or replaced. Ships of between 24 metres and 400 gross tonnage must carry a Declaration on Anti-fouling Systems along with evidence of the source of the paint.[56]

---

[51] FUND Convention, Article 10 'Annual contributions to the Fund shall be made in respect of each contracting State by any person who, in the calendar year referred to in Article 12, paragraph 2(a) or (b), has received in total quantities exceeding 150,000 tons...'

[52] CLC Convention, Article XI 'The provisions of this Convention shall not apply to warships or other ships owned or operated by a State and used, for the time being, only on government non-commercial service'

[53] FUND Convention, Article 4(2) 'the Fund shall incur no obligation under the preceding paragraph if: (a) It proves that the pollution damage resulted from an act of war, hostilities, civil war or insurrection or was caused by oil which has escaped from or been discharged from a warship or other ship owned or operated by a State and used, at the time of the incident, only on Government non-commercial service'

[54] For example, New Zealand has not granted immunity to warships from liability or compensation arising from oil pollution through the following provision in the Maritime Transport Act (MTA) 1994, Section 4, 'The provisions of sections 344 to 353, 361 and 368 of this Act shall apply to warships of the New Zealand Defence Force and, for the purpose of determining liability under any of those provisions in respect of any act or omission involving a warship, any provision of this Act, the Resource Management Act 1991, or any other enactment referred to in those provisions shall be deemed to apply to warships.(MTA, ss. 344 – 354 *Liability for Pollution from Ships*. S.361 *General Provisions in Respect of Liability for Pollution Damage* and s.368 *Miscellaneous*.)

[55] AFS Convention, Article 3(2), *Application*

[56] IMO website http://www.imo.org/About/Conventions/ListOfConventions/Pages/International-Convention-on-the-Control-of-Harmful-Anti-fouling-Systems-on-Ships-(AFS).aspx accessed 8 August 2011

BALLAST WATER MANAGEMENT

The third of the three principal environmental protection Conventions is the International Convention for the Control and Management of Ship's Ballast Water and Sediments (BWM Convention), 2004. The BWM Convention has wording identical to that of MARPOL and AFS in its encouragement for application to warships, but, unlike those two Conventions, has yet to receive enough support to enter into force.[57] The BWM Convention aims to prevent the spread of harmful aquatic organisms through the carriage and discharge of ship's ballast water. Ships are required to be surveyed and certificated and develop a Ballast Water and Sediments Management Plan, with its implementation recorded in a Ballast Water Record Book.

REGULATORY REQUIREMENTS FOR AUSTRALASIAN NAVIES

The Australian Government is party to all six MARPOL Annexes and has waived immunity for ships 'owned or operated by the Commonwealth and used, for the time being, on government non-commercial service'. This waiver is brought into effect through the application of a number of Marine Orders[58] made pursuant to the Protection of the Sea (Prevention of Pollution from Ships) Act 1983 (Cwth) and the Protection of the Sea (Harmful Anti-fouling Systems) Act 2006 (Cwth). New Zealand has, to date, taken a selective approach to the ratification of MARPOL and its application to warships. All Annexes except Annex VI have been brought into effect through a number of Marine Protection Rules[59] made pursuant to the Maritime Transport Act 1994. Annexes I, II and V apply to 'warships and any other ship of the New Zealand Defence Force'. Annex III applies to 'New Zealand ships anywhere, except ships of the New Zealand Defence Force' and Annex IV only applies *inter alia* to warships and other ships of the New Zealand Defence Force sailing in the Antarctic Treaty area'.

CONCLUSION

There are international standards in place in the LOSC to prevent pollution of the marine environment by ships. These 'generally accepted standards' are brought into effect for merchant ships through Conventions adopted by the 'competent international organisation'; the IMO. There is a very high level of support for these standards amongst the States that register merchant ships. Although warships are granted sovereign immunity from these standards States have the ability to continue this immunity through their national law or to waive the immunity and require warships to act in a manner consistent, so far as is reasonable and practicable, with these IMO environmental protection standards. It may not be possible, for operational reasons, for Governments to waive crown immunity for warships in all cases but this decision would need to be balanced against an increasing societal sensitivity about the marine environment and the existence of appropriate, and widely accepted, international environmental standards for ships in general. Many navies have demonstrated that it is reasonable and practicable to meet these international standards. If crown immunity is not granted there will be significant design, construction, survey and operational implications for warships.

---

[57] The BWM Convention, Article 17, *Entry into force* 'This Convention shall enter into force twelve months after the date on which not less than thirty States, the combined merchant fleets of which constitute not less than thirty-five percent of the gross tonnage of the world's merchant shipping, have either signed it...' As at 11 July 2011 28 States representing 26.37% of world tonnage had ratified the BWM Convention. *Lloyds List* 11 July 2011

[58] Marine Orders 91, 92, 93, 94, 95 and 96

[59] Marine Protection Rules 120, 140, 150, 160 and 170.

The Floating Harbour Transhipper - An Operationally Effective Solution For Military And Emergency Response Duties

R Ballantyne[1], G Macfarlane[2], S Ballantyne[1], T Lilienthal[2]

1 Sea Transport Corporation, Southport, Queensland, Australia
2 Australian Maritime College, University of Tasmania, Launceston, Tasmania, Australia

ross@seatransport.com

ABSTRACT

The Floating Harbour Transhipper (FHT) is a pioneering logistics solution that was designed to meet the growing demands for coastal transhipment in the mining sector as well as commercial port operations. Addressing the two key needs of military and disaster response capabilities, Sea Transport Logistics Pty Ltd. have incorporated the successful aspects of this design and developed two complimentary designs. This technical paper reviews the first of several planned series of experiments carried out at the Australian Maritime College that has provided strong evidence of the potential success in significantly reducing relative motions of vessels when transhipping in open seas – this being the cornerstone of any successful open water transhipment operation.

INTRODUCTION

Although the FHT was developed initially as a commercial solution, we are cognisant of the essential roles of commercial vessels in Military conflicts and Disaster Response. During the Falklands War for instance, 43 merchant ships were taken up from trade (STUFT) together with 24 Royal Fleet Auxiliary and Royal Maritime Auxiliary Service vessels, all manned by merchant navy personnel. Merchant ships outnumbered naval vessels in the campaign.

With innovative designs and a change of Federal Government thinking, Australia's presently dismal Merchant and Royal Navy can fundamentally become the new world leader in defence, disaster response and coastal shipping.

With more extreme weather patterns & natural disasters occurring, along with national security uncertainly, the nation's defence capability is struggling to keep up. The transhipping objective for Defence/Disaster Response and Mining is the same, which is, to transfer large volumes of cargo with minimal time and costs into remote areas with little or no infrastructure in all types of weather.

This technical paper reviews a case study on the development of the Floating Harbour Transhipper; a ground-breaking Australian design that can not only increase the nation's export and import capabilities, but strengthen the nation's Emergency Response & Military capabilities. For a relatively simple modification for defence use, the FHT can provide 95% of the military requirements for one-tenth of the cost.

The Floating Harbour Transhipper Concept

Traditionally, the transfer of bulk ore cargo from a feeder vessel to a moored 'mother' ship is conducted with the feeder vessel moored side-by-side to the mothership. A novel concept proposed by Sea Transport Logistics (STL) is to use an FHT, where the feeder vessel is moored inside the stern well dock of the FHT/mothership. The FHT is a covered floating storage vessel which incorporates a wet dock facility at the aft end to suit the feeder vessels. It also has its own bulk cargo handling equipment, not just for the transfer of material from the feeder vessel into its own stockpile, but also from this stockpile to an export Ocean Going Vessel (OGV) moored alongside. This system eliminates grab spillage and dust, common to other transhippers.

The concept is depicted in Figure 1, where a feeder vessel of up to 10,000 dwt capacity is moored inside the well dock of a Capesize (100,000 DWT) capacity FHT. The FHT only requires 60% storage of the export vessel capacity. In this case, the feeder vessel is a patented quad-screw Stern Landing Vessel (SLV), as described by Ballantyne and Ballantyne (2007) and shown in Figure 2. In this figure, it is illustrated that the feeder can also carry containers should it be utilised for disaster response duties. This feeder is typically carrying bulk ore product for mining transhipment options, however, Figure 3 highlights the flexibility to be utilised for defence purposes.

The FHT concept provides an environmentally & operationally compelling solution that can provide significant advantages for the country, community and mining companies. The key advantages for the FHT and SLV feeder system relevant to mining, defence & disaster response purposes are summarised below:

- Transhipment to and from the feeder in seas of up to 5m wave heights;
- Shallower draught feeder vessels can be used from very small ports or unprepared beaches, eliminating the need for dredging of sensitive areas;
- The FHT well dock arrangement eliminates stevedoring damage to feeder vessels and transhipment vessels;
- The feeder vessels can be secured within the well dock bow first to push the FHT to redeploy to disaster sites, cyclone moorings or dry dock.

It is acknowledged that the new Canberra Class Landing ship, Helicopter Dock (LHD) does incorporate a similar stern door access for loading smaller vessels, however the FHT system is unique in the sense it is a permanently open aft wet dock which is not restricted to small craft only. The wet dock in this instance has been designed to reduce the surge effects and dampen the roll, pitch and heave motions.

The intention of the FHT system is to reduce transhipment delays caused by inclement weather, by greatly reducing relative motions between the FHT and feeder vessel. The feeder is significantly sheltered when inside the FHT well dock, when compared to the more exposed location when a feeder is in a traditional alongside mooring arrangement. For example, it is common for alongside transhipment operations to typically be limited to significant wave heights of up to approximately 2.5 metres which can be seen in any operation currently in the market, whereas preliminary investigations indicate that feeder vessel operations can be handled without stopping in seas of up to 5 metres for the FHT concept. The main purpose of this paper is to present some of the results from the first of a series of planned physical model experiments which were conducted to investigate this issue.

Literature Review

To the author's knowledge, there are no publicly available studies that have investigated the relative motions of two vessels, of similar relative sizes as proposed in the FHT concept, where the smaller vessel is moored inside an aft well dock. There exist several published studies that have investigated the relative motions of two vessels in a side-by-side arrangement, with most studies dealing with the application of numerical codes to this problem, however the majority look to validate their predictions by undertaking a comparison against limited scale model experimental data. For example, see Kodan (1984), Buchner *et al.* (2001), and Koo and Kim (2006).

EXPERIMENTATION

The model experiments were carried out in the shallow water wave basin at the Australian Maritime College in Launceston, Tasmania. The water depth in the basin was simulated to represent a typical coastal region having a constant depth of 15 metres. A variety of incident wave conditions and vessel headings were investigated as part of a comprehensive test program. Experimental results presented within this paper concentrate on a series of head sea tests in regular sinusoidal waves having nominal wave heights corresponding to 2m, 4.5m and 7.7m. A wide range of typical wave periods were investigated.

The primary particulars of the FHT and feeder vessel, in both model (scale 1:44) and full scale, are provided in Table 1. Results presented here are for the case with the FHT in ballast condition and the feeder vessel at half load condition. These lighter conditions were tested to simulate a common worse case scenario. Full displacement conditions in general produce smaller motions; hence it is common practice for ships to ballast down during cyclonic conditions.

	PANAMAX FHT		**Feeder Vessel**	
	Ship	Model	Ship	Model
LOA (m)	220.000	5.000	90.000	2.045
Beam (m)	44.000	1.000	21.000	0.477
Draught (m)	7.000	0.159	2.420	0.055
Displacement (tonnes) – Mid Condition	47070	0.539	3660	0.043
Trim (degrees)	0.0	0.0	0.0	0.0
VCG (m)	10.000	0.227	5.500	0.125
LCG (m)	113.500	2.580	43.000	0.977
GM (m)	22.714	0.516	9.698	0.220
KM (m)	32.714	0.744	15.198	0.345
Pitch Radius of Gyration (m)	55.000	1.219	22.500	0.536
Roll Radius of Gyration (m)	15.400	0.338	7.350	0.173

Table 1

The motions of both vessels were measured using a non-contact optical tracking system based on advanced infrared cameras (supplied by Qualisys). Two specific cases were investigated; (a) with the feeder vessel alongside the FHT and; (b) with the feeder vessel located inside the well dock of the FHT.

For the study with the feeder alongside the FHT, the feeder was located on the portside of the FHT with the longitudinal location defined by lining up the midships of both vessels and the bows orientated in the same direction, as can be seen in the photograph shown in Figure 4. The starboard side of the feeder model was attached to the port side of the FHT model using a pair of vertical slides and universal joints, refer Figure 5. The aft vertical slide/universal joint also incorporated a short horizontal slide. This arrangement allowed freedom in heave, pitch and roll, whilst constraining the feeder model in surge, sway and yaw (relative to the FHT model).

In the case where the feeder was located within the well dock, the stern of the feeder model was attached to the internal end wall of the FHT well dock using a vertical slide and universal joint and the bows facing opposite directions. This allowed freedom in heave, pitch, roll and yaw, whilst constraining the feeder model in surge and sway (relative to the FHT model). The photograph in Figure 6 provides a general view of this set up. Fenders were attached to the inside walls of the FHT well dock near the entrance to limit the yaw movement of the feeder.

A simplified mooring system was adopted to ensure that the FHT model maintained the required nominal heading to the incident waves. This mooring system included a pair of mooring lines, one each from the bow and stern of the FHT model (connected at the still waterline). The stern mooring line incorporated a bridle so as to avoid contact with the feeder model.

RESULTS AND DISCUSSION

A comparison of the resultant pitch and roll motions for an incident wave height of 2 metres at a heading of 180 degrees is shown in Figures 7 and 8 respectively. A range of wave periods from 4 to 12 seconds were investigated with both the feeder inside the FHT well dock and the feeder alongside the FHT.

The pitch motions of the FHT did not vary appreciably between the cases when the feeder vessel was located alongside or inside the well dock (refer Figure 7). Interestingly, the pitch motions of the feeder vessel are quite similar to that of the FHT while it was located inside the well dock, with typical maximum values around 0.4 to 0.5 degrees. However, when the feeder was located alongside the FHT its pitch motions became significantly larger, by almost an order of magnitude, with the peak value approaching 4 degrees. In each case, the peak pitch angles occur at incident wave periods in the region of about 10s to 11s.

As might be expected, the roll motions of the FHT were generally relatively small for head sea conditions, as shown in Figure 8. However, in wave periods greater than 8 seconds, the FHT rolls more when the feeder is moored alongside than when it is inside the well dock, suggesting that the presence of the feeder alongside adversely affects the motions of the FHT. The roll motions of the feeder vessel whilst inside the well dock is similar to the roll motions of the FHT, but considered to be relatively small with values generally less than 0.2 degrees at all wave periods investigated. Of significant concern are the notable roll motions of the feeder vessel when moored alongside, which are found to exceed 3 degrees (around a wave period of 9s) which, similar to the pitch motions, is around an order of magnitude greater than found when the feeder is located inside the well dock.

In summary, the results obtained highlight the potential reduction in pitch and roll motions that can be achieved by 'sheltering' the feeder vessel within the aft well dock of an FHT. It is acknowledged that the motions of the feeder vessel when alongside the FHT could be controlled to some extent by the manner in which it is moored to the FHT, however, in certain circumstances this may be impractical.

Further experiments were conducted in head seas at greater nominal incident wave heights in order to determine the effect this has on the motions of both the FHT and feeder vessel. Cross-plots of the motions as a function of increasing wave height are provided in Figures 9 and 10 for the nominal wave period of 10s. This wave period was selected as the maximum motions were found to occur at or close to this wavelength in the results presented in Figures 7 and 8. It should be noted that tests on the case with the feeder vessel alongside the FHT were limited to incident waves of approximately 4.3m in height due to unacceptably high motions at higher wave heights.

The pitch motions of the FHT and the feeder vessel, when located inside the FHT well dock, are very similar over the entire range of wave heights investigated (Figure 9). However, when the feeder was located alongside the FHT the pitch motions are approximately eight times higher than the pitch motion of the feeder when inside the FHT well dock. The pitch motions of the FHT and feeder (when inside the well dock) can be seen to follow a general trend of increasing magnitude relative to the increasing incident wave height, which agrees with linear ship motion theory, Lloyd (1998).

Relatively small roll angles (<1 degree) were experienced by the feeder vessel when located inside the FHT well dock at all three wave heights (Figure 10). In contrast, when the feeder was located alongside the FHT the roll angles were found to exceed six times this level. The roll motions of the FHT were marginally greater when the feeder was side-by-side the FHT than inside the FHT well dock.

Both the rotational motions (pitch and roll) of the feeder vessel when moored inside the well dock are significantly less at all incident waves heights investigated (2.0m, 4.3m and 7.7m) than those measured at just 2.0m high incident waves for the alongside case. Practical operations handling container cargo are limited to six degrees of roll before the containers will jam in the guides, so the preliminary results indicate that the feeder motions inside the FHT well dock are acceptable. As previously mentioned, it may be possible to utilise alternative mooring arrangements to reduce the alongside motions, however, there will be a practical limit as to how effective and safe this will be. Further analysis of the experimental data is presently underway to investigate the relative motions between the two vessels, including the heave at the LCG and other critical locations of both vessels, as are investigations of the practicality of implementing a more substantial system to moor the feeder vessel within the well dock to further reduce the motions.

CONCLUDING REMARKS

The concept of a Floating Harbour Transhipper (FHT) for transferring bulk cargo offshore in open seas is outlined and discussed, including the use of smaller shallow-draught feeder vessels to transport bulk goods and equipment from small harbours or unprepared beaches to the FHT. A novel aspect of the FHT is the aft well dock in which the feeder vessels are moored during the transfer operation, thus taking advantage of the benefits of being sheltered from incident waves.

A series of physical scale model experiments were conducted to investigate the differences in relative motions of the feeder vessel when moored inside the FHT well dock compared to a more conventional side-by-side mooring arrangement. The well dock configuration was found to significantly reduce the motions of the feeder vessel, in some cases by an order of magnitude.

Realising the strategic importance of the marine industry, re-direction of Federal funding for many expensive non-redeployable military vessels under the defence spending umbrella should be focussed on supporting the local commercial marine industry which can come to the fore if needed. Australia's Defence support capability can be increased significantly at no extra capital cost by the 'dormant charter' use of modern cost-effective designs of a coastal shipping fleet. This is common practice for many European countries.

REFERENCES

Ballantyne, S. and Ballantyne, R., 2007, *Stern landing vessels*, Proceedings of the RINA International Conference on Military Support Ships, London, UK.

Buchner, B., van Diij, A. and de Wilde, J., 'Numerical Multiple Body Simulations of Side by Side Mooring to an FPSO', Proceedings of the 11th International Offshore and Polar Engineering Conference, Stavanger, Norway, pp 343 – 353, 2001.

Kodan, N., 'The Motions of Adjacent Floating Structures in Oblique Waves', Proceedings of 3rd Offshore Mechanics and Arctic Engineering, OMAE, New Orleans, Vol. 1, pp.206-213, 1984.

Koo, B.J. and Kim, M.H., 'Hydrodynamic Interactions and Relative Motions of Two Floating Platforms with Mooring Lines in Side-by-side Offloading Operation', Applied Ocean Research, Vol. 27, pp 292-310, 2006.

Lloyd, A.R.J.M, 1998, *Seakeeping – Ship behaviour in rough weather*, published by the author, available through the Royal Institution of Naval Architects, London.

ACKNOWLEDGEMENTS

The authors thank Albert Sedlmeyer and James Keegan at STS and Peter Tomic, Kirk Meyer and Liam Honeychurch of AMC for their invaluable assistance on this project.

Figure 1 – Floating Harbour Transhipper (mining). An SLV is in the well dock and a capesize vessel is moored alongside

Figure 2 – Stern Landing Vessel (Feeder Vessel) – Disaster Response

Figure 3 – Stern Landing Vessel (Feeder Vessel) – Military

Figure 4 – Model tests with the feeder vessel alongside the FHT

Figure 6 – a view of the model of the feeder vessel moored inside the well dock of the FHT model

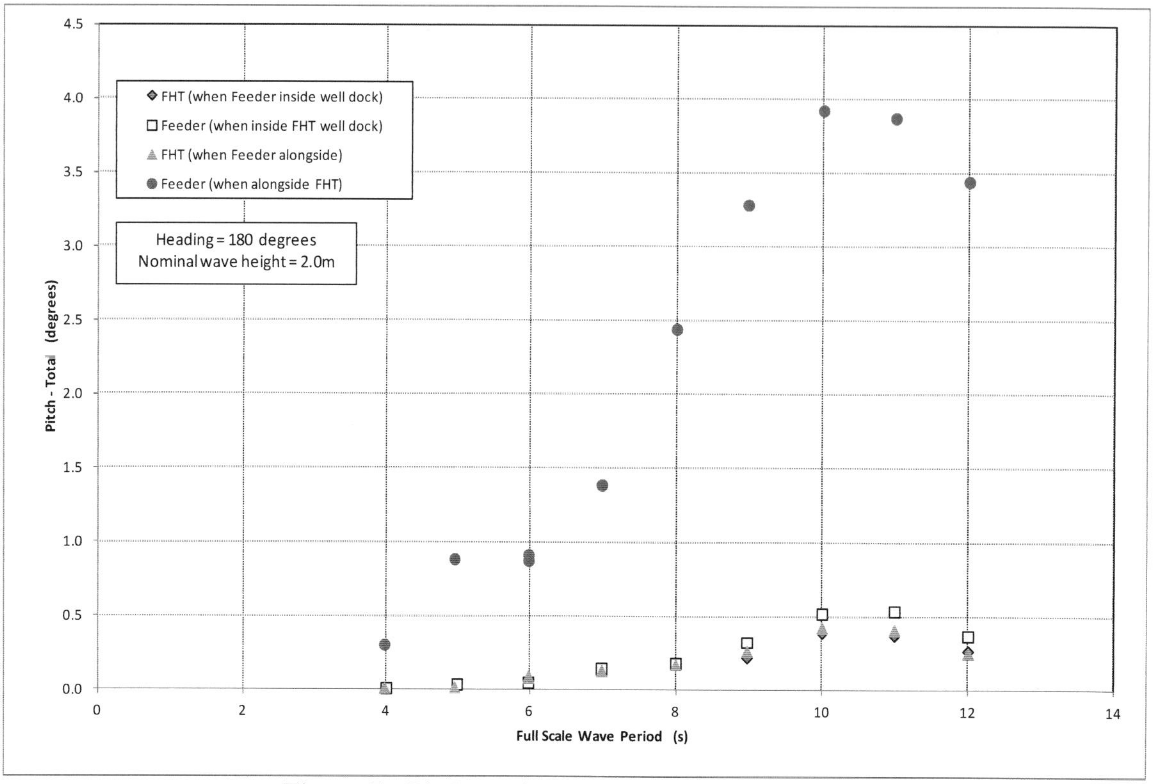

Figure 7 - Pitch motions as a function of wave period

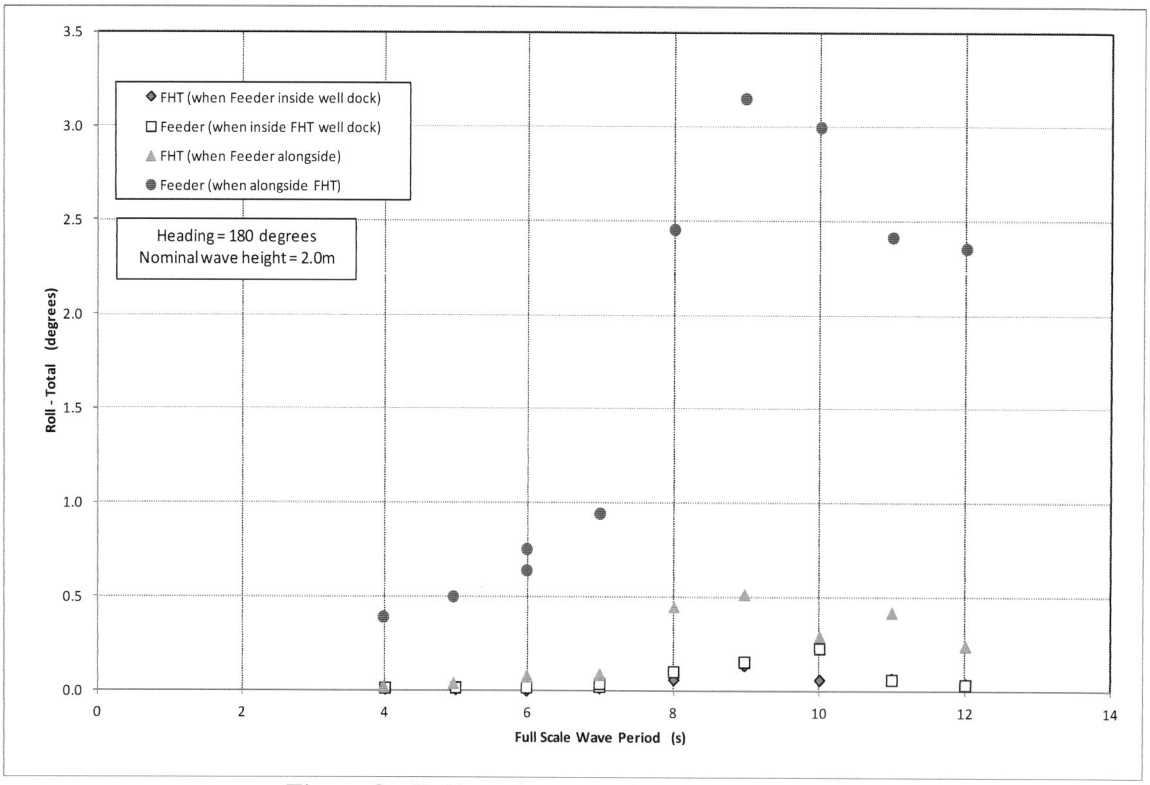

Figure 8 – Roll motions as a function of wave period

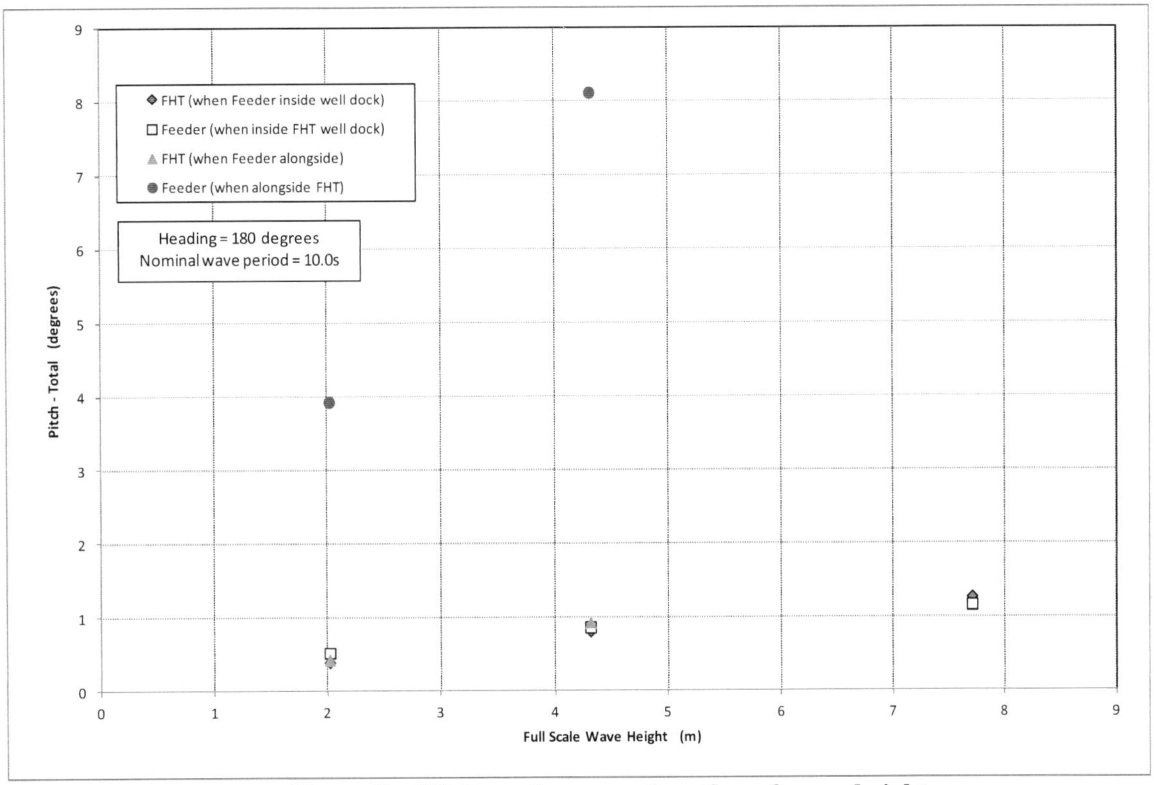

Figure 9 – Pitch motions as a function of wave height

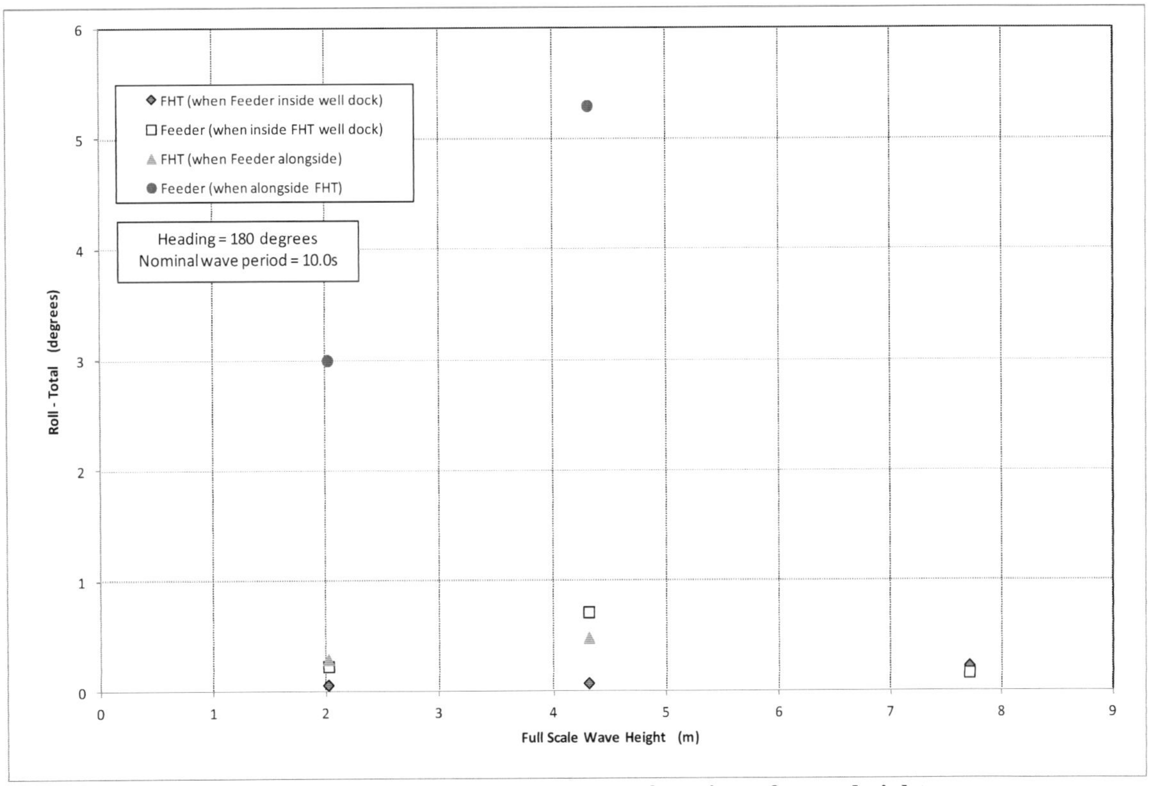

Figure 10 – Roll motions as a function of wave height

Predicting Berthed Ship – Passing Ship Interaction for Different Channel Geometries

Jonathan T. Duffy, Martin R. Renilson and Shaun P. Denehy
Australian Maritime College, Launceston, Australia
j.duffy@amc.edu.au

ABSTRACT

An investigation has been undertaken into the interaction forces and moments experienced by berthed ships due to passing ships. The aims were: (1) to quantify the influence of waterway geometry on the magnitude of the peaks and the form of the time domain interaction forces and moments; and (2) to quantify the influence of the form of the interaction forces and moments on berthed ship motions. The second aim was considered important, as in the past the peak values of the forces and moments were used, together with the assumption of an idealised form of the interaction forces, to predict the motions. It was felt necessary to verify this assumption, particularly for a range of different waterway geometries. Results were obtained from model scale experiments conducted in the Towing Tank at the Australian Maritime College to measure the time domain interaction forces and moments on a berthed ship due to a passing ship for different waterway geometries. The measured time domain interaction forces and moments are presented for three different channel geometries. The influence of the different channel geometries on the magnitude of the peaks and the form of the interaction forces and moments has been quantified. Measured and idealised interaction forces and moments were used as input to a time domain numerical simulation model to predict berthed ship motions. The influence of the form of the interaction forces and moments on the maximum berthed ship motions has been quantified for a typical mooring arrangement. From this the applicability of using an open water idealised form for cases with different channel widths and near bank geometries has been assessed.

Keywords: Ship interaction, Berthed ship motions

NOMENCLATURE

B	Ship beam	(m)
C_B	Block coefficient	
L_{PP}	Length between perpendiculars	(m)
N	Yaw moment	(Nm)
N''	Non-dimensional yaw moment	$\left[\frac{N}{0.5\rho U^2 L_{pp}^2 T}\right]$
T	Ship draught	(m)
U	Passing ship speed	(m/s)
X	Surge force	(N)
X''	Non-dimensional surge force	$\left[\frac{X}{0.5\rho U^2 BT}\right]$
Y	Sway force	(N)
Y''	Non-dimensional sway force	$\left[\frac{Y}{0.5\rho U^2 L_{pp} T}\right]$
ρ	Density of water	(kg/m^3)

1. INTRODUCTION

When a moving ship passes a berthed ship interaction forces are generated causing the berthed ship to move on its moorings. These motions should be minimised to prevent excessive mooring line loads and to reduce the likelihood of interruptions to loading/unloading procedures and damage to mooring and/or loading equipment.

Physical model scale experiments have been used in past studies to measure the interaction forces and moments on a berthed ship due to a passing ship [2,3,10,12].

Remery [10] used model scale experiments to investigate interaction forces and moments on tanker models and found that plotting the measured forces and moments to a base of the position of the passing ship relative to the berthed ship showed that the character of the curves was similar for the various sizes and passing distances of the passing ship.

Dand [2] conducted physical model scale experiments to measure the interaction forces and moments on a berthed ship due to a passing ship and developed formulae to predict the forces and moments.

Theoretical techniques have also been developed in previous work to predict the interaction forces and moments on a berthed ship due to a passing ship [5,8,9,13].

Using the results from Remery [10], Flory [4] found that if the interaction surge force, sway force and yaw moment were non-dimensionalised the various plots for different tanker sizes and separation distances were essentially the same, allowing one set of graphs to represent many cases. Flory [4] developed empirical formulae to predict the peaks of the time domain interaction forces and moments for cases where the passing ship path is parallel to the longitudinal axis of the berthed ship and the waterway width is infinite (open water).

Once the interaction forces and moments on the berthed ship are obtained, numerical simulation techniques can be used to predict berthed ship motions and mooring loads in the time domain. This information is important to ship operators and port authorities to ensure that the berthed ship motions and mooring line loads satisfy operational criteria [6,7]. Berthed ship motions due to external forces have been predicted using numerical simulations in previous studies [1,3,13,14].

In the present study interaction forces and moments on a berthed ship due to a passing ship were measured from physical model scale experiments for different channel geometries with varying near bank slopes and channel widths. The results were used to quantify the influence of the different channel geometries on the magnitude of the peaks and the form of the time domain interaction forces and moments.

The interaction forces and moments were then used as input to a time domain numerical simulation model to predict berthed ship motions and mooring loads. The influence of the form of the interaction forces and moments on the maximum berthed ship motions was quantified for a typical mooring arrangement. From this, the applicability of using the generic form of the forces and moments for open water cases proposed by Flory [4] was assessed when applied to cases with different near bank slopes and channel widths.

2. SCALE MODEL EXPERIMENTS TO MEASURE INTERACTION FORCES AND MOMENTS

Knowledge of how channel geometry influences the forces and moments acting on a berthed ship due to a passing ship is important to enable realistic real-time simulation of the behaviour of a berthed ship. An effective method for determining such knowledge is through the use of model scale experiments [2,3,10,12].

In the present study results have been obtained from physical model scale experiments conducted in the Towing Tank at the Australian Maritime College to measure the interaction forces and moments on a berthed ship due to a passing ship. The tests were conducted with a constrained model to simulate the berthed ship, and a towed model to simulate the passing ship. Interaction forces and moments were measured on the berthed ship model, which was constrained in surge, sway and yaw and free to heave, pitch and roll. Interaction forces and moments were not measured on the passing ship model.

2.1 Ship model details

All tests were carried out using two MarAd L series bulk carrier models [11]. The passing ship model was fitted with the standard cylindrical bow, whilst the berthed ship model was fitted with a bulbous bow. Principal particulars of each of the models are given in Table 1.

Table 1: Principal particulars of the models

Model	L_{PP}/B	B/T	C_B
Passing ship: Cylindrical bow bulk carrier	5.00	4.05	0.85
Berthed ship: Bulbous bow bulk carrier	5.00	4.05	0.85

2.2 Test program

Forces and moments on the berthed ship due to the passing ship were measured for three different channel geometries. The channel cross section along the Towing Tank was constant for each of the conditions. The passing ship path was parallel to the longitudinal axis of the berthed ship. The separation between the passing and berthed ship centrelines was constant for all conditions. The channel width for Case A represented the full width of the Towing Tank and is shown in Figure 1. Case B had a sloping near bank on the starboard side of the berthed model and is shown in Figure 2. For Case C the starboard bank was the same as Case B, but the distance between the passing ship model and the port bank was reduced, as illustrated in Figure 3. The water depth to draught ratio was 1.07 for the passing and berthed ship model in each case. The direction of travel of the passing ship is shown in Figure 4. The results in the present paper are for a passing ship speed corresponding to a Froude length number of 0.074 for each case. The speed of the passing ship model was constant for the duration of the passing manoeuvre.

2.3 Test procedure

The passing ship model was towed along the centreline of the Towing Tank. The passing ship model reached the desired constant speed well before it reached the berthed ship model. To maintain consistency in the tank test section and to synchronise the beginning of each measurement, the data acquisition was commenced at a common location for each test run. Sufficient time was allowed between consecutive runs to achieve calm water conditions.

The experiments yielded a time domain trace of interaction surge force, sway force and yaw moment. Examples of these traces are given in Section 4.

3. NUMERICAL SIMULATION OF BERTHED SHIP MOTIONS

The time domain interaction forces and moments obtained from physical model experiments were used as input to numerical simulations using Termsim II (a time domain numerical simulation program suite developed by MARIN [14]) to predict the motions of the berthed ship and associated mooring loads.

3.1 Berthed ship details

The berthed ship was modelled as a 250m L_{PP} bulk carrier with other principal particulars as per Table 1.

3.2 Mooring legs

A generalised mooring arrangement was used for the present study consisting of 6 mooring legs, as shown in Figure 4. Each mooring leg consisted of a number of mooring lines. The properties of each mooring line were based on Nylon double braided lines, 88.9mm in diameter with breaking strength of 1638kN (167t). The pre-tension on each mooring line was set at 147kN (15t), which is a typical value for such mooring arrangements.

3.3 Fenders

The fender reaction characteristics were modelled in the numerical simulations. Bridgestone C3000H cell fenders were modelled.

3.4 Berthed ship motions

Time domain traces of the motions of the berthed ship and the associated mooring loads were obtained from the numerical simulations. The surge, sway and yaw motions of the berthed ship are considered in the present paper.

Each numerical simulation was run for 200 seconds (full scale) before the passing ship force was applied to allow the berthed ship to obtain an equilibrium position. Examples of time domain berthed ship motions are presented in Section 4.

4. RESULTS

In Section 4.1 the measured interaction forces and moments on the berthed ship are presented for each case and compared to the idealised force and moment traces from Flory [4]. In each case the magnitude of the largest peak of the idealised force trace has been matched to the corresponding peak for the measured force trace. The results are presented such that the maximum negative surge force occurs at the same time for both the measured and idealised force traces.

The influence of channel geometry on the magnitude of the peaks and the form of the force and moment traces has been identified.

The sign convention is given in Figure 4 with the origin at midships. The midship location of the berthed and passing ships coincided at approximately 175 seconds.

4.1 Interaction forces and moments

4.1.1 Surge force

The non-dimensional surge force on the berthed ship for Cases A, B and C is presented in Figure 5. The general form of the measured surge force traces is similar for all cases.

From Figure 5 it can be seen that the measured surge force was initially in the opposite direction to the passing ship direction (negative) followed by a positive surge force after the midship locations of the ships coincided until the end of the pass. Dand [2] measured interaction surge forces with similar general form. It can be seen that the maximum interaction surge force was greatest for the case with a sloping near bank and narrow channel width (Case C).

The idealised surge force traces deviated from the measured surge force traces in the early stages of the pass, as it predicted a positive surge force. The magnitude of the negative peak of the idealised surge force has been matched to the measured value. The magnitude of the positive peak at approximately 200 seconds was overestimated by the idealised curve. The measured surge force was positive in the latter stages of the pass; however the idealised trace predicted a negative surge force after approximately 250 seconds. It can be seen that, in general, the times at which the peak forces and moments occurred for the idealised traces correlated well with the measured traces.

4.1.2 Sway force

The non-dimensional interaction sway force on the berthed ship for Cases A, B & C is presented in Figure 6. The general form of the force and moment time domain traces is similar for all cases. The measured sway force was towards the fenders as the passing ship approached, was away from the fenders when the passing ship was abeam the berthed ship and was towards the fenders in the latter stages of the pass.

It can be seen that when the sloping bank was alongside the berthed ship (Cases B and C), the magnitude of the peak sway force was less than that for the wider channel case (Case A). Pinkster and Ruijiter [9] found a similar trend when comparing forces for a vertical quay and open water.

The magnitude of the idealised positive sway force peak was matched to the measured value. The magnitude of the negative peaks was generally underestimated by the idealised curves. It can be seen that, in general, the times at which the peak forces and moments occurred for the idealised traces showed reasonable correlation with the measured traces.

4.1.3 Yaw moment

The non-dimensional interaction yaw moment on the berthed ship for Cases A, B and C is presented in Figure 7. The general form of the moment time domain traces is similar for all cases.

The yaw moment was generally bow away from the wharf as the passing ship approached. Between approximately 120 and 170 seconds the yaw moment became bow in to the wharf. The yaw moment was bow away from the wharf as the passing ship moved away, followed by a small bow in yaw moment.

The magnitude of the idealised negative yaw moment peak at approximately 150 seconds was matched to the measured value. The magnitude of the positive yaw moment peaks was underestimated by the idealised traces for Cases B and C, as can be seen in Figure 7. From Figure 7 it can be seen that, in general, the times at which the peak forces and moments occurred for the idealised traces showed reasonable correlation with the measured traces.

4.2 Predicted berthed ship motions and loads

4.2.1 Time domain berthed ship motions

The measured and idealised time domain interaction forces and moments on the berthed ship were used as input to Termsim II to perform numerical simulations of the berthed ship behaviour. Predicted berthed ship surge, sway and yaw motions are presented in Figures 8, 9 and 10, respectively. The non zero position at zero time represents the equilibrium position of the berthed ship. The midships of the passing and berthed ships coincided at approximately 175 seconds.

4.2.2 Maximum berthed ship motions

The predicted maximum berthed ship motions are presented for each case using the measured and idealised interaction force and moment traces. The magnitude of the predictions using the different force and moment inputs was assessed to determine the applicability of the idealised interaction force and moment traces for the cases tested.

The peak to peak surge, zero to peak sway and peak to peak yaw motions for Cases A, B and C are presented in Figures 11, 12 and 13, respectively.

As expected, the surge motion was greatest for the cases with the sloping near bank and narrow channel (Case C). The largest sway and yaw motion was experienced for the wide channel case (Case A).

The predicted surge motion was marginally larger when using the idealised force and moment traces for all cases.

The difference between the predicted berthed ship motions using the measured force and moment input and the idealised force and moment input for each case is given in Table 2. For the cases tested, there was only a marginal difference in the predicted maximum berthed ship motions when using the idealised force and moment traces that are based on open water data. The differences in the predicted values are generally within experimental uncertainty.

From Table 2 it can be seen that, in general, the difference in berthed ship motions did not increase for the cases with narrow channel width, despite the idealised forces and moments being based on the situation in open water.

From the present analysis the maximum predicted berthed ship motions using the idealised forces and moments were similar to those predicted with the measured forces and moments. Hence, within experimental uncertainty, the idealised force and moment curves appear to be suitable to predict the maximum berthed ship motions for early design work for the cases tested, in the absence of measured data, provided the values of the peak forces and moments can be predicted.

Analysis of further cases with varying degrees of lateral channel restriction is being undertaken to assess the applicability of using the idealised interaction force and moment traces to predict the berthed ship motions for more general geometries.

Table 2 Percentage difference in berthed ship motions predicted with measured and idealised forces and moments

Case	**Surge (%)**	**Sway (%)**	**Yaw (%)**

A	6	2	13
B	5	-11	-3
C	10	-4	9

4.2.3 Maximum Mooring Leg Loads

The predicted maximum mooring leg loads for Case A using the measured and idealised interaction force and moment traces is shown in Figure 14. As can be seen the breast lines experienced the greatest load. The same was found for Cases B and C. The maximum loads in the forward breast and aft breast lines for each case are shown in Figures 15 and 16, respectively.

The difference in the predicted maximum mooring leg loads when using the idealised and the measured force and moment traces is shown in Table 3. For Case A (widest channel) the largest difference was 16% for the forward breast lines.

The maximum difference in mooring leg load for Case B was 24% on the aft spring lines and for Case C it was 19% on the forward spring lines. In both these cases the absolute difference in load between the idealised and measured force and moment inputs was small, however as the maximum load on the spring lines was not very large this gave a large percentage difference. The percentage difference in the breast legs for Cases B and C (which were the critical legs, with large forces) was less than 10% in both cases.

It can be seen that the difference between the maximum predicted mooring leg loads was generally higher than the difference in predicted maximum berthed ship motions. However, the differences for the critical legs are generally within experimental uncertainty when measuring the interaction forces and moments. Hence, within experimental uncertainty, the idealised force and moment curves may be suitable to estimate the limiting situations in the early stages of berth design for the cases tested, in the absence of measured data. Analysis of further cases is being undertaken to assess the applicability of using the idealised interaction force and moment traces to predict berthed ship motions and mooring loads for cases where there is varying degrees of lateral channel restriction.

Table 3: Percentage difference in berthed ship mooring leg loads with measured and idealised forces and moments

Case	Mooring leg					
	Head (%)	**Fwd breast (%)**	**Fwd spring (%)**	**Aft spring (%)**	**Aft breast (%)**	**Stern (%)**
A	-11	16	11	11	8	6
B	1	2	9	24	-1	5
C	-7	7	19	10	9	10

5. CONCLUDING REMARKS

An investigation has been undertaken into the interaction forces and moments experienced by a berthed ship due to a passing ship. The influence of waterway geometry on the magnitude of the peaks and the form of the time domain interaction forces and moments has been quantified for a selection of test cases. Berthed ship motions have been predicted for cases with varying lateral channel restriction using both interaction force and moment data from physical model experiments and idealised forces and moments based on open water data. The influence of the form of the interaction forces and moments on maximum berthed ship motions has been quantified for the test cases.

From the results it was found that:

- the general form of the interaction forces and moments on the berthed ship were similar for the test cases with varying degrees of lateral channel restriction;
- the magnitude of the peak interaction surge force was largest for the narrow channel configurations for constant ship passing distance; and
- the greatest peak interaction sway force and yaw moment was experienced with the widest channel configuration for a constant ship passing distance.

From the present analysis the maximum berthed ship motions and maximum mooring leg loads predicted using the idealised forces and moments were slightly different to those predicted with the measured forces and moments. However, the differences are generally within experimental uncertainty when measuring the interaction forces and moments. Hence, within experimental uncertainty the idealised force and moment curves may be suitable to predict the limiting situations in the early stages of berth design for the cases tested, in the absence of measured data, provided that the maximum values of the forces and moments can be predicted.

Further analysis is being undertaken to assess the applicability of the idealised interaction force and moment traces for different channel geometries.

6. REFERENCES

1. Dand, I.W. (1981), Simulation of the behaviour of a moored ship when passed by other ships, *National Maritime Institute*, Report NMI R111.

2. Dand, I.W. (1981), Some measurements of interaction between ship models passing on parallel courses, *National Maritime Institute*, Report NMI R108.

3. Duffy, J.T., Webb, G. (2003), Berthed ship-passing ship interaction: a case study for the Port of Newcastle, *Coasts and Ports 2003,* Auckland, 9-12 September.

4. Flory, J.F. (2002), The effect of passing ships on moored ships, *Proceedings of Prevention First 2002 Symposium,* Long Beach, 10-11 September.

5. Muga, B.J., Fang, S.T. (1975), Passing ship effects from theory and experiment, OTC 2368, *Proceedings of 1975 Offshore Technology Conference*, Texas.

6. OCIMF, (1997), *Mooring equipment guidelines, 2nd Edition*, Witherby & Co. Ltd., London.

7. PIANC, (1995), *Criteria for movements of moored ships in harbours, a practical guide s1, Report of Working Group no. 24 of the Permanent Technical Committee II,* Supplement Bulletin no. 88n, 1995.

8. Pinkster, J.A., Naaijen, P. (2003), Predicting the effect of passing ships, *Proceedings of the 18th International Workshop on Water Waves and Floating Bodies*, Le Croisic, France.

9. Pinkster, J.A., Ruijiter, M. (2004), The influence of passing ships on ships moored in restricted waters, *Offshore Technology Conference*, Texas, 3-6 May.

10. Remery, G.F.M. (1974), Mooring forces induced by passing ships, OTC 2066, *Proceedings of 1974 Offshore Technology Conference,* Texas.

11. Roseman, D.P. (ed.) (1987), The MarAd systematic series of full form ship models, *SNAME Publications*.

12. Simmons, T., Isaacs, L.T., Renilson, M.R. (1992), Bank effects on vessel interaction forces, *Proceedings of the 11th Australasian Fluid Mechanics Conference*, Hobart, 14-18 December.

13. Spencer, J.M.A., McBride, M.W., Beresford, P.J., Goldberg, D.G. (1993), Modelling the effects of passing ships, *International Colloquium on Computer Application in Coastal and Offshore Engineering*, Kuala Lumpa.

14. Van Der Molen, W., Ligteringen, H., Van Der Lem, J.C., de Waal, J.C.M. (2003), Behaviour of a moored LNG ship in swell waves, *Journal of Waterway, Port, Coastal and Ocean Engineering*, January/February.

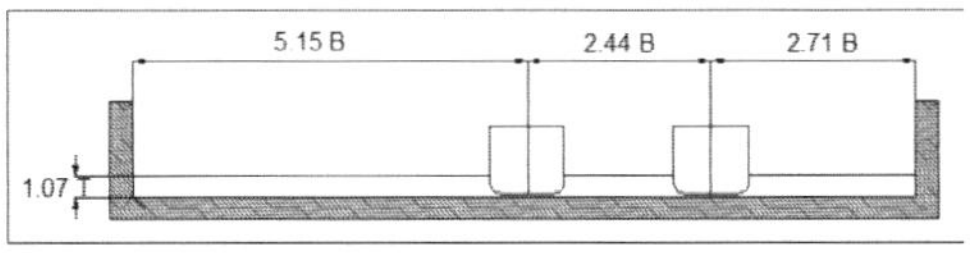

Figure 1: Channel cross section for Case A

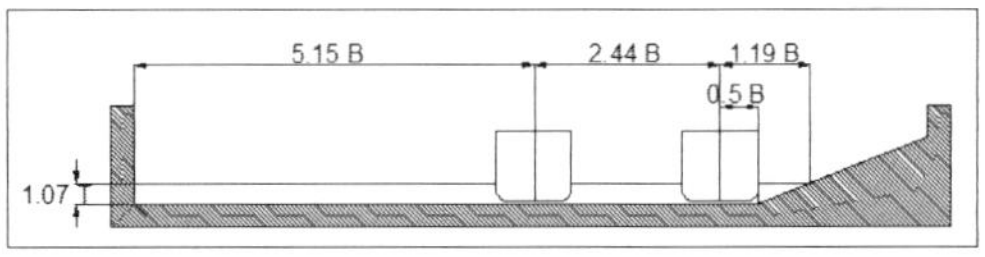

Figure 2: Channel cross section for Case B

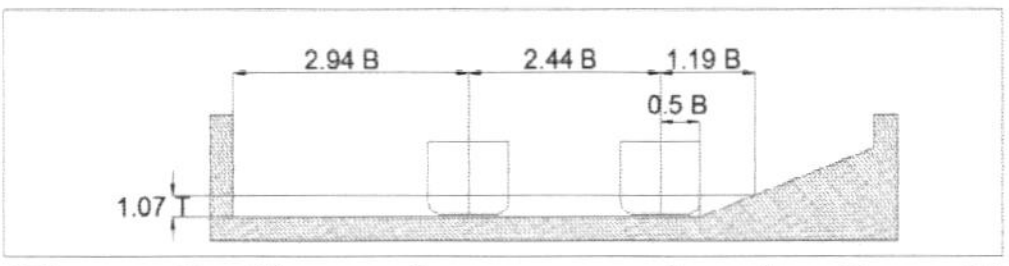

Figure 3: Channel cross section for Case C

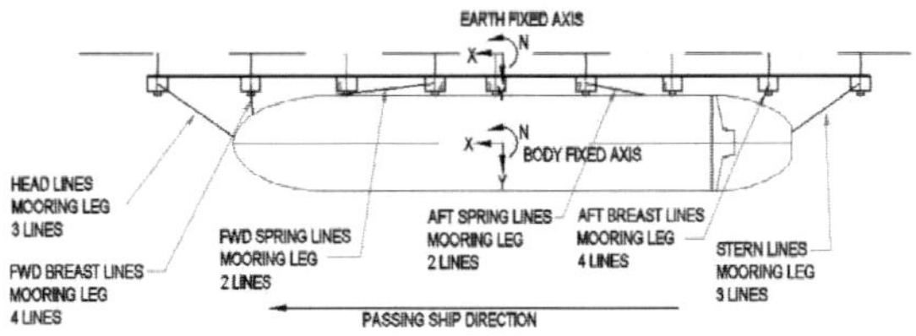

Figure 4: Schematic of the mooring arrangement and sign convention

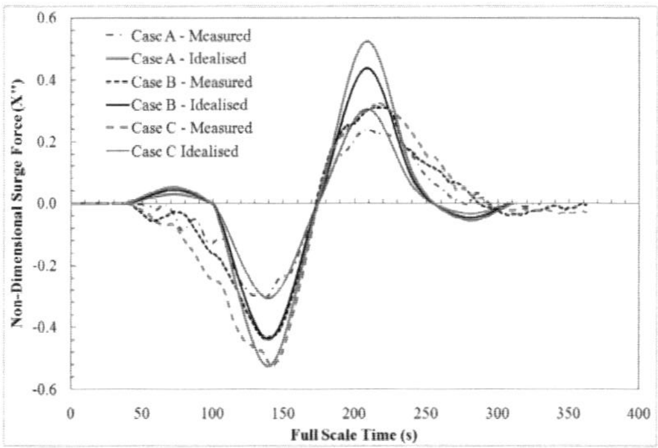

Figure 5: Non-dimensional interaction surge force for Cases A, B & C

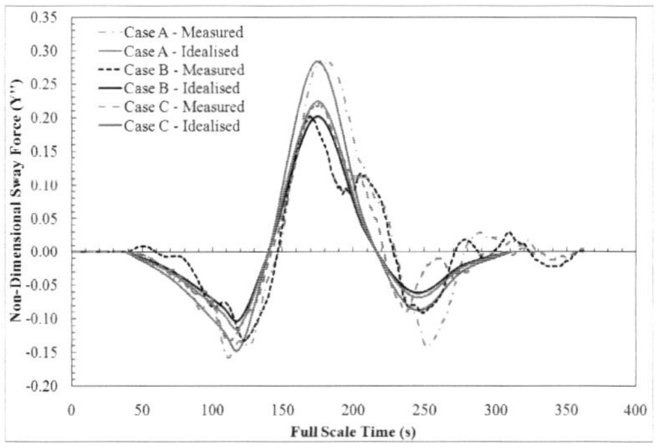

Figure 6: Non-dimensional interaction sway force for Cases A, B & C

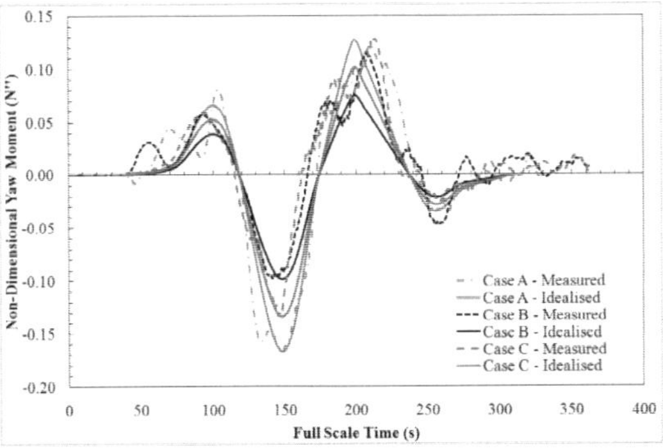

Figure 7: Non-dimensional interaction yaw moment for Cases A, B & C

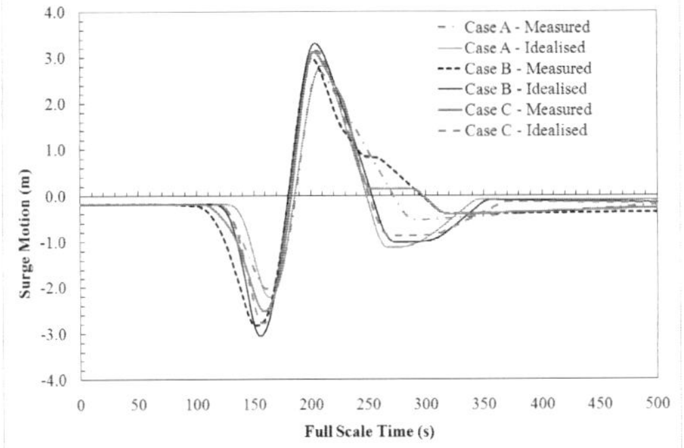

Figure 8: Predicted surge motion

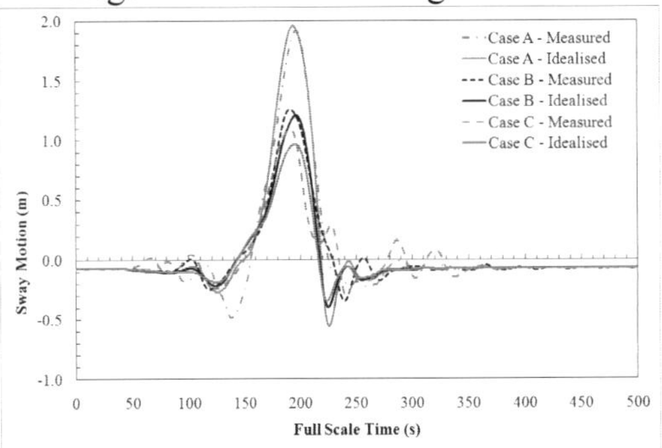

Figure 9: Predicted sway motion

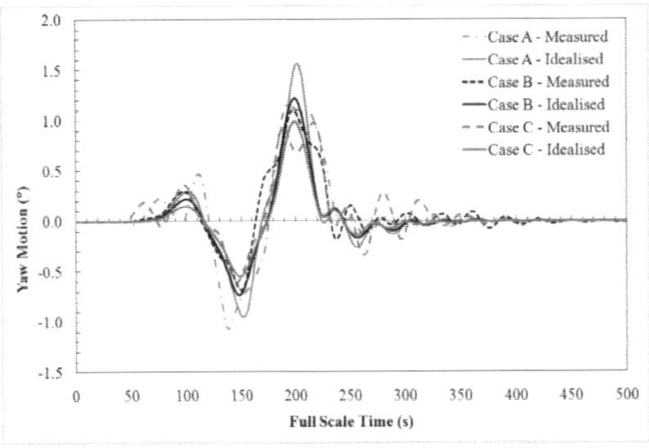

Figure 10: Predicted yaw motion

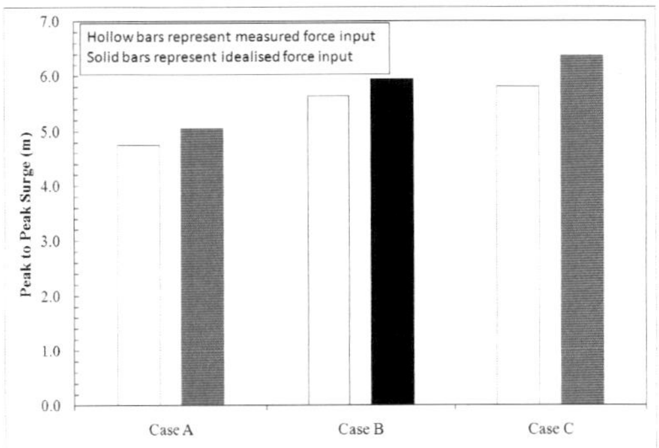

Figure 11: Predicted maximum surge motion

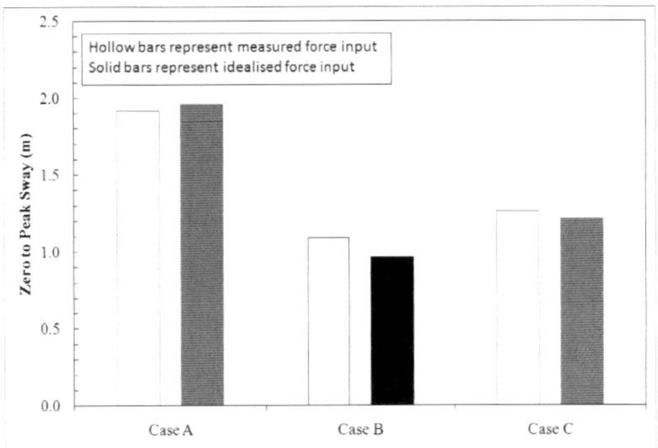

Figure 12: Predicted maximum sway motion

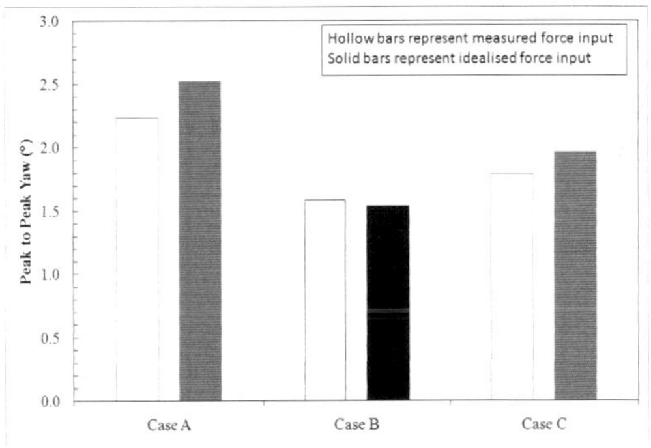

Figure 13: Predicted maximum yaw motion

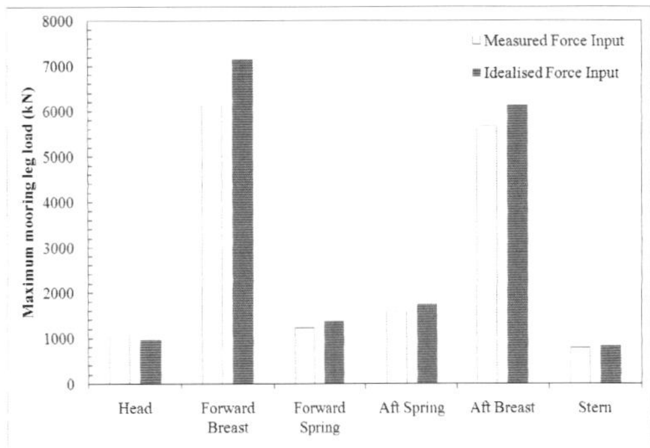

Figure 14: Predicted maximum mooring leg loads for Case A

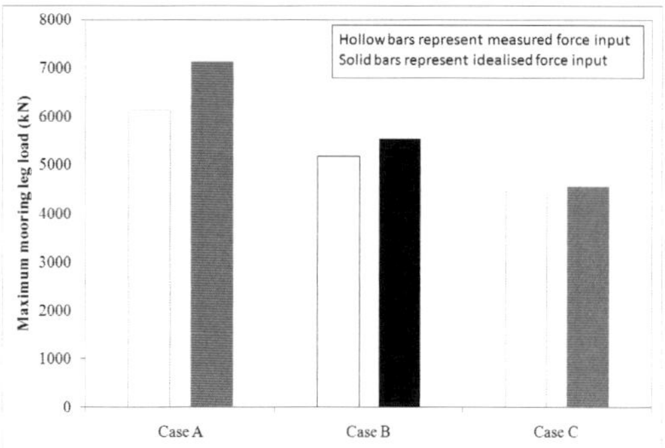

Figure 15: Predicted maximum load in the forward breast leg

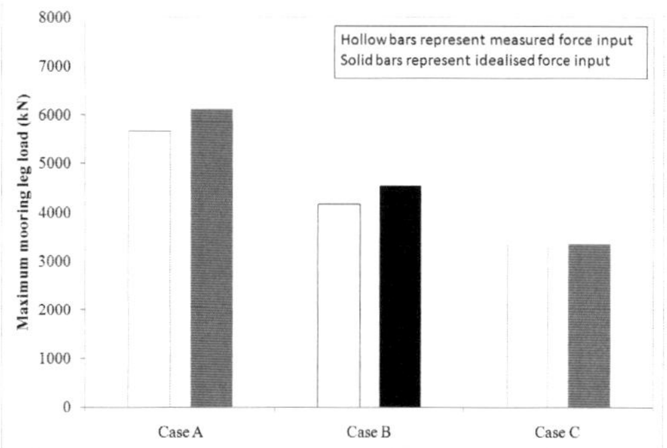

Figure 16: Predicted maximum load in the aft breast leg

Particulate Emission Characteristics from Medium-Speed Marine Diesel Engines

Sergey Ushakov, Norwegian University of Science and Technology, Norway
sergey.ushakov@ntnu.no
Harald Valland, Norwegian University of Science and Technology, Norway
Jørgen B. Nielsen, MARINTEK AS, Norway
Erik Hennie, MARINTEK AS, Norway

Particulate matter (PM) emissions from diesel engines are of major concern due to their adverse health effects. While PM emissions from the automotive sector have been regulated for decades, there have been no direct regulations on PM from sea-going vessels. The engines used in maritime sector are significantly different from automotive engines not only regarding their size, but also in regard to power output and fuels used.

This study presents results from exhaust particle emission measurements from two different medium-speed turbocharged marine diesel engines: 2-stroke Wätrsilä WX28B and 4-stroke Rolls-Royce KR3. The engines were tested according to ISO 8178-4 E2 and E3 cycles, using two typical marine fuels: marine gas oil (MGO) and heavy fuel oil (HFO). The size distributions were obtained using Scanning Mobility Particle Sizer (SMPS) and Electrical Low Pressure Impactor (ELPI), operated simultaneously.

The obtained number distributions for 2-stroke engine are mainly bimodal with the pronounced nucleation mode, representing mainly non-volatile species originating from lubrication oil, and found below 20 nm and accumulation mode at 20-60 nm, referred to carbonaceous agglomerates, while for 4-stroke engine only accumulation mode was observed, and appeared to be insensitive to changes in load conditions. Additionally, results revealed a surprisingly good agreement both in total particle number and mean count median diameter values between 4-stroke Rolls-Royce KR3 and 4-stroke heavy-duty diesel engine for MGO fuel, while results from a 2-stroke Wätrsilä WX28B engine were very different.

1. INTRODUCTION

During the last decades the particulate matter emissions have become a topic of significant interest. A number of toxicological and epidemiological studies (e.g. Dockery et al., 1993; Pope, 2000; Pope at al., 2002) confirmed the existence of statistical correlation between the daily mortality and morbidity and the exposure to small particles. Diesel engine PM emissions are known as a major source of fine particles in the environment with the majority of them being in nanoparticle (i.e. below 50 nm) size range (Kittelson, 1998). Modern medical investigations showed the possibility of nanoparticles to penetrate into the human body cells (Suzuki et al., 2007) causing structural damage and can have a potential to cross the blood-brain barrier (Oberdörster et al., 2004). So the awareness of the probable adverse health effects associated with aerosol particles became the main driving force to limit diesel PM mass emissions. The regulations originally restricted the total suspended particulate (TSP) mass, but later shifted their main attention to mass of particles having aerodynamic diameter smaller than 10 μm and 2.5 μm respectevely, resulting in corresponding PM10 and PM2.5 standards, as more strong evidences of smaller particle effect on human health have been found (e.g. Goo and Kim, 2003). Although, not all researchers agree that PM mass is the best appropriate metric to assess the negative effects of diesel PM on human health. Other suggestions are, for example, surface area (e.g. Oberdörster et al., 1996) or/and particle number concentration (Ferin et al., 1992) to be more appropriate.

A number of studies comprehensively investigated PM mass, number concentrations and size distributions from automotive light- and heavy-duty diesel engines (e.g., Kittelson, 1998; Kittelson et al., 2006). The diesel PM emissions normally range from 3 nm to approximately 1 μm in diameter with characteristic bimodal-shape size distribution with two distinctive modes. The nucleation mode representing the particles smaller 50 nm is believed to be semivolatile and consisting mainly of hydrocarbons and sulphuric acid PM formed during the dilution process (Kittelson et al., 1999), but could also contain non-volatile species originated from lubrication oil (e.g. Tobias et al., 2001; Sakurai et al., 2003), while accumulation mode is usually associated with carbonaceous agglomerates larger than 50 nm in size with volatile compounds adsorbed on the surfaces of these solid particles.

Despite the large number of studies concerning automotive engine PM, there were only few ones (e.g. Lyyränen et al., 1999; Kasper et al., 2007) related to particle emissions from marine diesel engines (MDE), which are believed to be as significant PM source as road traffic (Eyring et al., 2005) with corresponding adverse health effects resulting in number of serious diseases and decrease of life expectancy (e.g. Corbett et al., 2007). And while the particle emissions from in-land sources continuously decrease due to stricter environmental standards applied, the amount of PM emitted by sea-shipping can either decrease as the permittable sulfur content in marine fuels will be reduced and a wider use of LNG fuel is expected or it can increase because of significantly increasing ship traffic and absence of any direct regulation of particle emissions from marine diesel engines. This is still a hot topic for debates and up to now none of these theories has been validated.

There are several principal differences between automotive and marine diesel engines, which should be pointed out here. Firstly, they operate on highly different types of fuel with cars and trucks using a well-refined, high-quality products, and MDEs operating mostly on heavy fuels. It means the latter type of engines has to burn fuels with relatively large content of sulphur, ash and other residual contaminants, i.e. the compounds that are avoided in auto diesel fuel. Secondly, engines' geometrical dimensions and size also differ significantly, which may have effect on engine thermodynamical performance substantively. Thirdly, operating characteristics are not the same for automotive engines and MDEs. So taking into account all the mentioned differences, it can be expected that there will be also a pronounced difference in PM emission characteristics.

In this study the formation of particles, their physical properties and morphology were investigated from two very different marine diesel engines, operating on marine gas oil and low-grade residual fuel oil. Essentially the same instrumentation and measurement techniques as used for automotive PM characterization were applied here, allowing intercomparison of the results. Additionally, particle morphology and composition were assessed using scanning electron microscopy (SEM) with additional energy-dispersive X-ray (EDX) spectroscopy.

2. EXPERIMENTAL

2.1. Test engines, fuels and operating conditions

Two engines were evaluated in the current study: a 2-stroke Wätrsilä WX 28B and a 4-stroke Rolls-Royce KR3. They are both turbocharged medium-speed, direct-injection (DI) marine diesel engines with the detailed technical specification provided in Table 1. For relative comparison, the main specifications for a turbocharged 4-stroke DI heavy-duty engine are shown in the right column highlighting the engine differences discussed before. Both engines during the experiment were operated on two totally different types of fuel: low-sulphur marine gas oil (MGO), which can be considered as the cleanest fuel used in maritime industry, and heavy fuel oil (HFO) that is pure or nearly pure residual fuel oil. The major fuel specifications are presented in Table 2. In fact, HFO is the main fuel used in sea-shipping, since it is much cheaper comparing to other fuel types, but at the same time contains such undesirable residuals like ash, asphaltenes and metal sediments. The sulphur content of marine fuels is much more significant, reaching up to 4.5% by mass, than in well-refined fuels used in on-road diesel vehicles that normally use diesel fuel with around 10 ppm of sulphur in most European counties. The high sulphur content could lead to the formation of large amounts of sulfate particles, as well as to high SO_x emission concentrations. Additional factor influencing diesel PM level and composition is the content of sulphur, ash and metal compounds in lubrication oil used (Kittelson et al., 2008).

Table 1 Specifications of the test engines.

Engine type/specification	Loop-scavenged 2-stroke Wätrsilä WX 28B	4-stroke Rolls-Royce KR3	4-stroke Scania DC1102
Application	Ship propulsion	Ship propulsion/ Electricity generation	Truck engine
Number of cylinders	1	3	6
Cylinder bore, mm	280	250	127
Piston stroke, mm	360	300	140
Displ. per cylinder, l	22.2	14.7	1.8
Power, kW	300	500	280
Speed, rpm	600	750	1800
MEP, bar	13.5	18.1	15.4

Table 2 Major fuel specifications.

Fuel type/parameters	Marine Gas Oil	Heavy Fuel Oil
Viscousity@50 °C, cSt	2,5	380
Density@15 °C, kg/m^3	845	991
Sulphur content, % by mass	0,05	1,5
Vanadium, mg/kg	<1	300
Aluminum + silicium, mg/kg	<1	80
Zinc, mg/kg	<1	15
Phosphor, mg/kg	<1	15
Calcium, mg/kg	<1	30
Ash content, % by mass	<0,01	0,15
Water content, % by volume	0,01	0,5

There are currently no standard test cycle for marine diesel engines when measuring PM emissions. And the legislative body, International Maritime Organization (IMO), is limiting only NO_x emissions and allowed fuel sulphur content. Therefore, the same fixed operating conditions as used for NOx emission certification in accordance with the rules of IMO NO_x Technical Code (as stated in Annex VI of the MARPOL 73/78 convention) were adopted for current study to obtain PM size distributions. Due to different applications, the investigated MDEs were operated on different test cycles with 2-stroke Wätrsilä WX 28B tested over ISO 8178-4 E3 cycle for propeller-law-operated engines and ISO 8178-4 E2 test cycle for constant-speed main propulsion application used for a 4-stroke Rolls-Royce KR3 engine. Load, speed and corresponding exhaust gas temperatures (after turbocharger) are listed in Table 3. It should be noted that due to the PM data acquisition system problem the PM emission results from a 2-stroke engine operated on HFO at maximum load were lost and there were no possibility to repeat the experiment due to a turbocharger change.

Table 3 Test points with corresponding measured exhaust gas temperature.

Engine type/spec.		2-stroke Wätrsilä WX 28B				4-stroke Rolls-Royce KR3			
Load, %		100	75	50	25	100	75	50	25
MGO	Speed, rpm	659	600	523	415	750	750	750	750
	Exhaust gas temp., ° C	258	267	249	236	367	353	330	277
HFO	Speed, rpm	659	600	523	415	750	750	750	750
	Exhaust gas temp., ° C	261	270	253	240	326	321	317	287

2.1. Particulate matter sampling and measurement set-up

All exhaust gas samples during this study were extracted from the engine tailpipe with the help of "chinese hat" probe fitted near the centerline of the exhaust pipe. This special type of sampling probe was chosen to prevent the sampling system from overloading especially by big particles. After been sampled the extracted gas was transported to the dilution system by means of 1.5 m-long sampling line that was kept heated up to 400 °C. The heating was applied as it firstly prevents volatile species in the gas stream from condensation and, secondly eliminates the PM losses caused by thermophoretic forces.

Schematic representation of PM sampling-measurement set-up used in the study is provided in Figure 1. A commercially available Dekati two-stage ejector dilutor was used to dilute the sampled gas with properly conditioned particle-free dry air. This process roughly simulates the actual physical sub-processes and chemical reactions that occur when hot combustion products mix with much cooler atmospheric air. It was decided to study only solid particles, so barely "hot" dilution with applied primary dilution temperature (PDT) of as high as 400 °C was used. In reality the exhaust gas is diluted with much cooler ambient air which causes condensation of vapour-phase semi-volatile hydrocarbons and sulphuric acid via homogeneous or heterogeneous nucleation (Khalek et al., 2000; Kittelson et al., 1999). The total dilution ratio (DR) was kept in the range of 50-80 during the experiment for all test points. Its value was calculated as the ratio of NO_x concentration in the raw exhaust to that in the diluted exhaust. The applied method normally gives a more precise value of the dilution ratio than the one where values of DR is calculated based on the measured diluent and inlet flow rates (De Filippo and Maricq, 2008).

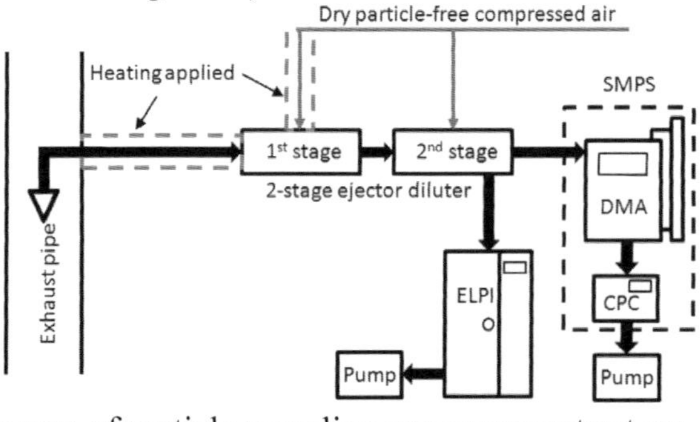

Figure 1 Schematic diagram of particle sampling-measurement set-up.

Scanning mobility particle sizer (SMPS) and electrical low-pressure impactor (ELPI) were used to obtain the required particle number and size distributions in the study. They were arranged to work in tandem increasing the measurement system capability. SMPS spectrometer (Knudson and Whitby, 1975) provided the main number size distribution results as it has a very high particle size resolution, but on the other hand its slower response makes it applicable only for steady-state measurements. The studied particle size range was approximately 6-215 nm, which was dictated by instrument flow rate settings. A 12-channel ELPI (Keskinen et al., 1992), as all cascade impactors, has a drawback of a much coarser size resolution, but at the same time has several great advantages like high time-resolution of 2-3 sec, and also can cover a very wide size range from 17 nm to 10 μm. It can be also used to collect PM mass for gravimetric and/or composition analysis. In this study, 47-mm thin aluminium foils, greased with Apiezon L vacuum grease to avoid particle bouncing, were installed on ELPI collection plates were used to collect the particles for further scanning electron microscope (SEM) and energy-dispersive X-ray (EDX) analysis. Several selected films, containing PM mass, were later analyzed with SEM/EDX techniques by Jeol JXA-8500F instrument, which is a high-performance thermal field emission electron probe micro analyser combining high SEM resolution with high quality X-ray analysis of submicron areas.

3. RESULTS AND DISCUSSION

3.1. PM emission characteristics for various fuels and operating conditions

Particle size distributions from a 2-stroke Wätrsilä WX 28B and a 4-stroke Rolls-Royce KR3 engines operated on MGO and HFO were obtained for various operating conditions using SMPS in rather narrow size range of 6-215 nm. The results were plotted on log-log scale and are presented on Figure 2. All the results were acquired with the "hot" dilution settings with corresponding PDT=400 °C, meaning that they reflect the distributions of mainly solid, non-volatile particles in diesel exhaust. The "hot" dilution allowed to keep the volatile species, normally comprised of light semi-volatile hydrocarbons and sulphuric acid (Kittelson, 1998; Tobias et al., 2001), in the vapour phase as they were beyond the main scope of the study. At the same time, authors want to highlight the significant contribution of the mentioned volatile components to PM emission, especially in terms of particle number, but were comprehensively studied elsewhere (e.g. Khalek et al., 2000; Kittelson et al., 2006).

The size number distributions from the 2-stroke Wätrsilä WX 28B engine showed a pronounced bimodal shape both on MGO and HFO for all operating conditions, except the maximum load where the distribution shape was unimodal (only accumulation mode). As expected and seen on Figure 4, the total particle concentration was higher when operating on residual HFO than on MGO with the highest difference of around an order of magnitude occurred at low load. The marked distinction occurred in the size range of the smallest particles that contributed from approximately 65% to 90% of total PM on all the fuels as seen in Figure 5, while there was particularly no difference in number concentrations above 100 nm due to fuel quality at any of the test points. The mean CMD values shown on Figure 3 increase from 18.3 nm to 19.9 nm and from 30.8 nm to 42.6 nm at 25% and 75% load respectively when shifting from MGO to HFO, which is in accordance to observations from other researchers (Ristimaki et al., 2010). All this can be probably explained by the significant difference in fuel composition with special emphasis on ash content between MGO and HFO (Lyyränen et al., 1999), but not the decaying nature of the nucleation mode with the load, especially when applying "hot" dilution. The earlier observations from this engine indicated the high lub oil consumption (especially below 50% load) which can result in additional production of non-volatile nucleation mode particles (Sakurai et al., 2003). The high lub oil consumption is more likely caused by specific engine design characterized by 'valveless' cylinder cover and exhaust port located in the bottom part of the cylinder, which is more typical for V-type engines. So authors do believe that observed nuclei mode on both fuels is composed primarily from metal compounds originated from lubrication oil (Sakurai et al., 2003). The accumulation mode formed when running on MGO is believed to compose mainly of carbonaceous agglomerates due to incomplete fuel combustion, while on HFO it is composed of carbon particles together with incombustible species originated from fuel ash.

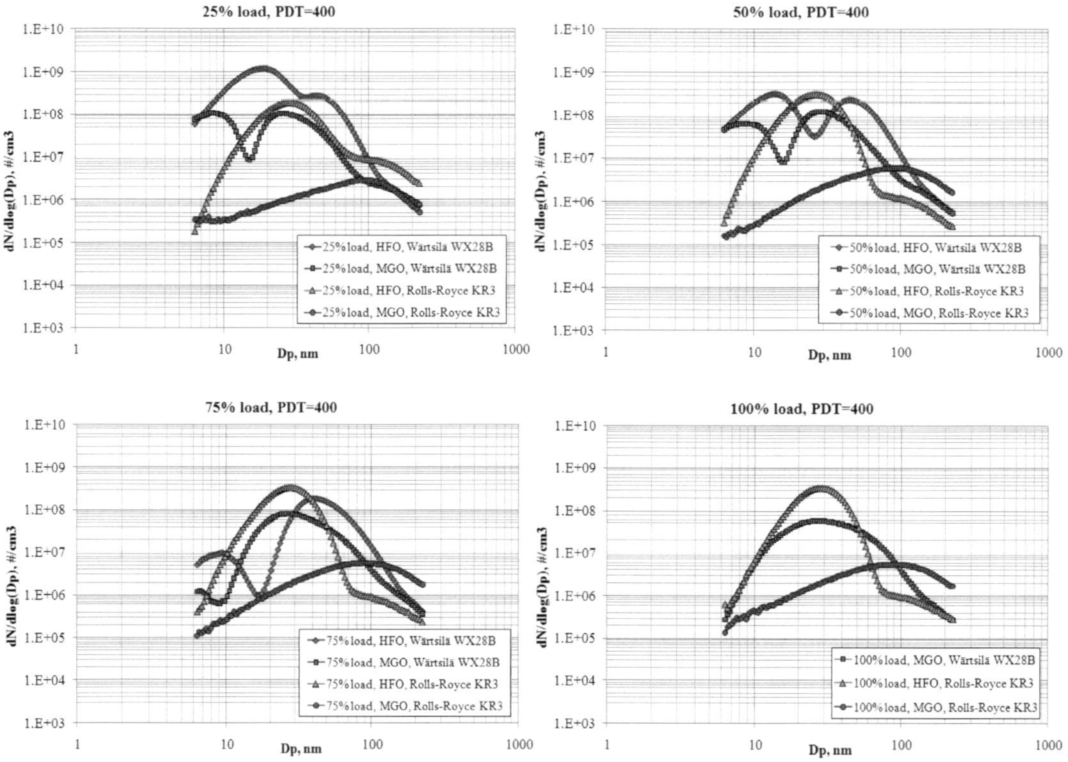

Figure 2 Particle number distribution characteristics at various operating conditions for 2-stroke Wätrsilä WX 28B and 4-stroke Rolls-Royce KR3 operated on MGO and HFO.

For all load conditions on both MGO and HFO the particle size distributions from a 4-stroke Rolls-Royce KR3 engine showed only a unimodal-shape without the lubrication oil-dependent nuclei mode as observed for Wätrsilä WX 28B engine. But the difference in the total number of particles between the fuels was dramatic and found to be around two orders of magnitude as shown on Figure 4. The total number of particles showed a slight increase for all tested fuels while increasing the load. This can probably be explained by the small increase in carbonaceous particle number due to increased amount of injected fuel and associated more locally rich combustion zones inside the cylinders (Tobias et al., 2001). At the same time the mean CMD values showed no dependence to load conditions and was around 25-30 nm and 60-70 nm for HFO and MGO respectively. Contribution of nanoparticles to the total PM number was above 90% while operating on HFO, but on MGO it was only around 25-35% as shown on Figure 5, which is more typical for automotive heavy-duty diesel engines. Results suggest that accumulation mode on MGO is mainly consist of carbonaceous fraction, while on HFO it is comprised of elemental (black) carbon aggregate particles and PM originated from the incombustible fuel ash species (Lyyränen et al., 1999; Kasper et al., 2007), which are mainly composed of vanadium, aluminum, calcium, silicium, asphaltenes and other sediments as given in Table 2 with heavy fuel oil specification.

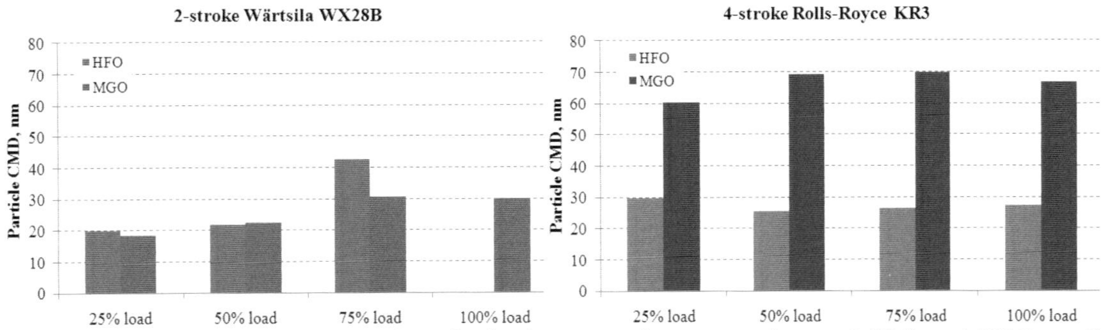

Figure 3 Particle mean CMD values for both test engines operating on MGO and HFO at all operational modes.

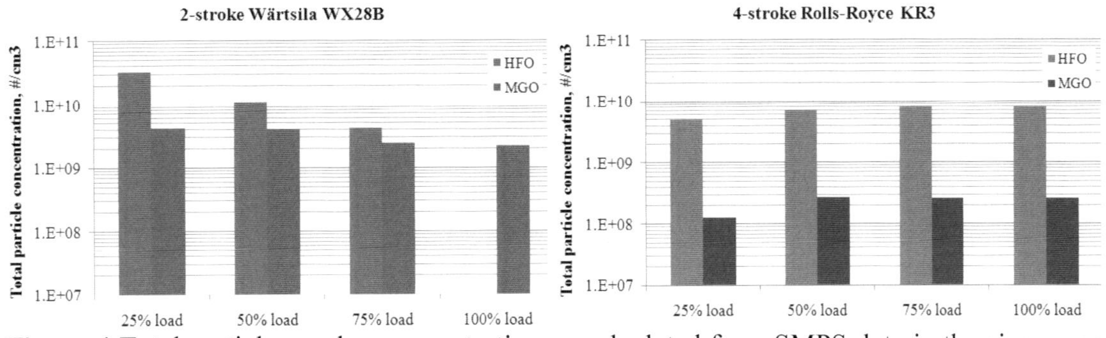

Figure 4 Total particle number concentration as calculated from SMPS data in the size range of 6-215 nm.

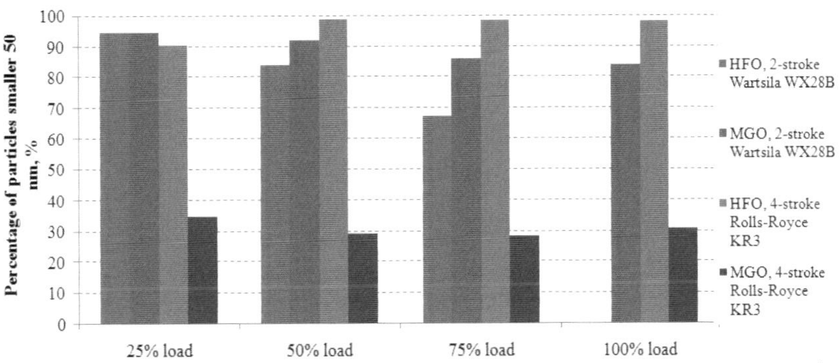

Figure 5 Mean percentage of particles smaller 50 nm in the SMPS scans at each operational mode with PDT of 400 °C.

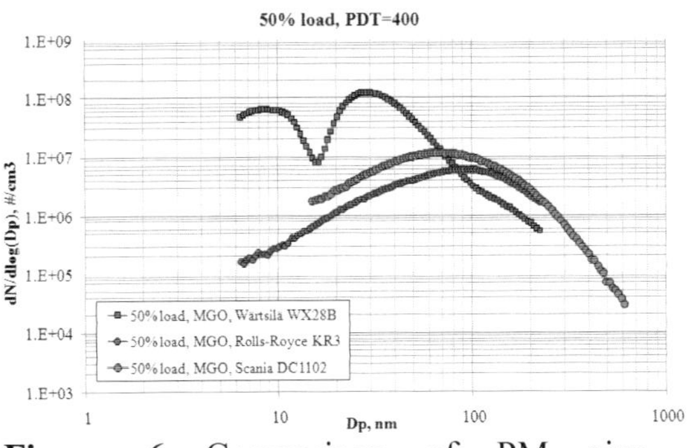

Figure 6 Comparison of PM size distributions from MDEs and heavy-duty engine at 50% load on marine gas oil fuel.

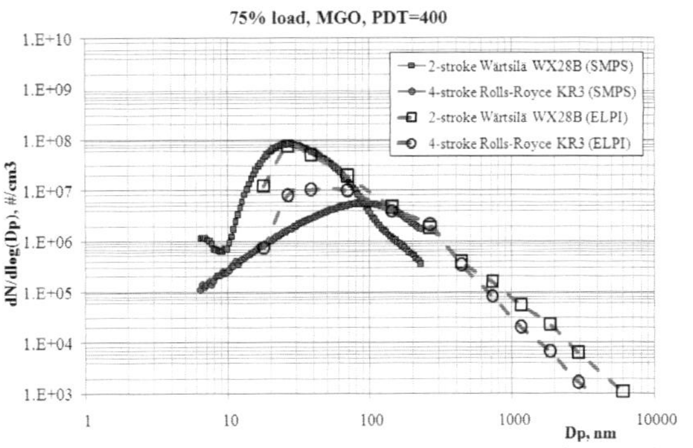

Figure 7 Comparison of number size distributions measured by SMPS and ELPI at 75% load while operating on MGO fuel.

As particle emission characteristics from a 4-stroke Rolls-Royce KR3 while operating on MGO showed an interesting behavior, very different from other results, it was considered to compare obtained PM characteristics with ones from a 4-stroke heavy-duty DI diesel engine. The results of comparison are shown on Figure 6 for all engines run on MGO at 50% load and suggest that total particle number emitted by a 2-stroke MDE is significantly higher than one from automotive engine, even if not take into account the nuclei PM associated with engine lub oil. This and also the fact that PM size distribution emitted by MDE shows a distinct accumulation mode at around 30 nm, which is much smaller than is known from automotive engines (50-150 nm) are in accordance to the findings of other researchers (Kasper et al., 2007). At the same time, results revealed an interesting feature, i.e. that PM size distribution from a 4-stroke MDE was almost identical to one from a 4-stroke HD engine in the range of result overlap (as different instrument settings were used). Small difference in PM number was registered for particles below 100 nm in size, while almost perfect coincidence was observed for sizes above 100 nm and even means CMD values were similar between MDE and HD engine with corresponding 69.1 nm and 66.3 nm respectively. This good agreement is difficult to explain, especially if the significant difference between the engines in terms of geometrical dimensions and operating conditions is taken into account.

A nice agreement between ELPI and SMPS PM size distributions were obtained for all tested fuels and operating conditions, as shown on Figure 7 for 75% load, MGO fuel case. It also showed that a significant number of particles exist above 1 μm. They are denoted as coarse particles and consist from accumulation mode ones that have been deposited on cylinder and exhaust system surfaces and later reentrained. Nevertheless, one should not ignore the fact that ELPI and SMPS operate on very different working principles and even the size axis has a different meaning for them. For SMPS, where particles are segregated in electrical field, the distributions are a function of electrical mobility diameter, while for ELPI, where particles are separated by impaction, they are a function of an aerodynamic diameter.

3.2. Particle morphology and composition analysis

The scanning electron microscopy and energy-dispersive X-ray analysis were performed for the PM samples collected with the help of ELPI, which cascade impactor configuration allows it obtaining size-selected particle samples. Due to its complexity and time consumption the detailed morphology analysis was performed only for particles collected from a 4-stroke Rolls-Royce KR3 when operating on MGO at 75% load conditions.

The particles were classified into several groups based on morphology and elemental composition. The majority of observed particles were carbon agglomerates, or aggregates, exhibited a complex, branched morphology as seen on Figure 8a, and consisted of spherical carbon particles with mean diameters between 30 and 90 nm. These agglomerates were observed in all sizes, containing approximately from ten to as many as several thousands of primary carbon particles (Kocbach et al., 2005). Some carbonaceous particles were also found as separate spherules in the smallest particle size range together with irregular-shaped ones (sharp-edged particles as seen on Figure 8b), which were mainly composed of Ca, Na, Fe, K and Si as revealed the EDX spectroscopy (Figure 8c) and are believed to originate from lubrication oil. Some particles formed from fuel oil droplet residue and the particles indicating mechanical abrasion were found in the coarse mode size. It should be noticed here that PM morphology results from a 4-stroke Rolls-Royce KR3 MDE appeared to be very similar to results from automotive diesel PM morphology and composition studies as found in literature (e.g. Berube et al., 1999; Kocbach et al., 2005).

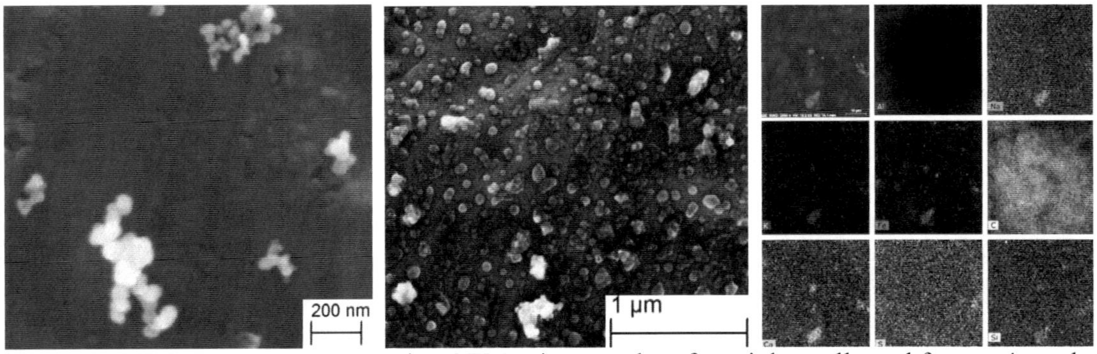

Figure 8 (a-b) Representative SEM micrographs of particles collected from a 4-stroke Rolls-Royce KR3 MDE operating on MGO at 75% load. (c) Elemental analysis of metallic particles with the help of EDX spectroscopy.

4. CONCLUSIONS

The combined ELPI-SMPS set-up showed a good stability and repeatability during the experiments performed under "hot" dilution conditions which allowed to study only solid, non-volatile particles. For a 2-stroke Wätrsilä WX 28B the obtained distributions for both MGO and HFO were bimodal, comprised of a pronounced load-independent accumulation mode and a nucleation mode which decayed with the load increase. The total particle number concentrations for both fuels were dominated by nanoparticles and decreased with the load, but was considerably higher for HFO in comparison to MGO with the biggest difference occurred at low load conditions. The observed load-dependent, for both the fuels, nuclei mode is believe to compose primarily of metal compounds and originated from high lub oil emissions, which are more likely caused by a specific engine design. At the same time, the accumulation mode on MGO is represented almost entirely by carbonaceous agglomerates formed due to incomplete combustion, while on HFO the mentioned carbon particles are found together with considerable amounts of incombustible residual fuel ash species. For all load conditions on both tested fuels the particle size distributions from a 4-stroke Rolls-Royce KR3 marine diesel engine were barely unimodal with pronounced accumulation mode. The maximal difference in total PM number between HFO and MGO was around two orders of magnitude. The total particle number increased with load, which can be explained due to increased amount of fuel injected and associated more locally rich combustion zones formed inside the cylinder. The calculated mean CMD values were fairly constant at all loads and were around 25-30 nm for HFO and 60-70 nm for MGO respectively. Contribution of nanoparticles to total PM number was above 90% on HFO, but only 25-35% on MGO fuel.
The surprisingly good agreement not only in overall shape, but also in terms of total PM number and mean CMD values were found between automotive and 4-stroke marine engine. This interesting finding is hard to explain and it probably needs separate, more detailed investigation. At the same time, total particle number concentration from a 2-stroke MDE was considerably higher than from automotive HD engine. In addition, Wätrsilä WX 28B engine showed a distinct accumulation mode at around 30 nm, which is much smaller than is known from automotive engines (50-150 nm), but is completely in accordance to the findings from the studies of MDE particulate emissions.

Several distinct particle classes were observed during PM morphology analysis: the carbonaceous agglomerates consisted of spherical carbon particles with mean diameters between 30 and 90 nm were observed in a very wide size ranges. Among the smallest particles the separate spherical carbon particles together with irregular-shaped metal ones, which could originate from lubrication oil, were found. Some supermicron particles formed from fuel oil droplet residue and the particle indicating mechanical abrasion were also observed. Hence, three main origins of PM as fuel, lubrication oil and mechanical wear due to friction of moving parts have been confirmed.

REFERENCES

Berube, K.A., Jones, T.P., Williamson, B.J. et al. (1999) Physicochemical characterization of diesel exhaust particles: factors for assessing biological activity, Atmos. Environ. 22, 1599–1614.

Corbett, J.J., Winebrake, J.J., Green, E.H., Kasibhatla, P., Eyring, V., Lauer, A. (2007) Mortality from Ship Emissions: A Global Assessment. Environ. Sci. Technol. 41, 8512-8518.

De Filippo, A. and Maricq, M.M. (2008) Diesel nucleation mode particles: semivolatile or solid? Environ. Sci. Technol. 42, 7957-7962.

Dockery, D., Pope, C.A., Xu, X., Spengler, J., Ware, J. et al. (1993) An association between air pollution and mortality in six U.S. cities. New England Journal of Medicine 329, 1753-1759.

Ferin, J., Oberdörster, G., Penney, D.P. (1992). Pulmonary retention of ultrafine and fine particles in rats. Am. J. Respir. Cell Mol. Biol. 6, 535-542.

Goo, J.H. and Kim, C.S. (2003) Theoretical analysis of particle deposition in human lungs considering stochastic variations of airway morphology. J. Aerosol Sci. 34, 585-602.

Kasper, A., Aufdenblatten, S., Forss, A.,Mohr, M., Burtscher, H. (2007) Particulate Emissions from a Low Speed Marine Diesel Engines. Aerosol Sci. Technol. 41, 24-32.

Keskinen, J., Pietarinen, K., Lehtimäki, M. (1992) Electrical low pressure impactor. J. Aerosol Sci. 23, 353-360.

Khalek, I.A., Kittelson, D.B., Brear, F. (2000) Nanoparticle growth during dilution and cooling of diesel exhaust: experimental investigation and theoretical assessment. SAE Paper 2000-01-0515.

Kittelson, D.B. (1998) Engines and nanoparticles: a review. J. Aerosol Sci. 29, 575-588.

Kittelson, D.B., Arnold, M., Watts, W.F. (1999) Review of diesel particulate matter sampling methods. Final Report. University of Minnesota: Minneapolis, MN.

Kittelson, D. B., Watts, W. F. et al. (2008) Effect of Fuel and Lub Oil Sulfur on the Performance of a Diesel Exhaust Gas Continuously Regenerating Trap. Environ. Sci. Technol. 42, 9276-9282.

Kittelson, D.B., Watts, W.F., Johnson, J. (2006) On-road and laboratory evaluation of combustion aerosols – Part 1: Summary of diesel engine results. J. Aerosol Sci. 37, 913-930.

Knudson, E.O. and Whitby, K.T. (1975) Aerosol classification by electric mobility: apparatus, theory, and application. J. Aerosol Sci. 6, 443-451.

Kocbach, A., Johansen, B.V. et al. (2005) Analytical electron microscopy of combustion particles: a comparison of vehicle exhaust and residential wood smoke. Sci. Total Environ. 346, 231–243.

Lyyränen, J., Jokiniemi, J., Kauppinen, E.I., Joutsensaari, J. (1999) Aerosol Characterization in Medium-Speed Diesel Engines Operating with Heavy-Fuel Oils. J. Aerosol Sci. 30, 771-784.

Oberdörster, G., Sharp, Z., Atudorei, V., Elder, A., Gelein, R., Kreyling, W., Cox, C. (2004) Translocation of inhaled ultrafine particles to the brain. Inhal. Toxicol. 16, 437-445.

Oberdörster G., Finkelstein J., Ferin J. et al. (1996) Ultrafine particles as a potential environmental health hazard. Studies with model particles. Chest 109 (Suppl. 3), 68–9.

Pope III, C.A. (2000) Epidemiology of fine particulate air pollution and human health: Biologic mechanisms and who's at risk? Environ. Health Perspect. 108 (Suppl.), 713–723.

Pope, C.A. III., Burnett, R.T., Thun, M.J., Calle, E.E. et al. (2002) Lung cancer, cardiopulmonary mortality, and long-term exposure to fine particulate matter. JAMA. 287, 1132-1141.

Ristimaki, J., G., Hellen, et al. (2010) Chemical and physical characterization of exhaust particulate matter from a marine medium speed diesel engine. CIMAC Congress, Paper # 73.

Sakurai, H., Tobias, H.J., Park, K., Zarling, D. et al. (2003) On-line measurements of diesel nanoparticle composition, volatility, and hygroscopicity. Atm. Env., 37, 1199-1210.

Suzuki, H., Toyooka, T., Ibuki, Y. (2007) Simple method to evaluate uptake potential of nanoparticles in mammalian cells using a flow cytometric light scatter analysis. Env. Sci. Tech., 41, 3018-3024.

Tobias, H.J., Beving, D.E. et al. (2002) Chemical analysis of diesel engine nanoparticles using a nano-DMA/thermal desorption particle beam mass spectrometer. Environ. Sci. Tech. 35, 2233-2243.

Extending Submarine Equipment Beyond its Service Life

Dave Groocock, Babcock Pty Ltd, dave.groocock@babcock.com.au
Brian Menzies, Babcock Pty Ltd, brian.menzies@babcock.com.au
Dr Carl Howard, The University of Adelaide, carl.howard@adelaide.edu.au

ABSTRACT

The original design service life of existing submarine platforms is frequently extended to cover capability gaps caused by program delays for introducing replacement platforms. The effects on equipment must be assessed prior to any approval of an extension to the operating life of the existing in-service vessels.

While the majority of equipment has design margins and safety factors, the components degrade through normal operating loading, wear and corrosion, leading to the design loads being closer to the design capability or strength of the equipment as it gets older. Many platforms are used differently to the original design specifications, thereby creating variances from the design life. If equipment is to be extended past the original design life, information must be gained on the current condition and its historical use. An investigation into the usage history of the equipment will enable engineers to determine the margin remaining in the design and/or the remaining life of equipment and therefore determine if the service life can be extended safely.

If this analysis shows that equipment must be replaced, this may provide opportunities to the designer, builder, maintainer and operator of either the current platform and/or future platform. These opportunities may involve replacing a system on the existing in-service submarines to prove a new technology prior to installing it on a future platform. This Technology Demonstrator approach must be tempered with the requirement to meet availability targets and on-going operations of the existing platform.

This paper discusses the development work being conducted by Babcock in conjunction with The University of Adelaide to determine the stakeholder requirements and parameters to be reviewed in order to properly assess equipment for extension of service life.

BACKGROUND

The service life of submarine equipment can be defined in a number of ways, with time duration and operating parameters that impact on the equipment to be taken into account when determining the service life. The main aspects and sub-aspects of the submarine platform life as a whole that can be considered to affect equipment service life can be broken down to:

- Non-operational time
  - Time in the water, but not operational (alongside)
  - Time out of the water (docked)
- Operational time and usage
  - Cycles to depth
  - Number of firings or other operational functions

For instance, regarding corrosion protection coatings such as paint, the service life can be considered to be the complete life of the submarine platform from the time that the paint was applied to the components, as it is providing protection to the components at all times. Conversely, the service life of cathodic protection systems can be considered to be only the time that the submarine spends in the water. Water provides the cathodic cell, therefore when the submarine is out of the water docked for maintenance the anodes are not corroding and losing material and subsequently are retaining the existing length of service life remaining, or are corroding through other mechanisms at much slower rates. When reviewing the service life for a given component, it is important to be aware of and clearly define the aspects or combination of aspects of the above that affect the component.

Yule & Woolner states that HMAS Collins was initially launched in 1993, sea trials commenced in 1994, with an initial into service date of 1996 (Yule & Woolner, 2008). The other five submarines in the Collins class followed, with HMAS Rankin being the last submarine to be introduced into service with the Royal Australian Navy (RAN) in early 2003 (Yule & Woolner, 2008). Since this time, there have been a significant amount of maintenance and availability issues related to the submarines, resulting in the appointment of John Coles to carry out a review into the Australian submarine sustainment industry. The Coles Review is aiming to provide recommendations for improving the sustainment of the submarines regarding value for money, infrastructure needs and management options, as outlined in the Defence Minister, Stephen Smith's media release dated 24th August 2011 (Smith & Clare, 2011). The history of the submarines offers scope for significant improvement in the way the submarines are managed and maintained in order to give the RAN the submarine capability that is required. This reform process is aiming at improving the Collins Class availability to better support the RAN through enhanced maintenance regimes, asset management and other similar techniques resulting in the provision of more at-sea days for the Collins Class submarines. It's expected that the reform process will be an on-going program of improvements to be made throughout the sustainment industry that will likely extend throughout the remaining Collins Class life and into the sustainment of the SEA 1000 Class of submarines.

HMAS Collins is currently due to be retired around 2025, dependent on docking duration and operational deployment lengths, confirmed by Minister Chris Evans on behalf of the Minister for Defence Stephen Smith, in recent Senate debates, question asked on 29th April 2011 and answered on the 13th October 2011 (Evans, 2011). Based on current predictions HMAS Collins will be out of service by 2025, with the remainder of the Collins Class fleet to be retired over the following years. HMAS Rankin is expected to be retired in the early 2030's based on the gap between the in service dates of the first (HMAS Collins) and last (HMAS Rankin) vessels of the class. The replacement submarine, designated SEA 1000 was initially scheduled to be in service by 2025; however in the August 2011 update of the Defence Capability Plan (DCP) Public version, there is no specified in service date (DMO and DoD, 2011). Even based on a project initialisation date of mid 2012, an optimistic 15 year period from initialisation to in service, Collins was a relatively quick project at 13 years, results in the First of Class (FOC) SEA 1000 submarine in service towards 2027 (Yule & Woolner, 2008). This timing will likely result in a capability gap where the Australian submarine fleet will be reduced to two submarines in the fleet as the Collins Class are retired to be replaced by the SEA 1000 vessels as shown in Figure 1. This is based on an assumption that the FOC SEA 1000 submarine will be delivered into service with the RAN in 2030. The major aspect that will impact on this is the decision of the submarine design for SEA 1000, the current options being Military-Off-The-Shelf (MOTS), evolved MOTS/Collins or a bespoke/indigenous design, with the decision to be made by the Defence Materiel Organisation (DMO) and Australian Government.

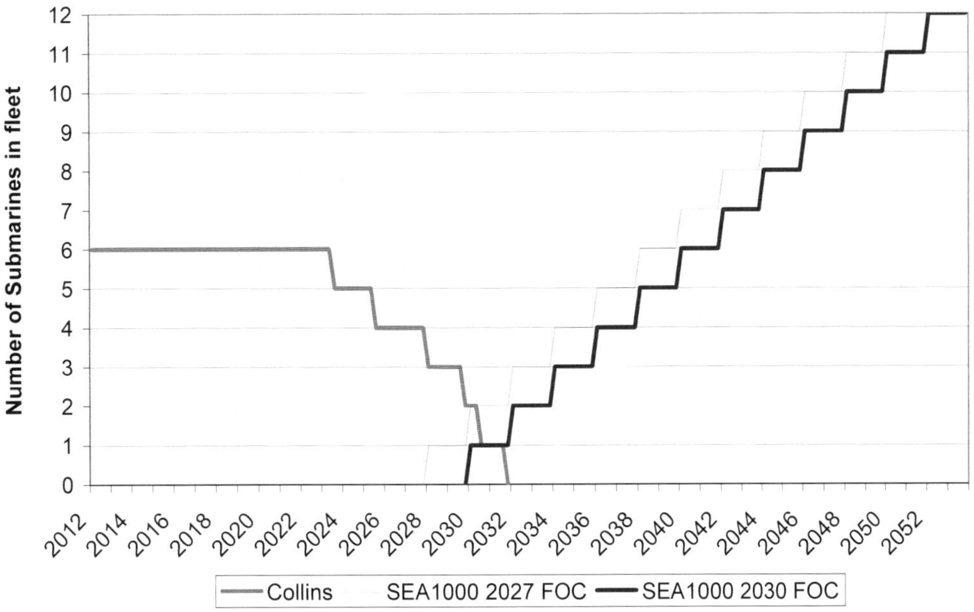

Figure 1: Expected Submarine Availability Current Collins Class retirement

In order to maintain a submarine force, the surest option is to extend the service life of some or all of the Collins Class submarines beyond their current expected retirement dates if possible. This study has investigated the above-described life extension and made judgements on the number of submarines that might be required to undergo a life extension and the duration of the extension in order to deliver a submarine capability to the RAN that is adequate to meet the requirements outlined in the DCP (DMO and DoD, 2011) and the Defence White Paper (Pezzullo *et al*, 2009). Following the above analysis of the likely scope of the Collins Class life extension, an investigation has been carried out into the components within the Weapons Discharge System (WDS) and the effects of the above-described life extension on these components. The aim was to identify the areas of the equipment that need to be reviewed, assessed and approved in order to confirm that the WDS is suitable for extension of the service life and will continue to operate correctly and safely, conforming with the requirements of the submarine platform, the WDS has equipment of the type that principles of assessment can be extended to other systems across the submarine.

EQUIPMENT

The equipment that makes up the Collins Class WDS can be broken down into categories for assessment and analysis. Some equipment will fit a number of the categories below, depending on what aspect of the equipment is reviewed.

- Pressure Exposed equipment – those components that are exposed to pressure, either due to depth pressure, stored pressure or operational pressure such as during firing torpedoes. These are likely to be the most safety critical components;
- External equipment – those components that are outside the submarine pressure hull and therefore exposed to the environment, pressure, temperature, seawater and potentially sunlight, air temperature and humidity when surfaced;
- Internal equipment – components that are housed within the submarine pressure hull and therefore are not exposed to the outside environment. This equipment can potentially still be exposed to seawater, moisture, and temperature;
- Mechanical and Rotating equipment – components that perform some kind of mechanical role such as opening, closing or securing doors, or providing movement to

other equipment. This category also includes machinery that is rotating on a regular, on-going or temporary basis;

- Hydraulic equipment – equipment that is performing a hydraulic role, such as hydraulic actuators, motors, etc. These can be internal or external components depending on where they are located and what systems they are operating within; and
- Electrical and Electronic equipment – components that comprise electrical or electronic components, involved in the control of equipment, power distribution, navigation, sensors, communications and the combat system.

The major components of the WDS have been investigated and assessed for their suitability for life extension. Issues have been highlighted and areas of further investigation have been included in order to capture what cannot be done with the information available. These major components are discussed below.

Torpedo Tubes
The torpedo tubes on the Collins Class fit within the Pressure Exposed equipment and External equipment categories as outlined above. During normal operating conditions they are usually held at the atmospheric conditions of the submarine internal environment and therefore the part of the tube that is external to the pressure hull is subjected to depth pressure as the submarine changes depth. Any failure to be able to operate the tubes is extremely concerning as they are the major offensive weapon and defensive strength of the platform. The inability to fire weapons could be catastrophic for the submarine.

Due to the issues of the reduced availability of Collins Class submarines, it is likely that the torpedo tubes have had less usage than expected, resulting in fewer cycles to depth and fewer firings of weapons than originally designed for. This may mean that there is a significant amount of service life remaining in the tubes, such that the life extension of the tubes could be approved relatively easily.

Since the tubes are external components, corrosion and the protection scheme needs to be considered. Whether this is through paint schemes, sacrificial anodes or an impressed current cathodic protection system, the tubes must be fully assessed and analysed to ensure corrosion hasn't caused a reduction in cross-sectional area or localised pitting that would cause strength issues during service, firing or a shock event that would then result in a failure or leak path for water to enter the submarine.

External equipment
External components within the WDS include a wide suite of equipment, including actuators, bow caps (the forward, external door to the torpedo tube), sensors, etc. Some of these are covered below; however the common theme between them is that protection must be provided from the environment. Whether this is corrosion protection, pressure protection or mechanical damage protection or a combination largely depends on the equipment considered.

Regarding corrosion protection, improvements have been made in the performance of paint schemes with new additives allowing protection to be provided without the use of lead and other elements that are harmful to the environment and personnel. The increase in awareness of pollution in the maritime environment has required the use of environmentally friendly materials, which have required the reconsideration of paint protection schemes applied to submarines. There has also been development in the technologies involved in the paint, including the introduction of schemes that only require the application of a single coat of paint that offers the same, or increased, level of protection that historically was only available with multiple coats of paint with individual layers, while being easier to apply either in-situ or when the component is removed from the submarine. The single coat paint schemes can also be made more flexible and durable, allowing protection from impacts and wear that occur in service.

Revisions to sacrificial anode protection systems have been identified to utilise different materials, such as aluminium or magnesium that also provides cost and weight savings and potentially performance improvements over existing zinc anodes. Analysis must be carried out to ensure there are no adverse impacts on surrounding components from changing the materials of the anodes if this option is pursued. Impressed current cathodic protection systems can also offer enhanced corrosion protection to external components. The impact on the electro-magnetic signature of the platform must be considered and known prior to implementation of a system like this.

Internal equipment
There is a significant amount of internal equipment related to the WDS on the Collins Class submarine. This includes hydraulic components, mechanical components and electrical and electronic components. The service life of the various internal components is related to the number of cycles through use rather than depth cycles and the elapsed time since introduction. Since there is a low incidence of the equipment being exposed to seawater, it is predominately related to the time since the components were installed, particularly with the electrical and electronic components. A companion to the WDS is the Weapons Handling and Stowage System that stores the weapons in the Weapons Stowage Compartment (WSC) and assists in loading the weapons into the torpedo tubes. The service life of some of these components will be related to the number of weapons stored in the submarine, the movement of these weapons within the WSC and the duration that they were stored.

Hydraulic equipment
The hydraulic equipment within the Collins Class submarines is located both internally and externally to the pressure hull. The external hydraulic system has externally mounted equipment operating the bow caps, bow shutters, inlet duct door and the various latches and interlocks among other equipment mounted in the forward free flood space. The internal equipment for the external hydraulic system is mounted within the pressure hull but operates the above external components, such as control packs and selectors, linked through pressure hull penetrations. There is also an internal hydraulic system in the Collins Class submarines that operates within the submarine without interaction with external components that is not considered here. It is expected however that the issues relevant to this system will be similar to those affecting the external hydraulics, without the external pressure issues.

The components within the hydraulic system must be maintained correctly in order to allow the systems and components to have their service life extended. This includes ensuring a full strip down of all components and replacement of all rubber O-rings is to be carried out where not already occurring during maintenance periods, requiring additional maintenance effort. The storage of components and valves while not in service is also key to ensuring the continued reliable service of the hydraulic components. Temperature controlled storage, along with UV protection of rubber components, will ensure the shelf life and subsequent in-service life of the O-rings themselves is maximised. This should be introduced at all stages of the maintenance procedure if possible to ensure the longest possible life for rubber components. The flushing of all systems must be managed closely to ensure impurities are removed to reduce the chance of any damage to equipment or components.

Electronic & Electrical equipment

The Control System contains the majority of the electronic and electrical equipment within the WDS. These electronic components are located internally to the submarine and therefore are exposed to atmospheric conditions, with the additional risk of seawater exposure depending on the location of the equipment. Additional electronic components are located externally to the pressure hull, with proximity sensors being mounted in various locations within the free flood space on the WDS, sensing the positions of doors, latches and valves to feed into the control system as a level of redundancy with the status of the equipment. These sensors are linked with the control system through pressure hull penetrations with a cable feeding in the signal.

The service life of the electronics components is related predominately to the elapsed time since design/build and introduction, while the external components service life is also related to the number of cycles to depth that the submarine has performed due to the effects of the pressure on the sealing and hull penetrations.

Open architecture control systems are a potential solution to enable easy upgrades of components, hardware and software aimed at preventing obsolescence issues. While this may be the case, care must be taken to ensure backwards and forwards compatibility with all systems and equipment. For example, an electronic printed circuit board may be upgraded, but if the latest software loaded on the board is not backwards compatible with the existing operating system there will be issues with the integration of the card. Moving too fast towards an open-architecture philosophy if the advantages and disadvantages associated with the new design philosophy are not properly assessed has the potential to cause significant issues with the platform when in-service. The Collins Class life extension may be able to upgrade some parts of the system, particularly if trying to prove some technologies for SEA 1000. Care must be taken to ensure that the system is safe, reliable and maintainable for the duration of the life extension.

Mechanical and rotating equipment

The mechanical and rotating equipment within the Collins Class WDS can be mounted both internally and externally to the pressure hull. Items such as the Air Turbine Pump (ATP) cross the boundary between internal and external by being mounted on the pressure hull and having rotating components on both sides. Items such as the Top Stop & Rear Catch, actuators and guide rails are external and are therefore can be considered as external equipment while also performing a mechanical role.

The service life for mechanical and rotating equipment can be determined by a mixture of all aspects previously discussed. The service life of the majority of the equipment is impacted by the operational aspects, such as cycles to depth and number of firings, with time elapsed since installation effecting mechanical and rotating equipment mounted externally. Similar to above, the Collins Class availability issues therefore may be beneficial when considering equipment for a life extension, with a reduced at sea/in-service time potentially allowing components to be extended.

Some components such as the ATP are subject to ongoing development with improved systems available currently compared with the original equipment fitted to the Collins Class.

Spares are becoming obsolete for older equipment; therefore alternatives can be sought at the component or assembly level, potentially offering capability enhancements to the submarine platform. The introduction of upgraded equipment into the Collins Class submarines at the time of life extension may give an enhanced capability to the RAN, while producing operational data under Australian conditions to improve upon and lead the SEA 1000 design and maintenance aspects. This enhanced capability within the Collins Class, could potentially be the land attack capability proposed in the Defence White Paper (Pezzullo *et al*, 2009). If this capability is desired by the Australian Government for the SEA 1000 submarines, then introduction and proving within the Collins Class may be beneficial.

Off The Shelf Equipment

The use of Commercial, or Military, Off The Shelf (COTS or MOTS) equipment is being considered for a variety of equipment within the defence arena, including the submarine platform itself for SEA 1000. Other equipment can include electrical and electronic components, communication equipment, sensors and equipment such as Reverse Osmosis Units that generate potable water from seawater. COTS equipment has the advantage of potentially being significantly cheaper to procure than either MOTS or bespoke equipment. There is also often significant background and in-service information available for a piece of COTS equipment depending on its use, some of which is directly applicable to a military application. While it can be advantageous to utilise COTS equipment, shock protection is often required for equipment that isn't designed with a military use in mind, leading to either significant re-design efforts, or sophisticated mounting arrangements to minimise the forces imparted on the equipment.

TECHNICAL ADVANCES

Condition Based Monitoring

Since the introduction of the Collins Class, Condition Based Monitoring (CBM) systems have become more prevalent in a wide variety of industries. CBM gives the ability to monitor the performance and degradation of equipment throughout the submarine. This maintenance method is particularly useful for external components that cannot be easily inspected without the submarine being removed from the water at docking periods. A correctly designed CBM system can be monitored real-time by a control system on-board the submarine or by ship staff or post-processing of the data can be performed by land based staff. It will allow analysis of how equipment is being operated and wearing and then will aid in determining when the equipment will fail. The advantage of real-time analysis is that degradation of components between post-processing periods can be captured prior to failure, thereby preventing a larger scale and greater impact failure of the system. If this analysis is to occur on-board the submarine, significant training and development of the personnel who are to be monitoring the data output needs to occur, and there is likely to be an increase in the amount of equipment needed to be installed into the submarine.

On the Collins Class submarines, it may be possible to design a CBM system that will store data on-board the submarine, enabling the analysis to be performed when the submarine is alongside for planned maintenance. If it is shown that processing of the data is required to be performed more regularly, there may be a potential for the submarine to transmit data when at sea via the existing or an upgraded communication system to allow processing and analysis at an external facility, with any issues transmitted back to the submarine during the next communication period (if urgent). This may require enhanced communication ability such as communication at speed and depth to that already available on the Collins Class vessels, requiring upgrades to be installed as a Technology Demonstrator for SEA 1000. This type of approach is typical in non-defence industries.

There are significant advantages and improvements a CBM system can introduce into the maintenance planning and docking periods. Maintenance may be carried out on components in response to condition rather than at scheduled periods, with components that are seeing degradation in condition in-service able to be identified and replacements or repairs ready to be applied to the submarine immediately when it docks rather than having to go through the assessment process during the docking period.

There are a wide variety of sensors and other equipment able to be implemented into a Collins Class based CBM system to assist in the life extension of the submarines. These are briefly explained below:

- Acoustic monitoring of pressure vessels, pressure exposed equipment and structural components – this will allow any residual defects in the equipment that is exposed to pressure to be monitored each and every time the equipment gets exposed to pressure through the submarine diving, torpedo firing, or other operational functions. Sensors can be equally spaced on torpedo tubes, tanks, bow caps, rear doors, etc in the WDS and a wide variety of equipment throughout the submarine in order to ensure any defects present within the equipment are known and monitored.
- Monitoring of hydraulic oil quality, water, and other impurities – sufficient sensor placement (upstream and downstream of major equipment) will give an indication of any impurities within the oil, and therefore give an early indication of any failure or pending failure of the equipment within the system. This will potentially allow either the components or system where the impurities have been identified to be isolated depending on system design, preventing any damage to the wider system, allowing the system to continue to be utilised where sufficient redundancy is available.
- Acoustic & Vibration monitoring of bearings and other rotating equipment – this is widespread in the monitoring of wind turbine gearboxes for example. The system allows the condition of gearboxes and shafts to be monitored through placement of accelerometers and acoustic sensors to pick up any increase in vibration or noise that would indicate bearing failure or performance degradation. This has widespread application in a submarine, with the ATP within the WDS, and propeller shafts, diesel generators, main propulsion motor, etc. in addition to linear actuators.
- Thermal monitoring of electrical and electronic components – this will identify if there are hotspots within electrical cabinets, giving the ability to provide more adequate ventilation and cooling, or to shutdown the area affected (if there is sufficient redundancy in the system) to prevent a fire breaking out with potentially catastrophic consequences or damage occurring to the overheating component causing failure.
- Discharge monitoring for electrical components – where the current or voltage can be monitored enabling identification of a fault case where there is damage to any insulation or breakdown in insulation that could result in a short circuit and loss of the electrical supply.

Any CBM system that can identify a potential point of failure within the WDS may allow that area of the system to be isolated and not used. For the WDS, this could be vital, if one tube or the equipment operating this tube becomes unusable due to damage or degradation but it is able to be isolated, the remainder of the WDS can be used. Even if the isolation can be reduced to a bank of three port or starboard torpedo tubes, the impact on the operational capability of the submarine is minimised, with the platform retaining the ability to launch weapons in an offensive or defensive capacity.

A CBM system can identify and distinguish between various levels of equipment, with critical equipment, essential equipment and general purpose equipment able to be assessed and analysed separately with differing levels of importance on any failure. Critical equipment will vary between mission critical and safety critical equipment, while essential equipment will cover equipment necessary to operate the submarine. General purpose equipment will include equipment that aids in the operation but are not vital, such as automatic or remote operating equipment where local or manual operation is available. Priority can then be placed on maintaining equipment critical to support submarine operations.

Composites

Utilising the wrapping of components in composite materials to a greater extent than already used can also provide enhanced protection to external components that are susceptible to corrosion, fatigue or damage. Where components are damaged or corroded, wrapping them in a composite material with strengthening such as carbon or glass fibre can extend the service life of the components. Resins and composite strengthening materials can be specified to allow existing equipment to be repaired and approved for use for the extended service life for the Collins Class.

THE WAY FORWARD

An appropriate program for the life extension of the Collins Class submarines will involve a historical analysis in order to define a baseline for the existing equipment. This will require gathering operational data from the Collins Class submarines, including depth excursions, firings from each tube in each submarine, the time spent alongside, the time spent at sea, the time spent out of the water under maintenance and other relevant data. The above aspects of the different types of equipment will need to be assessed in order to clearly define the baseline for the existing equipment. This may have to be defined for individual components, such as tubes, or submarine platforms depending on the aspects of the service life that impact on the equipment, resulting in different analyses and maintenance/extension work being carried out on different submarines.

Any upgrades, Technology Demonstrators or enhanced capability desired in an extended Collins Class must be identified as early as possible in order to carry out the design tasks to introduce them at the earliest opportunity. This early implementation into the Collins Class will feed more information at an early stage into the SEA 1000 project to give the design team sufficient data and detail to make informed decisions about the capability and technology being tested and demonstrated. Without the data to make these informed decisions, the design team must make judgments on background information, extrapolation from overseas suppliers and assumptions. This could cause similar problems that were seen in the Collins Class initial sea trials whereby issues unique to Australian conditions cause carry on problems with the class that take time and money to resolve, with the resultant negative public perception that is still proving extremely difficult to turn around.

CONCLUDING REMARKS

The term ‘service life’ can have multiple interpretations when referring to submarine equipment. It can refer to the time elapsed since the component was designed, built, installed or commissioned, or it can refer to the usage of the particular component when in service, the depth profile history, the number of times exposed to pressure through weapon discharge, or a combination of the above. When reviewing a components' service life, a clear understanding must be gained of the aspects of the above that will impact on the service life of the component when compared to its original design life.

In assessing the baseline for the individual component, system or submarine for suitability for a service life extension, a significant amount of information will be assessed. This will include aspects related to existing service the equipment has been subjected to and the maintenance performed on the equipment. The service life extension analysis will determine the planned maintenance that is required, if this needs to change from that already planned. An appropriately designed and implemented CBM system may enable less maintenance to be performed on the equipment, while still giving the maintainer sufficient, if not enhanced, detail on the performance and degradation of the equipment in-service. This will enable a complete asset management plan to be determined and implemented across the submarine to ensure continued, safe operation of the platform.

A service life extension to some or all of the Collins Class vessels therefore appears to be the logical solution to cover the delay in SEA 1000. This allows the Australian Government, the DMO, the RAN and the Australian submarine industry to ensure the SEA 1000 project is planned correctly and not rushed with the inevitable problems that a hastily executed project would cause. The best decisions can be made regarding the submarine platform design to provide the capability that the RAN requires.

CLOSURE

The SEA 1000 project does not have a publically defined start date, or a planned in service date at this point in time. Planning for the Collins Class life extension needs to commence now in order to be able to firstly approve the extension and then design and implement any design upgrades or technology demonstrators in order to provide the RAN with the required submarine capability through the transition period from Collins Class to SEA 1000. The submarine operational logs must be examined in order to assess the historical usage of the equipment on the Collins Class submarine, with industry beginning to look at the various components and the impacts of an extension on systems, sub-systems and components.

REFERENCES

Defence Material Organisation and The Department of Defence (2011). *Defence Capability Plan – Public Version*. Capability Development Group. Canberra, Commonwealth of Australia. From http://www.defence.gov.au/dmo/id/dcp/dcp.cfm

Evans, C. (2011), *Defence: Submarines (Question No. 633) – Senate Debate*, Australian Senate. Canberra, From http://www.openaustralia.org/senate/?id=2011-10-13.543.1

Pezzullo, M. Babbage, R., Abigail, P., and Thompson, M. Department of Defence (2009). Defending Australia in the Asia Pacific Century: Force 2030 - The Defence White Paper 2009. Dept. of Defence. Canberra, Commonwealth of Australia. From http://www.defence.gov.au/dmo/id/dcp/dcp.cfm

Smith, S. and Clare, J. Defence, (2011). Release of Terms of Reference for Submarine Sustainment Review [Press Release]. From http://www.minister.defence.gov.au/2011/08/24/minister-for-defence-and-minister-for-defence-materiel-release-of-terms-of-reference-for-submarine-sustainment-review/

Yule, P. and D. Woolner (2008). *The Collins Class Submarine Story: Steel, Spies and Spin*. Melbourne, Cambridge University Press.

The Effect of Length to Diameter Ratio on the Hydrodynamic Coefficients of a Submarine Operating Near the Surface in Calm Water

D Ranmuthugala[1], D Fletcher[1], D Neulist[1], M Renilson[1] and B Anderson[2]

1. Australian Maritime College, University of Tasmania, Australia
2. Defence Science and Technology Organisation, Australia

ABSTRACT

Conventional submarines regularly operate near the surface to recharge batteries and to meet operational requirements. During these operations, a submarine will generate surface waves which can significantly increase its resistance and thus the required propulsive power. In addition, forces and moments are induced in the vertical plane that can adversely affect the trajectory of the submarine.

A number of design and operational factors influence the wavemaking resistance and the vertical forces and moments. These include the proximity to the free surface, forward speed, size and configuration of the appendages, and the length to diameter (L/D) ratio. These effects are dependent on the interaction of the various hull-borne wave systems, which in turn are highly dependent on the Froude number.

In order to quantify these effects, a series of captive model experiments on a 1.556m model of the DARPA SUBOFF submarine has been conducted in the Australian Maritime College towing tank supplemented by Reynolds-Averaged Navier-Stokes (RANS) Computational Fluid Dynamic (CFD) simulations of the model under varying conditions and configurations. This paper describes the RANS CFD modelling of the SUBOFF geometry at different L/D ratios and Froude numbers to determine the influence of these parameters on resistance.

1. INTRODUCTION

When a submarine is travelling close to the surface the pressure disturbance around the hull causes waves to be formed on the water surface, and the consequent wavemaking resistance results in an increase in the total resistance compared to the case when the boat is deeply submerged. As this additional drag component will be dependent on the hull L/D it is possible that the optimum L/D value may be different for a submarine close to the surface, compared to one operating deeply submerged.

Conventional submarines (SSKs) are regularly required to operate near the surface to run diesel engines to recharge batteries, and to fulfil other operational requirements. The impact of the near surface operation on its range and speed need careful consideration. Hence, an understanding of the flow behaviour, and resulting increase in resistance, due to these operations is necessary.

In order to better understand the additional power required when operating close to the surface, and the resulting effect on the design and operation of a submarine, the Australian Defence Science and Technology Organisation (DSTO) is conducting experimental and numerical studies in conjunction with the Australian Maritime College (AMC).

2. OBJECTIVE

Previous experimental work by the authors has investigated the additional power required when close to the surface compared to that when operating deeply submerged (Dawson *et al.* 2010, and Renilson *et al.* 2011). The influence of the sail on the additional power when close to the surface was investigated, and a preliminary numerical investigation into the effect of the L/D ratio was undertaken.

From this preliminary work it was considered possible that the optimum L/D value for a submarine when operating close to the surface may be different to that when operating deeply submerged. Hence, the objective of the work presented here is to determine the additional power required when operating close to the surface as a function of L/D.

3. BACKGROUND

3.1 SSK design considerations

For the purpose of this study a full scale SSK with a length of 80m was used as being representative of a large SSK. A non-dimensional ratio of L/D = 8.6 was chosen, as it also corresponded to the geometry of the axisymmetric SUBOFF submarine hull form for an SSN developed by the Defence Advanced Research Projects Agency (DARPA) - Model 5470 described in Groves *et al.* (1989). As can be seen from figure 3.1, this corresponds to the principal dimensions (length and L/D ratio) for a typical large SSK.

3.2 Basis hull form

To enable the results to be compared with those from other research projects, much of the work has been based on the axisymmetric SUBOFF submarine geometry. This is an internationally accepted benchmark model and serves as a reliable source of publicly available experimental and CFD data. A detailed description of the SUBOFF geometry is presented by Groves *et al.* (1989).

For the purpose of this work the presence of the sail was neglected, thereby simplifying the numerical approach considerably.

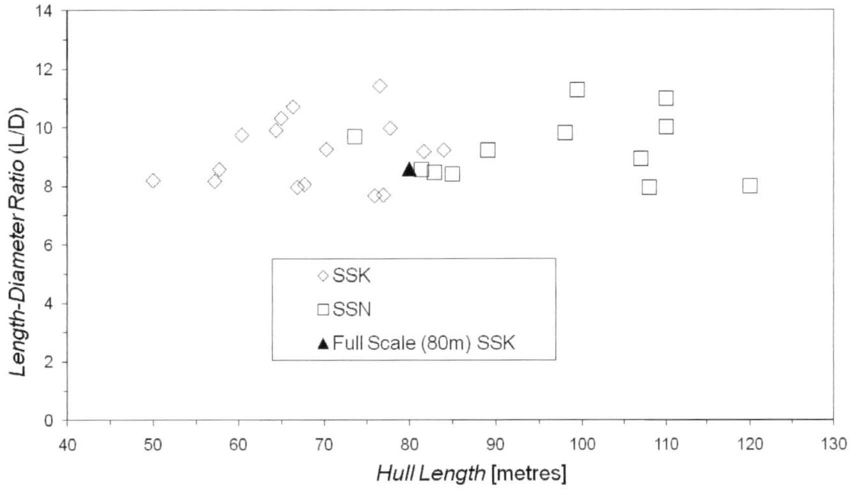

Figure 3.1 L/D as a function of L for existing vessels

3.3 Configurations investigated

The drag was obtained for three different length to diameter (L/D) ratios operating at a non-dimensional depth (defined as H* = H/D) of H* = 1.1 (*i.e.* near the surface, see figure 3.2) and at depth (*i.e.* without the effect of the free surface) at four different Froude numbers. The various simulation configurations in equivalent full scale dimensions are presented in table 3.1, although it should be noted that all the calculations were conducted at model scale, using a scale ratio of 1/51.4 which corresponded to the model experiments reported in Dawson *et al.* (2010), Wilson-Haffenden *et al.* (2010) and Renilson *et al.* (2011).

Table 3.1 Simulation Configurations (Full Scale)

Diameter (m)	Length (L)	Wetted surface (m^2)	L/D	Speeds (knots)
9.306	70	1,724	7.52	6, 8, 10, 12
9.306	80	2,018	8.60	6, 8, 10, 12
9.306	90	2,311	9.67	6, 8, 10, 12

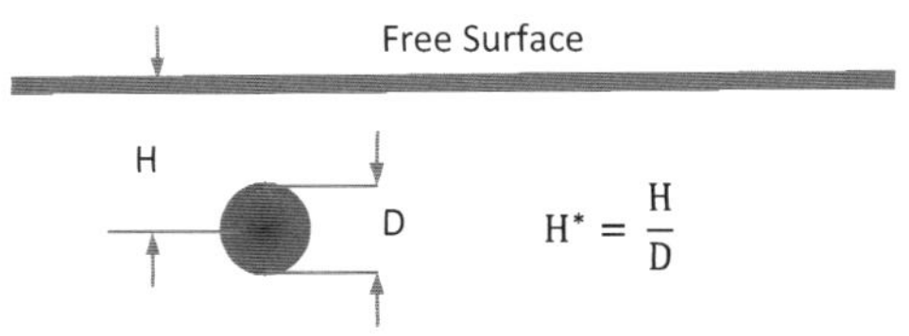

Figure 3.2 Definition of non-dimensional depth

4 NUMERICAL APPROACH

In order to assist the validation of the CFD results and to enable comparisons with the experimental data, the computational domain was given the dimensions of the AMC's towing tank. The computations were conducted at model scale Reynolds Numbers using ANSYS CFX software.

The computational domain and the boundary conditions for the CFD simulations are shown in figure 4.1. The symmetry of the geometry about the x-y plane enabled the domain to be reduced in half.

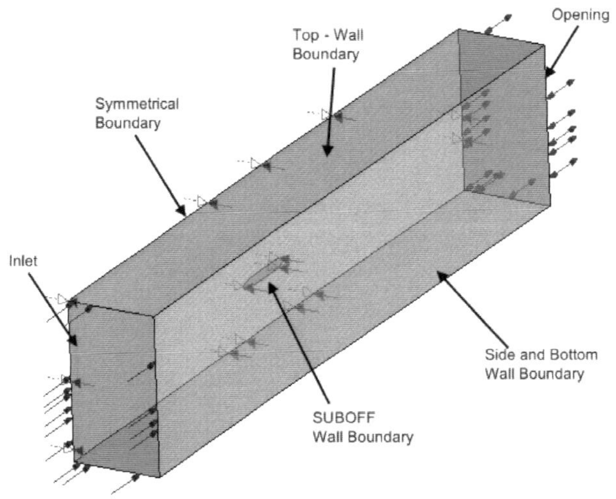

Figure 4.1 Computational domain

The computational domain consisted of a block structured grid refined around the free surface to capture the undulations in water level due to the waves generated by the body. The free surface was modelled using a Eularian technique, employing a variation of the Volume of Fluid (VOF) method based on a fluid volume fraction. In order to adequately capture the free surface it was essential to have a fine grid around the interface to minimise the smearing effects due to numerical diffusion. The homogenous free surface model, in which all fluids share the same local velocities and pressures, was used as no significant mixing or entrapment between the phases was expected. The free surface height was defined using the water volume fraction at the inlet, while the outlet consisted of the associated hydrostatic pressure at the boundary.

The mesh was constructed using ANSYS-ICEM-CFD software, utilising a combination of O, C, and H topologies as shown in figure 4.2. As the objective was to determine the effect of L/D ratio on the drag force of a submarine operating near the surface at constant speeds, the investigation consisted of steady-state numerical simulations with the submarine well submerged (without the free surface) and at a depth of $H^* = 1.1$. An issue when conducting CFD simulations with a free surface is that the surface waves cause numerical fluctuations, thus causing difficulties with convergence, especially at low Froude numbers. Therefore, it is usually advisable to attempt transient solutions when simulating free surface conditions in order to capture the surface fluctuations. However, the current work concentrated on steady-state simulation to run for sufficient iterations until a converged solution was achieved. As a result, the flow regime close to the submarine and the effects of the free surface were captured sufficiently to be confident that the results predicted the correct trends.

All simulations were carried out using the Shear Stress Transport (SST) turbulence model, which is able to adequately model both the near wall and free stream conditions. The maximum y^+ for the various simulation runs, which is dependent on the first cell height and the Reynolds numbers and hence the model length and velocity, spanned from 2 to 6.

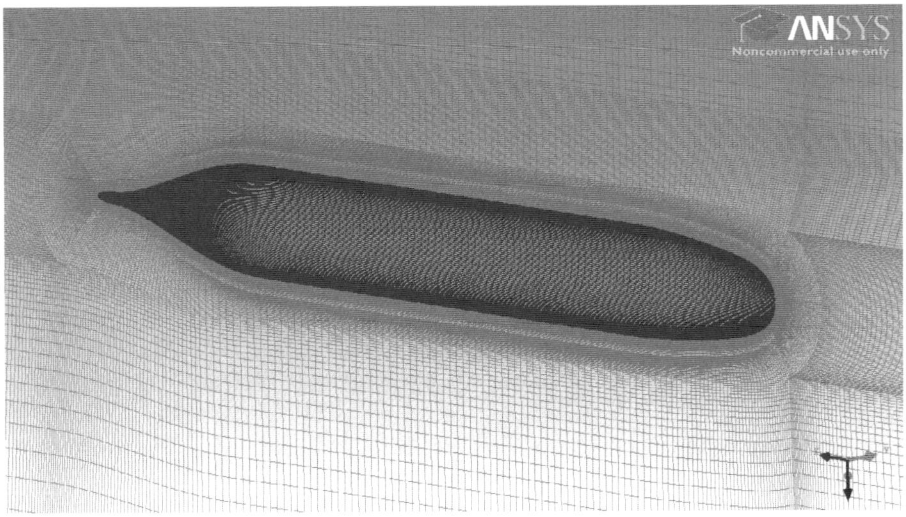

Figure 4.2 Mesh domain

5 GRID INDEPENDENCE STUDY

A grid independence study was conducted on the simulation mesh in order to determine the appropriate grid resolution. The 1.556m length hull form (L/D = 8.60) at a non-dimensional depth of H^*= 1.1, a Froude number of 0.18 with a Reynolds number of 1.25×10^6 was chosen for the study as this provided one of the more complex combinations for the flow simulations. The mesh density was systematically increased from around 2.5×10^6 to 7×10^6 elements, with the drag coefficient (C_t) monitored in order to identify the optimal mesh size.

Figure 5.1 shows the percentage error of C_t (in respect to the largest mesh *i.e.* 7×10^6 elements) as a function of the mesh size for the near surface condition described above. It is seen that above around 5×10^6 elements the C_t is relatively independent of the mesh size. Hence, a mesh configuration of 5×10^6 elements was used for the remainder of the simulations.

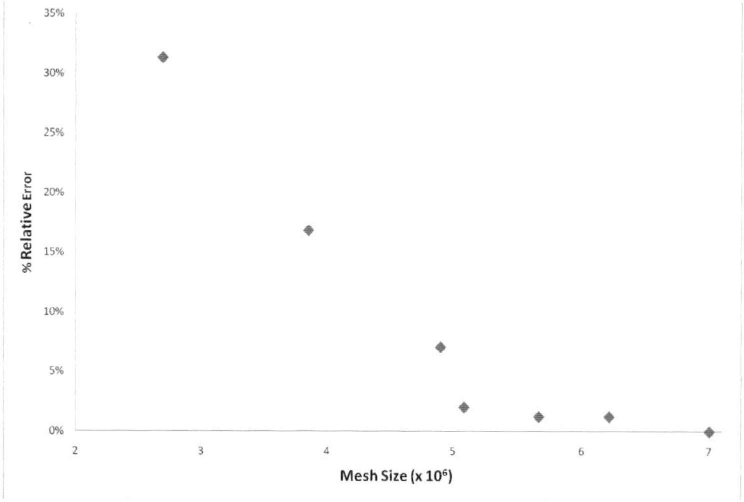

Figure 5.1 Percentage relative error of C_t as a function of mesh size compared to 7×10^6 element mesh

6 COMPARISON WITH EXPERIMENTAL DATA

6.1 Pressure distribution

In order to validate predictions made using the CFD technique discussed above, runs were conducted with the submarine well submerged (*i.e.* no free surface) with no sail at a Reynolds number of 1.2×10^7 to compare the C_t and the surface pressure coefficient (C_p), and skin friction coefficient (C_f) against experimental data for the un-appended SUBOFF geometry published by Huang *et al.* (1992). The agreement was generally very good, as can be seen from figure 6.1, which is a plot of C_p along the upper surface of the SUBOFF geometry for the selected mesh at a Reynolds number of 1.2×10^7 with the hull well submerged.

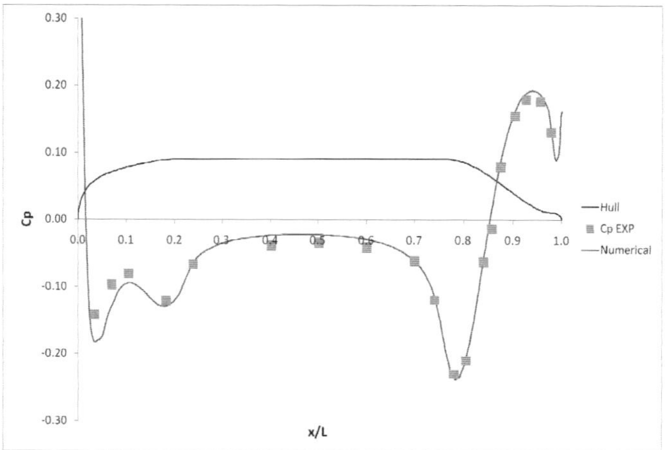

Figure 6.1 Distribution of C_p on the upper surface for the SUBOFF fully submerged at Re = 1.2×10^7

(Numerical results are for $y^+ = 40.3$, and a grid size of 5.1×10^6 elements)

This provides further confidence in the ability of the mesh to be used to accurately capture the flow. The predicted C_t value of the SUBOFF when deeply submerged was 3.23×10^{-3}, showing good agreement with the experimental value of 3.30×10^{-3}.

6.2 Froude Number

To give further confidence to the results of the CFD, model experiments were conducted in the AMC towing tank. Details of these experiments are given in references 1, 4 & 5. A model with the principal particulars given in table 6.1 was used for this work. The model was held at various fixed distances below the water surface as given in table 6.2. It was sting mounted from the rear as shown in figures 6.2 and 6.3 to reduce the effect of interference between the supporting mechanism and the water surface. Tests were conducted at the range of speeds given in table 6.3. Note that Froude scaling was used to convert from model to full scale.

Table 6.1: Principal particulars of model

Hull	
Length (m)	1.436
Diameter (m)	0.181
Wetted area (m^2)	0.753
Sail	
Chord (m)	0.131
Span (m)	0.079
Wetted area (m^2)	0.026

Table 6.2: Test depths

Non-Dimensional Depth (H*)	Model CL Depth (m)	Full scale CL Depth (m)
1.1	0.200	10.23
1.3	0.235	12.09
2.2	0.400	20.46
3.3	0.600	30.69
4.4	0.800	40.92
5.5	1.000	51.15

Table 6.3: Speeds used in test program

Model scale (m/s)	Full scale (knots)	Froude number (based on L = 80m)
0.5	7.08	0.133
1.0	13.61	0.266
1.5	20.69	0.400
2.0	27.77	0.533
2.6	35.94	0.693

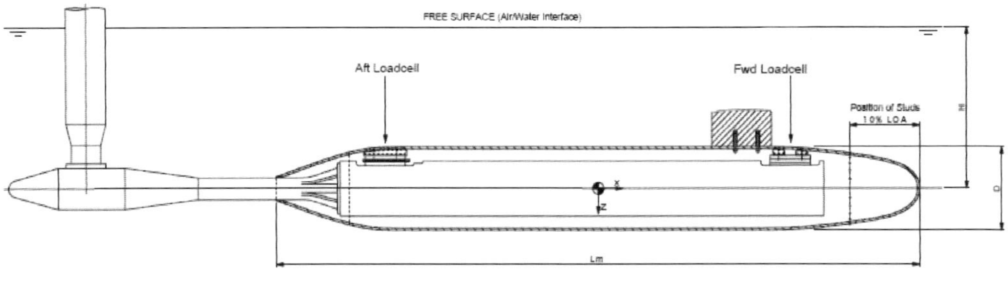

Figure 6.2 Drawing of model showing sting arrangement

The drag forces obtained both numerically and experimentally at a non-dimensional depth (H*) of 1.1 and well submerged (without the effects of the free surface) in the relevant Froude number range are presented in Figure 6.4. As can be seen, this shows good correlation between the numerical predictions and the experimental measurements.

Figure 6.3 photograph of model under test

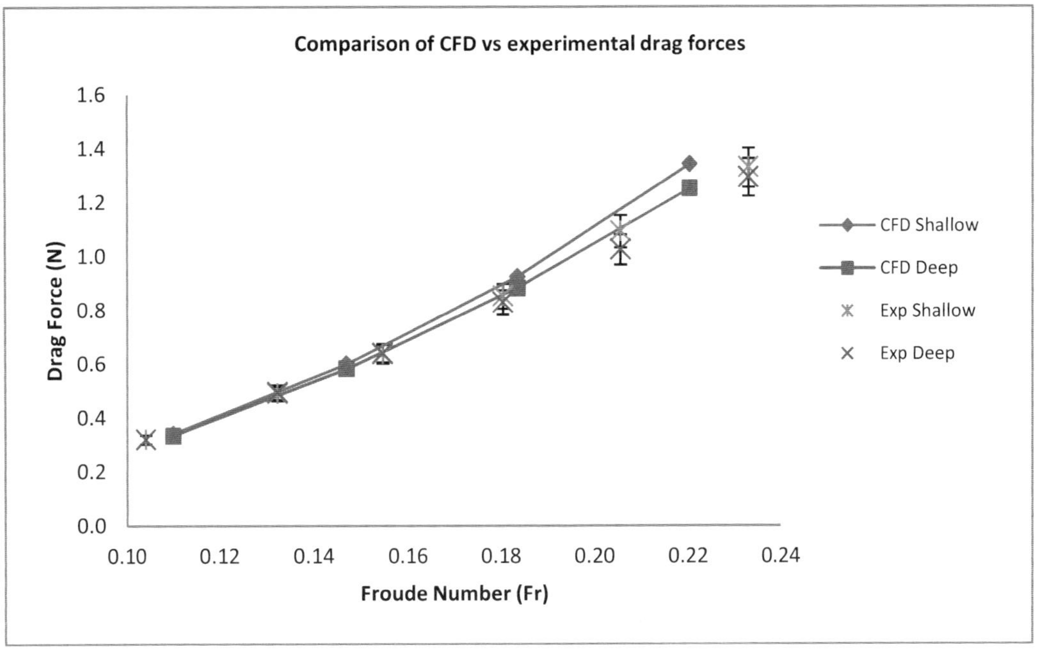

Figure 6.4 Comparison of predicted (CFD) and measured (Exp) drag

7 RESULTS AND DISCUSSION

In order to determine the effect of the free surface on a full scale boat the predicted drag on the model was then scaled to obtain the full scale C_t values, using Froude scaling with no form factor and the ITTC'57 correlation line for the skin friction correction. From this, the percentage increase in power required when operating close to the surface, compared to when operating deeply submerged, was calculated for each L/D value, and hence the influence of the presence of the free surface on the optimum L/D obtained.

The scaled results are presented in figure 7.1 as functions of full scale speed. Note that full scale speed was used rather than either Reynolds number or Froude number because the different length for each of the hull forms, as given in table 7.1, would result in potentially misleading results. However, it is also important to note that when obtaining the C_t values the full scale wetted surface for each of the hulls was used, and that this is different for each hull form as given in table 3.1.

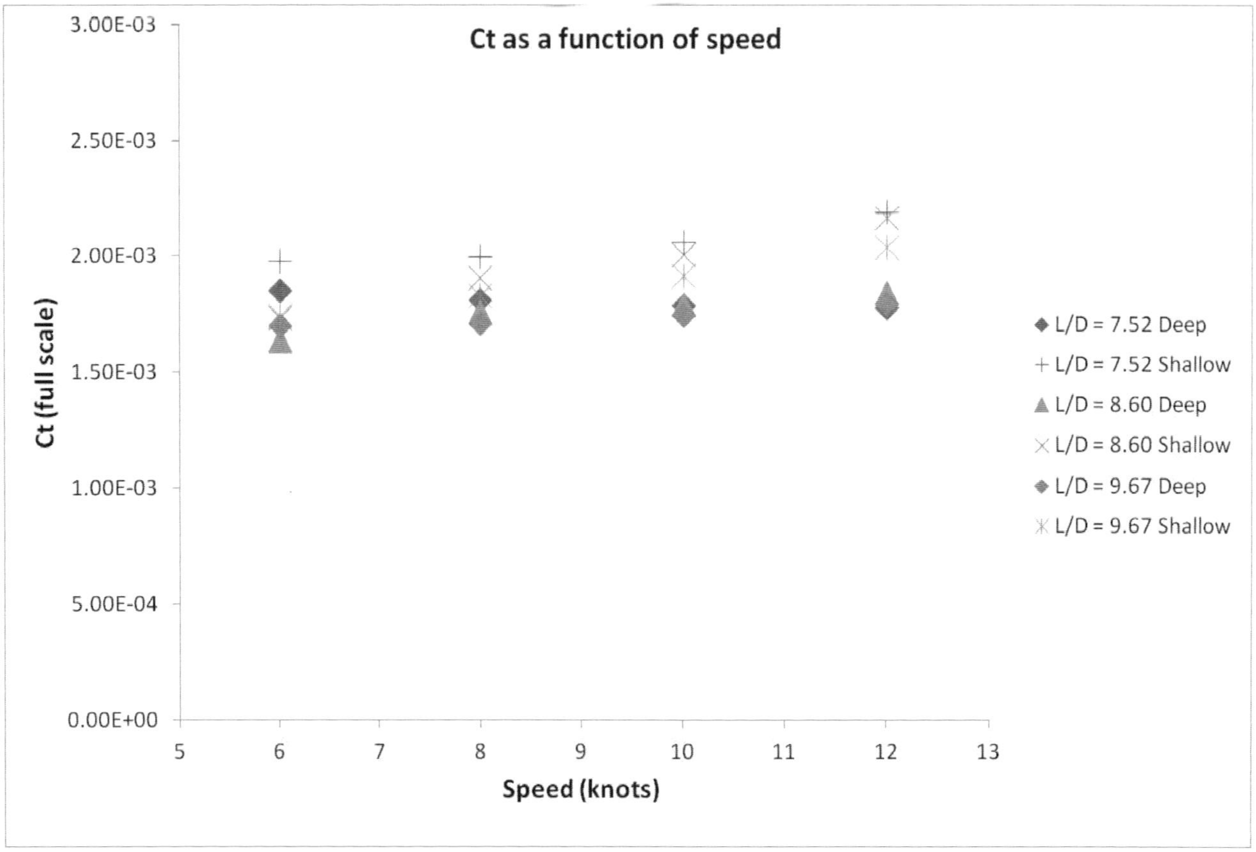

Figure 7.1 Full scale C_t as a function of full scale speed

Table 7.1 Froude number as a function of speed for the three L/D ratios

Speed (knots)	Froude Number		
	L/D = 7.52	L/D = 8.60	L/D = 9.67
6	0.118	0.110	0.104
8	0.157	0.147	0.138
10	0.196	0.183	0.173
12	0.236	0.220	0.208

As can be seen from figure 7.1 the C_t values for the shallow cases, where the hull is close to the free surface, are greater than for the deep cases, where the hull is deeply submerged and there is no free surface. The difference is an indication of the wavemaking drag caused by the proximity to the free surface.

Figure 7.2 is a plot of the percentage increase in power required due to the proximity of the free surface as a function of speed. As can be seen from this figure the percentage increase in power required due to the proximity to the free surface is considerable. In addition, the increase is greater for lower values of L/D.

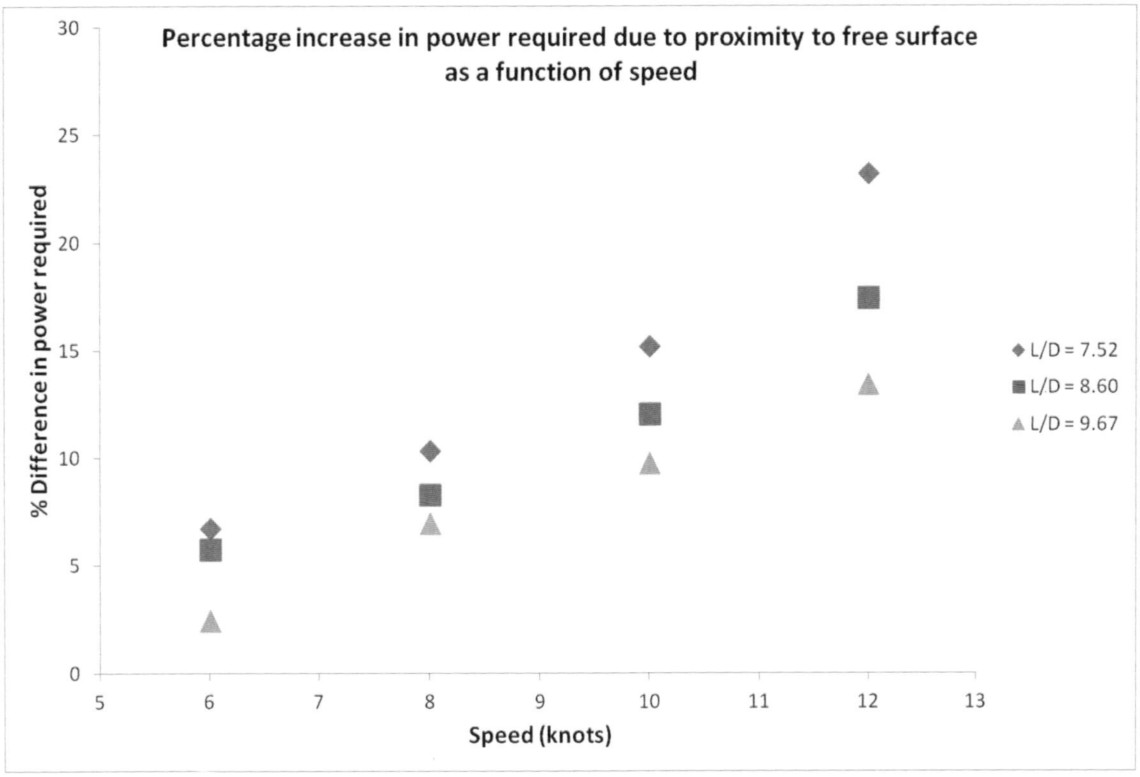

Figure 7.2 Percentage increase in power required due to proximity to free surface

The effect of L/D on the increase in power required is easier to see in figure 7.3, which is a plot of the increase in power required as a function of L/D for the four speeds. As can be seen in this figure, the lower the L/D value the greater the influence of free surface.

As this applies at all the speeds investigated, it is not thought to be purely a Froude number effect, but is likely to be due to the increased wavemaking caused by the smaller L/D value. This may suggest that the optimum L/D value for a boat which is designed to frequently operate close to the surface may be higher than the value for a boat which is designed only to operate well below the surface, where wavemaking does not have an influence.

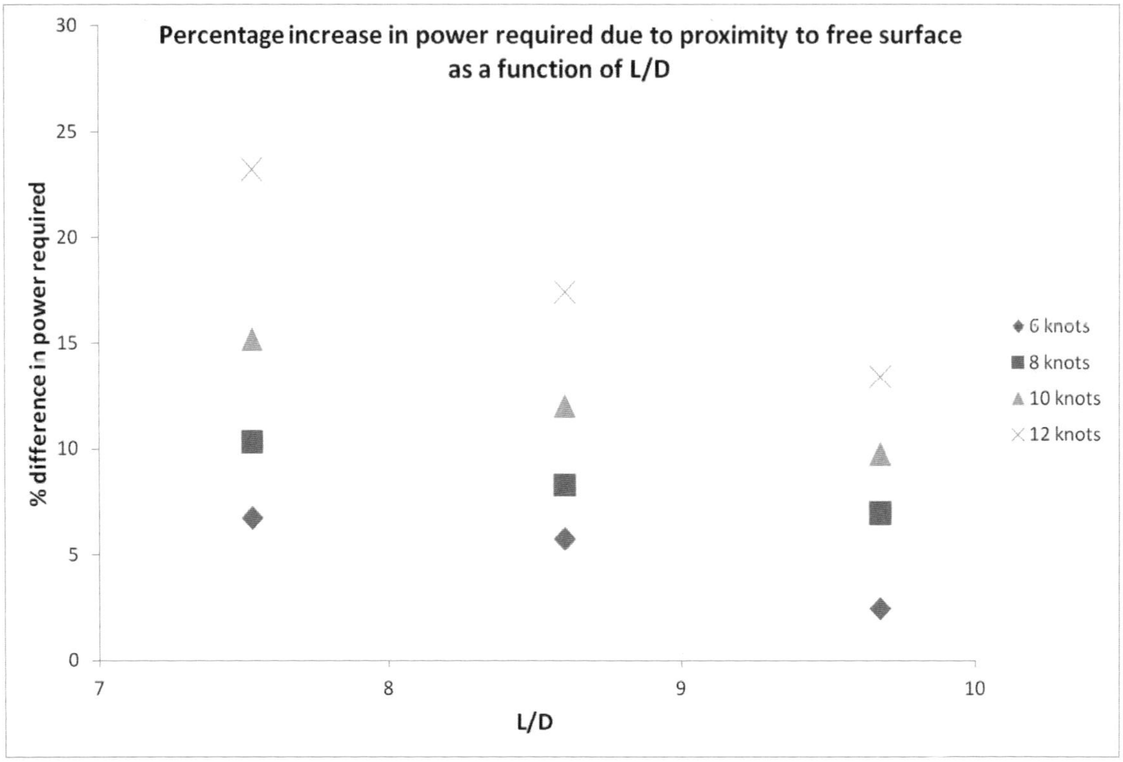

Figure 7.3 Percentage increase in power required due to proximity to free surface

8 CONCLUDING REMARKS

When a submarine is operating close to the surface it will generate surface waves, which will result in wavemaking resistance, and hence an increase in the power required to achieve a given speed.

From the preliminary results it appears that this wavemaking resistance is a function of L/D, and is smaller for larger values of L/D. This suggests that the optimum L/D ratio for a submarine designed to operate near the surface is likely to be larger than for one optimised to operate deeply submerged.

9 NOTATION

C_f Non-dimensional friction coefficient, defined as $\frac{2R_f}{\rho S U^2}$

C_p Non-dimensional pressure coefficient, defined as $\frac{2(p - p_0)}{\rho U^2}$

C_t Non-dimensional drag coefficient, defined as $\frac{2R}{\rho S U^2}$

D Hull diameter

H Distance of centreline below calm water level (figure 3.2)

H* Non-dimensional distance of centreline below calm water level, defined as $\frac{H}{D}$ (figure 3.2)

L Hull length

R Total resistance

R_f Frictional resistance

S Wetted surface

p Total pressure

p_o Atmospheric pressure

U Axial velocity

ρ Water density

10 REFERENCES

1. Dawson, E., Anderson, B., Van Steel, S, Renilson, M., and Ranmuthugala, D., (2010) 'An experimental investigation into the effects of near-surface operation on the wave-making resistance of SSK type submarines', *Maritime Systems and Technology Conference, Nov 2010, Rome*, 2010.

2. Groves, N.C., Huang, T.T., and Chang, M.S., (1989) 'Geometric Characteristics of DARPA SUBOFF Models (DTRC Model Nos 5470 and 5471)' *David Taylor Research Centre, Bethesda, Maryland, USA*, 1989.

3. Huang, T., Liu, H., Groves, N., Forlini, J., Blanton, J., & Gowing, S. (1992) 'Measurements of Flows Over an Axisymmetric Body with Various Appendages in a Wind Tunnel: the DARPA Suboff Experimental Program'. *USA: David Taylor Model Basin.*

4. Renilson, M.R., Ranmuthugala, D., Dawson, E., Anderson, B., Van Steel, S., and Wilson-Haffenden, S. (2011) 'Hydrodynamic design implications for a submarine operating near the surface', *Proceedings of Warship 2011: Naval Submarines and UUVs, Royal Institution of Naval Architects, 29 – 30 June 2011.*

5. Wilson-Haffenden, S., Renilson, M., Ranmuthugala, D., & Dawson, E. (2010) 'An Investigation into the Wavemaking Resistance of a Submarine Travelling Below the Free Surface'. *Sydney NSW: Proceedings of the Pacific 2010 International Maritime Conference, Sydney, Australia, January 2010.*

Research into the Circular Wave Pool – a new method of generating controlled breaking waves

Jonathan R. Binns, Martin Renilson, Giles Thomas, Gregor Macfarlane
Australian Maritime College, University of Tasmania
jrbinns@amc.edu.au

Steven Schmied
Webber Wave Pools

Rene Huijsmans
Delft Technical University

Surfing as a sport is increasing in popularity worldwide. However, as it depends on being able to access wave conditions which are not always available, surfers have been trying to make indoor pools which recreate the surfing experience in a controlled manner. One of the more promising developments in this area is a circular wave pool, patented by Greg Webber in 2005 [9].

The original idea was to produce crystal clear perfect surfable waves in a large circular pool, breaking on an inner island. Although never designed to replace surfing, this phenomena would complement the surfing industry perfectly, opening up new markets and cementing old ones. Greg Webber and Steven Schmied took the idea to TU Delft and then to the Australian Maritime College for further research.

6 years later the research team has been analysing shapes, testing and predicting their wave-making performance in straight lines and finally producing perfect scaled waves in a circular test facility. This paper reports on this research and on how it has been used to further the design of the wave pool. Appropriate surfing speeds have been established from observations of real surfers. Scale model tests were used to determine the optimum Froude depth and length numbers. Finally a set of working beach locations and slopes have been estimated from the experimental study.

In addition to being able to be used to create perfect waves for surfing, the circular wave pool can also be used for a range of research activities, for example to generate controlled waves to investigate the loads generated by breaking waves on coastal structures and to investigate waves generated by manoeuvring vessels.

INTRODUCTION

Wave pool designs have progressed steadily in the last 80 years, however a commercially viable design still appears elusive [7]. In order to find the solution to the commercial difficulties of wave pool designs, a novel idea to produce continuous breaking waves has been patented by Liquid Time Pty Ltd [9], whereby one or more immersed bodies are rotated within an annular wave pool; see Figure 1 for the concept design. An immersed body here could be any object that disrupts the water surface, and most commonly this would be a ship hull, a submerged body (submarine), or even a wave dozer [4].

The inner ring of the annulus has a sloping bathymetry to induce wave breaking from the wake of the moving body, with the break point following the circular path around the island at a given depth proportional to the wave height. The aim is to be capable of creating waves suitable for surfers from beginner to expert level. As an added benefit, by providing a safe learning environment, the overall surfing ability of the participants should be improved. The concept was proven, at least for a linear track, using a fishing vessel generating waves in a river estuary, where the vessel travelled in a straight line close to the bank. Figure 2 shows that one of the smaller moving-body (specifically a fishing vessel) generated waves can still be consistently surfed.

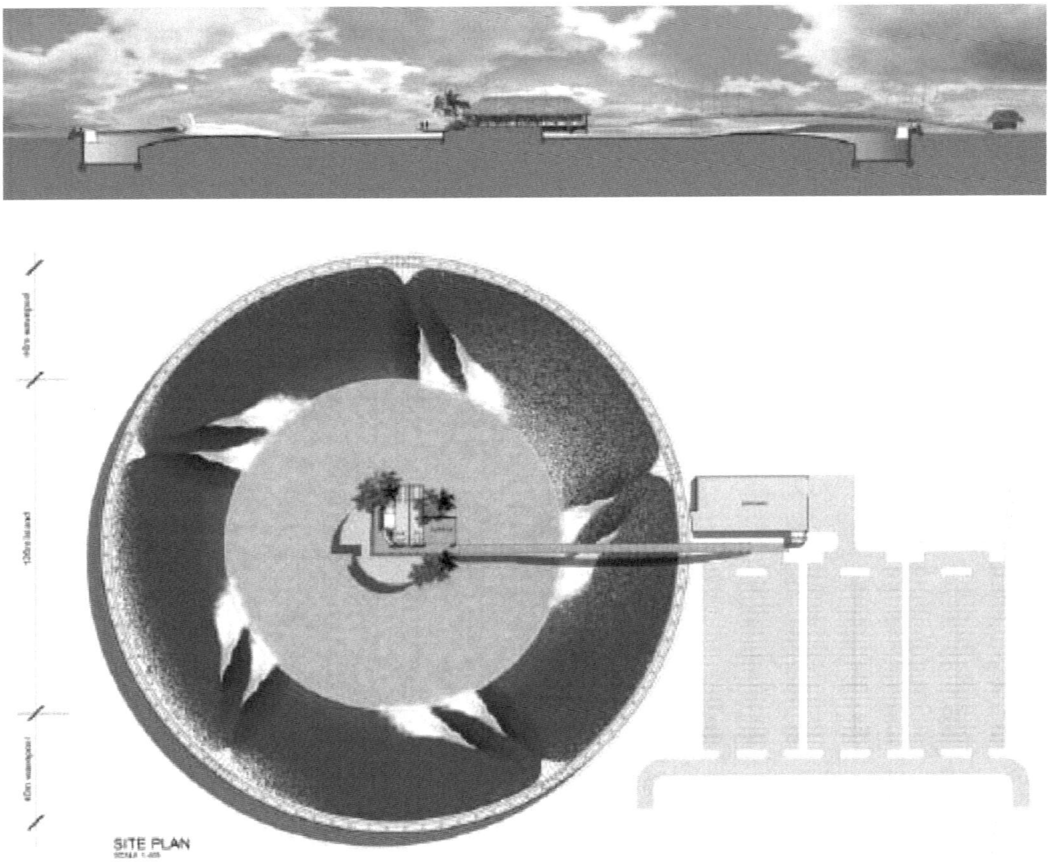

Figure 1 Proposed wave pool design by Webber Wave Pools and Liquid Time Pty Ltd. (reproduced with permission of Liquid Time Pty Ltd)

Figure 2 River testing such as that shown in this figure has proven that even the smallest of hull generated waves can be consistently surfed

The design space available to develop a wave pool is relatively large in comparison to waterborne vessel design. The main aim in designing a wave pool is to produce high quality surfable waves with the longest duration possible. The main constraint on the design is to produce the waves in the smallest space possible with a minimum amount of energy. In early linear tank testing of boat-like shapes it became apparent that increasing wave heights by simply increasing vessel beam meant that the limit of wave height creation due to a breaking bow wave was soon reached [8]. Also, from these early studies a clear advantage was seen in the use of "wavedozer" hulls which are essentially a flat plate driven at an angle of attack through the water. These wavedozer shapes form the basis of all work described within this paper. The initial wavedozer design is shown in Figure 3.

Figure 3 The first wavedozer shape tested, direction of travel is from left to right

OPTIMUM SURFING SPEEDS

As an initial approximation, the wave speed at the break point (the surfer's path) was assumed to be approximately the same as the moving body speed (V). An initial aim of the project was to estimate such speeds by considering two questions:

1. What is the optimum speed for a surfing wave?
2. What is the minimum speed for a wave to be surfable?

Field observations of surfing waves were conducted at Lorne Point, Victoria [7]. Lorne point was chosen as the waves break parallel with the shoreline, and with a desirable shape at small (less than 1m) wave heights. Thus, Lorne Point is considered a close representation of waves desired for the final wave pool.

From observations at Lorne Point, the approximate minimum water depth was found to be 1m, giving a minimum wave speed of 3m/s, using the shallow water wave dispersion relationship given by Dean & Dalrymple [3]. This observation was supported by Dally [2] and Hutt *et al.* [5], who determined that the minimum mean wave speed was as low as 2.4m/s. The maximum water depth for Lorne Point was estimated at 3m, thus giving a maximum wave speed of 5.4m/s. Combining the observations of Dally and Hutt, a maximum wave speed of 6.0m/s was obtained. Therefore the wave speed variation for good surfing was estimated to lie between 3 to 6m/s.

Viewing the wave pool as a hydrodynamic optimisation problem, the first parameter has been set. The absolute speed of the wave crests must lie in the range of 3 to 6m/s. Considering then that the crest line will be between 60° to 90° to the line of travel, the body creating the waves must travel at 3 to 12m/s, the upper limit being confined to expert surfers only. For the purpose of this study a full scale speed of 6.8m/s has been used as the base for further optimisation.

OPTIMUM FROUDE NUMBERS

There are two Froude number spatial scales that have a large influence on the size and slope of waves produced by an immersed body. The first scale of interest is the length of the body and the second is the water depth. Considering that the velocity of the body has been set, as per the previous section, these two spatial scales can be used to create the length and depth based Froude numbers for optimum creation of surfable waves.

The length based Froude number to create the largest waves possible has been well documented to lie between 0.45 to 0.5 ([6], [1]). If the absolute value of the speed has been set by desired surfer speeds, then setting the length Froude number has the effect of prescribing the length scale of the body. It then remains to determine what the length scale of the body actually is.

For thin ship forms determining the length scale of the hull is relatively easy, being the waterline length, however for the shapes considered here (a wavedozer) this estimation is quite difficult. For example, a study was conducted whereby the angle of attack of the wavedozer was varied whilst maintaining aft corner depth, thus altering its waterline length considerably. Although potential flow numerical results predicted a large variation in wave heights as a function of angle of attack, the experimental results shown in Figure 4 failed to discern a logical pattern with respect to Froude length number variation for these bodies varied in this way. The result presented in this figure are for wave height normalised by water depth (H^*) as a function of model scale speed (V). The experiments were conducted at a model scale of 1:20 in a model scale water depth of 250 mm, wave dozer immersion was 50 mm and the wave heights were measured at a distance of 375 mm from the sailing line in a pool of diameter of 5 m. At this point none of the waves were fully breaking, all of the waves were steady and the highest Froude depth number was 0.99, thus the wave patterns retained a familiar Kelvin wave pattern in plan view. This is certainly the subject of further investigation.

Due to the non-dispersive nature of shallow water waves, wave heights at a Froude depth number of 1.0 are much larger than at lower speeds. That is, the energy is concentrated in fewer waves at this Froude depth number, thus making larger wave heights. In this critical region a soliton may be pushed forward from the vessel, such as that depicted in the image series of Figure 5. These waves can be large in height (although they have no preceding trough), but they are also unstable and likely impossible to surf. Therefore, an aim is to set the surfing pool design with operation just below a Froude depth number of 1.0, without actually producing such disastrous solitons.

With the speed set from surfing data, the length of the wavedozer and depth of water set from experiments, the major parameters of the wave pool have largely been determined.

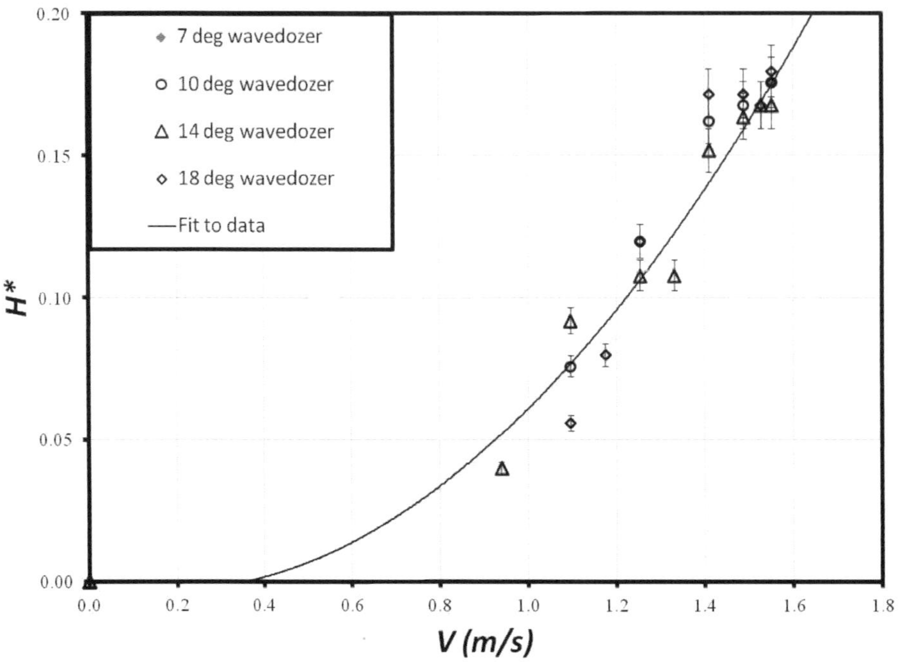

Figure 4 Variation in wave height (difference between trough and crest) normalised by water depth as a function of velocity for changes in wave dozer angle of attack, wave dozers are all variations of the geometry shown in Figure 3.

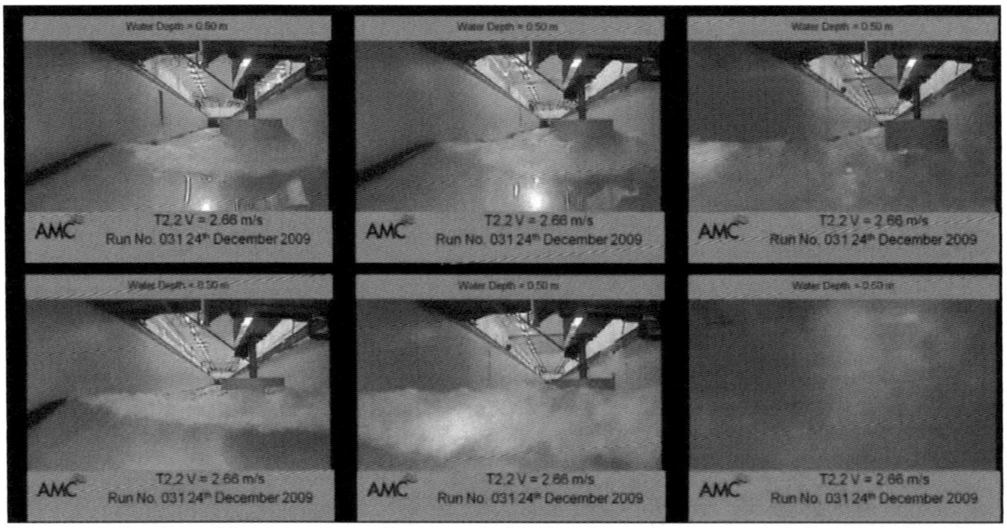

Figure 5 Series of images (starting top left, finishing bottom right), showing the development of a large soliton from a wide body at a critical Froude depth number, The carriage began to slow after the fourth frame and was completely stopped before the final frame shown. The final picture is obscured because the large wave completed swamped the camera, even though the camera at this height had never been wet for all runs at lower speeds. Model used was a parabolic Wigley hull with a length to beam ratio of 1.17 in a water depth of 500 mm in a 3.55 m wide towing tank.

OPTIMUM BEACH LOCATION

The location and bathymetry of the beach is still being examined in detail. However, as the exact break point of a surfable wave is controlled by shoaling bathymetry, it is clear that the bathymetry for breaking waves needs to coincide with the wave train from the body, producing a maximum wave height which is stable in space and time. To achieve this, a location has been selected outside of the near field, such that the wave pattern is not dominated by the local effects around the moving body, indicated by a constant wave height with respect to transverse distance.

For the purpose of finding the best location for initiating wave breaking by tailoring the beach bathymetry, experiments were performed on a circular track model with no internal beach at a diameter of 10m. The test conditions were set by the speed and Froude number limits discussed above. The resulting measured wave height variation with distance from line of travel normalized by the track radius is shown in Figure 6. In this figure changes of lateral distance of between 0.125 and 0.225 times the radius can be seen to have a small change in wave height. Therefore, it is proposed that the best location for the beach will be in this area, as changes in wave height could be controlled solely by changes in bathymetry, permitting the creation of the best surfing wave.

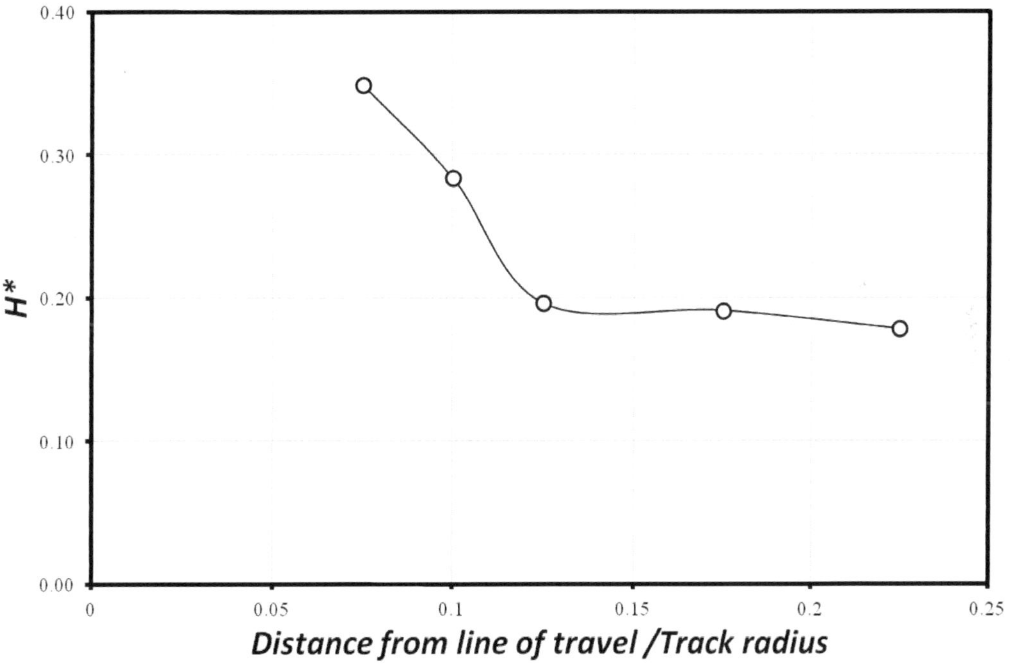

Figure 6 Variation in wave height normalised by water depth with distance from body line of travel for the 14° angle of attack wave dozer. Scale was 1:20, transom depth of 50 mm, measurement distance of 375 mm from line of travel for a 5 m radius pool.

CONCLUDING REMARKS

The design of a circular wave pool concept has been produced by Webber Wave Pools and patented within Australia and Internationally. This design shows great promise to not only produce a unique facility for expanding the surfing industry but also to conduct significant research into repeatable breaking waves.

The optimisation of the surfing wave pool begins with the desired full scale speed required for surfing. This is due to the small window placed on practical surfing speeds. Depth and length Froude numbers for the moving body which generates the waves can then be controlled for optimum wave creation. The design process then proceeds by maximising wave height whilst maintaining a non-breaking wave at a distance from the travel line of the pressure source. The wave used for surfing can then be formed by altering the bathymetry at the correct location.

Although there is still significant research and design work to be completed, the last three years have yielded some extremely valuable quantitative results. The promise of making the perfect repeatable-surfable wave seems to be coming true.

ACKNOWLEDGEMENTS

The authors would like to acknowledge the support of our industry partner Webber Wave Pools and the Australian Maritime College technical team.

The authors would also like to acknowledge the support of the Australian Research Council through their award of our Linkage Project grant LP0990307.

REFERENCES

1. Claughton, A., (1998), Balance of air and water forces, in *Sailing Yacht Design - Theory*, A. Claughton, J. Wellicome, and J. Shenoi, Editors. Addison Wesley Longman Limited: Edinburgh, UK. p. 3-13.
2. Dally, W.R., (2001), "*The maximum speed of surfers.*" Journal of Coastal Research, **SI 29**: p. 33-40.
3. Dean, R.G. & Dalrymple, R.A., (1984), Water Wave Mechanics for Engineers and Scientists. Advanced Series on Ocean Engineering. Vol. 2. Singapore: World Scientific.
4. Driscoll, A. & Renilson, M.R., *The wavedozer: a system of generating stationary waves in a circulating water channel*, in *AMTE(H)*. 1980.
5. Hutt, J.A., Black, K.P., & Mead, S.T., (2001), "*Classification of surf breaks in relation to surfing skill.*" Journal of Coastal Research, **SI 29**: p. 66-81.
6. Lewis, E.V., (1988), Principles of Naval Architecture, ed. E.V. Lewis. Vol. II - Resistance and Propulsion. New Jersey, USA: SNAME. vi + 327.
7. Schmied, S., Binns, J., Renilson, M., Thomas, G., Macfarlane, G., & Huijsmans, R., (2010), "*A novel method for generating continuously surfable waves.*" Marine Technology Society Journal, **44**(2): p. 7-12.
8. Schmied, S.A., Binns, J.R., Renilson, M.R., Thomas, G.A., Macfarlane, G.J., & Huijsmans, R., (2011), A novel method for generating continuously surfable waves - a comparison of predictions with experimental results. in *30th International Conference on Ocean, Offshore and Arctic Engineering*. Rotterdam, The Netherlands: ASME.
9. Webber, G., (2005), Wave generating apparatus, 20049070401, Intellectual Property Australia, Australia

AUTHOR INDEX

AUTHOR INDEX